A Career in Theoretical Physics

World Scientific Series in 20th Century Physics

Published

Vol. 1 Gauge Theories – Past and Future
        edited by R. Akhoury, B. de Wit and P. van Nieuwenhuizen

Vol. 2 Scientific Highlights in Memory of Léon van Hove
        edited by F. Nicodemi

Vol. 3 Selected Papers, with Commentary, of T. H. R. Skyrme
        edited by G. E. Brown

Vol. 4 Salamfestschrift
        edited by A. Ali, J. Ellis and S. Randjbar-Daemi

Vol. 5 Selected Papers of Abdus Salam (with Commentary)
        edited by A. Ali, C. Isham, T. Kibble and Riazuddin

Forthcoming

Vol. 6 Research on Particle Detectors
        by G. Charpak

World Scientific Series in 20th Century Physics – Vol. 7

A CAREER IN THEORETICAL PHYSICS

P W Anderson
Department of Physics
Princeton University

World Scientific
Singapore • New Jersey • London • Hong Kong

Published by
World Scientific Publishing Co. Pte. Ltd.
P O Box 128, Farrer Road, Singapore 9128
USA office: Suite 1B, 1060 Main Street, River Edge, NJ 07661
UK office: 73 Lynton Mead, Totteridge, London N20 8DH

While every effort has been made to contact the publishers of reprinted papers prior to publication, we have not been successful in a few cases. Where we could not contact the publishers, we have acknowledged the source of the materia. Proper credit will be given to these publishers in future editions of this work after permission is granted.

The author and publisher would like to thank the following publishers of the various journals and books for their assistance and permission to include the selected reprints found in this volume:

Academic Press
Addison-Wesley
American Assoc. Adv. Sci.
American Institute of Physics
American Physical Society
Elsevier Science Publishers
Gordon and Breach Science Publishers
IAEA

IDSET
Marcel Dekker
MIT Press
The Nobel Foundation
Pergamon Press
Plenum Publishing
Società Italiana di Fisica
Taylor and Francis

Library of Congress Cataloging-in-Publication
Anderson, P. W. (Philip W.), 1923–
    A career in theoretical physics / P. W. Anderson.
      p.   cm. -- (World Scientific series in 20th century physics ;
    vol. 7)
    ISBN 981021717X. -- ISBN 9810217188 (pbk.)
    1. Mathematical physics. I. Title. II. Series.
QC20.A55   1994
530.1--dc20   94-13919
                                                                          CIP

Copyright © 1994 by World Scientific Publishing Co. Pte. Ltd.

All rights reserved. This book, or parts thereof, may not be reproduced in any form or by any means, electronic or mechanical, including photocopying, recording or any information storage and retrieval system now known or to be invented, without written permission from the Publisher.

For photocopying of material in this volume, please pay a copying fee through the Copyright Clearance Center, Inc., 27 Congress Street, Salem, MA 01970, USA.

INTRODUCTORY ESSAY

It was in the winter of 1966–7 that I first tried to express my philosophy of what was important in science — or at least, of what was important that was possible for me. I had been invited to spend a month in the pleasant climate of La Jolla as the Regents' lecturer at the UCSD, visiting a department staffed by many of my old friends from Bell Labs, such as Bernd Matthias, Harry Suhl, and George Feher, who had been recruited by Walter Kohn, who was also by courtesy a Bell alumnus. One of the lectures I gave was called "More is Different". The original version, now lost, went over with the audience like a lead balloon, at least according to that portion of the audience most closely related to me. But a cleaned up version, which was published in *Science* in 1971, has been surprisingly influential. I have chosen to reprint it, out of chronological order, as an additional introduction to this collection.

I already had this overall vision, if only very vaguely, as I completed my thesis and other early work on the breadths of spectral lines (see paper 2 in the book). This work was among the first to approach what was normally thought of as a dissipative process and dealt with by Boltzmann equations for the microscopic variables, in terms of the macroscopic moments of the system as a whole in my own mind called it the "one big molecule" method. I was beginning to see that somehow, starting from the fundamental atomic-level laws of quantum theory, and without any more or less mystical appeals to the difference between the atomic domain and the macroscopic domain where deterministic Newtonian physics held way, one could really understand many of the properties of matter in that macroscopic domain. I also experienced two other insights: that these properties were often extraordinary; and that this understanding was often intellectually challenging and deep, at a level which was for me at least more exciting and worthwhile than any other purely intellectual exercise.

Thus I was launched into the search for the "Aufbauprinzips" of the everyday world. I admired Fröhlich's book on dielectric theory because it took the same point of view; I encountered it in my early work on ferroelectrics (papers 1 and 3). Much more important, I came to see the crucial importance of the concept of broken symmetry, which has been a lifelong interest (papers 6, 7, 9, 10, 11, 15, 22, 26, 32, 39). Broken symmetry is the clearest instance of the process of emergence which lies behind "More is different". My interest began with the antiferromagnetic ground state work in 1952–3, from which I developed the concept of the Goldstone boson. In applying similar thinking to superconductivity, the Higgs phenomenon and the concept of Generalized Rigidity arose, the latter referring to the similarities between superfluidity, superconductivity, and elastic rigidity of all sorts. The topological theory of defects and dissipation, of which papers (7), (11) and

(27) are examples, is another facet of the conceptual structure of broken symmetry. Other examples were omitted for reasons of space.

The first natural direction in which to seek for understanding of the real world is the explicit quantitative calculation of properties of real materials. This enterprise is particularly rewarding when the basic principles which govern these properties are new, or when the effects are subtle. This has been the mainspring of much of my work, from the early work on magnetic resonance and other spectra, to superexchange (paper 8), the microscopic theory of superconductivity (papers 4, 5, 9, 14, 18, 30, 33), and even work on the electronic structure of solids such as paper 21, foreshadowing the corrections to local density theory, and paper 29 where the idea is to develop a method which explains and exploits the strong locality of the chemical bond, perhaps the most fundamental observation in the field of chemistry, which nonetheless seems to interest quantum chemists not at all.

In 1956, stimulated by a pioneering paper on percolation by Hammersley and by some puzzling experiments, I started to look at the next level of complexity, those phenomena which are intrinsic to disorder. The first result was the theory of localisation of waves (papers 25, 28, 16). This was slow of acceptance, and, perhaps discouraged by this fact, it was not until 1969 that I returned to it and also began to look at other disorder phenomena such as glass and spin glass (papers 23, 24, 34). Spin glass stimulated a very fertile period of worldwide activity, first in the development of totally novel methodology, such as a new statistical mechanis of nonergodic systems, and then in applications of the new methodology to a "cornucopia" of new ideas and concepts and problems: neural networks, computer algorithms, rugged evolutionary landscapes, even immunology. This has left me involved in such fields as economics and the origin of life. The interest in complex, emergent phenomena made it natural for me to participate at an early stage in the Santa Fe Institute, founded on the twin principles of interdisciplinarity and of the importance and tractability of genuinely complex systems. There has not been much space for these ideas; only papers 24, 31, 34 and 39 touch on them.

Left behind, in my old field of quantum condensed matter physics, was a worry about the validity and completeness of the theories of many-body quantum systems which had occupied that world in the 1950's and 60's. My work on the Anderson model of magnetic impurities (paper 12) had led on to the "infrared catastrophe" in Fermi systems which was the key to my work on the Kondo phenomenon (papers 13, 20) which solved that isolated many-body system but only at the cost of going beyond the conventional perturbation theory. I was also disturbed by a class of strongly interacting superconducting materials (30), as well as our notable lack of success in dealing with the field of quantum valence fluctuations, i.e. "heavy electrons" (35, 36, 37). With the appearance of the high T_c cuprates in 1987, my apprehensions became manifestly justified, and I began to formulate theories of metals based on new principles (papers 40, 41, 42, 43). This is the area which now most absorbs me, and where my efforts have been rewarded with the discovery of several new and fascinating phenomena. I reach my 70th year still involved in two major, very different types of problem. Although there is no easy way to round this one off, I have tried to give a flavor of the ideas involved.

My original vision seems to me to have been sound. The problem of understanding the "here and now" world seems to show no sign of being exhausted. The contribution of physics is the method of dealing correctly both with the substrate from which emergence takes place, and with the emergent phenomenon itself. Examples are legion: for one, superconductivity cannot be properly understood simply as a phenomenology without understanding electrons and their interactions, nor on the other hand as a property of a small number of electrons without taking into account the macroscopic system. Ever newer insights into the nature of the world around us will continually arise from this style of doing science. There will be physics even in a world without the SSC.

Regarding the choice of articles in this book: the sheer volume of either a complete selection of research articles or, even more, a complete set of reviews would have been much too great. I have given preference to materials which are for one reason or another not easy of access: lecture notes and out of print sources, for example. I have tried but failed to provide a selection of all major topics; the book is long enough. I have included a few items on history as I saw it, and on general aspects of science.

Oxford, December 2, 1993

CONTENTS

Introductory Essay	v
More is Different	1
1. Theory of Ferroelectric Behaviour of Barium Titanate	5
2. Use of Stochastic Methods in Line-Broadening Problems	13
3. Qualitative Considerations on the Statistics of the Phase Transition in $BaTiO_3$-type Ferroelectrics	35
4. Theory of Dirty Superconductors	45
5. Calculation of the Superconducting State Parameters with Retarded Electron-Phonon Interaction (with P. Morel)	51
6. Plasmons, Gauge Invariance, and Mass	61
7. Hard Superconductors	67
Hard Superconductivity: Theory of the Motion of Abrikosov Flux Lines (with Y. B. Kim)	107
8. Exchange in Magnetic Insulators	113
9. Superconductivity (with B. T. Matthias)	131
10. Coherent Matter Field Phenomena in Superfluids	143
11. Considerations on the Flow of Superfluid Helium	165
12. Multiple-Scattering Theory and Resonances in Transition Metals (with W. L. McMillan)	179
13. The Kondo Effect I	217
The Kondo Effect II	224
Kondo Effect III: The Wilderness — Mainly Theoretical	232
Kondo Effect IV: Out of the Wilderness	238
14. Superconductivity in the Past and Future	245

15. Macroscopic Coherence and Superfluidity	263
16. The Fermi Glass: Theory and Experiment	273
17. Space-Time and Scaling Techniques in the Kondo Problem	281
18. Comments on the Maximum Superconducting Transition Temperature	287
Comment on 'Model for an Exciton Mechanism of Superconductivity' (with J. C. Inkson)	299
19. Conference Summary	303
20. Asymptotically Exact Methods in the Kondo Problem (with G. Yuval)	311
21. Many-Body Effects at Surfaces	333
22. Uses of Solid State Analogies in Elementary Particle Theory	363
23. Possible Consequences of Negative U Centers in Amorphous Materials	389
24. Survey of Theories of Spin Glass	395
25. Disorder: A Frontier of Theoretical Physics	413
26. Some General Thoughts about Broken Symmetry	419
27. The Rheology of Neutron Stars: Vortex Line Pinning in the Crust Superfluid (with M. A. Alpar, D. Pines and J. Shaham)	431
28. Localization Redux	445
29. Chemical Pseudopotentials	453
30. Some Remarks on Strong Electron-Phonon Coupling Metals (with C. C. Yu)	463
31. Spin Glass Hamiltonians: A Bridge between Biology, Statistical Mechanics and Computer Science	495
32. Measurement in Quantum Theory and the Problem of Complex Systems	501
33. It's Not Over Till the Fat Lady Sings	515
34. Spin Glass I: A Scaling Law Rescued	525
Spin Glass II: Is There a Phase Transition?	528
Spin Glass III: Theory Raises its Head	529

	Spin Glass IV: Glimmerings of Trouble	531
	Spin Glass V: Real Power Brought to Bear	533
	Spin Glass VI: Spin Glass as Cornucopia	535
	Spin Glass VII: Spin Glass as Paradigm	537
35.	Valence Instabilities and Related Narrow-Band Phenomena	539
36.	Present Status of Theory: 1/N Approach	548
37.	The Problem of Fluctuating Valence in f-Electron Metals	562
38.	Some Ideas on the Aesthetics of Science	569
39.	Theoretical Paradigms for the Sciences of Complexity	584
40.	50 Years of the Mott Phenomenon: Insulators, Magnets, Solids, and Superconductors as Aspects of Strong-Repulsion Theory	595
41.	The "Central Dogmas"	637
42.	The "Infrared Catastrophe": When Does it Trash Fermi Liquid Theory?	657
43.	Experimental Constraints on the Theory of High-T_c Superconductivity	673

CONTENTS
(in Sections)

Introductory Essay		v
More is Different		1

Ferroelectricity, Soft Modes

[1]	Theory of Ferroelectric Behaviour of Barium Titanate	5
[3]	Qualitative Considerations on the Statistics of the Phase Transition in $BaTiO_3$-type Ferroelectrics	35

Line Broadening, Correlation Function

[2]	Use of Stochastic Methods in Line-Broadening Problems	13

Superconductivity: BCS

[4]	Theory of Dirty Superconductors	45
[5]	Calculation of the Superconducting State Parameters with Retarded Electron-Phonon Interaction	51
[7]	Hard Superconductors	67
	Hard Superconductivity: Theory of the Motion of Abrikosov Flux Lines	107
[9]	Superconductivity	131
[14]	Superconductivity in the Past and Future	245
[15]	Macroscopic Coherence and Superfluidity	263
[18]	Comments on the Maximum Superconducting Transition Temperature	287
	Comment on 'Model for an Exciton Mechanism of Superconductivity'	299
[30]	Some Remarks on Strong Electron-Phonon Coupling Metals	463

Broken Symmetry

[6]	Plasmons, Gauge Invariance, and Mass	61
[10]	Coherent Matter Field Phenomena in Superfluids	143

[11]	Considerations on the Flow of Superfluid Helium	165
[22]	Uses of Solid State Analogies in Elementary Particle Theory	363
[26]	Some General Thoughts about Broken Symmetry	419

Magnetic State, Mott Insulators

| [8] | Exchange in Magnetic Insulators | 113 |

Superfluidity: $He^3 + He^4$

[11]	Considerations on the Flow of Superfluid Helium	165
[15]	Macroscopic Coherence and Superfluidity	263
[27]	The Rheology of Neutron Stars: Vortex Line Pinning in the Crust Superfluid	431

Local Moments and Kondo Effect

[12]	Multiple-Scattering Theory and Resonances in Transition Metals	179
[13]	The Kondo Effect I	217
	The Kondo Effect II	224
	Kondo Effect III: The Wilderness — Mainly Theoretical	232
	Kondo Effect IV: Out of the Wilderness	238
[17]	Space-Time and Scaling Techniques in the Kondo Problem	281
[20]	Asymptotically Exact Methods in the Kondo Problem	311

Localization

[16]	The Fermi Glass: Theory and Experiment	273
[23]	Possible Consequences of Negative U Centers in Amorphous Materials	389
[25]	Disorder: A Frontier of Theoretical Physics	413
[28]	Localization Redux	445

General

| [19] | Conference Summary | 303 |
| [27] | The Rheology of Neutron Stars: Vortex Line Pinning in the Crust Superfluid | 431 |

[31]	Spin Glass Hamiltonians: A Bridge between Biology, Statistical Mechanics and Computer Science	495
[32]	Measurement in Quantum Theory and the Problem of Complex Systems	501
[33]	It's Not Over Till the Fat Lady Sings	515
[38]	Some Ideas on the Aesthetics of Science	569
[39]	Theoretical Paradigms for the Sciences of Complexity	584

Electronic Structure

[12]	Multiple-Scattering Theory and Resonances in Transition Metals	179
[21]	Many-Body Effects at Surfaces	333
[29]	Chemical Pseudopotentials	453

Spin Glass and Non-Ergodic Systems

[23]	Possible Consequences of Negative U Centers in Amorphous Materials	389
[24]	Survey of Theories of Spin Glass	395
[34]	Spin Glass I: A Scaling Law Rescued	525
	Spin Glass II: Is There a Phase Transition?	528
	Spin Glass III: Theory Raises its Head	529
	Spin Glass IV: Glimmerings of Trouble	531
	Spin Glass V: Real Power Brought to Bear	533
	Spin Glass VI: Spin Glass as Cornucopia	535
	Spin Glass VII: Spin Glass as Paradigm	537

Mixed Valence

[35]	Valence Instabilities and Related Narrow-Band Phenomena	539
[36]	Present Status of Theory: 1/N Approach	548
[37]	The Problem of Fluctuating Valence in f-Electron Metals	562

High-T_c and the New Physics

[40]	50 Years of the Mott Phenomenon: Insulators, Magnets, Solids, and Superconductors as Aspects of Strong-Repulsion Theory	595

[41]	The "Central Dogmas"	637
[42]	The "Infrared Catastrophe": When Does it Trash Fermi Liquid Theory?	657
[43]	Experimental Constraints on the Theory of High-T_c Superconductivity	673

A Career in
Theoretical Physics

More Is Different

Broken symmetry and the nature of the hierarchical structure of science.

P. W. Anderson

The reductionist hypothesis may still be a topic for controversy among philosophers, but among the great majority of active scientists I think it is accepted without question. The workings of our minds and bodies, and of all the animate or inanimate matter of which we have any detailed knowledge, are assumed to be controlled by the same set of fundamental laws, which except under certain extreme conditions we feel we know pretty well.

It seems inevitable to go on uncritically to what appears at first sight to be an obvious corollary of reductionism: that if everything obeys the same fundamental laws, then the only scientists who are studying anything really fundamental are those who are working on those laws. In practice, that amounts to some astrophysicists, some elementary particle physicists, some logicians and other mathematicians, and few others. This point of view, which it is the main purpose of this article to oppose, is expressed in a rather well-known passage by Weisskopf (1):

Looking at the development of science in the Twentieth Century one can distinguish two trends, which I will call "intensive" and "extensive" research, lacking a better terminology. In short: intensive research goes for the fundamental laws, extensive research goes for the explanation of phenomena in terms of known fundamental laws. As always, distinctions of this kind are not unambiguous, but they are clear in most cases. Solid state physics, plasma physics, and perhaps also biology are extensive. High energy physics and a good part of nuclear physics are intensive. There is always much less intensive research going on than extensive. Once new fundamental laws are discovered, a large and ever increasing activity begins in order to apply the discoveries to hitherto unexplained phenomena. Thus, there are two dimensions to basic research. The frontier of science extends all along a long line from the newest and most modern intensive research, over the extensive research recently spawned by the intensive research of yesterday, to the broad and well developed web of extensive research activities based on intensive research of past decades.

The effectiveness of this message may be indicated by the fact that I heard it quoted recently by a leader in the field of materials science, who urged the participants at a meeting dedicated to "fundamental problems in condensed matter physics" to accept that there were few or no such problems and that nothing was left but extensive science, which he seemed to equate with device engineering.

The main fallacy in this kind of thinking is that the reductionist hypothesis does not by any means imply a "constructionist" one: The ability to reduce everything to simple fundamental laws does not imply the ability to start from those laws and reconstruct the universe. In fact, the more the elementary particle physicists tell us about the nature of the fundamental laws, the less relevance they seem to have to the very real problems of the rest of science, much less to those of society.

The constructionist hypothesis breaks down when confronted with the twin difficulties of scale and complexity. The behavior of large and complex aggregates of elementary particles, it turns out, is not to be understood in terms of a simple extrapolation of the properties of a few particles. Instead, at each level of complexity entirely new properties appear, and the understanding of the new behaviors requires research which I think is as fundamental in its nature as any other. That is, it seems to me that one may array the sciences roughly linearly in a hierarchy, according to the idea: The elementary entities of science X obey the laws of science Y.

X	Y
solid state or many-body physics	elementary particle physics
chemistry	many-body physics
molecular biology	chemistry
cell biology	molecular biology
.	.
.	.
.	.
psychology	physiology
social sciences	psychology

But this hierarchy does not imply that science X is "just applied Y." At each stage entirely new laws, concepts, and generalizations are necessary, requiring inspiration and creativity to just as great a degree as in the previous one. Psychology is not applied biology, nor is biology applied chemistry.

In my own field of many-body physics, we are, perhaps, closer to our fundamental, intensive underpinnings than in any other science in which nontrivial complexities occur, and as a result we have begun to formulate a general theory of just how this shift from quantitative to qualitative differentiation takes place. This formulation, called the theory of "broken symmetry," may be of help in making more generally clear the breakdown of the constructionist converse of reductionism. I will give an elementary and incomplete explanation of these ideas, and then go on to some more general speculative comments about analogies at

---

The author is a member of the technical staff of the Bell Telephone Laboratories, Murray Hill, New Jersey 07974, and visiting professor of theoretical physics at Cavendish Laboratory, Cambridge, England. This article is an expanded version of a Regents' Lecture given in 1967 at the University of California, La Jolla.

other levels and about similar phenomena.

Before beginning this I wish to sort out two possible sources of misunderstanding. First, when I speak of scale change causing fundamental change I do not mean the rather well-understood idea that phenomena at a new scale may obey actually different fundamental laws—as, for example, general relativity is required on the cosmological scale and quantum mechanics on the atomic. I think it will be accepted that all ordinary matter obeys simple electrodynamics and quantum theory, and that really covers most of what I shall discuss. (As I said, we must all start with reductionism, which I fully accept.) A second source of confusion may be the fact that the concept of broken symmetry has been borrowed by the elementary particle physicists, but their use of the term is strictly an analogy, whether a deep or a specious one remaining to be understood.

Let me then start my discussion with an example on the simplest possible level, a natural one for me because I worked with it when I was a graduate student: the ammonia molecule. At that time everyone knew about ammonia and used it to calibrate his theory or his apparatus, and I was no exception. The chemists will tell you that ammonia "is" a triangular pyramid

$$N(-) \quad H(+) \quad \downarrow \mu$$
$$(+)H \quad H(+)$$

with the nitrogen negatively charged and the hydrogens positively charged, so that it has an electric dipole moment (μ), negative toward the apex of the pyramid. Now this seemed very strange to me, because I was just being taught that nothing has an electric dipole moment. The professor was really proving that no nucleus has a dipole moment, because he was teaching nuclear physics, but as his arguments were based on the symmetry of space and time they should have been correct in general.

I soon learned that, in fact, they were correct (or perhaps it would be more accurate to say not incorrect) because he had been careful to say that no stationary state of a system (that is, one which does not change in time) has an electric dipole moment. If ammonia starts out from the above unsymmetrical state, it will not stay in it very long. By means of quantum mechanical tunneling, the nitrogen can leak through the triangle of hydrogens to the other side, turning the pyramid inside out, and, in fact, it can do so very rapidly. This is the so-called "inversion," which occurs at a frequency of about 3×10^{10} per second. A truly stationary state can only be an equal superposition of the unsymmetrical pyramid and its inverse. That mixture does not have a dipole moment. (I warn the reader again that I am greatly oversimplifying and refer him to the textbooks for details.)

I will not go through the proof, but the result is that the state of the system, if it is to be stationary, must always have the same symmetry as the laws of motion which govern it. A reason may be put very simply: In quantum mechanics there is always a way, unless symmetry forbids, to get from one state to another. Thus, if we start from any one unsymmetrical state, the system will make transitions to others, so only by adding up all the possible unsymmetrical states in a symmetrical way can we get a stationary state. The symmetry involved in the case of ammonia is parity, the equivalence of left- and right-handed ways of looking at things. (The elementary particle experimentalists' discovery of certain violations of parity is not relevant to this question; those effects are too weak to affect ordinary matter.)

Having seen how the ammonia molecule satisfies our theorem that there is no dipole moment, we may look into other cases and, in particular, study progressively bigger systems to see whether the state and the symmetry are always related. There are other similar pyramidal molecules, made of heavier atoms. Hydrogen phosphide, PH_3, which is twice as heavy as ammonia, inverts, but at one-tenth the ammonia frequency. Phosphorus trifluoride, PF_3, in which the much heavier fluorine is substituted for hydrogen, is not observed to invert at a measurable rate, although theoretically one can be sure that a state prepared in one orientation would invert in a reasonable time.

We may then go on to more complicated molecules, such as sugar, with about 40 atoms. For these it no longer makes any sense to expect the molecule to invert itself. Every sugar molecule made by a living organism is spiral in the same sense, and they never invert, either by quantum mechanical tunneling or even under thermal agitation at normal temperatures. At this point we must forget about the possibility of inversion and ignore the parity symmetry: the symmetry laws have been, not repealed, but broken.

If, on the other hand, we synthesize our sugar molecules by a chemical reaction more or less in thermal equilibrium, we will find that there are not, on the average, more left- than right-handed ones or vice versa. In the absence of anything more complicated than a collection of free molecules, the symmetry laws are never broken, on the average. We needed living matter to produce an actual unsymmetry in the populations.

In really large, but still inanimate, aggregates of atoms, quite a different kind of broken symmetry can occur, again leading to a net dipole moment or to a net optical rotating power, or both. Many crystals have a net dipole moment in each elementary unit cell (pyroelectricity), and in some this moment can be reversed by an electric field (ferroelectricity). This asymmetry is a spontaneous effect of the crystal's seeking its lowest energy state. Of course, the state with the opposite moment also exists and has, by symmetry, just the same energy, but the system is so large that no thermal or quantum mechanical force can cause a conversion of one to the other in a finite time compared to, say, the age of the universe.

There are at least three inferences to be drawn from this. One is that symmetry is of great importance in physics. By symmetry we mean the existence of different viewpoints from which the system appears the same. It is only slightly overstating the case to say that physics is the study of symmetry. The first demonstration of the power of this idea may have been by Newton, who may have asked himself the question: What if the matter here in my hand obeys the same laws as that up in the sky—that is, what if space and matter are homogeneous and isotropic?

The second inference is that the internal structure of a piece of matter need not be symmetrical even if the total state of it is. I would challenge you to start from the fundamental laws of quantum mechanics and predict the ammonia inversion and its easily observable properties without going through the stage of using the unsymmetrical pyramidal structure, even though no "state" ever has that structure. It is fascinating that it was not until a couple of decades ago (2) that nuclear physicists stopped thinking of the nucleus as a featureless, symmetrical little ball and realized that while it really never has a dipole moment, it can become football-shaped or plate-shaped. This has observable consequences in the reactions and excitation spectra that are studied in nuclear physics, even though it is much more difficult to demonstrate directly than the ammonia inversion. In my opinion, whether or not one calls this intensive research, it is as fundamental in nature as many things one might so label. But it needed no new knowledge of fundamental laws and would have been extremely difficult to derive synthetically from those laws; it was simply an inspiration, based, to be sure, on everyday intuition, which suddenly fitted everything together.

The basic reason why this result would have been difficult to derive is an important one for our further thinking. If the nucleus is sufficiently small there is no real way to define its shape rigorously: Three or four or ten particles whirling about each other do not define a rotating "plate" or "football." It is only as the nucleus is considered to be a many-body system—in what is often called the $N \rightarrow \infty$ limit—that such behavior is rigorously definable. We say to ourselves: A macroscopic body of that shape would have such-and-such a spectrum of rotational and vibrational excitations, completely different in nature from those which would characterize a featureless system. When we see such a spectrum, even not so separated, and somewhat imperfect, we recognize that the nucleus is, after all, not macroscopic; it is merely approaching macroscopic behavior. Starting with the fundamental laws and a computer, we would have to do two impossible things —solve a problem with infinitely many bodies, and then apply the result to a finite system—before we synthesized this behavior.

A third insight is that the state of a really big system does not at all have to have the symmetry of the laws which govern it; in fact, it usually has less symmetry. The outstanding example of this is the crystal: Built from a substrate of atoms and space according to laws which express the perfect homogeneity of space, the crystal suddenly and unpredictably displays an entirely new and very beautiful symmetry. The general rule, however, even in the case of the crystal, is that the large system is less symmetrical than the underlying structure would suggest: Symmetrical as it is, a crystal is less symmetrical than perfect homogeneity.

Perhaps in the case of crystals this appears to be merely an exercise in confusion. The regularity of crystals could be deduced semiempirically in the mid-19th century without any complicated reasoning at all. But sometimes, as in the case of superconductivity, the new symmetry—now called broken symmetry because the original symmetry is no longer evident—may be of an entirely unexpected kind and extremely difficult to visualize. In the case of superconductivity, 30 years elapsed between the time when physicists were in possession of every fundamental law necessary for explaining it and the time when it was actually done.

The phenomenon of superconductivity is the most spectacular example of the broken symmetries which ordinary macroscopic bodies undergo, but it is of course not the only one. Antiferromagnets, ferroelectrics, liquid crystals, and matter in many other states obey a certain rather general scheme of rules and ideas, which some many-body theorists refer to under the general heading of broken symmetry. I shall not further discuss the history, but give a bibliography at the end of this article (3).

The essential idea is that in the so-called $N \rightarrow \infty$ limit of large systems (on our own, macroscopic scale) it is not only convenient but essential to realize that matter will undergo mathematically sharp, singular "phase transitions" to states in which the microscopic symmetries, and even the microscopic equations of motion, are in a sense violated. The symmetry leaves behind as its expression only certain characteristic behaviors, for instance, long-wavelength vibrations, of which the familiar example is sound waves; or the unusual macroscopic conduction phenomena of the superconductor; or, in a very deep analogy, the very rigidity of crystal lattices, and thus of most solid matter. There is, of course, no question of the system's really violating, as opposed to breaking, the symmetry of space and time, but because its parts find it energetically more favorable to maintain certain fixed relationships with each other, the symmetry allows only the body as a whole to respond to external forces.

This leads to a "rigidity," which is also an apt description of superconductivity and superfluidity in spite of their apparent "fluid" behavior. [In the former case, London noted this aspect very early (4).] Actually, for a hypothetical gaseous but intelligent citizen of Jupiter or of a hydrogen cloud somewhere in the galactic center, the properties of ordinary crystals might well be a more baffling and intriguing puzzle than those of superfluid helium.

I do not mean to give the impression that all is settled. For instance, I think there are still fascinating questions of principle about glasses and other amorphous phases, which may reveal even more complex types of behavior. Nevertheless, the role of this type of broken symmetry in the properties of inert but macroscopic material bodies is now understood, at least in principle. In this case we can see how the whole becomes not only more than but very different from the sum of its parts.

The next order of business logically is to ask whether an even more complete destruction of the fundamental symmetries of space and time is possible and whether new phenomena then arise, intrinsically different from the "simple" phase transition representing a condensation into a less symmetric state.

We have already excluded the apparently unsymmetric cases of liquids, gases, and glasses. (In any real sense they are more symmetric.) It seems to me that the next stage is to consider the system which is regular but contains information. That is, it is regular in space in some sense so that it can be "read out," but it contains elements which can be varied from one "cell" to the next. An obvious example is DNA; in everyday life, a line of type or a movie film have the same structure. This type of "information-bearing crystallinity" seems to be essential to life. Whether the development of life requires any further breaking of symmetry is by no means clear.

Keeping on with the attempt to characterize types of broken symmetry which occur in living things, I find that at least one further phenomenon seems to be identifiable and either universal or remarkably common, namely, ordering (regularity or periodicity) in the time dimension. A number of theories of life processes have appeared in which regular pulsing in time plays an important role: theories of development, of growth and growth limitation, and of the memory. Temporal regularity is very commonly observed in living objects. It plays at least two kinds of roles. First, most methods of extracting energy from the environment in order to set up a continuing, quasi-stable process involve time-periodic machines, such as oscillators and generators, and the processes of life work in the same way. Second, temporal regularity is a means of handling information, similar to information-bearing spatial regularity. Human spoken language is an example, and it is noteworthy that all computing machines use temporal pulsing. A possible third role is suggested in some of the theories mentioned above: the use of phase relationships of temporal pulses to handle information and control the growth and development of cells and organisms (5).

In some sense, structure—functional structure in a teleological sense, as opposed to mere crystalline shape—must also be considered a stage, possibly intermediate between crystallinity and information strings, in the hierarchy of broken symmetries.

To pile speculation on speculation, I would say that the next stage could be hierarchy or specialization of function, or both. At some point we have to stop talking about decreasing symmetry and start calling it increasing complication. Thus, with increasing complication at each stage, we go on up the hierarchy of the sciences. We expect to encounter fascinating and, I believe, very fundamental questions at each stage in fitting together less complicated pieces into the more complicated system and understanding the basically new types of behavior which can result.

There may well be no useful parallel to be drawn between the way in which complexity appears in the simplest cases of many-body theory and chemistry and the way it appears in the truly complex cultural and biological ones, except perhaps to say that, in general, the relationship between the system and its parts is intellectually a one-way street. Synthesis is expected to be all but impossible; analysis, on the other hand, may be not only possible but fruitful in all kinds of ways: Without an understanding of the broken symmetry in superconductivity, for instance, Josephson would probably not have discovered his effect. [Another name for the Josephson effect is "macroscopic quantum-interference phenomena": interference effects observed between macroscopic wave functions of electrons in superconductors, or of helium atoms in superfluid liquid helium. These phenomena have already enormously extended the accuracy of electromagnetic measurements, and can be expected to play a great role in future computers, among other possibilities, so that in the long run they may lead to some of the major technological achievements of this decade (6).] For another example, biology has certainly taken on a whole new aspect from the reduction of genetics to biochemistry and biophysics, which will have untold consequences. So it is not true, as a recent article would have it (7), that we each should "cultivate our own valley, and not attempt to build roads over the mountain ranges . . . between the sciences." Rather, we should recognize that such roads, while often the quickest shortcut to another part of our own science, are not visible from the viewpoint of one science alone.

The arrogance of the particle physicist and his intensive research may be behind us (the discoverer of the positron said "the rest is chemistry"), but we have yet to recover from that of some molecular biologists, who seem determined to try to reduce everything about the human organism to "only" chemistry, from the common cold and all mental disease to the religious instinct. Surely there are more levels of organization between human ethology and DNA than there are between DNA and quantum electrodynamics, and each level can require a whole new conceptual structure.

In closing, I offer two examples from economics of what I hope to have said. Marx said that quantitative differences become qualitative ones, but a dialogue in Paris in the 1920's sums it up even more clearly:

FITZGERALD: The rich are different from us.

HEMINGWAY: Yes, they have more money.

References

1. V. F. Weisskopf, in *Brookhaven Nat. Lab. Publ.* 888T360 (1965). Also see *Nuovo Cimento Suppl. Ser I* **4**, 465 (1966); *Phys. Today* **20** (No. 5), 23 (1967).
2. A. Bohr and B. R. Mottelson, *Kgl. Dan. Vidensk. Selsk. Mat. Fys. Medd.* **27**, 16 (1953).
3. Broken symmetry and phase transitions: L. D. Landau, *Phys. Z. Sowjetunion* **11**, 26, 542 (1937). Broken symmetry and collective motion, general: J. Goldstone, A. Salam, S. Weinberg, *Phys. Rev.* **127**, 965 (1962); P. W. Anderson, *Concepts in Solids* (Benjamin, New York, 1963), pp. 175–182; B. D. Josephson, thesis, Trinity College, Cambridge University (1962). Special cases: antiferromagnetism, P. W. Anderson, *Phys. Rev.* **86**, 694 (1952); superconductivity, ———, *ibid.* **110**, 827 (1958); *ibid.* **112**, 1900 (1958); Y. Nambu, *ibid.* **117**, 648 (1960).
4. F. London, *Superfluids* (Wiley, New York, 1950), vol. 1.
5. M. H. Cohen, *J. Theor. Biol.* **31**, 101 (1971).
6. J. Clarke, *Amer. J. Phys.* **38**, 1075 (1969); P. W. Anderson, *Phys. Today* **23** (No. 11), 23 (1970).
7. A. B. Pippard, *Reconciling Physics with Reality* (Cambridge Univ. Press, London, 1972).

Theory of Ferroelectric Behavior of Barium Titanate
Ceramic Age **57**, 29 (1951)

This is my first review paper — a forerunner of the "soft mode" paper of 1958. Bill Shockley hired me at Bell to work on his ideas on ferroelectricity and I did, but soon became more interested in generalities, feeling that calculations were premature. He was unhappy with my work, which seemed to ignore his ideas for detailed calculations; but he taught me a lot of solid state physics and the background and the experience were both valuable to me.

Theory of
Ferroelectric Behavior of Barium Titanate

I — Review of Experimental Information

By P. W. ANDERSON
Bell Telephone Laboratories
Murray Hill, N. J.

THE purpose of this paper is, first, to give a review of certain typical experimental information on ferroelectrics, particularly on barium titanate and its sister-structures, which have absorbed most of the recent effort in this field; and second, to present a rough idea of the present theoretical picture of the atomic mechanisms underlying the phenomenon of ferroelectricity, again concentrating on barium titanate. Since it does not seem necessary to attempt to give complete references in a paper of this character, particularly because a quite complete bibliography of the subject is available in Von Hippel's excellent review article[1], it should be made clear at the start that the writer is borrowing material from various sources.**

The first and most obvious question to be answered is: what is a ferroelectric, and why is it so called? The name does not come from any relation to iron, but from the very close analogy between the phenomena of ferroelectricity and the better-known ferromagnetism. By definition, ferroelectricity is the dielectric analogue of ferromagnetism[2].

Fig. 1 shows the most striking point of this analogy: The ferroelectric hysteresis loop in barium titanate (BaTiO₃). Here we are plotting the dielectric induction D versus the applied field E, as can be done on a cathode-ray oscilloscope with certain rather simple circuit arrangements. The full scale of D on these diagrams roughly is 2×10^7 volts/cm, while the full scale in E is only about 4×10^3 volts/cm. Thus the mean or "true" dielectric constant $\epsilon = \dfrac{D}{E}$, corresponding to the slope of the loop as a whole, is about 10^4.

On the other hand, the second diagram in Fig. 1, in which the effect of applying a small extra voltage at various points of the loop is indicated, shows that the effective reversible dielectric constant ϵ_{rev} for small fields is only about 1000, at all points of the loop, and that for small fields there is little or no hysteresis loss. To those who are familiar with ferromagnetism, this will indicate that small fields are causing little or no domain motion, since ϵ_{rev} also is about equal to the saturation dielectric constant, which represents the dielectric constant of a sample in which all domains are completely aligned, and cannot be realigned by the measuring field.

On the ferroelectric hysteresis loop, one can define quantities entirely analogous to those which characterize

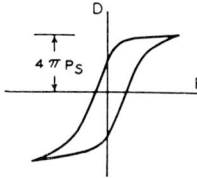

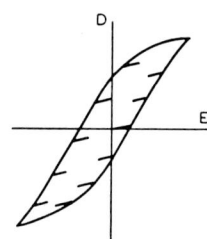

Fig. 1: Hysteresis loops in BaTiO₃.

magnetic loops: the coercive field E_c which is required to return the induction to zero, the remanent polarization $P_r = \dfrac{D_r}{4\pi}$ at zero applied field, and the saturation polarization, the polarization P_s of a hypothetical specimen with all domains parallel at no field. This latter quantity can be obtained by linearly extrapolating the saturation line at high fields back to the $E = 0$ axis: this is shown in Fig. 1. In BaTiO₃ the saturation polarization is larger than in the other classes of ferroelectrics, about 16×10^{-6} coul/cm².

The shapes of these hysteresis loops, as in ferromagnetism, are very sensitive to small amounts of impurity and even to the previous thermal and electrical history of a single specimen. Nevertheless, P_s and, usually, ϵ_{rev} roughly are constant for a given substance.

The second similarity between ferromagnetism and ferroelectricity lies in the Curie point phenomenon. Fig. 2 shows a plot of ϵ_{rev} versus the tem-

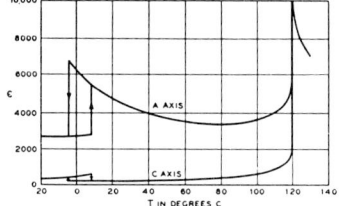

Fig. 2: Dielectric constant as a function of temperature for a single crystal of barium titanate.

perature for a single crystal of BaTiO₃. A very high peak occurs at the "Curie point" $T_c = 120°$; above this peak the substance behaves like a relatively normal dielectric of high dielectric constant, while below 120°, as shown in the Fig., there is complicated crystal-direction dependence for ϵ_{rev}. In this region we have also the field dependence of the dielectric properties indicated by the hysteresis loops of Fig. 1. Returning to the region above T_c, Fig. 3 shows the D versus E plot here. At temperatures only slightly higher than T_c, shown by the curve labelled $T_c + 2°$, the hysteresis loop has disappeared, flattening out to a single curve which, nonetheless, still shows some evidence of dielectric saturation in the curvature at high fields. At still higher temperatures the dielectric constant is lower, and the characteristic is perfectly linear as in an ordinary dielectric.

Another characteristic of the region above the Curie point which is strikingly similar to the behavior of ferromagnets is the Curie-Weiss law. This law states that the inverse of the sus-

---

* Presented at Tenth Symposium on Ceramic Dielectrics, School of Ceramics, Rutgers University, New Brunswick, N. J., November 17, 1950.
** The general rule followed in this paper is that if a development is covered in some detail by Von Hippel, no reference for such is noted.
[1] A. Von Hippel, Rev. Mod. Phys. 22, 221 (July 1950).
[2] Thus an electret, by this definition, is not a ferroelectric, since very few of the ferromagnetic properties have analogues in electrets.

(Reprinted from CERAMIC AGE, Newark, N. J., April, 1951)

ceptibility (magnetic or electric) is proportional to the temperature, i.e.

$$\frac{1}{\epsilon} = A + BT \quad (1)$$

and that actually the intercept at $\frac{1}{\epsilon} = 0$ is approximately T_c, i.e.

$$\epsilon \sim \frac{const}{-T-T_c} \quad (2)$$

The correctness of this law is demonstrated in Fig. 4. The small deviations near the Curie point are found also in ferromagnets, but are not quite the same in detail.

Thermally, the electric and magnetic Curie points are not quite so similar in structure. Although both show singularities in the specific heat at this point, the ferroelectric singularity is considerably smaller for the $BaTiO_3$ class than is the usual ferromagnetic specific heat singularity. Also, while most (not all) known ferromagnetic Curie points are thermodynamically classified as second-order, or λ-point, transitions, with no latent heat or discontinuity in volume or other "first order" properties, it is possible that the $BaTiO_3$ Curie point is a very slightly

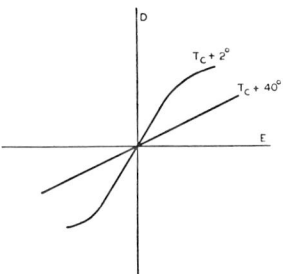

Fig. 3: D vs. E for $BaTiO_3$ above the Curie point.

first-order change with a true volume discontinuity. Further reference will be made later to some of these points.

A third striking experimental similarity between ferromagnetism and ferroelectricity lies in the domain structure. It is well-known that in order to explain the magnetic hysteresis loop Weiss and others postulated that a ferromagnetic specimen is divided into a system of regions, or *domains*, each polarized to saturation in a certain direction. At zero applied field, however, the magnetic domain polariza-

[3] For a review see C. Kittel, Rev. Mod. Phys. 21, 541 (1949).
[4] G. C. Danielson, unpublished.
W. Kaenzig, Phys. Rev. 80, 94 (1950).

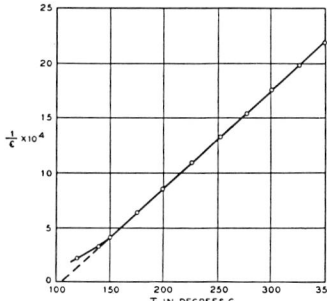

Fig. 4: 1_ϵ as a function of temperature in $BaTiO_3$.

tions add up to zero polarization for the specimen as a whole. Through beautiful theoretical and experimental work[3] the domains in ferromagnets have come in quite recent years to be understood and definitely "pinned down" theoretically, as well as made visible experimentally with the powder pattern technique. The opposite order is the rule in ferroelectrics: the domains are easily visible and, in some cases, have been the first observed evidence of ferro-electricity: it has been easier to identify a ferroelectric optically by looking for domains than electrically by finding a hysteresis loop — although the latter, it must be said, is a far surer test, since domain-like twinning is found to be surprisingly common.

To explain why the domains are so easily visible we must first go into the crystallography of $BaTiO_3$. Above the Curie point, this substance crystallizes in the cubic perovskite structure, as shown in Fig. 5. One may think of the large barium and oxygen ions as lying on the corners and face-centers of a cube, respectively, together making up a face-centered structure; the small titanium ions are then at the cube centers, surrounded by an octahedron of oxygen ions. At T_c and below this structure is found, by x-rays, to deform into a tetragonal unit cell, with one of the three cubic axes (arbitrarily of course) lengthening, the oth-

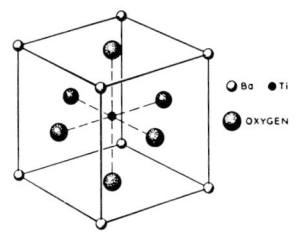

Fig. 5: Crystal structure of $BaTiO_3$.

er two shortening, to become the c and a axes respectively of the tetragonal structure. The ratio c to a is 1.01 at room temperature.

Recent x-ray results indicate that in addition to this general deformation the relative ionic positions shift slightly, the Ti ions moving .044 Å in the direction of the c-axis, the O ions moving very slightly the other way[4]. This is confirmed by the fact that the spontaneous polarization P_s has been found, by experiments on single crystals, to lie along the c-axis. It is fairly certain that the positively-charged Ti ions move in the direction of P_s and the negative O ions in the opposite direction.

One can think of the tetragonal deformation as an electrostrictive effect.

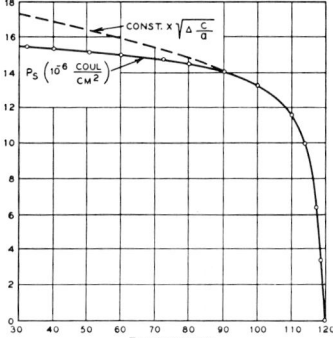

Fig. 6: Proportionality of strain and P_s^2.

It is found that an "electrostrictive" law

$$\Delta \left(\frac{c}{a} \right) = const \times P^2 \quad (3)$$

is obeyed if we insert for P the saturation polarization P_s. The same law, with the same constant, is found to hold *above* the Curie point for polarizations due to *applied* fields. There, of course, the effect is true electrostriction. Fig. 6 shows how equation (3) is obeyed. The deviations at lower temperatures are due not to the failure of the law but to inability to properly saturate the crystal when measuring P_s, even in the so-called "single-domain" crystals. The cubic magnetic materials also have this same slight tetragonal distortion, although in these the effect always is thought of as magnetostriction superposed upon a basically cubic structure, while the $BaTiO_3$ effect is so large that calling it electrostriction is rather a stretch of the imagination.

Because of this tetragonality a polarized piece of $BaTiO_3$ is optically anisotropic, with different indices of refraction in the two different directions; with a polarizing microscope, then, it is seen easily that a typical crystal of $BaTiO_3$ is subdivided into many domains, polarized in various directions, just as the existence of the hysteresis loop would lead one to suspect. A typical domain pattern is seen in Fig. 7.

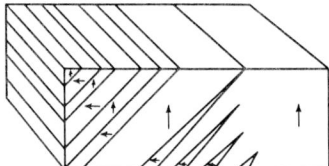

Fig. 7: A typical domain pattern in $BaTiO_3$.

Precisely as in ferromagnetism, it is found always that the component of polarization perpendicular to a domain wall is the same on each side of the wall, so that no free charge need form in the wall to compensate the surface charge density which would otherwise appear. Such a charge density would have a high electrostatic self-energy.

Two more lines of experimental work, not directly connected with the analogy of ferromagnetism and ferroelectricity the writer has been developing, should be mentioned. The first is the search for more ferroelectric materials. It is found that many compounds with structures similar to $BaTiO_3$ — the "perovskite structure" — are also ferroelectric.[5] Thus $BaTiO_3$ is not an isolated phenomenon as are Rochelle salt and KDP, the other two ferroelectric substances known; this, with its relatively simple structure, contributes to the interest in work on $BaTiO_3$.

Because of the practical importance of this subject, I should also say something about the piezoelectric effect in $BaTiO_3$. Some confusion has arisen as to whether this effect should be called electrostriction or a true piezo-effect. In the precise sense of the words, below the Curie point the strain is proportional to applied field in each single domain of a $BaTiO_3$ crystal, and thus we have to do with a pure piezo-effect; however, in two ways the phenomenon is related closely to electrostriction. First, the actual piezo-electric constant is readily derived from the electrostrictive equation (3) which we have already mentioned, as follows:

[5] Matthias, Phys. Rev. **76**, 175, 430, 1886, (1949)
Shirane et. al., Phys. Rev. **80**, 485, (1950)
Smolenskii, J. Tech. Phys. USSR **20**, 137, (1950)
[6] P. Weiss, Phys. Zeits. **9**, 358 (1908).

$$\text{strain} = \Delta\left(\frac{c}{a}\right) = \text{const} \times P^2 \quad (4)$$

$$\frac{d}{dE}\left(\Delta\left(\frac{c}{a}\right)\right) = \text{const} \times 2P\frac{dP}{dE}$$

Thus the change of strain with field involves the constant of electrostriction, the saturation polarization, and the dielectric susceptibility $\frac{dP}{dE}$; since the latter two quantities are large, the result is a large and useful effect.

In another sense, in an actual unpolarized crystal or ceramic the effect of a field is not at first to give a linear piezo-effect, because the deformation of domains polarized in one direction is cancelled completely by that of domains polarized in opposite directions. Thus, it is necessary to "pole" a specimen for piezo-electric work by the application of high fields, and accordingly to leave the specimen in an at least partially polarized state.

II — Review of the Theory of Ferroelectricity

In commencing the theoretical part of this paper, one can remark that the main task of the theoretician has been to try to de-emphasize the close analogy of ferromagnetism and ferroelectricity, at least as far as the actual basic mechanism is concerned.

The ideas involved in the present theoretical picture of the mechanism of ferroelectricity are not new and not complicated. We must go back to 1908, the year in which P. Weiss first proposed an acceptable, if somewhat ad hoc, theory of ferromagnetism.[6] In his proposal he cited the familiar formula with which the names of Lorentz, Lorenz, Clausius, Mosotti, Debye, and others have been connected in its various aspects:

$$F = \begin{cases} E + \dfrac{4\pi P}{3} & \text{(electric case)} \\[6pt] H + \dfrac{4\pi M}{3} & \text{(magnetic case)} \end{cases} \quad (5)$$

In this formula, F is the "internal" field, the field which actually acts to polarize the individual molecule inside the material one is discussing. The external applied field is E or H, the polarization of the substance per unit volume P or M, and the important point of the formula is that it *does* contain a contribution of P or M.

The derivation of this formula is a matter of simple electro- or magnetostatics. It can be best understood by referring to Fig. 8. We take the particular molecule on which we are trying to compute the acting field, and surround it with an imaginary sphere of macroscopic size. If the symmetry of the substance under consideration is as high as cubic (this includes isotropy, as in liquids) the field due to all the molecules inside this sphere can be shown rigorously to cancel out, if the polarization inside the sphere is uniform. The problem of finding the field due to all charges *outside* the sphere, then, is simply one of ordinary macroscopic electrostatics; it involves the fields of: (a) the surface charge on the sphere left by cutting it out; (b) the surface charges on the dielectric surfaces ((a) and (b) involve P); and (c) the charge on the condenser plates. The result of such a calculation is the Lorentz formula (5).

The term $\dfrac{4\pi}{3} P$ is, as von Hippel has called it, a "bootstraps" term, i.e. a term by which the polarization is capable of pulling itself up by its own bootstraps. In more conventional lan-

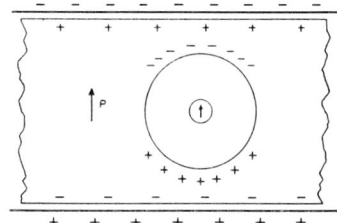

Fig. 8: Internal field in a dielectric.

guage, this is a "coöperative force" term. To show how this term acts, let us compute the polarization of a molecule in the dielectric under an external field:

$$P_{mol} = \alpha F \quad (6)$$

This equation defines the polarizability α of a molecule. Then let us find the total polarization of a cc of the substance containing n molecules, and finally insert for F equation (5):

$$P_{tot} = n\alpha F = n\alpha\left(E + \frac{4\pi P_{tot}}{3}\right)$$

or,

$$\chi = \frac{P}{E} = \frac{n\alpha}{1 - \dfrac{4\pi n\alpha}{3}} \quad (7)$$

The susceptibility χ is connected with the dielectric constant by

$$\epsilon = 1 + 4\pi\chi \qquad (8)$$

Now it is clear that without the "bootstraps" term we get the "naive" or Drude formula for susceptibility which is normally used in gases:

$$\chi_{F=E} = n\alpha \qquad (9)$$

Thus the $\dfrac{4\pi}{3}$ term has increased the susceptibility and the dielectric constant. In fact, if we imagine that $\dfrac{4\pi}{3} n\alpha$ is allowed to increase toward the value unity, we see that χ and ϵ increase without limit, and a little thought will convince one that if $\dfrac{4\pi}{3} n\alpha$ becomes greater than 1, the polarization will become infinite without the application of any external field at all. However, of course equation (6) breaks down when F and P become too large, and in such a region we will have merely a finite spontaneous polarization, just as is observed in ferroelectricity and ferromagnetism. This is the phenomenon which has been called the "$\dfrac{4\pi}{3}$ catastrophe".

In the case of magnetism, the actual fact is that all known substances at normal temperatures have $\dfrac{4\pi}{3} n\alpha \ll 1$; thus it was necessary for Weiss[6] to postulate boldly an additional cooperative force term N:

$$F = H + \left(\dfrac{4\pi}{3} + N\right) M$$

and N had to be taken of order of magnitude 10^4. This completely unsubstantiated hypothesis was remarkably successful, when combined with the domain idea, in explaining ferromagnetism in considerable detail, since the precise behavior of α, and of the high field equation corresponding to (6), was being discovered by Langevin and others at about the same time.

---

[7] W. Heisenberg, Zeits. fur Phys. **49**, 619 (1928).

[8] The best discussion of these general concepts is given by H. Fröhlich, **Theory of Dielectrics** (Oxford University Press, London, 1949). Also see J. Pirenne, Helv. Phys. Acta, **22**, 479 (1949).

[9] J. H. Van Vleck, J. Chem. Phys. **5**, 320 (1937).

[10] J. Pirenne, Physica **15**, 1019 (1949).

The explanation of the factor N, however, had to wait 20 years for Heisenberg, who in 1928 showed that a roughly similar force resulted from the quantum-mechanical exchange effect.[7]

In dielectric substances, on the other hand, there is no such limitation on $n\alpha$, as the existence of numerous substances with dielectric constants greater than 10 will testify. We see here the first major point of difference between ferroelectricity and ferromagnetism; no such esoteric cooperative force as the exchange force is necessary in the dielectric case, since the simple electrostatic "Lorentz force" seems always to be quite adequate.

On the other hand, in the dielectric case the shoe is almost too much on the other foot. Extrapolation from gas polarizabilities for dipolar substances by the Curie law led, for many polar liquids, to values of the polarizability which indicated that $n\alpha$ should be considerably greater than the ferroelectric value. That is why Van Vleck named the problem the "$\dfrac{4\pi}{3}$ catastrophe":

the theory seems to lead to the prediction, for example, that water at 100°C should be ferroelectric, and presumably therefore frozen or at least of considerably different physical character from the water we know.

The theoreticians were rescued from their theoretical catastrophe in the 1930's when Onsager showed that for polar liquids, at least, the original Lorentz field equation (5) is incorrect. The reason for this is that while on the *average* the polarization of the sphere is uniform, actually at any instant the permanent dipole of the central molecule affects the distribution in the sphere, leading to a situation in which the field due to the interior of the sphere does not cancel out when the order of taking averages is correctly handled. He concluded from his formula that the "catastrophe" could not occur at all for permanent dipoles; this may not be quite correct, but it is probable that the Curie temperatures would be so low that other interactions than the electrostatic ones would already have taken over the situation, and one should more properly speak of phase or order-disorder transitions, such as occur in the hydrogen halides, than of Lorentz catastrophes.[8] Van Vleck[9] has shown, however, that the problem of the Onsager inner field does not arise at all in the case of solids with "harmonic" or induced polarization, i.e. in substances in which there are no permanent dipoles, but only charges elastically bound to equilibrium positions. This demonstrates to us the second important difference between ferromagnetism and ferroelectricity: ferromagnetism always involves permanent magnetic dipoles (the field factor N is not of magnetostatic origin, and thus is not affected by the Onsager calculation) while ferroelectricity probably cannot involve permanent electric dipoles. It is possible that ferroelectricity can occur only in materials with large polarizabilities of the induced type.

Rochelle salt and KDP have long been thought to be dipolar-type ferroelectrics; however, Pirenne[10] has recently shown that the isotope effect in the latter substance cannot be explained on the dipolar hypothesis, and that probably KDP is at least an intermediate, more complicated case than either the harmonic (induced) or the dipolar.

A model of $BaTiO_3$ has been proposed which frequently is called the "rattling titanium" hypothesis. It is assumed that the titanium ion "rattles" among six equivalent equilibrium positions, one near to each of the oxygens

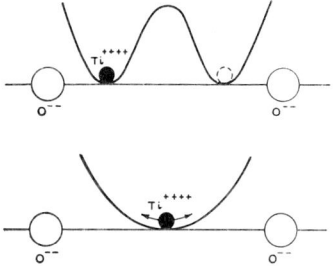

Fig. 9: Models for $BaTiO_3$.

in the octahedron. A schematic diagram of this situation is given in the first half of Fig. 9. This model leads to quite a high polarizability for the Ti ion, which as we shall see later is an unnecessary frill; on the other hand, so far as Onsager field effects are concerned, the "rattling" ion is simply equivalent to a free rotating dipole, which as we have seen is not likely to exhibit ferroelectricity. Most theoreticians interested in ferroelectricity now feel that while the "rattling" model is a convenient visualization aid, as used for instance in Von Hippel's article,[1] the second diagram, representing a pure "induced polarization" model, is far more likely to be correct.

Another aspect of ferroelectricity in the $BaTiO_3$ class which shows the superiority of the "induced" model is demonstrated by a quantitative consideration of the Curie-Weiss law above T_c. As we saw in equation (2) and Fig. 4, the susceptibility above the Curie point is given approximately by

$$\chi = \frac{K}{T-T_c} \quad (10)$$

Now above the Curie point $\frac{4\pi}{3} n\alpha$ is less than one, so that equation (7) holds:

$$\chi = \frac{n\alpha}{1 - \frac{4\pi}{3} n\alpha} \quad (7)$$

In all of the permanent dipole-like cases (square well, rattling titanium, or true rotating dipole) one and only one temperature dependence of the polarizability is to be expected, that given by the familiar dipolar "Curie law":

$$n\alpha = \frac{C}{T} \quad (11)$$

This can be inserted into equation (7), and it is found indeed that an expression of the form (10) results, as follows:

$$\chi = \frac{\frac{C}{T}}{1 - \frac{4\pi}{3}\frac{C}{T}} = \frac{C}{T - \frac{4\pi}{3}C}$$

but then

$$\frac{4\pi}{3} C = T_c$$

and we have, substituting for C,

$$\chi = \frac{\frac{3}{4\pi} T_c}{T - T_c} \quad (12)$$

It is a simple matter to test this equation in the case of $BaTiO_3$; T_c is known to be about 400° Kelvin, which gives us

$$K = \frac{3}{4\pi} T_c \simeq 100$$

---

[11] This discussion is similar to that given by Slater, Phys. Rev. 78, 748 (1950), but, as Slater points out, very similar conclusions were reached by Devonshire, Phil. Mag. 40, 1040 (1949), Ginsburg, J. Exp. Th. Phys. USSR, 12, 239 (1948), and Richardson and myself (unpublished). The two latter were more concerned with the size of the specific heat anomaly than with the Curie law; the two phenomena are, however, easily shown to be closely related by a simple thermodynamic argument, and in fact the specific heat anomaly is several orders of magnitude too small to be explained by a dipolar mechanism.

However, the measured value of the numerator of (10) is larger than 10,000. This means that the dipolar type of model is in error numerically by a factor of more than two orders of magnitude.[11]

On the other hand, the induced polarization model is capable of giving as large a value of K in (10) as we like. The reason for this is that the temperature coefficient of induced polarizability is generally quite small, of the order of $10^{-5}—10^{-6}$ per degree (the same order as coefficients of volume expansion); actually, this coefficient is due to the combination of the effects of change of n with temperature with those of a slight change in shape of the potential troughs in which the ions find themselves as the lattice changes size and the thermal agitation changes. Both of these effects are of the order of magnitude we have mentioned. Then if we postulate a small negative temperature coefficient δ for $n\alpha$, we get

$$n\alpha = (n\alpha)_{T=T_c}(1 - \delta(T-T_c))$$
$$= \frac{3}{4\pi}(1 - \delta(T-T_c)) \quad (13)$$

This, substituted into 7, gives

$$\chi = \frac{\frac{3}{4\pi}}{1-(1-\delta(T-T_c))} = \frac{\frac{3}{4\pi\delta}}{T-T_c} \quad (14)$$

Thus

$$K = \frac{3}{4\pi\delta} \quad (15)$$

and this tells us that by a choice of δ which is quite consistent with normal magnitudes we may easily explain the value of K which is observed. Thus we have a second very sound and convincing argument that the polarization in $BaTiO_3$ is of induced type.

We may now say a little about what this kind of thinking gives us in terms of a picture for $BaTiO_3$. In the first place, at optical frequencies this is a fairly normal material, having the slightly high index of refraction 2.4. The application of equation (7) to $n^2 = \epsilon_{optical}$ gives us

$$\frac{4\pi}{3} n\alpha_{optical} = .60 \quad (16)$$

This part of the polarizability, is, as is well-known, due entirely to the motions of the electrons, since the ionic motions cannot follow optical frequencies. Thus it is only necessary for us

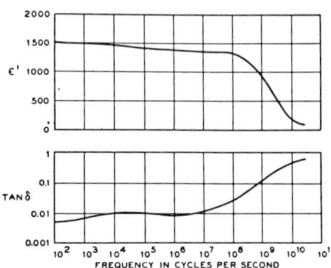

Fig. 10: Dispersion and loss in $BaTiO_3$ as a function of frequency.

to find an induced type of polarizability, coming from ionic motions, which is large enough so that added to (16) it gives 1:

$$\frac{4\pi}{3} n(\alpha_{optical} + \alpha_{ionic\ motion}) \geq 1 \quad (17)$$

Such polarizabilities, proceeding entirely from the harmonic binding of ions (we think particularly of the motion of the highly-charged titanium ion), are not at all surprising; LiF, for instance, has a value as high as this. Actually, recent computations by Slater[11] have shown that even this amount of α_{ionic} is not required, due to a peculiarity of the Lorentz field in the perovskite structure. (The original Lorentz calculation does not hold quite exactly for the O^{--} ions, but instead a larger Lorentz field is to be expected.)

A number of rather fantastic explanations have been suggested to explain the "surprising" phenomenon of ferroelectricity. As we have seen, it is not necessary to go to any great lengths to understand this phenomenon; what has always seemed to me a little surprising is that there are not *more* known ferroelectrics, since a combination of circumstances in which (17) is obeyed seems not to be a priori unusual. Perhaps, some have suggested, many well-known substances are ferroelectric, but do not have Curie points at measurable temperatures and have such large coercive fields that the polarization is to all intents and purposes frozen in and thus completely unobservable.

One final point on the two models of ferroelectricity, in relation to some recent experimental evidence, should be made. Many investigators have measured the frequency response of the dielectric constant of $BaTiO_3$ at room temperature, and all have found more or less the behavior of Fig. 10.[1] At first glance it appears that here is a typical relaxation spectrum, indicating that some kind of potential barrier must be surmounted in order for the dielectric polarization to change; the polarizable entities thus cannot follow rapid motion of the field. This seems at first a very strong argument for the "rattling titanium" model of Fig. 9, since this model provides an immediate visualization for this potential barrier. In fact, a reasonable size for the barrier between Ti positions gives a very good value for the relaxation frequency. However, recent measurements by Powles and Jackson[12] in England have thrown considerable doubt on this interpretation. These workers have measured the losses as a function of temperature, at frequencies near the relaxation frequency. Some of their results are shown in Fig. 11. We see that the losses drop off very sharply above the Curie point; other work, on (Ba-Sr)TiO_3 mixtures, indicates that even the remaining losses seem to be illusory, and thus that $BaTiO_3$ is essentially a lossless material above the Curie point.

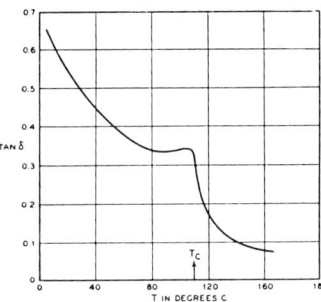

Fig. 11: High-frequency loss in $BaTiO_3$ as a function of temperature.

[12] Proc. Inst. Elec. Eng. part 3, p. 383 (1949).

This behavior cannot be explained on the relaxation idea; a relaxation frequency such as is given by the "rattling Ti" model cannot disappear so abruptly simply because of the disappearance of spontaneous polarization. The induced polarization model correctly predicts that there should be no large losses at any frequency lower than the fundamental infrared frequency, which should appear at a few tenths of a millimeter wavelength. The observed losses below the Curie point have been ascribed to two possible phenomena: (1) Slater's suggestion (made in informal discussion) is that these losses are acoustic radiation losses from the small particles of the ceramics upon which all these measurements have been made. This suggestion can be checked by making measurements at high frequencies on single crystals. (2) The second possibility is that the losses have something to do with domain effects; for instance, they may be domain wall motion relaxation such as is observed in some ferromagnets. This may be wrong in $BaTiO_3$, but is appearing more and more likely in some of the newer materials, in which something very like small-signal hysteresis losses seem to be present. Here also there are indications, however, that the losses are *not* ordinary relaxation losses.

In this second, theoretical, part of this paper I have tried to present the idea that $BaTiO_3$ and its sister materials are simply victims of the long-predicted and rather well-understood Lorentz "$\frac{4\pi}{3}$ catastrophe." The atomic mechanism leading to the ferroelectric phenomena in these crystals is thus accidental, in the sense that quantitative prediction is rather hopeless, but certainly not surprising. The electronic and ionic polarizability, added together, pass the "critical value" at the Curie point. The results of this occurrence are the various experimentally observed phenomena discussed in the first portion of the paper.

Use of Stochastic Methods in Line-Broadening Problems
Kyoto Lectures, March 1954

These lectures, given at the very young Yukawa Institute as a final thanks to Japan, for my Fulbright visit in 1953–54, are a summary of my approach to line-broadening problems, the field of my thesis and of much of my early work at Bell. During my stay with Kubo in Tokyo he began to go on with these ideas to develop the correlation function methods which became so important in condensed matter theory. (As did Lax at Bell, but not with so much effect.) My paper with Weiss in *Rev. Mod. Phys.*, in spite of its appearance in a review journal, was not a review. Several reviews appeared on my approach to pressure broadening but none were by me.

USE OF STOCHASTIC METHODS IN LINE-BROADENING PROBLEMS

P. W. ANDERSON

In recent years there has been a revival of interest in the old subject of the breadths of spectral lines. This revival is the result of the experimental opening up of the radio-frequency range of spectra: microwave gas spectroscopy, microwave magnetic resonance (magnetic resonance of electronic magnetic moments) and nuclear radio-frequency magnetic resonance. The radio-frequency range has the great advantage that monochromatic sources are available, so that the contour of a spectral line can be studied in complete detail insofar as the absorption of energy is greater than the background and noise effects.

You are all familiar with the basic problem of line-breadth theory. An isolated quantum-mechanical system — atom or molecule — always has discrete energy levels and radiates or absorbs discrete frequencies. In any macroscopic sample the spectrum will, however, not be discrete but essentially continuous, because the atoms or molecules interact. Observed line breadths contain this effect, which will be the subject of these lectures; they also contain at least three other effects: (1) The breadth due to the interaction with the radiation field, and (2) The instrumental breadths due to slit widths or noise-modulation of the source; these can be essentially completely removed in the radio range. (3) The Doppler breadth due to thermal motions of atoms: in gases this can be overwhelmed by increasing the density, in solids it is negligible.

There is then a new opportunity to study the interaction of atoms by quantitative comparison of theories with the wealth of new experiments. Naturally, the existing theories were not completely adequate and there has been a development of new theoretical methods; it is with a group of these methods, as well as with some of the older ones, that I wish to deal today. These methods all have in common the fact that they approach the problem as a problem in stochastic theory, that is, the theory of functions which vary in a random fashion in time. The interest in these problems from the standpoint of stochastic theory lies in the fact that they represent a rather unusual application of this theory which goes beyond the more normal type of problem. In most physical applications of stochastic theory, what is desired is the distribution of the function in question, given some kind of information about the process; for instance, consider the study of resistances or mobilities.

The question is what the final distribution of velocities is, given the transition probabilities; however, in the line-broadening problem, which is in its essence equivalent to the problem of relaxation, one must in addition find the distribution as a function of time in some way.

This will be much clearer expressed mathematically. What one might consider the most fundamental theorem of the line-broadening theory is the Fourier integral theorem. Suppose that we are studying magnetic or electric dipole radiation from some macroscopic quantum-mechanical system, whose Hamiltonian is H. We can define an operator $\mu(t)$ which is the Heisenberg representation of the total dipole moment operator of the system. The Fourier integral theorem states that the spectrum of the absorption or emission by this dipole moment is

$$I(\omega) = \text{const.} \times \text{Tr}\left\{\rho \left|\int_{-\infty}^{\infty} e^{-i\omega t}\mu(t)dt\right|^2\right\}. \tag{1}$$

ρ is the equilibrium density matrix.

This expression will appear self-evident to most of you, practically being the definition of the spectrum. However, since a simple proof can be given I shall give it: Label the eigenvalues of the macroscopic system E_i, E_j, etc. Now we all know that

$$I(\omega) = \text{const.} \times \sum_{E_j, E_i(E_i-E_j=\hbar\omega)} e^{-E_i/k_T} |(i|\mu|j)|^2.$$

The limitation on the sum may be removed by defining

$$(i|\mu(t)|j) = e^{\frac{i(E_i-E_j)}{\hbar}t}(i|\mu|j)$$

and selecting the appropriate frequency by using the δ-function expansion

$$\delta\left(\omega - \frac{E_i - E_j}{\hbar}\right) = \int_{-\infty}^{\infty} e^{i\frac{E_i-E_j}{\hbar}t - i\omega t} dt$$

so that

$$I(\omega) = \sum_{E_i, E_j} e^{-E_i/k_T} \int e^{-i\omega t}(i|\mu(t)|j)dt \int e^{i\omega t}(j|\mu^*(t)|i)dt$$

which, re-expressed as a trace, is exactly the above expression, since $\mathcal{H}\mu(t) - \mu(t)\mathcal{H} = i\hbar\dot{\mu}$ is the definition of $\mu(t)$ and leads to the above expression.

The above is still not obviously a problem in stochastic theory. Hoever, it becomes so when one physical assumption common to all the problems of which I speak is made. This is the assumption: that

$$\mathcal{H} = \mathcal{H}_0 + \mathcal{H}_i + \mathcal{H}_m \tag{2}$$

with the following conditions

(a) $$\mathcal{H}_m \gg \mathcal{H}_i \ ,$$

(b) $$[\mathcal{H}_m, \mathcal{H}_0] = 0 \ , \qquad (3)$$

(c) $$[\mathcal{H}_m, \mu] = 0 \ .$$

The meaning of the three parts is the following.

(a) $\mathcal{H}_0$ is the unperturbed Hamiltonian of the isolated, static atoms or molecules. In the gaseous pressure-broadening problem, it includes the internal degrees of freedom of the molecules. In the magnetic resonance problem it includes $-\mu \cdot H_0$, the Zeeman energy, as well as any of the fine and hyperfine structure terms we wish to include.

(b) $\mathcal{H}_i$ is that part of the intermolecular or interatomic interactions which is effective in broadening lines. Usually this is a relatively small part: for instance, in magnetic resonance only magnetic forces can broaden lines, while ordinary exhange repulsion, valence attraction, van der Waals' forces, etc., have no effect. In pressure broadening only *anisotropic* interactions can broaden microwave lines since these are only of the rotational type. In optical spectra there are also often special circumstances which allow the assumption of $\mathcal{H}_i$ to be small.

Finally, (c) $\mathcal{H}_m$ is the remainder of the Hamiltonian. Always it contains the translational kinetic energy of the atoms and molecules; also, usually the major portion of the interatomic interaction. The choice is made strictly on the basis of the problem at hand.

Now! Why does $\mathcal{H}_m$ commute with $\mathcal{H}_0$ and μ? The answer is different for the two cases. In pressure broadening, the degrees of freedom in $\mathcal{H}_0$ are internal ones, while $\mathcal{H}_m$ is the mutual translational motion as well as that major part of the interaction which does commute with μ and $\mathcal{H}_0$ (as most of the isotropic interactions will, for other than electronic transitions).

In magnetic resonance, $\mathcal{H}_m$ is entirely non-magnetic in character, $\mathcal{H}_0$ entirely magnetic, and the theorem is obvious.

There is one special case which we shall consider later: the famous case of "exchange narrowing" in which $\mathcal{H}_m$ is the energy of Heisenberg exchange interactions. I hope you will let us treat this when we come to it.

Finally, as to $\mathcal{H}_m \gg \mathcal{H}_i$. This may be justified on a purely experimental basis. Namely, since $\mathcal{H}_m$ is the total interaction energy it must have the value at least kT per degree of freedom or roughly so, at least when it can be treated classically as in most of our applications. Thus we may think of $\mathcal{H}_m$/atom as of order kT. But observed spectral line breadths are seldom more than 1 cm^{-1} or at worst a few cm^{-1}. The reason is easy to see: anything broader very usually is too broad and weak to be visible, and in any case in the radio range it is no longer a "line" but a band. But the interaction energies of interest are certainly of the order of the line breadth or smaller and not much larger. Thus $\mathcal{H}_m \gg H_i$ if T is larger than a few degrees.

This makes it possible now to discuss the problem in stochastic terms. First, we can see that $\mathcal{H}_0$ is easily disposed of. $\mathcal{H}_0$ can always be easily diagonalized — one must do this in order to know what are the spectral lines under discussion. Let its diagonal energies be E_j^0, E_k^0, etc., their differences are $\hbar\omega_{jk}^0$. Naturally E_j^0, E_k^0 are always highly degenerate, but they are distinct, discrete energies. We perform the canonical transformation on all operators $\mu, \mathcal{H}_i, \mathcal{H}_m$

$$(A(t))_{jk} = A'(t)_{jk} e^{i\omega_{jk} t} \tag{4}$$

and the time-dependence of the primed quantities due to $\mathcal{H}_0$ is eliminated;

$$i\hbar A' = [(\mathcal{H}'_i + \mathcal{H}'_m), A'] \tag{5}$$

while the spectrum becomes

$$I(\omega) = \text{const.} \times \text{Tr} \sum_{j,k} e^{E_i/kT} \int dt \int dt' e^{i(\omega-\omega_{jk})(t-t')} \rho'_j \mu'_{jk(t)} \mu^{*'}_{jk(t')}$$

$$= \sum_{j,k} I_{jk}(\omega) . \tag{6}$$

It is assumed that we have found a way to define $\mathcal{H}_0$ so that its eigenvalues are well separated, no matter how degenerate they may be — a very careful discussion of the handling of this separation and what to do if it fails would be out of place here but is contained in a paper by Kubo and Tomita for those who are dubious (also in my paper on pressure broadening for that case). ρ and μ are still operators but now their dimensionality is reduced to operating within subspaces which are split apart by the various eigenvalues of $\mathcal{H}_0$. This process of separation into lines is sometimes difficult but almost never impossible. We now have

$$I_{jk}(\omega) = \text{Tr} \, \rho_j \int dt \int dt' \mu'_{jk} \mu'^{*}_{jk} e^{i(\omega-\omega_{jk})(t-t')}$$

$$= \int e^{i(\omega-\omega_{jk}^0)\tau} \varphi_{jk}(\tau) d\tau \tag{7}$$

where

$$\varphi_{jk}(\tau) = \langle \text{Tr} \mu'_{jk}(0) \mu^{*'}_{jk}(\tau) \rangle_{\text{avg over time and statistics}} \tag{8}$$

is the well known correlation function.

Now we are well on the way. Since μ and $\mathcal{H}_m$ commute,

$$i\hbar\mu = [\mathcal{H}'_i(t), \mu] \tag{9}$$

while

$$i\hbar\mathcal{H}'_i = [\mathcal{H}'_m(t), \mathcal{H}'_i(t)] . \tag{10}$$

We now make the stochastic assumption: that $\mathcal{H}_m$, which essentially represents the thermal motion of the atoms as a whole, develops in time in an entirely random way relative to the interactions $\mathcal{H}'_i$ and the unperturbed motions $\mathcal{H}_0$. Thus the time-behavior of $\mathcal{H}'_i$ and of μ' in turn is random in the sense that it is controlled by the random thermal motion $\mathcal{H}'_m$.

A real justification of the randomness assumption for $\mathcal{H}_m$ would have to go deep into the fundamentals of statistical mechanics. Here let us say only that it is the identical assumption of randomness which is made in theories of diffusion, conductivity, etc., in calculating transition probabilities.

An interesting point is that we expect our approach to fail when $\mathcal{H}_i \sim \mathcal{H}_m$. This failure is identical with the failure of conductivity theory which leads to superconductivity: when the interactions are stronger than the thermal reservoir, normal irreversible theory breaks down and completely new methods are always required. No known line-breadth theory is valid in the case $h\Delta\nu > kT$. (Just as conductivity theory breaks down for $h/\tau \geq kT$.)

The problem then is: under the assumption that random thermal motion controls the time-dependence of $\mathcal{H}_i$, find the spectrum as defined above. I think that at this point it is appropriate to begin to specialize to actual cases. First I should like to discuss the pressure-broadening case, then later the magnetic resonance case.

The virtue of the pressure-broadening problem is that for a sufficiently rarified gas a nearly exact solution can be found in principle, satisfactory in most cases. Only one further assumption need be made, that the translational motion of the molecules is classical. This assumption is nearly always quite satisfactory, as may be easily verified from the uncertainty principle. One can easily show that wave-packets can be formed which have a negligible extent in both momentum and position. The effect of $\mathcal{H}_m$ on $\mathcal{H}_i$ is then easily calculated on the basis that the molecules follow classical paths relative to each other, so that $\mathcal{H}_i$ becomes simply a function of the internal degrees of freedom and the time. For instance, the interaction between two molecules passing at a reasonably large distance from each other is

$$\mathcal{H}'_i = F(r_{ij}, \text{internal degrees of freedom})$$
$$= F(\sqrt{b^2 + v^2 t^2}, \text{degrees of freedom})$$

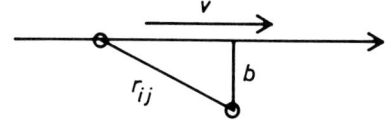

The technique which is most appropriate here is to use the method of time-development matrices. In the first place, since in the rarified gas all the molecules may be treated independently we focus our attention on a single radiating molecule and watch the development of its moment as a function of time. From the standpoint of the single molecule $\mathcal{H}'_i$ becomes a time-dependent Hamiltonian acting on its internal coordinates. For instance, thinking of a simple case of foreign-gas broadening by polarizable classical molecules of a dipolar molecule the interaction might be $\mu \cdot E$ (due to induced dipole), which gives something like

$$\frac{\mu^2 \cos^2 \theta(\mu, r_{ij})}{r^6} \sim \sum_{\text{coll}} \frac{\mu^2 \cos^2 \theta(\mu, r_{ij})}{\sqrt{b_{\text{coll}}^2 - v^2(t_0 - t_{\text{coll}})^2}}$$

and the part of the Hamiltonian acting on the internal coordinates of the molecule would be just $f(t) \cos^2 \theta$.

Now we introduce a quantity called the time development matrix $T(t)$ such that

$$\mu'(t) = T^{-1}(\mathbf{t})\mu'(0)T(t) \tag{11}$$

in fact for any operator $A'(t) = T^{-1}(\mathbf{t})A'(0)T(t)$ which satisfies the equations

$$i\hbar T = \mathcal{H}'_1(t)T \ .$$

and

$$T_{(0)} = 1 \ .$$

It is of course unitary. In the case in which there is no time-dependence and in addition $\mathcal{H}_0$ has not been transformed out of the problem, so that $\mathcal{H}_1 \neq f(t)$

$$T = e^{i\mathcal{H}_1 t/\hbar}$$

however, such an assumption in this case is erroneous — we find ourselves ignoring second and higher-order perturbation effects, which can lead to errors. However, often (as in the case of van der Waals forces) one may include the higher-order perturbations in an effective Hamiltonian, and then in such a case it is often allowable to set

$$T(t) = T e^{i/\hbar \int_0^t \mathcal{H}'_1(t')dt'} \ . \tag{12}$$

Still another matrix must be introduced for the problem we consider. This is the matrix $T(t)$ for a special condition: namely, that only one collision has occurred between $t = 0$ and $t = t'$. Namely, it is the solution for $t = +\infty$ of

$$i\hbar T = \mathcal{H}'_1 T$$

if $\mathcal{H}'_1$ represents only one collision. It is a function of the collision parameter b and any other variables (such as angles) which may be necessary to specify the collision. In the special case in which the T may be written as above

$$T(b) = e^{i\int_{-\infty}^{\infty} \mathcal{H}'_1(t')\sqrt{b^2+v^2t'^2}\,dt'}$$
$$= e^{ip(b)} \, . \tag{13}$$

This is a form I often use; of course it can always be written this way in any case. $T(b)$ is closely related to the usual collision matrix: it is its analogue under the assumption of classical paths but quantum-mechanical internal motions (of course in the interaction representation).

It is now possible to state our stochastic problem in a very definite form. Namely, it will be remembered that the original problem was to find

$$\varphi_{jk}(\tau) = \text{Tr}\,\langle \mu^*_{jk}(0)\mu_{jk}(\tau)\rangle_{\text{avg}} \tag{8}$$

but this may be written, for a sufficiently rarified gas that all collisions are independent, and in which the lengths of collisions may be neglected relative to the time between them; and if we introduce the degenerate indices m of the various operators,

$$\varphi_{jk}(\tau) = \sum \mu^{kj}_{mm'} \left(\prod_{\text{coll in }\tau} T^{-1\,j}\right)_{m'm''} \mu^{jk}_{m''m'''} \left(\prod_{\text{coll in }\tau} T^k\right)_{m'''m} \, .$$

I introduce a new notation involving operators with four indices operating on two indices quantities. In this notation

$$T^{-1}AT = (T^{-1} \times \tilde{T})A \, . \tag{14}$$

In indices

$$(T^{-1}AT)_{mm'} = \sum_{m''m'''} (T^{-1} \times \tilde{T})_{m,m';m'',m'''} A_{m''m'''}$$
$$= \sum_{m'',m'''} (T^{-1}_{mm''}\tilde{T}_{m'm'''}) A_{m''m'''} \, .$$

The quantity in brackets is the direct product of the operators T^{-1} and $\tilde{T}$, and except for the extra indices, operates entirely as though it were an ordinary linear operation. I like to call this operator U. This technique has then the advantage that instead of the rather inconvenient operation $T^{-1}\mu T$ we may use instead the simpler algebra of ordinary linear operators.

$\varphi_{jk}(\tau)$ becomes, then

$$\left\langle \mu_{jk}^* \cdot \prod_{\substack{\text{collisions} \\ u \text{ in } \tau}} (U^{jk}(u)) \cdot \mu_{jk} \right\rangle$$

where
$$U^{jk}(u) = T_{(j)}^{-1} \times \tilde{T}(k) . \tag{15}$$

The average may be taken by a simple differential-equation technique. We calculate

$$\langle \mu_{jk}(\tau + d\tau) \rangle - \langle \mu_{jk}(\tau) \rangle = d\mu_{jk}(\tau)$$

Now the rarified-gas assumption tells us that the averages over collisions in $d\tau$ and collisions in τ are independent, since the collisions are short and well-separated. We may then assume $d\tau$ to be short compared to the time between collisions, and long compared to the duration of a collision.

First we calculate the quantity (actually $\langle \mu_{jk}(\tau) \rangle$) multiplying $\mu_{jk}^*(0)$

$$\frac{d}{d\tau}(\mu_{jk}(\tau)) = \frac{1}{d\tau} \left\langle \prod_{\text{collision in } d\tau} U^{jk}(u) - 1 \right\rangle_{\text{avg}} \mu_{jk}(\tau) \tag{16}$$

(since the second average is just the quantity itself). Now let us consider the statistics of the collisions. The probability of a collision whose parameters fall within the limits $d\sigma$ about a given value — ($d\sigma = 2\pi b\, db \times$ (possibly) functions of angles — $\sin \vartheta\, d\vartheta\, d\varphi$ etc., functions of the state of the colliding molecule with appropriate Boltzmann factors, and so forth) is $n(v)v\,dv\,d\sigma\,d\tau$ where $n(v)$ is the number-density/cc with velocity v. It is usually an unimportant assumption quantitatively to assume $v = \bar{v}$, $\int n(v)dv = n$, the total density. Then the average becomes

$$\frac{d\mu_{jk}}{d\tau} = -nv \int d\sigma (1 - U^{jk}(d\sigma))\mu_{jk} . \tag{17}$$

This equation may of course be solved in general, and we get μ_{jk} as a sum of decaying exponentials;
$$\mu_{jk} = a_j e^{-nv\sigma_j \tau}$$

where σ_j are the eigenvalues of the constant operator

$$\sigma = \int d\sigma (1 - U^{jk}(d\sigma)) . \tag{18}$$

It turns out, however, that only one of the eigenvalues of this quantity enters in the final result, because of group theory. A very simple way of seeing this is to note that the transformation properties under rotation of $\mu_{jk}(\tau)$ are just exactly those of $\mu_{jk}(0)$, since the average over all types of collisions is rotationally isotropic — collisions occur from all directions with equal probability, so their effect is isotropic. But all of you know the theorem that matrix elements of a vector-component internal to a degenerate state are determined. Thus $\mu_{jk}(\tau) = \text{const.} \times \mu_{jk}(0)$ and

$$\frac{d\mu_{jk}}{d\tau} = \text{const.} \times \mu_{jk} .$$

Thus

$$\mu_{jk} = \mu_{jk}(0) e^{-nv\sigma_j \tau} \tag{19}$$

where σ_1 is a single eigenvalue of the σ matrix. A simple exercise in group theory, or actually the simple operation

$$\sigma_1 = \frac{\mu_{jk}^*(0) \cdot \sigma \cdot \mu_{jk}(0)}{|\mu_{jk}(0)|^2} \tag{20}$$

will suffice to compute this eigenvalue.

I do not want to bother you here with numerical details of results. The above treatment has actually given satisfactory results in some cases in which the interatomic forces are well known and the molecules simple.

During my stay in Japan I have been attempting to apply the above techniques to more general problems. Since the relaxation behavior of every matrix, not only μ, can be computed by them, in particular ρ, the density matrix, one expects that the technique really contains the solution of the problems of energy transfer and relaxation. However, the classical path method is not too well adapted to such problems because one can show that it does not give a quite satisfactory treatment of thermal equilibrium — there is, in fact, no differentiation between upward and downward energy transfers. I am sure that this snag can be surmounted in time by some simple new technique and the whole problem solved for sufficiently rarified gases.

I should like to discuss one more pressure-broadening problem before going on to magnetic resonance. This is again of stochastic theory interest because it stems from the very old method of Markoff for random walk distribution functions, which has itself been directly used by Holtzmark and later Margenau for special situations.

A simplification which is often extremely valuable and very seldom troublesome quantitatively of the basic problem of calculating (8) is to act as though the levels j and k were non-degenerate and transitions were not allowed: that is, the so-called "adiabatic assumption" that collisions take place adiabatically. Then the only thing happening during a collision is an adiabatic change in the energy levels E_j and E_k; and

$$\mu_{jk}(t) = \mu_{jk}(0) e^{i \int^t dt' \frac{(E_j - E_k)}{\hbar} t'}$$
$$= \mu_{jk}(0) e^{i \int^t \omega_{jk}(t') dt'}$$

ω_{jk} will, of course, be some function of the relative position of perturbing molecules and radiating molecules.

Good results can be obtained from this assumption even in cases in which it is obviously untrue, by substituting for $\omega_{jk}(t')$ some mean squared value for the substates calculated on a static basis, or some similar trick. A good fact to remember is that the adiabatic assumption never leads to large errors unless one is directly concerned in calculating transition probabilities.

Now it turns out that an additional, rather reasonable assumption removes the assumption of rarefaction of the gas completely; namely, that when a given molecule is interacting with more than one perturber, the frequency perturbation is the sum of those due to the separate ones

$$\omega_{jk}(1,\ldots,N) = \sum_{n=1}^{N} \omega_{jk}(R_n)$$

where R_n is the distance of the nth perturber from the initial molecule. For a sufficiently rarified gas this is a pretty good assumption because only one perturber at a time is near the given atom most of the time. Also, for second-order van der Waals' forces the additivity assumption is pretty good.

Then the correlation function becomes

$$\langle \mu_{jk}^*(\tau) \mu_{jk}(0) \rangle_{\text{avg}} = \mu^2 \left\langle e^{i \int_0^\tau \sum_n \omega_{jk}(R_n) d\tau} \right\rangle_{\text{avg}}$$
$$= \mu^2 \left\langle \prod_n e^{i \int_0^\tau \omega_{jk}(R_n) d\tau} \right\rangle_{\text{avg}}.$$

We assume that the substance is still in the gaseous state so that the distributions of the various perturbers are independent (in liquid state molecules knock against each other and have some correction). Then we may take a separate average for each term and get

$$\varphi(\tau) \sim \left(\left\langle e^{i \int_0^\tau \omega_{jk}(R) d\tau} \right\rangle_{\text{avg}} \right)^N$$

where N is the total number of perturbers. Eventually we want to set $N \to \infty$ but

$$N/V = \text{number density} = n = \text{const.}$$

How do we do the average? It is an average over all the possible paths $R(t)$ of a given perturbing particle. We can specify these by giving the positions $R(0)$ at $t = 0$ and the vectorial velocities, as well as the quantum state of the perturbing particle. For isotropic, classical perturbers this state is constant and the direction of approach immaterial; for such perturbers we may (again assuming all velocities $\bar{v}$ with little error) take all velocities in the same direction and we need average only over the initial position $R(0)$, but for more complicated situations one could easily find an averaging technique which would be appropriate. For our simple special case, however $\langle\ \rangle_{\text{avg}}$ then, $= \frac{1}{V}\int dV_{R(0)}$ (all space)

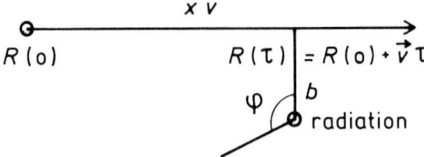

I like to express the position in the coordinate system shown above, in terms of φ, b, and x; then

$$|R(t)|^2 = b^2 + (x - vt)^2$$
$$\omega_{jk}(R(t)) = \omega_{jk}(\sqrt{b^2 + (x - vt)^2})$$

while

$$\int dV = \int_0^\infty b\,db \int_0^{2\pi} d\varphi \int_{-\infty}^\infty dx$$

so

$$\left\langle e^{i\int \omega(R)d\tau} \right\rangle_{\text{avg}} = \frac{1}{V}\int dV e^{i(\int_0^\tau \omega(\sqrt{b^2+(x-vt)^2})\,dt)} \ .$$

A trick is useful here. As $V \to \infty$ the above integral diverges as does the denominator. This may be corrected by doing

$$\frac{1}{V}\int dV(1 - (1 - e^{i\int_0^\tau \omega dt})) = 1 - \frac{V'(\tau)}{V}$$

where

$$V' = \int_0^\infty b\,db \int_0^{2\pi} d\varphi \int_{-\infty}^\infty dx \left(1 - e^{i\int_0^\tau \omega(\sqrt{b^2+(x-vt)^2})\,dt}\right) \ .$$

Now the limiting process is easy

$$\varphi(\tau) = \lim_{N,V\to\infty} \left(1 - \frac{V'(\tau)}{V}\right)^N = \lim_{V\to\infty} \left(1 - \frac{V'}{V}\right)^{nV}$$
$$= e^{-nV'(\tau)} .$$

And

$$I(\omega) = \int_{-\infty}^{\infty} e^{i\omega\tau - nV'(\tau)} d\tau .$$

Let me just calculate two limiting cases for V'. For small τ

$$\int_0^\tau \omega(\sqrt{b^2 + (x - vt)^2}) d\tau = \tau\omega(R_0)$$

and

$$V' = \int dV \left(1 - e^{i\omega(R_0)\tau}\right) .$$

This result is identical with that of the so-called "statistical" theory which makes a direct application of Markoff's method under the assumption that $I(\omega)$ is the distribution of momentary frequencies $\omega(R)$, with no effects of $d\omega/d\tau$ or any higher rates of change. The statement is often made that this approximation is valid on the wings of the line, or for sufficiently high pressures. Really the criterion is one of sufficiently small τ. This is roughly equivalent to the other two as we see from the expression for $I(\omega)$, but in one important case, namely that in which all the $\omega(R)$ are of the same sign, this necessarily gives only one wing of the line, the other vanishing completely, and even on the wing, at high densities, we must calculate a first correction. This is not too hard with the present technique and a method has been found. These calculations are fairly long and have not been published yet, so I can give you only the result: that the intensity of the "wrong" wing falls off exponentially as a power of ω close to the first:

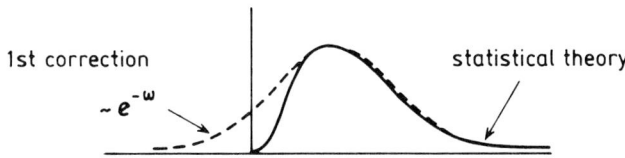

This result is entirely different from that obtained by Lindholm by a different method, which I am sure is incorrect. The correction terms on the "regular" wing agree with those published by Holstein.

The other limiting case is also easily derived: large τ. For τ large compared with the duration of the collision, only those $R_{(0)}$ need be considered which are such that a collision occurs

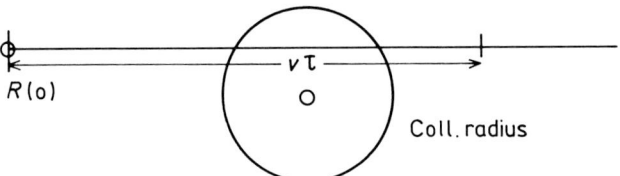

The picture is

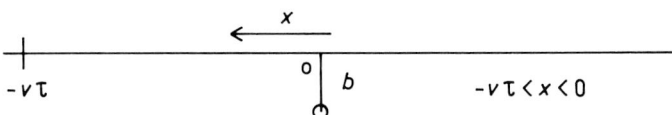

For all such x the integral $\int_0^\tau \omega(R) d\tau$ may be approximated by $\int_{-\infty}^{\infty} \omega(R) d\tau = \eta(b)$: a function of b alone, giving the "phase-shift" due to a single collision of parameter b, if $x < -v\tau$ or $x > 0$ no collision occurs:

$$\int_0^\tau \omega(R) d\tau = 0, \qquad 1 - e^{i\int} = 0 .$$

Then

$$V' = \int_0^\infty bdb \int_0^{2\pi} d\varphi \int_{-v\tau}^0 (1 - e^{i\eta(b)})$$
$$= 2\pi v\tau \int_0^\infty bdb(1 - e^{i\eta(b)})$$

for $\tau < 0, \eta \to -\eta$, we get $V'(-\tau) = V'(\tau)$.
This gives the impact theory limit:

$$\varphi(\tau) = e^{-mv\sigma\tau}$$
$$\sigma = 2\pi \int_0^\infty bdb(1 - e^{i\eta(b)}) .$$

This is the result our previous theory would have given if specialized to this case. It can be shown that the first correction to V' is a constant and that V' can be expanded as $a\tau + b + c/\tau + \ldots$ but b is a very complicated integral and I have not succeeded in evaluating it completely as yet. Only this last step prevents a numerically satisfactory solution for the entire problem, since two terms for V' for τ small and for τ large should be easy to fit together numerically giving the whole of V'.

Now I should like to turn to the discussion of magnetic resonance line breadths. The magnetic resonance case, while describable in the same stochastic terms, has a number of essential differences. One of the most vital is that the perturbation $\mathcal{H}'(t)$ is not intermittent as in the pressure broadening case, but continuously present. This destroys the simplifications we could use when we could separate the collisions. Also, a new phenomenon appears, that of narrowing, because of the following consideration. In pressure broadening the rate of collisions determines the breadth ($nv\sigma$; you remember) and during a collision is usually fairly large (it must be at least such that

$$\frac{\mathcal{H}'(t_{\text{coll}})}{\hbar} \geq 1, \qquad \frac{\mathcal{H}'}{\hbar} \geq \frac{1}{t_{\text{coll}}} \gg \frac{1}{t} \;).$$

Thus we are never confronted with a situation in which the time-rate of change of frequency is greater than $\mathcal{H}'/\hbar$. However, with $\mathcal{H}'$ constantly present there is no reason why its time-rate of change need be less than $\mathcal{H}'/\hbar$ itself, since there is no effective lower limit on $\mathcal{H}'$ relative to the rates of motion. When $\mathcal{H}'$ changes rapidly relative to its own magnitude (which determines the line-breadth, and therefore the important range of values τ in $\varphi(\tau)$) it is possible for $\mathcal{H}'$ to average to a small value over these intervals of time, and for the line to become *narrower* rather than broader as the motion becomes more rapid.

Explicit picture: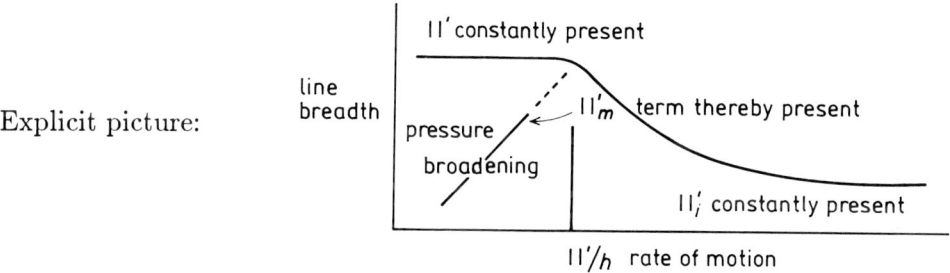

Interest at the present time is focused on this "narrowing" range. I should like to discuss two topics of interest in this range: Kubo's quantum mechanical theory for $\mathcal{H}'$ continuously distributed; and my own simplified theory for narrowing of fine structure. The latter may be less familiar to most of you so I shall start with it and go on to Kubo's theory if I have time.

In my theory I start with the same simplifying assumption that I used in the last pressure-broadening case: the assumption of adiabatic motion. This is *of course* invalid in magnetic resonance in most cases, because there are important off-diagonal elements of the perturbation Hamiltonian (not however for non-idential atoms — dilution problem) but in the same spirit as we discussed before the adiabatic technique may be modified by using the *correct* mean values of frequencies or mean squares including off-diagonal terms, and the resulting errors are usually small.

Then the problem is to calculate

$$\varphi(\tau) = \left\langle e^{i \int_0^\tau \omega(t) dt} \right\rangle_{\text{avg}}$$

where $\omega(t)$ is a stochastic function about which we may be able to know or guess something from the physical situation. For instance, in a crystalline solid in which $\mathcal{H}'$ is the dipolar interactions of local fields we guess that ω has a gaussian distribution with the well-known mean squared breadth as calculated by Van Vleck. My paper on exchange narrowing dealt with such Gaussian cases, but they are far better treated by Kubo's method so I won't discuss them here.

There are a large number of types of cases which are, however, not properly treated as yet by Kubo's method. Among these are all cases of fine structure, in which $\omega(t)$ may take on one or another of a set of discrete values, but the motion is such as to cause transitions among them (e.g., exchange between electrons with nuclear hyperfine structure, in which the nuclei are fixed in definite states but the electrons change places; diffusion in solids with some kind of crystal fine structure, etc.). Also, most cases in which the motion is the same kind of barrier-jumping, like diffusion in solids, are not well-approximated by Gaussian methods. At present the only attack on such problems is to make the opposite assumption that the random function $\omega(t)$ is Markoffian.

A Markoffian function $f(t)$ has the property that the probability $W(f_2, |f_1, t)$ that given f_1 at $t = 0$ the value is f_2 at time t is independent of the earlier history — the values at $t < 0$ no matter what the value of t is. Such a process can be thought of as proceeding in discrete jumps, whose probability is determined by the value only at the present time.

Let us present some of the basic mathematics of Markoffian functions. The entire process is determined by the second-order transition probabilities $W(f_2|f_1, \Delta t)$ which, if the process is to act sensibly at $\Delta t \to 0$, must be capable of being written

$$W(f_2|f_1, \Delta t) = \delta_{f_1, f_2} - \Pi(f_i, f_1) \Delta t$$

where now Π alone controls the process completely. The process can be expected to be stationary if it is a physical one: that is, there is a probability function $W_1(f)$ such that

$$\sum_{f_1} W_1(f_1) W(f_2|f_1, t) = W_1(f_2)$$

or,
$$\sum_{f_1} \Pi(f_2, f_1) W(f_1) = 0 ,$$

(the above is a form of the Smoluchowski equation: in general, the time-rate of change of a distribution is $\frac{dW'(f)}{dt} = \frac{W'(t+\Delta t) - W'(t)}{\Delta t} = \sum_{f_1} \Pi(f, f') W'(f')$ and this is it).

Another theorem is that the probability of taking on at least one value is unity:
$$\sum_{f_2} W(f_2|f_1, \Delta t) = 1 \longrightarrow \sum_{f_2} \Pi(f_2, f_1) = 0 .$$

This shows that Π is a non-symmetric matrix with eigenvectors $\begin{pmatrix} \cdot \\ \cdot \\ \cdot \end{pmatrix}$ on the left, W_1 on the right belonging to the eigenvalue zero.

The following simplified technique of calculating $\varphi(\tau)$ was suggested to me by Prof. Kubo. Consider the quantity to be averaged:
$$e^{i \int_0^\tau \omega(t) dt} = F(\tau) .$$

Let us consider the case in which the initial state is ω_1, the final state ω_2, and realize that eventually we must average over ω_1 with probability W_1 and over ω_2 with equal probability for all cases (i.e., mult. matrix $F(\tau)_{\omega_1, \omega_2}$ by the two eigenvectors of Π on the left and right)

$$F(\tau + \Delta \tau)_{\omega_1, \omega_3} = \sum_{\omega_2} F(\tau)_{\omega_1, \omega_2} W(\omega_3/\omega_2, \Delta \tau) e^{i\omega_2 \Delta \tau}$$
$$= F(\tau)_{\omega_1, \omega_3} e^{i\omega_3 \Delta \tau} - \sum_{\omega_2} \Pi(\omega_3, \omega_2) F(\tau)_{\omega_1, \omega_2} \Delta \tau$$

or,
$$\frac{d}{d\tau} (F(\tau))_{\omega_1, \omega_3} = i\omega_3 F(\tau)_{\omega_1, \omega_3} - \sum_{\omega_2} F(\tau)_{\omega_1, \omega_2} \Pi(\omega_3, \omega_2) .$$

In matrix language,
$$\frac{dF}{d\tau} = i\omega F - \Pi F$$
if we define a diagonal matrix $i\omega$ and Π, F in obvious ways.
$$\varphi(\tau) = 1 \cdot F \cdot W_1 .$$

Now clearly we have the complete solution if we can but work it out, since we need merely find the eigenvalues λ_j of $(i\omega - \Pi)_{\omega_1, \omega_2}$ and the appropriate coefficients of their eigenvectors relative to W_1 and then we shall have
$$\varphi(\tau) = \sum_j C_j e^{+\lambda_j \tau}$$

and $I(\omega)$, in turn, will be a sum of resonance distributions

$$\sum C_j \frac{\lambda_j}{(\omega - \lambda_j)^2 + \lambda_j^2}.$$

This general procedure, while clear and straightforward in principle, cannot always be carried out. For more than 4 ω_i's we get in general an insoluble secular equation. Therefore it pays to calculate first- and second-order approximations in the two cases
(a) $\qquad\qquad\qquad\qquad\qquad\qquad \pi \gg \omega$;
(b) $\qquad\qquad\qquad\qquad\qquad\qquad \omega \gg \pi$.

Case (b) is simple, ω is already a diagonal matrix $(\omega)_{jk} = \omega_j \delta_{jk}$. The first-order approximation is

$$\lambda_j = i\omega_j - \Pi_{jj}$$

Π_{jj} is the rate of transitions *away from* the given frequency $j-i$, the inverse lifetime $(\tau_1 = 1/\omega_j)$ of that frequency. The coefficients of these approximate eigenvectors are just $W_1(\omega_j)$: one gets in zeroth approximation for slow motion $\varphi(\tau) = \sum W_1(\omega_j) e^{i\omega_j \tau}$: The unperturbed distribution; and second

$$\varphi(\tau) = \sum W_1(\omega_j) e^{-i\omega_j \tau - \Pi_{jj} \tau}$$

a Lorentz distribution about each unperturbed eigenvalue.

Next we get the second-order perturbation:

$$\lambda_j^{(2)} = i\omega_j - \Pi_{jj} - \sum_k \frac{\Pi_{jk} \Pi_{kj}}{i(\omega_k - \omega_j)}$$

$$= -\Pi_{jj} + i\left(\omega_j - \sum_k \frac{\Pi_{jk} \Pi_{kj}}{\omega_k - \omega_j}\right).$$

The next term is thus a *shift* of the center frequencies *toward the center of the distribution* — since Π must lead more often to the center than the wings. This shift could have been predicted from the theorem that $\overline{\Delta\omega^2}$ is independent of the rate of motion $\mathcal{H}_m$ (provable direct from the expression for $\varphi(\tau)$); as the lines broaden by $\omega_e(j)$, the whole pattern must necessarily narrow as far as the line-centers are concerned.

The calculation from the other end is more difficult but can also be carried out in any particular case. The zeroth approximation is trivial: if we can neglect the $i\omega_j$, relative to Π, obviously $dF/d\tau = 0$ because ω_i is the eigenvector of Π belonging to the eigenvalue $\lambda = 0$. The first-order calculation merely introduces $i\bar{\omega}_j$: the average frequency: so here we get a single infinitely narrow line. To obtain the second order we must set $\lambda = i\bar{\omega}_j + \delta$ and neglect higher orders of δ. The calculation leads eventually to a term of order $-\omega_j^2/\omega_e$: the usual narrowed line-breadth.

For details of actual cases I should refer you to my paper to appear in the *Journal of the Physical Society of Japan*.

A most interesting case is the one in which the possible frequencies ω_j form a continuous distribution. In this case the perturbation techniques break down for the limit $\Pi \to 0$, so that (as Kubo has pointed out to me) the best solution is to use the time differential equation directly, which can be done in simple cases.

As a last demonstration of the stochastic techniques which can be applied in these problems I should like to discuss Kubo's technique for the so-called "Gaussian" case in which the distribution of $\mathcal{H}'$ is continuous and the motion not of the abrupt "Markoffian" type but more gradual.

Here it is possible to proceed without the initial assumption of adiabatic motion. The technique is closely related to the so-called "moment method". In the moment method we take the expression

$$\varphi(\tau) = \mathrm{Tr}\, \langle \mu(\tau) \mu^*(0) \rangle_{\mathrm{avg}} \Big/ \mathrm{Tr}\, \langle \mu^2 \rangle_{\mathrm{avg}}$$

and calculate $\mu(\tau)$ as a power series in τ. There are the well-known relationships

$$\frac{d^2\varphi}{d\tau^2} = -\overline{\omega^2}\; ; \qquad \frac{d^4\varphi}{d\tau^4} = \overline{\omega^4}$$

and so forth; and since in the absence of exchange or motion $\overline{\omega^4} \simeq 3(\overline{\omega^2})^2$ one may set approximately $I(\omega) \simeq e^{-\frac{\omega^2}{2\bar{\omega}^2}}$ Kubo has observed that this is equivalent to expanding

$$\varphi(\tau) = e^{-\frac{1}{2}\frac{d^2\varphi}{d\tau^2}\tau^2 + \cdots}$$

rather than the more obvious $1 - \frac{1}{2}\frac{d^2\varphi}{d\tau^2}\tau^2 + \ldots$ which is not useful because of the fact that $\varphi(\tau \to \infty)$ must approach zero.

In the case in which there is motion the second moment remains the same but the expansion in this form becomes invalid. This is because $\frac{d^2\varphi}{d\tau^2} = \frac{\langle \dot\mu^2 \rangle}{\langle \mu^2 \rangle} = \frac{\langle [\mathcal{H}',\mu]^2 \rangle}{\langle \mu^2 \rangle}$; but the averaging process after a finite interval τ has intervened causes this expression to fail rapidly for $\tau \neq 0$.

A more hopeful expansion can be obtained by calculating $\mu(\tau)$ by means of its *true* equation of motion

$$i\hbar \dot\mu = [\mathcal{H}'(t), \mu]$$

rather than

$$i\hbar \dot\mu = [\mathcal{H}'(0), \mu]\ .$$

The true equation may be solved by successive approximations:

$$\mu_0 = \mu(0)$$

$$\dot{\mu}_1 = \frac{1}{i\hbar}[\mathcal{H}'(t), \mu(0)]$$

$$\dot{\mu}_1(\tau) = \frac{1}{i\hbar} \int_0^\tau dt [\mathcal{H}'(t), \mu(0)]_{\text{avg}}$$

$$\dot{\mu}_1(\tau) = \frac{1}{(i\hbar)^2} \int_0^\tau dt \int_0^t dt' [\mathcal{H}'(t), [\mathcal{H}'(t'), \mu(0)]]$$

we expand not in terms of τ but of $\mathcal{H}'$.

We then set

$$\varphi(\tau) = e^{-\frac{\langle \mu_2(\tau)\mu(0)\rangle_{\text{avg}}}{\langle \mu^2\rangle}} + \ldots$$

again by the exponential assumption $\langle \mu_1 \rangle_{\text{avg}}$ vanishes on general principles).

Explicitly,

$$[\mathcal{H}'(t), [\mathcal{H}'(t'), \mu]] = \mathcal{H}'(t)\mathcal{H}'(t')\mu - \mathcal{H}'(t)\mu\mathcal{H}'(t')$$
$$- \mathcal{H}'(t')\mu\mathcal{H}'(t) + \mu\mathcal{H}'(t')\mathcal{H}'(t) .$$

Now one may set

$$\langle \mathcal{H}'(t)\mathcal{H}'(t') \rangle_{\text{avg}} = (\mathcal{H}'(t))^2 \varphi_{\mathcal{H}'}(t'-t)$$

where $\varphi_{\mathcal{H}'}$ is the correlation function of the stochastic function $\mathcal{H}'(t)$. The commutator $\langle [\mathcal{H}'(t), [\mathcal{H}'(t), \mu]] \rangle_{\text{avg}/\hbar^2}$ is just $\overline{\Delta\omega^2} \langle \mu^2 \rangle$: This is just the same as if there were no motion, since $\mu(0)$ and $\mu(t)$ are essentially equivalent.

Thus

$$\varphi(\tau) = e^{-\overline{\Delta\omega^2} \int_0^\tau dt \int_0^t dt' \varphi_{\mathcal{H}'}(t'-t)}$$

which can be shown to be

$$= e^{-\Delta\omega^2 \int_0^\tau (\tau-t)\varphi_{\mathcal{H}'}(t)dt} .$$

This is the identical answer which is obtained by the adiabatic method; but Kubo's technique gives an explicit technique for calculating $\varphi_{\mathcal{H}'}$ rather than allowing for guesswork. The two techniques turn out to coincide exactly in the one case, exchange narrowing, in which both can be carried out; but obviously Kubo's has a far sounder basis.

This expression leads to the narrowing phenomenon in well-known ways. For $\varphi_{\mathcal{H}'}(\tau)$ slowly-varying we may set it $\simeq 1$, and

$$\int_0^\tau (\tau - t) = \tau^2/2$$

$$\varphi(\tau) = e^{-\overline{\Delta\omega^2}\tau^2/2}$$

which leads to the unperturbed Gaussian distribution. For $\varphi_{\mathcal{H}'}$, rapidly-varying, we set

$$\int_0^\infty \varphi_{\mathcal{H}'} d\tau = \frac{1}{\omega_e}$$

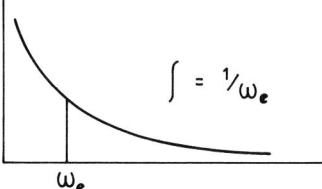

and $\int t\varphi_{\Delta\mathcal{H}} \sim 1/\omega_e^2$, negligible
so
$$\varphi(\tau) = e^{-\tau \frac{\overline{\Delta\omega^2}}{\omega_e}} .$$

The usual answer: one gets a Lorentz line-shape of breadth $\Delta\omega = \overline{\Delta\omega^2}/\omega_e$.

This concludes my survey of stochastic methods used in line-broadening problems. This by no means exhausts all the methods which have been or will be used, nor all the problems which they might be. The subject of line-shapes in metals of the electronic resonances, where one has electrons diffusing in and out of the skin depth region, is a very important one, which has been discussed on this basis by Kittel and his co-workers with the result that they have been the first to observe electronic paramagnetic resonance of really free electrons.

Kittel's theory of diffusion is

$$\dot{M}_+ = i\omega_0 M_+ - i\gamma M_z H_+$$
$$H_+ = e^{-kx+i\omega t} = H(x(t))e^{i\omega t} .$$

Solve for forced solution:

$$\frac{d}{dt}\left(M_+ e^{-i\omega_0 t}\right) = e^{i(\omega-\omega_0)t} \text{const.} \times H(t)$$
$$M_+ e^{-i\omega_0 t} = \int e^{i(\omega-\omega_0)t} \text{const.} \times H(t) dt .$$

Thus $M_+(\omega)$ is a Fourier component of $H(t)$ which can be found by Khinchine theory:

$$\langle H(0)H^*(\tau)\rangle_{\text{avg}} .$$

Electron diffusing in skin depth (k of order 1/skin) will break off when diffuses out of skin:
$$\frac{1}{\Delta\omega} \simeq \text{time to diffuse out } T, \qquad \lambda = mfp$$

skin depth $\qquad\qquad l = \sqrt{T/\tau}\lambda = \sqrt{Tv\lambda}$

so $\qquad\qquad\qquad T = l^2/v\lambda$

$$\Delta\omega = v\lambda/l^2 = \frac{v\lambda\sigma\omega}{c^2}$$

where $\qquad\qquad l^2 = c^2/\sigma\omega$

$$\sigma = ne^2\lambda/2mv$$

so $\qquad\qquad \dfrac{\Delta\omega}{\omega} = \dfrac{ne^2\lambda}{2mv}\dfrac{v\lambda}{c^2}$

$$= n\left(\frac{e^2}{2mc^2}\right)\lambda^2 = nr\lambda^2$$

(r is the classical electron radius).

Other stochastic methods have been used for nuclear resonances by Torrey, while Slichter and co-workers have discussed the effects of diffusion in an inhomogeneous external magnetic field in liquids.

One of the most important fields remains as yet almost untouched: that of relaxation processes as opposed to line-shapes themselves. This has been started by Bloembergen, Purcell and Pound for nuclear magnetic resonance, but they had the same trouble one has with the stochastic method in pressure-broadening relaxation — that the transition probabilities as computed by this method do not lead to the Boltzmann distribution: the energy distribution is not property managed. They seem to have satisfactorily solved this problem, but not to have carried out their investigations very far in various directions. With the new interest of Dicke, Hahn and others in transient techniques the relaxation problem has become much more important and more thought needs to be given to it.

With this I conclude these two lectures on stochastic methods in line-broadening problems. I hope that I have been able to interest some of you in carrying these studies further, or in applying the stochastic methods used here to other problems.

Qualitative Considerations on the Statistics of the Phase Transition in BaTiO$_3$-type Ferroelectrics

Conference Proceedings, Lebedev Physics Institute,
Academy of Sciences of the USSR,
Nov. 1958, *Fizika Dielektrikov*, p. 290, Moscow, 1960

This is the aforesaid soft mode paper. I went to Moscow to meet the Landau group and to talk with them and with Shirkov about superconductivity, but the invitation was to a dielectrics meeting. The work was from my notebooks of 1950–54, and I felt that in sending me Bell was paying me its "dielectric bill" for all that work under Shockley.

QUALITATIVE CONSIDERATIONS ON THE STATISTICS OF THE PHASE TRANSITION IN BaTiO$_3$-TYPE FERROELECTRICS

P. W. ANDERSON
Bell Telephone Laboratories
Murray Hill, New Jersey

The phase transition of BaTiO$_3$ is remarkably free of fluctuation effects relative to other similar cooperative transitions. This is related to the fact that the singularity which causes it involves only a very few of the modes of motion of the lattice, the remainder staying regular and harmonic. Various experimental consequences of these ideas are explored. The concept that the individual ions "rattle" in multiple-minimum potentials is shown to be misleading.

Barium titanate has occupied a special place in the minds of those interested in ferroelectrics because of its relatively simple structure and properties, and because of the technically important magnitude of some of its ferroelectric parameters. It is typical of a class of ionic ferroelectric crystals for which the theory of the phase transition is probably quite simple; it is our purpose to show why this is so, and some of the consequences of the theory.

I should apologize for the always qualitative, and sometimes trivial, nature of the reasoning; my excuse is that in the long history of the subject, and even in the many years since I first suggested some of these things, there have been no publications along these lines.

The theory of an abnormal ionic dielectric like BaTiO$_3$ is best approached by understanding a normal ionic dielectric of relatively high dielectric constant. For this the old methods of Debye[1] for studying thermal expansion due to anharmonic forces are adequate, although we shall re-express them in a more convenient form.

We shall simplify the dielectric by including only two branches of the vibration spectrum, the acoustic and the lowest — and presumably most completely polar-optical one. The zeroth approximation to the potential energy is given by the harmonic forces:

$$V_0 = \sum_k s^a(k) q^a_{-k} q^a_k + s^0(k) q^0_{-k} q^0_k. \tag{1}$$

q^a_k and q^0_k are the acoustical and optical mode coordinates respectively, and s^a_k and s^0_k the stiffnesses. The q_k's are linear combinations of displacements like x^n_j, j standing for the unit cell and n for the atom in that cell:

$$q_k^\alpha = N^{-1/2} \sum_{j,n} (a|n) e^{ik \cdot r_j} x_j^n$$

N being the total number of cells j.

The independent harmonic oscillators q_k are perturbed by the presence of anharmonic terms of various sorts, all resulting because the interatomic potential is steeper for closer approaches, less steep for distant ones, than a parabola. Typical terms are:

$$\begin{aligned} H' = &N^{-1/2} \sum_{\substack{k,k' \\ (\alpha,\beta,\gamma)=\overset{\circ}{a}}} \lambda_3^{\alpha\beta\gamma}(k) q_k^\alpha q_{k'}^\beta q_{-k-k'}^\gamma \\ &+ N^{-1} \sum_{k,k',Q} \lambda_4^{\alpha\beta\gamma\delta} q_k^\alpha q_{-k+Q}^\beta q_{-k'-Q}^\gamma q_{k'}^\delta \\ &+ \ldots \end{aligned} \qquad (2)$$

The basic approximation in a normal crystal is that the anharmonicity is small, so that in calculating the partition function or various averages it is correct to expand the exponential

$$e^{-\beta(H_0+H')} \sim e^{-\beta H_0}(1 - \beta H') \ .$$

(In most ferroelectrics classical theory is adequate, so for H_0 read V_0.) Thus all problems, to a certain order, reduce to finding averages of H', usually multiplied by polynomials in one or a few q_k's, using the unperturbed potential V_0. In calculating such averages we are helped by a theorem:

Only quantities having zero total momentum (i.e. uniform throughout the crystal) have finite averages. Such are $|q_k|^2, |q_k|^4, q_0, |q_0|^2 \ldots$.

Taking into account this rule only a few terms in H' can ever give contributions to the various averages.

I. Any powers of q_0's.
II. Any products of $|q_k|^2 |q_{k'}^2|$ etc.; the terms $|q_k|^4$ are N less numerous and negligible.
III. Powers of q_0's times $|q_k|^2$.

Using these rules we extract the useful part of H':

$$N^{-1/2} \sum_{k,\alpha\beta} \left(\lambda_3^{\alpha\beta\beta}(0) + 2\lambda_3^{\beta\alpha\beta}(k)\right) q_0^\alpha |q_k^\beta|^2 \qquad (3)$$

$$N^{-1} \sum_{k,k'} \left[2\lambda_4^{\alpha\alpha\beta\beta}(k-k') + \lambda_4^{\alpha\beta\beta\alpha}(0)\right] |q_k^\alpha|^2 |q_{k'}^\beta|^2 \qquad (4)$$

in which both sums include the values $k = 0$.

The final observation is that averages of q's for different k's are independent.

Using this concept it is easy to show quite generally that, to lowest order in the anharmonicity, all the thermal properties follow from a set of effective Hamiltonians, one for each mode, which are obtained by averaging (3) and (4) over all but the mode of interest. For instance,

$$H_{\text{eff}}(q_k^\alpha) = s_k^\alpha |q_k^\alpha|^2$$
$$+ N^{-1/2} |q_k^\alpha|^2 \sum_\beta \left(\lambda_3(0) + 2\lambda_3(k)\right) \langle q_0^\beta \rangle$$
$$+ N^{-1} |q_k^\alpha|^2 \sum_{k;\beta} \left(2\lambda_4(k-k') + \lambda_4(0)\right) \langle |q_{k'}^\beta|^2 \rangle , \qquad (5)$$

$$H_{\text{eff}}(q_0^\alpha) = s_0^\alpha |q_0^\alpha|^2 + N^{-1/2} \lambda_3(0)(q_0^{\alpha 3} + 2q_0^{\alpha 2} q_0^\beta)$$
$$+ q_0^\alpha \frac{1}{N^{1/2}} \sum_k \left(\lambda_3(0) + 2\lambda_3(k)\right) \langle |q_k^\beta|^2 \rangle$$
$$+ \frac{1}{N} |q_0^\alpha|^2 \sum_k \left(2\lambda_4(k') + \lambda_4(0)\right) \langle |q_{k'}^\beta|^2 \rangle . \qquad (6)$$

From (5) we see that the anharmonicity leads to a thermal variation of stiffness (and thus of frequency) for each mode; when this variation is large we know the method fails. The macroscopic properties, on the other hand, depend on the equations for the q_0's. $N^{1/2} q_0^\alpha$ are the dilatation and other uniform strains, so that as the vibrations produce, through the cubic anharmonic terms, a linear term in q_0^a, this forces a displacement of q_0^a, which is the cause of thermal expansion. Similarly the last term changes the compressibility: these results are like the old Debye Theory.[1]

The optical modes q_0^0, and especially the transverse ones for which there are no depolarization forces, control the dielectric properties. In fact, the ionic polarization of the crystal is

$$P = \sum_{j,n} e_{jn} x_{jn} ce N^{1/2} q_0^0 . \qquad (7)$$

The energy is an external field is $-E \cdot P$. Symmetry arguments show, when the crystal has a center of symmetry, that q_0^0 cannot enter in odd powers; thus the effective Hamiltonian may be written

$$H_{\text{eff}} = -Ee N^{1/2} q_0^0 + s_0^0 |q_0^0|^2 + 2N^{-1/2} \lambda_3(0) |q_0^0|^2 q_0^a$$
$$+ N^{-1} |q_0^0|^2 \sum_{k,\beta} \left(2\lambda_4(k) + \lambda_4(0)\right) \langle |q_k^\beta|^2 \rangle + 3N^{-1} \lambda(0) |q_0^0|^4 + \ldots$$
$$\simeq -Ee N^{1/2} q_0^0 + s_0^0(\text{eff}) |q_0^0|^2 . \qquad (8)$$

The dielectric constant, or more exactly the polarizability $(\varepsilon - 1)/4\pi$, can be derived from (7) and (8) using $\partial H/\partial q_0^0 = 0$; we get

$$\varepsilon - 1 = \frac{4\pi N e^2}{2s_0^0(\text{eff})} \ . \tag{9}$$

From (8) we see that there are two temperature-dependent effects on ε. The λ_3 term is the effect of thermal expansion, and increases ε with T because the more expanded lattice is softer. This is the normal behavior in low dielectric constant materials. The λ_4 term comes mostly from the optical branch and is the effect of the steep sides of the ions' potential wells. When the stiffness of the optical vibrations is lower (from (9), this is high ε) this will predominate, as is observed.

A last idea from the theory of simple dielectrics which is useful is that of the dipolar contribution to $s(k)$. Ignoring the charges on the ions there are certain short-range restoring forces due mostly to ionic repulsion, which would lead to a stiffness $s_{\text{rep}}(k)$. Then there are the purely electrostatic interactions of the dipoles caused by the displaced ions:

$$e^2 \sum_{i,j} r_{ij}^{-3} [x_i \cdot (1 - 3\hat{r}_{ij}\hat{r}_{ij}) \cdot x_j]$$

which, being quadratic in the displacements x, can also be transformed into a stiffness contribution

$$s_{\text{dip}}(k)|q_k^0|^2 = -\frac{1}{2}|q_k^0|^2(ne^2 L_k) \tag{10}$$

L_k here is the usual Lorentz local field parameter for $k = 0$ (i.e. for cubic sites $4\pi/3$ for transverse modes, $-8\pi/3$ for longitudinal ones[2]), and varies with k in the manner calculated by Heller and Marcus[3] and others. A sketch of a hypothetical k-dependence is given in Fig. 1. Then the optical branch stiffnesses may roughly be written

$$s^0(k) = s_{\text{rep}}(k) - \frac{1}{2}ne^2 L_k \ .$$

When the Lorentz correction nearly cancels the repulsive term we have a high-dielectric constant material; when it is slightly greater at $k = 0$ we get a ferroelectric.

Let us now go on to discuss the ferroelectric case. Starting at absolute zero, we have two choices: to discuss normal modes expanding about the actual ferroelectric state at $T = 0$, which leads to a highly complicated picture; or to expand about the hypothetical cubic state, assuming that deviations from it are not too large; this is the fruitful way.

In the absence of the dipolar interactions, we imagine the potential expanded about the cubic state to be very normal, with a normally positive stiffness for all optical modes, and reasonably small anharmonic terms. Because of the dipolar interactions, however, the effective stiffness at $T = 0$ of a relatively small range of optical modes near $k = 0$ is

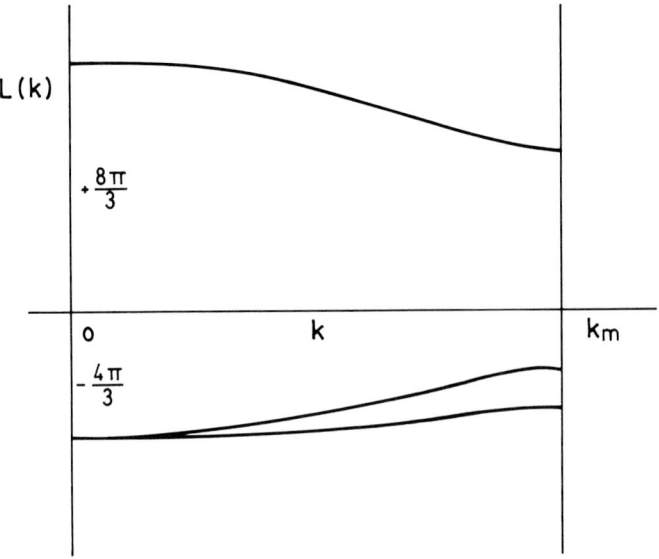

Fig. 1. Schematic dependence of Lorentz Sum $L(k)$ on k in cubic crystal.

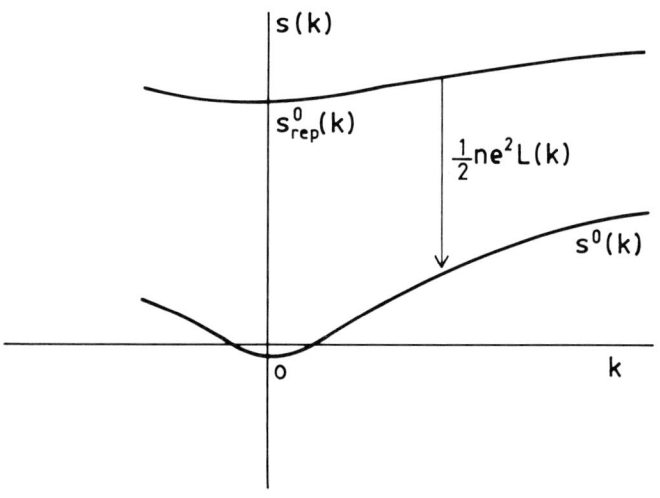

Fig. 2. Contributions to $s(k)$ in BaTiO$_3$.

negative, as shown in Fig. 2. Clearly the most unstable mode (with the largest negative stiffness) is q_0^0, which will have an effective potential in the above formalism of

$$V_{\text{eff}} = s_0^0(\text{eff})|q_0^0|^2 + 3N^{-1}\lambda(0)|q_0^0|^4 + \ldots . \tag{11}$$

q_0^0 has the famous double-minimum potential, and will displace to the position of one of the minima, with a value of order $N^{1/2}$. This is the ferroelectric polarization. From the above expressions we would expect the 4th-order term in the potential to limit the displacement. Slater and Devonshire[4] however, have shown that the positive λ_4 contribution is outweighed by a negative contribution involving the electro-strictive term $|q_0^0|^2 q_0^a$ in second order, so that the displacement is really limited by sixth order terms, a fact which is responsible for the, at first, rather puzzling abruptness of the ferroelectric transition. Nonetheless the total displacement of q_0^0 is rather small, in BaTiO$_3$ and, to a lesser extent, in several similar ferroelectrics such as KNbO$_3$, etc.

This displacement of q_0^0 has the effect, because of the term

$$N^{-1}|q_k^0|^2(2\lambda_4(k) + \lambda_4(0))\langle|q_0^0|^2\rangle$$

of increasing the stiffness of all the other modes with negative stiffnesses so that they all have positive stiffnesses. Nonetheless, we can expect these stiffnesses to be small and the anharmonic terms to be fairly large for these k's. The basic simplicity of the problem is this: that *these modes with peculiar behavior represent only a very tiny fraction of all modes*, and even inserting into $\Sigma\lambda|q_k|^2$ the worst possible values for their mean displacements, they do not contribute an appreciable amount. Thus it is perfectly satisfactory to calculate the effective Hamiltonians as though these modes were reasonably harmonic, having the effective stiffnesses we compute; they then contribute very little, and the temperature dependence of the effective stiffness remains perfectly normal except for the small correction due to the displacement of q_0^0. The situation is shown in Fig. 3, where the unperturbed stiffnesses, those at low T with q_0^0 displaced, and the stiffness near T_c are all shown.

This explains one of the rather surprising observed facts about BaTiO$_3$: that the quadratic term in the Devonshire free energy equation continues to vary smoothly and linearly as $A(T - T_0)$ near the transition, in spite of the large fluctuations implied by the high dielectric constant.[5] A second fact is that the thermal vibrations, as measured by x-ray or neutron scattering, should not change much at T_c, and in particular should remain spherical. This is being confirmed by the most recent measurements.[6]

Another interesting physical effect which has not been tested is also implied by Fig. 3. This is that near, but somewhat above, T_c the stiffness for transverse optical modes, and thus the reststrahl-frequency, should decrease to very low values. Rough estimates suggest $\lambda \sim$ 1–3 mm.

In summary, we have pointed out that under the special circumstances which may apply to barium titanate, the temperature variation of the parameters in the effective Hamiltonian, which is equivalent to Devonshire's free energy function, should be controlled by modes of motion which are not sensitive to any fluctuations caused by the ferroelectric transition, which explains the experimental success of simple analytic forms for this function. Essentially, not the whole of the substance, but only a few modes of very long

q_0^0 has the famous double-minimum potential, and will displace to the position of one of the minima, with a value of order $N^{1/2}$. This is the ferroelectric polarization. From the above expressions we would expect the 4th-order term in the potential to limit the displacement. Slater and Devonshire[4] however, have shown that the positive λ_4 contribution is outweighed by a negative contribution involving the electro-strictive term $|q_0^0|^2 q_0^a$ in second order, so that the displacement is really limited by sixth order terms, a fact which is responsible for the, at first, rather puzzling abruptness of the ferroelectric transition. Nonetheless the total displacement of q_0^0 is rather small, in $BaTiO_3$ and, to a lesser extent, in several similar ferroelectrics such as $KNbO_3$, etc.

This displacement of q_0^0 has the effect, because of the term

$$N^{-1}|q_k^0|^2 \big(2\lambda_4(k) + \lambda_4(0)\big) \langle |q_0^0|^2 \rangle$$

of increasing the stiffness of all the other modes with negative stiffnesses so that they all have positive stiffnesses. Nonetheless, we can expect these stiffnesses to be small and the anharmonic terms to be fairly large for these k's. The basic simplicity of the problem is this: that *these modes with peculiar behavior represent only a very tiny fraction of all modes*, and even inserting into $\Sigma \lambda |q_k|^2$ the worst possible values for their mean displacements, they do not contribute an appreciable amount. Thus it is perfectly satisfactory to calculate the effective Hamiltonians as though these modes were reasonably harmonic, having the effective stiffnesses we compute; they then contribute very little, and the temperature dependence of the effective stiffness remains perfectly normal except for the small correction due to the displacement of q_0^0. The situation is shown in Fig. 3, where the unperturbed stiffnesses, those at low T with q_0^0 displaced, and the stiffness near T_c are all shown.

This explains one of the rather surprising observed facts about $BaTiO_3$: that the quadratic term in the Devonshire free energy equation continues to vary smoothly and linearly as $A(T - T_0)$ near the transition, in spite of the large fluctuations implied by the high dielectric constant.[5] A second fact is that the thermal vibrations, as measured by x-ray or neutron scattering, should not change much at T_c, and in particular should remain spherical. This is being confirmed by the most recent measurements.[6]

Another interesting physical effect which has not been tested is also implied by Fig. 3. This is that near, but somewhat above, T_c the stiffness for transverse optical modes, and thus the reststrahl-frequency, should decrease to very low values. Rough estimates suggest $\lambda \sim 1\text{--}3$ mm.

In summary, we have pointed out that under the special circumstances which may apply to barium titanate, the temperature variation of the parameters in the effective Hamiltonian, which is equivalent to Devonshire's free energy function, should be controlled by modes of motion which are not sensitive to any fluctuations caused by the ferroelectric transition, which explains the experimental success of simple analytic forms for this function. Essentially, not the whole of the substance, but only a few modes of very long

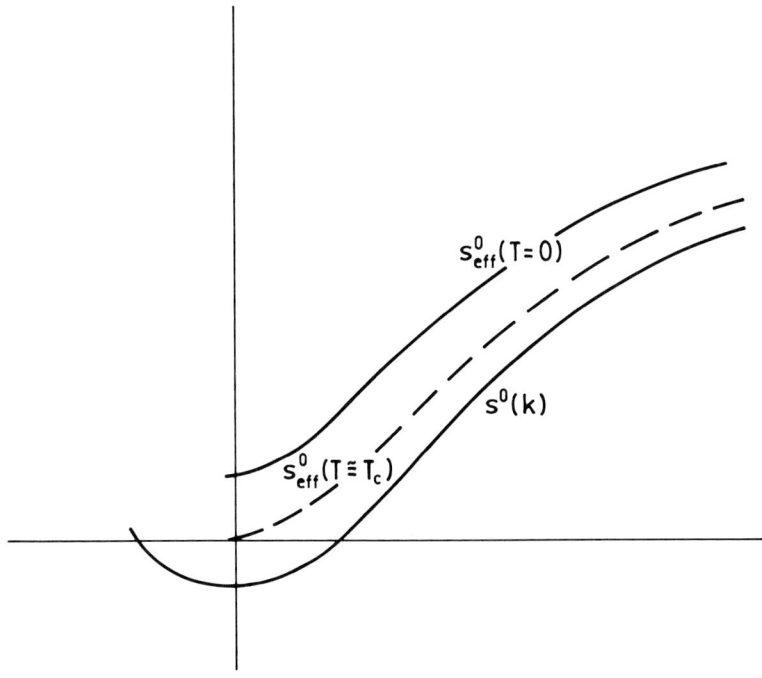

Fig. 3. Temperature dependence of $s(k)$ in $BaTiO_3$.

wavelength, experience the double-minimum potential, so that the physical theory is entirely different from that expected if each ion sees a strongly anharmonic, double-minimum potential.

References

1. P. Debye, in *Vorträge über die Kinetische Theory der Materie und Elektrizität*, M. Planck *et al.* (Teubner, Leipzig, 1914).
2. In noncubic materials, see G. Skanavi, *Doklady Akad. Nauk* **59**, 231 (1948).
3. W. R. Heller and A. Marcus, *Phys. Rev.* **81**, 809 (1951).
4. A. F. Devonshire, *Phil. Mag.* **60**, 1040 (1949); **62**, 1065 (1951); J. C. Slater, *Phys. Rev.* **78**, 748 (1950).
5. Drougard, Landauer and Young, *Phys. Rev.* **98**, 1010 (1955).
6. Frazer, Danner and Pepinsky, *Phys. Rev.* **100**, 745 (1955).

Theory of Dirty Superconductors
J. Phys. Chem. Solids **11**, 26 (1959) (appeared 1960)

This is not really a review, but it introduces the importance of time reversal to superconductivity and the scattered state representation which was used so ably by Tsuneto and by the de Gennes group.* The names "Dirty Superconductors" and "Dirty Limit" are used extensively to this day. The ideas were developed during a summer of enjoying Charlie Kittel's hospitality at Berkeley in 1958, and actually written up before (3), which was in winter '58; but the publication was delayed by problems at Pergamon Press. To me this paper and its implications were the conclusive experimental proof of the essential correctness of BCS, because superconductivity's insensitivity to most impurities is such a striking, universal fact, while its extreme sensitivity to magnetic impurities is equally striking. Many people seem not to understand the great logical force of such an argument.

A note on chronology: my paper on localization, though published in '58, was completed in summer '57. Late summer '57 through spring '58 was occupied primarily in working out the "Goldstone" and "Higgs" consequences of broken gauge invariance and hence in sorting out gauge invariance in the BCS theory. Paper (4), then, was my next logical step in superconductivity. That summer I also heard Orgel's lecture on transition metal complexes, which were an essential step in the superexchange work which I did over the winter '58–'59.

*My version is the answer to some questions first posed by Hal (H.W.) Lewis, who was at Bell en route from IAS to Santa Barbara.

THEORY OF DIRTY SUPERCONDUCTORS

P. W. ANDERSON

Bell Telephone Laboratories, Murray Hill, New Jersey

(*Received* 3 *March* 1959)

Abstract—A B.C.S. type of theory (see BARDEEN, COOPER and SCHREIFFER, *Phys. Rev.* **108**, 1175 (1957)) is sketched for very dirty superconductors, where elastic scattering from physical and chemical impurities is large compared with the energy gap. This theory is based on pairing each one-electron state with its exact time reverse, a generalization of the k up, $-k$ down pairing of the B.C.S. theory which is independent of such scattering. Such a theory has many qualitative and a few quantitative points of agreement with experiment, in particular with specific-heat data, energy-gap measurements, and transition-temperature versus impurity curves. Other types of pairing which have been suggested are not compatible with the existence of dirty superconductors.

ONE of the most striking experimental facts about superconductivity is that it is often insensitive to enormous amounts of physical and chemical impurities. For one example, several substances in essentially an amorphous state have been shown to be superconductors, such as bismuth and beryllium films laid down at liquid-helium temperatures.[1] As another example, there are disordered alloy systems with 20–50 per cent of chemical scattering centers, but with transition temperatures comparable with those of pure elements.[2] These quantities of crystal imperfections are large enough to scatter the electrons at an extremely rapid rate. In fact, if we were to take the mean free time before scattering for the electron as a measure of the electrons' uncertainty in energy, that uncertainty in energy is large compared not only with the energy gap ϵ_0, but with the Debye energy $\hbar\omega_D$. Plane-wave states for the electrons definitely have this very large degree of energy uncertainty.

On the other hand, the experiments of SERIN *et al.*[3] have shown that, starting with a pure single crystal of a superconducting material, there is usually a rather sharp initial drop in the superconducting transition temperature as the first small percentage of chemical imperfection is added. They show that this initial drop is proportional to the extra resistivity caused by these imperfections, and therefore proportional to the amount of scattering. If the impurities which are introduced are magnetic ions rather than ordinary chemical impurities, MATTHIAS *et al*[4] have shown that this initial sharp drop continues, and superconductivity is very soon destroyed. On the other hand, for ordinary impurities the sharp drop stops rather soon and is replaced by a more gradual behavior, which seems to be determined primarily by the fact that the impurity adds or subtracts electrons from the band, changes the density of states, and in various ways gradually varies the parameters of the free electrons. Thus we may divide superconductivity into two regions: (1) the region of relatively pure superconductors where scattering has a rather sharp effect on superconducting transition temperatures; and (2) the region of very imperfect superconductors, where additional scattering has very little effect. It is the purpose of the present paper to give a theory of this region of the "dirty" superconductor.

The fundamental assumption we will make is that in this region the problem of the electron wave functions is best solved by first diagonalizing the scattering interaction between the electrons and the impurities, and then calculating the phonon interactions between electrons. Finally, one calculates from this the superconducting properties. That is, we find a new set of one-electron wave functions for the electrons, and then solve the problem of the interactions of the electrons in terms of these, rather than in terms of ordinary plane-wave functions, such as are appropriate in the region of the pure superconductor. All of this, of course, assumes that the scattering is perfectly elastic, as it is from any form of chemical or physical imperfection. (At the low temperatures where superconductivity is important, inelastic phonon scattering does not play an important role.) So what we do is simply to assume that somebody has solved for us the extremely difficult problem of the wave functions of the electrons in the presence of the scatterers, and write down the resulting wave function $\psi_{n\sigma}$:

$$\psi_{n\sigma} = \sum_k (n|k)\phi_{k\sigma}, \quad (1)$$

where $\phi_{k\sigma}$ are the Bloch waves and $(n|k)$ the unitary transformation solving the scattering problem. (We give ψ_n a spin index σ; this may not be valid in heavy elements because of spin-orbit coupling, but the theory is still valid there.)

The basic observation which we make is that if $\psi_{n\sigma}$ is such an exact one-electron wave function in the presence of scatterers, and if the scatterers are nonmagnetic, the time-reversed wave function, $(\psi_{n\sigma})^*$, is also an exact wave function of the one-electron Hamiltonian. What is more, $(\psi_{n\sigma})^*$ has the same energy as $\psi_{n\sigma}$ itself. We shall call this energy E_n. The time-reversed wave function $(\psi_{n\sigma})^*$ is:

$$(\psi_{n\sigma})^* = \sum_k (n|k)^* \phi_{-k-\sigma}. \quad (2)$$

Now, having the correct one-electron wave functions, we proceed to derive the phonon interaction between these electrons. It is not at all correct simply to transform the B.C.S. interaction[5] directly into the interaction between these new wave functions $\psi_{n\sigma}$; the reason for this is that $\psi_{n\sigma}$ contains plane wave functions which have quite different energies, and in particular, in the presence of strong scattering contains wave functions from outside of the region where the B.C.S. electron interaction is attractive. Therefore, this method would lead us to the conclusion that the interaction between electrons is altered rather seriously. Actually, it is correct instead to write down the interaction between these new scattered-electron wave functions and the phonons, and then to do the second-order perturbation theory which

gives us the interaction between the electrons caused by the phonons. When the calculation is done this way, the energy denominators in the second-order perturbation theory contain the energies not of the initial plane-wave states, E_k, but rather the energies E_n of the scattered one-electron states. Whether or not the interaction is attractive depends primarily on what these energy denominators are. Therefore, we find that the attractiveness or not of the interaction is now a function not of E_k, the energy of the plane-wave states, but of E_n, the energy of the scattered-electron states. Without going into excessive detail, we simply write down the part of the interaction which corresponds to the B.C.S. truncated Hamiltonian:[5]

$$\mathscr{H}_{\text{red.}} = -\sum_{n,n'} V_{nn'} c_n^* c_{-n}^* c_{-n'} c_{n'},$$

$$V_{nn'} = \sum_{k,k'} \frac{|(n|k)|^2 |(n'|k')|^2 |M_{k-k'}|^2 \hbar\omega_{k-k'}}{(\hbar\omega_{k-k'})^2 - (E_n - E_{n'})^2}. \quad (3)$$

Here c_n^* and c_{-n}^* are creation and destruction operators for electrons in state $\psi_{n\sigma}$ and $(\psi_{n\sigma})^*$, $M_{k-k'}$ and $\hbar\omega_{k-k'}$ have the usual meaning, and $(n|k)$ is defined in equation (11).

This interaction is summed over all the plane wave functions which are contained in the scattered function n, with a coefficient which is given by the square of the amount of the state contained in the state n. Normalization requires the following equation:

$$\sum_k |(n|k)|^2 = 1. \quad (4)$$

Because of equation (4), if the parameters entering in the interaction were constants, as was assumed by B.C.S., the interaction would be exactly the same for the scattered state as it was for the plane-wave state. As it is, the interaction is not a constant, but at best only roughly so, and expression (3) picks out of the total interaction only the constant part. That is, the interaction (3) is strictly the average interaction over all the states going to make up the scattered state n. Since the states which make up this scattered state are, at least under conditions of strong scattering, taken more or less randomly from all the regions of the Fermi surface, we conclude that the interaction will be (aside from a smooth energy dependence) a constant, averaged over the entire Fermi surface.

The rest of the interaction, the part which was removed in the truncated Hamiltonian of B.C.S., is changed in a very radical way. In the pure substance the rest of the interaction can be thought of as an interaction between pairs which do not have exactly zero total momentum but rather some finite momentum. However, under conditions of strong scattering this part of the interaction does not take this form at all; each individual matrix element is smaller by a number of the order of the total number of electrons. This is because the momentum selection rule no longer holds, so that there is no reason why any individual matrix element should vanish. This increases the total number of matrix elements and must therefore decrease their magnitude. Thus, in the scattered state the B.C.S. part of the interaction, or rather the transformed B.C.S. part, plays a much more obviously unique role than it does in the pure superconductor.

One final comment about this interaction: obviously if the scattering is strong enough the procedure which we have followed, of first diagonalizing the one-electron Hamiltonian including scattering, and then introducing the electron–phonon interaction and the interaction between the electrons which results from it, is correct. When we ask the question: at what degree of scattering is this no longer the correct procedure? the first guess would be that one should find some average amount of electron–phonon interaction and when the scattering becomes less than that the procedure is no longer correct. However, it is fairly easy to convince oneself that this is not the correct way, but that the diagonalization of the one-electron part of the Hamiltonian is correct at very much smaller amounts of scattering. As a matter of fact, the procedure we have followed seems to be correct until the actual interaction *between* the electrons caused by the electron–phonon interaction begins to come in to play. That is, it is correct until $\hbar/\tau$ becomes comparable with the energy gap, which is a reasonable measure of the electron–electron interaction. This is, in fact, the experimental criterion for the transition from the region of strong scattering, as here defined, to the region of weak scattering.[3]

Now we shall draw some physical conclusions from these ideas. First we should observe that it is possible to solve the B.C.S. integral equation and derive a theory of superconductivity just as well as in the new situation with the new averaged interaction and the scattered wave functions ψ_n as it was in the old situation with the old interaction and the plane-wave functions ψ_k. A general result, which is fairly easy to prove, is that the energy gap, and therefore the transition temperature, will always be slightly smaller for the scattered states than they would be in the pure case, essentially because the average taken in the scattered case is not as favorable as one gets in the pure case. This explains why it is that in the pure superconductor region the transition temperature drops so radically, and yet stops dropping after one gets into the region of strong scattering; and at the same time it explains why this drop is relatively small, because one does not expect the difference between the two energy gaps to be very large. Thus, we see that these ideas explain fairly satisfactorily the general features of SERIN's results. It is also clear that when the time-reversal transformation cannot be made, that is, when the energy of the state ψ_n is not the same as the energy of ψ_{-n}, all this cannot be done, and the transition temperature will continue to drop as the degree of magnetic scattering increases.

An interesting question is what size of particles and at what degree of scattering will superconductivity actually cease. The first point is that, as long as the particle size remains fairly large, no quantity of scattering which leaves the substance a metal would seem to be capable of actually destroying superconductivity, because the average which is taken over the Fermi surface does not depend in any important way on the actual amount of scattering. On the other hand, on reducing the particle size, we will begin to get to the point at which the scattered wave functions ψ_n have energies E_n which are separated by discrete energy gaps. That is, their energies must extend over something like the total Fermi energy, which is about 10 eV; and if there were only about a thousand electrons, that would mean that the energy differences between the states would be of the order of 0·01 eV. With such energy differences among the E_n, superconductivity would no longer be possible; in fact, it is fairly easily seen that the energy differences must be less than the energy gap. That means that particles with fewer than about 10^4–10^5 electrons will begin to be affected. So far no experiments are reported on particles of this order of magnitude of size, although the particles used in REIF's nuclear-resonance experiments are beginning to approach this point.[6]

Another conclusion is that the B.C.S. theory in its original form, assuming a constant interaction, will be more nearly correct in the dirty superconductor region that it will be for pure superconductors; that is, in the impure superconductors the interaction is relatively a constant, and therefore the energy gap itself will be a constant and the thermal and other results of B.C.S. should be nearly exact. On the other hand, in pure single crystals of superconductors, one can very quickly show that the energy gap will be a strong function of the momentum vector on the Fermi surface, because most superconductors have fairly complicated Fermi surfaces, and it would be a miracle if the interactions were sufficiently constant to maintain a constant energy gap. There are two types of experiments which bear on this point. One type of experiment is the electronic specific heat and, in fact, various recent experiments[7] show deviations from the exponential specific-heat curve of the B.C.S. theory in the direction which would be expected if the energy gap were a function of position on the Fermi surface. Experiments on less perfect single crystals have shown less such deviations, as is to be expected. On the other hand, most of the direct experiments, in particular the experiments of TINKHAM and his co-workers[8] on the optical measurement of the energy gap and the experiments of HEBEL and SLICHTER[9] measuring the density of states by the relaxation time in nuclear resonance, are necessarily undertaken in the dirty superconductor region. In the case of TINKHAM's experiments, the measurement is necessarily made near a surface, while the other measurement is made on small particles. Therefore, neither of these types of experiments would have been expected to show any considerable anisotropy of the energy gap. The former do not; the structure observed in the latter experiments[10] is probably a collective excitation.[11] The question of the experimental investigation of the anisotropy is a fascinating one which remains open so far as I know. Our conclusion is that the recent experiments on the detailed investigation of the specific-heat curve must be considered as being more of a confirmation of the B.C.S. theory than vice versa, because they show that the expected anisotropy of the energy gap is actually there.

In conclusion I should like to make a number of acknowledgements and apologies. In the first place various notions about time reversal and superconductivity have appeared independently in a number of places. Most particularly, ABRAHAMS and WEISS at Rutgers have made similar calculations, and BARDEEN and MATTIS[12] have used a related wave function. In the second place, the idea of the anisotropy of the energy gap occurred independently to COOPER and to PIPPARD and HEINE[13]. Thus, the purpose of the present paper is merely to summarize in a physically consistent way all of these ideas, and to show that there is good agreement qualitatively with experiments. I should also acknowledge interesting discussions with C. HERRING and H. SUHL.

A final comment is that the nucleus is itself a dirty superconductor in the sense that the nucleus is a very fine particle with only a very small number of Fermi particles in it. Thus, it is not surprising to find that the theory of nuclei contains a "pairing" concept for $+m_1-m$ pairs[14] which shows very great similarity to the discussion in this paper.

It is hard to see how any explanation other than time-reversal invariance will explain all these facts; in particular, the suggestions of inexact pairings or pairings of parallel-spin electrons advanced relative to the explanation of Knight shift results[15] certainly fail of agreement with the facts about dirty superconductors. It seems to the author that these considerations represent the strongest arguments that the theory of superconductivity must have some relation to zero-momentum, opposite-spin pairs.

REFERENCES

1. LAZAREV B. G., SUDOVTSEV A. I. and SMIRNOV A. P., *Zh. Eksper. Teor. Fiz.* **33**, 1059 (1958); (translation) *Sov. Phys.* (*J.E.T.P.*) **6**, 816 (1958).
2. MATTHIAS B. T., WOOD E. A., CORENZWIT E. and BALA V. B., *J. Phys. Chem. Solids* **1**, 188 (1956).
3. LYNTON E. A., SERIN B. and ZUCKER M., *J. Phys. Chem. Solids* **3**, 165 (1957).
4. MATTHIAS B. T., SUHL H. and CORENZWIT E., *Phys. Rev. Letters* **1**, 92 (1958).
5. BARDEEN J., COOPER L. N. and SCHREIFFER J. R., *Phys. Rev.* **108**, 1175 (1957).

6. REIF F., *Phys. Rev.* **106,** 208 (1957).
7. PHILLIPS N. E., *Phys. Rev. Letters* **1,** 363 (1958). BOORSE H. A., *Phys. Rev. Letters* **2,** 391 (1959).
8. RICHARDS P. L. and TINKHAM M., *Phys. Rev. Letters* **1,** 318 (1958).
9. HEBEL L. C. and SLICHTER C. P., *Phys. Rev.* **107,** 901 (1957).
10. TINKHAM M., RICHARDS P. L. and GINSBURG D., private communication.
11. ANDERSON P. W., *Phys. Rev.* **112,** 1900 (1958).
12. BARDEEN J. and MATTIS D. C., *Phys. Rev.* **111,** 412 (1958).
13. PIPPARD A. B. and HEINE V., *Conference on Properties of Metals at Low Temperature*, Geneva, N.Y. (1958).
14. BOHR A. and MOTTELSON B. R., private communication; M. G. MAYER, *Phys. Rev.* **78,** 22 (1950).
15. HEINE V. and PIPPARD A. B., *Phil. Mag.* **3,** 1046 (1958).

Calculation of the Superconducting State Parameters with Retarded Electron-Phonon Interaction

(with P. Morel)

Phys. Rev. **125**, 1263 (1962)

This paper mentions that there will be structure in the energy gap function associated with phonons: when I brought this out in a talk at Birmingham in late '61, R. Peierls asked how it might be measured. At most a few weeks later I heard from J.M. Rowell at Bell Labs. that it had been, in the tunneling characteristic. The next few years left BCS with a sound quantitative basis. The paper owed much to Schrieffer's discussions with me and with Pierre Morel, my student inherited from D. Pines; Pierre was simultaneously serving as science attache at the French embassy.

Calculation of the Superconducting State Parameters with Retarded Electron-Phonon Interaction

P. Morel
French Embassy, New York, New York

AND

P. W. Anderson
Bell Telephone Laboratories, Murray Hill, New Jersey
(Received October 10, 1961)

> The energy gap and other parameters of the superconducting state are calculated from the Bardeen-Cooper-Schrieffer theory in Gor'kov-Eliashberg form, using a realistic retarded electron-electron interaction via phonons and including the Coulomb repulsion. The solution is facilitated by observing that only the local phonon interaction, mediated entirely by short-wavelength phonons, is important, and that a good approximation for the phonon spectrum is therefore an Einstein model rather than Debye model. The resulting equation is solved by an approximate iteration procedure. The results are similar to earlier gap equations but the derivation gives a precise meaning to the interaction and cutoff parameters of earlier theories. The numerical results are in good order-of-magnitude agreement with the observed transition temperatures but lead to an isotope effect at least 15% less than the accepted $-\frac{1}{2}$ exponent (T_c proportional to $M^{-\frac{1}{2}}$). Also, the present theory predicts that all metals should be superconductors, although those not observed to do so would have remarkably low transition temperatures.

INTRODUCTION

ALTHOUGH the original Bardeen, Cooper, and Schrieffer (BCS) theory of superconductivity[1] has gone far in explaining the basic processes which account for the condensation of fermion systems, it must still be considered as a phenomenological theory with respect to the use which is made of a rather nonphysical "effective potential" to describe the complex Coulomb and phonon-induced interactions between the electrons in a metal. We may recall that the BCS effective potential is instantaneous and displays a strongly oscillating behavior in coordinate space (since its Fourier transform is sharply cut off in momentum space). Since the strength V of this effective interaction appears only as an adjustable parameter, the BCS model is still adequate to describe most properties of superconductors; it is difficult, however, to see how this parameter V could be related to the retarded, time-dependent electron-phonon interaction and the essentially instantaneous Coulomb repulsion or a combination thereof.

On the contrary, several physical reasons lead us to the conclusion that the coordinate space dependence of the interaction potential plays a very subsidiary role whereas its retardation in time is indeed of major importance in determining the various cutoff phenomena. Firstly, it appears that the quite long range part (in space) of the phonon-induced interaction results primarily from the emission and absorption of very long wavelength phonons. These phonons can enter either through "direct" processes, in which the dielectric screening of the metal is nearly complete, or through "umklapp" processes. These latter processes have been supposed in the past[2,3] to play a major role, but this appears very much in doubt in view of the deformation potential theorem which requires that the effective potential acting to cause scattering by any phonon be, to first order, proportional to the strain produced by the phonon and not to the displacement. In this perspective, an Umklapp process may be viewed as the excitation of a true phonon corresponding to an actual deformation of the lattice plus a stationary wave corresponding to a translation of the lattice as a whole. Since the electron wave functions follow adiabatically a translation of the lattice, only the actual deformation causes any scattering. Consequently, Umklapp processes are no more effective than direct ones. The net result, then, is that the phonon-induced interaction is mediated primarily through short-wavelength phonons. It is well known experimentally and theoretically that the short-wavelength part of the phonon spectrum is rather sharply peaked about a few definite frequencies, so that it is a quite good approximation to think in terms of an Einstein model of a few groups of single-frequency, nonpropagating phonons. And of course, phonons which do not propagate in space lead directly to a localized interaction.

A second line of reasoning leading to the same conclusion stems from the particular form of the Gor'kov-Eliashberg energy gap equation.[4,5] In these works, the energy gap function in coordinate space

---

[1] J. Bardeen, L. N. Cooper, and J. R. Schrieffer, Phys. Rev. **108**, 1175 (1957); hereafter referred to as BCS.

[2] D. Pines, Phys. Rev. **109**, 280 (1958).
[3] P. Morel, J. Phys. Chem. Solids **10**, 277 (1959).
[4] L. P. Gor'kov, J. Exptl. Theoret. Phys. (U.S.S.R.) **34**, 735 (1958). [Translation: Soviet Phys.—JETP **7**, 505 (1958).] G. M. Eliashberg J. Exptl. Theoret. Phys. (U.S.S.R.) **38**, 966 (1960). [Translation: Soviet Phys.—JETP **11**, 696 (1960).]
[5] P. W. Anderson, *Proceedings of the Seventh International Conference on Low-Temperature Physics* (University of Toronto Press, Toronto, 1960).

and time is

$$\Sigma(\mathbf{r}-\mathbf{r}', t-t') = F(\mathbf{r}-\mathbf{r}', t-t') V(\mathbf{r}-\mathbf{r}', t-t'),$$

where V is the retarded potential caused by phonons and F is the Gor'kov pair Green's function. F has a long range in space and oscillates with a wavelength of the order of $2k_0^{-1}$ (k_0 is the Fermi momentum), while the spatial dependence of V is only related to the particular features of the phonon spectrum and may display an oscillating behavior with wavelength of the order of the inverse of the Debye momentum. Destructive interference makes the product of F and V small for all but quite small spatial distances $(\mathbf{r}-\mathbf{r}')$. Thus, we find once more that only the short-range part of the interaction potential is important.

It is therefore desirable that a treatment actually taking into account the time dependence of the electron-phonon interaction be developed, with the hope that the main features of the BCS model, and particularly the isotope effect, may be retrieved by this approach. Fortunately, a new and powerful analysis of the condensation phenomenon in terms of particle propagators has been derived by several authors,[4] thereby providing us with a suitable formalism for our purpose. In the first section, we shall briefly set up our notation and write the Dyson equation for the problem. In Sec. II, we reduce this equation to a single integral equation after integrating over all spatial variables; we then solve this equation approximately by a suitable iteration process in the case of electron-phonon interaction only and within the frame of the Einstein model; we find that the corresponding gap equation is identical to the BCS equation. It has been known since the work of Bogoliubov[6] that the screened Coulomb interaction acts to a degree like a hard core, at least in the instantaneous interaction model previously used, and so is quite ineffective in weakening the phonon-induced attraction. In Sec. III, we find that this result does indeed hold in our time-dependent treatment; adding an essentially instantaneous repulsive potential to the retarded phonon-induced potential, we solve the integral gap equation to find identically Bogoliubov's result. All the above results are derived in the case where the phonon spectrum reduces to a single frequency. We investigate in Sec. IV the effect of the spreading of this spectrum about one central frequency and we find closely similar results (identical in the weak-coupling limit) thereby justifying *a posteriori* our use of the Einstein model. Finally (Sec. V), we derive the expression for the isotope effect and we compute the transition temperatures for nontransition metals. Unfortunately, as in the calculation of Swihart,[7] the exponent in the isotope effect is found to differ appreciably (10 to 20%) from the ideal value $\frac{1}{2}$, more in fact than the quoted

errors of the present experimental data. On the other hand, we find a general order-of-magnitude agreement between the computed and measured transition temperatures better than that of Morel.[3]

I. THE DYSON EQUATION

Following closely the notation of Gor'kov and Eliashberg,[4] we define the Green's function $G(x-x')$ describing the propagation of an ordinary single particle:

$$G(x-x') = i\langle T\psi_+(x)\psi_+^*(x')\rangle = i\langle T\psi_-(x)\psi_-^*(x')\rangle, \quad (1)$$

as well as the Green's function $F(x-x')$ describing the merging of two particles into the condensed phase or the inverse process (i.e., the creation or the destruction of a "ground pair"):

$$F(x-x') = ie^{i\mu(t+t')}\langle T\psi_+(x)\psi_-(x')\rangle$$
$$= ie^{-i\mu(t+t')}\langle T\psi_-^*(x)\psi_+^*(x')\rangle, \quad (2)$$

where the last equality is obtained by time reversal (provided we use a representation in which the state vectors are invariant under this transformation). Here x stands for the four-vector $\mathbf{x}, t$ and T is the usual time ordering operator (see reference 4). The average in (1) is taken in the ground state $|\varphi_0(N)\rangle$ corresponding to a total number N of particles; the averages in (2) are taken between the ground states $|\varphi_0(N)\rangle$ and $|\varphi_0(N\pm 2)\rangle$. Lumping together in ϵ_k the kinetic energy of a single particle and the mass correction due to the self-energy in the normal fluid, we may write the Dyson equations in momentum space:

$$G(\mathbf{k},\omega) = -\frac{\omega+\epsilon_k}{\omega^2-\epsilon_k^2-\Sigma^2(\mathbf{k},\omega)+i\delta},$$
$$F(\mathbf{k},\omega) = \frac{\Sigma(\mathbf{k},\omega)}{\omega^2-\epsilon_k^2-\Sigma^2(\mathbf{k},\omega)+i\delta}, \quad (3)$$

where $\Sigma(\mathbf{k},\omega)$ is the self-energy corresponding to the processes in which two particles either merge into the condensed phase or emerge from it. This self-energy plays the part of the "energy gap" of BCS theory since the energy spectrum of the individual excitations of the condensed fluid is now given, in the low-energy limit, by

$$E_k = [\epsilon_k^2 + \Sigma^2(\mathbf{k},0)]^{\frac{1}{2}}. \quad (4)$$

$\Sigma(\mathbf{k},\omega)$ is related to the propagator $F(\mathbf{k},\omega)$ by

$$\Sigma(\mathbf{k},\omega) = \frac{i}{(2\pi)^4}\int \Gamma(\mathbf{k}-\mathbf{k}', \omega-\omega')F(\mathbf{k}',\omega')d^3k'd\omega', \quad (5)$$

where Γ is the exact vertex part, i.e., the sum of the contributions of all possible interaction processes or combinations of elementary electron-phonon inter-

---

[6] N. N. Bogoliubov, V. V. Tolmachev, and D. V. Shirkov, *A New Method in the Theory of Superconductivity* (1958) (translation: Consultants Bureau, Inc., New York, 1959).
[7] J. C. Swihart, Phys. Rev. **116**, 45 (1959).

action:

$$H_{\text{el-ph}} = -i \sum_{\mathbf{kk'}} \left(\frac{N}{M}\right)^{\frac{1}{2}} \frac{ZqV(q)}{[2\omega_q]^{\frac{1}{2}}} (b_q + b_{-q}^*) c_{\mathbf{k'}}^* c_{\mathbf{k}}$$

$$= i \sum_{\mathbf{kk'}} \alpha_q (b_q + b_{-q}^*) c_{\mathbf{k'}}^* c_{\mathbf{k}}, \quad (6)$$

and direct Coulomb interaction:

$$H_{\text{Coul}} = \sum_{\mathbf{k},\mathbf{k'},\mathbf{K}} V(\mathbf{k'}-\mathbf{k}) c_{\mathbf{k'}}^* c_{\mathbf{K}-\mathbf{k'}}^* c_{\mathbf{K}-\mathbf{k}} c_{\mathbf{k}}. \quad (7)$$

In the above expressions, N, M, and Z are, respectively, the number of ions per unit volume, the mass, and the valency of the ions; b_q (b_q^*) and $c_{\mathbf{k}}$ ($c_{\mathbf{k}}^*$) are the usual annihilation (creation) operators of the phonon and electron fields, respectively; $V(q)$ is the Fourier transform of the screened electrostatic potential:

$$V(q) = \frac{4\pi e^2}{q^2 + k_s^2} = \frac{4\pi e^2}{q^2 + 4\pi e^2 N_0}, \quad (8)$$

where k_s^{-1} is the screening radius (for the Fermi-Thomas model, k_s^2 is proportional to the density of states N_0 on the Fermi surface). Finally, let us remark that the phonon momentum $\mathbf{q}$ appearing in Eq. (6) is equal to $(\mathbf{k'}-\mathbf{k})$ only if this vector happens to be in the first Brillouin zone (normal process). In general, $\mathbf{q}$ is a function of the vector $(\mathbf{k'}-\mathbf{k})$ given by the momentum conservation relation:

$$\mathbf{q} = \mathbf{k'} - \mathbf{k} + \mathbf{K}_N, \quad (9)$$

where $\mathbf{K}_N$ is a suitable vector of the reciprocal lattice. It must be clearly stated that the summation in (6) extends over all $\mathbf{k}$ and $\mathbf{k'}$ near the Fermi surface, since, because of the periodicity of the crystal, the deformation potential caused by the phonon of momentum $\mathbf{q}$ has nonvanishing matrix elements not only for scattering from the electron state $\mathbf{k}$ to the state $\mathbf{k'}$ such that:

$$\mathbf{k'} - \mathbf{k} = \mathbf{q},$$

but also to the states $\mathbf{k'}$ satisfying the more general condition (9). Because of the deformation potential theorem, the matrix element is approximately the same whether the scattering is a normal or an "umklapp" process, for a given $\mathbf{q}$.

Now, we shall restrict the vertex part Γ to its lowest order terms only, i.e., the sum of

$$D_0(\mathbf{q},\omega) = \alpha_q^2 \left[\frac{1}{\omega_q - \omega - \eta_q} + \frac{1}{\omega_q + \omega - \eta_q} \right] = \frac{2\alpha_q^2}{\omega_q} u_q(\omega), \quad (10)$$

and $-V(q)$ for the phonon induced and Coulomb interactions respectively. Here ω_q and η_q are the real and imaginary parts of the frequency of the phonon of momentum $\mathbf{q}$. Since the phonon damping is rather small (η_q is usually of the order of $10^{-2}\omega_q$), the frequency dependent factor $u_q(\omega)$ may be written:

$$u_q(\omega) = P\left[\frac{\omega_q^2}{\omega_q^2 - \omega^2}\right] + \frac{i\pi}{2}\omega_q[\delta(\omega + \omega_q) - \delta(\omega - \omega_q)]. \quad (11)$$

It has been shown by Migdal[8] that this simplification is acceptable for the phonon-electron interaction since higher-order terms are of the order of $(M)^{-\frac{1}{2}}$ or smaller. We have approximated the exact Coulomb interaction by an instantaneous potential, neglecting high-order retarded polarization terms: This is perfectly acceptable since dispersion occurs only at rather high frequencies of the order of the plasma frequency and is completely negligible in the small energy range we are considering here. We are also neglecting mixed terms corresponding to a combination of phonon exchange and Coulomb interactions: These terms are certainly smaller than $(M)^{-\frac{1}{2}}$. The gap equation (5) together with (3) and (10) constitutes therefore the mathematical formulation of the problem. We shall now be concerned with simplifying and solving this equation in the perspective outlined in the introduction.

II. SOLUTION OF THE GAP EQUATION FOR ELECTRON PHONON COUPLING ONLY

We shall first neglect the Coulomb repulsion altogether and consider only the contribution $D_0(q,\omega)$ to the vertex part Γ. The gap equation then becomes

$$\Sigma(\mathbf{k},\omega) = \frac{i}{(2\pi)^4} \int \frac{2\alpha_q^2}{\omega_q} u_q(\omega - \omega') F(\mathbf{k'},\omega') k'^2 \\ \times \sin\theta' d\theta' d\varphi' dk' d\omega', \quad (12)$$

where $F(\mathbf{k},\omega)$ depends only upon the kinetic energy:

$$\epsilon_k = v_0(k - k_0). \quad (13)$$

(v_0 is the velocity of the electrons on the Fermi level) and $\mathbf{q}$ is given by (9). Since F vanishes rapidly[9] when $\mathbf{k'}$ is allowed to depart from the Fermi surface, we may replace the integration over $\mathbf{k'}$ by integration over the energy and the angular variables. Moreover, we see from (6) that

$$\frac{2\alpha_q^2}{\omega_q} = \frac{NZ^2}{Mc^2}\left[\frac{4\pi e^2}{k_s^2 + q^2}\right]^2 = \frac{1}{N_0}\left[\frac{k_s^2}{k_s^2 + q^2}\right]^2. \quad (14)$$

We have made use of the standard expression for the velocity of sound (see reference 3). Note that this factor is almost independent of q, so that we may replace q^2 by a mean value of the order of the square or the Debye momentum (more precisely $\frac{3}{5}q_D^2$). The factor

---

[8] A. B. Migdal, J. Exptl. Theoret. Phys. (U.S.S.R) **34**, 1438 (1958). [Translation: Soviet Phys.—JETP **7**, 996 (1958).]

[9] The q^{-2} dependence of $\Gamma(q,\omega)$ insures the convergence of the original equation (5). Replacing the integration over k' by integration over ϵ' may, however, introduce an artificial divergence in the case of an instantaneous interaction. We shall in this case cut off the energy integration at the Fermi energy.

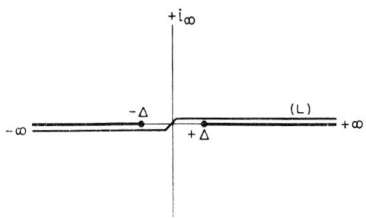

FIG. 1. Contour of integration (L) in the complex (z) plane for Eq. (18). Note that (L) does not cross the branch line of the function $D(z)$ on the real axis (heavy line).

$u_q(\omega-\omega')$ is strongly dependent upon the phonon frequency ω_q. For polyvalent metals, however, the relation between ω_q and the vector $(\mathbf{k}'-\mathbf{k})$ is quite complicated, so that the angular average of (12) requires a complicated procedure taking into account the structure of the crystal (see Morel, reference 3). On the other hand, since we expect $\Sigma(\mathbf{k},\omega)$ to be essentially independent of $\mathbf{k}$, we may average (12) over all $\mathbf{k}$ in the energy shell. This is equivalent to averaging $u_q(\omega-\omega')$ over the phonon spectrum (all phonon momenta are roughly equiprobable). This approximation, which is akin to the spirit of BCS theory, is of course self-consistent since it removes all $\mathbf{k}$ dependence in the right-hand side of (12); it amounts simply to assuming that the gap is isotropic, which has been shown to be approximately correct from an experimental point of view.[10] We obtain then

$$\Sigma(\omega) = \frac{i\lambda}{2\pi} \int d\epsilon' \int d\omega' F(\epsilon',\omega') U(\omega-\omega'), \quad (15)$$

where λ is a parameter which plays the role of the "$N_0 V$" of BCS:

$$\lambda = \tfrac{1}{2}\left[\frac{k_s^2}{k_s^2 + \tfrac{3}{5} q_D^2}\right]^2, \quad (16)$$

and $U(\omega)$ is the average of the phonon propagator $u_q(\omega)$ over the phonon spectrum, represented by the distribution function $g(\omega_q)$:

$$U(\omega) = \int_0^{\omega_{\max}} u_q(\omega) g(\omega_q) d\omega_q. \quad (17)$$

It is worthwhile to pause at this point and remark that λ is not directly proportional to the density of states on the Fermi surface N_0 as the analogy with BCS's expression $N_0 V$ tends to suggest. The dependence of λ upon N_0 enters only through k_s^2 and we find particularly that however large N_0 may be or however complicated the structure of the crystal may be $[(q^2)_{\mathrm{av}} \ll q_D^2]$, the value of this parameter cannot exceed $\tfrac{1}{2}$.

Restricting ourselves for the time being to the Einstein model, i.e., to the case where the phonon spectral distribution function g reduces to a single δ

[10] P. W. Anderson

function centered at $\omega_q = \omega_1$, we shall proceed to integrate (15) over the energy. Before doing so, however, we write this equation in the equivalent form:

$$\Sigma(\omega) = \frac{i\lambda}{2\pi} \int_{(L)} dz \int_{-\infty}^{+\infty} d\epsilon' \frac{\Sigma(z)}{D^2(z) - \epsilon^2} U(\omega-z), \quad (18)$$

$$D(z) = [z^2 - \Sigma^2(z)]^{\frac{1}{2}}. \quad (19)$$

Note that $D(z)$ retains the same determination when z follows the contour (L) represented in Fig. 1. If we choose the determination in the upper half of the complex plane, we obtain in a straightforward fashion:

$$\Sigma(\omega) = \lambda \int_{(L)} \frac{\omega_1}{2}\left[\frac{1}{\omega_1 - z + \omega - i\eta} + \frac{1}{\omega_1 + z - \omega - i\eta}\right] \frac{\Sigma(z)}{2D(z)} dz. \quad (20)$$

As we expect from the time-reversal invariance of F and V, this equation is compatible with an even solution on the real axis. On the other hand, the detailed features of the solution do not appear clearly. It is, therefore, illuminating to go back temporarily to the time representation of relations (3) and (18):

$$F(\mathbf{k},t) \simeq [\Sigma(E_\mathbf{k})/2E_\mathbf{k}] e^{iE_\mathbf{k} t},$$

$$\Sigma(t) = \lambda U(t) C(t) = \lambda U(t) \int_{-\infty}^{+\infty} F(\mathbf{k}',t) d\epsilon'. \quad (21)$$

The oscillating time dependence of the integrand is smeared out by the integration and $C(t)$ is a rather slowly varying function of time (see Sec. IV) so that the resulting $\Sigma(t)$ behaves essentially like $U(t)$. Consequently, we expect that the solution of the integral equation (20), may be approximately proportional to $U(\omega)$. In order to test this possibility, we shall try to solve it by iteration, starting from the trial function:

$$\Sigma_1(\omega) = \Delta U(\omega),$$

$$\Sigma_2(\omega) = \lambda \Delta \int_0^\Delta \frac{U(z) U(\omega-z) dz}{i[\Delta^2 - z^2]^{\frac{1}{2}}} + \int_\Delta^\infty \frac{U(z) U(\omega-z) dz}{[z^2 - \Delta^2]^{\frac{1}{2}}}, \quad (22)$$

where we have taken advantage of the smallness of $\Sigma(\omega)$ to replace it by the constant $\Sigma(0) = \Delta$ in the expression for $D(z)$. These integrals cannot be computed with any accuracy near the singularity $\omega \simeq \omega_1$; however, in the regions both above and below this singularity, the integration can be carried out approximately and one finds:

$$\Sigma_2(\omega) = \lambda\Delta[\ln(2\omega_1/\Delta)U(\omega) + \tfrac{3}{4}(1-i\pi/2)\omega/\omega_1 + O(\omega^3)],$$
$$\omega \ll \omega_1,$$
$$\Sigma_2(\omega) = \lambda\Delta[\ln(2\omega_1/\Delta)U(\omega) - i\pi(\omega_1^3/\omega^3) + O(1/\omega^4)], \quad (23)$$
$$\omega \gg \omega_1.$$

This expression is reasonably similar to the trial function $\Sigma_1(\omega)$ and, indeed, converges toward Σ_1 in the weak coupling limit ($\lambda \ll 1$) if the value of the gap $\Sigma(0) = \Delta$ satisfies the following condition:

$$1/\lambda = \ln(2\omega_1/\Delta), \quad (24)$$

identical to the BCS gap equation. Finally, let us note, for further reference, that $\Sigma_2(0)$ is real and also that $\Sigma_2(\omega)$ has no singularity at $\omega = 2\omega_1$ although its complete analytical expression includes the term:

$$\frac{\lambda\Delta\omega_1^3}{2\omega(\omega - 2\omega_1 + 2i\eta)}\left[i\pi\left(\frac{1}{\omega-\omega_1} - \frac{1}{\omega_1}\right)\right.$$
$$\left. + \frac{1}{\omega-\omega_1}\ln\left|\frac{\omega-\omega_1}{\Delta}\right| - \frac{1}{\omega_1}\ln\left(\frac{\omega_1}{\Delta}\right)\right],$$

which corresponds to a pole with a vanishing residue.

III. SOLUTION OF THE GAP EQUATION INCLUDING THE COULOMB INTERACTION

On account of the reasonable success of the above scheme, we wish to extend it to solve the gap equation including the Coulomb repulsion $-V(q)$. Since Eq. (5) is linear with respect to the vertex part Γ, the introduction of the essentially instantaneous Coulomb interaction is equivalent to adding a constant to the frequency-dependent potential, i.e., replace $\lambda U(\omega)$ by

$$U'(\omega) = \lambda U(\omega) - \mu, \quad (25)$$

where μ is the angular average of $V(q)$:

$$\mu = \frac{1}{4\pi^2 v_0}\int_0^{2k_0}\frac{4\pi e^2}{k_s^2 + q^2}qdq = \frac{k_s^2}{8k_0^2}\ln\left[\frac{k_s^2 + 4k_0^2}{k_s^2}\right]. \quad (26)$$

We are looking for a solution displaying the general behavior of $U'(\omega)$ and more precisely, we shall start the iteration process with the trial function:

$$\Sigma_1'(\omega) = \Delta[(1+\xi)U(\omega) - \xi], \quad (27)$$

with two adjustable parameters Δ and ξ. We have then

$$\Sigma_2'(\omega) = \Delta\int_0^\Delta \frac{[(1+\xi)U(z) - \xi][\lambda U(\omega-z) - \mu]}{i[\Delta^2 - z^2]^{\frac{1}{2}}}dz$$
$$+ \int_\Delta^{\epsilon_F}\frac{[(1+\xi)U(z) - \xi][\lambda U(\omega-z) - \mu]}{[z^2 - \Delta^2]^{\frac{1}{2}}}dz. \quad (28)$$

Note that we have introduced a cutoff of the order of the Fermi energy in the last integral in order to prevent the logarithmic divergence due to the constant term $\xi\mu$ in the numerator of the integrand (see footnote 9). As before, these integrals cannot be computed near the singularity $\omega \approx \omega_1$ but we obtain the following expressions for $\Sigma_2'(\omega)$ in the low- and high-frequency limits, respectively:

$$\Sigma_2' = \Delta[\lambda\ln(2\omega_1/\Delta) - (1+\xi)\mu\ln(2\omega_1/\Delta) + \xi\mu\{\ln(2\epsilon_F/\Delta)$$
$$-i\pi/2\} + (\omega/\omega_1)\{\tfrac{3}{4}\lambda(1+\xi)(1-i\pi/2)$$
$$+(i\pi/2)\xi\lambda\} + O(\omega^2)], \quad \omega \ll \omega_1, \quad (29)$$
$$\Sigma_2' = -\Delta[\mu(1+\xi)\ln(2\omega_1/\Delta) - \xi\mu\{\ln(2\epsilon_F/\Delta) - i\pi/2\}$$
$$+ i\pi\lambda\xi\omega_1/\omega + O(1/\omega^2)], \quad \omega \gg \omega_1.$$

Although (27) and (29) are not as accurately self-consistent as (22) and (23) in the previous section,[11] Σ_2' does indeed converge towards Σ_1' in the weak-coupling limit ($\lambda, \mu \ll 1$). In this limit, the adjustable parameters Δ and ξ must satisfy

$$(\lambda - \mu)\ln(2\omega_1/\Delta) + \mu\xi\ln(\epsilon_F/\omega_1) = 1,$$
$$\mu\ln(2\omega_1/\Delta) - \mu\xi\ln(\epsilon_F/\omega_1) = \xi. \quad (30)$$

Hence,

$$\ln\left(\frac{2\omega_1}{\Delta}\right) = \left[\lambda - \frac{\mu}{1+\mu\ln(\epsilon_F/\omega_1)}\right]^{-1}. \quad (31)$$

This relation is identical to the equation found by Bogoliubov and coworkers[6] using a similar model, with the difference that the cutoff ω_1 of the phonon-induced interaction and the cutoff ϵ_F of the instantaneous Coulomb repulsion appear now as the consequence of the frequency dependence of these interactions rather than as arbitrary cutoffs in momentum space of an effective instantaneous interaction. The effect of the Coulomb repulsion is indeed a reduction of the parameter "N_0V" as expected but we find also that the Coulomb repulsion is somewhat less effective than the phonon-induced attraction on account of its instantaneous character {μ appears in reference 6, Eq. (3.7) reduced by the factor $[1+\mu\ln(\epsilon_F/\omega_1)]^{-1}$ of the order of 0.4}.

IV. EFFECT OF THE FINITE RANGE OF THE PHONON SPECTRUM

The actual phonon spectrum of a solid extends over a finite range, the width $2\omega_2$ of which may be a significant fraction of the Debye frequency (of the order of 20% for example). Accordingly, we see from (17) that the singularities of the average phonon induced interaction $U(\omega)$ are spread over the same range and therefore, $U(\omega)$ is much smoother than its components $u_q(\omega)$. Equivalently, $U(t)$ has only a rather short range in time, of the order of ω_2^{-1}, since the components $u_q(t)$ interfere destructively if the time interval t is of the

---

[11] The most striking deficiency is the appearance of an imaginary term $-i(\pi/2)\xi\mu$ in the zero-order term of the expansion (29). It is likely that this imaginary term (negligible in the weak-coupling limit) is spurious and due to our introducing an artificially sharp cutoff of the term proportional to $\xi\mu$ at the Fermi energy.

order of ω_2^{-1} or larger. This may be best demonstrated by taking a simple model for the phonon spectrum, namely a Lorentz distribution:

$$g(\omega_q) = \frac{1}{\pi} \frac{\omega_2}{(\omega_q - \omega_1)^2 + \omega_2^2}. \quad (32)$$

Note that this mathematically simple model is still a reasonably accurate approximation of the "well-behaved" phonon spectra found for a rather large class of metals. It would not be a satisfactory approximation for multi-peaked spectra observed in some instances. The corresponding phonon-induced interaction is found immediately to be

$$U(t) = \tfrac{1}{2}(\omega_1 - i\omega_2) e^{-i\omega_1|t|} e^{-\omega_2|t|},$$

$$U(\omega) = \tfrac{1}{2}(\omega_1 - i\omega_2) \left[\frac{1}{\omega_1 + \omega - i\omega_2} + \frac{1}{\omega_1 - \omega - i\omega_2} \right]. \quad (33)$$

Note the rapid damping due to the large imaginary part of the pseudo-phonon frequency $(\omega_1 - i\omega_2)$. Because $U(\omega)$ is actually smoother than $u_q(\omega)$, one sees physically that the actual solution of (20) should be smoother than (27). However, the computation scheme used in the previous sections is not suited for this generalization because it relies upon the sharpness of the singularity of $u_q(\omega)$. We shall, therefore, take a different approach to the problem, involving the transformation of Eq. (20) to the time representation (21).

Our iteration procedure now consists in the following steps. Firstly, we choose a trial function

$$\Sigma_1(\omega) = \Delta[(1 + \xi - \zeta)U(\omega) + \zeta A(\omega) - \xi], \quad (34)$$

where Δ, ξ, and ζ are adjustable parameters and $A(\omega)$ an even function of the frequency, smoothly decreasing from 1 at $\omega = 0$ to zero at $\omega = \pm\omega_1$. Secondly, we carry (34) into the expression for $C(t)$:

$$C(t) = C(-t) = \int_{(L)} \frac{\Sigma_1(z)}{2D(z)} e^{-izt} dz. \quad (35)$$

Thirdly, we compute $\Sigma_2(t)$ according to relation (21):

$$\Sigma_2(t) = [\lambda U(t) - \mu \delta(t)] C(t), \quad (36)$$

and finally, we shall Fourier-transform the resulting expression for Σ_2 in order to compare it to the initial Σ_1. In the course of this program, however, we shall need to make some approximations, the most important of which is described below.

Since expression (35) is practically linear with respect to Σ [we may replace $\Sigma(z)$ by $\Sigma(0) = \Delta$ in the expression for $D(z)$], the three terms of (34) contribute, respectively,

$$(1 + \xi - \zeta)C_1(t), \quad \zeta C_2(t), \quad -\xi C_3(t),$$

to the final expession of $C(t)$. The most important term is the first one and is found to be (for positive t):

$$C_1(t) = \Delta[\ln 2 - \text{Ci}(\Delta t)] + \Delta[\cos(\omega_1 t) \text{Ci}(\omega_1 t) \\
+ \sin(\omega_1 t)\{\text{Si}(\omega_1 t) - \pi/2\}] - i(\pi/2)\Delta \\
+ i(\pi/2)\Delta e^{-i(\omega_1 - i\omega_2)t}, \quad (37)$$

where $\text{Si}(z)$ and $\text{Ci}(z)$ are the sine and cosine integral functions. In spite of its appearance, the second bracket is a perfectly smooth function of t, increasing monotonically from $-\infty$ at $t = 0$ [logarithmic divergence, of the order of $\ln(\omega_1 t)$] and approaching zero like $-(\omega_1 t)^{-2}$ for large t. Consequently, this function is very accurately approximated by

$$\ln(e^\gamma \omega_1 t) - \tfrac{1}{2}\ln(1 + e^{2\gamma}\omega_1^2 t^2).$$

Both the third and the fourth terms are imaginary and we notice that the latter introduces a correction proportional to $\exp[-2i(\omega_1 - i\omega_2)t]$ in the expression for $\Sigma_2(t)$ or equivalently a pole at $2(\omega_1 - i\omega_2)$ in the corresponding expression for $\Sigma_2(\omega)$. Moreover, this extra term will in turn bring about corrections at $3\omega_1$, $4\omega_1$, etc., upon iteration. Our intent is, of course, to neglect these high-energy corrections (which are quite small in the weak-coupling limit) but we must not overlook the fact that the corrections at $2\omega_1$, $4\omega_1$, $6\omega_1$, $\cdots$ bring small imaginary contributions which ultimately cancel the imaginary contribution of the third term $-i(\pi/2)\Delta$ of (37) [for $\Sigma_2(\omega)$ is the average of (29) with respect to the phonon frequency and, therefore, $\Sigma_2(0)$ must be real]. In order to be consistent, we shall therefore neglect both imaginary terms of (37) and retain only

$$C_1(t) \simeq \Delta[\ln(2\omega_1/\Delta) - \tfrac{1}{2}\ln(1 + e^{2\gamma}\omega_1^2 t^2)], \quad 0 < t < \Delta^{-1}. \quad (38)$$

Here γ is the Euler constant. Similarly, the second term of (34) is cut off at a frequency of the order of $\tfrac{1}{2}\omega_1$ and therefore

$$C_2(t) \simeq \Delta[\ln 2 - \text{Ci}(\Delta t) + \text{Ci}(\tfrac{1}{2}\omega_1 t)] \\
\simeq \Delta[\ln(\omega_1/\Delta) - \tfrac{1}{2}\ln(1 + \tfrac{1}{4}e^{2\gamma}\omega_1^2 t^2)]. \quad (39)$$

Finally, the last term of (34) is cut off at a frequency of the order of the Fermi energy; thus:

$$C_3(t) \simeq \Delta[\ln 2 - \text{Ci}(\Delta t) + \text{Ci}(\epsilon_F t)] \\
\simeq \Delta[\ln(2\epsilon_F/\Delta) - \tfrac{1}{2}\ln(1 + e^{2\gamma}\epsilon_F^2 t^2)]. \quad (40)$$

Collecting terms and performing the required Fourier transformation, we obtain the expression for $\Sigma_2(\omega)$ in the three regions: $\omega \simeq 0$, $\omega \simeq \omega_1$, and $\omega \gg \omega_1$, respectively:

$$\Sigma_2''(\omega)\big|_{\omega \approx 0} = \lambda\Delta[\ln(2\omega_1/\Delta) + 0.23(1 + \xi) + 0.14\zeta] \\
- \mu\Delta[\ln(2\omega_1/\Delta) - \zeta \ln 2 - \xi \ln(\epsilon_F/\omega_1)],$$

$$\Sigma_2''(\omega)\big|_{\omega \approx \omega_1} = \lambda\Delta U(\omega)[\ln(2\omega_2/\Delta) + 0.28 + 0.12\xi + 0.12\zeta] \\
- \mu\Delta[\ln(2\omega_1/\Delta) - \zeta \ln 2 - \xi \ln(\epsilon_F/\omega_1)], \quad (41)$$

$$\Sigma_2''(\omega)\big|_{\omega \to \infty} = -\mu\Delta[\ln(2\omega_1/\Delta) - \zeta \ln 2 - \xi \ln(\epsilon_F/\omega_1)].$$

Adjusting Δ, ξ, and ζ to fit Σ_2 and Σ_1 in these regions, we find the following self-consistency condition:

$$\ln\left(\frac{2\omega_1}{\Delta}\right) = \left[\frac{\lambda}{1-0.23\lambda} - \frac{\mu}{1+\mu \ln(\epsilon_F/\omega_1)}\right]^{-1}, \quad (42)$$

almost identical to (31). Note that the phonon-induced interaction (attraction) is enhanced by the factor $(1-0.23\lambda)^{-1}$ typically of the order of 1.05 to 1.1. This close similarity with the results of Sec. III justifies *a posteriori* our use of the simplified Einstein model; this also encourages us to place actual numbers on the parameters λ and μ and compute the corresponding transition temperature. We have plotted the self-consistent "gap-function" $\Sigma_2(\omega)$ for the typical case $\lambda = 0.3$, $\mu = 0.25$, $\omega_2 = \omega_1/5$ and $\ln(\epsilon_F/\omega_1) = 6$ (Fig. 2). Note the striking similarity with the simple model of BCS and Bogoliubov, i.e., a constant gap, cutoff at $\omega = \omega_1$.

V. TRANSITION TEMPERATURE AND ISOTOPE EFFECT

It is interesting at this point to use our model to evaluate the critical temperature (or rather the corresponding "$N_0 V$") for some well-behaved metals and compare our predictions with experimental data. Firstly, we notice that the two most important parameters characterizing the electronic behavior of metals, i.e., the electron density (or Fermi momentum) and the density of individual states on the Fermi level, enter the expressions for λ and μ only through the combination:

$$a^2 = k_s^2/4k_0^2 = 4\pi e^2 N_0/4k_0^2. \quad (43)$$

We may compute both k_0 and N_0 for the Fermi-Thomas model from the interelectron spacing (obtained from basic crystallographic data). Now, the actual density of states on the Fermi surface is the Fermi-Thomas value corrected for the effective mass (which in turn is estimated from specific-heat measurements). From (26), it is clear that

$$\mu = \tfrac{1}{2} \ln[(1+a^2)/a^2]. \quad (44)$$

On the other hand, we have not accurately computed the angular average of the strength of the phonon-induced interaction because of the complication brought by the existence of Umklapp processes; we have, rigorously,

$$\lambda = \int_0^1 \left[\frac{a^2}{a^2 + q^2/4k_0^2}\right]^2 x \, dx, \quad (45)$$

where $x = |\mathbf{k} - \mathbf{k}'|/2k_0$ is equal to $q/2k_0$ only for normal processes. For alkali metals, it is indeed a fair approximation to neglect the effect of umklapp processes and take

$$\lambda \simeq \int_0^1 \left[\frac{a^2}{a^2 + x^2}\right]^2 x \, dx = \frac{1}{2} \frac{a^2}{1+a^2}. \quad (46)$$

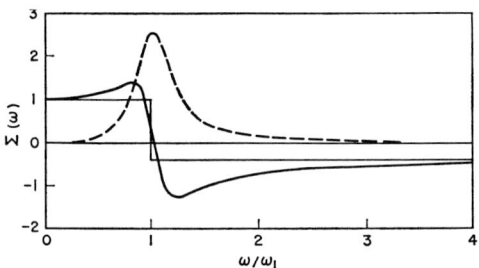

FIG. 2. Plot of the self-consistent energy-dependent "gap function" $\Sigma(\omega)$ for typical values of the parameter. The full line and the dashed line represent the real and imaginary part of $\Sigma(\omega)$, respectively. The step function represents the gap function used by Bogoliubov (equal to 1 from 0 to ω_1 and to $-\xi$ from ω_1 to the Fermi energy).

For polyvalent metals, however, the umklapp processes play an important role and it is preferable to replace $q^2/4k_0^2$ by the mean value over the first Brillouin zone. For metals with a simple structure (quasi-spherical zone), this mean value is of the order of $\tfrac{3}{5}(4Z)^{-\frac{2}{3}}$; then

$$\lambda \simeq \frac{1}{2}\left[\frac{a^2}{a^2 + \tfrac{3}{5}(4Z)^{-\frac{2}{3}}}\right]^2. \quad (47)$$

Now, it must be emphasized that the above expression is no more than an order of magnitude estimate of λ; for, we have not taken into account any effect of the crystalline structure although it is clear that it determines the mean value of $q^2/4k_0^2$. Also, (47) is based on the assumption that the screening radius of the electron-ion interaction is the same as the screening radius of the direct Coulomb interaction between electrons and may be estimated on the basis of the Fermi-Thomas model. This assumption is very much open to doubt, particularly for the heaviest ions. In any case, it may be seen from Table I that this simple model leads to a fair order of magnitude agreement between the "$N_0 V$" estimated from experimental data using the BCS expression for the critical temperature and our parameter:

$$\lambda - \mu^* = \lambda - \frac{\mu}{1 + \mu \ln(\epsilon_F/\omega_1)}, \quad (48)$$

computed from basic crystallographic and thermal data.

We have also plotted both parameters λ and μ^* versus a^2 for monovalent, bivalent, and tetravalent metals [using a typical value $\ln(\epsilon_F/\omega_1) = 6$]. Since a^2 is always of the order of 0.3 or larger (0.8 to 1 for alkali metals), this plot indicates that most if not all metals should be superconducting (see Fig. 3). On the other hand, the computed critical temperature is exceedingly low if $(\lambda - \mu^*)$ is smaller than 0.15, say. For example, the critical temperature of sodium would be of the order of 10^{-3} °K; it is then safe to assume that even if perfectly pure sodium were superconducting at such low temperatures in zero magnetic field,

TABLE I. In this table, a^2 is computed from crystallographic data and m^*/m is estimated from the electronic specific heat. These data, as well as the Debye temperature and the critical temperature (columns 3 and 4) are taken from the "*American Institute of Physics Handbook*," McGraw-Hill, 1957. N_0V (column 8) is estimated from the critical temperature with the help of BCS equation. The exponent of the isotope effect (last column) is derived from this experimental value of N_0V with the help of relation (50).

	a^2	m^*/m	Θ_D(°K)	T_c(°K)	λ	μ^*	$\lambda - \mu^*$	N_0V_{exp}	$-(d\Delta/\Delta)(M/dM)$
Na	0.67	1.6	160	...	0.25	0.12	0.13	...	...
K	0.83	...	100	...	0.25	0.12	0.13	...	...
Cu	0.45	1.15	343	...	0.20	0.10	0.10	...	...
Au	0.51	1.1	164	...	0.18	0.10	0.08	...	...
Mg	0.45	1.3	342	...	0.32	0.12	0.20	...	...
Ca	0.55	0.75	220	...	0.27	0.11	0.16	...	...
Zn	0.39	0.9	235	0.9	0.25	0.09	0.16	0.18	0.35
Cd	0.43	0.75	164	0.56	0.23	0.09	0.14	0.175	0.34
Hg	0.43	2	70	4.16	0.37	0.10	0.27	0.35	0.46
Al	0.35	1.5	375	1.2	0.33	0.10	0.23	0.175	0.34
In	0.40	1.35	109	3.4	0.34	0.10	0.24	0.29	0.44
Tl	0.415	1.15	100	2.4	0.32	0.09	0.23	0.27	0.43
Sn	0.37	1.2	195	3.75	0.34	0.10	0.24	0.25	0.42
Pb	0.38	2.1	96	7.22	0.40	0.10	0.30	0.39	0.47
Ti	0.32	≈ 3	430	0.4	0.41	0.11	0.30	0.14	0.25
Zr	0.51	1.1	265	0.55	0.37	0.11	0.26	0.16	0.30
V	0.28	≈ 9	338	4.9	0.47	0.12	0.35	0.24	0.41
Nb	0.29	≈ 8	320	8.8	0.47	0.12	0.35	0.32	0.45
Ta	0.29	≈ 5	230	4.4	0.45	0.11	0.34	0.25	0.42
Mo	0.27	1.9	360	...	0.38	0.10	0.28	...	...
U	0.30	8	200	1.1	0.47	0.12	0.35	0.19	0.36

the least residual field (magnetic impurities) would exceed the very small critical field and prevent superconductivity. This argument practically excludes all monovalent metals.

Finally, we notice that all experimental values of "N_0V" for metals appear to be smaller than 0.4 (see Table I) in good agreement with our prediction since μ^* is practically 0.10 in all cases and λ cannot exceed 0.5 (to the first order of the expansion in the electron-phonon interaction).

Isotope Effect

The isotope effect consists in the dependence of the energy gap Δ upon the phonon frequency ω_1 (the phonon frequency itself is proportional to $M^{-\frac{1}{2}}$, everything else being equal). Differentiating (31) with respect to the phonon frequency, we obtain in a straightforward fashion:

$$d\Delta/\Delta = -\tfrac{1}{2}(dM/M)\{1 - [\lambda \ln(2\omega_1/\Delta) - 1]^2\}. \quad (49)$$

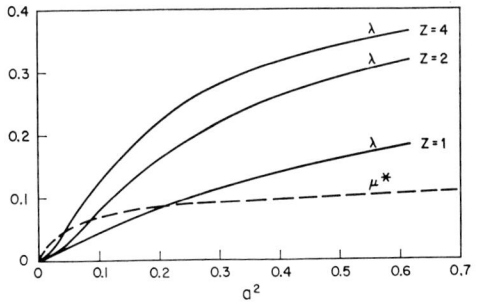

FIG. 3. Plot of the parameters λ and μ^* vs a^2 [see expressions (44) to (48)] for different valencies $Z = 1, 2,$ and 4.

Using the property $\mu^* = 0.1$, we can derive from (49) a semiempirical relation:

$$d\Delta/\Delta = -\tfrac{1}{2}(dM/M)[1 - 0.01(N_0V)^{-2}], \quad (50)$$

indicating a significant discrepancy from the "ideal" value $-\tfrac{1}{2}(dM/M)$ postulated by BCS. Note that the discrepancy is larger for low temperature superconductors (see Table I).

VI. CONCLUSION

The positive aspect of our results is that they confirm the approximations to the effective potential made in the past, particularly Bogoliubov's approximation, and give a precise meaning to each parameter which can, therefore, in principle, be computed from the basic crystallographic data. Also, the orders of magnitude of the computed transition temperatures are generally satisfactory.

The negative features are, firstly, the isotope effect which is particularly in conflict with the results of Geballe and Matthias on zinc, secondly, the prediction that all metals should superconduct at sufficiently low temperatures, and thirdly, the fact that it is not obvious from the theory why no material with T_c higher than 18°K have yet been found. This last point is not really too disturbing since the drastic simplifications on which this theory is based could only be expected to be reasonable in the weak-coupling limit; in the strong-coupling case of high-transition-temperature superconductors, such effects as the phonon scattering lifetime and, in general, the complex diagrams which have been omitted from our Eq. (12), should be incorporated in the theory.

On the other hand, the first and second difficulties are of a more serious nature for they do occur in the weak-coupling limit. Because our results are essentially insensitive to the actual computation scheme [compare Eqs. (31) and (42), and also Eq. (6.21) of reference 6], we do not feel that our solution of the basic equation (5) could be in serious error, at least in the weak-coupling limit. Moreover, the predicted transition temperatures are not remarkably sensitive to the errors, or approximations, which do remain in our treatment, with the possible exception of the cases in which zone boundary perturbations are too extensive at the Fermi surface (such as Bi, Sb, $\cdots$), so that the free-electron Fermi sphere is a very poor approximation. This suggests strongly that the results derived here are exactly the logical conclusions of the original assumptions of the BCS theory. Consequently, what serious difficulties, such as the isotope effect, are to be found, seem to us to require either a new look at the basic assumptions of the theory, possibly using a different interaction taking into account more complex diagrams than the lowest order diagrams implicitly contained in the BCS's treatment, or re-examination of the experimental evidence.

ACKNOWLEDGMENT

It is a pleasure to thank Professor J. R. Schrieffer and Professor P. Nozières for helpful discussions of the Green's function formalism of Sec. II.

Plasmons, Gauge Invariance, and Mass
Phys. Rev. **130**, 439 (1963)

The physics behind the Higgs phenomenon is not, in spite of the belatedly added reference to Schwinger's opaque remarks, in any way beholden to Schwinger in its inception, since it was a direct transcription into particle language of my work (referred to above) on broken gauge invariance in superconductivity. My ideas on broken symmetry in both vacua and in condensed matter systems did receive great impetus from some brief contacts with Y. Nambu: his phrase "independent orthogonal Hilbert spaces" was very liberating and expressed what I had been groping towards since 1953. I also brushed shoulders, at least, with Steve Weinberg at the Cavendish; and as I acknowledge, J.C. Taylor was at Bell for a summer and we talked at length.

Plasmons, Gauge Invariance, and Mass

P. W. ANDERSON
Bell Telephone Laboratories, Murray Hill, New Jersey
(Received 8 November 1962)

Schwinger has pointed out that the Yang-Mills vector boson implied by associating a generalized gauge transformation with a conservation law (of baryonic charge, for instance) does not necessarily have zero mass, if a certain criterion on the vacuum fluctuations of the generalized current is satisfied. We show that the theory of plasma oscillations is a simple nonrelativistic example exhibiting all of the features of Schwinger's idea. It is also shown that Schwinger's criterion that the vector field $m \neq 0$ implies that the matter spectrum before including the Yang-Mills interaction contains $m=0$, but that the example of superconductivity illustrates that the physical spectrum need not. Some comments on the relationship between these ideas and the zero-mass difficulty in theories with broken symmetries are given.

RECENTLY, Schwinger[1] has given an argument strongly suggesting that associating a gauge transformation with a local conservation law does not necessarily require the existence of a zero-mass vector boson. For instance, it had previously seemed impossible to describe the conservation of baryons in such a manner because of the absence of a zero-mass boson and of the accompanying long-range forces.[2] The problem of the mass of the bosons represents the major stumbling block in Sakurai's attempt to treat the dynamics of strongly interacting particles in terms of the Yang-Mills gauge fields which seem to be required to accompany the known conserved currents of baryon number and hypercharge.[3] (We use the term "Yang-Mills" in Sakurai's sense, to denote any generalized gauge field accompanying a local conservation law.)

The purpose of this article is to point out that the familiar plasmon theory of the free-electron gas exemplifies Schwinger's theory in a very straightforward manner. In the plasma, transverse electromagnetic waves do not propagate below the "plasma frequency," which is usually thought of as the frequency of long-wavelength longitudinal oscillation of the electron gas. At and above this frequency, three modes exist, in close analogy (except for problems of Galilean invariance implied by the inequivalent dispersion of longitudinal and transverse modes) with the massive vector boson mentioned by Schwinger. The plasma frequency is equivalent to the mass, while the finite density of electrons leading to divergent "vacuum" current fluctuations resembles the strong renormalized coupling of Schwinger's theory. In spite of the absence of low-frequency photons, gauge invariance and particle conservation are clearly satisfied in the plasma.

In fact, one can draw a direct parallel between the dielectric constant treatment of plasmon theory[4] and Schwinger's argument. Schwinger comments that the commutation relations for the gauge field A give us one sum rule for the vacuum fluctuations of A, while those for the matter field give a completely independent value for the fluctuations of matter current j. Since j is the source for A and the two are connected by field equations, the two sum rules are normally incompatible unless there is a contribution to the A rule from a free, homogeneous, weakly interacting, massless solution of the field equations. If, however, the source term is large enough, there can be no such contribution and the massless solutions cannot exist.

The usual theory of the plasmon does not treat the electromagnetic field quantum-mechanically or discuss vacuum fluctuations; yet there is a close relationship between the two arguments, and we, therefore, show that the quantum nature of the gauge field is irrelevant. Our argument is as follows:

The equation for the electromagnetic field is

$$p^2 A_\mu = (k^2 - \omega^2) A_\mu(\mathbf{k},\omega) = 4\pi j_\mu(\mathbf{k},\omega).$$

---

[1] J. Schwinger, Phys. Rev. **125**, 397 (1962).
[2] T. D. Lee and C. N. Yang, Phys. Rev. **98**, 1501 (1955).
[3] J. J. Sakurai, Ann. Phys. (N. Y.) **11**, 1 (1961).
[4] P. Nozières and D. Pines, Phys. Rev. **109**, 741 (1958).

A given distribution of current j_μ will, therefore, lead to a response A_μ given by

$$A_\mu = \frac{4\pi}{k^2 - \omega^2} j_\mu = \frac{4\pi}{p^2} j_\mu. \quad (1)$$

(1) is merely the statement that only the electromagnetic current can be a source of the field; it is required for general gauge invariance and charge conservation according to the usual arguments.

The dynamics of the matter system—of the plasma in that case, of the vacuum in the elementary particle problem—determine a second response function, the response of the current to a given electromagnetic or Yang-Mills field. Let us call this response function

$$j_\mu(\mathbf{k},\omega) = -K_{\mu\nu}(\mathbf{k},\omega) A_\nu(\mathbf{k},\omega). \quad (2)$$

By well-known arguments of gauge invariance, $K_{\mu\nu}$ must have a certain form: Schwinger points out that in the relativistic case it must be proportional to $p_\mu p_\nu - g_{\mu\nu} p^2$, and equivalent arguments give one the same form in superconductivity.[5] It will be convenient to consider, for simplicity, only the gauge

$$p_\mu A_\mu = 0. \quad (3)$$

Then the response is diagonal: $K_{\mu\nu} = -g_{\mu\nu} K$. For a plasma with n carriers of charge e and mass M it is simply (in the limit $p \to 0$)

$$K = ne^2/M. \quad (4)$$

In an insulator the response is not relativistically invariant. If the insulator has magnetic polarizability α_m and electric α_e, the response equations may be written, in the gauge (3),

$j_\mu = -\alpha_e p^2 A_\mu$ (longitudinal and time components),

$\mathbf{j} = -\alpha_m p^2 \mathbf{A}$ (transverse components).

In a truly relativistic situation such as our normal picture of a vacuum, we expect

$$j_\mu = -\alpha p^2 A_\mu \quad (5)$$

to describe normal polarizable behavior.

Since we cannot turn off the interactions, we do not actually observe the responses (1), (2), or (5). If we insert a test particle, its field A_μ^e induces a current j_μ which in turn acts as the source for an internal field A_μ^i:

$$j_\mu = -K(A_\mu^i + A_\mu^e), \quad A_\mu^i = +4\pi j_\mu/p^2,$$

or, the total field is modified to

$$A_\mu = [p^2/(p^2 + 4\pi K)] A_\mu^e. \quad (6)$$

The pole at which A propagates freely occurs at a mass (frequency)

$$m^2 = -p^2 = 4\pi K, \quad (7)$$

---

[5] M. R. Schafroth, Helv. Phys. Acta **24**, 645 (1951).

which in a conductor is

$$m^2 = \omega^2 - k^2 = \omega_p^2. \quad (8)$$

ω_p is the usual plasma frequency $(4\pi n e^2/M)^{1/2}$.

It is not necessary here to go in detail into the relationship between longitudinal and transverse behavior of the plasmon. In the limit $p \to 0$ both waves propagate according to (8). The longitudinal plasmon is generally thought of as entirely an attribute of the plasma, while the transverse ones are considered to result from modification of the propagation of real photons by the medium. This is reasonable in the classical case because the longitudinal plasmon disappears at a certain cutoff energy and has a different dispersion law; but in a Lorentz-covariant theory of the vacuum it would be indistinguishable from the third component of a massive vector boson of which the transverse photons are the two transverse components.

How, then, if we were confined to the plasma as we are to the vacuum and could only measure renormalized quantities, might we try to determine whether, before turning on the effects of electromagnetic interaction, A had been a massless gauge field and K had been finite? As far as we can see, this is not possible; it is, nonetheless, interesting to see what the criterion is in terms of the actual current response function to a perturbation in the Lagrangian

$$\delta L = j_\mu \delta A_\mu. \quad (9)$$

This will turn out to be identical to Schwinger's criterion. The original "bare" response function was K:

$$j_\mu = -K_{\mu\nu} \delta A_\mu.$$

Taking into account the interaction, however, we must correct for the induced fields and currents, and we get

$$j_\mu = -K' \delta A_\mu^e = -K[p^2/(p^2 + 4\pi K)] \delta A_\mu^e \to$$
$$-(p^2/4\pi) \delta A_\mu^e, \quad p^2 \to 0. \quad (10)$$

Thus, the new response to an applied perturbing field (9) is very like that of an ordinary polarizable medium. The only difference from an ordinary polarizable "vacuum" with bare response (5) is that in that case as $p \to 0$

$$K' \to -[\alpha/(1+4\pi\alpha)] p^2, \quad (11)$$

so that the coefficient of $p^2/4\pi$ is less than unity.

This criterion is precisely the same as Schwinger's criterion

$$\int B_1(m^2) dm^2 = 1,$$

where $B_1 m^2$ is the weight function for the current vacuum fluctuations. This can be shown by a simple dispersion argument. Schwinger expresses the unordered product expectation value of the current as

$$\langle j_\mu(x) j_\nu(x') \rangle = \int dm^2 \, m^2 B_1(m^2) \int \frac{dp}{(2\pi)^3} e^{ip(x-x')}$$
$$\times \eta_+(p) \delta(p^2+m^2)(p_\mu p_\nu - g_{\mu\nu} p^2).$$

The Fourier transform of the corresponding retarded Green's function is our response function:

$$K'(p) = \int \frac{dm^2 \, m^2 B_1(m^2)}{p^2 - m^2} [p_\mu p_\nu - g_{\mu\nu} p^2],$$

and

$$\lim_{p \to 0} K'(p) = (p_\mu p_\nu - g_{\mu\nu} p^2) \int dm^2 \, B_1(m^2).$$

Thus, (aside from a factor 4π which Schwinger has not used in his field equation) his criterion is also that the polarizability α', here expressed in terms of a dispersion integral, have its maximum possible value, 1.

The polarizability of the vacuum is not generally considered to be observable[6] except in its p dependence (terms of order p^4 or higher in K). In fact, we can remove (11) entirely by the conventional renormalization of the field and charge

$$A_r = AZ^{-1/2}, \quad e_r = eZ^{1/2}, \quad j_r = jZ^{1/2}.$$

Z, here, can be shown to be precisely

$$Z = 1 - 4\pi\alpha' = 1 - \int_0^\infty dm^2 \, B_1(m^2).$$

Thus, the renormalization procedure is possible for any merely polarizable "vacuum," but not for the special case of the conducting "plasma" type of vacuum. In this case, no net true charge remains localized in the region of the dressed particle; all of the charge is carried "at infinity" corresponding to the fact, well known in the theory of metals, that all the charge carried by a quasi-particle in a plasma is actually on the surface. Nonetheless, conservation of particles, if not of bare charge, is strictly maintained. Note that the situation does not resemble the case of "infinite" charge renormalization because the infinity in the vacuum polarizability need only occur at $p^2 = 0$.

Either in the case of the polarizable vacuum or of the "conducting" one, no low-energy experiment, and even possibly no high-energy one, seems capable of directly testing the value of the vacuum polarizability prior to renormalization. Thus, we conclude that the plasmon is a physical example demonstrating Schwinger's contention that under some circumstances the Yang-Mills type of vector boson need not have zero mass. In addition, aside from the short range of forces and the finite mass, which we might interpret without resorting to Yang-Mills, it is not obvious how to characterize such a case mathematically in terms of observable, renormalized quantities.

We can, on the other hand, try to turn the problem around and see what other conclusions we can draw about possible Yang-Mills models of strong interactions from the solid-state analogs. What properties of the vacuum are needed for it to have the analog of a conducting response to the Yang-Mills field?

Certainly the fact that the polarizability of the "matter" system, without taking into account the interaction with the gauge field, is infinite need not bother us, since that is unobservable. In physical conductors we can see it, but only because we can get outside them and apply to them true electromagnetic fields, not only internal test charges.

More serious is the implication—obviously physically from the fact that α has a pole at $p^2 = 0$—that the "matter" spectrum, at least for the "undressed" matter system, must extend all the way to $m^2 = 0$. In the normal plasma even the final spectrum extends to zero frequency, the coupling rather than the spectrum being affected by the screening. Is this necessarily always the case? The answer is no, obviously, since the superconducting electron gas has no zero-mass excitations whatever. In that case, the fermion mass is finite because of the energy gap, while the boson which appears as a result of the theorem of Goldstone[7,8] and has zero unrenormalized mass is converted into a finite-mass plasmon by interaction with the appropriate gauge field, which is the electromagnetic field. The same is true of the charged Bose gas.

It is likely, then, considering the superconducting analog, that the way is now open for a degenerate-vacuum theory of the Nambu type[9] without any difficulties involving either zero-mass Yang-Mills gauge bosons or zero-mass Goldstone bosons. These two types of bosons seem capable of "canceling each other out" and leaving finite mass bosons only. It is not at all clear that the way for a Sakurai[3] theory is equally uncluttered. The only mechanism suggested by the present work (of course, we have not discussed non-Abelian gauge groups) for giving the gauge field mass is the degenerate vacuum type of theory, in which the original symmetry is not manifest in the observable domain. Therefore, it needs to be demonstrated that the necessary conservation laws can be maintained.

I should like to close with one final remark on the Goldstone theorem. This theorem was initially conjectured, one presumes, because of the solid-state analogs, via the work of Nambu[10] and of Anderson.[11] The theorem states, essentially, that if the Lagrangian

---

[6] We follow here, as elsewhere, the viewpoint of W. Thirring, *Principles of Quantum Electrodynamics* (Academic Press Inc., New York, 1958), Chap. 14.

[7] J. Goldstone, Nuovo Cimento **19**, 154 (1961).
[8] J. Goldstone, A. Salam, and S. Weinberg, Phys. Rev. **127**, 965 (1962).
[9] Y. Nambu and G. Jona-Lasinio, Phys. Rev. **122**, 345 (1961).
[10] Y. Nambu, Phys. Rev. **117**, 648 (1960).
[11] P. W. Anderson, Phys. Rev. **110**, 827 (1958).

possesses a continuous symmetry group under which the ground or vacuum state is not invariant, that state is, therefore, degenerate with other ground states. This implies a zero-mass boson. Thus, the solid crystal violates translational and rotational invariance, and possesses phonons; liquid helium violates (in a certain sense only, of course) gauge invariance, and possesses a longitudinal phonon; ferro-magnetism violates spin rotation symmetry, and possesses spin waves; superconductivity violates gauge invariance, and would have a zero-mass collective mode in the absence of long-range Coulomb forces.

It is noteworthy that in most of these cases, upon closer examination, the Goldstone bosons do indeed become tangled up with Yang-Mills gauge bosons and, thus, do not in any true sense really have zero mass. Superconductivity is a familiar example, but a similar phenomenon happens with phonons; when the phonon frequency is as low as the gravitational plasma frequency, $(4\pi G\rho)^{1/2}$ (wavelength $\sim 10^4$ km in normal matter) there is a phonon-graviton interaction: in that case, because of the peculiar sign of the gravitational interaction, leading to instability rather than finite mass.[12] Utiyama[13] and Feynman have pointed out that gravity is also a Yang-Mills field. It is an amusing observation that the three phonons plus two gravitons are just enough components to make up the appropriate tensor particle which would be required for a finite-mass graviton.

Spin waves also are known to interact strongly with magnetostatic forces at very long wavelengths,[14] for rather more obscure and less satisfactory reasons. We conclude, then, that the Goldstone zero-mass difficulty is not a serious one, because we can probably cancel it off against an equal Yang-Mills zero-mass problem. What is not clear yet, on the other hand, is whether it is possible to describe a truly strong conservation law such as that of baryons with a gauge group and a Yang-Mills field having finite mass.

I should like to thank Dr. John R. Klauder for valuable conversations and, particularly, for correcting some serious misapprehensions on my part, and Dr. John G. Taylor for calling my attention to Schwinger's work.

[12] J. H. Jeans, Phil. Trans. Roy. Soc. London **101**, 157 (1903).
[13] R. Utiyama, Phys. Rev. **101**, 1597 (1956); R. P. Feynman (unpublished).
[14] L. R. Walker, Phys. Rev. **105**, 390 (1957).

Hard Superconductors
Lectures at the 1963 Seminar of the Theoretical Physics Division of the
Canadian Association of Physicists, University of Alberta

Hard Superconductivity: Theory of the Motion of Abrikosov Flux Lines
(with Y.B. Kim)
Rev. Mod. Phys. **36**, 39 (1964)

The point here was a demonstration theory that flux lines do move and cause dissipation, which was a totally new and deep concept; not the specific assumptions of the model, which have been occasionally set up as a straw man in subsequent work. Kim noticed the slow decay of magnetization in the "critical state" and I proposed a logarithmic fit, on the basis of a "creep" mechanism like that of dislocations. Later we observed flux "flow", as well as noisy "flux jumps". But the exciting idea was the generalization of old ideas of dissipation by dislocation motion in solids to the general concept of defect motion in ordered systems: the genesis of one of the main lines of the theory of broken symmetry. We thought this work very important, and were unhappy about Bell management's decision to oppose publicity for it, on the basis of Matthias' and Geballe's objections that it negated their discovery of high-field superconductors.

HARD SUPERCONDUCTORS

P. W. ANDERSON

These are notes on lectures presented at the Banff Summer School of Theoretical Physics. They were transcribed mostly by Dr. J. M. Vail, directly from tapes of the lectures as given, and my rough notes. After a background discussion of superconductivity, they review the microscopic derivation of the Ginzburg–Landau equations and the fundamental paper of Abrikosov on type II superconductors.

LECTURE 1

I shall begin by discussing some general features of the theory and experiments surrounding superconductivity. The most prominent features of superconductivity are persistent currents in a ring and the Meissner effect. I shall discuss these below before returning to the real properties of hard superconductors.

I hope that many of you are free of prejudice on the subject and so I can follow some of the definitions which I prefer (some of which might not be generally accepted). Probably most pure metals are soft superconductors, i.e., they obey the London equations. This implies that, in equilibrium, in simply connected regions the magnetic field (H) is excluded from the material. Since it costs energy to expel the magnetic field, it follows that the material pays an energy $H^2/8\pi$ to remain superconducting. This energy must be overcome by the condensation energy, since the superconductor must have a lower free energy than the normal metal. The maximum field that the material can afford to expel is

$$\frac{H_{\text{crit}}^2}{8\pi} = F_n(H=0) - F_s(H=0) \,.$$

This critical field is deducible from calorimetric measurements and, for good soft superconductors, the observed critical field is in excellent agreement with the calorimetric measurements. There are papers of Mapother (Illinois) on this subject. All the above is valid for simply connected regions. Now we shall consider a ring. Suppose a ring is made superconducting first, and then an external magnetic field is applied. Here $H = 0$ inside the ring. Hence we must have a current flowing with no external source — a persistent current. If on the other hand the field was applied first, then $H \neq 0$ inside the hole, i.e., the magnetic field is trapped. This suggests a practical application. Namely if one drives a current through a coil one may generate a magnetic field without any loss of energy. Unfortunately this will not work for field $H > H_{\text{BC}}$, H_{BC} being the bulk critical field which is less than 2000 gauss. So this is not useful from a practical viewpoint.

Another feature of soft superconductors which is closely related to the expulsion of the magnetic field is the so-called intermediate-state phenomenon. This state can be thought of as arising from the increase in flux near the edges of the superconducting sample.

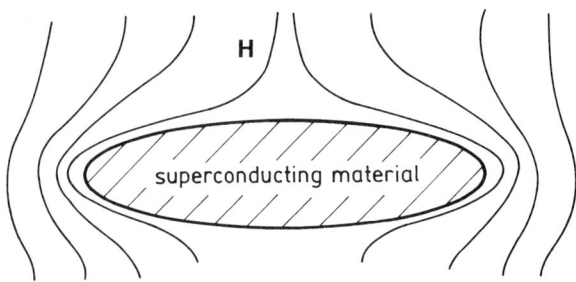

When the field at the edges is greater than H_{CB} the edges will leave the superconducting state and become normal. This state is not magnetostatically stable, and it is convenient to think of the sample as having superconducting (S) and normal (N) regions.

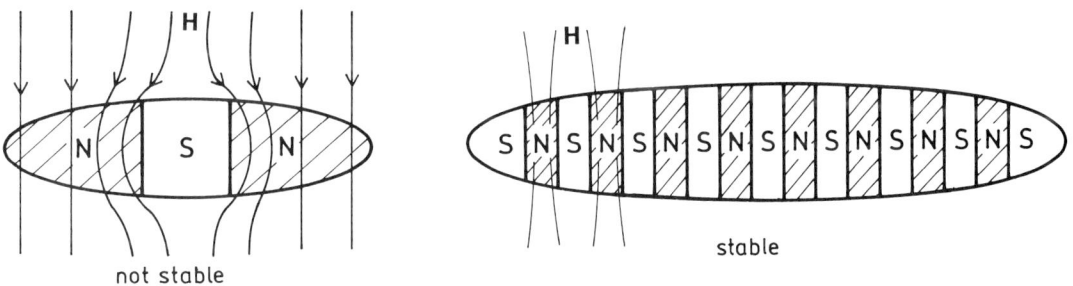

This state can be observed by covering the sample with magnetic powder. The powder will stay on the normal regions. The above described intermediate state tends to obscure the Meissner effect. Usually one measures a small percentage of H in the sample depending on its geometry, because the domains of S and N are caught by some imperfections and hence are not completely mobile. Only a perfect sample gives a perfect Meissner effect. The reason that the domains are macroscopic is that it costs energy to have a surface between the N and S domains.

I shall now define a hard superconductor as a material which remains a superconductor above H_{CB}, in other words a material in which H penetrates the sample on a microscopic basis. Thus a hard superconductor, because of the penetration of H into the sample, does not lose the free energy $H^2/8\pi$ but rather

$$\frac{H^2 \text{ applied}}{8\pi} - \frac{H^2 \text{ material}}{8\pi}.$$

Hence the sample can remain superconducting above H_{CB}. I would like to remark here that most alloys are hard superconductors.

It is convenient to subdivide hard superconductors into two classes:
(a) Abrikosov's type II superconductors
(b) "True" hard superconductors

The B vs. H curve and the magnetization (M) vs. H curves for the Abrikosov hard superconductors are

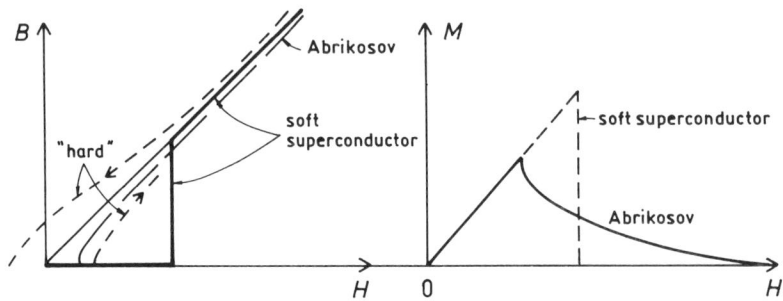

The magnetization is approximately unique and reversible. The current $\mathbf{J} = \nabla \times \mathbf{M}$ flows on the surface. This is much like the Meissner effect in soft superconductors where $\mathbf{M} = -4\pi\mathbf{H}$ and hence the current flows only on the surface. It should be remarked that in practice the Abrikosov superconductors show some hysteresis, bulk currents and other irreversible effects, thus shading imperceptibly into the "true" hard superconductors which withstand large fields and bulk currents.

This completes my survey of the experimental side of the picture and from here on I shall concentrate on the theory and outline some of the historical background. Two years after the discovery of the Meissner effect there already existed some experimental evidence for departure from it. This led Mendelssohn to propose the so-called "Sponge model". In this model one visualizes the material being meshed with filaments of superconducting material of dimensions less than the penetration depth. This is possible because of the prediction of the London theory that superconductors of small dimensions compared to the penetration depth are capable of sustaining a larger critical field. These filaments were supposed to be associated with dislocations in the specimen.

London in his book (1950) proposed that soft superconductors require a positive surface energy to form domains of S and N in the sample. Thus soft superconductors are characterized by a finite number of domains. Pippard and — in a much more complete manner — Abrikosov generalized London's idea and suggested that in hard superconductors the domains are microscopic in size because in a sense the surface energy is no longer positive.

At this point I would like to give an outline of the course.
(1) The Phenomenological Equations of London (1934) and Landau–Ginzburg (1950).
(2) The derivation of the above equations from microscopic theory (Gor'kov).
(3) Abrikosov phenomenological theory of hard superconductors. I should remark that Abrikosov essentially solved the problem of hard superconductors in (1957) prior to the appearance of the microscopic theory of superconductors.
(4) Application to real superconductors. Here I shall consider the dynamics of superconductors, i.e., how currents decay.

It is interesting to note that all but (4) above grew out of the work of essentially five people before a microscopic theory was available. These people are London, Landau and Ginzburg, Pippard and Abrikosov. Their work is an important basis for both the theoretical and experimental investigation in this field.

A. *The London Equation*

For ordinary mortals it is convenient to arrive at Landau's equations via the London equations: I shall follow this scheme.

The current due to one electron is $\mathbf{j} = \frac{e}{m}(\mathbf{p} - \frac{e}{c}\mathbf{A})$ where the average is over the wave function of the electron. The total current is then

$$\mathbf{J} = \left\langle \sum_{j=1}^{n} \frac{e}{m} \frac{i\hbar}{2} (\psi_j^* \nabla \psi_j - \psi_j \nabla \psi_j^*) - \frac{e^2}{mc} \sum_{j=1}^{N} \psi_j^* \mathbf{A} \psi_j \right\rangle .$$

If we assume that ψ_i remains fixed upon the application of $\mathbf{A}$, i.e., that the Fermi surface is rigid, then we get for the diamagnetic current

$$\mathbf{J}_D = \frac{e^2}{mc} n \mathbf{A} .$$

In a normal metal the wave function does change and the paramagnetic current (the first term in J) essentially cancels the diamagnetic current. However when the superconductivity occurs London assumed that the electrons in the metal condense in some sense, and this makes the wave function rigid, thus preventing repopulation. We now know this assumption to be true, although it was an ad hoc hypothesis at the time. It gives the famous London equation

$$\mathbf{J} = -K\mathbf{A} = -\frac{n_s e^2}{mc} \mathbf{A} .$$

This equation is clearly not gauge invariant. The gauge invariant form is (1937)
$$\nabla \times \mathbf{J} = -K\mathbf{H} = -K\nabla \times \mathbf{A} \ .$$
This equation, together with ordinary electrodynamics, leads to the Meissner effect, and the usual London penetration phenomena, as follows:
$$\nabla \times \nabla \times \mathbf{J} = -K\nabla \times \mathbf{H} = -K\frac{4\pi}{c}\mathbf{J} \ .$$
Thus
$$\nabla \times \nabla \times \mathbf{J} = -\frac{1}{\lambda_L^2}\mathbf{J}$$
where
$$\lambda_L^2 = \frac{mc^2}{4\pi ne^2} \sim (500 \text{ Å})^2 \ .$$
λ_L is the penetration depth. This equation has the solution
$$\mathbf{J} \propto e^{-x/\lambda_L} \ .$$
This implies that a weak field at the surface of a superconductor decays exponentially at a rate given roughly by λ_L.

Another phenomenon of superconductivity is the persistent current. This phenomenon means that if we, by some means, induce a current in a ring which is superconducting it remains indefinitely. A typical experimental way of doing thus is to take a superconducting ring above its transition point, put it in a magnetic field, cool it down through the transition point and then remove the external magnetic field, which produces a current flowing around the ring. Doing it this way we see that it is equally possible to describe the persistent current phenomenon as an inability of the flux through a multiply connected sample of a superconductor to decay once it is trapped. This phenomenon, as London himself pointed out, is not necessarily explained by his original equations, especially the case that the ring is not thick compared to the London penetration depth. This case is often observed. Studies of it have been made at IBM and by Mercereau at Ford. One also often finds trapped flux in simply connected soft superconductors, but this is always ascribable to freezing of the intermediate state domains.

To discuss this phenomenon London introduced the concept of the fluxoid. He pointed out that his equation did lead to the notion of the conservation of the fluxoid through a hole in a superconductor in the sense that it predicts:
$$\nabla \times \left(\mathbf{A} + \frac{\mathbf{J}}{K}\right) = 0$$
$$\oint_C \left(\mathbf{A} + \frac{\mathbf{J}}{K}\right) \cdot d\mathbf{s} = 0 \quad \text{(Stokes' theorem)}$$
$$\int_S \mathbf{H} \cdot \mathbf{n}\, dS + \oint \frac{\mathbf{J}}{K} \cdot d\mathbf{s} = 0$$

where **n** is the unit outward normal to the surface S spanning the closed curve C, whose line element is ds. These equations are true for any circuit passing entirely through a superconductor and which includes in the surface S only a superconductor, simply because the London equation holds only in the superconductor. If we now apply these equations to the multiply-connected superconductor we cannot pass our circuit around a hole and have these equations to be true, but we can pass two circuits around a hole:

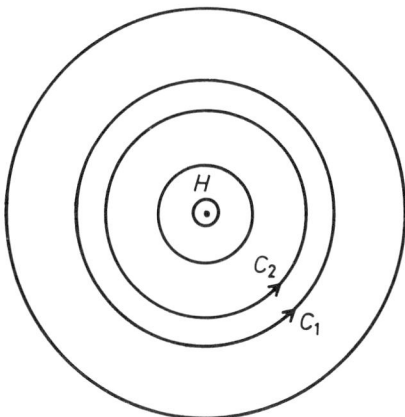

$$\int_{S_1} \mathbf{H} \cdot \mathbf{n}\, dS_1 + \oint_{C_1} \frac{\mathbf{J}}{K} \cdot d\mathbf{s}_1 = \text{same as integrals 2 so fluxoid is conserved.}$$

Now this conservation of fluxoid is clearly associated, in some sense, with the persistence of current but in fact the persistence of current seems to involve something extra. Persistent currents have been known to last as much as a month, and yet it is quite obvious that with the persistent current the state has higher energy than without it.

Therefore one would expect that some kind of dissipative process should lead one from a state with the given persistent current to one with less. Now London's equation tells one also that a superconductor cannot sustain a constant electric field because $d\mathbf{j}/dt \sim d\mathbf{A}/dt = \mathbf{E}$, and one gets constant acceleration of the current, but that cannot and does not exclude the existence of all dissipative processes, as we know from microwave skin effect measurements, for instance, or from the occurrence of domain motions in the intermediate state. The additional fact that fluxoid is conserved in time as well as in space is not a direct or obvious consequence of London's picture. One may quibble about this point, but I think all of this is irrelevant in the sense that physically it is pretty obvious, in the first place, that London's equations are phenomenological equations and must not be taken quite so seriously, and secondly that there is something extra physically going on in the phenomenon of the persistent current.

London went on at this time to suggest that the fluxoid through the hole in the superconductor was not only a constant of the motion in space and time, but that it would be quantized. We will discuss the quantization more completely when we get to the Landau–Ginzburg equations but we can see immediately that quantization of the fluxoid would make more reasonable the existence of persistent currents in that, if it were quantized, there presumably would be some kind of macroscopic free energy barrier over which the system would have to climb in order to get from one value of the quantized flux to another.

As you will remember, London's equation $\mathbf{J} = -\frac{n_s e^2}{mc}\mathbf{A}$ comes from an analogy with a term in the current of a single electron with wave function ψ which we write of course as follows:
$$\mathbf{J} = \frac{i\hbar e}{2m}(\psi^*\nabla\psi - \psi\nabla\psi^*) - \frac{e^2}{mc}\mathbf{A}|\psi|^2 .$$
It was from this equation that Landau went on to create the famous Landau–Ginzburg phenomenological theory of superconductivity. I say this was the reasoning, but in fact, only Ginzburg and God really know what the direction of Landau's mind was: I am guessing. His basic contribution was to suggest that this equation be taken seriously as a manifestation of a macroscopic quantum behavior of the coupled electron gas as a whole. His reasoning started (this he states) from the idea that every phase transition must have some order parameter, which is zero in the normal state and nonvanishing in the condensed state. Because a superconductor seemed to exhibit in some sense macroscopic quantum phenomena he suggested that the appropriate order parameter, in the case of superconductivity, should be a kind of macroscopic electronic wave function ψ. The most important new feature of this ψ is that it is complex, which I think is a necessary and brilliant idea if one is to have true transport current as in a wire, not just $\nabla \times \mathbf{M}$ currents as in ferromagnetism. Then we use the full equation for the current in the superconductor, that is we must take as our current the following equation:
$$\mathbf{J} = \frac{ie\hbar}{2m}(\psi^*\nabla\psi - \psi\nabla\psi^*) - \frac{e^2}{mc}\mathbf{A}|\psi|^2 .$$
In this equation we retain not only the so-called diamagnetic term containing $\mathbf{A}$, but terms containing the gradient of the wave function itself. Notice that this equation is properly gauge invariant as London's equation in its original form was not; that is, this equation is gauge invariant if we make the appropriate gauge transformation both on the wave function ψ and on the vector potential $\mathbf{A}$. This seems a much more natural way of making an equation gauge invariant than the ad hoc subtraction of the appropriate terms, or the expression of it only in terms of fields. This equation is then the famous second Landau–Ginzburg equation.

This equation, incidentally, already implies that the flux is quantized through a ring. The Russians did not state this directly in any place I know of, but it certainly is implicitly said in Abrikosov's work in 1957 by analogy with helium II. This quantity ψ is a physical order parameter, or so we have assumed, and therefore it must be single valued around a ring. Let us then draw a picture of our ring

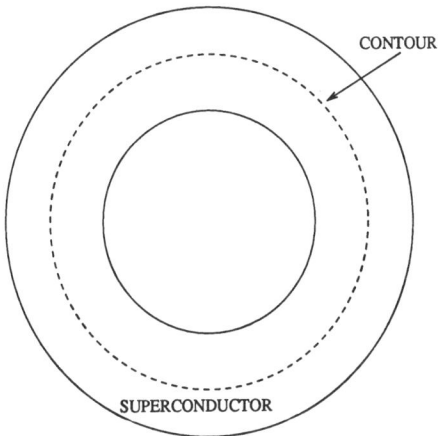

and suppose that, as we would certainly expect in any macroscopic specimen, the magnitude of the order parameter around this ring is constant. Since we have decided that the magnitude of ψ must be constant we are left only with a phase angle, and so around this macroscopic ring we may allow ψ to have the following form:

$$\psi(\mathbf{r}) = f(\mathbf{r})e^{i\varphi} .$$

In this equation φ is a function which presumably we might as well take to be linear in the distance around the ring, which must increase by 2π, or any multiple of 2π (including 0, of course) as we go around the ring. If ψ has this functional form, $\mathbf{J}$ the current, will be given by the following equation:

$$\mathbf{J} = \frac{e\hbar}{m}|f|^2\left(\boldsymbol{\nabla}\varphi - \frac{e}{\hbar c}\mathbf{A}\right) .$$

In a large specimen, in the interior of the specimen, the Meissner effect tells us that the current must be zero; thus the two terms of this equation must cancel when they are taken around a circuit through the interior of the specimen, and this tells us that the flux must be quantized in units $2\pi\hbar c/e$, as follows:

$$\oint_C d\mathbf{s}\cdot\left(\boldsymbol{\nabla}\varphi - \frac{e}{\hbar c}\mathbf{A}\right) = 0 ,$$

$$\Delta\varphi = \frac{e}{\hbar c}\oint_C \mathbf{A}\cdot d\mathbf{s} = \frac{e}{\hbar c}\int_S \mathbf{H}\cdot\mathbf{n}\,ds = \frac{e}{\hbar c}\Phi$$

where Φ is the total flux through the circuit c, and $\Delta\varphi$ is the change of φ in one circuit of the ring c, i.e., $2n\pi$. Thus

$$2n\pi = \frac{e}{\hbar c}\Phi \quad \text{or} \quad \Phi = \frac{hc}{e}.$$

LECTURE 2

At this point it is worthwhile to go back and discuss the parameters that enter into this equation for the current. Since this is purely a phenomenological theory, in any true sense these parameters are of course just numbers. There are essentially two different significant parameters. The first is the London penetration depth, which is proportional to the coefficient $\frac{e^2}{mc}|f|^2$ outside of the whole business. London estimated the penetration depth by simply taking the number of superconducting electrons to be equal to the total number of electrons, and giving e, m and c their standard values. When this is done the London penetration depth turns out to be about the observed experimental value. One important fact about flux quantization is the magnitude of this London penetration depth, that is the magnitude of the coefficient outside the whole business. The magnitude of this coefficient is such that if the variation of the phase does not, even in a quite small sample, exactly cancel the vector potential, one would predict enormously large currents. One may put this point on a quantitative basis by observing that this coefficient has the dimensions, as you know, of an inverse length squared. The total current flowing through a given ring will then be proportional to the ratio of the cross-sectional area of the ring to the square of the London penetration depth, in dimensionless units. Such a current would in turn produce a very large magnetic field which would strongly modify the vector potential **A**. Therefore the only self-consistent solution of these equations is a solution in which the two terms on the right-hand side cancel out each other almost exactly. This is true for any sample whose area is large compared to the square of the London penetration depth, whether it is so thin that the magnetic field actually penetrates it or not. One may ask what happens if one tries to apply a magnetic field configuration to such a sample which does not give one proper quantization of the flux through the ring. The answer can be one of two things — either the ring produces very large shielding currents in order to bring the flux through the ring to the right value, leading to an increase in the kinetic energy of the current and therefore to a higher free energy; or the wave function no longer has the same simple form that we have assumed here. This situation has been exhaustively discussed by Douglass.

The second interesting parameter is essentially the charge e which enters in the ratio between the part of the current which comes from the phase variation, and the part of the current which comes from the vector potential **A**. The other constants here are $\hbar$ and c and they of course may be presumed to have their usual values. It was first pointed out by Ginzburg that the data on thin films already in 1955 indicated experimentally that this constant here should not be e but an effective charge e^*. From this data Ginzburg estimated a value of somewhere between 1-1/2 and 2 times the usual electronic charge. He remarks in a footnote that Landau says that the value cannot be a nonintegral number of charges because of gauge invariance and draws no conclusion from this. The obvious conclusion which one can draw now is that the value of e^* should have been assumed to be twice the electronic charge, and in a sense it was at this point that the quantization of flux in units of $hc/2e$ first could have been suggested theoretically.

I should comment at this point that of course the remarks of London about the fluxoid rather than the flux being the quantity which is conserved and also quantized, are still exactly true in the Landau–Ginzburg theory. What is exactly quantized is the fluxoid and in a sample whose cross-sectional area is small compared to the square of the London penetration depth the actual quantum of flux will vary from the value $hc/2e$.

This idea of the quantization of flux leads to a very natural and physical explanation of the phenomenon of persistent currents. We recognize that a state in which this quantum number n is greater than 0 is obviously not the lowest quantum state of the system because it contains a current and a resulting magnetic field, and the magnetic field energy is positive. Therefore, Bloch's theorem about the lowest quantum state of any system being a state in which no current is flowing is perfectly well satisfied in this theory. On the other hand, a state in which this quantum number n is not equal to 0 differs from the ground state, in which n is equal to 0, throughout the material, in the value of a macroscopic parameter, the wave function ψ. Furthermore, in order to get from the state in which n is not equal to 0 to the state in which it is equal to 0 it is not at all obvious what we should do — in fact most of the rest of this course will be concerned with discussing what we should do — but it is obvious that we are going to have to make a rather modification in the wave function and that it is very likely that this large modification of the order parameter in the intermediate states between the two quantum states is likely to be a modification which requires a fairly large free energy. Thus we can understand that it is likely that there is a *macroscopic*, or at least large, free energy barrier between the different quantum states and therefore that any one of the quantum states can be quite stable, although in principle it is still only a metastable state.

We have then the second of the Landau–Ginzburg equations, which I have presented here first because it is this second equation which is most directly related to the London equation, and also because it leads to all of the really interesting properties of superconductors. In fact it is fair to say that in order to establish that a theory of superconductivity leads to the phenomenon we call superconductivity, it is only necessary to establish that this theory of superconductivity leads to the second Landau–Ginzburg equation. In fact, it needs only lead to it in the limit of very long wavelengths; that is, of very slow variations of the wave function ψ and of magnetic fields. On the other hand, of course, it is necessary that it leads to it at all temperatures. N. R. Werthamer at Bell Laboratories and also

Ludwig Tewordt have recently established that the present theory of superconductivity, having its origin in the BCS theory, does in fact lead to the second Landau–Ginzburg equation at all temperatures, under the conditions appropriate to a macroscopic ring. From the second Landau–Ginzburg equation one calculates the three characteristic superconducting effects: the Meissner effect, the persistent current effect and flux quantization.

Nevertheless this equation is still incomplete because it contains a completely unknown and unspecified order parameter Ψ. In order to complete a set of phenomenological equations for describing superconductivity, it is obviously necessary to have a second equation which will determine Ψ under given circumstances. In other words, it is necessary to have a constitutive equation for the order parameter in the presence of a given kind of sample, temperature and distribution of magnetic field. It is possible to arrive at a general form for such an equation by very simple reasoning. In the first place, in the absence of a magnetic field we can presume that Ψ will simply have a constant thermal equilibrium value at any given temperature throughout a homogeneous sample. We know that this thermal equilibrium value must go to zero at the transition temperature and above it, and since the superconducting transition is a second order phase transition, we know that Ψ must rise continuously from zero as we go below the transition temperature. A well-known example of such a phase transition is ferromagnetism for which the magnetization M is an approximate order parameter. The procedure in determining the transition temperature would then be to evaluate the free energy F as a function of M, and minimize it at various temperatures. When a minimum occurs at a value of $M \neq 0$, we have the ferromagnetic state. The situation is illustrated schematically in the following three diagrams for $F(M)$ at various temperatures. In the third diagram, the magnetization $M(T)$ for which the curve is a minimum, would be the value physically realized at that temperature. We now return to a consideration of the superconducting phase transition.

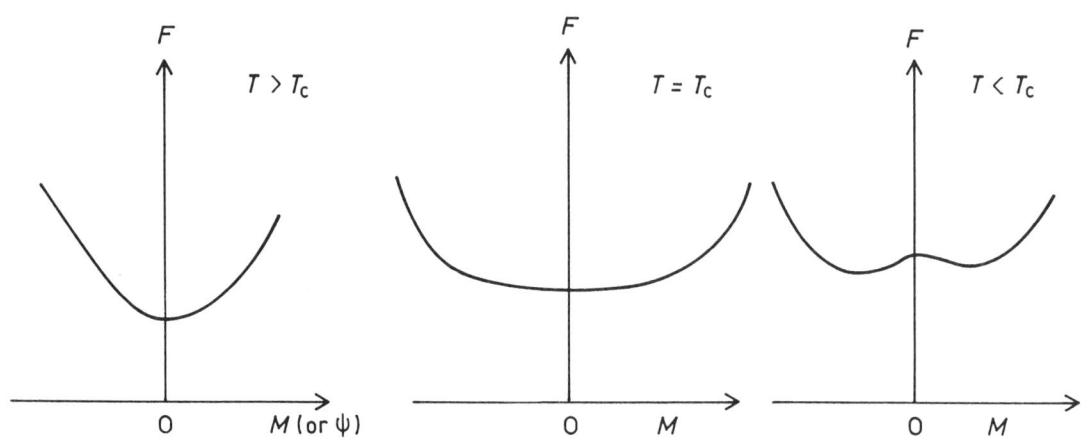

In keeping with Landau's basic philosophy about second order transitions, what we can do then is to expand things around the phase transition point because we know that the order parameter is going to be in some sense small at that point. The equation which is to determine the order parameter is obviously that the derivative of the free energy with respect to the order parameter at a given temperature must be zero — that is the definition of thermal equilibrium. We can get an order parameter which obeys the above conditions if we assume that the free energy as a function of the order parameter, if the order parameter is constant in space, is given by the following equation:

$$F = \alpha(T)|\Psi|^2 + \beta(T)|\Psi|^4 + \cdots .$$

If we take as coefficients in this equation a coefficient of the $|\Psi|^2$ term which is negative and vanishes at the transition temperature, and a roughly constant coefficient for the fourth power terms as follows:

$$F = \text{const.} \times \left[\left(\frac{T}{T_c} - 1 \right) |\Psi|^2 + \frac{|\Psi|^4}{N} \right],$$

we see that the derivative with respect to Ψ gives us the following equation determining Ψ as a function of temperature:

$$\text{const.} \times \left[\left(\frac{T}{T_c} - 1 \right) \Psi + \frac{2|\Psi|^2}{N} \Psi \right] = 0 .$$

Clearly, this kind of expansion must fail at lower temperatures, but it is reasonably obvious that it will only fail in the form which we have assumed for the functional dependence of F on the order parameter Ψ.

In the presence of magnetic fields, normal-superconducting boundaries, and so forth we must have more terms in this expression. We follow the clue that what we are really looking for is the free energy as a function of Ψ, because what we can do then to get Ψ is to take the variational derivative with respect to it. We can find the kinetic energy and field terms in the free energy from the following general principle: namely the current is, by general reasoning, given by the variational derivative of the mean energy with respect to the vector potential $\mathbf{A}$. $\mathbf{J} = e \frac{\delta F}{\delta \mathbf{A}}$. We have the expression for the current, and therefore we can immediately write down a kinetic energy term which will give this expression

$$T = \frac{1}{2m} \left(-i\hbar \boldsymbol{\nabla} \Psi - \frac{e}{c} \mathbf{A} \Psi \right)^2 .$$

It seems at this point very likely, and in fact will become obvious when we come to talk about the microscopic theory, that this kinetic energy expression is the first term in an expansion in terms of $\nabla\Psi$. The free energy expression we used was a series expansion in the order parameter Ψ and therefore the theory is limited to temperatures very near to the temperature at which Ψ vanishes under any given condition. This idea of the expansion in terms of $\nabla\Psi$ tells us that a second limitation is that we must confine ourselves to situations in which $\mathbf{A}$ is relatively small and the wave function Ψ is rather slowly varying.

Now we can write the first of the Ginzburg–Landau equations by writing down the variational equation we obtained by minimizing the sum of these two energies. This first Ginzburg–Landau equation is the following:

$$\left\{ \frac{1}{2m}\left(-i\hbar\nabla - \frac{e}{c}\mathbf{A}\right)^2 - C\left[\left(1 - \frac{T}{T_c}\right) - 2\frac{|\Psi|^2}{N}\right] \right\}\Psi = 0$$

where C is a constant. We see that it looks very like a wave equation for the wave function, or order parameter, Ψ, except that where we would normally have a term giving the potential energy as a function of position, instead we have a term which comes from the energy as a function of temperature and of the wave function Ψ.

Before going on I would like to make one remark pointing out that these equations are not in fact exactly valid in precisely the region in which one would at first think that they should be most nearly exact, namely precisely at the transition temperature. It has now been recognized even by the Landau school that the Landau theory of second order critical points is not in fact exact because of the possibility of critical fluctuations near the transition point. We have no reason to expect that superconductivity will be an exception. However, the phenomenon of superconductivity is probably the best case for this expansion technique of Landau, because as it turns out the condensation energy of any given electron depends on its interaction with a very large number of other electrons — in most cases, all those within the so-called coherence length, to be discussed later, a distance of the order of 100 to 10^4 Å, involving the order of 10^6 to 10^{12} electrons. Fluctuation phenomena therefore tend to be fairly well averaged out even near T_c for a superconductor.

The most important property of the first Landau–Ginzburg equation is that it also contains a natural unit of length, with an important physical meaning. Consider a system subject to some spatially-dependent perturbation: write $\Psi = \Psi_0 + \delta\Psi$ where Ψ_0, the equilibrium order parameter, may be assumed spatially uniform. If no magnetic field is present the first Landau–Ginzburg equation reduces to:

$$\nabla^2(\delta\Psi) = -\frac{1}{\xi_0^2}\left\{\left(1 - \frac{T}{T_c}\right) - 6\frac{|\Psi_0|^2}{N}\right\}\delta\Psi$$

where we have replaced $2mC/\hbar^2$ by $1/\xi_0^2$, and have used the fact that Ψ_0 satisfies the first Landau–Ginzburg equation, and assumed that the perturbation $\delta\Psi$ is small in evaluating the term in $|\Psi|^2$ on the right-hand side. Since Ψ_0 is uniform, the first Landau–Ginzburg

equation reduces to:
$$2\frac{|\Psi_0|^2}{N} = \left(1 - \frac{T}{T_c}\right).$$

When this is substituted in the preceding equation we obtain:
$$\nabla^2 \delta\Psi = \frac{2}{\xi_0^2}\left(1 - \frac{T}{T_c}\right)\delta\Psi .$$

In one dimension this equation has physically meaningful solutions of the form $\exp(-x/\xi)$ where
$$\xi(T) = \frac{\xi}{\sqrt{2\left(1 - \dfrac{T}{T_c}\right)}}$$

is called the "coherence length": it is a representative scale for the possible space-variations in the superconducting wave function Ψ. The same distance for damping of a weak perturbation occurs in the two- and three-dimensional cases as derived above. An important dimensionless parameter of the theory is the ratio of the Landau penetration depth to the coherence length,
$$K = \frac{\lambda_L(T)}{\xi_0(T)} = \sqrt{\frac{mc^2}{\pi N \xi_0^2 e^2}}$$

where N is the density of superconducting electrons. In the Landau–Ginzburg theory the ratio is approximately temperature independent, as given above.

The Landau–Ginzburg theory was developed primarily to solve the problem of interfacial energy between regions of normal and superconducting phase in the intermediate state. In both the London and Landau theories, these regions were considered to be of macroscopic dimensions. We consider the free energy in a magnetic field:

$$F = \frac{1}{2m}\left[\left(-i\hbar\nabla - \frac{e\mathbf{A}}{c}\right)\Psi\right]^2 + \frac{1}{\xi_0^2}\left[\left(\frac{T}{T_c} - 1\right)\Psi^2 + \frac{|\Psi|^4}{N}\right] + \frac{1}{8\pi}(H-H_{\text{applied}})^2 .$$

We examine this in the vicinity of a normal-superconducting interface for two cases $K \gg 1$ and $K \ll 1$, using the concepts of magnetic field penetration depth and order parameter coherence length. We plot schematically H and Ψ in the vicinity of the interface.

For $K \ll 1$, i.e., $\xi \gg \lambda_L$, the order parameter rises slowly in the superconducting phase, relative to the quick drop of magnetic field. We compare the free energy density in normal and superconducting phase, defining a surface free energy density:

$$E_s = F_n - F_s$$

where n and s refer to normal and superconducting phases respectively. By the definition of H_c, the free energy density is the same deep inside the normal and superconducting regions; otherwise, of course, the boundary cannot be stable. For a distance of order ξ_0 into the superconducting phase, the field energy is as unfavorable as in the pure superconducting phase, while the compensating $|\Psi|^2$ term has not risen to its full value. Some field energy is regained in the field penetration region but very little. The gradient term can only add energy, so we get a positive surface free energy density of the order of $\xi_0 \frac{H_c^2}{8\pi}$ for $K \ll 1$. As London pointed out, this positive surface free energy is essential to stabilize the macroscopic domains which occur in the "intermediate state" of soft superconductors. The size of the domains are large in order to reduce the total area of domain boundaries and save surface energy; surface energies — thus ξ_0 — can be measured by using this fact and finding that domain size which minimizes the combination of magnetostatic and surface energy.

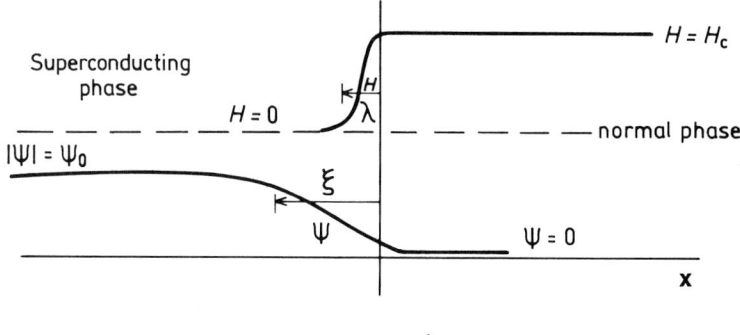

Case 1: $K = \frac{\lambda_L}{\xi} \ll 1$.

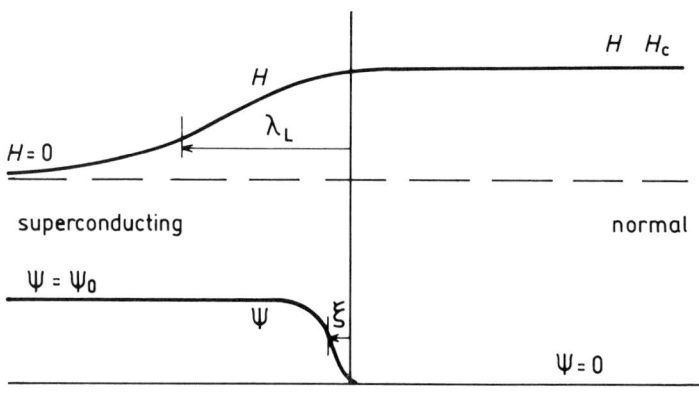

Case 2: $K = \frac{\lambda_L}{\xi} \gg 1$.

For $K \gg 1$, we have the following situation, in which $\xi \ll \lambda_L$.

In this case, over a depth of order λ_L, Ψ has achieved the value necessary to minimize the $|\Psi|^2$ terms, while the field energy is more favorable than in the pure superconductor. Thus the surface energy is effectively negative of order $-\lambda_L \frac{H_c^2}{8\pi}$. It is important to realize that this negative energy is only a free surface energy in a limited sense; it depends on the presence of the applied magnetic field and is really only a regained field energy. The surface energy at $H = 0$ must be positive on general thermodynamic grounds. In this case of negative surface free energy, the subdivision of the intermediate state into normal and superconducting regions would be expected, from energetic considerations, to proceed to the microscopic scale, thus producing a maximum area of interface. We conjecture that this is the case in hard superconductors: The mathematical criterion is that $\lambda_L > \xi_0$. This conjecture, and much of the knowledge about the actual values of λ_L and ξ_0, has been the result of Pippard's application of microwaves in the study of superconductivity; we will discuss his results later.

LECTURE 3
DERIVATION OF THE LANDAU–GINZBURG EQUATIONS

Since everything which follows will be based entirely on the Landau–Ginzburg equations, or possibly more general but similar equations, we shall now sketch the relationship between them and the BCS microscopic theory of superconductivity. This derivation was first given by Gor'kov in a Green's function formulation of the theory of superconductivity. A simpler version which is directly relevant to the hard superconductors is given by de Gennes *et al.*[a] This work is strongly recommended as collateral reading, for those who wish to follow the reasoning in detail. The general structure of the argument is clearer, however, in the Green's function formulation of Gor'kov.

The Green's function approach is based on the "generalized Hartree–Fock" method in the theory of superconductivity. We begin with the Heisenberg equations of motion for the electron field operator $\psi^\dagger(r, t)$:

$$i\hbar \frac{\partial}{\partial t} \psi^\dagger(\mathbf{r}, t) = \frac{1}{2m} \left(\mathbf{p} - \frac{e\mathbf{A}}{c} \right)^2 \psi^\dagger + V(\mathbf{r})\psi^\dagger$$
$$+ \int dt' \int d^3r' \, V(\mathbf{r} - \mathbf{r}', t - t') n(\mathbf{r}', t') \psi^\dagger(\mathbf{r}, t)$$

where time dependence in the many-body interaction term allows for the necessarily retarded electron-electron interaction via the phonon field, and the operator $n(\mathbf{r}, t) = \psi^\dagger(\mathbf{r}, t)\psi(\mathbf{r}, t)$. The Hartree and Hartree–Fock terms follow in the usual way from linearizing this equation by taking account of the mean density $\langle n(r, t) \rangle$ and the mean exchange

[a] Ref. 4.

charge $\langle\psi^\dagger(r,t)\psi(r',t')\rangle$. The potential $V(\mathbf{r})$ refers to any background potential, such as the static periodic lattice potential, and we shall assume that the Hartree and Hartree–Fock terms may be absorbed into $V(r)$ and we shall concentrate on a third term which is responsible for superconductivity.

$$\int dt' \, d^3r' \, \Delta^\dagger(\mathbf{r},\mathbf{r}',\,t-t')\psi(\mathbf{r}',t')$$

where

$$\Delta^\dagger = V(\mathbf{r}-\mathbf{r}',\,t-t')\langle\psi^\dagger(\mathbf{r}',t')\psi^\dagger(\mathbf{r},t)\rangle\,.$$

This term represents a "mean potential for pair formation". It corresponds to a diagram of the following form:

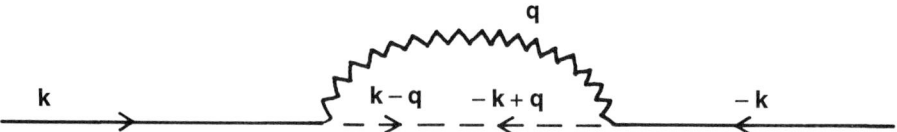

We think of such a diagram as representing the following physical process: a free hole of momentum $\mathbf{k}$ emits a phonon and is scattered into the sea of bound pairs. The corresponding electron of the bound pair $\mathbf{k}-\mathbf{q}$, $-\mathbf{k}+\mathbf{q}$ absorbs the phonon and appears at $-\mathbf{k}$. Thus the presence of the bound pair sea allows for a kind of "off-diagonal self-energy part" which changes an electron into a hole of the same momentum and spin. Such terms can only be generated self-consistently: they are not derivable from perturbation theory.

We now derive the equations of motion for the Green's functions F and G, defined by

$$G = \langle T\psi^\dagger(\mathbf{r},t)\psi(\mathbf{r}',t')\rangle$$
$$F^\dagger = \langle\psi^\dagger(\mathbf{r},t)\psi^\dagger(\mathbf{r}',t')\rangle\,.$$

Operating on the equation of motion for $\psi^\dagger$ by multiplying by $\psi(\mathbf{r}'',t'')$ or $\psi^\dagger(\mathbf{r}'',t'')$, subtracting a term $\mu\psi^\dagger$, thus measuring energies relative to the Fermi energy ($\mu =$ chemical potential), we obtain:

$$i\hbar\frac{\partial}{\partial t}G(\mathbf{r},t,\mathbf{r}',t') = H_0 G + \int d^3r''dt''\Delta^\dagger(\mathbf{r},\mathbf{r}'',\,t-t'')F(\mathbf{r}'',\mathbf{r}',\,t''-t') + \delta(\mathbf{r}-\mathbf{r}',\,t-t')$$

where we have replaced the interaction term by $\Delta^\dagger$: the superconductivity term.

This system of equations is highly nonlinear, and will have to be treated by some approximation procedure. Gor'kov's procedure is to solve the equations schematically for the Green's function on the left-hand side by multiplying by the "inverse" of the operator $\left(i\hbar \frac{\partial}{\partial t} - H_0\right)$. This gives an expression for F in terms of Δ and G; this expression is then used in the definition of Δ to solve for Δ in terms of Δ and G.

We shall begin by outlining the simplified form of solution which corresponds to the BCS theory of a pure superconductor with $\mathbf{A} = 0$. If we Fourier analyze F, G, and Δ in space and time, we obtain equations which are schematically of the form:

$$\hbar \omega F^\dagger(\mathbf{k}, \omega) = \varepsilon_k F^\dagger(\mathbf{k}, \omega) + \Delta(\mathbf{k}, \omega) G(\mathbf{k}, \omega)$$

and

$$-1 + (\hbar \omega - \varepsilon_k) G(\mathbf{k}, \omega) = \Delta(\mathbf{k}, \omega) F(\mathbf{k}, \omega) = \frac{[\Delta(\mathbf{k}, \omega)]^2 G(\mathbf{k}, \omega)}{(\hbar \omega + \varepsilon_k)}$$

whence

$$\left\{(\hbar \omega)^2 - \varepsilon_k^2 - \Delta^2(\mathbf{k}, \omega)\right\} G(\mathbf{k}, \omega) = \hbar \omega + \varepsilon_k .$$

In the above expressions, we have Fourier analyzed in terms of eigenfunctions of $H_0 = \frac{p^2}{2m} - \mu + V(\mathbf{r})$; the corresponding eigenvalues are ε_k. Similarly we obtain:

$$F(\mathbf{k}, \omega) = \frac{\Delta(\mathbf{k}, \omega)}{(\hbar \omega)^2 - \varepsilon_k^2 - \Delta^2(\mathbf{k}, \omega)} .$$

In all this, we assume that the system is uniform in space and time, so that the double time and spatial variables in $G(\mathbf{r}, \mathbf{r}'; t, t')$ are really involved only in differences; i.e., G has the form $G(\mathbf{r} - \mathbf{r}'; t - t')$. We shall see that this solution is similar to that of BCS, but with slightly greater generality. The identification of $E_k^2 = \varepsilon_k^2 + \Delta_k^2$ (suppressing ω dependence of Δ) and of Δ_k with the energy gap, will be seen to be appropriate, since E_k is then the energy of the pole in the Green's function.

The connection with BCS can be seen by comparing the expression:

$$\Delta(\mathbf{k}, \omega) = \int d\omega' \sum_{k'} V(\mathbf{k} - \mathbf{k}', \omega - \omega') \langle F^\dagger(\omega', k') \rangle$$

with the BCS expression for the (temperature dependent) energy gap:

$$\Delta_{\mathbf{k}} = \sum_{k'} V_{\mathbf{k}\mathbf{k}'} \langle C^\dagger_{\mathbf{k}'} C^\dagger_{-\mathbf{k}'} \rangle_{\text{thermal average}}$$

$$= \sum_{k'} V_{\mathbf{k}\mathbf{k}'} \frac{\Delta_{\mathbf{k}'}}{\sqrt{\varepsilon_{\mathbf{k}'}^2 + \Delta_{\mathbf{k}'}^2}} \tanh\left(\frac{\beta E_{k'}}{2}\right) .$$

In the preceding paragraph, we have a form for $F(\mathbf{k}, \omega)$ which is defined by adding a term $i\varepsilon$ in the denominator. Then $\text{Im } F(\mathbf{k}, \omega)$, and $\langle C^\dagger_k C^\dagger_{-k} \rangle$ play similar roles: a Dirac δ-function in the energy of the denominator occurs, from Im $F(\mathbf{k}, \omega)$, and must be multiplied by a factor appropriate to the as yet undefined thermal averaging procedure. Because of the δ-function, one might guess that $\tanh\left(\frac{\beta E_k}{2}\right)$ could be replaced by $\tanh\left(\frac{\beta \hbar \omega}{2}\right)$, and that one could then write:

$$\Delta(\mathbf{k}, \omega) = \int d\omega' \sum_{\mathbf{k}'} V(\mathbf{k} - \mathbf{k}', \omega - \omega') F(\mathbf{k}', \omega') \tanh\left(\frac{\beta \hbar \omega}{2}\right).$$

Now $\tanh\left(\frac{\beta \hbar \omega'}{2}\right)$ has poles at imaginary values of ω', namely:

$$\omega'_n = i\pi kT(2n+1).$$

Integrating over the contour shown below then permits one to evaluate $\Delta(\mathbf{k}, \omega)$ by the residue theorem: Fourier transforming back to spatial variables then gives the result:

$$\Delta(\mathbf{r}, \mathbf{r}', \omega) = kT \sum_{n'} V(\mathbf{r}, \mathbf{r}', \omega - \omega_{n'}) F(\mathbf{r}, \mathbf{r}', \omega_{n'}).$$

The above discussion has been heuristic: however, when thermal averaging is performed according to the rigorously correct prescription, the same final result is obtained. It is not obvious how to define the time Fourier transformed $V(\mathbf{r}, \mathbf{r}', \omega)$ for complex ω, but it will turn out not to be relevant in this connection. This formalism is presented as a substitute for BCS, but is in fact considerably more accurate for the physical case which occurs, in which Δ is primarily a function of ω rather than of $\mathbf{k}$; that is, when the quasi-particle description of the system is rather a poor starting point.

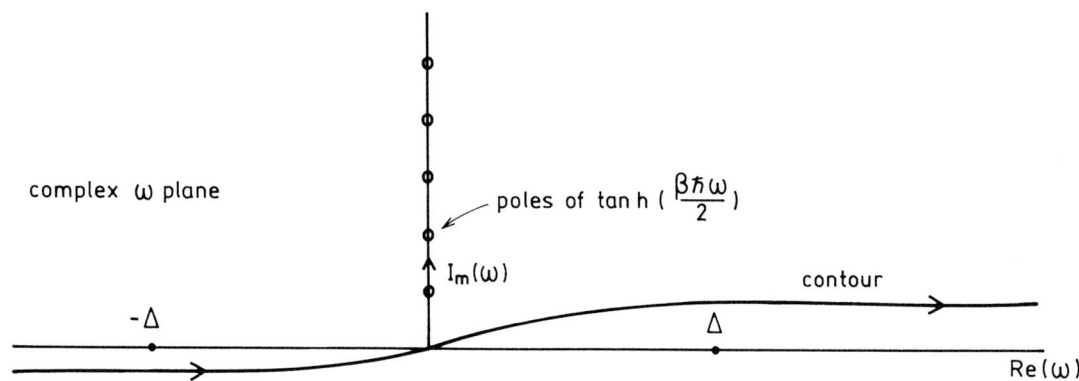

We now return to a systematic treatment of the spatially inhomogeneous Green's function equations. We Fourier transform in time to obtain:

$$\Delta^\dagger(\omega) = \int d\omega' V(\omega - \omega') F^\dagger(\omega')$$

where dependence of Δ, V, and F on spatial variables is understood above and in all that follows. No assumption about spatial homogeneity will be needed, but approximate uniformity in time (i.e., slowly varying fields) has been assumed: the time dependent generalization of the Landau–Ginzburg equations has not yet been obtained. The equations for F and G on time Fourier transformation are:

$$(\hbar\omega - H_0)G_\omega(\mathbf{r},\mathbf{r}') = \delta(r - r') + \int d^3r'' \Delta^\dagger_\omega(\mathbf{r},\mathbf{r}'') F_\omega(\mathbf{r}'',\mathbf{r}')$$

and

$$(\hbar\omega - H_0)F^\dagger_\omega = \int \Delta^\dagger_\omega G_\omega$$

where the notations G_ω and $G(\omega)$ refer to the same quantity exactly. Now, for the normal metal, in which the superconducting interaction Δ does not operate, we have: (G^0_ω is the normal metal Green's function)

$$\{\hbar\omega - H_0(\mathbf{r})\}G^0_\omega(\mathbf{r},\mathbf{r}') = \delta(\mathbf{r} - \mathbf{r}') .$$

We therefore see that G^0_ω is in a sense the inverse of $(\hbar\omega - H_0)$. If we multiply the equation for $F^\dagger_\omega$ by $G^0_{\omega^*}$, and integrate, we obtain:

$$\int d^3 r\, G^0_{\omega^*}(\mathbf{r},\mathbf{r}_1) \left\{ \hbar\omega - \frac{(\mathbf{p} - \frac{e\mathbf{A}}{c})^2}{2m} - V(\mathbf{r}) \right\} F^\dagger_\omega(\mathbf{r},\mathbf{r}')$$

$$= \int d^3 r \int d^3 r''\, G^0_{\omega^*}(\mathbf{r},\mathbf{r}_1) \Delta^\dagger_\omega(\mathbf{r},\mathbf{r}'') G_\omega(\mathbf{r}'',\mathbf{r}') .$$

Now $(\hbar\omega - H_0)$ is an Hermitian operator if ω is real, otherwise the Hermitian conjugate involves ω^*. We wish to apply the operator $(\hbar\omega - H_0)$ to the left on $G^0_{\omega^*}$ to make use of the resulting δ-function. Because of our choice of arguments this is permitted. Note however that $G(\mathbf{r},\mathbf{r}_1)$ is not symmetrical in the arguments: $G(\mathbf{r},\mathbf{r}_1)$ obeys the same equation in $\mathbf{r}_1$ as in $\mathbf{r}$ except that $\partial/\partial r$ must be replaced by $-\partial/\partial r_1$. We thus have:

$$F^\dagger_\omega(\mathbf{r}_1,\mathbf{r}') = \int d^3 r \int d^3 r''\, G^0_{\omega^*}(\mathbf{r},\mathbf{r}_1) G_\omega(\mathbf{r}'',r') \Delta^\dagger_\omega(\mathbf{r},\mathbf{r}'') .$$

Finally, we insert this expression into the equation for $\Delta^\dagger_\omega$

$$\Delta^\dagger_\omega(\mathbf{r},\mathbf{r}') = \int d\omega' V(\omega - \omega',\mathbf{r},\mathbf{r}') F^\dagger_\omega(\mathbf{r},\mathbf{r}')$$

$$= \int d\omega' V(\omega - \omega',\mathbf{r},\mathbf{r}') \int d^3 r'' \int d^3 r'''\, G^0_{\omega^*}(\mathbf{r}'',\mathbf{r}') G_{\omega'}(\mathbf{r}''',\mathbf{r}') \Delta^\dagger_\omega(r'',r''') .$$

This is the self-consistent equation in Δ which must be satisfied. It may be described as quasi-linear: it is linear except for the dependence of G_ω on Δ. The quantity Δ is referred to as the energy gap *function*, because in the simple case of a pure superconductor, the excitation energy of a quasi-particle is given by $E_k^2 = \varepsilon_k^2 + \Delta_k^2$ where Δ_k depends implicitly on E_k, and Δ_k is the Fourier transform of $\Delta(\mathbf{r} - \mathbf{r}')$. In more complicated cases the quantity Δ defined as VF is more fundamental than any actual energy gap. In general, it is not necessary that there can even be an energy gap, because it is not always true that the quasi-particle energy is given by $E_k^2 = \varepsilon_k^2 + \Delta_k^2$, in fact it will certainly not be so if Δ varies in space.

Thus far, the equation for Δ is completely general except for assuming static or quasi-static time dependence. We now specialize the equation in a completely inessential way, by assuming, first, that $V(\mathbf{r} - \mathbf{r}')$ is very short range *in space*: this is very generally true because of the screening effect of the electrons, which is effective beyond distances of the order of 10^{-8} cm, whereas the Landau–Ginzburg equation deals with distances of the order of the coherence length $\xi_0 \sim 100 - 10^4$ Å; thus $\mathbf{r}$ and $\mathbf{r}'$ must be very close together for a significant contribution to $\Delta(\mathbf{r}, \mathbf{r}')$. We thus approximate the spatial dependence by $\Delta(\mathbf{r})\delta(r - r')$ where $\Delta(\mathbf{r})$ is of the form $\int d^3 r' V(\mathbf{r} - \mathbf{r}') F(\mathbf{r}, \mathbf{r})$: i.e., we use an average value of V. Secondly, almost all important effects in superconductivity, and certainly all of importance for hard superconductors, occur over relatively long time intervals, that is, near $\omega = 0$: Thus the gap function at $\omega = 0$ is the physically important parameter in the theory. Thus Δ will be assumed independent of ω in the low-energy range. We then approximate $\Delta_\omega(\mathbf{r}, \mathbf{r}')$ by $\Delta(\mathbf{r})$ as above, and identify it with the Landau–Ginzburg order parameter ψ. The equation for $\Delta^\dagger$ now simplifies to:

$$\psi^*(\mathbf{r}) = \int d\omega' \, \langle V(\omega') \rangle \int d^3 r' \, G^0_{\omega'}(\mathbf{r}', \mathbf{r}) G_{\omega'}(\mathbf{r}', \mathbf{r}) \psi^*(\mathbf{r}') \,,$$

where $\langle V(\omega') \rangle$ is assumed to equal the approximately constant value of $V(\omega - \omega')$ in ω, averaged over spatial variables as above, and cut off at a value of $\omega' \sim k_B \theta_D$, corresponding to our idea that short time effects are to be neglected. An assumption that $\Delta(\omega_{\text{finite}})$ is proportional — not necessarily equal — to $\Delta(\omega = 0)$ has been made here. This can only be justified by referring to our experience with the actual solution of gap equations — see particularly Schrieffer, Scalapino and Wilkins and Morel and Anderson — in which this turned out to be really true. This equation for ψ^* is nonlinear because of the occurrence of the actual Green's function $G_{\omega'}(\mathbf{r}', \mathbf{r})$, which depends on Δ, or ψ. However, the main point of the theory is now clear, namely a relationship between ψ at $\mathbf{r}$ and its values at neighboring points $\mathbf{r}'$, giving the spatial coupling in the theory. I would like to emphasize that so far the equation is exact for the situations of interest.

We now proceed to eliminate $G_{\omega'}$, in order to obtain the constitutive equation for ψ, in three steps of approximation. We shall introduce the temperature dependence as previously outlined, by replacing the integral over ω' by a discrete sum over residues at the poles of $\tanh\left(\frac{\beta\hbar\omega'}{2}\right)$. In treating G_ω we note that its equation has the form:

$$(\hbar\omega - H_0)G_\omega = \delta(\mathbf{r} - \mathbf{r}') + \int \Delta_\omega^\dagger F_\omega$$

and, since $\Delta_\omega^\dagger$ is identified with ψ^*, and, expanding everything in powers of ψ we will also have F_ω starting with a linear term in ψ, we see that in order to get the linear terms of the constitutive equation, we may replace G_ω by G_ω^0. Thus, the linear approximation to the Landau–Ginzburg equation is:

$$\psi^*(\mathbf{r}) = k_\mathrm{B}T \sum_n \langle V(\omega_n)\rangle \int d^3 r' \, G^0_{\omega_n^*}(\mathbf{r}', \mathbf{r}) G^0_{\omega_n}(\mathbf{r}', \mathbf{r}) \psi^*(\mathbf{r}') \;.$$

The first approximation then is to expand F_ω and G_ω in power series in ψ, as Landau did. The second approximation is to assume weak magnetic fields. We have:

$$(\hbar\omega - \mu)G_\omega^0(\mathbf{r}, \mathbf{r}') - \frac{\hbar^2}{2m}\left(\nabla_\mathbf{r} + \frac{ie}{\hbar c}\mathbf{A}(\mathbf{r})\right)^2 G_\omega^0(\mathbf{r}, \mathbf{r}') = \delta(\mathbf{r} - \mathbf{r}') \;.$$

The critical step, due to Landau, is to try the construction

$$G_\omega^0 = G_\omega^{00}(\mathbf{r}, \mathbf{r}')e^{iS} \;.$$

Then G_ω^{00} satisfies:

$$\left\{(\hbar\omega - \mu) - \frac{\hbar^2}{2m}\nabla^2\right\} G_\omega^{00}(\mathbf{r}, \mathbf{r}') = \delta(\mathbf{r} - \mathbf{r}')$$

(i.e., G_ω^{00} is the Green's function for the normal metal in the *absence* of magnetic field) provided:

$$S = -\frac{e}{\hbar c}\mathbf{A}(\mathbf{r})\cdot(\mathbf{r} - \mathbf{r}') \;.$$

The equation for G_ω^{00} is valid in the following sense: consider

$$\left\{\nabla_\mathbf{r} + \frac{ie}{\hbar c}\mathbf{A}(\mathbf{r})\right\} G_\omega^{00}(\mathbf{r}, \mathbf{r}') \exp\left\{-\frac{ie}{\hbar c}\mathbf{A}(\mathbf{r})\cdot(\mathbf{r} - \mathbf{r}')\right\}$$

$$= \left\{\nabla_\mathbf{r} G^{00}\right\}e^{iS} + \left\{\frac{ie}{\hbar c}\mathbf{A}(\mathbf{r}) - \frac{ie}{\hbar c}\mathbf{A}(\mathbf{r})\right\} G_\omega^{00} e^{iS}$$

$$+ \text{ terms of order } \left\{\frac{e}{\hbar c}|(\mathbf{r} - \mathbf{r}')|B G_\omega^{00}(\mathbf{r} - \mathbf{r}')e^{iS}\right\}$$

where the magnetic induction B enters from differentiation of $\mathbf{A}$. The second term is zero, and the assumption of weak magnetic fields makes the third term negligible compared to the first, as we shall discuss in a moment. Application of $\left(\nabla - \frac{ie}{\hbar c}\mathbf{A}\right)$ again gives the term in $\nabla^2 G_\omega^{00}$ and other terms of higher order. If the last term above is to be negligible compared to the first, we must have:

$$\frac{e\lambda B}{\hbar c} < \frac{1}{\xi}$$

since $|(\mathbf{r} - \mathbf{r}')|$ will be of the order of the penetration depth λ when it occurs multiplying B, and $\nabla_r G_\omega^{00}(\mathbf{r},\mathbf{r}')$ is of order ξG_ω^{00}, where the coherence length ξ represents the effective range of the Green's function. Thus is we denote $\lambda\xi$ by R^2, we have the criterion:

$$B < \frac{\hbar c}{eR^2}$$

which is a weak field approximation. The meaning of this approximation has not been made clear in the literature. It is really a WKB approximation, and the criterion for validity is basically that the amplitude G_ω^{00} not vary due to $\mathbf{A}$ over the relevant ranges. This will be a good approximation in impure materials, but not so good generally. It is interesting to rewrite the criterion in the form $R^2 < \Phi_0/B$, where Φ_0 is the basic quantum of fluxoid: a limitation on the range $R = (\lambda\xi)^{\frac{1}{2}}$ for validity of the solution. Using this approximate form for G_ω^0, we obtain, for ψ:

$$\psi^*(\mathbf{r}) = \sum_n k_B T \langle V(\omega_n) \rangle \int d^3 r'\, G_{\omega_n^*}^{00}(\mathbf{r},\mathbf{r}') G_{\omega_n}^{00}(\mathbf{r},\mathbf{r}')$$

$$\times \exp\left\{-\frac{2ie}{\hbar c}\mathbf{A}(\mathbf{r}) \cdot (\mathbf{r} - \mathbf{r}')\right\} \psi^*(\mathbf{r}') \, .$$

This then is the result of the second major approximation, which we have seen may be characterized as weak $\mathbf{B}$ field. We note at this point the factor 2 occurring with e, giving an effective charge $e^* = 2e$ in the Landau–Ginzburg equation.

Finally, the third major approximation is to assume that ψ varies slowly over distances of the order of ξ, the range of the Green's function. We then expand the exponential and ψ about the point $\mathbf{r}$:

$$\psi^*(\mathbf{r}) = k_B T \sum_n \langle V(\omega_n) \rangle \int d^3 r'\, G_{\omega_n^*}^{00} G_{\omega_n}^{00}$$

$$\cdot \left\{1 - \frac{2ie}{\hbar c}\mathbf{A}(\mathbf{r}) \cdot (\mathbf{r} - \mathbf{r}') - \frac{2e^2}{\hbar^2 c^2}[A(r) \cdot (r - r')]^2 + \ldots\right\}$$

$$\cdot \left\{\psi^*(\mathbf{r}) - (\mathbf{r} - \mathbf{r}') \cdot \nabla_r \psi^*(\mathbf{r}) + \frac{(\mathbf{r} - \mathbf{r}')^2}{2}\nabla_r^2 \psi^*(\mathbf{r}) + \ldots\right\} \, .$$

The first order term will vanish on integration, by symmetry, so that the final equation has the form:

$$\psi^*(\mathbf{r}) = k_B T \sum_n \langle V(\omega_n) \rangle \int d^3 r'\, G^{00}_{\omega_n^*} G^{00}_{\omega_n} \left\{ 1 + \frac{(\mathbf{r}-\mathbf{r}')^2}{2}\left(\nabla - \frac{2ie}{\hbar c}\mathbf{A}(\mathbf{r})\right)^2 \right\} \psi^*(\mathbf{r}).$$

This then has the following form:

$$\psi^*(\mathbf{r}) = f(T) \left\{ 1 + \frac{\bar{\ell}^2}{2}\left(\nabla - \frac{2ie}{\hbar c}\mathbf{A}(\mathbf{r})\right)^2 \right\} \psi^*(\mathbf{r})$$

having replaced $|r-r'|^2$ by a mean squared length $\bar{\ell}^2$. This is the correct form for the linear term of the Landau–Ginzburg equation when $f(T)$ is evaluated near the transition point. For exact expressions, see Gor'kov and de Gennes. However, even from the present expression we can get some information. Firstly, one can easily convince oneself that when impurities are present, the range of the Green's function G is limited by the mean free path ℓ. Thus $\bar{\ell}^2$ is about equal to ℓ^2. We also know something about $f(T)$: since T_c is the point at which a solution is just possible with $\mathbf{A} = 0$, we have $f(T_c) = 1$, and $f(T)$ probably varies roughly linearly, with $f(T) < 1$ for $T > T_c$, so we may estimate:

$$\left(1 - \frac{T_c}{T}\right)\psi^*(\mathbf{r}) = \frac{\bar{\ell}^2}{2}\frac{T_c}{T}\left(\nabla - \frac{2ie}{\hbar c}\mathbf{A}(\mathbf{r})\right)^2 \psi^*(\mathbf{r})$$

or, more generally

$$\left(\frac{T}{T_c} - 1\right)\psi^* + \text{nonlinear terms} = \frac{\bar{\ell}^2}{2}\left(\nabla - \frac{2ie}{\hbar c}\mathbf{A}\right)^2 \psi^*.$$

This then evaluates the length parameter of the first Landau–Ginzburg equation directly and simply.

The current equation can also be obtained by similar Green's function manipulations. Its definition $J \sim \lim_{r \to r'}\left(\frac{\partial}{\partial r} - \frac{\partial}{\partial r'}\right) G(\mathbf{r},\mathbf{r}')$ involves G. The equation is then solved in terms of F and thus Δ. A second way is to reconstruct the free energy functional from the first equation and take the derivative with respect to $\mathbf{A}$. It emerges from these arguments that the limitation of T near T_c was not the most severe. The most restrictive limitation was $\frac{R^2 B e}{\hbar c} \ll 1$: nothing can be done to relax this. However, N. R. Werthamer and L. Tewordt have managed to apply the $e^{i\mathbf{A}(\mathbf{r})\cdot(\mathbf{r}-\mathbf{r}')}$ construction for most values of T in the limit $R \to 0$, thus obtaining generalized Landau–Ginzburg equations in this, the so-called London limit. Note that this limit is the one giving flux quantization, since $\mathbf{A}$ is small and Δ is slowly varying in a ring. Finally, we write down a standardized form of the Landau–Ginzburg equation:

$$\xi_0^2\left(i\nabla - \frac{2e}{\hbar c}\mathbf{A}\right)^2 \psi = \left(1 - \frac{T}{T_c}\right)\psi - \frac{2|\psi|^2}{N}\psi.$$

LECTURE 4

The next major contribution in understanding superconductivity was made by Pippard. His experiments were essentially based on measurements of the skin effect in normal and superconducting metals. The energy absorption is related to the penetration depth associated with **B**, and its associated electric field **E** which is not present in superconducting material at low frequencies. As higher frequencies are used, as in normal metals, the skin depth, corresponding to shorter wavelengths, becomes smaller. This affords the opportunity to measure not only the local electromagnetic response, but any possible nonlocal effects. Pippard did in fact find that the response function K was nonlocal, whence the local London expression

$$\mathbf{J}(\mathbf{r}) = -K\mathbf{A}(\mathbf{r})$$

had to be modified to a form:

$$\mathbf{J}(\mathbf{r}) = \int d^3r' \, K(\mathbf{r} - \mathbf{r}')\mathbf{A}(\mathbf{r}')$$

or, in Fourier transformed language:

$$\mathbf{J}(\mathbf{q}) = \mathbf{K}(\mathbf{q})\mathbf{A}(\mathbf{q}) \, .$$

The function K was found to be roughly an exponential function in space, and its range in pure metals did *not* correspond to the penetration depth, but to a larger length which Pippard called the *coherence length* ξ. A second fact, which he first suggested, was that the coherence length is *reduced* by scattering, while the skin depth is *increased*. It is not obvious, and as far as I know, only the microscopic theory shows, that Pippard's coherence length, which is a length associated with the current equation only, and Landau's length ξ, which enters his constitutive equation, are actually roughly the same. It is reasonable, because each of them presumably represents the range of the appropriate kernel of an integral equation involving some combination of Green's functions, but the assumption that they are always the same seems to have been made from the start. Each may now be calculated from fundamental theory in certain limits, but this assumption has not yet been verified in detail, essentially because the Landau–Ginzburg theory has never been generalized to the point where nonlocal electromagnetic response is an essential ingredient of the theory. It is rather easy to get a qualitative understanding of the linear electromagnetic responses. Consider first the normal metal: when we apply an electromagnetic field, represented by $\mathbf{A}(\mathbf{q})$, a current response results from scattering electrons from a state $\mathbf{k}$ to $\mathbf{k}+\mathbf{q}$, and we measure a current $\mathbf{J}(\mathbf{q})$ corresponding to $\mathbf{k}+\mathbf{q} \to \mathbf{k}$: thus the response function due to the paramagnetic term $\mathbf{A} \cdot \mathbf{J}$ is:

$$K_p(\mathbf{q}) = \sum_{k'} \frac{|\langle \mathbf{k}|J|\mathbf{k}+\mathbf{q}\rangle|^2}{(\varepsilon_{\mathbf{k}+\mathbf{q}} - \varepsilon_{\mathbf{k}})} \, .$$

There is, however, in addition, the London diamagnetic current, corresponding to the response $K_D = ne^2 A/mc$, which is almost completely cancelled by K_p in the normal phase. However, in the superconducting phase, if q is very small, i.e.: $k^2 k_f q/m \ll \Delta$, the energy gap, then the paramagnetic part of the response is blocked out by the gap, and only K_D occurs. If q is larger, then some of the paramagnetic response is regained: we expect K_p to begin to reappear at $q \sim \Delta/\hbar v_F$. Under scattering, however, the states to which the energy gap applies are scattered states with a range of $\mathbf{k}$ values. The scattering $\mathbf{k} \to \mathbf{k} + \mathbf{q}$ contains components of both low and high energy, so the matrix elements $\langle \mathbf{k}|\mathbf{J}|\mathbf{k}+q\rangle$ involve not only energy denominators $(\varepsilon_{\mathbf{k+q}} - \varepsilon_{\mathbf{q}})$, but a spread of energies of the order of $\hbar/\tau$. Thus K_p may have an appreciable value even as $q \to 0$, namely $K_p \sim \hbar/\tau\Delta$. At the same time its wavelength dependence is spread over $q \sim 1/\ell$, not $1/\xi_0$, so the nonlocality is reduced by impurity scattering, for example.

Abrikosov's Theory for Hard Superconductors

We now wish to discuss the fundamental paper by Abrikosov dealing with the case $K = \lambda/\xi \gg 1$ (*Sov. Phys. JETP* **5**, 1174 (1957)). The theory is based on the Landau–Ginzburg equations, although in fact some of the results are independent of them. Abrikosov's contribution was to recognize that scattering processes might have the effect of producing large values of K, and to work out the theory for this case. As mentioned in the introductory lecture, Pippard, Landau, and Ginzburg had made brief suggestions along these lines.

Abrikosov treats two important regions of the theory, in each of which some of the mathematics is complicated, but I believe that all of the physics can be presented without excessive difficulty. We shall henceforth work strictly within the Landau–Ginzburg theory.

Consider then the Landau–Ginzburg equations:

$$\xi_0^2 \left(i\boldsymbol{\nabla} + \frac{2e}{\hbar c}\mathbf{A}\right)^2 \psi = \left(1 - \frac{T}{T_c}\right)\psi - \frac{2\psi^3}{N}$$

$$\frac{4\pi J}{c} = \boldsymbol{\nabla} \times \boldsymbol{\nabla} \times \mathbf{A} = \frac{2}{\lambda_0^2 N}\left\{\frac{\hbar c}{4ie}(\psi^*\boldsymbol{\nabla}\psi - \psi\boldsymbol{\nabla}\psi^*) - \mathbf{A}|\psi|^2\right\}.$$

At a given temperature in zero field the equilibrium value of ψ is given by:

$$|\psi|^2 = \frac{N}{2}\left(1 - \frac{T}{T_c}\right).$$

We may therefore define the effective coherence length and effective penetration depth as:

$$\xi^2 = \frac{\xi_0^2}{(1 - T/T_c)}, \quad \lambda^2 = \frac{\lambda_0^2}{(1 - T/T_c)}.$$

The free energy per unit volume depends on the normalization of ψ, determined by the constant N. We determine ψ so that $|\psi|^2 = n_s$, London's number of superelectrons. Then the free energy is:

$$F = \psi^* \left\{ \frac{\hbar^2}{2m} \left(i\nabla + \frac{2e}{\hbar c} \mathbf{A} \right)^2 \right\} \psi - \frac{1}{\xi_0^2} \frac{\hbar^2}{m} \left(1 - \frac{T}{T_c} \right) |\psi|^2 + \frac{|\psi|^4}{2N} \frac{\hbar^2}{n\xi_0^2}$$

plus field energy terms and therefore, in the absence of fields, the free energy at equilibrium, relative to that of the normal state, gives the relation:

$$\frac{H_{\mathrm{CB}}^2}{8\pi} = \frac{\hbar^2}{m\xi_0^2} \frac{N}{2} \left\{ \left(1 - \frac{T}{T_c}\right)^2 - \frac{1}{2}\left(1 - \frac{T}{T_c}\right)^2 \right\} = \frac{\hbar^2 N}{4m\xi_0^2} \left(1 - \frac{T}{T_c}\right)^2.$$

It is convenient to introduce dimensionless units. Abrikosov uses a dimensionless system of temperature-dependent units which are quite annoying, and which were probably responsible for his missing the flux quantization in units of hc/e^* implicit in his theory. Nonetheless, in order to follow his paper, we shall introduce his units. We begin by normalizing ψ to the field-free value:

$$\psi_0^2 = \frac{N}{2}\left(1 - \frac{T}{T_c}\right).$$

Then $\psi' = \psi/\psi_0$: we henceforth omit the prime, obtaining:

$$\frac{\xi_0^2}{(1 - T/T_c)} \left(i\nabla - \frac{2e}{\hbar c} \mathbf{A} \right)^2 \psi = \psi - \psi^3$$

$$\frac{\lambda_0^2}{(1 - T/T_c)} (\nabla \times \nabla \times \mathbf{A}) = -|\psi|^2 \mathbf{A} + \frac{\hbar c}{4ie} (\psi^* \nabla \psi - \psi \nabla \psi^*).$$

We must measure distances in terms of the penetration depth λ, so that $r' = r\sqrt{1 - T/T_c}/\lambda_0$, and again omitting primes:

$$\nabla \times \nabla \times \mathbf{A} = -|\psi|^2 \mathbf{A} + \frac{\hbar c}{4ie\lambda_c} \sqrt{1 - \frac{T}{T_c}} (\psi^* \nabla \psi - \psi \nabla \psi^*).$$

Finally, fields are measured in units of $1/\sqrt{2}H_{\mathrm{CB}}$, which, from our previously derived form for H_{CB} gives:

$$H' = \frac{\sqrt{2}H}{H_c} = \frac{\xi_0}{\hbar}\sqrt{\frac{m}{\pi N}}\frac{H}{(1 - T/T_c)}$$

so that **A**, which is $H \times$ distance, is measured in units of

$$H_c\lambda = \frac{\hbar}{\xi_0}\sqrt{\frac{\pi N}{m}}\left(1 - \frac{T}{T_c}\right) \times \left(1 - \frac{T}{T_c}\right)^{-1/2}\sqrt{\frac{mc^2}{4\pi Ne^2}} = \frac{\hbar c}{2\xi_0 e}\sqrt{1 - \frac{T}{T_c}}.$$

Thus we finally obtain the dimensionless equations:

$$\nabla \times \nabla \times \mathbf{A} = -|\psi|^2 \mathbf{A} + \frac{1}{2iK}\left(\psi^*\nabla\psi - \psi\nabla\psi^*\right)$$

and

$$\left(i\frac{\nabla}{K} + \mathbf{A}\right)^2 \psi = \psi - \psi^3 .$$

It is convenient to recognize that a gradient applied to ψ implies a coherence effect and thus a factor $1/K$ in these units.

The first question to which Abrikosov applies these equations is the following: suppose we start at very high fields. Then at high enough fields there will be no superconductivity, but there will be a field H_{c2} at which a stable solution with very small ψ is obtained, i.e., at which the linearized equations have a solution. This solution may occur for $H_{c2} > H_{\mathrm{CB}}$ or for $H_{c2} < H_{\mathrm{CB}}$. If it is the latter, it is unstable, representing a supercooled state, which is not a solution of immediate interest; in that case clearly we have the familiar first-order thermal transition. However, if $H_{c2} > H_{\mathrm{CB}}$, it is clearly important, since it indicates that superconductivity will occur in the system above the bulk critical field.

In linearizing the solution (neglecting terms in $|\psi|^2$), we immediately see that $\nabla \times \mathbf{H} = 0$, which has a solution $H = \text{constant}$, and if we take $A_y = H_0 x$, we shall see that H_0 is the external field strength. The linearized approximation assumes that the transition is a second order phase transition, with the order parameter ψ rising continuously from zero. With the above choice for **A**, the other equation takes the form (linearized);

$$\left\{-\frac{\partial^2}{\partial x^2} - \frac{\partial^2}{\partial x^2} - \left(\frac{\partial}{\partial y} - iKH_0 x\right)^2\right\}\psi = K^2\psi .$$

This is the wave equation for a free particle in a magnetic field, familiar from the theory of Landau diamagnetism.

We are looking for the lowest and most nearly constant solution we can get, so we consider ψ independent of Z, the direction of **H**. The rest of the equation is of the familiar two-dimensional harmonic oscillator form, in which all solutions with different y dependences are degenerate. Thus consider $\psi = \psi(x)$ only:

$$\frac{d^2}{dx^2}\psi + K^2(1 - H_0^2 x^2)\psi = 0 .$$

This is backwards from the usual form, since we consider H_0, and not K, as the eigenvalue. However, for given K, the highest value of H_0, (or alternatively the lowest value of K for given H_0) has the solution:
$$\psi = e^{-K^2 x^2/2}$$
where $H_0 = K$ (unstable solutions also occur for smaller values of H_0 equal to $K/(2n+1)$ where n is an integer). Since, in dimensionless units $H_0 = 1$ corresponds to a field $H_{CB}/\sqrt{2}$, we see that $K > 1/\sqrt{2}$ gives the above stable solution with $H_{c2} > H_{CB}$. Thus $K > 1/\sqrt{2}$ is the basic criterion for hard, or type II, superconductivity.

If we include the y-dependence, the equation for ψ is:

$$\left\{\frac{\partial^2}{\partial x^2} + \left(\frac{\partial}{\partial y} - iKH_0 x\right)^2\right\}\psi = -K^2\psi\ .$$

Then a solution degenerate with that given above has $\psi \propto e^{iky}$, with x shifted by k/KH_0, where $H_0 = K$. Thus we have solutions:

$$\psi = \exp\left\{iky - \frac{K^2}{2}\left(x - \frac{k}{K^2}\right)^2\right\}$$

corresponding to functions centered about different points in the x-direction. Packets could be built up from these looking like cylindrical orbitals, but this might be easier in a different gauge.

At this point some remarks about the filamentary model of hard superconductors would be appropriate. The above solutions, which, like cyclotron orbits, can be roughly localized in space, in this case to dimensions $\sim \xi_0$, may be thought of as "filaments of superconductivity" running along the z-direction. Abrikosov's theory shows that $H_{c2} = KH_{CB}$ is the highest field at which such filaments may exist in thermal equilibrium. But clearly, below H_{c2} the bulk of the sample can be superconducting. Above H_{c2}, one could hope that dislocations or grain boundaries might remain superconducting after the bulk is no longer capable of it. However, our equation tells us that ξ must still be the scale of such filaments, and it seems difficult to find an energy of the order of $(H^2 - H_{c2}^2)/8\pi\,\xi^2$ per unit length in the small perturbations associated with dislocations, for example. In other words, the fallacy of the filamentary model is the assumption that the filament, whatever it is made of, can be localized sharply in a usually superconducting matrix just because a field is present to make the matrix normal. A field does not produce such an effect in the Landau–Ginzburg equations. Superconducting inclusions could presumably occur, but I do not believe that small, highly localized inhomogeneities can produce the effect. Two other sources of energy which the filaments must overcome are the kinetic energy of localization in the z-direction, and the excess kinetic energy of whatever current they carry.

Thus if filaments exist, they will probably lead only to small currents, and only slightly increased critical fields. Such effects are observed in cases where H_{c2} can be accurately identified. (A theory of such effects on the surfaces of cavities or insulating inclusions has been presented by St. James and de Gennes (*Phys. Lett.*, 1964) and its application to the sponge model suggested by Kim *et al.* (*Phys. Rev. Lett.*, 1964). Note added in proof.)

Thus Abrikosov's model emphasizes the idea that flux penetrates an otherwise superconducting bulk, while Mendelssohn's idea is that superconducting filaments penetrate a normal bulk. But we have shown that superconductivity can only penetrate a reasonably homogeneous sample below H_{c2} plus a small increment depending on inhomogeneity. Of course this does not apply to very inhomogeneous samples such as metals dispersed in glass.

Having found a basic set of solutions of the linearized equations, Abrikosov goes on to satisfy them in the nonlinear region $H \simeq K$, where an expansion in the difference of K from H would appear to be possible. In this region the equations may be expanded using $\mathbf{A} = \mathbf{A}_0 + \delta \mathbf{A}$ and $\psi = \psi_0 + \delta \psi$, where ψ_0 must be a solution of the linearized equation, with a constant multiplier to be determined. Only ψ_0 is necessary in the $\mathbf{A}$-equation:

$$-\nabla \times \nabla \times (\delta \mathbf{A}) = |\psi_0|^2 \mathbf{A}_0 + \frac{i}{2K}(\psi_0^* \nabla \psi_0 - \psi_0 \nabla \psi_0^*)$$

for if δA is of order $|\Psi_0|^2$, this is consistent, the terms in $\psi_0^* \nabla(\delta \psi)$ and $\delta \psi \nabla \psi_0$ being of higher order in ψ_0, as we shall see. The equation for $\delta \psi$ is:

$$\left(\frac{i\nabla}{K} + A_0\right)\delta\psi + 2\delta A \cdot \left(\frac{i\nabla}{K} + A_0\right)\psi_0 = \delta\psi - |\psi_0|^3 \; .$$

If, as we have assumed, $\delta \mathbf{A} \sim |\psi_0|^2$, then it is consistent to choose $\delta \psi \sim |\psi_0|^3$. These equations for $\delta \psi$ and $\delta \mathbf{A}$ are very complicated. At this point Abrikosov exhibits some mathematical virtuosity and feats of intuition in obtaining a solution using a particular form which he chooses for ψ_0. This solution is two-dimensional, which is reasonable, periodic, which may be reasonable, and is constructed as a linear combination of the solutions obtained in the linearized case. His ansatz is:

$$\psi_0 = \sum_{n=-\infty}^{+\infty} c_n e^{ikny} \exp\left\{-\left[K^2/2\left(x - \frac{kn}{2K^2}\right)^2\right]\right\}$$

where k and the coefficients c_n are arbitrary. From this form for ψ_0 he obtains the identity (which I and other have been unable to derive in any direct way, though it can be verified from the equations, using the form given for ψ_0):

$$H = \frac{\partial A}{\partial x} = H_0 - \frac{|\psi|^2}{2K}$$

whence
$$A = H_0 x - \frac{1}{2K} \int_0^x |\psi(x')|^2 dx'$$
where H_0 is a constant to be determined. If this solution for A is substituted into the equation for $\delta\psi$, an inhomogeneous linear form results. Use is now made of the theorems that the inhomogeneous term in an equation involving an Hermitian operator in the homogeneous form, must be orthogonal to all solutions of the homogeneous equation. Thus, multiplying by ψ_0^* and integrating, and making use of specific properties of the form chosen above for ψ_0, Abrikosov obtains the following result:

$$\frac{(K-H_0)}{K} \overline{|\psi_0|^2} + \left(\frac{1}{2K^2} - 1\right) \overline{|\psi_0|^4} = 0 \tag{a}$$

where the bar indicates integration over all of the space. This may be checked for $\psi_0 = e^{-k^2 x^2/2}$, leading to complicated, tricky integrations: the ansatz actually used for ψ_0 is an amazing guess, if it is a guess. Now if **B** is viewed as the mean value of H, we obtain:

$$B = \mathbf{H} = H_0 - \frac{\overline{|\psi_0|^2}}{2K}$$

which we rewrite:

$$B = H_0 - \frac{1}{2K} \frac{(\overline{|\psi_0|^2})^2}{\overline{|\psi_0|^4}} \times \frac{\overline{|\psi_0|^4}}{\overline{|\psi_0|^2}}.$$

The quantity $\beta = \overline{|\psi_0|^4}/(\overline{|\psi_0|^2})^2$ gives a measure of the mean variation of $|\psi_0|^2$ and is independent of the magnitude of ψ_0 and of H_0. Thus finally, we write:

$$B = H_0 - \frac{(K-H_0)}{(2K^2-1)\beta}$$

having used the equation (a) above relating $\overline{|\psi_0|^2}$ to $\overline{|\psi_0|^4}$ in the ratio:

$$\frac{(2K^2-1)}{2K(K-H_0)}.$$

LECTURE 5

Abrikosov now proceeds to minimize the total free energy difference between the superconducting state with $H \neq 0$ and $H = 0$, with respect to variations of β. Using the Landau–Ginzburg equations, the total free energy, which is an integral over all space, may be expressed in terms of $\overline{|\psi|^2}$ and $\overline{|\psi|^4}$, whence the following expression in terms of β and B is obtained:

$$\frac{F_{\text{SH}} - F_{\text{SO}}}{(H_{\text{CB}}^2/4\pi)} = \frac{1}{2} + B^2 - \frac{(K-B)^2}{1 + (2K^2-1)\beta}.$$

The process of minimizing with respect to $\beta = \overline{|\psi_0|^4}/(\overline{|\psi_0|^2})^2$ is reminiscent of the treatment of the BCS theory for the quasi-linearized approximation for complicated cases such as $\ell = 2$ bound states, where it is also necessary to minimize with respect to $\overline{\Delta^4}$.

The coefficients C_n are determined by assuming that the periodicity has the same wavelength in the y and x directions, i.e., $C_n = C_{n+1}$. The minimum value of β then occurs for $k = \sqrt{2\pi}K$, whence $\beta_{\min} = 1.18$, and:

$$\psi_0 = Ce^{-K^2x^2/2}\theta_3\left[\sqrt{2\pi}Ki(x+iy); 1\right].$$

For the square lattice periodicity assumed here, the mapping of contour lines $|\psi_0|^2 = $ constant is given in the figure on page 1177 of Abrikosov's paper. Nodes occur in the "wave function" ψ at points $x, y = (2n+1) \times K\sqrt{\pi/2}$.

If we evaluate the flux per node in the material, we see that, for $B = K$,

$$\int_{\text{node}} \mathbf{B} \cdot \mathbf{n}\, ds \sim \frac{2\pi}{K^2}K = \frac{2\pi}{K}$$

since the square area containing one node has side of length $(2\pi)^{\frac{1}{2}}/K$. In real units, this flux is $2\pi H_{CB}\lambda\xi$, and since the units of $H_{CB}\lambda$ are $\frac{\hbar c}{2\xi_0 e}\left(1-\frac{T}{T_c}\right)^{1/2}$ and since $\xi = \xi_0\left(1-\frac{T}{T_c}\right)^{-1/2}$, the flux per node is $hc/2e$: there is one flux quantum (with effective charge $e^* = 2e$) per node. Abrikosov apparently did not notice the cancellation of temperature and material-dependent terms, because of the system of units. Abrikosov states that a triangular lattice solution is less stable than the square one, but recently this result has come into question.

It remains to obtain physically measurable quantities from the theory. The magnetic field varies from its maximum value H_0, which occurs at the nodes, to the minimum value, $H_0 - \frac{(K-H_0)\sqrt{2}}{(2K^2-1)}$. This field strength outside the sample is given by $4\pi\frac{\partial \bar{F}}{\partial B} = H_0$. This can be checked directly, but it also follows from the form $H_0 = H_0 - \frac{|\psi_0|^2}{2K}$ at the surface. The magnetization $M = B - H/4\pi$ reduces, for maximum field $H = H_0$, to the form $-M = K - H_0/1.18(2K^2 - 1)$. Thus the curve M vs. H_0 will have a small negative slope for $K > 1/\sqrt{2}$ in this region, as shown on the diagram.

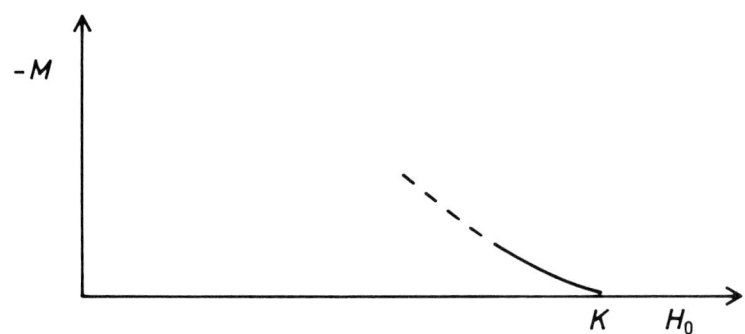

The above expression for M illustrates an important fact for practical applications, namely that, since $\nabla \times \mathbf{M} \neq 0$ at the surface, the Abrikosov state superconducting carries a surface current, as do ordinary superconductors. Note also that the maximum field variation is relatively small for large K. For example with $H_0 = K/2$,

$$H_{\min} = \left\{ \left(\frac{K}{2}\right) - \frac{\left(K - \frac{K}{2}\right)\sqrt{2}}{(2K^2 - 1)} \right\}$$
$$\approx \left(\frac{K}{2} - \frac{1}{2\sqrt{2K}} \right)$$

which deviates from the maximum value of $K/2$ only by a term of order $(2\sqrt{2}K)^{-1}$.

The general picture that emerges is of great importance, namely little variation of H, but permeation of the sample throughout for $H < H_{c2} \equiv K$. The real situation in hard superconductors is probably much less regular then indicated in Abrikosov's contour map of ψ, but the concept which emerges, of a thermal equilibrium *mixed* state of a superconductor with field penetration is vitally important.

A historical note: the existence of such a state, the applicability of thermodynamics to it, and the two critical fields were all proposed on experimental grounds by Schubnikov in 1936. This work played a vital role in stimulating the Landau–Ginzburg–Abrikosov work.

I believe that Abrikosov's second part is even more vital and enlightening from a general standpoint. Here he attacks the other end of the magnetic field range. Start with the perfect superconducting state: all superconductors will be perfect, with ψ = constant, at $H = 0$. There will be a second critical field H_{c1} at which penetration first occurs as H is increased. How does H_{c1} occur? Here he introduces the key idea of quantized flux lines. He mentions the analogy with quantized vortices in HeII, but not what would seem to be by far the more useful analogy with dislocations in crystals. These quantized flux lines are the central concept of the whole subject.

Briefly, the idea is as follows: as we apply more and more field, eventually the field will begin to penetrate. However, it does not penetrate continuously, but in quantized flux lines, since only integral quanta of flux can be surrounded by superconductivity; or better, it penetrates as a nodal line of $\psi = \Delta(!)$. The magnetic energy per unit length of a flux line outside the superconductor relative to that inside comes from integrating the Lorentz force

$$\int \mathbf{F} \cdot \mathbf{n}\, dV = \int \frac{\mathbf{J}}{c} \times \mathbf{\Phi}_0 \cdot \mathbf{n}\, ds = \int \frac{1}{4\pi} (\nabla \times H) \times \mathbf{\Phi}_0 \cdot \mathbf{n}\, ds$$

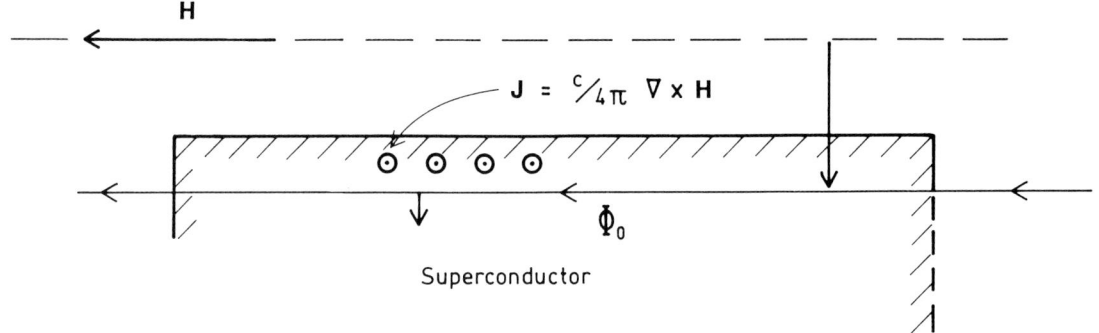

When this energy overbalances the energy ε of creation of a line per unit length, the flux line will move into the superconductor. We therefore wish to solve the equations to find ε, and the interaction of flux lines. We write:

$$\psi = f(r)e^{i\theta},$$

where

$$f(\mathbf{r} \to 0) = 0$$

and

$$f(\mathbf{r} \to \infty) = 1.$$

We introduce the quantity

$$Q = \left|\mathbf{A} - \frac{\nabla \theta}{r}\right| = A_\theta - \frac{1}{Kr}$$

where A_θ is the transverse (as opposed to radial) component of $\mathbf{A}$. As an ansatz we take $A_r = 0$. In these terms, the Landau–Ginzburg equations reduce to:

$$-\frac{1}{K^2 r}\frac{d}{dr}\left(r\frac{df}{dr}\right) + Q^2 f = f - f^3$$

$$\frac{d}{dr}\left[\frac{1}{r}\frac{d}{dr}(rQ)\right] = Qf^2$$

also

$$\mathbf{H} = \nabla \times \mathbf{A} = \nabla \times \mathbf{Q} \to -\frac{1}{r}\frac{d}{dr}(rQ) \text{ in magnitude}.$$

The formulation requires that as $r \to \infty$, $f \to 1$, $H \to 0$ and $Q \to 0$. Since H is finite, $Q \to 1/Kr$ as $r \to 0$. The above equations with the given boundary conditions are not generally solvable. However, we may consider $K \gg 1$. Then the variation of f is confined to a range of order $\xi_0 \ll 1$ (in Abrikosov's units), while H and Q extend much farther, We then have, approximately:

$$\frac{d}{dr}\left(\frac{1}{r}\frac{d}{dr}(rQ)\right) = Q$$

whence the solution $Q = K_1(r)/K$ where K_1 is a Hankel function. This is just a solution of London's equation at the radius r. We may now assume $Q \sim 1/Kr$ for small r in the equation for f, which then reduces to:

$$-\frac{1}{K^2}\frac{1}{r}\frac{d}{dr}\left(\frac{(r\,df)}{dr}\right) + \frac{f}{K^2 r^2} = f^3 - f \ .$$

The solutions of this equation are, for r large: $f \approx 1 - \frac{1}{K^2 r^2} + \ldots$ and for r small $f \approx cr + \ldots$, and good numerical solutions can be obtained in the intermediate region, connecting these two asymptotic forms. The free energy per unit length of a filament of flux may likewise be simply evaluated, since only the London region is involved, because for $K \gg 1$, the volume for $1/K$ to unity is much greater than that from zero to $1/K$ (again referring to Abrikosov's units). Thus

$$\varepsilon = 2\pi \int_0^\infty \left\{H^2 + \frac{(1-f^4)}{2}\right\} r\, dr \ .$$

Now $H = -\frac{1}{r}\frac{d}{dr}(rQ)$ and $Q = K_1(r)/K$ gives $H = \frac{1}{K}K_0(r)$. Thus, using an approximate form for $\int K_0^2(r) \cdot r\, dr$, and numerical integration, Abrikosov obtains:

$$\varepsilon \approx \frac{2\pi}{K^2}(\ln K + 0.081 + \ldots) \ .$$

We note that there appears to be an error in the second line of Abrikosov's equation (31), which we think should read:

$$\varepsilon = \pi \int_0^\infty \{(1-f^4)r - 2r f^2 Q^2\} dr \ .$$

At this point, Abrikosov points out that the flux associated with one filament is $\Phi_0 = \oint \mathbf{A} \cdot d\mathbf{s} = 2\pi/K$: the system of units obscures from the reader that this is just an amount equal to $hc/2e$. If we return to our expression for the magnetic energy of a flux line outside the superconductor, with $H = H_{cl}$, the field at which penetration begins, then, *per unit length*, this energy is:

$$H_{cl} \frac{\Phi_0}{4\pi} \quad \text{in conventional, non-Abrikosov units}$$

and penetration begins when this is equal to ε, namely when:

$$H_{cl} \cdot \frac{\Phi_0}{4\pi} \to \frac{4\pi H_{cl}}{K} = \frac{2\pi}{K^2}(\ln K + 0.081)$$

in Abrikosov's units, or:

$$H_{cl} = \frac{1}{2K}(\ln K + 0.08).$$

This is the final form for the field at which penetration begins.

We now consider the interaction between filaments. This is also relatively easy to obtain because it contains contributions mainly from the London regions of the filaments. Considering only these outer regions, we have, from the Landau–Ginzburg equation:

$$-\nabla \times H = |\psi|^2 \mathbf{A} + \frac{1}{2K}(\psi^* \nabla \psi - \psi \nabla \psi^*).$$

We now take account of the fact that $|\psi|^2 \approx 1$ in the superconducting phase, and obtain:

$$-\nabla \times \nabla \times \mathbf{H} = \mathbf{H} + \frac{1}{2K} \nabla \times (\psi^* \nabla \psi - \psi \nabla \psi^*).$$

Again, near each filament $\psi \approx e^{i\theta}$, so $\nabla \psi = i\theta/r$, and the curl gives a two-dimensional δ-function. Thus

$$-\nabla \times \nabla \times \mathbf{H} = H - \frac{2\pi}{K} \sum_m \delta(r - r_m)$$

where m labels the flux filaments, or finally

$$\nabla^2 H - H = -\frac{2\pi}{K} \sum_m \delta(r - r_m).$$

This linear equation may be solved to give

$$H = \frac{1}{K} \sum_m K_0(|r - r_m|)$$

where K_0 is the zeroth order Hankel function. Actually, H obeys the same equation which applies to the potential of a system of equally charged rods in an electrostatistically screening medium, such as a metal, for K_0 is the logarithmic Coulomb potential in two dimensions, screened off at the radius λ_L.

In order to use the London limit, for which $|\psi|^2 \approx 1$, in discussing the free energy, for which $|\psi|^2 = 1$ gives an extremum, we must rewrite F, given by:

$$F = H^2 + \left(\frac{\psi^4}{2} - \psi^2 + \frac{1}{2}\right) + \left(\frac{i\boldsymbol{\nabla}\psi^*}{K} - \mathbf{A}\psi^*\right) \cdot \left(\frac{i\boldsymbol{\nabla}\psi}{K} - \mathbf{A}\psi\right).$$

With the approximation $|\psi|^2 = 1$, the first bracketed term is approximately zero, and, inserting $|\psi|^2$ into the last term gives:

$$F \approx H^2 + \left(\frac{i\boldsymbol{\nabla}\psi^*}{K} - \mathbf{A}\psi^*\right)\psi\psi^*\left(\frac{i\boldsymbol{\nabla}\psi}{K} - \mathbf{A}\psi\right).$$

This last term is what London calls $|j_s|^2$: the supercurrent. It can be rewritten, but for a possible factor $(4\pi)^2/c$:

$$F \approx H^2 + (\boldsymbol{\nabla}_{2\text{ dimension}} H)^2.$$

For evaluating the average value of F (spatial integral), partial integration yields:

$$\bar{F} \approx H(H - \nabla^2 H)$$

and using the source equation for H this becomes:

$$\bar{F} = \frac{2\pi}{K^2} \sum_{i,j} K_0(r_{ij}).$$

We thus see that the average free energy corresponds to lines interacting in pairs via the "screened" potential K_0. This property is of great importance in the theory of flux creep, to be discussed later.

Abrikosov's further discussion is mainly concerned with the lattice symmetry which minimizes F. The energy differences are so small compared to the pinning energies to be discussed below, that regular lattices probably do not occur. A final point of interest arises from using the asymptotic form of K_0:

$$\bar{F}_{\text{int}}(r_{ij} \gg \lambda) \approx \text{const. } B^{5/4} \exp\left\{-\sqrt{\frac{4\pi}{\sqrt{3}KB}}\right\}$$

where B refers to the density of flux lines: $B = 2\pi n/K$. Thus for small B, that is, at the beginning of penetration, $\bar{F}_{\text{int}} \to 0$ very fast, with a vertical tangent in B vs. $\partial F/\partial B = H$. This enables us to complete the by now famous magnetization curve for Abrikosov type II superconductors:

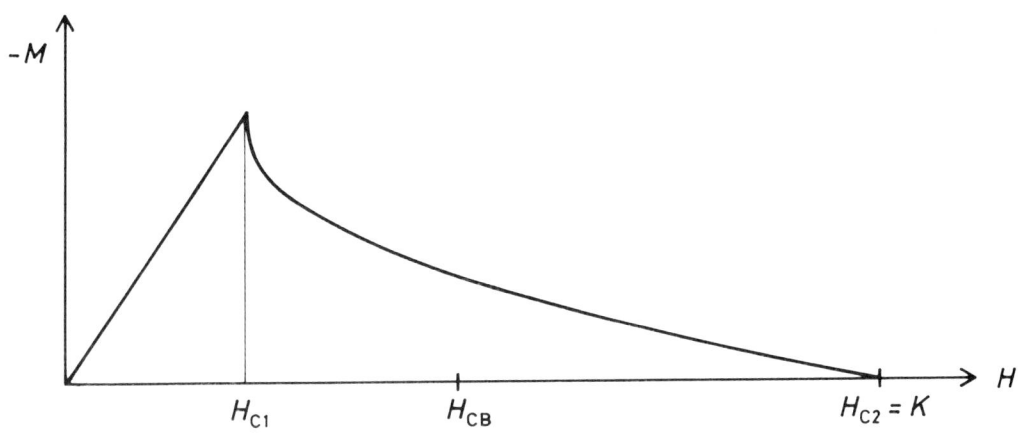

Experimental curves, notably for niobium with very small percentages of impurities, closely approximate this curve, with slight irreversibility. The irreversibility is quite large in many other hard superconductors. However, I believe that, more important than this curve, are the two basic physical concepts elucidated by Abrikosov's work, namely: (1) the thermodynamically stable mixed state between H_{c1} and H_{c2}, in which region there is considerable magnetic penetration ($B \neq 0$); and (2) the concept of quantized lines of flux. The latter concept is absolutely vital to the theory of superconductivity, especially for hard superconductors, to the same extent that dislocations are to the theory of the strength of materials. Actually, there is a remarkably close analogy between the two, which we shall exploit in describing the hard superconductors. Here, given flux quantization, anyone familiar with vortices or dislocations will see the possibilities of the flux line concept, which has the advantage of mobility over the usual filamentary mesh model. The whole problem of persistent currents becomes the problem of the rigidity of the structure against the passage of flux lines through it. Although Abrikosov mentions the importance of inhomogeneities in a last paragraph, the first serious thought about the problem was sparked by the experimentalists, specifically Kunzler and co-workers at Bell Telephone Laboratories, who discovered that they could pass considerable currents through Nb_3Sn in the presence of 10^5 gauss, without destroying superconductivity. This led to elaboration of what now seem the obvious consequences of Abrikosov's theory by Yntema, Gorter and others.

LECTURE 6
HARD SUPERCONDUCTIVITY: THEORY OF THE MOTION OF ABRIKOSOV FLUX LINES

NOTE: The final section of these notes is substantially the text of a paper presented by the author, with Y. B. Kim, to the *International Conference on the Science of Superconductivity*, held August 26–29, 1963, at Colgate University, Hamilton, New York. The conference proceedings will be published in *Reviews of Modern Physics*. The material is available as MM-63-1111-17 and therefore we do not reproduce it here.

Acknowledgements

I would like to thank the Theoretical Physics Division of the Canadian Association of Physicists for their hospitality at Banff, and I would like to extend particular thanks to Dr. J. M. Vail and Dr. M. Revzen who gave much of their free time at Banff in order to transcribe these notes.

In addition, thanks are due to Dr. M. Razavy and Dr. E. J. Woods for the organization of the note-talking, and for some proofreading.

P. W. ANDERSON

Hard Superconductivity: Theory of the Motion of Abrikosov Flux Lines

P. W. ANDERSON and Y. B. KIM

Bell Telephone Laboratories, Inc., Murray Hill, New Jersey

The central concept in the theory of what one might call the "critical phenomena" of hard superconductors—critical currents, critical fields, decay of persistent currents, "excess" voltages, etc.—is clearly Abrikosov's notion of the quantized flux line.[1] This is made almost obvious by the remark that, because of the now universally accepted validity of the quantization of flux through superconductors, the smallest possible breakdown of superconductivity is the motion of a single quantum of magnetic flux through the wire or ring. Thus, in all cases so far conceived, the lowest activation energy for any critical breakdown is that for the motion—and creation, if necessary—of single Abrikosov flux lines. This statement is independent of whether the mechanism for hard superconductivity is the GLAG one or the Mendelssohn sponge theory, although we assume the former to be valid in most cases. Even the decay of currents in true soft superconductors under α-particle bombardment[2] is probably best explained by the threading of Abrikosov lines through normal holes punched by the α particles.[3]

The purpose of this paper is to see how many of the phenomena of hard superconductivity we can understand qualitatively in terms of the thermally activated motion of Abrikosov lines past pinning centers, without going into unnecessary detail on the nature of the pinning centers—whether they are dislocations, cavities, precipitates, etc.—or the precise internal structure of the superconductor. Our task, then, is to study the process—presumably thermally activated barrier penetration—by which flux lines move.

Let us then suppose that we have a superconductor penetrated by a magnetic field H and carrying a bulk current, for simplicity $\perp H$, $J = c\nabla \times \mathbf{H}/4\pi$. The magnetic field will penetrate in the form of Abrikosov lines; their density is clearly not uniform because of J, and we expect their arrangement is to some extent irregular. The magnetic energy per unit volume is $H^2/8\pi$; we can think of this as a magnetic pressure exerted by the flux lines on each other, and in the absence of pinning centers this pressure would have to be equalized by a rearrangement of the lines, leading to $J = 0$. Examination of Abrikosov's theory shows that actually at all but low fields the internal and external fields are nearly the same, so that we usually assume $B = H$, a minor simplification of which Friedel *et al.* have considered the errors.[4]

In finding the rate of the activation process we need to know two things: the driving force exerted by the magnetic pressure, and the nature of the barriers. The former is more available to us theoret-

---

[1] A. A. Abrikosov, Zh. Eksperim i Teor. Fiz. **32**, 1442 (1957) [English transl.: Soviet Phys.—JETP **5**, 1174 (1957)].
[2] P. de Feo and G. Sacerdoti, Phys. Letters **2**, 264 (1962).
[3] N. Cabibbo and S. Doniach, Phys. Letters **4**, 29 (1963). We have proposed a slightly different mechanism.
[4] J. Friedel, P. G. de Gennes, and J. Matricon, Appl. Phys. Letters **2**, 119 (1963).

ically but, on the other hand, more complex, entirely because it is really transmitted only through the interactions between the flux lines themselves. We can only guess at the correct results before more detailed computations of the behavior of nonuniform distributions of flux lines become available.

In Abrikosov's paper, in the appropriate case for hard superconductors ($\kappa \gg 1$, $H_{c2} > H > H_{c1}$) the interaction free energy of the lines is written in two equivalent ways:

$$F_{\text{int}} = (1/8\pi)\overline{\mathbf{H}\cdot(\mathbf{H} - \delta^2\nabla^2\mathbf{H})} \quad (1)$$

$$= \frac{H_c^2}{4\kappa^2}\sum_{i,j} K_0\left(\frac{|\mathbf{r}_i - \mathbf{r}_j|}{\delta}\right), \quad (2)$$

κ being the famous Landau–Ginsburg parameter δ_0/ξ_0, δ the penetration depth, and H_c the thermodynamic critical field. The two-dimensional position vectors of the lines are the $\mathbf{r}_i$. In the case most usual in hard superconductors, the distances $|\mathbf{r}_i - \mathbf{r}_j|$ are small compared to δ, $\overline{\nabla^2 H}$ is relatively small, and so the interaction free energy is practically given by the naive expression for the magnetic energy. This would mean that the force per unit volume would be given by $\nabla F = -\mathbf{H} \times \nabla \times \mathbf{H}/4\pi = \mathbf{J} \times \mathbf{H}/c$, the Lorentz force; and the force per flux line is then $\mathbf{J} \times \mathbf{\Phi}_0/c$ per unit length, where the magnitude of $\mathbf{\Phi}_0$ is the flux unit $hc/2e$ and it is directed along the flux line.

It is very useful to notice that (2) is formally the same as the electrostatic interaction between lines of electric charge, except that it is screened away at the penetration depth δ—i.e., K_0 has the logarithmic nature of two-dimensional charge interactions as $r \to 0$, while it is $\sim e^{-r/\delta}$ as $r \to \infty$. This tells us that, because of this relatively long-range interaction, local perturbations of the line density are very unfavorable energetically—for instance, simply putting in locally one extra flux line costs an energy of the order of $H\Phi_0$ per unit length, which is much greater than the energy available from any reasonable pinning centers. Thus the arrangement can be irregular *only* on a scale *greater than* δ—locally the density must be uniform, and any local variation must be only a slight increase in the local density spread out over a region of radius $>\delta$.

This is the idea behind the concept of "flux bundles"[5]—that in fact, while it is probably the individual flux line's internal structure (of size $\sim\xi_0$) which is caught by a pinning center, that line individually cannot jump over the barrier alone, because it would get badly out of equilibrium with the local density in its neighborhood; but rather a whole bundle of lines, of radius $\sim\delta$, must move simultaneously. Therefore, of course, it is the force on the total bundle which acts against the pinning barrier. On the other hand, while the line *density* must be very uniform, the K_0 function is actually a very slowly varying one near $r \to 0$ so that the arrangement of lines need not be crystallographically regular, and the bundles can slide past each other reasonably easliy.

The free energy of a bundle in the region of a barrier, then, can now be written down. The force is about $JH\delta^2l/c = J\Phi_0 n_b l/c$, where l is the effective length of line over which the force acts, presumably the distance between pinning centers, and n_b is the number of lines in a bundle; as a function of position x of the bundle we have

$$F_{\text{force}} = JH\delta^2lx/c\,.$$

The size of the barrier is presumably about ξ_0, and the appropriate scale factor for its energy is $(H_c^2/8\pi)\xi_0^3$; suppose a fraction p of this is effective. The total barrier free energy then—realizing that we have both p and l as undetermined parameters so that we need not specify the determinable ones more precisely—is

$$F_b = (pH_c^2\xi_0^3/8\pi) - (JH\delta^2l\xi_0/c)\,. \quad (3)$$

It is perhaps barely worthwhile to make a guess at the pre-exponential factors in the rate equation, even though almost all the results are controlled entirely by (3). A particular barrier will allow one of the lines in the bundle through at a rate/sec of

$$R = \omega_0 e^{-F_b/kT} \quad (4)$$

ω_0 is a vibration frequency of the bundle, $\sim 10^5 - 10^{10}$/sec. Since the bundles are of width δ, the rate per unit area is obtained by dividing by δ. The equation for diffusion of flux density $|B|$ is given by finding the rate at which flux enters and leaves a small element of volume:

$$d|B|/dt = -\boldsymbol{\nabla}\cdot(\Phi_0 \mathbf{R}/\delta)\,, \quad (5)$$

where the gradient is two-dimensional, Φ_0 is the flux unit, and $\mathbf{R}$ is of magnitude (4) and directed in the direction of the gradient of magnetic pressure

$$\boldsymbol{\alpha} = \boldsymbol{\nabla} p = \boldsymbol{\nabla}(H^2/8\pi) = \mathbf{J} \times \mathbf{H}/c\,. \quad (6)$$

We shall often use the notation α for the appropriate combination in the force term even when JH is not correct.

There are two cases in which we would obviously expect the above reasoning to fail: the two extremes of the Abrikosov state, near H_{c2} and H_{c1}, the upper

---

[5] P. W. Anderson, Phys. Rev. Letters **9**, 309 (1962).

and lower critical fields, respectively. Near the lower critical field the lines are more than a distance δ apart; also the differences between B and H, and therefore the complications due to surface currents, etc., are much greater. Qualitatively, we expect that as the number $n_b \to 1$, the value of the field will matter less and less, and the effective force will reduce to that on a single line, proportional to J alone. This is indeed the qualitative behavior in many materials, e.g., those in which we find the force term as $J(H + B_0)$, with B_0 of the order of H_{c1}.[6] But no justification for this particular form has appeared.

As $H \to H_{c2}$, the lines will be forced together until the forces between them are no longer the long-range, smooth electromagnetic forces, varying as $\ln(r_i - r_j)$, but are the much more steeply varying forces which ensue when the regions in which $\Psi \neq$ const overlap. One would expect the bundle concept to fail completely, and the "hard core" interaction between lines to lead to new effects. One suggestion one might make would be that the lattice of lines may become *rigid*, so that the bundles can no longer slide independently past each other. In that case the rate might be expected to become much slower, a possible explanation for the "peak effect."[7]

The two immediate conclusions which were drawn from (5) were the critical current curve and the flux creep rate equation.[5] The "critical current" came from supposing that the critical parameters as measured in most cases—notably Kim's experiments[6]—simply represented a point at which the rate R became immeasurably slow. Call this rate R_c. Then we have

$$kT \ln (R_c/\omega_0) = -(F_b)_{crit}$$

or

$$\alpha_{crit} = \frac{(JH)_{crit}}{c} = \frac{pH_c^2}{8\pi} \frac{\xi_0^2}{\delta^2 l} - \frac{kT}{\delta^2 l \xi_0} \ln \frac{R_c}{\omega_0}. \quad (7)$$

A similar expression was found to agree reasonably well with the temperature dependence of α_{crit} in Refs. 5 and 6. This expression is somewhat better qualitatively because one has the most structure-sensitive parameter l in both terms.

The flux creep rate equation was also treated approximately in Ref. 5. Equation (5) may be put in a more useful form by writing it as an equation for the pressure gradient α itself (for simplicity we specialize to a one-dimensional situation as in a tube wall):

[6] Y. B. Kim, C. F. Hempstead, and A. R. Strnad, Phys. Rev. Letters **9**, 306 (1962).
[7] For instance, S. H. Autler, E. S. Rosenblum, and K. H. Gooen, Phys. Rev. Letters **9**, 489 (1962).

$$\frac{\partial \alpha}{\partial t} = -\frac{\partial}{\partial t}\left(\frac{H}{4\pi}\frac{\partial H}{\partial x}\right)$$
$$= \frac{\partial}{\partial x}\frac{H\Phi_0}{4\pi\delta}\frac{\partial R(\alpha)}{\partial x}.$$

Usually, R will depend exponentially on α. H, of course, will depend only roughly linearly on α. Also, we really do not quite understand the pre-exponential factors in the rate equation anyhow, and an extra factor H could not easily be checked experimentally in most cases. Thus, it is easiest and within errors of the theory to neglect the derivative of H relative to that of R, and we obtain

$$\frac{\partial \alpha}{\partial t} = \frac{H\Phi_0 \omega_0}{4\pi\delta} e^{-F_0/kT} \frac{\partial^2}{\partial x^2} e^{\alpha/\alpha_1} = K_0 \frac{\partial^2}{\partial x^2} e^{\alpha/\alpha_1}, \quad (8)$$

where

$$\alpha_1 = \frac{kT}{\delta^2 l \xi_0} = \frac{\alpha_{crit} - F_0/\delta^2 l \xi_0}{\ln(\omega_0/R_c)}. \quad (9)$$

α_1 is usually very small, of the order of $10^{-2}\,\alpha_{crit}$, which means that the barriers are indeed high compared to kT. K_0 is defined by this equation and F_0 is the force-free barrier height $F_b(\alpha = 0)$.

A steadily decaying solution of (8) is

$$e^{\alpha/\alpha_1} = (-\alpha_1 x^2 + bx + c)/2K_0 t$$
$$\alpha = \alpha_1\{\ln[(-\alpha_1 x^2 + bx + c)/2K_0] - \ln t\}. \quad (10)$$

This logarithmic time dependence has been repeatedly observed.[6,8]

Another solution of the nonlinear diffusion equation (8) is the steady-state one, such as one would obtain physically by supplying power from an external source to maintain a current through a tube or plate sample:

$$(\partial/\partial x)e^{\alpha/\alpha_1} = c$$
$$\alpha = \alpha_1(\ln c + \ln x). \quad (11)$$

Because α_1 is so small c will be enormously large in all cases, and thus α will be essentially constant throughout the sample, as is physically obvious from the exponential rate dependence on α.

We have made no progress in studying the transient solutions which are relevant when one quickly applies external fields or currents to a hard superconductor. Clearly, the effect will be to create local concentrations of magnetic pressure which could diffuse away but may well not be able to do so stably—i.e., in such a way as to decay continuously into a steady-state solution. We are currently studying this

[8] Y. B. Kim, C. F. Hempstead, and A. R. Strnad, Phys. Rev. **131**, 2486 (1963).

problem with the idea that possible instabilities in Eq. (8) may well be related to the phenomena of magnet instability.

A whole range of further applications of these ideas are suggested by the observation that the existence of flux creep clearly implies power dissipation in all hard superconductors.[8] That is to say not only that apparently perfectly superconducting samples below the so-called "critical curve" are still dissipating power—and thus offering resistance to current flow—at a finite rate, but that apparently nonsuperconducting, resistive samples far *above* the usually accepted critical conditions are also often truly superconducting in the thermodynamic sense, especially under transient conditions before thermal or flux diffusion instabilities have had time to occur.

A simple way to derive the resistive power dissipation is to start from (5) and Maxwell's equation

$$\nabla \times \mathbf{E} = -\dot{\mathbf{B}}/c$$

in a simple one-dimensional case where we assume we have B in the z direction, J and E in the y direction, and flux creeping in the x direction. Then,

$$\frac{dE_y}{dx} = -\frac{\dot{B}_z}{c} = \frac{d}{dx}\left[\frac{\Phi_0 \omega_0}{\delta c} \exp\left(-\frac{F_0}{kT} + \frac{\alpha}{\alpha_1}\right)\right];$$

i.e., to maintain the flux we have to apply an E field

$$E_y = \frac{\Phi_0}{dc} \exp\left(-\frac{F_0}{kT} + \frac{\alpha}{\alpha_1}\right). \quad (12)$$

There is, then, clearly a power dissipation

$$P = E_y J_y = \frac{J\Phi_0}{\delta c} \exp\left(\frac{\alpha}{\alpha_1} - \frac{F_0}{kT}\right) \quad (13)$$

in the material.

This immediately suggests the possibility of severe thermal instabilities in hard superconductors. Let us write down the equation for the heat content of the material:

$$\frac{dQ}{dt} = c_{sp}\frac{dT}{dt} = -\kappa \nabla^2 T + P.$$

κ is the heat conductivity. Suppose a small temperature fluctuation $\delta T(r,t)$ occurs. Whether it grows or decays is determined by the equation

$$c\frac{d\delta T}{dt} = -\kappa \nabla^2(\delta T) + \frac{\partial P}{\partial T}\delta T. \quad (14)$$

Suppose the fluctuation occurs on a scale of size r. Then for stability we must have

$$\frac{\kappa}{r^2} > \frac{\partial P}{\partial T} = \frac{P}{T}\left(\frac{F_0}{kT} - \frac{\alpha}{\alpha_1} - \frac{1}{kT}\frac{dF_0}{dT}\right).$$

By differentiating Eq. (7) we may obtain an expression for the dimensionless ratio in parentheses:

$$\frac{\kappa}{r^2} > \frac{P}{T}\left.\frac{T}{\alpha_1}\left(\frac{\partial \alpha}{\partial T}\right)\right|_R, \quad (15)$$

where $(\partial \alpha/\partial T)_R$ means that we fix the rate R. Experimental data tell us that this ratio is of the order $10^2 - 10^3$, which means that thermal instability is a severe problem. That is, the increase in temperature ΔT of the given region over the surroundings caused by the power input P is of the order

$$\Delta T \simeq Pr^2/\kappa$$

so this tells us that

$$\frac{\Delta T}{T} < \left|\frac{T}{\alpha_1}\left(\frac{\partial \alpha}{\partial T}\right)_R\right|^{-1} \simeq 10^{-2} - 10^{-3} \quad (16)$$

is the stability requirement: a very tiny rise in temperature can presage a complete thermal breakdown. In particular, if, because of excessively rapid current or field changes or of "weak spots," the stress α becomes concentrated in a small region, P will be exponentially larger locally while κ/r^2 only increases as the square of the size, so that local thermal instability can be a problem. The obvious practical morals are three: first, that magnet configurations allowing good thermal conduction are vital; second, the rather discouraging remark that so far good hard superconductors are also bad thermal conductors for obvious reasons, so that the better the magnet the worse the stability problem will be; and third, because of the factor $1/r^2$ very big magnets will have extra stability problems.

The final remark we would like to make is that the existence of this effective resistivity caused by flux creep allows us to investigate experimentally the creep rate over a much wider range of α than was possible with the original flux decay measurements, by measuring the resistivity of wire samples. In particular, one can go to far higher stresses—the "resistive state" of superconductors.[8] In this region Kim et al. have found that the exponential law $R \propto e^{\alpha/\alpha_1}$ begins to fail and is replaced by a roughly linear relation $R \propto \alpha$. This occurs when the effective barrier $F_0 - (\alpha/\alpha_1)kT$ is no longer large compared to kT; we would then expect a viscous resistance to flux line motion, a process of "flux flow" rather than "flux creep," analogous to the motion of magnetic domain walls above the coercive field. Kim has suggested a very useful semi-empirical formula for the velocity of flux lines:

$$v = [(\text{const} \times \alpha)^{-1} + (\text{const} \times e^{\alpha/\alpha_1})^{-1}]^{-1}. \quad (17)$$

In summary, the general qualitative ideas of flux creep theory appear to be soundly based and to allow for the qualitative and semiquantitative understanding of a wide range of phenomena. Many problems remain, both for detailed quantitative study and even for better qualitative understanding. To list a few of these:

(1) More detailed understanding of flux line interactions, in particular a sounder basis for the "bundle" concept and an understanding of B_0 and of the peak effect.

(2) The peculiar transient pulses observed by Kim et al.[8] These support the creep idea qualitatively, but are too large to be individual lines or bundles and too small to be instabilities. Are they avalanches?

(3) The nonlinear diffusion equation: when is it unstable?

(4) The viscous state. This is a completely new theoretical problem and I know of no obvious way even of approaching it from fundamental theory.

Exchange in Magnetic Insulators
Transition Metal Compounds, Buhl Int. Conf. on Materials, Pittsburgh, ed. E.R. Schatz
(Gordon and Breach, New York, 1964), pp. 17–28

In the Buhl lecture of 1963, I talked about the "magnetic state" which characterizes all strongly repulsive systems, an insight still very valuable and unavailable to those trained in the school which nearly monopolizes U.S. solid-state education, and which emphasizes band calculations and one-electron theory. This latter school is very much beholden to the influence of J.C. Slater and his students. The motivation for the Buhl lecture was really Bob Shulman's simultaneous infatuation with his Jaguar and with food and wine; the former took us to Pittsburgh to partake of the latter unforgettably, with Bob Heikes and his wife. Heikes ran the workshop and, before his distinguished management career, contributed interesting ideas in this area.

The "magnetic state" concept was a first crude vision of the dichotomy which becomes increasingly clear as the years go on, between systems based on the "one-electron" physics of weakly interacting electrons, which can be well treated by "LDA" methods and direct computation, as opposed to magnetic systems which are dominated by the electronic repulsion and persist in defying even cleverly modified versions of those methods.

EXCHANGE IN MAGNETIC INSULATORS

PHILIP W. ANDERSON

It's been my experience that giving a talk on material to which my own contribution was mostly completed a number of years ago is a very frustrating thing. In the first place, the advances in the field which have been made recently were by other people and it's always a difficult thing to describe properly other people's work. More serious is the fact that when one has completed a reasonably large job, one is always completely convinced that nothing further can be done and the theory has a beautiful, completed, closed appearance. A couple of years later one finds that this is faded somewhat and one is tempted to start off down the various new avenues of thinking which has opened up since the last time one looked at the subject. So I will admit that the longest part of my talk, which will be about my own work, is really in the nature of a very long introduction and, from my point of view, I am going through it only to reach the more interesting new things which have been opened up by the work of Toru Moriya, Bob Shulman and Conyers Herring. I should mention that practically everything I'm talking about was done in collaboration with Shulman, Sugano, and Moriya.

The contributions I made to this field a while back, were based on a name and the conceptual background to which this name led one. The name, the idea, is the idea of the Magnetic State of an insulating material. The concept which is implied is that there is a specific kind of state of the gas of electrons in a substance which one should call the magnetic state, which is distinguishable from other kinds of states of the electron gas, and which one should understand before one really starts thinking about such properties as exchange. I can best define what I mean by state by giving some examples: The metallic state, the insulating or semi-conducting state, or the super conducting state. Perhaps the most rigorous way of defining this concept of state is in terms of the low lying elementary excitations of the system. For instance, in the metallic state, these low lying elementary excitations are very numerous and have energies which go right down to zero; they are near a surface in momentum space, and they carry charge. In an insulator or semi-conductor or the other hand, the low lying elementary excitations have an energy gap, are near isolated points in momentum space and carry charge. The basic point that I would make about this idea of state is that one can't really hope to make any progress, for example, in understanding the electronic behavior of the metal, the super-conductor, or the semi-conductor unless one has firmly in mind the basic distinguishing characteristics which make it a metal as opposed to a super-conductor, say, and roughly why. For instance, start with a finite number of individual sodium atoms far apart, and start bringing them closer together. We could know everything we liked about the properties of each of these individual sodium atoms or even about combinations of a few of them but nothing about them is going to tell us anything about the properties of metallic sodium, i.e., of a macroscopic piece of sodium metal in which we have put a macroscopic number of sodium atoms together. Nothing is going to tell us until we actually try it out that the electrons are going to come off the individual atoms and take on free electron properties. I don't mean to imply that the science of magnetism, before the magnetic state came along, was really in such a miserable position as that one would be in if one would try to understand the idea of a metal without the idea of the free electron, but I do mean that there is a distinguishable magnetic state and that the reason for its occurrence is rather uniform in all cases. Using the idea that there is such a state simplifies and clarifies thinking about magnetism. The reason why, for so many years, one could talk reasonably intelligently about magnetism, whereas one couldn't possibly talk about the metallic state without realizing it was a distinguishable state of matter, is that the case in which magnetism is favored is the case in which the individual atoms are reasonably far apart. In a zeroth rough approximation they act like the individual isolated atoms would act. For instance an isolated atom of iron or manganese or gadolinium or sodium itself is magnetic. If we keep the atoms far enough apart, we have examples of magnetic substances. So it's a kind of zeroth approximation to treat magnetism as though the atoms or ions were just individual separated atoms and treat their interactions as perturbations. But it still is important to understand that when we put the atoms together into a solid there is a real problem as to whether the substance will take on the magnetic or metallic or some other state. As I said if we bring sodium atoms together they do not go into the magnetic state, they go into a state in which the free electrons have come off the sodium atoms, paired their spins and the resultant metal may be diamagnetic or paramagnetic but certainly doesn't have the Curie–Weiss behavior that is connected with the magnetic state. We bring the atoms of gadolinium together and we find that they do stay magnetic.

For this audience, actually, I hardly need to go in very great detail into the background of what is really happening in the magnetic state. You all know that the man who first explained what is going on was Mott in 1949. Mott imagined that we had some lattice of ions of the kind which we normally expect to be magnetic, say Ni^{++} ions (see Fig. 1) In order to make the system neutral, of course, we have to make an ionic crystal out of it and we have to put something in between. We can put in oxide ions or we can talk about a really dilute substance like nickel sulfate. Then we can compare the energy of a hypothetical magnetic state with a non-magnetic state. First, suppose that each Ni^{++} ion has the configuration (d^8) and a moment of two Bohr magnetons. You can imagine that each nickel ion has precisely its own 8 d electrons; and then the spins are free and the electrons are localized. Then one can take an electron from one of these Ni^{++} ions making it an Ni^{+++} and put that electron on a very distant Ni^{++} ion, and you will have an Ni^{+}

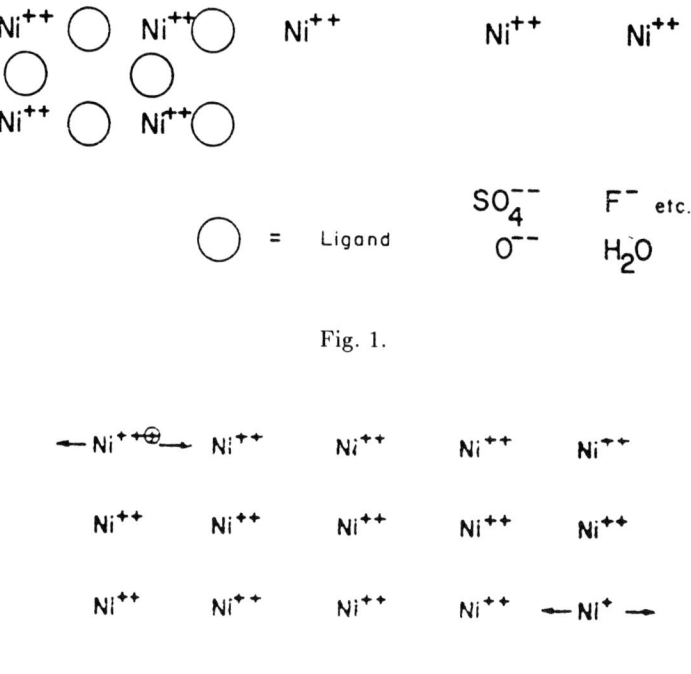

Fig. 1.

Fig. 2.

ion with 9 d electrons. The Ni^{+++} has only 7. (See Fig. 2.) Two energy changes will have occurred. The first energy change is quite obvious. When we took the electron from one ion and put it on another, we were taking it from an ion where there were only 7 other d electrons and were putting on an ion where there are 8. Thus the extra electron is repelled by the other extra electron on that ion by an amount which to zeroth approximation is given by the Coulomb repulsion integral between two d electrons: we can call it U and estimate it by the integral.

$$U = e^2 \int d^2(r_1) \frac{1}{r_{12}} d^2(r_2) dr_1 dr_2$$

This is the repulsive potential energy between two d electrons in the same orbital d. When one directly evaluates that by integration, one doesn't come out with a very reasonable value; for actually doing the problem, it is sensible to take the difference in energies between an Ni^{+++} and a Ni^{+} ion as compared to a pair of Ni^{++} ions from tables. One comes out with numbers of the order of 10 to 15 electron volts. So that's quite a large number and quite a large amount of energy that we've lost. We've regained, however, a certain amount of energy because now the hole on the Ni^{+++} ion and the extra electron on the Ni^{+} ion can hop back and forth from one atom to another, and so they are essentially free electrons now instead of localized electrons. If we imagine that there is a d band in the substance,

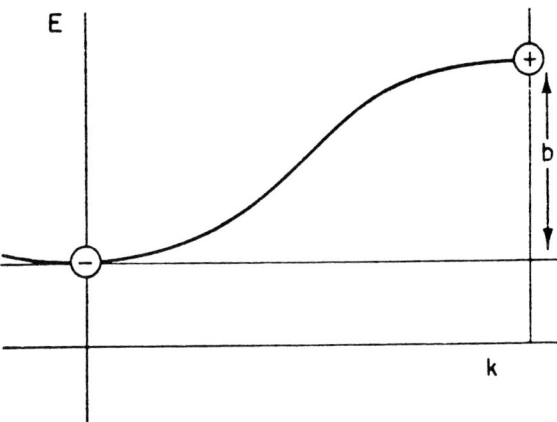

Fig. 3.

(see Fig. 3) the hole can occupy the very highest position in the band and the electron, if it likes, can occupy the very lowest position in the band. Say that the band width is b; we've gained back an energy of order b because the localized states of course are equal mixtures of all the states in the band. Now, again writing down an integral which does not really represent the most exact way of going at it but is what it would be in this highly oversimplified uncorrelated kind of picture, the band width will be roughly given by the hopping integral between the two local sites:

$$b_{jk} = \int dr \, d(r - R_j) \left(\frac{p^2}{2m} + v \right) d(r - R_k)$$

R_j and R_k are the positions of neighboring sites.

We find then that if the energy U, the Coulomb repulsive energy, is much greater than b, then we haven't regained enough energy so that the electrons really will continue to separate themselves and become free electrons; therefore they will stay in the localized magnetic state. If U is of order b, (actually it turns out that because of various numerical factors and because of screening the relationship is not symmetrical in U and b) one goes into the metallic state, the electrons separate and free themselves from their sites and one has a metal. It is actually found that in certain oxides of the lower d series, there probably is a thermal transition, between the two possible states, magnetic and metallic.

The great advantage of these ideas in thinking about problems having to do with magnetic materials is that they give one, in the first place, a zeroth order state, from which one can apply perturbation theory. This zeroth order state (or rather the zeroth order states) are then those in which all the electrons stay on their individual sites. The first excitations are those in which the electrons hop from one site to a nearby site, thereby losing an energy U; then one can go on from that to a perturbation theory in principle and hope that that perturbation theory converges. If it converges then one has an elementary excitation or quasi-particle way of thinking about magnetic problems. In a metal, it is useful to think of one's elementary excitations as being very like free moving single electrons; in the zeroth approximation the wave function of a single electron is just a plane wave, the first approximation is a Bloch wave in which one has modified the wave function for the effects of the background lattice, but it still is a single electron function and the last approximation is the quasi-particle. And one can think on any one of these three levels depending on how accurately one wants to solve the problem. To this, in the magnetic state one thinks of one's zeroth order of approximation as the electrons in single, isolated atom electronic states, and then one has to take into account in the first approximation the fact that the electron is in a periodic lattice containing ions of Ni^{++} and all kinds of other things, especially the nearly ligands. So from this, one must go to ligand field theory states which have a certain amount of the appropriate ligand functions mixed into the atomic functions, and possibly even the d functions are changed. Finally one comes to a really accurate quasi-particle theory in which one thinks of the electron as completely renormalized for all the possible many kind of electron excitations that should surround it as it sits on its individual site with its given spin. For instance, a certain amount of spin polarization of the background lattice will be carried with it of course. So the lowest energy excitations are the rotations of the spins of the electrons on the localized sites; to this zeroth approximation there is no interaction between the spin directions on the different sites. To a higher approximation, there will be such interactions and it's these interactions that we're interested in finding; but in general the low lying excitations are spin reorientations without real motion of electrons from one site to another, only virtual motions of the electrons between the sites. Then one has real charge carrying excitations, the lowest charge carrying excitations in the zeroth approximation having an energy U. We can correct U then for admixture, for polarization, etc., and then there is presumably some real physical renormalized U which represents the lowest lying energy for excitation of a free electron in the material. So from this point of view we have a certain number of parameters which can be thought of as either integrals over the atomic functions or on a different level as renormalized parameters. The important ones then are basically the quantitities b and U; and a number of others we shall talk about shortly.

Two important points I would like to make before I go any further. One is that the biggest effect, the most interesting effect of the background lattice on this magnetic state problem, is the admixture of the background lattice wave functions into the one electron wave function. Once one has taken that into account it is possible, though of course can't be proved, that the further perturbations due to the background lattice are not really very big. A second point is that it is never correct under the circumstances in which the magnetic state is the correct state, to talk as though one's basic wave functions were eigenfunctions of the one electron problem because the one electron problem is periodic, and the eigen functions of a periodic hamiltonian are running waves. We see here that the starting wave functions, because of the size of U as opposed to b, must necessarily be localized wave functions, and the only possible zeroth approximation is to make them localized. Thus we must be very careful not to make the standard assumption, that one always makes in all other many body problems, that the one electron hamiltonian is diagonalized by the starting wave functions.

Now starting from this point of view one can identify terms in the interactions of the spins on two neighboring sites. The biggest term is something which I have called kinetic exchange. Let's start from our one-electron orbitals on the two neighboring Ni^{++} ions. The zeroth approximation is the atomic function

$$U_{Ni}(r - R_j).$$

We have said that we have to admix into these orbitals a certain amount of the p function on the ligand — for instance an F^- function with an admixture parameter λ

$$\phi = U_{Ni}(r - R_j) - \lambda U_p.$$

Then if we make the same kind of wave function on a second Ni^{++} we do not come out with an orthogonal function, and it is very convenient to work with orthogonal wave functions so we orthogonalize. Then our basic wave function also contains a certain amount of U_{Ni} on the other atom, i.e.

$$\varphi_d(R_j) = U_{Ni}(r - R_j) - \lambda U_p + \gamma U_{Ni}(r - R_k)$$

is second order in λ, very small.

Let us study the energy as a function by the relative spin of an electron in $\varphi_d(R_j)$ and $\varphi_d(R_k)$. Now if we start from a zeroth order state in which our two electrons are parallel, that's the full story. We know that the two electrons, because their spins are parallel, must be in orthogonal wave functions because of the exclusion principle. Another way of putting it is that the wave function which contains one parallel spin electron in each of two orthogonal functions cannot be changed by further admixture between the two states. So, the energy of that state is not changed by any possible tendency of the electron to hop from one site to another. That's by no means the case if the two spins are anti-parallel. Then the electron can go from one wave function to the other virtually. The energy of the excited state it goes into is of course U higher because it's an extra electron on this atom. Actually, it's U plus corrections for the proximity of the two atoms, but it's relatived to our basic parameter U of the problem. The tendency for the electron to hop from R_j to R_k virtually is given by the matrix element b_{jk}, the admixture will be given by b_{jk}/U. So the real wave function will become

$$\varphi_d(R_j) + b_{jk}/U\,\varphi_d(R_k).$$

The energy improvement that it gets by doing that will be $-bjk^2/U$, but the electron on R_k can do the same thing so we get an energy contribution, if the spins are anti-parallel, $-2\,bjk^2/U$. This is always anti-ferromagnetic, and I contend it is the largest spin coupling energy in the problem. Therefore in general the spins want to be anti-parallel. That conclusion is true only when bjk does not vanish; if the wave function on Rj, for instance, is a $d\pi$ with a node pointing along the line toward the ligand whereas the wave function on Rk is a $d\sigma$, bjk vanishes by symmetry. In summary, if there is a b_{jk} kinetic exchange is probably the largest contribution.

The second biggest contribution is what I call potential exchange and that comes about because the space wave functions of the two electrons, though orthogonal, do overlap. When the two electrons are anti-parallel as I said that has no particular effect, but when the two electrons are parallel we know that there will be what is called the exchange hole in the overlap region. This occurs because the two electrons really can never be at the same point at the same time. Thus the parallel state is favored because the electrons are not quite as close together when they are parallel as we would think. The correction relative to just taking a straightforward Coulomb interaction integral between the two wave functions is the potential exchange integral, given by

$$\int \varphi_d(r_1 - R_j)\varphi_d(r_1 - R_k)\frac{\varepsilon^2}{(r_1 - r_2)}.$$

We see that this exchange is the self energy of the overlap charge distribution, by

$$\rho_{jk}(r) = \varphi_d(r - R_j)\varphi_d(r - R_k).$$

One of the properties of ρ_{jk} is that because we chose orthogonal wave functions to begin with the integral of ρ over all space is equal to zero. $\int \rho_{jk}(r)dr = 0$. That's just the orthogonality integral. This says that the overlap charge distribution is positive in some places, negative in others, and on the average it vanishes; therefore we should expect that it will not be very large and that the contributions to this integral from different places will tend to cancel. That is basically the reason why the potential exchange integral is usually considerably smaller than the kinetic integral.

That makes two kinds of exchange: the third is everything else, by which I mean all the specifically many body effects which can occur. The magnetic electrons can exchange polarize the electrons in the atom cores and the p band of the fluorine or oxygen. Such effects are probably the largest in the rare earths. They are certainly not the largest in the d band substances which we are interested in here. There are various of these effects and probably none of them are really very big.

Now we come to the question of practically calculating the exchange in a real substance. One of the things this general way of thinking leads to is a set of general rules as to when exchange will be positive, when negative, when it will be relatively large and when it will be relatively small, in terms of the covalency parameters and the crystal field parameters. These rules tend to work out very nicely in most cases, in fact I don't know of any cases which really contradict them. On the other hand, the really crucial test is to try to calculate something from first principles and really get a value for one exchange integral. There is a particularly fortunate case that one can calculate, the case of $KNiF_3$. The most fortunate thing about it is that Shulman and Sugano have already done the ligand field theory. One can think of the ligand field theory as the theory of the isolated elementary excitation of the electron gas, and the exchange theory is a theory of the interacting electrons. We have in this case, a calculation of the isolated electron problem which leads to reasonable values for various parameters and particularly for the crystal field parameter. The second nice thing about $KNiF_3$ is that it's particularly conveniently set up for the exchange problem. See Fig. 4.

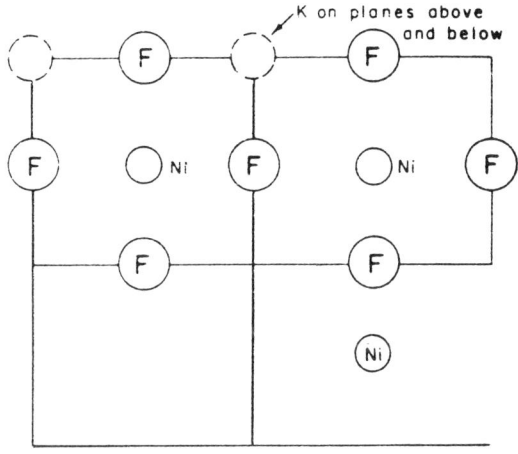

Fig. 4.

One main point is that the next nearest neighbor nickels are practically completely out of touch with each other. One really can treat the classical problem here of two nickel ions on opposite sides of the fluorine ion. It is even true that the only magnetic electrons on the nickel are in d_{z^2} orbitals which point toward the fluorine and so one can have the classical picture of two d_{z^2} orbitals interacting through a $P\sigma$. As I said, Shulman and Sugano have calculated the crystal field splitting $10Dq$. A calculation which leads to $10Dq$ is certain to give a value for b_{jk} because the two processes which lead to the major contributions to these two things are very similar. A major contribution to $10Dq$ comes from the interaction between the d sigma orbital and the p sigma orbital on the fluorine reacting back again, essentially, on the Ni^{++} ion. The major contribution to b_{jk} presumably comes from the same admixture and interacting with the other nickel. It is really practically the same process.

Thus from Shulman and Sugano's numbers, we calculated a very satisfactory answer for the quantity $2b_{jk}^2/U$. U now is something one gets from atomic tables and corrects with various corrections which seem reasonable and accurate to within any kind of accuracy that we are talking about here, which is not much better than 50%. We came out with a value for Jjk. One can calculate the Curie point, knowing this; we came out with a Curie point which was 75%, as I remember, too high. We were very happy about that because we knew that the potential exchange interaction was of the opposite sign to this, and we expected it to be somewhat smaller, and so it looked as though this was about the right answer. Now, however, with Moriya doing most of the work, we have calculated the potential exchange integral. We didn't do it before simply because the potential exchange integral involves a lot of three center integrals which are very hard to do. Using the expression

$$\varphi_d(r - R_j) = U_{\text{Ni}}(r - P_j) - \lambda V_p - \gamma U_{\text{Ni}}(r - R_k)$$

and putting in an s term also, and pumping that into the potential exchange integral, one can thereby divide it up into integrals involving all the various possible exchange integrals between d on atom j, p or s on the fluorine and d on atom k. There turn out to be something like 18 different 2 and 3 center integrals. Most of them had been done by Shulman and Sugano but a number of the three center ones had to be estimated by Moriya in a way which I think is fairly satisfactory. The way he estimated them was this: for instance, suppose you want to do

$$\int (U_{\text{Ni}}(r - R_j))^2 \frac{\varepsilon^2}{(r - r')} V_p(r') U_{\text{Ni}}(r' - R_k) dr\, dr';$$

that is a three center integral but you can, from looking at values of 2 center integrals, make a fairly good estimate of where the center of charge of the overlap between p and d on atom R_k is. One knows that the center of the charge distribution $(U_{\text{Ni}})^2$ is right on atom Rj; then the integral is obviously pretty close to the overlap between p and d divided by the distance between the centers of charge. One can make a number of estimates like that, I'm sure they're close enough to right for the time being and so one can calculate the potential exchange integral. The result was of the same size as the kinetic exchange integral so we came out with a transition temperature of zero, essentially. Now this was surprising because it represented an unusually large failure of what is called MeWeeny's rule, which is the rule I mentioned before, that exchange integrals are always considerably smaller than direct Coulomb integrals. The biggest Coulomb integral involved is pretty obviously the Coulomb integral on the fluorine, times the admixture of the fluorine orbital into Rj twice and the admixture twice into R_k. That admixture gives one a term of order λ_p^4 (Coulomb integral of Vp). It turns out that the exchange integral was of the order of 1/3 of that term; in other words, one has a cancellation only to 1/3. If one looks at typical potential exchange integrals in other cases, one finds that there is usually a reduction factor of from 1/5 to 1/35. So, we're in worse shape here than one usually is with regard to McWeeny's rule, we don't quite understand why it seems to come out that way. This may really not be as bad as all that because these calculations are extremely sensitive to the values of the various parameters one chooses. One really doesn't know U to within better than 20% or maybe even 30 or 40%; on the other hand to make things correct again, one would have had to change U by about 80% and we didn't want to do that, that seemed unlikely to be correct; it's about twice our expected error. Another parameter one could change, is λ_p. This integral is more sensitive to λ_p than b_{jk}/U and the sensitivity of λ_p is such that we would only have to change it by 30 or 40%, to get it back into agreement; but we believe we know λ_p better than that. For one thing, it's an experimental parameter coming from nuclear resonance results; and for another thing the experimental and theoretical values check and all the theoretical values involving λ_p for the properties of the isolated ion check, and why should then this one property convince us that we must use a λ_p 30% smaller. That's somewhat outside, but definitely outside, the error we would like to assign to it.

Now, in view of the fact that there are 20 or 30% effects involved here we went back to our original calculation and found that we had probably not done it much better than 20 or 30% even from a purely theoretical point of view; so Shulman, Sugano and myself are presently involved, with the help of M. Karplus, in redoing the original calculation of b_{jk}. It turns out that b_{jk} does involve at least two rather nasty three center integrals, which we had approximated by assuming Phillips' rule that wave functions which are properly orthogonalized do not see very large contributions from the core potentials. We used this rule to approximate to certain integrals, particularly the integral involving the nickel j, nickel k, and the one electron potential on the fluorine. That may or may not be right so we are going back and recalculating that.

My own personal bias on the other hand is that this and the other possibilities already mentioned are not likely to be the major source of the difficulty. The reason for this is clear from a number of things, perhaps most directly from thinking about the recent work of Conyers Herring on exchange. Herring's approach is quite different from the one here: his approach starts from a more purely theoretical point of view. He starts from the idea that, as I said, the magnetic state is the state of atoms basically at large distances from each other; he says to himself, then, let's start with our atoms infinitely far apart and calculate always the very first term in their interaction energy, which depends on the relative orientations of their spins. He was able to show, for instance, in this limit, that we'll call the limit r goes to infinity, that the Heisenberg Hamiltonian

$$\sum_{ij} J_{ij} S_i S_j$$

is a complete description of the spin interaction problem. It is correct to order e^{-2R/a_H} where a_H is the radius of the typical orbits on the different atoms. In the limit that r goes to infinity, that goes to zero very rapidly. Thus his approach was to attack the problem from the point of view that he was going to expand in this exponential. From that point of view, one simply has only to consider pairs and consider the interaction of pairs in the order of e^{-2R/a_H}. It all sounds very simple, but the first thing one finds out when one looks at it is that even the simplest of all the exchange problems of this sort, simply an isolated pair of hydrogen atoms, the problem of H_2, has never been solved. One thinks of H_2, the hydrogen molecule, as completely solved by numerical methods but numerical methods don't help you much when you get into a mathematical limit because the term you're looking for is exponentially smaller than the terms that the computing machine is giving you. So it's simply impossible to have the computing machine do this "easy" problem. It can't be done.

Herring did in fact solve the problem of H_2 finally, for the first time, in the course of this work but that is not the major result that we're interested in now. What I'd like to talk about is certain very general features of his work. There were three qualitative results on the problem of the exchange between two atoms very far apart. Point No. 1 is that the intermediate region between the two atoms is very much larger than the region in which the electrons are near their separate atoms. Specifically, since the configuration space of this problem is 6 dimensional, (there are 2 electrons and each one of them has 3 coordinates) the amount of space in which both electrons are quite far from both atoms is much larger than the amount of space in which they are close to the atoms. In general almost all the exchange integral of the lowest order comes from the intermediate region, where all the particles in the problem are very far apart from each other, and where the free atom wave functions are not good, so it's not very useful to use an atomic wave function kind of theory. This point No. 1 is not likely to be as important in our kind of problem because we've got an F^- in between the two atoms. It has a core potential which is very deep and quite attractive to the electrons that are involved in the exchange, and therefore it is very likely that the electron, in getting back and forth between the two atoms, will usually go by way of the fluorine atom. We expect to get a fair approximation by confining our considerations to atomic wave functions, although looking at Herring's work without thinking about the presence of the intermediate atom, we might think we would not be correct in doing so.

Point No. 2 — he finds in general that in his problem, in H_2 specifically, the potential and kinetic terms very nearly exactly cancel. One can write out the exchange integral in various ways in his theory and one of them is to write a rigorous wave function $\chi(r_1, r_2)$ derived from rigorous solutions of the problem. This rigorous 2 particle wave function has electron one centered around atom j and electron two centered around atom k. There is also a wave function one calls $P\chi$ which is the permuted wave function and then one computes the exchange integral by

$$J = \int \chi(r_1, r_2) H(r_1, r_2) P\chi(r_1, r_2).$$

This Hamiltonian H of course contains kinetic energy terms, core potential terms, and e^2/r_{12}. The term in e^2/r_{12} is a kind of a general potential exchange, that in $p^2/2m + v_{\text{core}}$ is a generalized kinetic exchange. He found that these two terms in the hydrogen molecule problem very nearly cancelled. Now at first you would think this cancellation is analogous with our cancellation; again I think that is not a correct conclusion. The point is that in the isolated atom problem so far one is working with neutral atoms and therefore if electron one is near atom j and electron two is somewhere in all the rest of space, the potentials of the core and of the other electron very nearly cancel and so, to a zeroth approximation, the other electron doesn't see the first atom; and that is the basic reason why one gets this cancellation. That is simply not true if they are positive ions instead of atoms. So, the kinetic term should be larger in our problem because we're talking not only not about hydrogen atoms but about atoms which have rather deep core potentials which are very much stronger than the potentials involved in H_2, and also about atoms which have a charge. Thus we have Conclusion No. 1, exchange comes from all space, not true in our case or not necessarily true although it's yet to be shown. Conclusion No. 2 is that potential and kinetic terms cancel; that can't be true in our case in general, in any case involving ions or in any case involving core potentials because our potentials are just very much larger. Conclusion No. 3 which probably is true is that correlation particularly in the e^2/r_{12} potential exchange term is very important. In fact, the only difference between the Heitler–London exchange answer for the H_2 problem and the true exchange answer is that at really large distance the Heitler–London answer changes sign and the potential exchange term becomes larger than the kinetic exchange term. There is a general theorem that says that can't be true and the only way one can satisfy this general theorem is by concluding that the correlation between the two electrons keeps them farther apart than the straightforward one electron theory would tell you.

Thus the conclusion, which I think is probably true for us also, is that our problem is correlation in the potential exchange term. To see that can have a big effect, one can make a very simple calculation. One of the many integrals of Shulman and Sugano which we used in our calculation of the potential exchange term was the integral involving the Coulomb interaction of two p electrons on the fluorine ion

$$\int V_p^2(r_1) e^2/r_{12} V_p^2(r_2) dr_1 dr_2.$$

If you calculate that straightforwardly from just the wave functions, this Coulomb interaction is 23 electron volts. This is just calculated. There's another way to estimate it and that is to go to tables of ionization potentials and ask yourself what is the difference in energy between a fluorine (−) ion and a fluorine (+) ion. That is roughly 4 volts, the electron affinity of fluorine; and the ionization energy of fluorine, $-E$ of F^+, that is 17 volts roughly. Now if we subtract these two, we get the difference between an F^- and an F^+ relative to a pair of F atoms. Again, we're calculating a U straight from the atomic tables by saying we are going to calculate the process of $F + F$ goes to $F^+ + F^-$ and that difference is if we assume that the wave functions do not change in the interim, that difference is the Coulomb interaction of two electrons. That difference then is 13 eV experimentally, this is actually an overestimate because in this case we are working, on the average, with fluorine wave functions which are considerably more compressed than F^- wave functions presumably are. We really should calculate the problem $F^- + F^-$ going to $F + F^{--}$, but we can't calculate that because it doesn't have the appropriate electrons in it and besides nobody has ever seen F^{--}; this presumably would have a still smaller Coulomb interaction effect. So, we can conclude that we have seriously overestimated the Coulomb interaction term in the potential exchange interaction, that probably the correct potential exchange interaction can be reduced by a factor something like a half for correlation effect and we are in reasonable agreement. That is the best we can do at this point. So in conclusion we must concede that exchange in d band ionic crystals looks like a more subtle and difficult problem than it looked 2 or 3 years ago. I still believe, personally, that we have calculated the major component and that we are working in the right ballpark. If one wanted actually to do more problems one would get roughly the right value for the exchange integrals. That this is an achievement can be pointed out by the fact that as far as I know, no magnetic curie point has ever been calculated from anything like first principles to within a factor of 10. As far as I know very few second order transition points of any kind, in fact no second order transitions, except possibly the superconducting transition of lead, have ever been calculated to within 100% accuracy. So, even 100% accuracy for a second order transition point is still a remarkably satisfactory result.

Discussion

W. J. Carr Jr., Westinghouse: Isn't it true that also in simpler problems, say in two atom problems, what you call the kinetic exchange, although it's larger, is approximately equal to the potential exchange? I'm not sure of that but I believe it's correct.

Dr. Philip W. Anderson: Yes, in all cases the kinetic exchange is somewhat bigger than the potential exchange, it has to be unless there is actual orthogonality between the two wave functions.

W. J. Carr Jr.: But they are still the same order of magnitude aren't they?

Dr. Philip W. Anderson: Yes, that is what I said in the table. In the only case which has been really calculated rigorously, H_2, they are of very much the same order of magnitude. I said however, that I don't think that is a general theorem that we can apply in every other case.

J. Appel, General Atomic: I would like to ask Dr. Anderson whether he can make any comments on how large the correlation correction to the b_{jk} could be, in a particular case, say nickel oxide.

Dr. Philip W. Anderson: It is very hard to say. I have not thought much about it. I have a feeling that the correlation correction to b_{jk} is less than it is to U or to this potential exchange parameter. I don't know how much less but I think considerably so. For one of the important things to realize about b is that although you write it using the whole one electron potential H, there are ways of rewriting it with only the kinetic energy and it is really basically an off-diagonal kinetic energy matrix element. One can approximately take out the dependence on the core potentials and in those circumstances it should be less sensitive to correlation effects.

Dr. Frederic Keffer, University of Pittsburgh: Can you get any estimate out of your theory of the biquadratic exchange which is now believe to be important in manganese oxides?

Dr. Philip W. Anderson: Yes, you can get estimates. It is just a higher term in the kinetic exchange but presumably there are also potential terms. One can get the order of magnitude very easily. It is got to be the order I think b^4/u^3 but there may be a factor 10 in the coefficient because there are many more terms possible here than there are otherwise.

R. F. Wood, Oak Ridge: In your wave function you mix in F^- orbitals in a molecular orbital type of calculation. You're trying to get the description of the electron in the vicinity of the fluorine. Do you feel that this approach really allows enough flexibility into the wave function or should you perhaps mix in higher excited states on the fluorine atom also. You say that you can measure the covalency parameter through NMR experiments. But if you did allow higher excited fluorine functions into the calculation, what then would you be comparing your experimental results on covalency to theoretically?

Dr. Philip W. Anderson: Well the second part is not mine to answer. I think that's really directed to Bob Shulman and the resonance experimentalists. The first part I would say yes if one makes a correlation correction to the part of the potential exchange that comes from the F^-, a considerable part of that is coming from excitations to higher states on the F^-. You can describe an expansion or a contraction of one of the F$^-$ orbitals such as one would have if another electron goes in as an excitation to a lot of higher states. I think it will be there and I don't know, something like 10% is certainly kind of a minimum order of magnitude for what it would be.

R. F. Wood: I really wonder if this covalency parameter is very well defined except, perhaps, on this somewhat simplified model. There's only one parameter entering into the wave function that you are using now, as I understand.

Dr. Philip W. Anderson: Yes.

R. G. Shulman, Bell Labs: I'll try to answer the second part of the question. Certainly if you could mix in excited states of fluorine — or any number of other excited states of the isolated atoms in the crystal, one should be able to get a better wave function and a better solution. There is no disagreement about it. But the question is how far can you go by mixing in the first most obvious excited state which is the state starting with the fluorine electron in the 2s 2p orbitals. So you go that far. You do your experiments and you interpret your experiments in terms of a certain amount of that mixture of 2p 2s. You calculate also, assuming that this is the only degree of departure from the free ion of the nickel plus 2. And you get pretty good agreement with whatever parameters of the d electrons that you are able to calculate which are fairly extensive. In fact, at that first go around, the agreement was good enough in the calculation that we didn't feel justified in increasing the flexibility of the wave function. I mean you could always make it a better wave function but what would the criterion be. The criterion should be that it would give better agreement with experimental results. Well we had pretty good agreement with experimental results considering the accuracy of the measurements, and the accuracy of the calculation so that we felt that there was no justification in invoking some new physical mechanism, which say would be an excited state of one of the two atoms involved, so that what you say is certainly true in the ideal case. Now the consequences of that upon this second case, the more complicated case of how things interact through the fluorine has been answered.

Dr. Philip W. Anderson: Perhaps I did not make it clear enough when I was talking about 10 to 20 percent accuracy, I don't think from a one electron point of view you would want to put in that much of some higher state. I think the calculations in terms of just the states we're using are probably very accurate from a one electron point of view. They are probably the best you can do, and you will get relatively minor corrections by including higher states. What I meant when I said there are probably a lot of higher states around is that in the correlation effects there will be a lot of excitations and a lot of polarization effects involving higher excited states, probably.

R. F. Wood: This says that you think you have a fairly good approximation to the Hartree-Fock one electron state.

Dr. Philip W. Anderson: Yes.

R. G. Shulman: There too I would like to point out that the approximation is quite good to the antibonding state because we are concerned with an electron which is primarily a d electron. So the d electron involves a small amount of fluorine mixture and you minimize the energy of the mixture. But the properties that we were concerned with were the properties of a magnetic electron, primarily a d electron, and we set up the Hamiltonian with a nickel surrounded by six fluorines in order to calculate the properties of an electron which was centered there and detached it from the crystal. I don't think we would want to make claims about knowing very much about the p electrons on the fluorines because they are out toward the edge of our problem and did not have the magnetic properties that we were generally concerned with. The problems we were concerned with were the electrons in an orbital which was predominately d in character with just a small admixture of the p.

John B. Goodenough: Lincoln Lab, M.I.T.: In your estimate of the b_{ji} that entered the kinetic energy term, you did not make clear how much of the splitting $10Dq$ was assumed to be due to the admixture of wave functions from the fluorine ions. There is also a large electrostatic portion to $10Dq$. What do you feel are the relative magnitudes of the electrostatic and the covalent contributions to this splitting.

Dr. Philip W. Anderson: Well that again is a question for Bob but I know the answer. The answer is there are lots and lots of contributions. I said that's the largest one and it is perhaps the largest one but there are many contributions. For instance, when one is thinking about these things qualitatively one does not think exchange contributions are going to be very large — they are, they're quite large. The Shulman–Sugano calculation of the crystal field theory has all of the various kinds of terms in and they all turn out to be roughly of the same order of magnitude. You can regroup them in certain ways and perhaps after regrouping, a large fraction comes from the covalency, but not by any means all. There's an electrostatic contribution, an s contribution, a p contribution, etc., it's a mess. Incidentally, they add up in an entirely different way in b_{jk} so that agreement between the two is strictly coincidental.

R. G. Shulman: In answer to your question about the point charge contributions to the crystal field splitting, we did publish it but it is a very long thing to read and it goes like this. It depends upon what you mean by point charge contribution but let us take the semiclassical model of the point charge and calculate it in a quantum-mechanical way not in a semiclassical way. You see the semiclassical way of calculating the point charge model is to assume that you have D electrons in the metal in quantum-mechanical orbitals and then you have point charges out here although they are really atoms and they are not really quantum-mechanical, they are point charges. Now of course this point charge calculation gives a certain contribution to $10Dq$ of the correct sign, and in the case of potassium nickel fluoride it gives a contribution of about 1200 wave numbers whereas the major part of the calculated value of 6,000 wave numbers or so came from the off diagonal matrix element which Anderson mentioned.

Superconductivity (Two Opinions)
(with B.T. Matthias)
Science **144**, 373 (1964)

A relic of my (mostly) love (some) hate relationship with Bernd. I came to be sure, even before 1987, that the full story of many superconductors included something new, while he became increasingly depressed by the evidence that phonons were right. Bernd's great strength was his refusal to play by other people's rules. With the metallurgists and chemists who were his natural competitors he exploited the narrowness of vision which led them to synthesize and characterize compounds in traditional ways, without recognizing that modern solid-state physics had opened up a cornucopia of useful and interesting classifications such as ferro-electricity, ferromagnetism ("pro" or "anti") and superconductivity, which could be recognized with essentially trivial measurement techniques. With theorists, it was our focus on simple models and simplest cases which was the Achilles' heel he found and, to be sure, used to the hilt to mystify and confuse us. The theorists loved to make elaborate theories of complicated effects, requiring beautiful precision measurements on the simplest, cleanest possible materials. Their success in doing this convinced them that they had "solved superconductivity". But Bernd realized that the average nonspecialist had no idea of the intellectual structure of the field and that, to him, the proper role of theory is to predict whether substance A or substance B is superconducting at all, and if so what is its critical temperature. (Reporters are the ultimate nonspecialists, but even physicists in other fields fall easily into this mode of thinking, and in fact one completely ignores it to one's peril.) Such questions are extraordinarily difficult, in large part because the answer comes mostly from negative results: substance A contains atoms Z, X, and Y which can combine in thousands of different ways, all of which must be less stable than A for it to exist at all in the given chemical state; it must then choose the physical state (crystal structure, magnetic structure, electronic structure) which favors superconductivity. So one is asking the theorist to do thousands of (to him) irrelevant calculations, most of them impossibly difficult, before he starts. To make the situation clearer, perhaps as many as a couple of dozen crystal structures, and no melting points, have been calculated accurately from first principles; and T_c is in principle (though not in fact) very much more complicated than either.

Thus Bernd pooh-poohed the predictive power of theory very convincingly for many years, and even entrapped many theorists into too literal interpretations of the parameters they fed into model theories, hence straying unsuccessfully into his bailiwick of chemical intuition and empirical rules. But with the increasing sophistication of the genuine microscopic theory especially exploited by Rowell, Dynes and McMillan, he seemed to lose his enthusiasm for the bizarre speculations he was wont to tease us with. As you will see later, I, on the other hand, was increasingly disturbed by the evidence that there were persistent "bad actors" among the higher T_c superconductors which did not obey the rules and had generally mysterious properties. Almost certainly, Bernd's claim to immortality will be that he dragged solid-state physics, much against its will, into taking seriously the many, many substances which did not fit our neat, logical — but probably incorrect — categories.

Bernd and I could both have used a touch of the philosophy of General Semantics: superconductor A is not identical with superconductor B until proven so.

24 April 1964, Volume 144, Number 3617

Superconductivity

Two physicists, P. W. Anderson and B. T. Matthias, approach this important phenomenon from different points of view.

I. A Theoretical Approach

The conditions under which this article is being written are unusual. With the other side of the coin being ably presented by my colleague, B. T. Matthias, I will not have to qualify my statements or judiciously distribute credits and concessions, but can flatly state my opinions, for what they are worth. I suspect that I will be proved wrong in some measure; I hope the fact of my stating these opinions will stimulate other physicists to try to prove me wrong.

The fact is that the theory of superconductivity—to which Bardeen, Cooper, and Schrieffer (1) made the most important contribution, which was announced almost exactly 7 years ago (2)—has had unprecedented and amazing success in changing this phenomenon from one of the most obscure to one of the simplest and best understood of all the phenomena relating to the properties of matter.

Since this statement appears to contradict what Matthias says in the discussion that follows, and since much—not all—of the disagreement is a matter of semantics and of philosophy, I would like to discuss briefly some of the elementary facts about the nature of scientific theories and of scientific proof which we all think we understand but which, on a little deeper examination, we find are not precisely the way we imagined.

Nature of Scientific Theories

The scientific method advances—we all learned—by induction from experimental facts. The scientist experiments, evolves hypotheses, tests them against more experiment, applies Ockham's razor, and has learned something. That is of course understood, but I suspect many physicists do not think carefully about how the process really goes in a mature science such as the physics of matter. The experiments against which a theory must be tested are not merely those under direct consideration but the ones carried out over the past 50 to 100 years which have given us all-but-absolute, unshakable confidence in a certain structure of fundamental laws: quantum mechanics, relativity, statistical mechanics, the consequences of symmetry, the regular nature of crystals, the band theory, and so on. I could give you tens of examples of the following general rule: a theory which contradicts some of these accepted principles and agrees with experiment is usually wrong; one which is consistent with them but disagrees with experiment is often not wrong, for we often find that experimental results change, and then the results fit the theory.

Here is a simple example: some time ago Heisenberg and Koppe proposed a theory of superconductivity (3) which predicted correctly that no transition temperature would be higher than 19°K. It was not a bad theory at the time, but Koppe is the first to agree now that it really was not a very reasonable approach, and that the theory is in disagreement with several now well known principles of solid-state theory. I suspect he would also agree that a theory which was consistent with all the a priori known facts of solid-state theory but did not predict a definite value for the upper limit on the transition temperature T_c would certainly have been more valid than his.

Thus the first—many would say the main—task of a solid-state theorist confronted with a phenomenon (like superconductivity, ferromagnetism, or—an example I shall continue to use—semiconductivity) is to find a way in which the special behavior of matter under consideration can be accounted for in a way that is merely *consistent* with all the things we already know about solids and may already know about the phenomenon. Actually, this is usually, for practical purposes, the end of the story: there is usually only one sensible way to account for the facts about the phenomenon consistently with known principles of physics.

Band Theory of Semiconductivity

Let me take an example which may be familiar to many readers: semiconductivity. The band theory of semiconductivity is mostly attributable to the work of Wilson (4) in 1932, and by 1940 or so, Mott and Gurney could say flatly (5), "The accepted theory of semiconductors is due to Wilson." If we examine the experimental data up to that time we find that this acceptance rested on little else than the plausibility of the theory; no quantitative, and very

---

P. W. Anderson is affiliated with Bell Telephone Laboratories, Murray Hill, N.J.
B. T. Matthias is on the staff of the department of physics, University of California, La Jolla, and the Bell Telephone Laboratories, Murray Hill, N.J.

373

little qualitative, data existed until after World War II. The theory had certainly predicted nothing quantitatively. The thermoelectric effect, the Hall effect, and the conductivity were found to behave in a manner made plausible by Wilson's theory, and that was about it.

From 1945 to 1950, such detailed *qualitative* achievements as Pearson and Bardeen's wide-ranging experimental study of silicon (6) and the various experiments associated with transistor action verified the theory in qualitative detail; but still the quantitative basis of the phenomena was practically untouched: the band structure in all its detail, on which all of the energetics and much of the transport theory depended, and the precise coupling constants which controlled the scattering and thus controlled all transport phenomena. Yet one could hardly find anyone aware of the evidence who would question the basic rightness of the theory. In the final stage of the advances of the past decade or so, two developments typical of an absolutely mature (perhaps even a dying) science have occurred: the correlation of detailed experiment with detailed quantum chemical calculations of band structures, and the beginning of the development of a "rigorous" mathematical theory. (I use quotes because of course the "proofs" in such theories always depend on the intrinsically unprovable convergence of the extraordinarily complicated infinite series which result from perturbation theory as well as on major—and unstated—idealizations of the physical world, as when all crystalline imperfections are neglected.)

Semiconductivity theory is one of the few instances of a theory in which the quantitative aspect—the third stage—has been handled with a reasonable degree of completeness. I have an idea that even though superconductivity theory is so young, it is already entering its third stage—the stage of quantitative energy and transition-temperature calculations, the stage in which there are some pretensions to mathematical rigor.

The BCS Theory

From the very first the Bardeen-Cooper-Schrieffer (BCS) theory of superconductivity was remarkably successful in its qualitative and semiquantitative aspects. Perhaps this is as good a time as any to give a brief account of it. The theory stems from an idea of Leon Cooper's, published in late 1956 (7). He showed that if the electrons in a metal have an attractive interaction, the commonly accepted state, the so-called "Fermi sea," is not stable, for the reason that two electrons taken from the top of the Fermi sea—the so-called Fermi level—can then always form a bound-pair state, thereby lowering their energy relative to the Fermi level. Since any two electrons may do this, a macroscopic number of bound pairs may form, changing the state qualitatively.

A wave function for a bound pair of electrons may be written

$$\psi(r_1,r_2) = \phi(r_1 - r_2)\exp[iK \cdot (r_1 + r_2)]$$

where $\hbar K$ is the momentum of the center of mass and ϕ is the wave function of the bound pair. Fourier analysis gives

$$\phi(r_1 - r_2) = \sum_k \exp[ik \cdot r_1 - ik \cdot r_2]f(k).$$

Thus, ψ is a superposition of states in which the two electrons simultaneously occupy one-electron states with momenta $(-k + K)$ and $(k + K)$. The most reasonable, and in fact the correct, value of K is zero, hence this is the origin of the famous $k, -k$ pairing. It turns out, also, that the spins of the two electrons are opposite too: the pairs bind in a state whose spin, momentum, and angular momentum are zero.

It was the momentum condition which gave Bardeen, Cooper, and Schrieffer the clue to go farther. They picked out that part of the interaction in which bound pairs of fixed total momentum zero occurred and found that a mathematically tractable—in fact exact—theory of that part could be developed. The resulting theory can be described in physical terms by saying that the entire gas of free electrons condenses into a single state of bound pairs. (One recognizes at once that the occupied electron states deep in the Fermi sea are almost always occupied in $k, -k$ pairs even in the normal state, so that they actually do not change much in this condensation; the properties of both normal and superconducting metals are entirely determined by states only near the Fermi level. This is the main way in which the exclusion principle enters the problem.) To unbind a pair and create an unbound particle or "normal quasi-particle" costs the binding energy of the pair and thus gives rise to the famous "energy gap." But this extra particle is neither a hole nor an electron, because it keeps emitting and absorbing bound pairs into the zero-momentum state, changing from electron to hole when it emits, and vice versa. This is the brilliant and rather tricky new feature of this theory which makes it so successful: the essential discarding, in a sense, of the detailed and exact conservation of particle number.

Electron pairs may be added to or subtracted from the sea of bound pairs; when one metal is in contact with another metal, this is the process which determines the Fermi level. Because these electron pairs occupy the whole metal—their momentum is fixed at zero, usually—they may be added or subtracted anywhere, and the result is infinite conductivity—the superflow. Since the pair state is occupied by a macroscopic number of electrons all acting in concert, any attempt to change the momentum of a single pair will effectively unbind that pair, whereas a change in the momentum of the whole electron gas is a macroscopic process involving macroscopic energy barriers; the "rigidity" of the bound pairs leads to quantization of flux.

The analogy with rigidity of a solid is close and enlightening. Though a bar of steel is full of vacancies, phonons, and other defects, we still are not surprised to find that it transmits a force from one end to the other without loss. That force is transmitted by the lattice of condensed atoms as a whole, the random motions of individual atoms being too small to disturb it. Similarly, supercurrents are transmitted by the condensed electronic state as a whole.

Schafroth, Blatt, Butler Theory

This picture of the superconducting state is rather similar to a proposal made by Schafroth, Blatt, and Butler (8), before the BCS theory was proposed: that superconductivity is caused by a Bose condensation of bound pairs of electrons. Just as helium atoms are a bound state of six fermions but obey Bose statistics, they argued, pairs of electrons can act like bosons and condense into the lowest energy state, $K = 0$, as it is suspected helium atoms do in the superfluid. Several years later, in fact, Blatt and others completed the proof that all the results predicted by the BCS theory can be made, by dint of great effort, to follow from such a picture. I think it is fair to say that this equivalence, and the importance of the ideas of Blatt, Butler, and Schafroth, have been overlooked to some extent.

Why do I then give the lion's share of credit to Bardeen, Cooper, and Schrieffer? Why isn't the BCS theory just a highly convenient working approximation within the Blatt-Butler-Schafroth theory? There are three reasons.

1) A very practical one: it is not obvious that, without the hints of the BCS theory, the mathematical results which correspond so closely to experiment would ever have been obtainable by Blatt and his associates from their very difficult and complicated theory. In any case, Bardeen, Cooper, and Schrieffer were certainly the first, by some years, to give any mathematical results for comparison with experimental results; in the event, the agreement was amazingly good. Before the BCS theory was advanced, electron pairs were merely a plausible hypothesis among other hypotheses.

2) A difference in emphasis which is practically one of principle: according to the BCS theory the thermal breakdown of superconductivity is dominated by the breakup of pairs—the energy gap—and not by their evaporating *as pairs* from the condensate of pairs for which $K = 0$. In metals, pairs in which $K \neq 0$ practically do not exist, because of electromagnetic interaction effects, in contrast to the situation for ⁴He (9). This physical fact was not recognized in the Blatt-Butler-Schafroth theory until much later.

3) The most important difference, one of principle: actually the BCS theory is a better and more useful one than the corresponding theories pertaining to helium II that existed at the time Blatt *et al.* constructed their theory, even better than theories held until very recently, when the usefulness of the BCS type of theory began to be recognized (10). A theory of the BCS type, to be workable, had perforce to introduce breakdown of the assumption of exact conservation of particle number, which now turns out to be by far the best way of describing superfluidity as well as superconductivity. Let me make this important point another way.

Photons are bosons. In dealing with microscopic quantum processes we find it useful to deal with photons as particles, and to work with states which we describe as containing a fixed small *number* of photons. In such a state the electric field fluctuates—particularly in phase, and phase and photon number are conjugate variables.

In dealing with "classical"—wavelike—electromagnetic phenomena on a macroscopic scale, on the other hand, we are familiar with the procedure of treating directly with the electric field as a fixed "*c*-number." We know, then, that the number of photons must be uncertain, but that disturbs no one. For macroscopic coherence and interference phenomena it is far better to use eigenstates of the field operator E, not the number operator n. We can of course make transformations from one to the other, because the wave and particle pictures are dual and equivalent; but for classical interference phenomena the description in terms of waves—fields—is far better.

Bardeen, Cooper, and Schrieffer had perforce—though they tried very hard not to—to use states emphasizing the *field* of electron pairs, rather than the *number*, in this same way. Basically, the phenomenon of superfluidity *is* best described in terms of the electron *field*. This choice of coherent states (*11*) in the particle field led directly to the gauge-invariance difficulties which troubled some critics of the theory in the first year or so after it was proposed, but by about 1959 that problem had been cleared up (9). Now, on the other hand, it is clear that in such phenomena as the Josephson tunneling effect or flux quantization the phase of the electron wave field has real physical meaning (*12*), and thus a strict interpretation of the requirements of number conservation and gauge invariance, such as was insisted upon by the Australian school, cannot be maintained.

Shortly after the BCS theory was proposed, a number of other theories based on similar ideas appeared. All of these in the end turned out to be essentially equivalent to BCS, although for a computation of a particular kind, any of the four (to my knowledge) others [Bogolyubov and Valatin's quasiparticles (*13*); Koppe (*14*), Suhl, and Anderson's (9) pseudospin approach; Nambu's spinor self-energy (*15*); and Gor'kov's Green-function theory (*16*)] might be superior. As I said, Blatt's theory is also equivalent to BCS, but I know of no applications in which it is preferable, and the correct version appeared only much later.

Predictions and Experiment

Let us return to the mainstream of the discussion. The BCS theory led immediately to a number of predictions about the properties of superconductors, which depended on a small number of semiempirical parameters: the transition temperature or the energy gap at $T = 0$; the density of states; the velocity at the Fermi surface. The properties considered fell into two classes, and in both of them the predictions were, on the whole, in excellent agreement with experiment. In the first class were phenomena which depended primarily on the existence of an energy gap and its temperature dependence: the specific heat, the Meissner effect, optical absorption. In the second were a number of phenomena which depended critically on the "coherence factors" of the theory. It was fashionable for a number of years to say knowledgeably that, after all, the BCS theory was not "proved"—that only in the matter of the energy gap was there agreement between experiment and theory. This was not true. In the original paper it was pointed out that the striking difference between nuclear relaxation, which peaks at T_c and then drops, and ultrasonic relaxation, which drops at T_c, was a coherence effect depending delicately on the form of the theory.

This type of coherence effect can be explained in a very simple way. As we pointed out, the BCS assumption is that the pairs have zero spin, zero angular momentum, and zero linear momentum. Another way of saying this is that the paired electrons exemplify time reversal: they are in states in which the momentum and spin of one are reflections of those of the other.

Many perturbing effects—scattering from a surface or a chemical impurity, or from an acoustic wave—are static in nature—that is, "time-reversal-invariant." Such a perturbation does not affect the bound pairs in a superconductor, in the BCS approximation (*17*). A magnetic perturbation, on the other hand—for example, the magnetic field of a nucleus or of a magnetic impurity—has just the opposite nature. This is the source of the two contrasting kinds of effects. One of the most striking predictions of this type to be supported by experiment is that ordinary chemical impurities do *not* lower T_c much and *do* narrow the distribution of energy gaps (*18*), whereas magnetic impurities *do* sharply lower T_c and "smear out" the gap distribution—effects leading in extreme cases to "gapless superconductivity." The effect on T_c was measured before it was arrived at theoretically, but the effect on the gap was predicted before it was measured *(19)*.

According to the BCS concept, these impurity-scattering phenomena are diagnostic for pairing of electrons of the simple type assumed in the theory. So far all superconductors investigated show them. (Because impurities are universally present it is very hard *not* to observe them!)

The one effect which appeared to violate the coherence-factor predictions was the Knight shift. To my mind the violation is more apparent than real, reflecting a failure of communication, in that the theorists have been discussing an idealized, pure system and the experiments are made on some of the most impure and inhomogeneous specimens of metals ever prepared. In investigations of the Knight shift in pure, bulk specimens, experimental results have not yet contradicted theory.

A second qualitative triumph of the BCS theory began in Russia. In 1959 Gor'kov announced that his version of the theory led to the Landau-Ginsburg phenomenological equations of superconductivity, with an effective charge $e^* = 2e$ *(20)*. The importance of this discovery was not at that time widely appreciated in the West, mainly because we did not appreciate a 1957 paper of Abrikosov's *(21)* which derived from these equations the entire theoretical apparatus necessary for understanding hard superconductivity—the technologically important superconducting magnets, in particular—and, rather obscurely, demonstrated the quantization of flux through a superconductor in units of $hc/e^* = hc/2e$. As a result of this failure of perception on the part of Western theorists, both of these phenomena were first clearly brought out by the experimentalists *(22)*. I am glad to give the experimentalists the credit of discovery if I may at the same time point out that, like a marginal note of Fermat's or a 17th-century anagram, Abrikosov had hidden away the truth, for us all to recognize afterward: the theory was all right, it was just that *we* were stupid.

It is clear from all this that by 1960 or 1961 the experimental support for the theory of superconductivity was overwhelmingly convincing, more so than that supporting, say, Wilson's equally plausible theory of semiconductors at the time the transistor was developed. For some reason—perhaps primarily psychological—physicists were, on the whole, much less ready to accept it than they have been to accept new theories in other, similar cases, but speculation on the precise reasons is hardly appropriate here.

The Third Stage

The first two stages of the development of the theory of superconductivity had, then, been compressed into 4 or 5 years. What about the much more difficult third stage: the quantitative prediction of transition temperatures, for instance, or of the condensation energy? I must again remind you that this kind of question has not been solved in most other cases: we don't have any good theory of melting points, of ferromagnetic or antiferromagnetic Curie points, or of ferroelectric transitions. But we are beginning to understand the superconducting transition quantitatively.

The approximation for the interaction suggested in the original Bardeen-Cooper-Schrieffer paper was admittedly extremely crude. It is known that as an electron passes through the lattice of ion cores in a metal it displaces the ions—polarizes them—in such a way as to attract other electrons. At the same time, its own negative charge repels other electrons. Bardeen and his co-workers assumed that these two mechanisms were independent (they are not, of course, since the electron polarizes the ion cores by electrostatic interaction also) and that the criterion for superconductivity was that the lattice polarization, or "phonon" effect, was the larger. Since that effect was attractive only for electrons differing in energy by less than a largest phonon energy $\hbar \omega_D$, an artificial cutoff (above which no interaction at all took place) was introduced at that energy. The result was the famous equation

$$3.5\, kT_c = \epsilon_g = \hbar \omega_D \cdot \exp[-1/N(0)(V_{phon} - V_{coul})].$$

This equation gives correctly the order of magnitude of the transition temperatures, since it is expected that the coupling constant $N(0)V$ should be considerably less than unity, and it gives the right isotope effect for many metals—the dependence of ϵ_g, and of T_c which is proportional to it, on $\hbar \omega_D$ and thus on the isotopic mass of the ions as $M^{-\frac{1}{2}}$. One could hardly expect that such an admittedly crude approximation could be relevant to the real criterion for superconductivity, and the very first more detailed discussion, by Swihart *(23)*, showed that the isotope effect was an artifact of the cutoff assumption. Nonetheless, for a few years this equation seems to have been taken seriously, even to some extent by its authors.

Bogolyubov had already briefly, and Swihart more thoughtfully, approached the problem from a more realistic point of view; but the advance which made the quantitative approach possible was made by Eliashberg *(24)*, whose ideas were amplified and clarified by Morel and Anderson *(25)*, by Schrieffer, Scalapino, and Wilkins *(26, 27)*, and by others.

The new feature brought in by these people was the idea that it was not only more correct but also more useful to emphasize the *retarded* nature of the electron-electron interaction in superconductors. This can be understood very simply if we think about the actual physical interaction between two electrons in a metal.

We know that after an electron has moved rapidly through the metal, the first thing that occurs is a reaction by the other electrons which screens off the electrostatic potential at a very short distance, less than a unit cell radius in most metals. This screening is instantaneous (or may be considered so for our purposes), and so practically no particle beyond the screening radius ever experiences any appreciable potential energy of interaction. Thus, the second electron must be very close in space *and* time to see this part of it.

The second reaction is the slow response of the metal ions which are attracted briefly to the electron's negative charge. The ions move toward the region where the electron was and, in a time equivalent to one lattice-vibration period, return to their unperturbed positions, executing a damped vibration. Their initial, largest swing is such that an attractive potential builds up in the region where the electron had been, but the swing occurs about a lattice-vibration period after the electron, moving with the Fermi velocity, has left the region. Again, most of this attractive, retarded part of the potential is felt only in a volume of space very close to the actual path of the electron. Thus we see that the whole interaction potential between electrons

376

has a very short range in space but is of a complicated nature in time, with rather long-range parts. The repulsive part is nearly instantaneous, the attractive part is retarded. The partner of the given electron, then, should avoid the space-time region where the potential is repulsive and stay in that where it is attractive.

Eliashberg generalized the equations of the BCS–Gor'kov theory to allow one to use a potential that depended upon both space differences and time differences. The essential result is a single integral equation in space and time —or, when transformed through Fourier analysis, in momentum and energy variables—for an "energy gap" which is a function of momentum k and energy E. The obvious approximation, which turns out to be even better founded than the foregoing discussion indicates, is to neglect the k-dependence—that is, the space-dependence—of this function and to use only the very much simpler equation in one variable, the energy, which results. This single, one-dimensional integral equation is manageable even for very complicated systems; what is more, as Schrieffer has pointed out, this theory is nearly rigorous, in the sense that it includes most of the complicated renormalization effects which might otherwise be large quantitative corrections. I like to call it the "E.A.S.Y." energy-gap equation, for Eliashberg, Anderson, Schrieffer, and Y. Nambu, all of whom contributed—among others—to its development and proper use.

At the same time it happened that developments in experiments on tunneling between superconductors have made it possible to measure the energy-gap function $\Delta(E)$—or at least quantities very closely related to it—in all its detail. The most detailed experiments have been performed on lead by Rowell et al. (28). By correlating these experiments with the theory we have made the following two advances: (i) it was demonstrated that the tunneling data could result from a gap equation of the given form, and this, in view of the complexity of these data, all but verifies the form of the equation (29); (ii) we then showed that the strength of the electron-phonon coupling given from the tunneling data is consistent both with the transition temperature of lead and with the electron-phonon resistivity of lead (30). In this way we have realized, in a rather backward fashion, the decade-old hope of Fröhlich and Bardeen that T_c could be calculated from the resistivity; rather, we have calculated both T_c and the resistivity from the much more detailed knowledge of the phonons and the electron-phonon coupling furnished us by the tunneling experiments (with, I should mention, a strong assist from measurements of the phonon spectrum made by means of neutron scattering).

Questions and Answers

A number of questions immediately arise. (i) Can we do the same thing with other metals? Answer: Yes, slowly, when good tunneling data are available, as they are not yet for most other metals; when we know more about their phonon spectra than we now do; and if the spectrum is as easily analyzed as that of lead. (ii) Is there any case in which we know as much from other sources as the tunneling data tell us in the case of lead? There is one: in many semiconductors we have good, detailed information about electron-phonon scattering, band structures, dielectric constants, and so on; and in degenerate semiconductors we can often vary some of the parameters—notably the number of electrons—at will. Cohen has modified the BCS–Gor'kov theory in a manner appropriate to the case of the degenerate semiconductor and has successfully predicted—to everyone's surprise—in what special circumstances degenerate semiconductors might become superconducting (31).

A third crucial test for the quantitative theory of the interactions which cause superconductivity is that of the isotope effect. In the absence of detailed information about the electron-phonon coupling one might hope to use the transition temperature as a parameter for determining this coupling and at least predict the isotope effect. In the retarded-interaction theory mentioned earlier, the source of the isotope effect is in the retarded part of the interaction caused by ion displacement, since this part will inevitably scale in time accurately with the lattice vibration period and thus with the isotopic mass. That is to say, the only effect of a change of isotopic mass will be to change the vibration period of the ion. Any deviation from $M^{-\frac{1}{2}}$ must measure the effectiveness of the Coulomb part of the interactions which is instantaneous.

It is well known in the theory of nuclear interactions that particles interacting by way of a "hard core"—that is, a strong, short-range repulsive interaction—can to a great extent avoid its effects by modifying their wave functions in the region of the hard core. This modification brings in very-high-momentum states but does not cost very much energy.

In the case of superconductivity theory we have a "hard core" in time, not in space, but the same principle operates: the electrons can avoid each other in this region by bringing very-high-energy components into their wave functions. But in a solid this may or may not be possible, depending on whether the bands are narrow or wide. When they are wide, as in most simple metals, the avoidance is nearly complete, and the isotope effect is nearly $-\frac{1}{2}$; but in narrow-band metals such as the transition metals the theory, as worked out by Garland (32) (after Anderson and Morel, who did not compute the problem accurately enough numerically or include the effect of variable band width), can lead to intermediate, zero, or even, in extreme cases, positive values for the isotope effect. On this issue theory and experiment have fought to a standoff: theory predicted intermediate values—even, fortuitously, the right value for molybdenum!—first (23, 25), but experiment found near-zero values before the theory was accurately enough developed to provide a basis for understanding them (33).

Matthias brings up some exciting open questions in these areas, which, at the risk of finally ruining my reputation for accuracy, I shall answer here to the best of my ability. (i) Will other mechanisms occur? (ii) Have they occurred? To the latter question, the answer is very probably no. In most of the more complicated mechanisms, electrons seem to be paired in anisotropic or otherwise unusual states, which are broken up by impurity scattering. We must look to very pure metals for anything really new. Will other mechanisms occur? Probably. The requirement for the formation of pairs is that the interaction be attractive not everywhere but simply in some, not necessarily very large, region of space and time. Many interactions—perhaps even screened Coulomb ones—have this property, but the transition temperature may be exponentially low. It is amusing to note that the original BCS criterion—that the interaction be attractive in momentum space—is not necessary or even, in many cases, satisfied; the vital thing is the real space-time

377

interaction, not its Fourier transform.

Are all metals superconducting? Probably most are; many theorists have accepted for some time the idea that most of them must be. Sodium has an almost vanishingly weak interaction in much of the relevant region, so its T_c may be very, very low; magnetic metals, to be superconducting, must be extremely pure, since their pair states are, perforce, not BCS pairs and will be sensitive to scattering, and the T_c of these metals may be very low. But I can see no reason why *some* component of the interaction cannot be attractive *somewhere* in every case.

P. W. ANDERSON

References and Notes

1. J. Bardeen, L. N. Cooper, J. R. Schrieffer, *Phys. Rev.* **108**, 1175 (1957).
2. J. R. Schrieffer, paper presented at the Am. Phys. Soc. meeting, 22 Mar. 1957.
3. W. Heisenberg, *Z. Naturforsch.* **2a**, 185 (1947); H. Koppe, *ibid.* **3a**, 1 (1948).
4. A. H. Wilson, *Proc. Roy. Soc. London* **A133**, 458 (1931); **A134**, 277 (1932).
5. N. F. Mott and R. W. Gurney, *Electronic Processes in Ionic Crystals* (Oxford Univ. Press, New York, 1940).
6. G. L. Pearson and J. Bardeen, *Phys. Rev.* **75**, 865 (1949).
7. L. N. Cooper, *ibid.* **104**, 1189 (1956).
8. M. R. Schafroth, S. T. Butler, J. M. Blatt, *Helv. Phys. Acta* **30**, 93 (1957). A full and very readable, if one-sided, account may be found in Blatt's forthcoming book (Academic Press, New York). The proof is mostly in J. M. Blatt, *Progr. Theoret. Phys. Kyoto* **23**, 447 (1960) and T. Matsubara and J. M. Blatt, *ibid.*, p. 451.
9. P. W. Anderson, *Phys. Rev.* **110**, 985 (1958); ———, *ibid.*, p. 827; ———, **112**, 1900 (1958); N. N. Bogolyubov, *Usp. Fiz. Nauk* **67**, 549 (1959).
10. E. P. Gross, *J. Math. Phys.* **4**, 195 (1963); P. C. Martin, *ibid.* **10**, 208 (1963).
11. I use the phrase "coherent state" here in the sense in which it is used by R. J. Glauber, *Phys. Rev.* **131**, 2766 (1963).
12. P. W. Anderson, "Weak superconductivity," in *Ravello 1963 Spring School Notes*," E. R. Caianello, Ed. (Academic Press, in press).
13. N. N. Bogolyubov, *Zh. Eksperim. i Teor. Fiz.* **34**, 58 (1958); J. G. Valatin, *Nuovo Cimento* **7**, 843 (1958).
14. H. Koppe and B. Muhlschlegel, *Z. Physik* **151**, 613 (1958).
15. Y. Nambu, *Phys. Rev.* **117**, 648 (1960).
16. L. P. Gor'kov, *Zh. Eksperim. i Teor. Fiz.* **34**, 735 (1958).
17. The importance of time reversal here was recognized by D. C. Mattis and J. Bardeen, *Phys. Rev.* **111**, 817 (1958); a more complete discussion was given by P. W. Anderson, *Phys. Chem. Solids* **11**, 26 (1959); also see P. W. Anderson, *Proc. Intern. Conf. Low Temp. Phys. 7th, Toronto, Ont.* (1961).
18. For a discussion of chemical impurities on T_c, see E. A. Lynton, B. Serin, M. Zucker, *Phys. Chem. Solids* **3**, 165 (1957); the theory is discussed in a host of papers, published after the papers cited in *17*; of these later papers, that of D. Markowitz and L. P. Kadanoff, *Phys. Rev.* **131**, 563 (1963), is the clearest. For discussion of the gap distribution from the standpoint of experiments, see Y. Masuda, *Phys. Rev.* **126**, 1221 (1962); from the standpoint of theory, see L. Gruenberg, in preparation.
19. Concerning the effect of magnetic impurities on T_c, early experimental and theoretical work by H. Suhl and B. T. Matthias [*Phys. Rev.* **114**, 977 (1959)] and by others, was followed by the more accurate theory of A. A. Abrikosov and L. P. Gor'kov [*Zh. Eksperim. i Teor. Fiz.* **36**, 319 (1959)]. This paper also predicted the effect on the gap seen by F. Reif and M. A. Woolf [*Phys. Rev. Letters* **9**, 315 (1962)].
20. L. P. Gor'kov, *Soviet Phys. JETP (English Transl.)* **9**, 1364 (1959).
21. A. A. Abrikosov, *Zh. Eksperim. i Teor. Fiz.* **32**, 1442 (1957).
22. B. S. Deaver and W. M. Fairbanks, *Phys. Rev. Letters* **7**, 43 (1961); R. Doll and M. Nabauer, *ibid.*, p. 51.
23. J. C. Swihart, *Phys. Rev.* **116**, 45 (1959).
24. G. M. Eliashberg, *Zh. Eksperim. i Teor. Fiz.* **38**, 966 (1960).
25. P. Morel and P. W. Anderson, *Phys. Rev.* **125**, 1263 (1962).
26. J. R. Schrieffer, D. J. Scalapino, J. M. Wilkins, *Phys. Rev. Letters* **10**, 336 (1963).
27. ———, in preparation.
28. J. M. Rowell, P. W. Anderson, D. E. Thomas, *Phys. Rev. Letters* **10**, 334 (1963).
29. This striking agreement may be seen in Fig. 2 of J. R. Schrieffer *et al.* (*26*); the staggering amount of detail available from tunneling curves is shown in Fig. 1 of J. M. Rowell *et al.* (*28*), and its detailed interpretation is discussed by Scalapino and Anderson, *Phys. Rev.*, in press.
30. Y. Wada, *Rev. Mod. Phys.*, in press.
31. M. L. Cohen, *Phys. Rev.*, in press.
32. J. M. Garland, *Phys. Rev. Letters* **11**, 111 (1963); *ibid.*, p. 114.
33. T. H. Geballe, B. T. Matthias, G. W. Hull, E. Corenzwit, *ibid.* **6**, 275 (1961).

II. The Facts

Written discussions between experimentalists and theorists are not too frequent these days, and if they are presented at all, the authors are trying very hard either to agree or to ignore each other. The field of superconductivity is no exception. It is particularly difficult to follow the two sides of an argument when the interval between presentations extends over weeks or months. We hope the argument can be followed more easily in a two-part discussion such as this.

I am indeed fortunate in having the present theoretical aspect of the BCS approach to superconductivity presented by my friend P. W. Anderson, who has been so instrumental and essential in this field. [Another theoretical point of view was given 2 years ago by H. Fröhlich (*1*). While he did not make any detailed quantitative statements, his general approach seemed to reflect the occurrence of superconductivity in a truer way.] For once I will not have the feeling of being unfair when I say that even today the theory cannot in any way predict the occurrence of superconductivity, much less the transition temperature. Reading Anderson's article can hardly lead one to another conclusion. For a long time experiments have indicated the almost general occurrence (*2*) of superconductivity in most metals that were sufficiently pure and cold. We have hoped it could be shown theoretically that a condensation of one kind or another in a Fermi gas should always occur under ideal conditions. Experimentally the answer seems to be in the affirmative. It will be up to the reader to decide whether this answer has now been given theoretically as well.

Theoretical Development

During the last 13 years the theory of superconductivity presented by Fröhlich and Bardeen (*3*) was based on an electron-phonon interaction. Right from the outset in 1950 we were told that this was the final solution to the last unsolved problem in solid-state physics. Since then we seem to have gone through quite a number of "final" stages in the theoretical development. The present one is the BCS theory, named after its three authors, Bardeen, Cooper, and Schrieffer (*4*), and presented in 1957. There has also been a more general theory by Schafroth, Butler, and Blatt (*5*), formulated in 1954, which is based on the Bose-Einstein condensation of electron pairs. As a rule these theories have been preceded by experimental results, and they have in general been unable to predict results. However there are some notable exceptions. Fröhlich predicted a gap in the electron spectrum on the top of the Fermi surface in the superconducting state for a one-dimensional model. This gap was indeed detected optically, by microwave spectroscopy (*6*) as well as by specific heat measurements (*7*). He also predicted the isotope effect, according to which the transition temperature T_0 varies with the inverse square root of the atomic mass, thus clearly indicating a correlation with lattice vibrations. In 1957, Bardeen, Cooper, and Schrieffer presented another electron-phonon theory (*4*), which explained the existence of the gap by means of pairs of electrons with opposite spins. They correctly predicted the behavior of nuclear relaxation in the superconducting state. They also predicted the relation between the gap width and the transition temperature for the elements. In 1961 Onsager (*8*) predicted, on the basis of the Schafroth theory, the existence of quantized flux of only half the value of that calculated by London. The factor 2 was, again, due to the formation of pairs of electrons with twice the charge of a single electron. All these theoretical predictions have been verified experimentally.

However, there have been just as many theoretical predictions that have not been verified by experiment. These include the prediction that the Knight shift in superconductors vanishes, that higher transition temperatures exist, and that the isotope effect, with a value of $M^{-\frac{1}{2}}$ is general; in addition, the theory has also incorrectly predicted the electronic heat conductivity.

Unfortunately, one question remained almost totally ignored in most theories and experiments; namely, What are the critical conditions for the occurrence of superconductivity itself? Derivations of a criterion were first attempted by Fröhlich and Bardeen, and later by Bardeen, Cooper, and Schrieffer. The latter group actually gave an equation for the transition temperature itself; this equation, however, contained an interaction constant that cannot be calculated at present. Apart from this difficulty, the critical conditions for superconductivity could not be predicted by this equation either. For example, according to the equation, yttrium and lanthanum should have the same transition temperature, that of yttrium being possibly a little higher, since both have the same $N(0)$ and almost the same V and Debye temperature. However, yttrium is not superconducting down to $0.07°K$ and α-lanthanum is superconducting at about $5°K$. This difference is discussed later. Moreover, this formula, cited by Anderson, is not only crude, as he says, but also incorrect for the transition elements, since the dependence of T_c on $N(0)$ is in most cases the exact opposite of that stated in the formula. For example, the T_c of yttrium, rhodium, and platinum decreases with an increase in $N(0)$. Since the formula was proposed it seems to have been discarded completely because it does not present the criteria for the occurrence of superconductivity, which, on the other hand, are easily given by a simple empirical rule (9).

Theory versus Description

Let me deviate for a moment and explain why I think the word *theory* is really inappropriate. A *theory*, according to the current *Oxford English Dictionary* is defined as "a conception or mental scheme of something to be done, or of the method of doing it. A systematic statement of rules or principles to be followed." According to the same source, a *description* is defined as "the combination of qualities or features that marks out or serves to describe a particular class."

Clearly, since the present "theories" are unable to state the rules for the occurrence of superconductivity, which, in my opinion, are essential in any explanation of the phenomenon, they should really be considered descriptions or at best models of superconductivity. It is, therefore, not surprising that the acceptance of the generality of the phenomenon, which had been considered rather limited until recently, is the result of the *empirical* approach of finding a great number of superconductors. The recent theoretical conjectures of Morel and Anderson (10), Casimir (11), and Fröhlich (12) support the experimentally arrived at conclusion that most metals, if sufficiently pure and cold, will eventually undergo a condensation of one kind or another. In the overwhelming number of cases this will be the onset of superconductivity, but there are rare exceptions, such as ferromagnetism. In the latter case the electron spins, instead of being antiparallel, are all aligned. It is also possible that different and still unknown transitions or condensations will occur in the millidegree temperature range. These might include phenomena such as nuclear ferromagnetism, predicted 25 years ago by Fröhlich and Nabarro, or other types of alignment.

Occurrence of Superconductivity

The conclusion that superconductivity is a rather widespread phenomenon resulted from the discovery, in recent years, of almost 1000 superconductors as compared with the 30 to 40 superconductors known about 13 years ago. Quite early in the experimental studies it was possible to state a simple rule (9) for predicting the transition temperature of a given metal. If the metal is a nontransition element, or a combination of nontransition elements, superconductivity almost invariably occurs above $0.3°K$ and the transition temperature generally increases slightly with n, the number of valence electrons per atom. *All* electrons outside of filled shells are considered valence electrons. While the chemical valence is sometimes related to this number, it has no meaning for $n > 7$ or 8, from a chemical viewpoint. The transition temperatures in systems of nontransition elements are not very high; they reach a maximum of $8.8°K$ in the lead–bismuth system. All nontransition elements investigated to date show the isotope effect predicted by Fröhlich and Bardeen and later formulated in the BCS theory. Quite frequently the combination of nontransition elements results in semiconductors, in which case an insulator rather than a superconductor is expected as the temperature is decreased. However, even some of the semiconductors, such as doped germanium telluride, become superconducting below $0.3°K$, as recently reported by J. K. Hulm at the International Conference on the Science of Superconductivity held at Colgate University.

The transition elements form the second group of metals in the periodic system. The transition temperatures at which these elements and their compounds (with either transition or nontransition elements) become superconducting are strongly dependent on the number of valence electrons per atom for the element and on the average number of valence electrons for the composite systems. This is in sharp contrast to findings for the nontransition-metal group. For the transition elements the maxima in transition temperature correspond to values near 5 and 7 for number of valence electrons per atom, are very pronounced (Fig. 1), and are rather independent of crystal structure. Hamilton and Jensen (13) recently found that the distribution of the transition temperature is quite *symmetric* with respect to column VI of the periodic system if lanthanum and uranium are excluded. This means, as far as the d-electrons are concerned, that the symmetry is with respect to a *half-filled* d-shell. Although the points shown in Fig. 1 are values for the elements only, the curve was actually traced through many intermediate points not shown in the figure. These points represent the transition temperatures of solid solutions of the elements in each other. The shape of the curve also persists for intermetallic compounds, but for these the height and precise location of the maximum transition temperatures depend somewhat on the crystal structure. No symmetry or regularity of this kind has ever been observed for the nontransition elements. Rather high transition temperatures occur frequently in systems containing a transition element. The maximum transition temperature known today is $18.05°K$ for the compound Nb_3Sn, which has the rather complicated β-wolfram type crystal structure and a ratio for number of valence electrons per atom of 4.75.

379

None of the transition elements or their compounds has ever shown an isotope effect proportional to $M^{-\frac{1}{2}}$, the value reported for the five nontransition elements measured to date. In ruthenium (14) there is no observable effect at all, and in molybdenum (15) the effect is proportional to $M^{-\frac{1}{4}}$. The variations in transition temperature, together with this entirely different isotope effect, would suggest to an unprejudiced observer that there is a drastic difference in the mechanisms causing superconductivity in the two groups of metals. Another difference between the two groups is noticed when their maximum attainable transition temperatures are compared. For alloys of the transition elements this temperature is 18°K, about twice the value for the nontransition elements.

Lanthanum and uranium are exceptions with respect to the symmetrical distribution of transition temperatures in the periodic system. A reason for this has recently been postulated (13). Both are the beginning elements of the $4f$ and $5f$ series, without having any occupied f levels. However, these f levels cannot be far from being occupied, and thus may act as virtual levels. The existence of these virtual levels had previously been proposed as a possible cause of superconductivity by means of a magnetic interaction (16), rather than by means of the usual phonon mechanism.

Superconductivity and Ferromagnetism

Many years ago the close relationship between superconductivity and ferromagnetism became apparent in investigations of these phenomena in isomorphous compounds (17). Since then we have found many examples which indicate not only that this close relationship exists but also that very often magnetic interactions must be responsible for the occurrence of superconductivity, and the reverse may even be true. This can be illustrated by a few examples: (i) U_6Fe is the superconducting compound of uranium with the highest transition temperature, higher than that of U_6Co or U_6Mn; (ii) iron in titanium raises the transition temperature of titanium faster than any other element does; (iii) however, if iron *lowers* the transition temperature by being *localized*, as, for instance, in molybdenum and some of its alloys, it does this also much more drastically than either cobalt or manganese does.

Scandium is not superconducting above 0.08°K, but solid solutions of chromium in scandium (and, so far as we now know, *only* of chromium) are superconducting near 3° to 4°K. This effect of chromium is presumably due to the magnetic interactions, similar to those of iron in titanium. Iron in scandium, in turn, has a magnetic moment and leads to antiferromagnetism.

The compound $ZrZn_2$ crystallizes in a cubic structure (type C15) which results in superconductivity in most other zirconium compounds. However, $ZrZn_2$ becomes ferromagnetic at 35°K. La_3In is the lanthanum compound with the highest superconducting transition temperature, but the analogous compound with scandium, Sc_3In, becomes ferromagnetic. So far this kind of free electron ferromagnetism has been observed only in these two compounds, which were found among 6000 intermetallic phases. The general rule is that superconductivity will occur.

There never has been any theoretical attempt to link the two phenomena. Yet our results indicate that the mechanisms leading to superconductivity are very often closely related to those causing ferromagnetism. Recently members of the Russian and Japanese schools have presented a number of papers describing magnetic interactions which lead to superconductivity (18). Strangely enough, these papers have been ignored thus far. Could it be that history is repeating itself, and that communications on magnetic interaction are receiving the same treatment from Western theorists that the Gor'kov and Abrikosov discoveries received?

Since none of the existing theories can either give criteria for the phenomenon of superconductivity or permit prediction of the transition temperature, it is impossible, for me at least, to understand recent attempts to enforce another consolidation of theory and experiment at any price. Let me illustrate this by the following example. Before the absence of an isotope effect in ruthenium was discovered experimentally, none of the existing theories had ever considered the possibility of the absence of this effect. In their 1961 review article, Bardeen and Schrieffer (19), referring to a different Coulomb cutoff in Bogolyubov's theory of superconductivity, stated, "If this calculation were valid, there would be two serious difficulties: (1) the exponent of the iso-

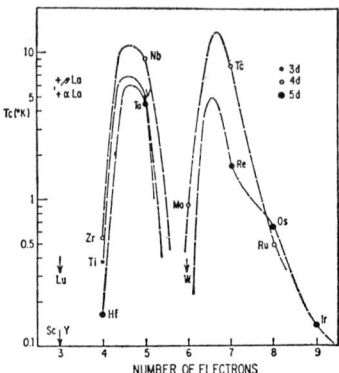

Fig. 1. Variation in superconductivity of the transition elements in the periodic system.

tope effect would be expected to depart significantly from ½, contrary to experiment, and (2) the effect of the Coulomb interactions would be reduced so much that nearly all metals would be expected to be superconducting." Since then we have shown experimentally the absence of an isotope effect in ruthenium and also the deviation from the factor ½. It has also been shown experimentally that most metals are superconducting.

It is not clear to me why suddenly these serious difficulties should no longer exist, as Garland has recently claimed they no longer do (20). On the other hand, the different mechanisms for the occurrence of superconductivity, as well as the general existence of the phenomenon itself, have long since been deduced, though purely empirically.

Summary

In summarizing, let me say again that some of the theoretical conclusions have been verified experimentally. These include the gap in the electron spectrum, the nuclear relaxation in the superconducting state, and the lattice-electron interaction for the nontransition elements, as well as the fact that pairs are involved in the condensation, as shown by quantized flux measurements. However, the predictions of the theory, especially those with respect to the Knight shift in superconductors and the isotope effect for the transition elements, have not been borne out. The theory has since been modified to account for the non-vanishing Knight shift found by experiment, and also for the experimentally determined isotope effect of the transition elements. However, there is no word at all on when and where to find a superconductor in the transition metals.

Unfortunately, the validity of the electron-phonon interaction has been, in my opinion, overstated to the point that a formula for the transition temperature was given which has now failed under the weight of the periodic system. Another result of the BCS formulation of electron-phonon hypothesis is the conclusion that superconductivity is the result of a more or less delicate balance between the electron-phonon interaction and the Coulomb repulsion. For this reason most theories have considered superconductivity an accident rather than the normal ground state of most metals. However, experimental results indicate that superconductivity is indeed a very general phenomenon. In contrast to the BCS formulation, Fröhlich, in paragraph 7 of his review article (1), has proposed that the occurrence of superconductivity as well as that of superfluidity might be understood in a very general manner as a result of an appropriate quantum hydrodynamics.

The difference in the transition temperatures of yttrium and lanthanum cannot be understood on the basis of an electron-phonon interaction, but it has been explained on the basis of a magnetic interaction.

The transition elements have many properties which can be symmetrically arranged according to the elements' position in the periodic system. The behavior of the transition temperature at which superconductivity occurs is another such symmetric function—a fact which gives added credence to the generality of the phenomenon. Since the rule that the maximum transition temperatures should occur when the number of valence electrons per atom is around 5 or 7 is fairly simple, it would seem that the correct underlying explanation should be equally simple. I believe that it has yet to be given. Should it ever be stated, we might then expect an answer to our second question: Why has it been relatively easy, within the last 10 years, to reach transition temperatures of 17° to 18°K in many intermetallic systems and impossible to raise this value even by as little as half a degree? It is not that we have not tried. More than 6000 metals have been checked; 18°K has been reached very often but never exceeded.

Another result has emerged from the experimental investigation of all these metals. As stated before, we found ferromagnetism in two compounds which had been expected to be superconducting—namely, $ZrZn_2$, a compound composed entirely of superconducting elements, and Sc_3In, wherein, to date, only indium has been found to be superconducting. The extreme rarity of these occurrences indicates that ferromagnetism is a much less likely kind of condensation than superconductivity. While there are many superconducting compounds formed entirely of nonsuperconducting elements, all ferromagnetic compounds, with the exception of the two mentioned, contain either chromium, manganese, iron, cobalt, or nickel or an f-electron element. Superconductivity can be predicted empirically on a routine basis today with very few exceptions. However, prediction of ferromagnetism is not possible at present, since the incidence of 2 in 6000 is rather unfavorable. Perhaps in this case we have not been looking in the right direction, and have thereby given the theory a chance to precede the results of experimental search.

Where do we stand with respect to cooperative phenomena in general? As I have tried to show, ferromagnetism is a rare and restricted phenomenon in comparison with superconductivity. In the field of ferroelectricity there was a superstition for 20 years that the occurrence was due to one specific mechanism, that of the hydrogen bond. Today we know that ferroelectricity is a general phenomenon. We have a great number of ferroelectrics, and the mechanisms responsible for the effect range from order-disorder among hydrogen bonds to the polarization of sulfur. Thus, I think ferroelectricity is better understood theoretically than superconductivity is.

To summarize in the form of a final question, to which the ultimate and final answer must come only from the theory: Will the electrons in any and every metal that is sufficiently pure and cold always undergo a condensation? If so, in a particular metal, at what temperature will condensation occur and of what type will it be? Clearly, we do not yet expect any quantitative answer to the problems of superconductivity. As Anderson mentions, not even melting points can be calculated today. Melting, however, is a universal phenomenon, and every material, when heated above 4000°C, will eventually either melt or sublime. It seems that a similar answer should also be given for electrical conductivity as a general phenomenon. When sufficiently cold and pure, what will a metal do? I think it will usually become superconducting.

B. T. MATTHIAS

References and Notes

1. H. Fröhlich, *Rep. Progr. Phys.* **24**, 1 (1961).
2. B. T. Matthias, T. H. Geballe, V. B. Compton, *Rev. Mod. Phys.* **35**, 1 (1963)
3. J. Bardeen, *Phys. Rev.* **79**, 167 (1950); ———, *ibid.* **80**, 567 (1950); H. Fröhlich, *ibid.* **79**, 845 (1950); ———, *Proc. Phys. Soc. London* **A63**, 778 (1950).
4. J. Bardeen, L. N. Cooper, J. R. Schrieffer, *Phys. Rev.* **108**, 1175 (1957).
5. M. R. Schafroth, *ibid.* **96**, 1442 (1954); M. R. Schafroth, S. T. Butler, J. M. Blatt, *Helv. Phys. Acta* **30**, 93 (1957).
6. H. Fröhlich, *Proc. Roy. Soc. London* **A223**, 296 (1954); M. A. Biondi and M. P. Garfunkel, *Phys. Rev.* **116**, 853 (1959); P. L. Richards and M. Tinkham, *ibid.* **119**, 575 (1960).
7. A. Brown, M. W. Zemansky, H. A. Boorse, *Phys. Rev.* **92**, 52 (1953).
8. L. Onsager, *Phys. Rev. Letters* **7**, 50 (1961).
9. B. T. Matthias, *Phys. Rev.* **97**, 74 (1955).
10. P. Morel and P. W. Anderson, *ibid.* **125**, 1263 (1962).
11. H. B. G. Casimir, *Z. Physik* **171**, 246 (1963).
12. H. Fröhlich, *Phys. Letters* **7**, 346 (1963).
13. D. C. Hamilton and M. J. Jensen, *Phys. Rev. Letters* **11**, 205 (1963).
14. T. H. Geballe, B. T. Matthias, G. W. Hull, Jr., E. Corenzwit, *ibid.* **6**, 275 (1961).
15. B. T. Matthias, T. H. Geballe, E. Corenzwit, G. W. Hull, Jr., *Phys. Rev.* **129**, 1025 (1963).
16. B. T. Matthias, V. B. Compton, H. Suhl, E. Corenzwit, *ibid.* **115**, 1597 (1959).
17. B. T. Matthias, E. Corenzwit, W. H. Zachariasen, *ibid.* **112**, 89 (1958).
18. A. I. Akhiezer and I. Y. Pomeranchuk, *Zh. Eksperim. i Teor. Fiz.* **36**, 859 (1959); A. I. and I. A. Akhiezer, *ibid.* **43**, 2208 (1962); I. A. Privorotskii, *ibid.*, p. 2235, J. Kondo, *Progr. Theoret. Phys. Kyoto* **29**, 1 (1963).
19. J. Bardeen and J. R. Schrieffer, in *Progress in Low Temperature Physics*, C. J. Gorter, Ed. (North-Holland, Amsterdam, 1961), vol. 3, p. 203.
20. J. W. Garland, Jr., *Phys. Rev. Letters* **11**, 111 (1963); *ibid.*, p. 114.
21. I thank T. H. Geballe for his many good criticisms and stimulating discussions, and, last but not least, for being with me in a field that is not greatly illuminated at present by the gray world of theory.

Coherent Matter Field Phenomena in Superfluids
Some Recent Definitions in the Basic Sciences
(Belfer Graduate School of Sciences, Yeshiva Univ., New York, 1969),
ed. A. Gelbart, Vol. 2, 1965–66.

A first shot at a general theory of broken symmetry and at the implications of the Josephson effect; as such much shorter than my article in Gorter's *Progress in Low Temp. Physics* of 1967. The discussion remarks are important: they refer back to a paper (not in my bibliography) taking issue with a remark of Gorkov and Galitskii, and reflect my growing conviction that broken symmetry has a lot to say about the fundamentals of quantum mechanics and vice versa.

COHERENT MATTER FIELD PHENOMENA IN SUPERFLUIDS

P. W. Anderson

BELL TELEPHONE LABORATORIES, MURRAY HILL, NEW JERSEY

It is ironic that during the same years that the elementary-particle physicists, at the highest energy end of the spectrum, have been doing their best to discard as many as possible of the properties of the quantum field operator, at the very lowest energy end of the spectrum, in the two low temperature phenomena of superconductivity and superfluidity, we have been provided with a series of very direct demonstrations which seem to bring the quantum particle field almost into the ordinary, tangible, macroscopic realm. This will be the burden of what I will be trying to say today. In these phenomena the quantum particle field plays a role very similar to the roles, with which we are familiar, of the classical fields, the electromagnetic and gravitational fields, in our ordinary macroscopic experience. This has been shown by a series of experiments of various kinds on coherence in quantum fluids. I would like here to emphasize more the basic meaning of these experiments than to describe in detail their results and their experimental mechanisms.

It is interesting that nowhere in this discussion will I have to treat in very much detail the microscopic theory of either of these two superfluids, either the superfluid, helium, or the superconducting electron gas in metals. In fact, the foundations of the subject of quantum coherence were laid long before there were any acceptable microscopic theories of these phenomena. One could even argue that there is no acceptable microscopic theory of helium to this day. These foundations were laid by Landau, Penrose, Onsager, London and many others. Even the two types of macroscopic quantization, the quantization of vorticity and the quantization of magnetic flux, were proposed by 1949, long before

21

there were any microscopic theories. The first approach to the properties of superfluids and superconductors which contained the possibility of understanding these phenomena formally and theoretically was, however, proposed in 1951 and 1956 by Penrose and Onsager [1]. This is still, in a sense, equivalent to the accepted way of discussing these phenomena, and since it is historically the first and possibly the best known, it seems to be a good starting point for our discussion today.

The proposal they made was that one could essentially define superfluidity of a Bose system such as liquid He as a state in which the density matrix of the system factorized in a certain special way. They said that the density matrix $\rho = \rho(r, r') = \langle \psi^*(r)\psi(r') \rangle$, which is defined as the average of the product of the field creation operator at point r and the destruction operator at point r', could be factorized into $f^*(r)f(r')$, a function of r times a function of r' plus small terms. The density matrix $\rho(r, r')$ of any ordinary system tends to vanish as the points r and r' attain macroscopic distances from each other. The "small terms" are normal, i.e., go to zero as $|r - r'|$ goes to infinity. The term which has no dependence on the relative positions of r and r' is characteristic of superfluidity. So the statement is that this is equivalent to superfluidity; I hope I will be able to make that point clear in the rest of the talk. I should say, of course, that at absolute zero the average is taken in some hypothetical ground state of the system; on the other hand, at finite temperatures one does not use ground states but a thermal average.

Beliaev [2] extended this concept of factorization to the time-dependent Green's functions. He observed that the density matrix, which is the average of the time independent field operators, is a special case of the general Green's function, which is the average of the Heisenberg field operators $\langle \psi^*(r, t)\psi(r', t') \rangle$. He assumed that that could be factorized into $f^*(r, t)f(r', t')$ + small terms. Almost immediately and in a practically succeeding issue of the *Journal of Experimental and Theoretical Physics*, Gor'kov [3] observed that the same kind of theory could be generalized to apply to the electrons in superconductivity. It was not possible to have this kind of theory for just the single field operators because of the exclusion principle, but if one substituted instead the two-particle Green's function, which has a pair of field operators ψ^*, and a pair of ψ's, one could do the same thing for the fermions. One could talk about the fermion gas in the metal in terms of pairs of electrons which then have sufficiently similar commutation relations to bosons so that one can carry out the same kind of thing.

So Gor'kov introduced a function $F^*(X_1, X_2)$ of two variables X_1 and X_2 (including both time and space variables) and factorized his two-particle Green's function, a function of four variables, into pairs of two variable functions:

$$G(X_1, X_2, X_3, X_4) = \langle \psi^*(X_1)\psi^*(X_2)\psi(X_3)\psi(X_4)\rangle$$
$$= F^*(X_1, X_2)F(X_3, X_4) + \text{small terms}$$

Again the important term is not correlated between variables of the two different F's, but now one has a strong correlation between X_1 and X_2. One thinks of F^* as a kind of a two-particle wave function, and the two bound electrons in such a wave function should remain close together. Gor'kov showed that this theory, if one took F to be homogeneous in space, was completely equivalent to the then just developed Bardeen–Cooper–Schrieffer (BCS) theory of superconductivity, which now, of course, has come to be recognized as the correct microscopic theory of superconductivity [4]. As far as the superfluid (helium) version is concerned all reasonably useful microscopic theories that have been proposed have led to this kind of behavior but, as I say, no really satisfactory microscopic theory yet exists.

In superconductivity there already existed a theory of Ginsburg and Landau [5], which seems to describe superconductivity from a phenomenological point of view very satisfactorily. This theory describes the properties of superconductors in terms of a function which they called the order parameter function $\Psi(r)$, but the basic equations of the theory were remarkably similar to the equations for ordinary one-particle quantum mechanics in terms of a single-particle wave function $\Psi(r)$. So this function came to be known as the wave function although its meaning was not at that time understood. Gor'kov very shortly showed that in the appropriate limiting cases he could derive the Landau–Ginsburg theory from his theory [6], in which he identified $\Psi(r)$ with the function $F(X, X)$ at identical values of the variables. So he gave the function Ψ a physical meaning. It was only later that the corresponding theory for superfluidity, which is not quite as useful, was proposed by Gross and Pitaevskii [7].

Coherence phenomena can be treated perfectly satisfactorily using this Penrose–Onsager "off-diagonal long-range order" scheme. Unfortunately, all of those who until recently have chosen to work with this theory have made an additional assumption which is not necessary, and which makes the amount of work one has to go through in order to derive results considerably greater. This assumption is, that one should work always with states (in making the averages with which one forms either the density matrices or the Green's functions), with fixed definite total particle number N, i.e., total number of helium atoms or total number of pairs of electrons. In the case where one does the thermal average over a grand canonical ensemble, it is sometimes convenient, of course, to use many values of the particle number, but even there it was assumed that there was no coherence between states of different particle number. Now, so long as one is talking about a single isolated superfluid system, a single hunk, say, of superconductor or a single bucket of superfluid with a cover on it so that nothing evaporates, then, of course, the number of helium atoms or the number of electron pairs is perfectly fixed and the Hamiltonian must commute with the number operator. It seems extremely reasonable that, since the number of particles is a perfectly good quantum number, we should insist, in taking our average, on keeping the total number of particles fixed.

Unfortunately, it begins to be a little more difficult if we start trying to do interference experiments. In doing the kind of experiment we are going to talk about, we will be thinking very characteristically of systems which consist of two separate parts between which we have some kind of connection, and the basic feature of that connection is that it can transfer particles coherently from one side of the system, or one piece of the system, to the other. We will, in other words, be talking about systems such as those in which we have two regions of space separated by "slits" as you do in an electromagnetic interference experiment. What is more, it will be very hard, in discussing these coherence experiments, as it would be very hard in discussing for example electromagnetic interference experiments, to keep the slits open at all times. You would often like to compare what happens when you close off a slit with what happens when the slits are open. You would like to be able to turn off or on switches, superconducting switches or superfluid switches, here and there in your system. And under those circumstances, there is no reason whatever to expect that you are going to get satisfactory results by working with states in which the numbers of particles on the two sides are absolutely fixed and in which there are not allowed to be coherent superpositions of states in which the particles are differently partitioned between the two sides.

Let me suppose I have a system composed of two pieces connected by such a switch, and that I start with each piece having ODLRO, i.e., being described by

$$G(x, x') = f^*(x)f(x'),$$

but each having just $N/2$ particles. It turns out that there is no way that I can make up from this description the correct description of the system with the switch open, i.e., with ODLRO embracing the full composite system, in which x or x' may range over both halves. Thus no separate piece of a superfluid system is adequately described by a single state with ODLRO. Of course, it is possible to do things properly, by starting from a supermacroscopic system which has definite fixed particle number, and which contains every piece that I am ever going to attach to my system, i.e., every superconductor that I may ever bring into contact with my system or every reservoir of liquid helium that I may be interested in. But that is just inconvenient.

It is much easier to do it in a different way. The different way is to satisfy our original requirement of factorization by taking as our basis states (confirming afterward that what we do is all right), states in which there is an average of the particle operator itself.

$$\langle \Psi | \psi^*(r, t) | \Psi \rangle = \langle \psi^*(r, t) \rangle = f^*(r, t)$$

Clearly, when we do this our original assumption of ODLRO is satisfied:

$$\langle \psi^* \psi \rangle = \langle \psi^* \rangle \langle \psi \rangle + \text{other terms},$$

but equally clearly, the reverse is not necessarily true.

In other words, this implies that I still have my original assumption but I have made an extra assumption in addition. Clearly also these are states which do not have a fixed number of particles because I have destroyed a particle between one Ψ and the other, so at least the state has components with N and with $N - 1$ particles, i.e., with two different numbers of particles. These states that I am going to use are very similar to, except for certain special features, the coherent states of the electromagnetic field with which people have begun to realize it is best to work in quantum optics. On the other hand, the states that we are confronted with in superfluidity and superconductivity do not have the simple complete coherence that one can get in the electromagnetic analogy. The maximum value that the mean value of Ψ can have is the square root of the density, because $\langle \psi^* \psi \rangle = \rho$, but in the helium case, it was shown in the original paper of Penrose and Onsager (and the estimate has been refined recently by MacMillan) that one gets of the order of $0.1\sqrt{\rho}$ for the absolute value. In the superconductivity case, the situation is even less coherent, one gets of the order of 10^{-4} of the maximum possible value. Nonetheless, these numbers are clearly recognizable as numbers or order unity rather than numbers of order 10^{-23}, and therefore the coherence is macroscopic in magnitude rather than microscopic.

Another way of putting it is that we all recognize that when we deal with macroscopic systems such as tables, chairs, solar systems and so on, it is extremely inconvenient to work with eigenstates of the conserved quantities such as total momentum, total parity, or anything like that;

it is much more convenient to work with wave packets of the eigenstates which have the property of localizing reasonably closely the macroscopic variables of the system, and that is exactly what we are doing here. One can make wave-packet transformations back and forth and demonstrate that we are working with the same kind of system.

The only condition we have is that someone sometime has to sit down with the equations and decide that the zero-point motion that we are ignoring in forming our wave packet, the diffusion of the wave packet, is quantitatively negligible compared with other fluctuations in the system. You have to decide, when you are talking about the solar system, that the uncertainty in the position of the sun is utterly negligible. In this case, you have to decide when you are talking about a macroscopic superfluid that the uncertainty in definition of the wave function of the quantum field is essentially negligible and it will not diffuse. I have done that calculation [8] and it does indeed turn out that even in the worst possible cases the zero-point motion is negligible compared to thermal fluctuations.

Now after a long prologue we come to the basic idea of what I am going to do today, which is to deal with the mean value of the quantum particle field as though it were one of the macroscopic thermodynamic and dynamical variables of the system, just as in, say, elasticity of crystals we deal with the position of a particular part of the crystal as a macroscopic thermodynamical variable, or in ferromagnetism we deal with the magnetization as a macroscopic variable. Both are, in fact, variables which destroy symmetry principles but are well known by our daily experience to be usable as thermodynamic variables. As in these cases, it is perfectly possible to define the variable not only for the total system but also for individual pieces of the system, small cells macroscopic from the atomic point of view but microscopic from our point of view. In other words, one can deal with coarse-grained averages as well as with macroscopic averages.

It is quite important that f, the mean value of the wave field, is a complex order parameter. It contains an absolute value and a phase:

$$f = |f| e^{i\varphi}$$

The two real parameters $|f|$ and φ turn out to be two different kinds of thermodynamic variables. In the thermodynamics of order–disorder systems, for example, we have a kind of thermodynamic variable, the long-range order in that case, which is characterized by the fact that while it is extremely convenient in discussing the thermodynamics and in keeping our concepts straight, there is no way we can get in there with a force and act on that particular variable. It is hard to find a force that pulls differently on copper and gold atoms. Similarly in this case, the magnitude of the mean value of the quantum field is not a variable that we can get in and manipulate. So this is not of much value to us in discussing macroscopic dynamics. The phase φ of the mean field, on the other hand, has the opposite property. It is a real dynamical variable such as the ones I gave as examples previously, the magnetization or the position of a piece of crystal. It is a variable corresponding to which there are forces, and thus is simultaneously a dynamical and a thermodynamical variable. The fact that it is a dynamical variable follows from the fact that the number operator N_i which corresponds to any one of our partial systems is the conjugate dynamical variable to the phase operator φ_i for the corresponding system. That is, the commutator of these two operators is the standard commutation relationship of conjugate dynamical variables, just like the P and Q of particle dynamics: $[N, \varphi] = i$. This is easily verified in a number of ways. Perhaps the simplest possible way is to ask oneself how one would go about forming, from wave functions with fixed numbers of particles, wave packets with fixed phase. That is fairly easy: we take our wave functions with fixed number, we sum over them in order to form a wave packet, we multiply by some real coefficient a_N which has a peak for some value of N and which is non-vanishing over a sufficiently broad region in order not to get into serious trouble with the uncertainty principle, and then we multiply by the coefficient $e^{i\varphi N}$:

$$\Psi(\varphi) = \sum_N a_N e^{i\varphi N} \Psi(N).$$

It is clear that in forming the mean value

$$\langle \psi \rangle = (\Psi(\varphi), \psi \Psi(\varphi)),$$

the term on the left with phase $e^{-i\varphi(N-1)}$ connects only with that on the right with phase $e^{i\varphi N}$, so that

$$\langle \psi \rangle = |f| e^{i\varphi}.$$

Thus the transformation matrix between phase and number operators is $e^{i\varphi N}$, just as for p and q. This implies the commutator, and also that we may write the operator equivalences

$$N = i\, \partial/\partial \varphi \qquad \varphi = -i\, \partial/\partial N.$$

There are limitations on the usability of these commutation relations but those are relevant only for very small systems, and we are talking specifically about a large system, so that we can really say this is a precise relationship. In addition this gives us an illustration of how to form wave packets for our system.

Now the essential dynamical equations for the whole phenomenon of superfluidity follow from these two ideas: that we are going to take states in which there are mean values of the particle field, and that the phase and the number operators are conjugate dynamical variables. The equations those lead to are, of course, the equations of motion of these two dynamical variables. We can write down the equation of motion of the number operator for one of our pieces of a dynamical system, connected by a superfluid connection to another system. Let the number and phase operators of the two systems be N_1, N_2 and φ_1, φ_2.

The equation of motion of the number N_1 is given by the standard quantum mechanical equation of motion, which, using the fact that these are conjugate dynamical variables, can be written

$$i\hbar \dot{N}_1 = [\mathcal{H}, N_1] = -i\, \partial \mathcal{H}/\partial \varphi_1,$$

and correspondingly we have the other Hamilton's equation for this pair of conjugate variables

$$i\hbar \dot{\varphi}_1 = [\mathcal{H}, \varphi_1] = i\, \partial \mathcal{H}/\partial N_1.$$

These equations are essentially equivalent to two equations which are familiar in the theory of superconductivity. The first is more or less equivalent to the London–Ginsburg–Landau current equation ([5], [9]), the equation which determines the supercurrent in a superfluid system. The second is an equation which has had a shorter history, although a similar equation has occurred in both the theory of superfluidity and the theory of superconductivity. It is equivalent to London's acceleration equation [9], telling how the system responds if we apply forces. As I have written them down, these are equations for operators, but they hold, of course, for mean values; so these are equally good equations for the mean value of the number operator and for the mean value of the phase in our wave packets.

The current equation is historically the first, and the most straightforward of the interference experiments that I promised to talk about are based on it. Of course, for an isolated system, as I said, the Hamiltonian commutes with the number operator, i.e., the Hamiltonian must obey gauge invariance, and be independent of the total phase. That just says, of course, that the number of particles moving into or out of an isolated system is zero. The situation is quite different if we talk now about two separate pieces of a superfluid system. Now imagine that we have two separate pieces of a superfluid system connected by a superfluid bridge (we must always remember that when I say superfluid I mean either helium or superconductors, in which case I am talking about pairs of electrons). Our very definition of superfluidity has said that we have a mean value of the quantum field operator in this whole system. If the energy were independent of the relative phase of the quantum field operator on the two sides I would not have such a mean value because on the average the two phases would fluctuate completely independently. So by my very definition of superfluidity I have said that there must be a dependence of the total energy on the relative phase on the two sides; there must be an energy which must be a minimum where the phases are equal, and must increase on each side of that minimum. Incidentally, of course, this tells me why superfluidity and superconductivity are low-temperature phenomena, because as thermal fluctuations increase, eventually at high enough temperatures the system is going to overcome this energy (which can be traced in every case to the quantum zero-point energy of the individual particles).

If, then, there is a dependence of the energy on $(\varphi_1 - \varphi_2)$:

$$U = U(\varphi_1 - \varphi_2),$$

we can have supercurrents between the two subsystems:

$$\frac{dN_1}{dt} = -\frac{dN_2}{dt} = \frac{1}{\hbar} \frac{\partial U(\varphi_1 - \varphi_2)}{\partial (\varphi_1 - \varphi_2)}.$$

We can equally well apply this to two small, semimacroscopic neighboring cells within an individual bucket of superfluid and we have again the same equation but in that case we get that the current will be the variational derivative of the energy with respect to the gradient of the phase:

$$J_s = \frac{1}{\hbar} \frac{\delta U}{\delta(\nabla\varphi)}.$$

This is the course-grained equivalent of the previous equation. Again we expect that the energy as a function of the gradient of the phase must be a minimum when the gradient is zero, and that it will be a quadratic function for small enough values of the gradient:

$$U = n_s(T)(\hbar^2/2m)(\nabla\varphi)^2.$$

As the coefficient it is convenient to put in $\hbar^2/2m$ times a perfectly arbitrary function $n_s(T)$ of the temperature and the internal state of the system, and then the current is the variational derivative of this with respect to the gradient of the phase and so may be written

$$J_s = n_s(T)(\hbar/m)\nabla\varphi.$$

Well, it is clear now why I have written the coefficient the way I did. I wanted to get a quantity which looked as though it was a velocity, a superfluid velocity $v_s = \hbar/m \, \nabla\varphi$ multiplying a superfluid number of particles. At this point, this is just a definition of the superfluid velocity. At absolute zero, in a perfectly homogeneous system, we can prove, in fact, that this is the velocity of the superfluid and that $n_s(T)$ is equal to the total number of particles or of electron pairs in the system. You prove that very simply by taking the perfectly homogeneous system, and translating it at a certain velocity, which must then be the superfluid velocity. I am aware of no proofs that v_s defined in this way means anything else in any real system, either superconducting or superfluid. In any real bucket of helium, any experimentalist like myself who has worked with it will tell you that there are always counterflows of normal and superfluid particles, and in any real superconductor there are impurities. Even at absolute zero in the presence of impurities n_s is not identically equal to N. Thus this is an idealization and one's reasons for calling this v_s are reasons basically of convenience rather than necessity. It is sometimes but not always a good approximation of the actual particle velocity in the system. But it is the current equation which is the basic equation.

In the case of superconductivity, another somewhat more familiar route may be taken to this equation. In the case of superconductivity you can recognize immediately that I have told you a whopper when I said that the energy is a functional of the gradient of the phase of the mean value of the wave field operator because that, of course, is not a gauge invariant statement. In the case of superconductivity my particles are charged; there is an electromagnetic field present and I can change the phase of the wave function arbitrarily if I at the same time change the value of the vector potential of that electromagnetic field. So in order to maintain gauge invariance I have to write instead that U is a functional of $(\hbar \nabla\varphi - 2e\mathbf{A}/c)$ where $\mathbf{A}$ = vector potential. $2e$ occurs in the factor rather than e because φ is the phase of a two-particle wave function. Then, of course, I take the well-known identity that the current is equal to $1/c$ times the variational derivative of the energy with respect to the vector potential and that gives me the electromagnetic current in this case.

At this point I would like to make two things clear. The first is that this discussion of the supercurrent already makes it very plausible (and I believe probably it can be made more than plausible by someone more patient than I) that superfluidity as we understand it, the existence of superflows in the absence of driving forces, and this type of coherence within the wave function or the off-diagonal long-range order concept to which it is equivalent, are both necessarily and sufficiently related to each other. Thus we suggest it is impossible to have supercurrents without off-diagonal long-range order and vice versa. The one seems plausible because there is only one variable, A, or the gradient of the phase, to which the current operator is conjugate, and therefore it does not seem that we can get supercurrents unless we have a dependence of the energy on the appropriate variable. Conversely, if we have a dependence of the energy on this variable we must have supercurrents. Now this seems absolutely trivial except that there are papers in the literature during the past year which violate both directions of the implication. There are papers producing hypothetical phases with off-diagonal long-range order which do not have any supercurrents, and there are papers producing hypothetical superfluid phases which have no off-diagonal long-range order, so that it may be worth pointing out that they seem to be necessarily and sufficiently connected.

The second importance of this part of the discussion is that this is the point at which macroscopic quantization begins to enter the picture. The first point at which macroscopic quantization enters is that the $\langle \psi \rangle$ is to be taken as a macroscopic variable of the system, and therefore it is a physical quantity. Thus it is not by any means to be taken as multiple valued. Therefore, as you follow this quantity around some circuit within which you have, say, two bridges between two superconducting samples (i.e., some multiply connected circuit) the value of that quantity must come back to the same value when we return to the same point. The phase, however, need not return to the same value: it can return to the same value $+2\pi$ or 4π or any number of integers times 2π. So we immediately get that there is a kind of quantization of a circuit integral. As we go around any closed curve within our superfluid, the integral of the gradient of the phase must be equal to $2n\pi$. And that is the basic quantization statement we have to make. Now in the case of superfluidity, as I have said, we often take the superfluid velocity to be $\hbar/m\, \nabla\varphi$. So we immediately get that the line integral of the superfluid velocity around a closed curve is equal to nh/m, where n is an integer. So we get quantization of the circulation of the superfluid velocity. The question as to whether actual physical velocities are quantized in this way is a question of how good this assumption about the superfluid velocity is, and it turns out to be a pretty good one. But as far as I know there is no reason to say that it is exact.

The quantization of the magnetic flux in a superconductor comes about in a similar way but not exactly the same. We again take the formula for J_s which, using the similar quadratic formula for the energy, is $J_s = n_s e v_s$ and $v_s = \hbar/m\, \nabla\varphi - 2e/mc \mathbf{A}$. Realizing that in most macroscopic superconducting circuits the superfluid velocity inside the material is practically zero because of electromagnetic shielding (i.e., the London penetration effect) we can usually set it equal to zero. Then we go through our circuit integration again and we find that the line integral

$\oint A\,dl$ is equal to $hc/2e$. This line integral in turn is easily shown to be the magnetic flux. Again this is an approximation because we have had to use a circuit in which there is no supercurrent and in many instances, which have actually been checked out experimentally, the sample can be made so small that there is supercurrent at every point. That inexactness in the macroscopic quantization is thus one which can actually be proved experimentally. So it is important to remember that the quantization in the sense of the line integral of $\nabla\varphi$ is, of course, exact but the quantization of the usual quantities that are discussed is rather inexact.

Now, let us return to talk about the actual interference experiments. In order to do this, we have to go back from these more sophisticated continuum results involving U as a functional of the gradient of the phase to the trivial question of what happens when we have two superfluid or superconducting samples connected by a superconducting bridge, which can be represented by an energy which depends on the relative phase on the two sides. By general principles you can see that the dependence of the energy on the relative phase must be periodic because there is no way the system can tell whether the phase difference is zero, 2π, etc. as long as there is only the single bridge. Therefore, in this case, we expect the function $U(\varphi_1 - \varphi_2)$ to be periodic. Now there are two ways in which this periodic function can be constructed. The way in which ordinary macroscopic samples behave is very disappointing. There is one energy parabola $U \sim (\varphi_1 - \varphi_2)^2$ at the origin, another centered at $\varphi_1 - \varphi_2 = 2\pi$, etc.:

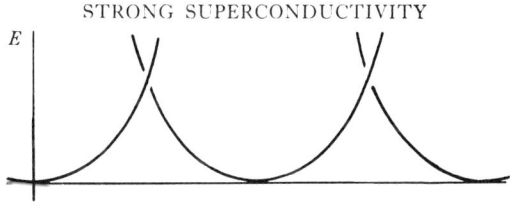

Figure 1

For reasons which are easy enough to understand but that I do not want to go into here, when you actually start working with this system, it tends almost always to pass through the intersections between the parabolas without paying any attention to the other parabola, so that it is not possible to get from one of these states to another easily. The tremendous contribution that Josephson [10] made to the subject is the fact that he discovered a way in which this periodic energy curve could be made to be simply sinusoidal. It did not then have any of these annoying metastable states up at high energy so that the system would follow the sinusoidal behavior as one changed the phase. The original Josephson junction was simply a tunneling barrier between two pieces of superconductor separated by an extremely thin film of insulator, but now it is realized that simply extremely thin wires or extremely thin bits of superconducting film behave in much the same way. Almost all the macroscopic interference experiments make use of this kind of junction between two pieces of superfluid, except for the original macroscopic flux quantization experiment by Doll–Näbauer [11] and Deaver–Fairbank [11].

What is the simplest way in which this could be used? We have two pieces of superconductor connected by an insulating film, and in order to arrive at this energy curve as a function of phase, I have had to assume that the phase difference is a certain fixed value everywhere in the junction. That need not be so, however, if I insert a magnetic field through the insulating barrier and parallel to it. Then the field is given by the curl of the vector potential so I can say that there is a vector potential perpendicular to the barrier.

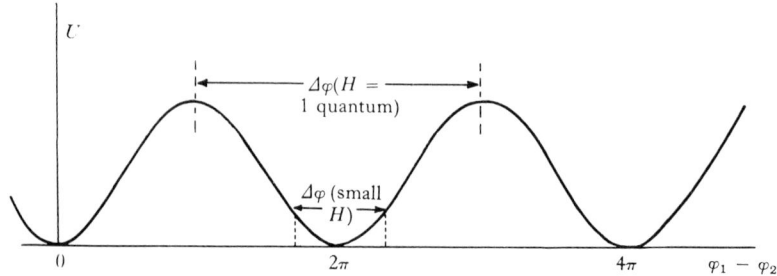

Figure 2

Thus the phase difference is a linear function of x. (Figure 3) Then if we draw the sinusoidal phase-energy curve we see that for small enough $H (\ll 4\lambda we/hc$, it turns out, where w is the width of the junction and λ the penetration depth) the phase difference is smeared only over a small portion of a cycle and the total energy is still sinusoidal; but when H is big enough to cover a full cycle, there is no longer any total dependence of energy on phase; it encompasses a whole 2π of the phase axis and then the energy clearly does not depend on where I am along the curve. (see figure 2) Therefore, since the total current is the derivative of the energy with respect to the phase, the total current will be exactly zero. In fact you can follow through the argument and convince yourself that the total current follows a perfect oneslit interference pattern. This experiment was the first check by Rowell and myself [12] that we were really seeing this kind of junction.

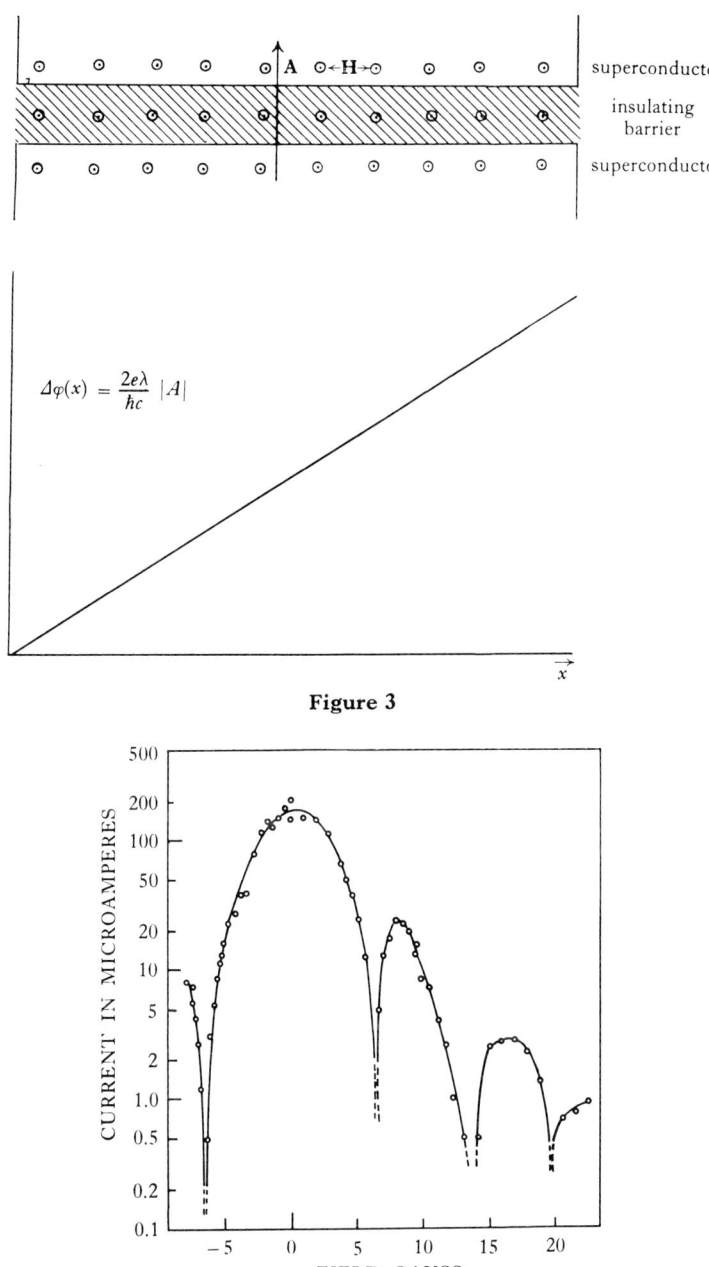

Figure 3

$$\Delta\varphi(x) = \frac{2e\lambda}{\hbar c}|A|$$

Fig. 4. The field dependence of the Josephson current in a Pb–I–Pb junction at 1.3°K.

Once you are convinced that you can do that then you can immediately see—we should have but Mercereau did—the next experiment to do. The next experiment to do is to make two of these things and separate them by a hole. In other words, I connect two superconducting reservoirs by two Josephson junctions and have a hole in the middle through which I pass a magnetic field; of course, the magnetic field also

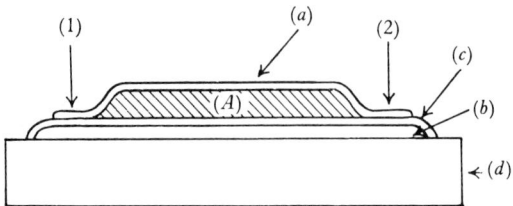

Fig. 5. Cross-section of a Josephson junction pair vacuum-deposited on a quartz substrate (d). A thin oxide layer (c) separates thin (~ 1000 Å) tin films (a) and (b). The junctions (1) and (2) are connected in parallel by superconducting thin film links forming an enclosed area (A) between junctions. Current flow is measured between films (a) and (b).

had to be in the two Josephson junctions. In that case, I get the original pattern for each of the junctions but the phase at one is correlated with the phase at the other depending on how much of the vector potential comes from the magnetic field in the hole. So as I increase the magnetic field, since the hole is much bigger than the junction I get a two-slit interference pattern with a one-slit interference pattern as its envelope. This was checked by Mercereau et al. [13].

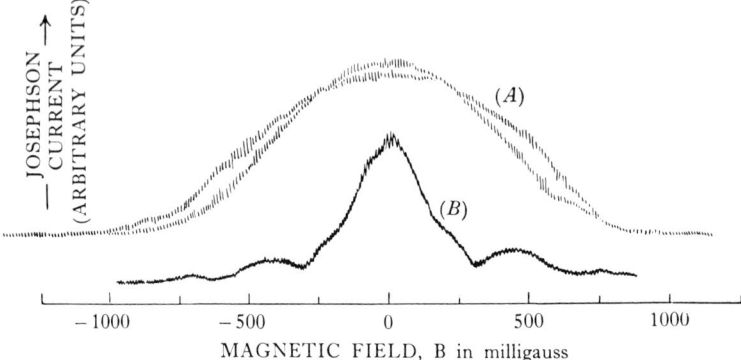

Fig. 6. Josephson current versus magnetic field for two junctions in parallel showing interference effects. Magnetic field applied normal to the area between junctions. Curve (A) shows interference maxima spaced at $\Delta B = 8.7 \times 10^{-3}$ G, curve (B) spacing $\Delta B = 4.8 \times 10^{-3}$ G. Maximum Josephson current indicated here is approximately 10^{-3} A.

The next experiment, also carried out by Mercereau *et al.*, was to check the idea of Bohm and Aharonov that the vector potential plays a special role in quantum mechanics relative to classical mechanics. They wanted to check that one could get an interference phenomenon without there being any actual magnetic field acting on the electrons so they made the same device but now enclosed the magnetic field in a solenoid, and insofar as possible there was no magnetic field from the solenoid in any other parts of the apparatus. Sure enough, now the one-slit pattern is gone, but one gets a simple diffraction pattern up to as large magnetic fields as they used [14]. Now, finally, Mercereau and his co-workers at Ford used the same thing as a very sensitive ammeter [15]. They carried out the final possibility, which is to make two Josephson junctions but now one connects the two sides by wire which is wound noninductively. So now they tested the supercurrent of the junctions in parallel in order to test the relative phase of the two junctions. This then is testing the Josephson current as a function of the superfluid velocity, i.e., as a function of $\nabla\varphi$, where $\nabla\varphi$ is caused by a superfluid velocity rather than by a vector potential. Sure enough one gets a nice interference pattern, in fact one can get several orders of magnitude higher sensitivity as a microammeter than with any other device known to man.

It is now interesting to go on to the other of these two equations, the acceleration equation. So far I have not said anything about what really makes the currents flow. How do I accelerate currents in the superfluid? Of course, the answer must be this equation. What is the first thing this equation tells us? Let us again suppose that we are talking about a set of neighboring microscopic cells in the uniform sample. Let us take the gradient of this equation, and a macroscopic, thermodynamic mean value:

$$\nabla \langle \hbar \, d\varphi/dt \rangle = \nabla \langle \partial \mathscr{H}/\partial N \rangle$$
$$m \, d/dt \, (\hbar/m \, \nabla\varphi) = \nabla\mu$$

Well, in the first place $\langle \partial \mathscr{H}/\partial N \rangle$ is just precisely the definition that we normally use of the chemical potential. In any system, no matter how dreadfully off equilibrium it may be, this is the chemical potential. The gradient of the chemical potential is the accepted definition of the force on the system, so what this tells you is that the superfluid velocity obeys the equation of motion $m\dot{v}_s = F$. That is why I call it the acceleration equation. There is, then, only one way in which we can have an actual difference in potential across a superfluid, and that is to allow acceleration to take place. We can do that very directly: for instance, suppose I had two buckets of liquid helium with an opening between them, I could make the difference in head on the two sides finite but I would know then that unless something else happened, the liquid helium would accelerate and one would get just U-tube oscillations; one would never get any dissipation.

Now fortunately, or perhaps unfortunately, that is not the only way in which acceleration can take place.

The other possibility is the relatively new idea of phase slippage by vortex motion [16]. In order to have a phase difference, a change in the phase around a circuit, I have to have a hole in the middle of it, otherwise I am going to have a point which is a branch point of the macroscopic particle field. I do not like that, but I have to assume that there may be situations in which one has to introduce branch lines of the quantum particle field. In fact, the simplest such arrangement I can introduce in order for the phase not to be single valued within a single macroscopic sample of superfluid is a line at which you take the mean value of the particle field to be zero, and then around that branch line I can allow the phase to have a variation of 2π: around that branch line I have $\oint \nabla \varphi \, dl = 2\pi$. This object is called a vortex core and occurs both in superconductivity and superfluidity. In superconductivity there is often a quantum unit of flux associated with it [17]. In superfluidity it is the well-known quantized vortex [18].

Now the concept of phase slippage comes from the fact that I can get a time-dependent difference in phase from one point in a sample of superfluid to another by the following procedure. Let us take two points in the superfluid, point 1 and point 2, and a path between them. If I have a quantized vortex line on one side of the path, the phase difference from 1 to 2 may have one value. Now I move my vortex line to the other side of the path; and because I have a branch line of the phase, I now know that my phase integral,

$$\int_1^2 \nabla \varphi \, dl = \varphi_2 - \varphi_1$$

is increased by 2π. If now I am fortunate enough to have another vortex that I can pass through the path, I can increase the phase by 2π again and so on. In other words I can keep the phase varying continuously if I am willing to pass vortices indefinitely through the path between the two points; so that I can have a chemical potential difference between point 1 and point 2 given by

$$\mu_1 - \mu_2 = h \langle dn/dt \rangle,$$

where $\langle dn/dt \rangle$ is the average rate of passage of vortices through the path.

Now, this tells us two things. One: it tells us the conditions under which dissipation can occur in superfluids: it can occur only if there are vortex lines. It turns out that in liquid helium it is very easy to create vortex lines and so one almost always has dissipation of some sort. There are two kinds of superconductors. The first is the so-called soft superconductors, which are very resistant to the presence of vortex lines, and pass supercurrents without any dissipation whatever. The second is the hard superconductors, which contain vortex lines and always have an associated dissipation [19].

Second, and more interesting from the point of view of interference experiments, is the idea of using this expression to connect a potential difference with a frequency. You can see this is like Schrödinger's equation $i\hbar\, \partial\psi/\partial t = H\psi$, i.e., $(i\hbar/\psi)\, \partial\psi/\partial t$ equal to an energy. So I can associate a frequency with an energy difference simply by Einstein's frequency condition, which when it is applied in superconductivity is called Josephson's frequency condition. A typical simplified experiment, not far from the experiment that we actually carried out, is the following. I can take two buckets of helium with a partition between them, with a little hole in the partition, and then I can make a level difference so that the liquid is trying to flow through the hole. As the liquid flows through the hole it turns out that it probably blows smoke rings. In any case it creates vortices and these vortices move continuously out of the hole so if I make a path from the liquid surface in one bucket to the other, vortices pass across that path at a rate determined by the head difference. How can I tell whether they are passing at that rate? I can tell by introducing an ultrasonic transducer which modulates the flow through the hole at the appropriate rate, i.e., I can introduce an a.c. flow and synchronize the motion of those vortices. When the a.c. flow is synchronized with the height difference I would expect a singularity to occur in the flow and that indeed is exactly what we have observed [20]. This is interesting because it is the only one of these interference experiments so far which have been carried out in a precisely analogous way in both superconductivity and superfluidity [21].

What is the conclusion to be drawn from all this? In the first place, probably the presently most interesting conclusion is that our ideas about superconductivity and superfluidity seem to be right. That is particularly interesting in liquid helium, where the microscopic theory is not in as strong condition as the microscopic theory is in superconductivity. In the second place, something which probably is much more important for the more distant future is that there are here the germs of a series of devices which act on an entirely different principle from anything which has appeared on the macroscopic level previously. It is so different, in fact, and so exciting, that, to anyone who is familiar with what is going on, it is intensely frustrating that we have not been able to make more use of them than we have. One of the reasons why I think we have not is that somehow these things are so different from what we are used to in our technology that we do not even yet have the sense to realize what we need them for, but I am sure that somewhere we need them.

QUESTION. Would it be possible for several types of off-diagonal long-range order, several types of condensation to occur simultaneously?

ANSWER. Well, that is an interesting question. That was proposed by Gor'kov and Galitskii. The question is very simple. What they suggested is essentially that you might also have a second term in the factorization, $\langle \psi(r)\psi(r')\rangle = f(r)f(r') + g(r)g(r')$. It does not seem too important in the superfluid case but in the superconducting case $F(r, r')$, which is a function of two variables, might be a d state and the corresponding G might be an s state and you might have a superposition of d and s, say. There was quite a hassle about whether this was possible. It never was completely resolved except by recourse to actually directly soluble models. It is also interesting because it is a question of whether the whole use of the coherence concept the way I did it is correct or whether it might be possible to have a ring with simultaneously two types of long-range order: two values of quantized flux, say. I think the answer is no. There are only two ways in which this could occur. You can, first, have not this but $[f + g(r)][f + g(r')]$. That is what happens if you make a coherent superposition, but that is not interesting because that is just a new kind of coherence. It is a different coherence, not a superposition of the original ones but a different coherent state. The other possibility is, you might actually make a wave function which is a linear superposition of the two kinds of order and that is the one that is harder to shoot down. I know of no mathematical proof that such a thing can be shot down but I know the physics of it. The physics is that really these two wave functions are very far apart from each other in Hilbert space. One has a microscopic number of particles quantized in one state, the other has a microscopic number of particles quantized in another. Therefore you have to change the states of an infinite number of particles to get from one to the other. Therefore you can never have any coherence between the two. This is very like Schrödinger's old paradox about the live cat and the dead cat. You can make a wave function which is a linear superposition of a live cat and a dead cat, but it does not mean anything.

REFERENCES

1. O. Penrose, *Phil. Mag.* **42**, 1373 (1951); O. Penrose and L. Onsager, *Phys. Rev.* **104**, 576 (1956). It has been pointed out to me that the concept of ODLRO is actually mentioned in the paper of V. L. Ginsburg and L. D. Landau, in 1950, which as we will shortly discuss gives the phenomenological theory of superconductivity. These authors failed to observe the limitation that fermions could not exhibit ODLRO except as pairs. It is clear, however, that by 1958, when the papers of Beliaev and Gor'kov to which we later refer appeared, the concept was clearly understood by the Russian group. It is unfortunate that the paper of Yang, *Rev. Mod. Phys.* **34**, 694 (1962), which has had great influence in this country, did not acknowledge the Russians' priority.
2. S. Beliaev, *J. exp. theor. Phys.* **34**, 417 (1958).
3. L. P. Gor'kov, *J. exp. theor. Phys.* **34**, 735 (1958).
4. J. Bardeen, L. N. Cooper, and J. R. Schrieffer, *Phys. Rev.* **108**, 1175 (1957).
5. V. L. Ginsburg and L. D. Landau, *J. exp. theor. Phys.* **20**, 1064 (1950).

6. L. P. Gor'kov, *J. exp. theor. Phys.* **36**, 1918 (1959).
7. E. P. Gross, *Nuovo Cim.* **20**, 454 (1961); L. P. Pitaevskii, *Sov. Phys. JETP* **13**, 451 451 (1961).
8. P. W. Anderson, in *Lectures on the Many-Body Problem* (E. R. Caianello, Ed.) Vol. 2, p. 113. Academic Press, New York, 1964.
9. F. London, *Proc. Roy. Soc.* **152A**, 24 (1935).
10. B. D. Josephson, *Physics Lett.* **1**, 251 (1962).
11. R. Doll and M. Näbauer, *Phys. Rev. Lett.* **7**, 43 (1961); B. S. Deaver and W. Fairbank, *Phys. Rev. Lett.* **7**, 43 (1961).
12. P. W. Anderson and J. M. Rowell, *Phys. Rev. Lett.* **10**, 230 (1963); J. M. Rowell, *Phys. Rev. Lett.* **11**, 200 (1963).
13. R. C. Jaklevic, J. J. Lambe, A. H. Silver, and J. E. Mercereau, *Phys. Rev. Lett.* **12**, 159 (1964).
14. R. C. Jaklevic, J. J. Lambe, A. H. Silver, and J. E. Mercereau, *Phys. Rev. Lett.* **12**, 274 (1964).
15. R. C. Jaklevic, J. J. Lambe. A. H. Silver, and J. E. Mercereau, *Phys. Rev.* **140**, A1628 (1965).
16. P. W. Anderson, *Rev. Mod. Phys.* (1966) to be published.
17. The theoretical discovery of the superconducting flux line is certainly due to A. A. Abrikosov, *J. exp. theor. Phys.* **32**, 1442 (1957).
18. R. P. Feynman, *Prog. of Low Temp. Physics* (C. J. Gorter Ed.), Vol. 1, Ch. II (1955).
19. P. W. Anderson, *Phys. Rev. Lett.* **9**, 309 (1962); Y. B. Kim, C. F. Hempstead, and A. R. Strnad, *Phys. Rev. Lett.* **9**, 306 (1962); *Phys. Rev.* **131**, 2486 (1963).
20. P. L. Richards and P. W. Anderson, *Phys. Rev. Lett.* **14**, 540 (1965).
21. P. W. Anderson and A. H. Dayem, *Phys. Rev. Lett.* **13**, 195 (1964).

Considerations on the Flow of Superfluid Helium
Rev. Mod. Phy. **38**, 298 (1966) (Sussex LT Meeting, 1965)

I feel this is the cleanest discussion of superfluidity available. Note that on many points this is contradictory or orthogonal to Landau orthodoxy as pronounced by Khalatnikov. Whether Landau would have agreed was never clarified because of his accident.

Considerations on the Flow of Superfluid Helium*

P. W. ANDERSON
Bell Telephone Laboratories, Murray Hill, New Jersey

First, we show that the most important equations of the dynamics of the two types of superfluids, He II and superconductors, follow quite directly from the simple assumption that the quantum field of the particles has a mean value which may be treated as a macroscopic variable. The background of this ansatz is also discussed. Second, we apply these equations to various physical situations in He II, notably the orifice geometry and the superfluid film, and show how they, and particularly the idea of phase slippage accompanying all dissipative processes, can be applied and what kinds of macroscopic interference phenomena may be expected. The effect of synchronization in the ac interference experiment is discussed.

I. INTRODUCTION

The material of the first part of this article covers really much the same areas of basic physics which are treated by Martin and by Nozieres in their articles at the same conference at which this was presented. Nonetheless the reader will find that the emphasis is sharply different. The striking macroscopic interference phenomena, the observation of which has been stimulated by Josephson's remarkable discovery,[1] call out for a description in terms of a definite wave function with a definable phase ϕ in every part of the system, while quantum hydrodynamics as pioneered by Landau[2] has tended to emphasize the superfluid velocity v_s and its equations of motion. The identification of $v_s = (\hbar/m)\nabla\phi$ has its limitations in relating these points of view, especially in Josephson junctions; while the opposite point of view, that ϕ exists, makes theorists uncomfortable because it breaks the gauge symmetry. Nonetheless I shall choose to take the latter path, assuming either that the reader has read such a discussion as Ref. 3 which gives the reason why this broken symmetry is possible, or that he will understand that the superfluid system is to be attached at some point to a large superfluid reservoir, with respect to which the phase is to be measured.

The idea of off-diagonal long-range order was introduced by Penrose and Onsager.[4] They suggested that superfluidity be described as a state in which the density matrix

$$\rho(r, r') = \langle \psi^*(r)\psi(r') \rangle$$

could be factorized:

$$\rho(r, r') = \psi^*(r)\psi(r') + \text{small terms}. \quad (1)$$

Beliaev[5] extended this for helium to a Green's function theory in which ψ was explicitly time-dependent, and first observed that the chemical potential determined the time dependence. Sortly thereafter Gor'kov[6] observed that superconductivity theory could be cast into the same form by substituting electron pair field operators $\psi\psi$ for the He atom bose field. In superconductivity there already existed a set of phenomenological equations proposed by Ginzburg and Landau[7] which dealt with an "order parameter" η, and it was soon recognized that this order parameter was the same as the factorized ψ of the Gor'kov theory.[8] It was only much later that Gross[9] and Pitaevskii[10] proposed similar sets of equations for liquid helium.

The notion that it was possible to regard the function which appears in these treatments as essentially the mean value of the quantum particle field has long been accepted in both helium[5] and superconductivity (there the first explicit discussion of the transformation between the "ODLRO" and mean field points of view was given by Anderson[11]) but only in the case of a homogeneous system; apparently the general case has not been discussed until recently even for superconductivity.[3] The basic idea is that it is as legitimate to treat the quantum field amplitude as a macroscopic dynamical variable as it is the position of a solid body; both represent a broken symmetry which, however, cannot be conveniently repaired until one gets to the stage of quantizing and studying the quantum fluctuations of the macroscopic behavior of the system.

Here we are going to discuss less the microscopic background of this ansatz than a number of its most important consequences for He II, many of which follow without further microscopic assumptions and are therefore of fundamental interest. Only some of what we will have to say is new, in the sense that many of the

---

* This paper was presented at the Brighton Symposium on Quantum Fluids at the University of Sussex, England, 18 August 1965. The full proceedings of this Conference will be published by the North-Holland Publishing Company in 1966, including the present article as well as those referred to above, in which a more conventional approach to superfluid dynamics is employed. Many of the other contributions have of course also been published in various journals.

[1] B. D. Josephson, Phys. Letters **1**, 251 (1962).
[2] L. D. Landau, J. Phys. USSR **5**, 71 (1941).
[3] P. W. Anderson, in *Lectures on the Many-Body Problem*, edited by E. R. Caianello (Academic Press Inc., New York, 1964), Vol. 2, p. 113.
[4] O. Penrose and L. Onsager, Phys. Rev. **104**, 576 (1956); see also O. Penrose, Phil. Mag. **42**, 1373 (1951).

[5] S. T. Beliaev, Zh. Eksperim. i Teor. Fiz. **34**, 417 (1958) [English transl.: Soviet Phys.—JETP **7**, 289 (1958)].
[6] L. P. Gor'kov, Zh. Eksperim. i Teor. Fiz. **34**, 735 (1958) [English transl.: Soviet Phys.—JETP **7**, 505 (1958)].
[7] V. L. Ginsburg and L. D. Landau, Zh. Eksperim. i Teor. Fiz. **20**, 1064 (1950).
[8] L. P. Gor'kov, Zh. Eksperim. i Teor. Fiz. **36**, 1918 (1959) [English transl.: Soviet Phys.—JETP **9**, 1364 (1958)].
[9] E. P. Gross, Nuovo Cimento **20**, 454 (1951).
[10] L. P. Pitaevskii, Zh. Eksperim. i Teor. Fiz. **40**, 646 (1961) [English transl.: Soviet Phys.—JETP **13**, 451 (1961)].
[11] P. W. Anderson, Phys. Rev. **112**, 1900 (1958).

appropriate equations have been written down (see the papers of Martin and Nozieres). The basic idea was previously stated in a Letter,[12] and some of the consequences have been explored in subsequent Letters by Donnelly[13] and by Zimmerman.[14] We will first discuss a particularly simple point of view on the derivation of the basic equations, following more closely than usually the corresponding ideas in superconductivity.[3,15] The emphasis will be on what can be shown to follow more or less rigorously from using this ansatz as a general semimicroscopic definition of superfluidity. Only the parameter values in the resulting equations require any other knowledge of the microscopic system. Then we will discuss the dynamical consequences for the orifice experimental geometry, as well as some more general situations such as film flow. In Appendix B we discuss briefly the interesting connection with classical ideal fluid hydrodynamics, where the basic "Josephson" theorem turns out to have a classical analogue which has not to our knowledge been clearly stated previously.

II. BASIC EQUATIONS

We take as our definition of a superfluid that it is a fluid in which the particle field operator ψ has a macroscopic mean value, in a sense which is defined shortly.

$$\langle \psi(r,t) \rangle = f(r,t) \exp[i\phi(r,t)]. \quad (2)$$

Here we allow slow (on the atomic scale) space and time variations; the essential point is that $\langle \psi \rangle$ has a mean value in the thermodynamic, quasi-equilibrium sense. One may think of the situation here as completely analogous to the definition of temperature in a nonequilibrium state; it is possible if there are equilibrating processes of shorter range in time or space than the coarse-grained scale of our averaging, which in turn is to occur over regions smaller than the macroscopic scale of physical measurements. Of course, in every system one can define a temperature and entropy if the average is allowed to extend over a long enough time, while only certain special systems will give a stable limiting value to $\langle \psi \rangle$. More explicitly, we visualize averaging ψ over some small region of space–time; if the region is small enough compared to the rates and ranges of microscopic fluctuations, we will obtain some finite value. As we increase the size ΔV of the region $\langle \psi \rangle$ will decrease to zero in a normal system very rapidly, in times of the order $h/(\text{mean kinetic energy})$ and ranges of order interparticle distances. In a superfluid system, on the other hand, we assume that at an intermediate, "coarse-grained" scale $\langle \psi \rangle$ approaches a limiting finite value, not changing until we reach a scale on which it varies because of the presence of macroscopic perturbations such as macroscopic fields and flows. (In Appendix A we discuss this definition a little more deeply.)

The whole problem of superfluid dynamics, then, reduces to the question of how to deal with $\langle \psi \rangle$ (often denoted the superfluid order parameter) as a thermodynamical variable. It is quite important that $\langle \psi \rangle$ is a *complex* order parameter; it has both an amplitude f and a phase ϕ. There is actually rather a sharp distinction between the two real thermodynamic variables f and ϕ, and it is the behavior with respect to ϕ which is responsible for specific superfluid properties. The distinction is that ϕ is coupled, like a polarization or a strain, to external forces, where f is merely an internal order parameter in the sense, for instance, of the original long-range order parameter of order–disorder systems, or of the antiferromagnetic sublattice magnetization. In principle a corresponding force might exist but in practice it does not, so f simply manifests itself as a convenient tool by which to describe the condensation process.

The point of ϕ, then, is that it is not only a thermodynamic but also a dynamical variable. The latter fact comes from its being the dynamically (not thermodynamically) canonically conjugate variable to N, the total number of particles in the system described by ϕ. (The limitations of this statement for systems of few particles[16] are irrelevant here.)

We illustrate this fact by forming wave packets in many-body wave-function space, just as in discussing the relationship of p and q it is useful to form wave packets. Let us write the wave function of one of our coarse-grained cells of volume ΔV as

$$\Psi(\Delta V) = \sum_N a_N \Psi_N(\Delta V) \quad (3)$$

where, since our cell is only a part of the superfluid, it is essential to realize that the state is a superposition of states Ψ_N with different numbers of particles N, with coefficients a_N large in some range of values $\Delta N \sim N^{\frac{1}{2}}$. An important simplifying assumption concealed in (3) is that the cell may be made big enough so that a_N and Ψ_N do not depend very much parametrically on the variables of the rest of the system: this is what is meant (see Appendix A) by a "satisfactory" local description.

We postulate that $\langle \psi \rangle$ has a limiting mean value, which must be of order $(N/\Delta V)^{\frac{1}{2}}$; we calculate this value from (3):

$$\frac{1}{\Delta V}\int_{\Delta V} d\tau \langle \psi(r) \rangle = \frac{1}{\Delta V}\sum_{N,N'} a_{N'}{}^* a_N \int (\Psi_{N'}, \psi(r)\Psi_N) \, d\tau$$

$$= \sum_N a_{N-1}{}^* a_N \int \frac{(\Psi_{N-1}, \psi(r)\Psi_N) \, d\tau}{\Delta V}. \quad (4)$$

---

[12] P. L. Richards and P. W. Anderson, Phys. Rev. Letters **14**, 540 (1965).
[13] R. J. Donnelley, Phys. Rev. Letters **14**, 939 (1965).
[14] W. Zimmermann, Phys. Rev. Letters **14**, 976 (1965).
[15] P. W. Anderson, N. R. Werthamer, and J. M. Luttinger, Phys. Rev. **138**, A1157 (1965).

[16] W. H. Louisell, Phys. Letters **7**, 60 (1963); L. Susskind and J. Glogower, Physics **1**, 49 (1964).

In the case of the Bose-condensed perfect gas, one may take

$$\Psi_N = (c_0^+)^N \Psi_{\text{vac}}, \qquad (5)$$

where

$$c_0 = \frac{1}{(\Delta V)^{\frac{1}{2}}} \int \psi(r) \, d\tau$$

$$c_0 \Psi_N = N^{\frac{1}{2}} \Psi_{N-1}, \qquad (6)$$

so that

$$\langle \psi \rangle = \sum_N a_{N-1}^* a_N (N/\Delta V)^{\frac{1}{2}}. \qquad (7)$$

This is the maximum possible value of the matrix element. (Here we have taken advantage, as we will always, of our freedom to choose the phase of Ψ_N to make the matrix element real in order to keep the phase factors—which of course are *not* irrelevant—in the a_N's.) In real superfluid systems and at finite temperatures (the case of superconductivity is of course quite similar if for a single Bose ψ or c we read Fermion pair operators) the matrix element does not take on its maximum possible value—in He, for instance, as McMillan has shown,[17] the value is about 11% of the maximum when ΔV is reasonably large. Even if

$$M = (\Psi_{N-1}, \bar{\psi}\Psi_N) \qquad (8)$$

is large, however, $\langle \psi \rangle$ will not be large unless the a_N preserve phase coherence. For example, we may form a wave packet using

$$a_N = (2\pi \Delta N)^{-\frac{1}{2}} \exp\left[-\tfrac{1}{2}(N-\bar{N})^2/(\Delta N)^2\right] \exp(i\phi N). \qquad (9)$$

In this case if $\Delta N \gg 1$,

$$\langle \psi \rangle \cong |M| \exp(i\phi) \qquad (10)$$

whereas if the phase factors are arbitrary $\langle \psi \rangle$ will be very much smaller, even if every individual Ψ_N represents a pure Bose-condensed state of the volume element ΔV. We will very shortly discuss the reason why wave packets like (9) actually occur in superfluid systems; first, however, let us dispose of a few formal preliminaries.

Any linear combination of the set of fixed-number states Ψ_N may be written as a linear combination of our basic wave packets (3)–(9) [which we will call $\Psi(\phi)$]. In particular, the number eigenfunction Ψ_{N_0} may itself be written

$$\Psi_{N_0} \propto \int_0^{2\pi} d\phi \exp(-i\phi N_0) \Psi(\phi) \qquad (11)$$

$$\left(= \sum_N a_N \, \delta(N, N_0) \Psi_N \right).$$

The operator $-i\partial/\partial\phi$ acting on $\Psi(\phi)$ has the same

---

[17] W. L. McMillan, Phys. Rev. **138**, A442 (1965).

effect as multiplying Ψ_{N_0} by N_0:

$$\int_0^{2\pi} d\phi \exp(-i\phi N_0)\left(-i\frac{\partial}{\partial\phi}\right)\Psi(\phi)$$

$$= N_0 \int_0^{2\pi} d\phi \exp(-i\phi N_0) \Psi(\phi) = N_0 \Psi_{N_0}$$

so that we may take

$$N \leftrightarrow -i(\partial/\partial\phi) \qquad (12\text{a})$$

and correspondingly it may be shown that

$$i(\partial/\partial N) \leftrightarrow \phi \qquad (12\text{b})$$

in the limit that N may be considered a continuous variable. Thus, as we stated, N and ϕ are conjugate dynamical variables.

In general, also, N and ΔN are sufficiently large that the system's dynamics may be treated reasonably well classically and the uncertainty in ϕ is not excessive—of course (12) implies $\Delta N \Delta \phi \sim 1$.

In any case the equations of motion of N and ϕ are

$$i\hbar \dot{N} = [\mathcal{H}, \dot{N}] = i(\partial\mathcal{H}/\partial\phi)$$

$$i\hbar \dot{\Phi} = [\mathcal{H}, \phi] = -i(\partial\mathcal{H}/\partial N).$$

Taking the mean values of these two equations and assuming that the wave packets are such that $\Delta N/N$ and $\Delta\phi$ are both small (quasi-classical case) is the source of the two equations which essentially characterize superfluidity: the equation for superfluid flow, corresponding to London's equation or the Ginsburg–Landau current equation in superconductivity, and the "Josephson" frequency equation, which is related to the acceleration equation for superfluid flow in both He and superconductivity. The second is somewhat simpler though less generally known: it is just

$$\hbar (d\phi/dt) = \partial E/\partial N = \mu. \qquad (13)$$

Here we have used the standard definition of the chemical potential μ; if the fluid is in motion so that the particles have kinetic energy $\tfrac{1}{2}mv_s^2$ that is to be included in μ as well as the internal energy and any external forces. Obviously the partial derivative holds S fixed. In an isothermal, assumed incompressible bath of liquid helium, with free surface at height h in a gravitational field g,

$$\mu = m(p/\rho) + mgh + \tfrac{1}{2}mv_s^2. \qquad (14)$$

In the nonisothermal case (14) should contain the thermomechanical term.

Equation (13) is of the utmost importance in understanding superfluidity. It says two things: first, that if the state of the superfluid is constant in time, because ϕ is a thermodynamic variable it will be constant and μ must be constant everywhere: there can be no potential differences in the truly steady state. Second, if there is any potential difference $\mu_1-\mu_2$ between two elements in the superfluid $\phi_1-\phi_2$ must change in time. This can happen in two ways. The simplest is acceleration. If we take the gradient of (13), we obtain

$$(d/dt)(\hbar\nabla\phi) = \mathbf{F}, \quad (15)$$

where $\mathbf{F}$ is the total force on the particles, and making the identification (which we shall discuss shortly) of

$$\hbar\nabla\phi = p_s = mv_s, \quad (16)$$

this is the statement that the superfluid may undergo acceleration without frictional damping by whatever external forces act upon it. A potential difference may lead to continuous acceleration, then.

Slightly more subtle and much more physically important is the concept of phase slippage by quantized vortex motion. As we have emphasized, ϕ is the phase of the thermodynamic variable $\langle\psi\rangle$, which is of course necessarily single-valued. ϕ, however, being a phase, need not be single-valued in a multiply connected system such as a toroid; it need merely return to its original value $\pm 2n\pi$ on traversing a path around a nonsuperfluid obstacle:

$$\oint \nabla\phi \cdot d\mathbf{s} = 2n\pi. \quad (17)$$

In terms of the superfluid momentum p_s this expresses the idea of the quantization of angular momentum in units of $\hbar$ or of vorticity $\boldsymbol{\omega}=\frac{1}{2}\nabla\times v_s$ through a closed curve in units of $h/2m$. As we shall see, however, the quantity v_s is not necessarily a measurable particle velocity and so (17) is somewhat more fundamental than the usual concept of vorticity quantization.

A bucket of superfluid may be made multiply connected not only by the presence of actual solid obstacles but by the introduction of one-dimensional regions of nonsuperfluidity within the liquid itself: lines where $\langle\psi(r)\rangle=0$. These are "vortex cores" and may of course move according to the usual laws of hydrodynamics along with the surrounding fluid. Around such a line there can be a circulation of one or, less usually, an integral number of quanta, according to (17).

Equation (17) shows that the integral of $\nabla\phi$ along a path on one side of a vortex core differs from that on the other by 2π (see Fig. 1). Thus when a vortex core moves across the line between points 1 and 2, this may cause a time rate of change of $\phi_1-\phi_2$. Mathematically, we may write

$$\langle\mu_1-\mu_2\rangle_{\mathrm{Av}} = T^{-1}\int_0^T dt\hbar\frac{d(\phi_1-\phi_2)}{dt} = \frac{\hbar}{T}\int_0^T dt\frac{d}{dt}\int_1^2 \nabla\phi \cdot dl,$$
$$(C)$$

where $\langle\ \rangle_{\mathrm{Av}}$ denotes time average, where we consider the limit T very long, and C is any path from 1 to 2. If the liquid is assumed to be in a reasonably steady

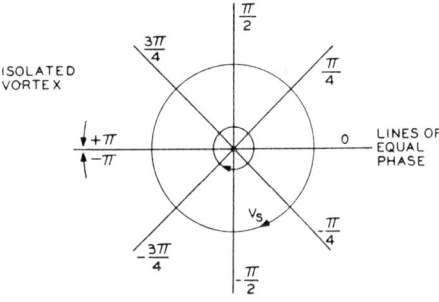

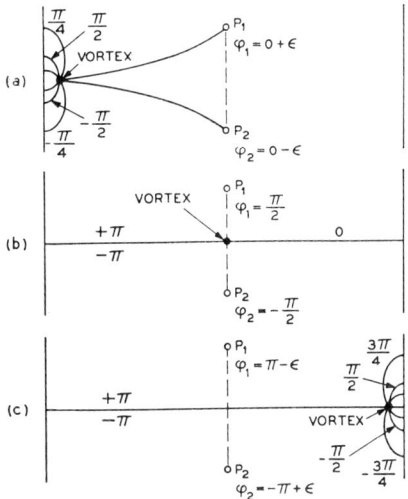

FIG. 1. Illustration of phase changes at two points P_1 and P_2 in a channel as a vortex moves between them. From a to b to c the vortex moves across from left to right; as it moves from one wall to the other the relative phase changes by 2π.

state, so that at 0 and T the positions of the various vortices do not differ very much, the integral is just equal to 2π times the number of vortices which have crossed C in this time interval. Thus we have

$$\langle\mu_1-\mu_2\rangle_{\mathrm{Av}} = h\langle dn/dt\rangle_{\mathrm{Av}}, \quad (18)$$

where $\langle dn/dt\rangle_{\mathrm{Av}}$ is the average rate of motion of vortices across a path from 1 to 2. This is the "phase slippage" concept which is used to explain the various "ac Josephson"-like experiments.[18,19] Of course, this is not incompatible with the acceleration equation (15), so in a sense this phenomenon too is merely a manifestation of the fact that potential differences occur only in an accelerated superfluid; but it is a point of view which had not previous to Ref. 12 found its way into the He literature, nor until recently, with the discovery

---

[18] S. Shapiro, Phys. Rev. Letters **11**, 80 (1963).
[19] P. W. Anderson and A. H. Dayem, Phys. Rev. Letters **13**, 195 (1964).

of the Josephson effect and of flux creep and flow, into that of superconductivity.

Let us now take up the current equation, the theory of which is somewhat more complex. There are two aspects to this. In the first place there is the quasi-rigorous Eq. (16) the meaning of which becomes a bit vague with more careful consideration. If we simply suppose that the state of a volume element ΔV moving with velocity v_s is obtained from that of a stationary element by a pure Galilean transformation, multiplying ψ by $\exp(imv_s x/\hbar)$, this equation is trivially valid. That is the usual derivation of it, in one form or another. I know of no acceptable proof of (16), however, in the sense of showing v_s to be a real particle velocity, in physically important situations such as counterflow of normal and superconducting fluids, or where v_s is varying reasonably rapidly, as near a vortex core. Since (16) is the statement which leads to vorticity quantization, this means that that concept, often claimed to be exact, is apparently not so.

The statement has been made in the literature[13,14] that the results—specifically the "ac Josephson effect"—which follow from (13) or the phase slippage concept can equally well be "derived" from vorticity quantization plus perfect fluid hydrodynamics for the superfluid. Neither of these latter ideas, however, has at the moment a very quantitative experimental background, while theoretically as we have just seen the phase has a much more secure theoretical meaning than v_s; it seems to us that (13) is a much more rigorous and complete theoretical statement than the hydrodynamic equations, which are derived via (15) and (16) from it. Normal fluid counterflow and dissipation do not affect it. In a perfect classical fluid, for instance, the vortices cannot move across stream lines, so clearly He II is not such a fluid.

Perhaps a more rigorous general reason for using (16) at least as a *definition* for v_s (other than that it allows us to use ϕ as a velocity potential for the superfluid flow) is (15), which shows that $m dv_s/dt$ does then give the rate of exchange of momentum per superfluid particle with external forces, an excellent operational definition of v_s.

The superflow, however, is best determined not from the expression for v_s but from the other of the two Hamilton equations for the conjugate variables N and ϕ:

$$\hbar (dN/dt) = \partial E/\partial \phi. \qquad (19)$$

Gauge invariance assures us that in fact the energy of an isolated bit of superfluid is independent of ϕ, so that in the absence of a coupling to its neighbors $dN/dt = 0$. Consider, on the other hand, a pair of neighboring volume elements ΔV_1 and ΔV_2. Our definition of superfluid implies that $\langle \psi \rangle$ has a tendency to constancy, so that the mean phases ϕ_1 and ϕ_2 of the two neighboring elements will be coupled by some energy which is a minimum when $\phi_1 - \phi_2 = 0$:

$$E = U(\phi_1 - \phi_2).$$

Assuming the effects of other neighboring elements can be treated independently (suitable arguments can be found for this step) we see that the flow across the boundary between ΔV_1 and ΔV_2 is given by

$$J_{\text{tot}} = \frac{dN_1}{dt} = -\frac{dN_2}{dt} = \frac{1}{\hbar} \frac{\partial U(\phi_1 - \phi_2)}{\partial (\phi_1 - \phi_2)}. \qquad (20)$$

This is the expression used in the theory of the Josephson current[1,3] across a barrier between two macroscopic pieces of superconductor; at this point we are carrying the same reasoning over to the continuous interior of a superfluid or superconductor. As in that case, we may pause now to point out that it is this coupling energy which enforces the phase coherence of each individual volume element of the superfluid. The kinetic energy matrix element which transfers particles across the boundary can cause transitions like

$$\{\Psi_1^N \to \Psi_1^{N-1}; \Psi_2^{N'} \to \Psi_2^{N'+1}\}, \qquad (21)$$

with a matrix element we may call M_{12}. (Here $\Psi_{1,2}$ is the many-body wave function of $\Delta V_{1,2}$.) If the wave packets (3) have coefficients a_N which, like (9), have coherent phase relationships, all transitions $N' \to N'+1$ can occur coherently with all transitions $N \to N-1$. Mathematically, the energy due to transitions like (21), inserting wave packets like (3), is

$$(\Psi_1\Psi_2, (\text{K.E.})\Psi_1\Psi_2) = \left(\sum_N a_N^* a_{N-1}\right)\left(\sum_{N'} a_{N'}^* a_{N'+1}\right) M_{12}.$$

$$(22)$$

As we see, this energy is orders of magnitude larger for coherent wave packets like (9), for which ϕ is determinate, than in the incoherent case.

In the interior of the superfluid, it is more convenient to go over to a continuum representation of the coupling energy which maintains phase coherence. We write U as a functional of the gradient of ϕ (as well as f, of course)

$$U = \int U[(f), \nabla \phi] \, d\tau$$

$$\simeq \int E(f, S)(\nabla \phi)^2 \, d\tau \qquad (23)$$

and then by considering a cell of wall area A and thickness d we find

$$\frac{J}{\text{unit area}} = \frac{dN_1}{dt} \bigg/ A = \frac{\delta U}{\hbar \delta \nabla \phi} = \hbar^{-1} E \nabla \phi. \qquad (24)$$

If we use the pseudo-identity

$$v_s = \hbar \nabla \phi / m,$$

we may write

$$J_s = n_s(f, S)v_s$$
$$= [\hbar n_s(T)/m]\nabla\phi, \quad (25a)$$

so that we can identify the parameter E of (24):

$$E = \delta^2 U/\delta(\nabla\phi)^2 = \hbar^2 n_s(T)/m. \quad (25b)$$

Thus the current equation contains a completely arbitrary parameter n_s. In pure, homogeneous systems at absolute zero Galilean invariance can be used to show that $n_s = n$, the total number of particles; but in impure systems at $T=0$, or any system at $T\neq 0$, no such identity holds, though in general n_s is of the same order of magnitude as n.

The supercurrent (24) exists even if the phase ϕ is completely stationary in time, which as we have shown implies the absence of accelerating forces. In fact, in the presence of accelerating forces and time-dependent ϕ we must expect additional quasi-particle currents, in general; the system will exhibit a two-fluid hydrodynamics, the complexities of which need not concern us here. So far as I know their proper treatment requires more knowledge of the actual physical system than we are assuming here.

One very important point about (24) taken together with (15) is that it makes it at least highly probable that there is both a necessary and a sufficient connection between the existence of supercurrents and our definition of superfluidity (1) (which is essentially equivalent to what Yang[20] has named ODLRO). Namely, $\langle\psi\rangle$ and therefore ϕ will not exist if the energy is not such as to maintain spatial coherence of ϕ, so $\delta^2 U/\delta(\nabla\phi)^2$ must exist and be positive in a superfluid in this sense, meaning necessarily supercurrents by (24). (15) shows that they flow with zero forces and are therefore supercurrents. Hypothetical phases with $\langle\psi\rangle$ but no supercurrents (Cohen[21]) seem to ignore this half of the argument.

Conversely, the only dynamically conjugate variable to N is the phase as we have defined it, so that the existence of a dN/dt in a stationary state implies a $\delta U/\delta\phi$, which implies that ϕ is a meaningful variable. Various hypothetical superconducting phases (e.g., Frohlich[22]) do not satisfy *this* half of the argument.

This concludes our general discussion of the basic equations of superfluidity. We restate the conclusions: the phase equation (13) and the corresponding equation of phase slippage are exact in the "integrated" sense that they give the phase difference between two distant points in undisturbed regions of superfluid. The existence of the order parameter alone guarantees the existence of quantized vortices, and according to (24) these are indeed vortices in that they contain a

[20] C. N. Yang, Rev. Mod. Phys. **34**, 694 (1962).
[21] M. H. Cohen, Phys. Rev. Letters **12**, 664 (1964).
[22] H. Frohlich and C. Terreaux, Proc. Phys. Soc. (London) **86**, 233 (1965).

superfluid circulation. However, the quantization of vorticity in any true sense is dependent on the imprecise assumption (15) that the phase is the velocity potential with fixed coefficient. This need not be true unless we treat (15) merely as a definition of v_s, for instance near a vortex core, just as the quantization of flux in superconductivity is not necessarily precise. Operationally, for example, the measurement of h/e by the ac Josephson effect, or of h/mg by the helium counterpart, is more precise in principle than by flux or vorticity quantization. In practice, of course, present-day methods are not capable yet of distinguishing these niceties, but it is of value to have a clear idea of the theoretical assumptions behind the various equations, since it is foreseeable that the most precise measurements of many important physical quantities will involve quantum coherence.

III. SOME DYNAMICAL CONSEQUENCES

The macroscopic quantum interference effects promised by the existence of $\langle\psi\psi\rangle$ in superconductivity have been relatively easy to observe for a number of reasons: the light mass of the electron, permitting weak superfluid connections to be made easily by use of the tunneling phenomenon, the coupling to the electromagnetic field which leads to flux quantization, and most particularly the fact that that coupling provides a second parameter, the penetration depth λ, which in screening out the current and magnetic field from the interior of the superconductor creates the Meissner effect, that is ensures that every superconductor exhibits a finite critical magnetic field H_{c_1} below which it contains no vortices and thus essentially exhibits constant ϕ. In He, $\lambda\to\infty$ so that the corresponding $\omega_{c_1}=0$; no rotation of a sufficiently large He sample is too small for vortices to be energetically favorable. Indeed, experiments show that few samples are ever free of vorticity. Even worse, the coherence length—the length given by the ratio of $\delta U/\delta|\psi|^2$ to $\delta U/\delta|\nabla\psi|^2$, which determines how rapidly the order parameter can vary and thus how large a vortex core is—is of order a few Å, so that no channel through which He can flow is too small to contain a vortex. All this means that the useful idealization in superconductivity of the "ideal Josephson junction," a weak link between two reservoirs having constant phase, is probably not relevant to the helium case. Any barrier in which the flow occurred only by quantum-mechanical tunneling would have to be of subatomic dimensions, especially in thickness, to permit any measurable current to flow, and could not be supported mechanically. Any attempt to replace the ideal junction by a thin channel, on the other hand, runs into exactly the same difficulties that are encountered with long thin bridges in the case of superconductivity,[23,19] namely, that the device acts like a

[23] R. D. Parks and J. M. Mochel, Rev. Mod. Phys. **36**, 284 (1964).

sequence of weak links in series and one is never sure exactly where the phase rigidity is breaking down. The closest approximation we could imagine to a single definable "weak superfluid junction" was the orifice geometry, which is analogous to the "short" thin film bridge of Anderson and Dayem[19] in superconductivity.

Let us analyze this system in some detail, as we did the Josephson junction previously.[3] First, a note as to the driving term to be inserted to represent any externally applied conditions. In the case of normal systems it is natural to place the ends of a specimen in contact with reservoirs at different chemical potential levels μ_1, μ_2, giving an "applied" potential gradient $\nabla\mu$, and to describe any situation in terms of solutions of the resulting applied field problem; we calculate J as a function of the pressure gradient or field, even though actually we may be driving the system with a constant current generator. The microscopic theory is done by inserting μN terms into the Hamiltonian, as is well-understood in the calculation of resistance, for instance, or thermoelectric power.

It is precisely the nature of superfluids that they cannot assume a stationary state under a field or pressure gradient, but will, as explained by Anderson, Werthamer, and Luttinger,[15] have to be described by phases with a time dependence obeying (13). This condition follows when we insert the appropriate μN terms, as there described; but that does not lead to a way to discuss the equally interesting case of an imposed current. One obvious technique is to impose a fixed phase difference by postulating an infinitely tight coupling to reservoirs consisting of large superfluids of fixed phase, but that is often quite unphysical. We have used without discussion[3] the technique of inserting a term

$$\Delta E = J_{12}(\phi_1 - \phi_2)$$

in the energy to describe the effect of a fixed supercurrent J_{12} between regions 1 and 2. In superconductivity a rather rough physical derivation of this can be given in terms of the electromagnetic interaction between J and the magnetic flux of a vortex line. It is analogous to, but more general than, the technique of inserting a $p \cdot v_s$ term sometimes used to derive the hydrodynamic equations.

In the case of He we may return very simply to Eq. (19). We showed that the particle accumulation rate dN_1/dt in a volume element ΔV_1 is given by:

$$dN_1/dt = \hbar^{-1}(\partial U/\partial \phi_1).$$

If no net accumulation is to occur, and if the current leaving ΔV_1 to neighboring elements is given in terms of the coupling energy between them by

$$J_{12} = dN_1/dt)_{\text{into superfluid}} = \hbar^{-1}(\partial U_{\text{coupling}}/\partial \phi_1)$$

there must be a compensating term

$$dN_1/dt)_{\text{current generator}} = -\hbar J_{12}\phi_1.$$

Similarly, to make the corresponding current leave element (2) we must have $dN_2/dt)_{\text{c.g.}} = \hbar J_{12}\phi_2$ so we may represent a constant current generator by

$$\mathcal{H}_{\text{gen}} = \hbar J_{12}(\phi_2 - \phi_1). \tag{26}$$

Using this let us discuss the orifice problem in the presence of a constant driving current. Current acceleration can be important only on the time-scale of the U-tube oscillations, which is usually longer than the time necessary to create a vortex or otherwise change the phase but can be included if desired. Also, it will greatly simplify one's thinking without falsifying any important physical features to assume $T \to 0$; i.e., only incompressible, superfluid flows.

Under these conditions $\hbar/m \nabla \phi = v_s$ and

$$\nabla \cdot v_s = 0$$

so

$$\nabla^2 \phi = 0.$$

Thus we can solve for the flow in the absence of vortices by a simple potential calculation. Let the radius of the orifice be a. The equipotentials and streamlines are along coordinate surfaces in a set of oblate spheroidal coordinates, defined in terms of cylindrical coordinates r, θ, z through the axis of the orifice by

$$(r^2/a^2 \cosh^2 u) + (z^2/a^2 \sinh^2 u) = 1;$$
$$(r^2/a^2 \sin^2 v) - (z^2/a^2 \cos^2 v) = 1;$$
$$\theta = \theta; \tag{27}$$
$$-\infty \leq u \leq \infty, \quad 0 \leq v \leq \pi/2, \quad 0 \leq \theta < 2\pi;$$

or

$$r = a \cosh u \sin v$$
$$z = a \sinh u \cos v. \tag{28}$$

The potential (i.e., phase) is

$$\phi(r, z) = C \int \frac{du}{\cosh u} = 2C \tan^{-1} e^u, \tag{29}$$

so that the total phase difference between the two large reservoirs separated by the orifice is

$$\phi_1 - \phi_2 = \pi C = \phi(+\infty) - \phi(-\infty). \tag{30}$$

The velocity is in the "u" direction and is

$$v_s = \frac{\hbar}{ma} \frac{C}{\cosh u (\sinh^2 u + \cos^2 v)^{\frac{1}{2}}}. \tag{31}$$

Some special values of the velocity field are:

Along the axis:

$$r = 0, \quad \sin v = 0, \quad z = a \sinh u$$
$$v_s = \frac{\hbar C}{ma}(\cosh u)^{-2} = \frac{\hbar C}{m} \frac{a}{a^2 + z^2}. \tag{32a}$$

In the orifice:
$$z=0, \quad r=a\sin v, \quad u=0$$
$$v_s = (\hbar/m)[C/(a^2-r^2)^{\frac{1}{2}}]. \quad (32\mathrm{b})$$

On the plane:
$$r=\pm a\cosh u, \quad \cos v=0$$
$$v_s = (\hbar C/mr)[a/(r^2-a^2)^{\frac{1}{2}}]. \quad (32\mathrm{c})$$

The total particle flow is
$$J = \frac{\rho}{m}\int v_s \cdot dS = \frac{2\pi\hbar C a\rho}{m^2} = \frac{2\hbar a\rho}{m^2}(\phi_1-\phi_2), \quad (33)$$

[using (30)] and the total kinetic energy is
$$K = \tfrac{1}{2}\rho\int v_s^2\, d\tau$$
$$= \tfrac{1}{2}\rho\frac{\hbar^2}{m^2}\int(\boldsymbol{\nabla}\phi)^2\, d\tau$$
$$= \tfrac{1}{2}\rho\frac{\hbar^2}{m^2}\Big(\int dS_{+\infty}\cdot - \int dS_{-\infty}\cdot\Big)\phi\boldsymbol{\nabla}\phi$$
$$K = \tfrac{1}{2}J\hbar(\phi_1-\phi_2) = (\rho a/m)[\hbar^2(\phi_1-\phi_2)^2/m]. \quad (34)$$

Thus this energy is a quadratic function of the phase difference $(\phi_1-\phi_2)$. However, it is essential to realize that this is only one branch of a multiple-valued function, because by gauge invariance K must be a periodic function of $\phi_1-\phi_2$ with period 2π. That is, if we were to pin down the phases in the two reservoirs by coupling to reservoirs of fixed relative phase $(\phi_1-\phi_2)_0$ we could satisfy the boundary conditions by any flow
$$J = (2\hbar a\rho/m^2)[(\phi_1-\phi_2)_0\pm 2n\pi]$$

and the corresponding kinetic energy
$$K = (\rho a/m)(\hbar^2/m)[(\phi_1-\phi_2)_0\pm 2n\pi]^2. \quad (34)$$

Equation (34) must also be used, then, in the presence of a current generator (27); thus the total energy is
$$E = (\rho a\hbar^2/m^2)[(\phi_1-\phi_2)\pm 2n\pi]^2 - \hbar J_{12}(\phi_1-\phi_2). \quad (35)$$

This energy considered as a function of $\phi_1-\phi_2$ has an infinity of points of metastable equilibrium where
$$J = J_{12}$$
$$(\phi_1-\phi_2) = 2n\pi + (m^2 J_{12}/2\hbar a\rho).$$
$$E = -J_{12}[n\hbar + (m^2 J_{12}/4a\rho)]. \quad (36)$$

The situation is diagrammed in Fig. 2. The first drawing assumes a small current generator, $J_{12} < \hbar a\rho/m^2$, which is the current value when $\phi_1-\phi_2 = \pi$, the value at which the parabolas cross. For this current, although none of the energy minima are truly stable each is at least the lowest energy state for a given fixed phase difference (this has not been proved but seems obvious).

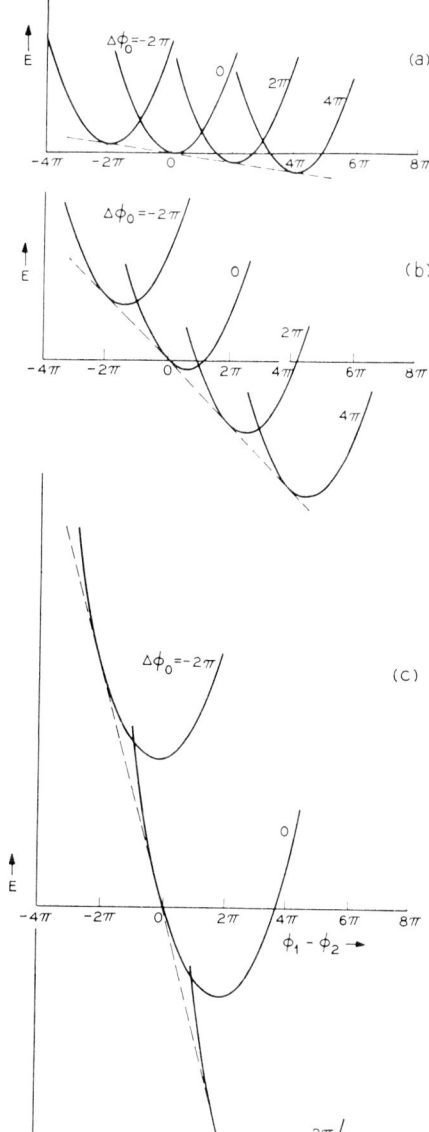

FIG. 2. Energy parabolas for potential flow in an orifice as a function of relative phase for different constant current generators J_{12}. $(\Delta\phi)_0$ is the relative phase "slippage" defined by $E(\Delta\phi_0) = J_{12}\Delta\phi_0$. (a) Low current; (b) $J_{12} \simeq v_c^0$; (c) J_{12} approaching observed critical current.

The system can still absorb energy from the current generator by "running downhill" in phase, but only if some fluctuation or external perturbation lifts it over the energy barriers between parabolas. In principle, if $J_{12} > \hbar a\rho/m^2$ the third drawing is correct, and there is no true energy minimum; this corresponds to a "zeroth-order" critical velocity (we take the mean value over the area of the orifice)

$$(v_s)^0 = 2\hbar/ma \qquad (37)$$

which is $\frac{1}{3}$ cm/sec for $a=10$ μ. Actually, the crossing points between parabolas do not represent possible transitions, because the two parabolas represent entirely different "sheets" of the energy connected only by passage of a vortex across the orifice, and we must consider the activation energy problem for creation of a vortex, as has been discussed by Vinen.[24] At the very least a length of vortex line of the order $2a$ must be created, which has energy of order

$$(E_{\text{vortex}})_{\min} = (\pi\rho\hbar^2/m^2) \ln(a/\xi) \times 2a,$$

where ξ is the coherence length, ~ 1 Å. This must be compared to the energy gained when a vortex line is halfway across the orifice, which is of order

$$\pi\hbar J_{12} = (\pi^2 \rho a^2 \hbar/m) v_s.$$

The result is another "critical" velocity

$$v_s^{(1)} = (2/\pi)(\hbar/ma) \ln(a/\xi)$$
$$\cong 7.5 \hbar/ma. \qquad (37')$$

This is also far smaller than observed critical velocities, indicating as discussed by Vinen that the great difficulty in forming vortices in most situations is probably nucleating them at the walls.

Yet another "critical" velocity may be estimated if we suppose that the mechanism for phase slippage is the most plausible one in a simple orifice geometry, that of blowing vortex rings out on the downstream side of the orifice, of approximately the size of the orifice itself. The energy of a vortex ring of radius a is

$$E_{\text{ring}} = (2\pi^2 \rho a \hbar^2/m^2) \ln(a/\xi)$$

which can be produced from an energetic point of view only if the energy available from the current source in each cycle, $2\pi\hbar J_{12}$, is equal to E_{ring}. From this we get

$$v_s^{(2)} = (\hbar/ma) \ln(a/\xi) \cong 11.5(\hbar/ma). \qquad (37'')$$

[Note that the momentum conservation equation, as opposed to these energy considerations, simply gives us the frequency condition (13) as expected from (15)].

The essential physical point here is that all of these "critical" velocities are much less than real observed superflow velocities, indicating that in all cases the generation and motion of vortices is controlled by large random fluctuations, presumably either in the generation near the walls or in the motion of vortices already present. In general, the working point of an orifice is found to be not near the lowest intersection of two energy curves, where

$$v_s = 2\hbar/ma \quad \text{and} \quad (\phi_2 - \phi_1)_0 = \pi,$$

but at a phase-difference of the order of 10π, where the energy available is much greater than that necessary to form a vortex and thus we may expect rather irregular and unstable behavior. When vortices are created under such conditions they are accelerated rather strongly and give up considerable energy to the normal excitations.

It is because of this large value of the phase difference that the orifice geometry—and, correspondingly, to a lesser extent superconducting thin film bridges—are more difficult to demonstrate spacial interference effects with than the Josephson tunnel junction, for which the system moves adiabatically from one energy minimum to the next, 2π away, whenever that is energetically possible. Incidentally, it is clear that since v_c is not very dependent on channel length, the total phase difference for a long channel at v_c is even greater than for a short one, and as a result even more randomness in the creation of vortices and even less sensitivity to the precise value of phase difference is to be expected for long channels.

It is probably for these reasons that of all the macroscopic quantum interference effects, only the driven ac experiment has as yet succeeded in He. This experiment depends on the principle of synchronization, in which a strong external ac signal is introduced which can override the internal fluctuations.

First let us consider a free-running orifice connecting two reservoirs with a height difference Z. This height difference will decrease at a rate

$$\frac{dZ}{dt} = \frac{J\rho}{mA} = v_s\left(\frac{\pi a^2}{A}\right), \qquad (38)$$

where A is the area of the surface in the smaller reservoir. J also determines the rate of generation of vortices; if J is greater than the observed J_c, vortices will be generated very rapidly, and conversely; but actually of course there is a functional dependence of the rate on the current:

$$dn/dt = f(v_s) \qquad (39)$$

which will be rather steep, as shown in Fig. 3(a), but at least finite at very low v_s. Finally, we have the Josephson relation (13):

$$mgZ = -h(dn/dt) = -hf[(A/\pi a^2)(dZ/dt)]. \qquad (40)$$

Equation (40) neglects the possibility of phase change by acceleration, i.e., it really contains a term in dJ/dt or d^2Z/dt^2, which will have no effect in the mean but does lead to the U-tube oscillations. If f were a step

---

[24] W. F. Vinen, *Proceedings of the International School of Physics "Fermi" (Varenna) 1961*, edited by G. Careri (Academic Press Inc., New York, 1963), p. 336.

function, dZ/dt would be fixed and the height difference would decay linearly, but presumably f is somewhat "soft," and as Z decreases the decay rate will do so also, but perhaps less slowly. Figure 3(b) shows a hypothetical decay curve which fits qualitatively with the vortex generation rate shown in 3(a).

Now suppose that as the height difference drops, we are causing an ac flow to be superimposed on the dc. During half the cycle we will be increasing the tendency to form vortices, during the other half decreasing it. When the height difference is such that $dn/dt = h\nu$, one vortex per cycle will be formed—presumably in the positive half-cycle, with quite high probability. When formed it uses up some considerable fraction of the available energy so another cannot be formed immediately; thus there will be a strong tendency for exactly one vortex per cycle to be formed, since the second half-cycle is not available. Because of (40) this will mean a tendency to fixed Z, i.e., a plateau in dn/dt [Fig. 3(a)] and thus in Z. Another way of putting it is that when the vortex formation is in a definite phase relationship with the ac signal, power can be transferred from the ac generator to the system as a whole, enough power to appreciably change the flow rate. If the alternating current is larger than the dc—as in our experiment[12]—clearly it is quite possible to stop the dc flow entirely, because we can control vortex formation wholly with the ac.

I have found a very simple mechanical analogy useful for understanding the ac Josephson effect. (See Fig. 4.) The relative phase of the two reservoirs I think of as the angular position coordinate ϕ of a set of locomotive wheels, and the velocity of the locomotive is the height difference Z; the ratio of position to phase is then the radius of the wheels corresponding to h/mg. The equation (40) for the rate of generation of vortices as a function of the flow (acceleration) may be inverted to give an effective nonlinear frictional force on the "locomotive"

$$\frac{A}{\pi a^2}\frac{dZ}{dt} = -f^{-1}\frac{mgZ}{h} = -f^{-1}\left(\frac{dn}{dt}\right). \qquad (41)$$

If f is a step function, this gives a constant frictional

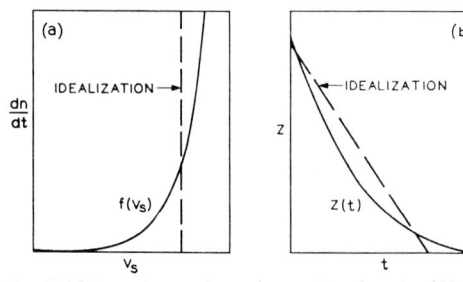

FIG. 3. (a) Rate of vortex formation as a function of v_s. "idealization" is the sharp critical current assumption; reality is probably more like $f(v_s)$ shown. (b) Decay of a helium head through an orifice or channel for critical current idealization and **real** situation.

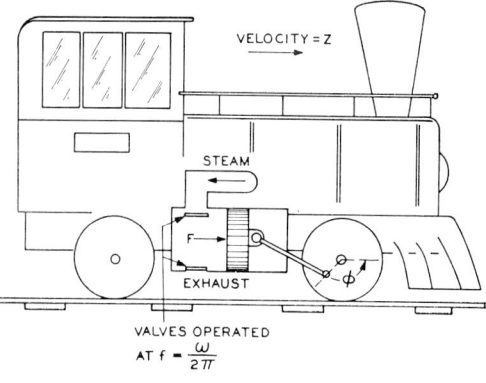

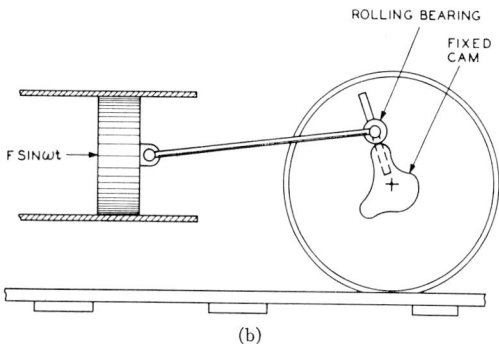

FIG. 4. (a) Illustration of "locomotive" mechanism of ac Josephson-type effect. (b) "Locomotive" system which could be driven at submultiple velocities.

force—i.e., the height (velocity) decreases linearly to zero. If f is as in Fig. 3, we get a "braking" action which leads to a decay between linear and exponential. It is an oversimplification to think of this as a constant force in time. Think of the locomotive as having square wheels and rusty bearings, so that the losses occur in some definite but not simple way during each cycle.

Now let us introduce the driven alternating current. This gives us an energy proportional to the phase coordinate, and alternating at frequency ω; it may be schematized by attaching a piston through a simple linkage to our locomotive wheels, and applying a force on the piston by (for instance) admitting steam intermittently at a rate $\omega/2\pi$:

$$F = F_0 \cos(\omega t + \phi_0). \qquad (42)$$

(See Fig. 4.)

As we very well know, such an alternating force is capable of keeping the locomotive going at a steady velocity Z if it is large enough, and if

$$\omega = 2\pi (dn/dt),$$

i.e., the force is applied once per revolution of the wheels. Also, as is less well-known but obvious, this mechanism can act as a brake or an accelerator at this

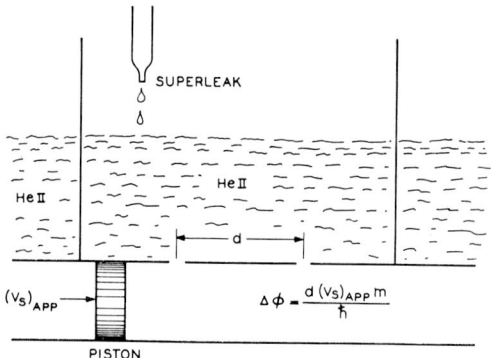

Fig. 5. Suggested space interference experiment with HeII.

velocity. The system will attempt to synchronize itself, even where the power available is not adequate to hold it in synchronization.

Another possibility is to run the wheels at n times the frequency of the valves which admit the steam. Clearly if the velocity of the locomotive is not perfectly uniform, or the valves do not regulate the steam harmonically, the two can run in synchronism and drive the locomotive. This kind of harmonic ($V=n\hbar\omega/2e$) is often observed in the true ac Josephson effect.

In helium and in the thin film bridges, the phase slippage takes place by means of vortices. This means that the motion to which the driving ac is coupled is highly anharmonic (square wheels). Another valid schematization of this is to make the piston linkage highly anharmonic—let it roll on a queer-shaped cam, for instance [see Fig. 4(b)]. Then not only can the wheels be driven faster than the frequency of the force, but also they can be driven at subharmonics, since, for example, the vortex may be formed only every nth cycle of the driving current. Thus we expect and observe both harmonics and subharmonics, in the Anderson–Dayem and Richards–Anderson experiments.[12,19]

Both of these experiments, for final quantitative description, must wait for quantitative theoretical treatments of the generation and motion of vortices. But the basic principle of phase synchronization of the external signal with the relative phase of the order parameter in the reservoirs is independent of detailed mechanism.

As for spacial interference experiments, the most promising seems to be the orifice analog of the Mercereau "current" experiment.[25] Here one drives a supercurrent through two orifices in parallel, and at the same time causes a superflow past the orifices on one side (see Fig. 5).

The second superflow enforces a phase difference $\Delta\phi$ at the two orifices on one side so that both orifices cannot simultaneously be at metastable minima of their energy curves (see Fig. 2), unless that phase difference is $2n\pi$. Another way of putting it is that the circulation

[25] R. C. Jaklevic, J. Lambe, J. E. Mercereau, and A. H. Silver, Phys. Rev. **140**, A1628 (1965).

through the two orifices must be quantized, leading to an additional circulating current which may aid breakdown (vortex creation) at one or both orifices. This effect would be periodic in the phase. Unfortunately, it is very sensitive to fluctuations and instabilities.

A phenomenon in which the idea of phase slippage must play an important role is the superfluid creep of films. It has been suggested that vortices form at the critical velocity with axes parallel to the film and perpendicular to the flow.[13] That is almost certainly correct—it gives dimensionally the correct critical velocity, which again is of order $\sim 10\hbar/md$. However, I would speculate somewhat differently on certain details.

First, the motion of the vortices. Examination of typical estimates indicates that the frictional forces on He vortices allow them to move with a velocity component parallel to the Magnus force of about 1% of the flow velocity. Thus vortices of the type postulated above will be generated at the solid surface and move out of the film into the free surface only a few thousand Å downstream (or vice versa, but this seems less likely). This will be the predominant dissipative mechanism if it occurs. It is hard to believe that vortex flow into the bulk fluids at either end can play an important role in a direct fashion.

There is, however, a somewhat more subtle question to be considered. If we are to take seriously the usual vortex creation critical velocity expression, h/md, it is not obvious that the smaller dimension of the film is really the value of d which must be considered. Why do not vortices form perpendicular to the film and move across it from one edge to the other? While at velocities $\sim h/mW$, where W is the width of the film ~ 1 cm (velocities $\sim 10^{-3}$ cm/sec) one would expect the formation of such vortices to be difficult dynamically, at velocities of 10^2–10^3 times that, still small compared with critical film velocities, there appears to be no such process at work.

It is suggested here that the predominant mechanism in film flow is the *pinning* of such vortices by surface flaws and thin spots in the film. Thus the film is a "hard superfluid": its flow is maintained by a pinning effect rather than by an absence of suitable vortices.

Vortices of the parallel type can also become pinned at either end and retard the motion of other vortices by their mutual repulsion. This is probably a mechanism which increases the critical velocity for rough substrates. Finally, the interaction of the pinned perpendicular vortices and the moving parallel vortices can lead to quite complicated effects: such things as the vortices reattaching themselves after a crossing in such a way that the pinned end attaches itself to the parallel vortex can occur, and become a mechanism for pinning of parallel vortices which may increase the pinning, and thus the critical velocity, as a function of the vorticity flowing into the film from the bulk liquids. Another mechanism which may play a role is motion of the free end of a pinned perpendicular vortex under the Magnus force until it becomes a pinned parallel vortex. This could be a copious source of vorticity.

In conclusion, then, the fundamental point made here is that in helium II. as in superconductivity, the Josephson equation and the associated concept of phase slippage are the most fundamental and exact consequences of our present theoretical understanding of superfluidity. Where phase slippage in superconductivity can occur in the absence of identifiable flux quanta, in helium with present technology it will always involve vortex lines because their core size is only a few Å, and no tunneling medium is available. Thus the crucial problem in helium flow is to find the vortex lines and study how they move across the flow path into the walls or disappear into the bulk. The complicated dynamics of vorticity is beyond the scope of this paper; we have merely speculated, with little quantitative study, in order to present concrete examples of the central ideas.

ACKNOWLEDGMENTS

I have benefitted throughout from the close collaboration of P. L. Richards. Discussions with P. C. Hohenberg, J. M. Luttinger, and W. L. McMillan were of value. A suggestion of D. J. Scalapino was reworked into the form of the space-interference experiment using orifices suggested in Sec. III. Questions asked by F. Reif and P. A. Wolff stimulated the first part of the work.

APPENDIX A. ODLRO VS MACROSCOPIC PARTICLE FIELDS

As explained in the introduction, Penrose's initial definition of the order parameter[4] in terms of a large eigenvalue of the density matrix was later extended by Beliaev,[5] by Gor'kov[6] (in terms of Green's functions), and a generalization conveniently named "off-diagonal long-range order" (ODLRO) by Yang.[20] That is, one writes

$$\langle 0_N | \psi^*(x)\psi(x') | 0_N \rangle = \sum_n \lambda_n f_n^*(x) f_n(x') \quad (A1)$$

and "ODLRO" is present when $\lambda_1 \sim N$, giving a contribution to the sum comparable to the sum of all others. One may then define a "ground state" $\langle 0_{N-1} |$ of the system with a different number of particles so that f_1 becomes a matrix element

$$[\lambda_1 f_1(x)]^{\frac{1}{2}} = \langle 0_{N-1} | \psi(x) | 0_N \rangle. \quad (A2)$$

In this way the necessity for dealing with states which are coherent mixtures of states with different numbers N of particles in the system is avoided, apparently, and for this reason most of the above authors prefer this scheme.

We argue that this approach is physically unnecessary, though valid, and occasionally inconvenient. For one, this definition does not permit convenient subdivision of a system. The over-all phase of f_1 is quite arbitrary—as it correctly should be for an isolated system with no particle exchanges permitted. On the other hand, once the phase at any space–time point is fixed the phase of the rest of the system is. Thus one cannot use the same description for any subdivision of the system; the λ and f for half of a bucket of He II simply do not describe it adequately. On the other hand, if one abandons the attempt to hold on to the broken gauge symmetry and ascribes a fixed, measurable phase to every superfluid system, recognizing that in principle the relative phase of any two may always be measured by a Josephson-type experiment, one immediately has a usable local description.

This is a satisfactory expedient unless the full generality of (A1) is meaningful—i.e., unless it is conceivable that more than one eigenvalue λ_1 is "large" and more than one intermediate state $| 0_{N-1} \rangle$ is involved. That is, we may ask whether the system may ever in any sense be a superposition of several distinct types of ODLRO. An attempt at such a theory was made by Gor'kov and Galitskii[26] for the d-state BCS theory, and proven invalid by various groups.[27] The question enters in many other cases—even, for example, in discussing flux quantization, one must be assured that one type of ODLRO only is present.

The most generally applicable argument here is that made by the author[28] in the Gor'kov–Galitskii case. It is that the two intermediate states $| 0_{N-1,1} \rangle$ and $| 0_{N-1,2} \rangle$ are demonstrably in entirely distinct Hilbert spaces in the limit $N \to \infty$, in the sense that $\sim N$ different particle states must be changed a finite amount to get from one to the other. Thus the $| 0_N \rangle$ state must be simply a superposition of a $| 0_{N,1} \rangle$ state communicating with $| 0_{N-1,1} \rangle$ and a $| 0_{N,2} \rangle$ state, and no interference effects whatever can connect the two types of states. In particular, every measurable quantity—energy, current, etc.—is the simple linear superposition of the two values. Then such a state is no more meaningful than Schrödinger's famous superposition of the quantum states of a dead cat and a live cat: a possible mathematical description of a physical absurdity.

APPENDIX B. A "NEW" COROLLARY IN CLASSICAL HYDRODYNAMICS?

Euler's equation of motion in a classical ideal fluid is

$$(\partial v/\partial t) + \nabla[(v^2/2) + \mu] = v \times \nabla \times v. \quad (B1)$$

μ is an appropriately defined chemical potential per unit mass. We now consider a general flow and draw a path C entirely inside the fluid—otherwise general—between two points P_1 and P_2 in the fluid. Points P_1 and P_2 are to be thought of eventually as being in reasonably quiet regions where the flow is steady over a long time T.

---

[26] L. P. Gor'kov and V. M. Galitskii, Zh. Eksperim. i Teor. Fiz. **40**, 1124 (1961) [English transl.: Soviet Phys.—JETP **13**, 792 (1961)].
[27] D. Hone, Phys. Rev. Letters **8**, 370 (1963); R. Balian, L. H. Nosanow, and N. R. Werthamer, *ibid*. **8**, 372 (1962).
[28] P. W. Anderson, Bull. Am. Phys. Soc. **7**, 465 (1962).

Let us now perform two integrations on (B1): first, along C from P_1 to P_2

$$\int_{P_1(C)}^{P_2} \frac{\partial v}{\partial t} d\mathbf{l} + (\tfrac{1}{2}v^2 + \mu)_{P_1}^{P_2} = \int_C (v \times \nabla \times v)\, d\mathbf{l}. \quad (B2)$$

[It was brought out in the discussions of the conference that (B2) is even more general than I had thought, in that most types of viscosity terms which might be added to (B1) involve gradients, so that if viscosity is not acting at P_1 and P_2 they cancel out.]

Second, we integrate over a very long time interval T and divide by T, thus taking a time mean value as is done in the virial theorem:

$$\int_{P_1}^{P_2} d\mathbf{l}\, T^{-1} \int_0^T \frac{\partial v}{\partial t} dt + \left\langle \left(\frac{v^2}{2} + \mu\right)_{P_2} \right\rangle_{Av} - \left\langle \left(\frac{v^2}{2} + \mu\right)_{P_1} \right\rangle_{Av}$$

$$= T^{-1} \int_0^T 2\int d\mathbf{l} \cdot (v \times \omega). \quad (B3)$$

We have defined ω, the vorticity, as $\tfrac{1}{2}\nabla \times v$ and written the time mean value at the points of steady flow in an obvious notation. The first term on the left-hand side is

$$T^{-1}\left[\left(\int_C v \cdot d\mathbf{l}\right)_T - \left(\int_C v \cdot d\mathbf{l}\right)_0 \right].$$

We define a "quasi-steady" flow as one in which this difference increases less rapidly than T; almost any turbulent flow one wishes to consider, or periodic flow, etc. will satisfy this condition. Then as far as time mean values are concerned we arrive at the basic corollary of Euler's equation:

$$\langle (\tfrac{1}{2}v^2 + \mu)_{P_2} \rangle_{Av} - \langle (\tfrac{1}{2}v^2 + \mu)_{P_1} \rangle_{Av} = \left\langle 2\int_{P_1}^{P_2} d\mathbf{l} \cdot (v \times \omega) \right\rangle_{Av}. \quad (B4)$$

It is easy to interpret the quantity on the right-hand side. Writing v as dr/dt, the particle velocity, this is

$$2\int_{P_1}^{P_2} (d\mathbf{l} \times dr/dt) \cdot \omega,$$

which is the *rate at which vorticity is being carried across the curve C* by the particle motion. Thus

$$\langle \Delta[(v^2/2) + \mu] \rangle_{Av} = \langle 2\text{``}(d\omega/dt)\text{''} \rangle_{Av}. \quad (B5)$$

We see immediately that this equation is far more important in a superfluid, where vorticity is conserved and quantized, than it is in ordinary fluids, where in a laminar flow, for instance, the right-hand side has little or no special significance. In helium, in fact, by turning to the integrated form of (B1) involving the potential we get the detailed Josephson equation without the special assumptions necessary here.

A number of somewhat surprising consequences immediately appear. One example is that the Pitot tube,[29] for instance, must involve transport of vorticity and thus motion of vortex lines in liquid He II. Ordinary aerodynamic lift and drag also would do so if the surface condition were $v=0$, but of course it is not; the vorticity there can be thought of as all in the surface layer outside the superfluid and thus not quantized.

I have tried at length to find a clear statement of (B4–5) in the classical literature, including the voluminous works of Rayleigh and Lamb, but have so far failed to find anything but corollaries and lemmas related to it. I will be pleased to hear from any reader who can find the theorem stated in this form; one can only assume that it was understood by the "classics" but is of no value in classical hydrodynamics so was never stated.

---

[29] J. R. Pellam, Phys. Rev. **78**, 818 (1950).

Multiple-Scattering Theory and Resonances in Transition Metals

(with W.L. McMillan)

Proceeding of the Int. School of Physics 'Enrico Fermi' XXXVII

(Academic Press, New York, 1967), pp. 50–86

At the Varenna School, 1966, there was a delightful gathering of the magnetism world organized by (eventually, Lord) Walter Marshall, in the heyday of the Varenna School when it was still possible to swim in Lake Como safely. Many contributions were good — I remember those of Volker Heine and Jim Phillips especially. There is much useful formalism in the discussion of Friedel's theorem and the Anderson model, but the actual calculation for "liquid Fe" is flawed by a serious program error discovered by J.M. Olson, a student of Kohn. Bill and I carefully never assigned the blame for this.

Multiple-Scattering Theory and Resonances in Transition Metals.

P. W. ANDERSON and W. L. MCMILLAN

Bell Telephone Laboratories Inc. - Murray Hill, N. J.

1. – Introduction.

It has long been known that the energy shift caused in a free-particle state of energy E and relative angular momentum l by a single spherical scattering potential is proportional to the partial-wave phase shift $\delta_l(E)$ [1]. FRIEDEL [2] pointed out that therefore the additional density of states near such a scatterer is given by

$$(1.1) \qquad \left(\frac{\mathrm{d}n}{\mathrm{d}E}\right) = \frac{1}{\pi} \sum_l (2l+1) \frac{\mathrm{d}\delta_l}{\mathrm{d}E},$$

and thus the total extra density of electrons in the neighborhood of the scatterer below E is

$$(1.2) \qquad \sum_l (2l+1) \frac{\delta_l(E)}{\pi}.$$

The conventional multiple-scattering theory for treating such problems as the electronic structure of random alloys or liquid metals [3] is based on correcting the self-energy of the electrons by an effective potential due to the individual scatterer which is proportional to the scattering matrix, either the T-matrix (which is diagonal for spherical scatterers in angular momentum representation),

$$T_{ll}(k) = \frac{(2l+1) \sin \delta_l \exp[i\delta_l]}{k},$$

or the reactance or K-matrix ($\propto \mathrm{tg}\,\delta_l$). Most naively, these contributions are simply added; in more sophisticated treatments the multiple-scattering corrections are more or less exactly taken into account, at the very least by calculating the T-matrix with the remainder of the scatterers corrected for by using the « medium » E vs. k relation to calculate k and δ. A characteristic form of the result, then is

$$(1.3) \qquad n(E) \propto \operatorname{Tr} \operatorname{Im} G(E) \propto \int d^3k \, \operatorname{Im} G(k, E),$$

$$(1.4) \qquad G(k, E) = \frac{1}{E - \hbar^2 k^2/2m - V_c(k, E)},$$

$$(1.5) \qquad V_c(k, E) = \frac{\pi \hbar^2 N}{m} \sum_i T_i(E, k) + \text{mult. scattering corrections}.$$

It is often stated that this theory is exact for low densities of scatterers, but the fact is that for very low density—a single scatterer or a few scatterers—it is not the same as the single-scatterer result (1), which is certainly exact in that case. The poles of G given by (4) are not at the actual shifted energies. This contradiction has been noticed by FUKUDA, NEWTON and DE WITT [4, 5] who have discussed the difficulties of the relationships among T, the « level shift » matrix K, and the true energy-level shift $\propto \delta_l(E)$. RIESENFELD and WATSON [6] attempted to construct a multiple-scattering theory around the use of the phase shift itself as an effective potential; but it is clear that such a procedure is necessarily only approximate, since as far as actual scattering effects are concerned, the phase shift is indeterminate to integral multiples of π. Thus as the phase shift goes through a resonance—e.g. from ~ 0 to $\sim \pi$ via $\pi/2$—the effective potential would show an unphysical shift, unphysical because $G(E, k)$ should be nearly the same for $\delta = 0$ as $\delta = \pi$, both representing zero scattering.

These questions are rather academic in most instances in which multiple-scattering theory has been used so far, such as nuclear matter, liquid non-transition metals or disordered alloys, transport theory in semiconductors etc., since in most cases either the phase shifts are small or only slowly-varying, or the density of states is not the quantity of major interest. But when we began to study the problem of random lattices of d-band metal atoms (with possible eventual applications to magnetic problems such as the properties of localized magnetic impurity states in transition metals or alloys, or eventually the Moriya theory of magnetism of transition metals) we soon realized that the essential characteristic of the so-called « $3d$ band » is that it is a true $l = 2$ scattering resonance in the midst of an essentially free electron continuum (the so-called $4s$-$4p$ band). This fact has recently been emphasized by ZIMAN [7]; it may also be noticed in the parametrizations of EHRENREICH and MUELLER [8] of the d-band problem in that the breadth of the d-bands is seen to come to a great extent from the parameter of admixture with free electron states, not mainly from the actual d-state overlap which is characteristic of tight-binding theory.

As a test problem on which to try out the methodology we chose what seemed to us the simplest, a crude model of « liquid iron » in which atomic positional correlations are neglected except to prevent core overlap. In this model we found the Fermi level right in the midst of an $l = 2$ resonance, and indeed when we attempted to use formulas like (3) to estimate the density of states we found that our results were off not just by a small amount but by an order of magnitude: such formulas do not include the resonant states. This was in spite of the fact that the use of the self-energy expression (2c) gave a very plausible rough fit to the E vs. k curves for solid iron in its two modifications [9], when we used phase-shifts computed from virtually the same potential used in the solid calculations [10]. (See the calculations of Sect. **3**).

In this paper we present a new method for calculating the density of states in energy in a multiple-scattering system, which has the advantage of being exactly valid—reducing to (1)— in the limit of low density and therefore avoids the difficulties of the usual multiple-scattering techniques. While—as with these earlier methods—in principle the starting formulas are exact, many approximations are necessary to do an actual calculation. Thus in order to demonstrate that these approximations do not do serious damage to physical reality we do an actual complete calculation, using the simplest possible approximations at every stage, on our test model of a random lattice of Fe atoms.

This is probably as demanding a calculation as could have been considered —very-high density of scatterers, a very strong, sharp resonance—so any reasonable result may be taken to represent success. In any case few previous calculations of density of states in a liquid metal has ever been carried so close to confronting reality, if not experiment.

The principle of our final method is to follow as closely as possible the correct calculation of the density of states for a single scatterer. Such a correct calculation can be done in the following way. We introduce a Greenian in ordinary space which satisfies the usual Green's function equation:

$$(1.6) \qquad \left[-\frac{\hbar^2 \nabla^2}{2m} + V(r) - E \right] G^+(r, r') = \delta(r - r'),$$

where $V(r)$ is the potential of the single scatterer. We choose the G^+ which satisfies outgoing wave boundary conditions:

$$(1.7) \qquad \lim_{r' \to \infty} G^+(r, r') \propto \sum_{l,m} c_l^m Y_l^m(\theta', \varphi') h_l(r'),$$

and the correct boundary conditions as either r or r' approaches the origin.

Now we calculate the density of states by

$$(1.8) \qquad \varrho(E) = \text{Tr Im } G(E) = \int^R d\mathbf{r} \text{ Im } G(r, r, E) \, .$$

By actually constructing G^+ in this case, we show that this gives a correct answer for the density of states in a sphere of radius R. We say « a » correct answer, because as pointed out by DeWitt [5, 4] to the order of the scattering effects the density of states depends on the boundary condition at R; the above gives a valid average over boundary conditions in the sense of DeWitt. In fact, it gives a valid result on integrating over any volume.

To do the multiple scattering calculation, we divide the volume up into cells in each one of which is a single scatterer; the potentials of no two scatterers are allowed to overlap. (This is the generalization of the « muffin-tin » potential approximation [11]). G^+ within such a cell may be expanded in spherical harmonics about its center and its dependence on $r_<$ (the lesser of the two radii) determined as for a single potential. The dependence on $r_>$ is determined by the problem of how the various spherical harmonics are reflected by the scatterers in all the other cells. Clearly we are looking towards a first approximation in which the cells are spherical and the reflections do not disturb spherical symmetry, so that the whole G^+ may be allowed to have spherical symmetry within the cell, and the density of states is a sum over angular-momentum values as in the single-scatterer case. But formally it turns out that the more complete calculation may be written down. Finally, the density of states is calculated by doing the integral (8) over each cell.

In this way the calculation is divided into two parts. Outside the cell we ask merely « how does the spherical wave of angular momentum l emanating from cell j propagate, and in particular how is it reflected back into that cell? » (*i.e.*, we put on outgoing boundary conditions and ask for the reflected wave). This is purely a scattering problem and for its solution clearly all the information we need is contained in the scattering matrix T of the individual potential, in principle. If we can, using all the wiles of multiple-scattering theory, succeed in replacing this scattering by that from an effective medium, well and good: that is correct. It is our contention that it is this and only this kind of question which a pure multiple-scattering theory is designed to answer, because its input simply does not contain enough information about the states inside the actual atomic spheres.

Inside the cell, on the other hand, we are using entirely a different piece of information, namely the explicit solution of the wave equation point by point for each angular momentum l. It turns out, indeed, that the same information is contained in the energy derivative of the phase shift, $d\delta_l/dE$, but *not* in the phase shift itself, as we will show. Once we know δ and $d\delta/dE$—*not* only the former— we may forget about the actual contents of the cell. In fact, the phase shift might be obtained from experiment, and many-body corrections inside the cell would then be included.

The order of presentation of the rest of the paper will be as follows.

First, we will have a rather long introductory discussion of the single-scattering problem, deriving the Friedel formula and showing its relationship to various other ways of treating the resonance problem. In the next Section we will present our unsuccessful attempt to do the liquid metal problem by purely multiple-scattering techniques, partly because the information acquired is used in the solution of the reflection problem, partly because the results are of physical interest, but mainly to show quite explicitly what the difficulty with that method is. Next we will present the formal background of our new method in a fairly simple form and motivate the approximations used in the following Section, where the actual calculation done for « liquid Fe » is described. In a concluding Section, we will discuss the results, which throw considerable light on the meaning of the electronic structure of the transition metals, and discuss other situations in which this method may be useful.

2. – Theory of the resonant single scatterer including some new or unfamiliar results.

One of the first things I would like to do in these lectures is to discuss the elementary theory of the single spherically symmetric scatterer and show how delicate a question it is and why it does not fit in well with the usual multiple-scattering techniques. I apologize for the fact that much of this is simple and familiar ground but as usual it is amazing how much insight a really soluble problem can give.

The elementary derivation of the Friedel formula goes as follows. We put our scatterer at the center of a large, spherically symmetrical box of radius R. In the absence of the scatterer, the solution of the radial wave equation for angular momentum l and energy $E = \varkappa^2$ which is okay at the origin is $j_l(\varkappa r)$ which asymptotically behaves like $(\sin(l\pi/2 + \varkappa r))/\varkappa r$. Thus the unperturbed energies are given by

$$E_0 = \varkappa_0^2 ; \qquad \sin\left(\frac{l\pi}{2} + \varkappa_0 R\right) = 0 ,$$

so

$$\varkappa_0 = \left(n + \frac{l}{2}\right)\frac{\pi}{R}, \qquad\qquad n = 0, 1, \ldots .$$

In the presence of the scatterer, the solution in any region outside the scattering potential, which we assume to be confined within a radius R_m, is the combination of spherical Bessel functions we will call $\mathscr{J}_l$:

(2.1) $$\mathscr{J}_l(\varkappa r) = \cos\delta_l j_l(\varkappa r) - \sin\delta_l n_l(\varkappa r), \qquad r < R_m.$$

and we hereafter extend the definition of $\mathscr{J}_l$ to be the correct solution of the radial wave equation at all radii. The phase-shift δ_l may be determined from the logarithmic derivation of $\mathscr{J}_l$ at any radius at which we may know it, by integration of the wave equation or otherwise:

(2.2) $$\begin{cases} \dfrac{\mathrm{d}\ln \mathscr{J}_l(\varkappa r)}{\mathrm{d}r} = \gamma_l(r, E), \\ = \dfrac{(\varkappa(\cos\delta_l j_l'(\varkappa r) - \sin\delta_l n_l'(\varkappa r))}{\cos\delta_l j_l - \sin\delta_l n_l}. \end{cases}$$

Since n_l asymptotically is

$$n_l(x) \sim -\frac{\cos(l\pi/2 + x)}{x},$$

$\mathscr{J}_l$ goes to

$$\mathscr{J}_l \sim (\varkappa r)^{-1} \sin\left(\varkappa r + \frac{l\pi}{2} + \delta_l\right),$$

so that the eigenenergy $E = \varkappa^2$ is determined by

$$\varkappa = \frac{(n + l/2)\pi - \delta_l}{R}.$$

The energy shift is

(2.3) $$\Delta E = 2\varkappa_0 \Delta\varkappa = \frac{2\varkappa_0 \delta_l}{R},$$

and self-evidently then it would seem that the change in total number of states would be given by:

(2.4) $$\begin{cases} \Delta n = \sum_l \dfrac{\mathrm{d}n_l}{\mathrm{d}\varkappa}\Delta\varkappa, \\ = \sum_l (2l+1)\dfrac{R}{\pi} \times \dfrac{\delta_l}{R} = \dfrac{1}{\pi}\sum_l (2l+1)\delta_l, \end{cases}$$

or

(2.5) $$\Delta\left(\frac{\mathrm{d}n}{\mathrm{d}E}\right) = \frac{1}{\pi}\sum_l (2l+1)\frac{\mathrm{d}\delta_l}{\mathrm{d}E}.$$

Now this formula is perfectly correct, in a certain sense, but in another sense it has a number of dubious features. One would think that it should represent a limit for fixed E as $R \to \infty$, but it does not; if we fix E and l, at any given instant either $\varkappa >$ or $\varkappa < ((n + l/2)\pi + \delta_l)/R$ and thus either $2l + 1$ states have been added or not. Since the total number of extra states is of order unity, the convergence is not uniform in any sense. As Fukuda and Newton showed, it is possible to find sequences R_n which completely falsify the density of states picture even in the limit $R_n \to \infty$. This is not unreasonable; since only ~ 1 state is affected by a resonance, the position of the box walls has an effect of the same order. Another way of putting it is that the energy shift (2.3) is of order $1/R$ or $N^{\frac{1}{3}}$ of the total number of states $\sim N$ in a box, i.e., of order (concentration n of perturbers)$^{\frac{1}{3}}$ rather than of order n. $n^{\frac{1}{3}}(R^{-2})$ of the states are affected, the remainder being affected much less than $\sim n$. This conclusion is, as Fukuda, Newton and Dewitt showed, probably independent of the spherical shape of the box and of the single-scatterer assumption since the degeneracy of the states in an arbitrarily shaped box will be high enough that an appropriate linear combination can be formed. This is a very serious situation because it indicates that it may be quite difficult to find a self-energy expression which will suitably reproduce the effect of a scatterer. In multiple-scattering theory, as we foresaw, we will want to write for the Greenian a function like

$$G(\omega, k) = \frac{1}{\omega - k^2 - \Sigma(k, \omega)},$$

where $\Sigma(k, E)$ is a self-energy which, if it is to reproduce perturbation theory at low densities, should be of order n. But then $G(\omega, k)$ will have no appreciable imaginary part at values of $\omega - k^2$ of the order $n^{\frac{1}{3}}$ where, as we have just seen, the most important scattering states are to be found.

There are a number of formal ways of showing that the Friedel formula (4) is nonetheless the correct result in any reasonable method of taking the limit and defining the density of states. The earliest use of a similar formula may have been by Wigner who showed that the density of states $(1/\pi)(\mathrm{d}\delta_l/\mathrm{d}E)$ which results from it may be thought of very precisely as the *lifetime* of the virtual state involved in the scattering process. If there is a true resonance, $\mathrm{d}\delta_l/\mathrm{d}E$ may be large and the resonant state long-lived; correspondingly, as we shall see, the state density or weight in Hilbert space may be rather high and difficult to relate to the scattering process.

The technique of de Witt for regularizing the $R \to \infty$ limit is rather formal and not relevant for our purposes. The principle was to find a really correct expression for the density of states in terms of the S-matrix, which could be generalized to slightly complex energies, and then one takes $R \to \infty$ first and $\mathrm{Im}(\omega) \to 0$ second.

Fundamental to the work I am going to talk about here is a way of doing the same calculation which, I believe, is simpler as well as more physical. It is more physical because it deals directly with the Green's function in space which is the quantity we really measure when we study a density of states. Almost any measurement reduces to introducing an electron at point r, time t, and measuring the resulting amplitude at point r', time t'; the mean in space of the imaginary part of the Fourier transform of $G(r, t; r, t') = \langle \psi(r, t')\psi^*(r, t)\rangle$ is what we mean by the density of states:

$$\varrho(\omega) = \operatorname{Im}\int d^3r\, G(r, r; \omega) = \operatorname{Im}\int d^3r \int dt\, \exp\left[(i\omega - \varepsilon sgnt)t\right] G(r, t; r, 0).$$

In general, the experiment we may be doing will not have box boundary conditions but will involve a hunk of metal with some kind of nonreflecting surroundings, so the boundary condition we *really* want is outgoing waves only: the electron introduced at time t should not return from outside some large radius R. What we really want, then is

$$\int_0^R dr\, G_+(r, r, E). \qquad G_+ \text{ has only outgoing waves}.$$

The equation for the Greenian in our units is

(2.6) $$\left(\nabla^2 + V(\mathbf{r}) - E\right) G(\mathbf{r}, \mathbf{r}') = \delta(\mathbf{r} - \mathbf{r}').$$

In the case of a single spherically symmetrical potential $V(r)$, the Greenian is also spherically symmetrical so that it may be expanded in spherical harmonics about the potential's center:

(2.7) $$G(\mathbf{r}, \mathbf{r}') = \sum_{l,m} Y_l^m(\hat{r}) Y_l^{-m}(\hat{r}') G_l(r, r') = \sum_l (2l+1) P_l(\cos\theta_{\hat{r}\hat{r}'}) G_l(r, r'),$$

where the $Y_l^m(\hat{r})$ are the spherical harmonics of the direction angles θ, φ of $\hat{r}$ and are normalized to unit solid angle;

(2.8) $$\int |Y_l^m(\hat{r})|^2 d\Omega = 4\pi,$$

G_l then obeys the appropriate radial equation

(2.9) $$-\frac{1}{r^2}\frac{d}{dr}\left(r^2 \frac{dG_l}{dr}\right) + \left(V(r) + \frac{l(l+1)}{r^2} - E\right) G_l(r, r') = \frac{1}{4\pi r^2}\delta(r - r').$$

It is very easily shown that the solution to (2.9) is

$$(2.10) \qquad G_l(r, r') = \frac{\varkappa}{4\pi} f_l(\varkappa r_>) \mathcal{J}_l(\varkappa r_<),$$

where both f_l and $\mathcal{J}_l$ are solutions of the radial wave equation which reduce in the region outside the « muffin tin » to certain spherical Bessel functions, $r_>$ and $r_<$ are the greater and lesser of r, r', respectively. $\mathcal{J}_l$ is in fact the function defined earlier which obeys the boundary condition at $r = 0$ and is therefore expressible outside the muffin tin as:

$$(2.11) \qquad \mathcal{J}_l(\varkappa r) = \cos\delta_l j_l(\varkappa r) - \sin\delta_l n_l(\varkappa r) \qquad r > R_m,$$

f_l is defined by two requirements: first, that its real part obey the condition which gives the correct normalization to G_l:

$$(2.12) \qquad W(\mathcal{J}_l, f_l) \equiv \mathcal{J}_l f_l' - f_l \mathcal{J}_l' = \frac{1}{\varkappa^2 r^2} \qquad (= W(j_l, n_l)).$$

This gives a discontinuity in G_l at $r = r'$ which is just such that (2.9) is satisfied:

$$\int_{r'-\varepsilon}^{r'+\varepsilon} \frac{\delta(r-r')}{4\pi r^2} \, dr = -\int_{r'-\varepsilon}^{r'+\varepsilon} \frac{1}{r^2} \frac{d}{dr}\left(r^2 \frac{dG_l}{dr}\right) dr + O(\varepsilon),$$

or

$$\frac{1}{4\pi} = -r^2 \frac{dG_l}{dr}\bigg|_{r'-\varepsilon}^{r'+\varepsilon} = -\frac{\varkappa^2 r^2}{4\pi}\left(f_l'(\varkappa r)\mathcal{J}_l(\varkappa r) - \mathcal{J}_l(\varkappa r) f_l'(\varkappa r)\right).$$

We may satisfy (2.12) by writing

$$(2.13) \qquad f_l = \mathcal{N}_l + a_l \mathcal{J}_l;$$

where $\mathcal{N}_l$ is the « other » solution of (2.1) which is 90° out of phase with $\mathcal{J}_l$ and is normalized like a spherical Bessel function. We may define $\mathcal{N}_l$ in terms of its behavior outside R_m, the muffin-tin radius, analogously to (2.11)

$$(2.14) \qquad \mathcal{N}_l(\varkappa r) = \cos\delta_l n_l(\varkappa r) + \sin\delta_l j_l(\varkappa r) \qquad r > R_m.$$

The second condition on f_l is that it obey whatever boundary condition we choose to set on the wave function at large distances. This determines a_l. If we set a real boundary condition—such as that f_l vanish at radius R—a_l will be real except at energies where $\mathcal{J}_l$ obeys the outer boundary condition, where a_l will have a pole. These are the eigenenergies.

As we have discussed already, we will use outgoing boundary conditions.

We cannot emphasize too strongly that the results even in the $n \to 0$ limit are not independent of boundary conditions, and that therefore the choice of boundary conditions is crucial to the method. Actually, having chosen this condition, in the real case random scattering removes the influence of the boundary, but only if we have handled it right in the first place. Incidentally, it has been called to my attention during the lecture that the following result and some of the derivations are identical with those of Friedel.

Outgoing waves result for a single scatterer if we set

$$(2.15) \qquad a_l = i; \; f_l = \mathcal{N}_l + i\mathcal{J}_l \to e^{i\delta_l} h_l(\varkappa r) \qquad r > R_m .$$

Here h_l is the spherical Hankel function.

We now calculate the density of states within a volume V by the orthodox procedure:

$$\varrho(E) = \frac{1}{\pi} \operatorname{Tr} \operatorname{Im} G(E) = \frac{1}{\pi} \int d\boldsymbol{r} \operatorname{Im} G(r, r, E) ,$$

$$(2.16) \qquad \varrho(E) = \frac{\varkappa}{\pi} \sum_l (2l+1) \int_0^R r^2 \, dr \, \mathcal{J}_l^2(\varkappa r) .$$

Here we have specialized to a spherical volume of radius R, within a large volume in which we apply outgoing boundary conditions. Approximately, this represents the density of states per atom in a low-density gas of scatterers of states per atom in a low-density gas of scatterers of density $3/4\pi R^3 = n$, where we neglect deviations from spherical symmetry and back-scattering corrections are at most of order n^2.

Now we demonstrate that (2.11) agrees with the phase-shift formula (2.1). First, however, observe that it passes the « empty-lattice » test; since $\mathcal{J}_l = j_l$ for an empty lattice ($V = 0$), the identity

$$\sum_l (2l+1) j_l^2 = 1 ,$$

leads to

$$(2.17) \qquad \varrho(E)|_{V=0} = \frac{\varkappa R^3}{3\pi} = \frac{\varkappa V}{4\pi^2} ,$$

which is correct.

A simple formula from perturbation theory suffices to demonstrate the check with (2.1). Though a similar formula is in the textbooks [17] for $l = 0$ at least, perhaps it will be easiest to derive. (Note: in fact, similar formulas were used by Friedel.) The equation satisfied by $\mathscr{I}_l$ is (2.9), (with r.h.s. = 0)

$$(2.9') \quad -\frac{1}{r^2}\frac{d}{dr}\left(r^2\frac{d\mathscr{I}_l}{dr}\right) + \left(V + \frac{l(l+1)}{r^2}\right)\mathscr{I}_l = E\mathscr{I}_l .$$

Let us perturb this equation by changing E to $E + dE$ out to a radius R, where $dE \ll R^3$. Then let us take the difference between the perturbed equation, the solution of which is assumed to be $\mathscr{I}_l + d\mathscr{I}_l$, and the original one:

$$-\frac{1}{r^2}\frac{d}{dr}\left(r^2\frac{d}{dr}(d\mathscr{I}_l)\right) + \left(V + \frac{l(l+1)}{r^2} - E\right) d\mathscr{I}_l = dE\,\mathscr{I}_l ,$$

to first order in dE. We now multiply by $r^2 \mathscr{I}_l/dE$ and integrate from 0 to R:

$$(2.18) \quad -\int_0^R dr\,\mathscr{I}_l \frac{d}{dr}\left(r^2 \frac{d}{dr}\left(\frac{d\mathscr{I}_l}{dE}\right)\right) + \int_0^R r^2 dr\left(V + \frac{l(l+1)}{r^2} - E\right)\frac{d\mathscr{I}_l}{dE} = \int_0^R r^2 \mathscr{I}_l^2(r)\,dr .$$

Now a series of partial integrations of the first term on the left transfers the differential operator into $\mathscr{I}_l$, which by virtue of (2.9') cancels out in the volume term. We are left with a surface integral:

$$(2.19) \quad R^2\left\{\frac{d\mathscr{I}_l}{dE}\frac{d}{dr}\mathscr{I}_l - \mathscr{I}_l\frac{d}{dr}\left(\frac{d\mathscr{I}_l}{dE}\right)\right\}\bigg|_R = \int_0^R r^2\mathscr{I}_l^2(r)\,dr ,$$

(the term at the origin vanishes because of the r^2 factor). Up to this point the l.h.s. could have been a surface integral of a derivative of a solution of the wave equation over any shape of boundary, but now we go specifically to spherical wave solutions. First, we observe that

$$\frac{d}{dE}(\gamma_l(R, E)) = \frac{1}{\mathscr{I}_l}\frac{d}{dE}\left(\frac{d\mathscr{I}_l}{dR}\right) - \frac{1}{\mathscr{I}_l^2}\frac{d\mathscr{I}_l}{dE}\frac{d\mathscr{I}_l}{dR} ,$$

so that

$$(2.20) \quad \int_0^R r^2\mathscr{I}_l^2(r)\,dr = R^2 \mathscr{I}_l^2(R)\frac{d\gamma_l(E, R)}{dE} ,$$

which is already in a useful form for computation, especially since continuity assures us that γ and $\mathscr{J}$ are the same at $R+\varepsilon$, slightly outside the sphere where we have changed the energy by dE, as they are inside, at $R-\varepsilon$, so that (2.20) contains quantities we can compute simply from tabulated values of γ_l. None the less it is edifying to manipulate further to show the relationship to (2.5). We simply take the derivative of γ_l using (2.2). That is a slightly delicate operation because the derivative referred to in (2.20) results from a change in energy dE *everywhere*, while our perturbation technique is only convergent if dE acts within a finite sphere. What we do is to observe that the energy derivative of γ_l at R is unaffected by the potential outside R, so that (2.20) gives that correctly. We then compute that directly in terms of δ and $\varkappa$ from the formal expression for γ_l,

$$(2.21) \quad \gamma_l(E) = \varkappa \left\{ \frac{\cos\delta_l j_l'(\varkappa R) - \sin\delta_l n_l'(\varkappa R)}{\cos\delta_l j_l - \sin\delta_l n_l} \right\},$$

by taking the complete derivative of this with respect to E, including its dependence on $\varkappa$ as well as $\delta_l(E)$,

$$(2.22) \quad \frac{d\gamma_l}{dE} = \frac{\partial\gamma_l}{\partial\delta_l}\frac{d\delta_l}{dE} + \frac{d\gamma_l}{d\varkappa}\frac{d\varkappa}{dE}.$$

The second term will turn out to give the unperturbed density of states in the weak-scattering limit; the first gives the Friedel term. To demonstrate the latter statement,

$$-R^2 \mathscr{J}_l^2 \frac{d\gamma_l}{d\delta_l} = -R^2\varkappa[(j\cos\delta - n\sin\delta)(-j'\sin\delta - n'\cos\delta) -$$
$$- (j'\cos\delta - n'\sin\delta)(-j'\sin\delta - n'\cos\delta)] = -R^2\varkappa(n_l j_l' - j_l n_l') = \frac{1}{\varkappa}.$$

Thus the « scattering » contribution is

$$(2.23) \quad \left(\int_0^R r^2 \mathscr{J}_l^2(r) dr\right)_{\text{scatt.}} = \frac{1}{\varkappa}\frac{d\delta_l}{dE}.$$

The « free » contribution does not simplify except asymptotically:

$$(2.24) \quad \frac{\partial\gamma_l}{\partial\varkappa} = \frac{\mathscr{J}_l'}{\mathscr{J}_l} + \frac{\varkappa R}{\mathscr{J}_l^2}(\mathscr{J}_l'' \mathscr{J}_l - (\mathscr{J}_l')^2).$$

Using the differential equation satisfied by $\mathscr{I}_l''$, this gives

$$(2.25) \quad -R^2 \mathscr{I}_l^2 \frac{\partial \gamma_l}{\partial \varkappa} = \left(\int_0^R r^2 \mathscr{I}_l^2(r)\,dr\right)_{\text{free}} = $$
$$= R^2 \left\{-\frac{\mathscr{I}_l \mathscr{I}_l'}{2} + \varkappa R[\mathscr{I}_l^2 + (\mathscr{I}_l')^2] - \frac{l(l+1)}{\varkappa R}\mathscr{I}_l^2\right\},$$

which, using the asymptotic behavior, goes asymptotically to

$$\left(\int_0^R r^2 \mathscr{I}_l^2(r)\,dr\right)_{\text{free}} \sim \frac{R^2}{\varkappa R} = \frac{R}{\varkappa}, \qquad (\varkappa R \gg l),$$

which is correct. As we have earlier remarked, the oscillatory dependence of the density for a given l on R in approximately the same order as the $d\delta/dE$ term is not incorrect but essential; it gives in fact the Friedel-R.K.Y. oscillations. The final usable formula for the density of states, then, is

$$(2.26) \quad \varrho(E) = \frac{1}{\pi}\sum_l (2l+1)\left(\frac{d\delta_l}{dE} - \frac{R^2}{2}\mathscr{I}_l^2(\varkappa R)\frac{\partial \gamma_l(\varkappa, \delta_l)}{\partial \varkappa}\right).$$

The second term on the right may be evaluated formally by (2.24), (2.25), or any of various other formulas using the special properties of the spherical Bessel functions, but none seem to gain much in clarity.

(2.26), then, is the correct formula in the very-low density case; the first term ensures that resonant states do not get lost, and represents the extent to which states « disappear » into the interiors of the scatterers, while the deviations of the second from the pure free-electron value presumably represent a first approximation to the effects of propagator modification which the usual theory gives and which would be present even if $\delta_l(E)$ were constant.

Before going on to make a rather long digression, I would like to try to convince you of the relevance of this formula and the further discussion of resonances to d-bands by showing that the essential nature of a d-band is a scattering resonance.

Band structures of transition metals are calculated—with remarkable, in fact rather surprising, success, I might add—by means of the A.P.W. method. The essential assumption of that method is the so-called « muffin-tin » potential: that the potential energy for an electron can be approximated by two parts: First, spherically symmetrical potentials confined to nonoverlapping « muffin-pan » regions around each atom, and second, a flat « muffin-tin » plateau between the atoms. In the muffin pans, the wave functions are expanded in terms of spherical harmonics; in the plateau regions, in terms of plane waves; and the matching condition at the muffin-tin radius R_m determines the eigenenergies and band structure. Unfortunately, the formulas, though easily computed, are not transparent and are extremely difficult to generalize to periodic structures.

To show that the d-band is not basically made up of overlapping *bound* d-states, we computed the $l = 2$ scattering phase shift for the Fe atomic potential used by L. MATTHEISS in his A.P.W. calculations (which is not seriously different from that of Wood for his more complete band structure.) Mattheiss was kind enough to lends us the Table of $\gamma_l(E, R_m)$ which is part of the APW. input.

It is easy enough to convert γ_l to a phase shift by means of the standard formula

$$(2.27) \qquad \operatorname{ctg} \delta_l = \frac{\varkappa n'_l(\varkappa r) - \gamma_l n_l(\varkappa r)}{\varkappa j'_l(\varkappa r) - \gamma_l j_l(\varkappa r)},$$

where n and j are the spherical Bessel functions.

The extra density of states is then given by the formula already referred to:

$$n(E) = \frac{N}{\pi} \sum_l (2l+1) \frac{\mathrm{d}\delta_l}{\mathrm{d}E}.$$

We plot the $l = 2$ contribution to this in Fig. 1. We see that the position of the d-band even calculated in this primitive fashion is very close to correct, and its width is about a factor two too small; by the nature of (1) it contains roughly the correct number of states. We emphasize that the obvious resonance is 0.7 Ryd, or 10 eV, above the top of the muffin tin: not a bound state in any, even remote, sense. Possibly the further complication of the d-bands may be thought of in terms, primarily, of a modification of the propagation of the plane waves which scatter off of one Fe atom by all the other atoms; this is the point of view we choose to take because it has some hope of being generalized to truly disordered systems such as liquids, alloys, molecules, paramagnetic magnetic metals, etc.

But first we have to really understand what may *not* be done, namely the usual kind of multiple-scattering theory. In order to understand that I have to lead you into the long digres-

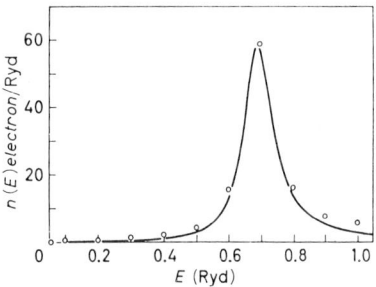

Fig. 1. – Friedel formula for «extra» density of states in d levels. Actual formula used includes all d states within R_s, *i.e.*, is equivalent to (37) for $l = 2$. Extra circled points are the result including the whole of (37).

sion I mentioned earlier; and this will in addition relate some of these ideas to things some of the other lecturers will be talking about.

What I will do is derive some relationships between the two good ways of doing the problem of a single scattering resonance; namely, the « extra-orbital » theory which I pioneered, and Friedel's phase-shift technique. The phase-shift technique has the advantage of being exact and general and certain of its results are quite simple; but the results are more difficult to visualize and to generalize. For instance, it is the extra-orbital theory which Hodges and Ehrenreich [8], Mueller [21], Phillips [22] and Heine [23] have been able to use as the basis not only of a pseudopotential theory of one-electron bands but of ferromagnetism, just as the extra-orbital theory allows a much neater treatment of the local-moment problem [24] than does the phase-shift theory. Much of this work, incidentally, has been stimulated by the ideas of the latter group and I would like to acknowledge my debt to them, though the point of view is different enough that I will not refer to their comparable ideas in detail.

First, I would like to put the extra-orbital theory in a formally satisfactory form, in a sense which will become clear shortly. The basic defect of the extra-orbital theory, from the point of view of trusting it on problems like the density of states which involve the normalization of the wave functions rather than the location of eigenenergies, is the fact that the set of wave functions used is overcomplete. The localized orbital one introduces to describe the resonance is expandable in plane waves, as is any function, so that the usual set

$$\varphi_{d,k} = \varphi_d, \frac{\exp[i\mathbf{k}\cdot\mathbf{r}]}{\sqrt{V}},$$

is not an orthonormal one: there is an overlap

(2.28) $$S_{dk} = \int \varphi_d^*(r) \exp[i\mathbf{k}\cdot\mathbf{r}] \frac{\mathrm{d}r}{\sqrt{V}}.$$

As discussed long ago by Cohen and Heine, this overcompleteness is a characteristic of pseudo-potential methods, and often can be turned to advantage—basically in that the choice of the set involves an arbitrariness which can be used to simplify the problem. The localized state problem is no exception.

Since the localized state problem is easiest done by Green's function methods, the essential desideratum is to have a Green's function technique suitable for overcomplete sets.

Let us suppose we have a set of states φ_α, $\varphi_{\alpha'}$ which are not orthonormal (overcomplete in this case) and that we also know the exact eigenstates φ_n of the same Hamiltonian. The true Greenian is of course

(2.29) $$G = \frac{1}{E - \mathscr{H}}; \qquad G_{nn'} = \delta_{nn'} \frac{1}{E - E_n},$$

and the true density of states

$$\varrho(E) = \frac{1}{\pi} \operatorname{Im} \operatorname{Tr} G(E) . \tag{2.30}$$

Formally, just as the correct secular equation for the nonorthonormal set is

$$|\mathscr{H} - E(1+S)| = |\mathscr{H}_{\alpha\alpha'} - E(\delta_{\alpha\alpha'} + S_{\alpha\alpha'})| = 0 \tag{2.31}$$

where

$$S_{\alpha\alpha'} = \int \varphi_\alpha^* \varphi_{\alpha'} \, dr - \delta_{\alpha\alpha'},$$

we may define a Greenian G' by

$$[E(1+S) - \mathscr{H}] G' = 1 , \tag{2.32}$$

which will be computed very analogously to the usual one. G', however, does not give the correct density of states using (2.30). We will show that the correct density of states is given by

$$\varrho(E) = \frac{1}{\pi} \operatorname{Im} \operatorname{Tr} (1+S) G' , \tag{2.33}$$

or, another way of putting it, that

$$G = (1+S) G' = [E - \mathscr{H}(1+S)^{-1}]^{-1} . \tag{2.34}$$

(2.33) is in fact almost self-evident except for the fact that (2.32) and the last equality in (2.34) are mathematically incorrect in the case of an overcomplete set of states. The reason for this is that both $[E(1+S) - \mathscr{H}]$ and $(1+S)$ considered as matrices in $\alpha\alpha'$ space have at least one zero eigenvalue in that case, so have no unique inverse. The simplest way to see that is to suppose we have made the set overcomplete by including two identical states φ_α. Then both have a pair of identical columns. But the set we use can be transformed into that case by a nonsingular transformation. Fortunately, (2.33) is still correct because the prefactor $(1 + S)$ is actually a projection operator which removes the objectionable state (I leave this as an exercise).

Neglecting, for this reason, the singular nature of the transformation, suppose there to exist a transformation

$$\varphi_\alpha = \sum_n U_{\alpha n} \varphi_n ,$$

then

$$(1+S)_{\alpha\alpha'} = \sum_n U^*_{\alpha n} U_{\alpha' n} = (U^* \tilde{U})_{\alpha\alpha'},$$

$$(1+S)^{-1} = \tilde{U}^{-1}(U^{-1})^*,$$

$$[(1+S)^{-1}\mathscr{H}]_{\alpha\alpha'} = \sum_{n\alpha''}(U^{-1})_{n\alpha}(U^{-1})^*_{n\alpha''}\int\sum_{n'n''}U^*_{\alpha''n'}\varphi^*_{n'}(r)\mathscr{H}(r)U_{\alpha'n''}\varphi_{n''}(r)\,\mathrm{d}r =$$

$$= \sum_{nn''}(U^{-1})_{n\alpha'}\mathscr{H}_{nn''}U_{\alpha n''} = (U\mathscr{H}U^{-1})_{\alpha\alpha'},$$

($\mathscr{H}$ is Hermitian). This is the appropriate quantity to enter the Green's function transformed to the α basis.

It is less simple to derive the correct expression for $G(r, r', E)$ from this overcomplete set, but with proper use of a modified metric, it is possible. Fortunately, it will not be necessary here. Another—and many may suggest simpler—way to do the whole job is to just orthogonalize the functions. The results are essentially identical.

Let us then use (2.30) and (2.31) to describe a localized resonant state such as we have just seen to be present at the level of the d-band in iron. The Hamiltonian in « α » representation is

(2.35) $$\mathscr{H} = \sum_k E_k n_k + \sum_m E_d n^d_m + \sum_{km} V^d_{mk}(c^*_k d_m) + \text{hermitian conj}.$$

Usually, for the sake of simplicity we will leave out the orientation index m since it will be fairly clear how it comes in. Nonetheless, it is important to realize that the fact that the resonance occurs for $l = 2$ and is degenerate is physically often vital; for one thing, the resonance only occurs because of the centrifugal barrier in the potential, $l(l+1)/r^2$. This is very large; for example, in Fe even at R_m it is 1 Ryd high (about $2 \times 3/(2.4)^2$), and even larger at smaller radii where V is still not very large or steep.

Our choice of the matrix element V_{dk} is to some extent a parametrization, since one can think of φ_d as simply defined by

$$V_{dk}\varphi_d \propto V(r) j_l(k, r).$$

This is similar to the use of core states in the pseudo-potential theory to exhaust the Hilbert space of the atomic potential.

φ_d as well as the plane wave states are chosen normalized; the only overlap is between the plane waves and k (see (2.28)). With (2.25) and (2.35), the same « Lee model » feature which allows the exact solution of the usual localized state problems still holds. The manipulations of ref. [24] (Anderson, L.S.) still may be carried out. Let us in fact calculate the density of states in this model. The Green's function equations are (using (2.32) and (2.34)):

(2.36)
$$\begin{cases} a) & (E - E_d) G_{dd} = 1 + \sum_k (V_{dk} - ES_{dk}) G_{dd}, \\ b) & (E - \varepsilon_k) G_{kd} = S_{kd} + (V_{kd} - ES_{kd}) G_{dd}, \\ c) & (E - \varepsilon_k) G_{kk} = 1 + (V_{kd} - ES_{kd}) G_{dk}, \\ d) & (E - \varepsilon_{k'}) G_{k'k} = (V_{k'd} - ES_{k'd}) G_{dk}, \\ e) & (E - E_d) G_{dk} = S_{dk} + \sum_{k'} (V_{dk'} - ES_{dk'}) G_{k'k}. \end{cases}$$

These equations may be much simplified by writing

(2.37) $$(V_{dk} - ES_{dk}) = V'_{dk}.$$

The first two are solved trivially by substituting (2.36 b) in (2.36 a) and we obtain

$$(E - E_d) G_{dd} - \left(\sum_k V'_{dk} \frac{1}{E - \varepsilon_k} V'_{kd} \right) G_{dd} = 1 + \Sigma_k V'_{dk} \frac{1}{E - \varepsilon_k} S_{kd}.$$

Define

(2.38) $$\Sigma_d = \sum_k V'_{dk} \frac{1}{E - \varepsilon_k} V'_{kd},$$

as the self-energy of the d-state, and we get simply

(2.39) $$G_{dd} = (E - E_d - \Sigma_d)^{-1} \Big(1 + \sum_k V'_{dk} (E - \varepsilon_k)^{-1} S_{kd} \Big).$$

The last three equations require a small amount of manipulation. We insert (2.36 d) into (2.36 e), adding and subtracting the term with $k = k'$ in the sum:

$$(E - E_d) G_{dk} = S_{dk} + \left(\sum_{k'} V'_{dk'} \frac{1}{E - \varepsilon_{k'}} V'_{k'd} \right) G_{dk} - V'_{dk} \frac{1}{E - \varepsilon_k} V'_{kd} G_{dk} + V'_{dk} G_{kk}.$$

Using (2.36 c) we observe that

$$V'_{dk} G_{kk} = \frac{V'_{dk}}{E - \varepsilon_k} + V'_{dk} \frac{1}{E - \varepsilon_k} V'_{kd} G_{dk},$$

and the last term cancels against the next-to-last above, giving

(2.40) $$G_{dk} = [E - E_d - \Sigma_d]^{-1} \left(S_{dk} + \frac{V'_{dk}}{E - \varepsilon_k} \right).$$

This then may be substituted into (2.36 c), giving

$$(2.41) \quad G_{kk} = \frac{1}{E-\varepsilon_k} + \frac{V'_{kd}}{E-\varepsilon_k}[E-E_d-\Sigma_d]^{-1}\frac{V'_{dk}}{E-\varepsilon_d} + \frac{V'_{kd}}{E-\varepsilon_k}S_{dk}[E-E_d-\Sigma_d]^{-1}.$$

We show that this technique gives us the correct density of states:

$$\varrho(E) = \frac{1}{\pi}\operatorname{Tr}\operatorname{Im} G = \frac{1}{\pi}\operatorname{Im}\left[\operatorname{Tr} G^0 + (E-E_d-\Sigma_d)^{-1} + \right.$$
$$\left. + \sum_k \frac{(V'_{kd}S_{dk} + V'_{dk}S_{kd})}{(E-\varepsilon_k)(E-E_d-\Sigma_d)} + \sum_k \frac{|V'_{kd}|^2}{(E-\varepsilon_k)^2}[E-E_d-\Sigma_d]^{-1}\right].$$

If we take into account the definition of Σ_d (2.38), and of V' (2.37), we observe that the last three terms may be written

$$\frac{d}{dE}\ln(E-E_d-\Sigma_d),$$

so that

$$(2.42) \quad \pi\varrho(E) = \pi\varrho_0(E) + \frac{d}{dE}(\text{phase angle of } \{E-E_d-\Sigma_d\}).$$

it is fascinating that this «Friedel Theorem» is completely independent of whether overlap is taken into account—see my earlier Nottingham paper [25]. φ_d could have been introduced from some external Hilbert space and it would still be true.

This demonstrates indirectly that

$$(2.43) \quad \operatorname{tg}^{-1}\frac{\operatorname{Im}\Sigma_d}{E-E_d-\operatorname{Re}\Sigma_d} = \delta_l;$$

we will show this directly immediately.

Before going on to the basic point of my digressing in this direction, I would like to make a few remarks about other implications of this formalism. One is that these techniques are very closely related to the pseudopotential-OPW schemes for simplifying and parametrizing band theory. Again there the scheme is to choose a best «core» function to represent the large effects of the potential; another way of thinking of that scheme is as an approximation that $\delta_l - n\pi$ is small and slowly varying. The use of a suitable set of V_{kd} to represent δ_l would be another reasonable technique of parametrization.

The Mueller-Phillips parametrization scheme [21], [22] for d-bands is a multiple-scattering version for perfectly periodic lattices of this extra-orbital theory as can actually be demonstrated; other ways of doing the same theory can be demonstrated using the above techniques, but so far I have been unable to develop a good way to reproduce the aperiodic system with this scheme even though I have tried quite hard—mostly because originally this seemed to be the only way to take into account the extra density of d-states engendered by the scattering resonance.

Now let me demonstrate the equality (2.43) more directly. We have already defined the phase shift in terms of the asymptotic behavior of the scattered wave functions. This may be restated in terms of the Greenian G_l in the asymptotic region as follows: by (2.10) and (2.15), asymptotically

$$G_l \sim \frac{\varkappa}{4\pi} \exp[i\delta_l] h_l(\varkappa r_>) [\cos \delta_l j_l(\varkappa r_<) - \sin \delta_l n_l(\varkappa r_<)] .$$

The unperturbed Greenian is

$$G_l^0 \sim \frac{\varkappa}{4\pi} h_l(\varkappa r_>) j_l(\varkappa r_<) .$$

Let us find the difference

$$\Delta G_l = G_l - G_l^0 = \frac{\varkappa}{4\pi} h_l(\varkappa r_>) [(\exp[i\delta]) \cos \delta - 1) j_l(\varkappa r_<) - \exp[i\delta] \sin \delta n_l(\varkappa r_<)] =$$

$$= \frac{\varkappa}{4\pi} h_l(\varkappa r_>) [(-\sin^2 \delta + i \sin \delta \cos \delta) j_l - \exp[i\delta] \sin \delta n_l] ,$$

(2.44) $$\Delta G \sim -\frac{\varkappa}{4\pi} \exp[i\delta_l] \sin \delta_l h_l(\varkappa r) h_l(\varkappa r') .$$

Since G_l^0 contains the requisite discontinuity, ΔG is continuous; it consists of outgoing (scattered) waves, with amplitudes given by

(2.45) $$T_{ll} = \frac{\varkappa}{4\pi} \exp[i\delta_l] \sin \delta_l ,$$

as we might well have expected.

What we now demonstrate is that the solutions $G_{kk'}$ of (2.36) have exactly the same form and that again the complex phase angle of T is δ_l. The dd and dk parts of the Greenian clearly do not contribute to its asymptotic value, so that we can merely look at G_{kk} and $G_{kk'}$; we write down directly the difference between $G_{kk'}$ and $G_{kk'}^0 = \delta_{kk'}/(E-\varepsilon_{\varkappa'})$:

(2.46) $$\Delta G_{kk'} = \frac{V'_{kd}}{E-\varepsilon_k} \frac{1}{(E-E_d-\Sigma_d)} \frac{V'_{dk'}}{E-\varepsilon_{k'}} .$$

(the overlap term is an artifact of the method and uniportant.) As noted in my Nottingham paper, this is of the form

$$\Delta G = G_0(k) T_{kk'} G_0(k')$$

where

(2.47) $$T_{kk'} = V'_{kd} \frac{1}{E - E_d - \Sigma_d} V'_{dk},$$

and it is pretty obvious that the phase angle of T will indeed turn out to be

$$\delta_l = \text{tg}^{-1} \frac{\text{Im}(\Sigma_d)}{E - E_d - \text{Re}(\Sigma_d)}.$$

Clearly (2.46) is of the form of two spherical waves multiplying an energy-dependent factor, like (2.44). In r-space one of the two spherical waves is

$$f(r) = \int d\boldsymbol{k} \exp[i\boldsymbol{k} \cdot \boldsymbol{r}] \frac{V_{kd}}{E - k^2}.$$

V_{kd} has the angular properties of the d-state — $l=2$ in an example — so the angular integration will leave us with:

$$f(\boldsymbol{r}) = Y_m^l(\hat{r}) \frac{1}{r} \int \frac{k \, dk \, V(|k|)}{E - k^2} \exp[ikr].$$

Here we simply have a choice as to how we go about handling the poles at $\varkappa = \pm \pi$ which contribute the only long-range part of this integral. This corresponds to satisfying the boundary conditions on G, and clearly we choose to have only outgoing waves, using the outgoing pole and including no contribution from the incoming one—i.e., closing our contour towards $+iE$. Then $f(r) = \text{const} \times Y_l^m(e^{i\varkappa r}/r) = \text{const} \times h_l(\varkappa r) Y_l^m$. Thus (2.46) is precisely equivalent to (2.44). It is interesting to note another check: if

$$\delta = \text{tg}^{-1} \frac{\text{Im} \Sigma}{E - E_d - \text{Re} \Sigma}; \quad \sin \delta \exp[i\delta] = \frac{\text{Im} \Sigma}{E - E_d - \Sigma},$$

and

$$\text{Im} \Sigma = \frac{1}{\pi} V_{kd}^2 \varrho(E),$$

which in fact is the numerator which results from doing the k integrations in (2.46).

3. – Multiple-scattering theory and why it doesn't work.

Now we come to the point of the digression. This is that in constructing a multiple-scattering theory, it is possible to deal accurately *only* with the asymptotic part of the scattering matrix T. While it is still possible to invent a T such that

$$(3.1) \qquad G = G_0 + G_0 T G_0$$

which includes the strictly internal structure involved in the dd part of the Green's function, a) this convention is rather arbitrary—as we see, G may even include contributions from a purely extraneous Hilbert space and give the right density of states; and b) it is only the asymptotic part of T which is treated properly in a multiple-scattering theory, which is a type of theory which deals only with how waves propagate as they scatter against many centers, not with the internal structure of the centers.

Formally, (3.1) is a correct statement of the answer to the single-scatterer problem. Now let us consider a collection of many scatterers α, β of density n. If the internal structures of these scatterers don't overlap, it is almost obviously true that the whole Greenian is given by

$$(3.2) \qquad G = \sum_\alpha G_0 T_\alpha G_0 + \sum_{\alpha \neq \beta} G_0 T_\alpha G_0 T_\beta G_0 + \sum_{\alpha \neq \beta \neq \gamma} G_0 T_\alpha G_0 T_\beta G_0 T_\gamma G_0 + \ldots .$$

If we could forget about the exclusions, (which are there because the T-matrix exactly solves the single-scatterer problem) it would be correct to write

$$G = \frac{1}{1 - \sum_\alpha G_0 T_\alpha} G_0 ,$$

and since

$$G_0 = [E - H_0]^{-1} ,$$

this gives

$$(3.3) \qquad G = \frac{1}{E - H_0 - \sum_\alpha T_\alpha} .$$

The standard dictum for dealing with this is very simple. It is pointed out that the off-diagonal matrix element $T^\alpha_{kk'}$ is multiplied by a factor $\exp[i(k - k') \cdot R_\alpha]$, and by Van Hove's remark in one form or another in a random lattice $\sum_\alpha \exp[i k \cdot R] \sim \sqrt{V}$.

Thus a remarkably good approximation in a truly random system is to take

$$(3.4) \qquad \left(\sum_\alpha T_\alpha\right)_{kk'} = n T_{kk} \delta(k-k') \ ;$$

we include only the forward scattering, and modify the propagation of waves for that.

There are three basic difficulties with this simple, and often remarkably accurate, procedure, the first two of which have been extensively discussed in the literature by such authors as Lax, Edwards, Brueckner, Watson, etc., the third of which is new.

1) The question of « off-the-energy-shell » propagation. The single-scattering problem is solved accurately by T, but the use of the asymptotic values and the usual definition of δ_l means that we know it really well only for $T_{kk}(E)$ where $E = k^2$. But the pole in G of eq. (3.3) where things are really happening is not near $E = k^2$. Thus somehow we must find a way of calculating T from whatever data we have with emphasis on accuracy not *on* the energy shell but near $E - k^2 = T$.

2) The question of the exclusions in the sum, as well as of taking into account the actual structure of the multiple-scattering medium.

These are actually related; both lead to correlations in the phases of the omitted terms which change the results radically for reasonably large densities of scatterers.

So far as I know, the technique we used for solving problems (1) and (2a) simultaneously (2b, the structure factor, we have not succeeded in handling as yet, and I believe it is very important) was new. This technique has not been justified formally as yet in terms of multiple scattering theory but I believe it can be.

The first guess one might make as to a right answer is as follows. We try

$$(3.5) \qquad G(k, E) = \frac{1}{E - k^2 - 2\pi N \sum_l ((2l+1)\exp[i\delta_l]\sin\delta_l)/\varkappa} \ ,$$

but it is very little effort to make the additional improvement of allowing the particles to move in the « medium » between scatterings, that is, to let them propagate with the medium wave vector k outside the atomic sphere rather than with the wave vector in the « muffin-tin plateau » potential. Then we modify (2.27) to read

$$(3.6) \qquad \operatorname{ctg} \delta_l(k, E)_{R_s} = \frac{k n'_l(kR_s) - \gamma_l n_l(kR_s)}{k j'_l(kR_s) - \gamma_l j_l(kR_s)} \ .$$

This phase shift depends on k and R_s, the radius of the sphere at which we assume the potential to be modified by the « effective potential » V_c.

Now, the interesting thing about (3.5) using (3.6) to calculate δ (and of course replacing $\varkappa$ by k in the denominator) is that as a rough description of propagation of plane waves in the medium of resonant scatterers, it seems to be reasonably good. However, in fact it is seriously incorrect, as can be verified simply by looking at the case where above all it should be valid, namely very weak interactions $\delta_l \ll \pi$. We can then express γ_l and therefore δ_l in terms of the mean potential matrix element $V_{kk} = \bar{V}$: in fact we have already given the appropriate formulas. When $T(\varkappa, E)$ is used, the energy shift is then correctly given; but when we modify to $T(k, E)$ the change in δ_l does not cancel against the change in k to lowest order (the arithmetic of this is tedious but simple in principle). Thus in fact when we tried (3.4) it did not fit to nearly free electrons at the bottom of the band.

Why is this? The reason is formally the exclusions in the sum (3.2). When we calculate T_α in the presence of the medium which has in it all the other scatterers, and which we approximate by $\bar{T}$, we must not include the multiple scattering off scatterer α twice. We must *remove* the part of the medium referring to α and replace it with the true scatterer. But this is just the same as saying, not that $E - k^2 = \bar{T}$, but that (T) of scatterer $= (T)$ of medium, *i.e.* that

$$(T)_{\text{scatterer less medium}} = 0 \;.$$

Thus the scheme we use is to calculate T using (2.6) and set it $= 0$:

$$(3.7) \qquad T_{kk} \propto \frac{1}{k} \sum_l (2l+1) \exp\left[i\delta_l(k, E)\right] \sin \delta_l(k, E) = 0 \;.$$

This gives us a (in general complex) relationship between k and E which is the dispersion relation for the random lattice. In general, for complex k we must use complex δ_l; this generalization to complex phase angles and Bessel functions may be shown to be perfectly valid, and did not complicate the equations much. Fortunately for the simplicity of this theory, we will soon see that the Greenian itself is essentially meaningless, at least in calculating the density of states; what we really want is a picture of how k runs as a function of real E. We do a straightforward machine iteration using the Table we already have available of γ_l as a function of real E, and using the complex k which gives us $T = 0$. Fig. 2 shows the result of this for « liquid Fe »; this plus the density of states, as we will discuss later, was calculated in less than a minute of IBM 7090 time. The shaded region lies within

$$k_r \pm k_i(E).$$

We did not wish to consider any specific crystal structure, since we were seeking a spherical average; thus it seemed most sensible to study essentially liquid Fe, and for simplicity we neglected correlations completely except to keep the atoms out of each other's immediate neighborhoods.

In recent years a number of calculations of band structures for transition metals have been carried out by the augmented plane-wave method in one of its varieties [9, 10]. One of the striking characteristics of the results of these calculations is that on a coarse scale, averaging over ~ 1 V intervals, the band structure is very approximately isotropic in k space. Examples are shown in Fig. 3 and 4. In Fig. 3 we plot the computed band structure of body-centered Fe from WOOD [9] in two very different crystal directions, and observe that the general level, the wide spread and relatively low energy value near the zone boundary, and several other features are common to both. This is precisely the opposite to what would be expected if the d bands were basically tight-binding atomic orbitals coupled by tight-binding atomic overlap-type matrix elements; instead we can only conclude that in some

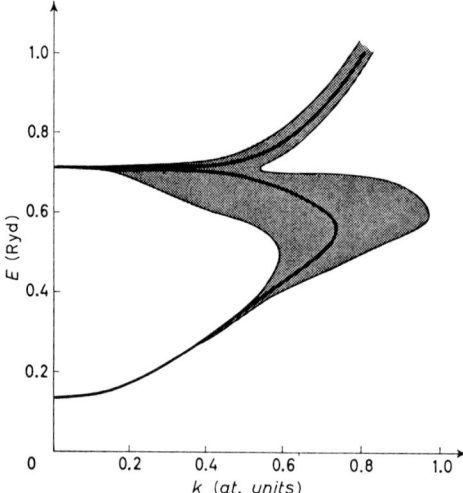

Fig. 2. – Result of multiple scattering calculation of $k(E)$. Line is $k_r(E)$, dashed area is within $k_r \pm k_i$.

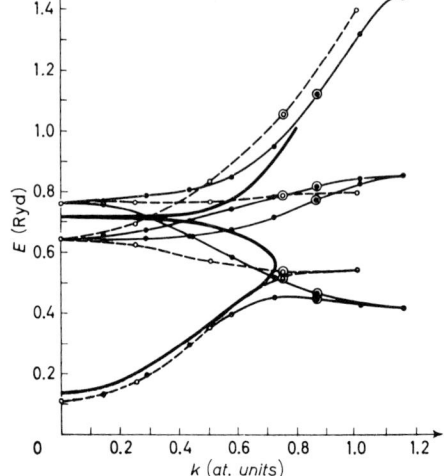

Fig. 3. – Band structure from J. H. Wood's calculations of b.c.c. Fe compared with crude multiple-scattering estimate. Dashed curves and points: 111 and 100 symmetry directions: ● (010); ○ (111). Points used in Fig. 4a circled. Solid curve: multiple scattering theory.

sense the d-band width and shape is contributed primarily by the interaction of each individual atomic d level with the essentially isotropic free plane-wave spectrum: in other words, that the d band is acting like a resonance or « virtual state » in the midst of a free-electron band.

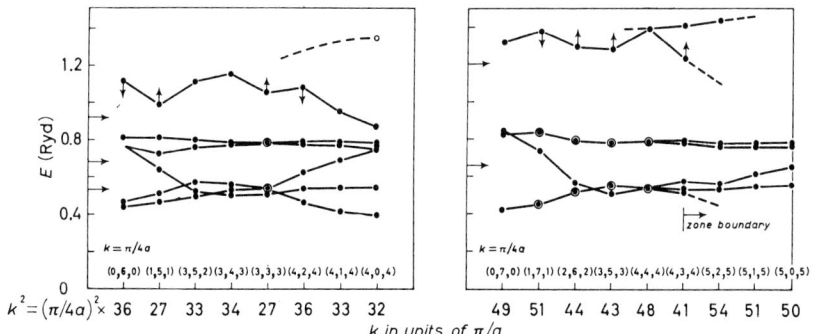

Fig. 4. – Wood's computed energy levels along roughly circular sections in k space. a) At (101) radius (largest entirely within first zone). Radii squared are given by lower abscissa scale; actual k-space points by upper. Arrows to left are theory at a mean radius. Circled points are degenerate. Arrows on points on upper branch indicate direction of shift which would result from correcting to mean radius. b) At (111) radius. Scales, etc. as in a). From $(\pi/a)(\frac{5}{4}, \frac{1}{2}, \frac{5}{4})$ on, points are outside first zone; therefore dashed branches are not relevant and have been omitted. Probably these points are only marginally significant.

In Fig. 4 we have demonstrated the same sort of information in another way. Here we have plotted a series of energy levels from Wood's published band structure around a rough great circle of a sphere in momentum space, keeping $|k|$ approximately fixed but moving in the 110 plane from (in Fig. 4a) $(\pi/a) \times (0, \frac{3}{2}, 0)$ through $(\pi/a)(\frac{3}{4}, \frac{3}{4}, \frac{3}{4})$ to $(\pi/a)(1, 0, 1)$, and Fig. 4b along a larger circle, on which the latter half is outside the first Brillouin zone so that certain of the levels which clearly belong to the smaller of the two equivalent k values have been omitted. The striking thing in each of these Figures is the rough isotropy of the band structure. There are some deviations near symmetry points, but the grouping into three roughly isotropic bands is obvious. Also fascinating is the fact that levels seem to jump back and forth between bands; it is clear that the « bands » need not containing integral numbers of levels.

Comparing the two Figures for k vs. E, we see that the « multiple-scattering plus resonance » theory does indeed give a very good qualitative description of the energy bands.

The reader may besomewhat disturbed by our comparing a theory meant to describe a perfectly random liquid with calculations which presumably give highly accurate results, but for the regular b.c.c. lattice of the same density. Our reason is twofold: first, after all the crystal is one of the infinity of structures the liquid might take up, so it should be expected to have its energy levels in at least the same neighborhoods as the liquid; and second, we hope to show that the solid's energy levels do not depend severely on its precise crystal structure and are roughly isotropic in k space, in order to show that it is not direct d-d overlap which plays the greatest role in d band behavior but interaction between the atomic resonance and the modified propagation of free plane waves. If that were so, then the whole field of the properties of d-band metals might be understood much more simply than heretofore.

Another attractive feature of Fig. 1 is the fact that the d band seems automatically to divide itself into two regions: the states below the resonance, which are shifted down by it and broadened (the breadth serving as a crude representation of the Bragg scattering which gives the anisotropy due to crystal structure in the solid) and the states nearer $k = 0$, which are above the resonance, and occupy a relatively narrower energy region. One can show that the former would be more mixed with free electron states, more sensitive to lattice effects, and generally less localized than the latter. All of this corresponds nicely to the actual behavior of the d band, i.e., superconductors in the lower half, magnets in the upper, and to our knowledge of the nature of the electronic states.

On the other hand, as far as the other use of the Greenian, as a measure of density of states by (1.3), is concerned, expressions like (3.4) are completely hopeless. They simply fail to indicate the existence of the 4 extra resonant states that are present in the peak of Fig. 1. We argue here that this dichotomy is not merely a consequence of the very approximate way in which we have handled the multiple-scattering corrections, but is intrinsic to the method. Essentially, the multiple-scattering Greenian tells us how plane waves propagate, so its region of singularity tells us the location of the energy states in the E, k plane, because every energy state will have some plane wave mixed into it. What it does not tell us is what goes on inside the atoms: how much d-state should be included to give a properly normalized wave-function in the interior of the atom?

This is why multiple-scattering methods seems to check satisfactorily against the Korringa-Kohn-Rostoker method (similar to A.P.W.) [13, 16] in the periodic case. In that case the mere existence of a pole of G may be used to presume a state and determine its normalization, because perfectly periodic system k-space also has a precisely definite number of states per cell *on a given branch*. Note that the accepted derivations give only the secular determinant; the proof that the normalization is correct, which is necessary to check the method, is missing and, we believe, fails.

In the limit of very low density the difficulty is really quite evident. The $\gamma_i(E_0)$ at a fixed energy E_0 are the only information available to us about the potential to determine the $T(E_0)$. Even the « off-energy-shell » scattering may be determined from γ_l, as we have shown; though in the limit of low density that will not be needed. But G is, in this theory, entirely known if we know T: and thus also ϱ by (1.3).

On the other hand, $\gamma_l(E_0)$ does not tell us very much about the potential. It is undoubtedly possible to construct potentials having almost any desired value of $(d\delta_l/dE)_{E_0}$ for a fixed $\gamma_l(E_0)$. In this limit, again—and almost certainly in all cases—it is essential to know enough about V to calculate $d\delta_l/dE$ in order to calculate the density of states correctly. The difficulty is, of course, that T does determine G near its pole in E and k space, but that this information is inadequate to the order of approximation needed.

That this inadequacy is fundamental is an opinion to which we are inclining more and more. Some indications from the study of the one-scatterer problem we went into in depth are very strong in this direction. First is that although in that problem it is possible to define a generalized T-matrix which is correct for all k and E the part we used in the partial summations of multiple-scattering theory was merely the asymptotic value; we could add in almost an arbitrary d-d part without seriously changing that. In fact, the Friedel theorem is found to hold even in the limit where the extra orbital comes from an entirely different, orthogonal Hilbert space. We can think of this as the limit of shrinking down the atomic sphere smaller and smaller, until the k-values describing the important part of the Green's function become completely inaccessible and unrelated to the single-pole type of approximation we are using.

Another way of putting it is that no type of random phase approximation, which at some point or another must be used to remove the residual « incoherent » scattering, can remove the contributions of order unity from within the atoms. Their root mean square is always $\sim n$, not actually $n^{\frac{1}{2}}$. This is related to the difficulties pointed out by De Witt et al., that the energy shifts are not $\sim n$, but $\sim n^{\frac{1}{2}}$.

Riesenfeld and Watson [6] attempted to remedy this difficulty by using for T the phase-shift itself multiplied by $(1/\pi)(2l+1)$ This seems to us an unacceptable procedure, although to lowest order in n it may be made to give the right answer. That is, however, only by making a perfectly *ad hoc* assumption that δ varies with k and E separately as though $E=k^2$ even where that is not valid. It is also hard to see how this definition may be made unitary, *i.e.*, can succeed in treating real scattering.

An interesting exercise we carried out was to actually Fourier-transform, approximately, the correct $G(r, r', E)$ which we derived in Sect. 3, to see what the real structure of $G(k, k, E)$ was. It became evident that the main difficulty is the single-pole approximation embodied in an expression like (3.3). That states that the important singularities of G are spread over a region of order

n, where actually—as remarked in ref. [4, 5]—some of the important poles spread over $\sim n^{\frac{1}{3}}$ for small n.

Thus our conclusion is that theoretically as well as by practical test, the multiple-scattering theory gives us a good description of the propagation of plane waves even in the presence of large numbers of strong, resonant scatterers; but it does not give us the density of states. (Note that this conclusion means that in so far as a Fermi surface can be defined in a random lattice, the famous theorem about the constancy of the Fermi-surface volume may well not be accurate.) We will not further belabor this lack of success but go on to the correct method.

4. – Formal description of method of calculation of density of states.

The essential point of the method is that it is a Green's function method which is none the less exact for a single isolated scatterer. The derivation of this result has already been given.

Of course, the d band in any transition metal, liquid or solid, is not the low-density case. We have already shown in Fig. 1 the result of a calculation using (2.26) (the second term doesn't differ much from free electrons) and an $R = R_s$ appropriate to solid Fe. This may be compared with the computed density of states of solid iron [18]. (See also Fig. 5). While the position of the d band was correct and its width only off by a factor 2 even in this most severe test, we should hope to do better, especially in view of the excellent agreement with the general trend of the E vs. k curves obtained by the multiple scattering theory in Fig. 3. We should emphasize again that the density of states calculated from the latter theory directly is so ridiculous we do not bother to present it. What Fig. 1 shows is that (2.26) is not a poor zeroth approximation: the biggest influence on the d band is the d-free electron interaction which produces the resonance, not direct d-d interaction; Fig. 3 shows that this resonance is also the largest contributor to the propagation modification of quasi-free electrons in the d band region. Is there a way to combine these two successes?

The thechnique for doing this was sketched out in the introduction. It starts with another exact observation: that the Greenian in any Wigner-Seitz cell around a given atom is determined if we know how the eigensolutions of a given energy reflect from the boundary of that cell. From the outset again we make one highly simplifying approximation, that of spherical symmetry. Then the above statement says merely: if we can calculate how the spherical wave $\mathcal{J}_l + i \mathcal{N}_l$ reflects from the contents of the system outside the boundary, and we already know δ_l and $\mathrm{d}\delta_l/\mathrm{d}E$ which essentially tell us all we need to know about how it propagates inside the cell, we can construct the entire Greenian. In fact, we have already written out the necessary expression, eq. (2.10) with

f_l defined by (2.13); the required reflection coefficient determines the single (for any l) parameter a_l. Using that fact, we can immediately write down the density of states, which is (2.15) modified by the factor $\operatorname{Im} a_l$ for each l:

$$(4.1) \quad \varrho(E) = \frac{\varkappa}{\pi} \sum_l (2l+1) \left[\int_0^{R_s} r^2 \, \mathrm{d}r \, \mathscr{J}_l^2(\varkappa r) \right] \operatorname{Im} a_l =$$

$$= \frac{1}{\pi} \sum_l (2l+1) \left[\frac{\mathrm{d}\delta_l}{\mathrm{d}E} - R_s^2 \mathscr{J}_l^2(\varkappa R_s) \frac{\partial \gamma_l(\varkappa, \delta_l, R_s)}{\partial \varkappa} \right] \operatorname{Im} a_l \,.$$

Then the final step in the problem is to determine a_l: what radial wave function

$$f_l = \mathscr{N}_l(\varkappa r) + a_l \mathscr{J}_l(\varkappa r)$$

in the region between the muffin tin radius R_m and the nearest other atom ($> R_s$) satisfies outgoing boundary conditions as $r \to \infty$ *when the other scatterers are taken into account?* This, now, *is* purely a multiple scattering problem; we have no difficulties of principle such as we had in establishing (4.1) but merely wish to answer the straightforward question of how spherical waves of energy $E = \varkappa^2$ propagate in the presence of the scattering medium.

The method we have chosen to make a first approximate attack on the calculation of a, in order to test the whole scheme, is the most straightforward one possible, consistent with the spirit of the multiple-scattering technique: we treat the system of scatterers outside R_s as a continuous dispersive medium in which the dispersion relation between E and k is that calculated in Sect. 3. That is, from R_s out we assume the medium to be characterized by a *complex* k which is given as a function of E by

$$(4.2) \quad T(k) = 0$$

where

$$(4.3) \quad T(k) = \frac{1}{k} \sum_l (2l+1) \exp[i\delta_l] \sin \delta_l(k, R_s) \,.$$

Note that this is different from the usual Lax-Edwards procedure of assuming real k and asking for the corresponding complex E, E_r giving the position and E_i the breadth of the band of states in energy which propagate with wave vector k. Here we are borrowing a suggestion from PHARISEAU and ZIMAN [15]; we inquire with what wave vector k_r, and decay constant in space k_i, waves of a fixed energy E propagate. We do this because we are forced to match to solutions of the wave equation internal to the atom *for real E*, complex E inside the atom being meaningless as well as inconvenient; our philosophy is to inquire for the reflection coefficient for a fixed E, not the propagation in time for a fixed k.

Replacing the medium by a continuum, then, the outgoing wave outside R_s is given by

$$(4.4) \qquad f_l(r)|_{r>R_s} = h_l(kr)$$

where for complex k we define h_l by its asymptotic behavior $\propto (e^{ikr}/r)$.

a_l is determined by matching the logarithmic derivatives of (4.4) and (2.13)

$$(4.5) \qquad \varkappa \left[\frac{\mathcal{N}'_l(\varkappa R_s) + a_l \mathcal{J}'_l(\varkappa R_s)}{\mathcal{N}_l + a_l \mathcal{J}_l} \right] = k \frac{h'_l(kR_s)}{h_l(kR_s)}.$$

While the basic computational formula for a_l must remain (4.5), considerable insight into its physical meaning is given by expressing it in terms of the reflection coefficient R_l for spherical waves of angu'ar momentum l. Let us write f in terms of ingoing and outgoing waves (h_l^* and h_l respectively):

$$(4.6) \qquad f_l(r)|_{r>R_s} = \text{const}\,(h_l(\varkappa r) \exp[i\delta_l] + R_l h_l^*(\varkappa r) \exp[-i\delta_l]).$$

(the phase factors define the reflection coefficient appropriate to the $\mathcal{J}_l$, $\mathcal{N}_l$ functions rather than j_l, n_l). By equating (4.6) to (2.13), we obtain the expressions:

$$\text{const.}\,(\mathcal{N} + i\mathcal{J} + R_l(\mathcal{N} - i\mathcal{J})) = \mathcal{N} + a\mathcal{J},$$

$$(4.7a) \qquad a_l = i\left(\frac{1-R_l}{1+R_l}\right),$$

$$(4.7b) \qquad \text{Im}\,a_l = \frac{1-|R_l|^2}{|1+R_l|^2}.$$

This exhibits several reasonable physical properties. 1) when $R = 0$, no reflection, $\text{Im}\,a_l = 1$: wherever the effect on k is relatively small (not the scattering: a small energy shift near $\varkappa = 0$ can make k pure imaginary) there is little correction. (Showing that we do have an expansion in n, incidentally.) 2) Where $|R| = 1$, $\text{Im}\,a = 0$: perfect reflection gives no contribution *i.e.*, the system must be in a « pass band » to have any density of states-except 3) where $R = -1$, which represents a *true bound state*: $\mathcal{J}_l$ satisfies the boundary condition alone. Near a resonance, where δ moves from 0 to π and $k - \varkappa$ varies wildly, we can expect to see regions with $a_l > 1$ and $a_l < 1$, and, in particular, since, as we saw in Sect. 3, $k = 0$ occurs, we will have one point where $a_l = 0$. The actual computation shows approximate bound states both above and below this point.

One final comment on the meaning of a_l. If $kR_s \gg 1$ the spherical functions are well approximated by plane waves, and in any case it is suggestive to write down the plane wave result for R. In the plane-wave approximation,

$$(4.8) \qquad R \simeq \exp[2i(\varkappa R + \delta)] \left(\frac{1 - k/\varkappa}{1 + k/\varkappa}\right).$$

We see immediately that this has the qualitative properties mentioned: for $k = 0$ or pure imaginary (a «stop band») $|R| = 1$ and $\text{Im } a = 0$ unless $(\varkappa R + \delta) = \pi/2$. For $k = \varkappa$, $R = 0$, $\text{Im } a = 1$. If k is real,

$$(4.9) \qquad \text{Im } a \simeq \frac{k\varkappa}{k^2 \cos^2(\varkappa R + \delta) + \varkappa^2 \sin^2(\varkappa R + \delta)},$$

(4.9) is useful in understanding the behavior of a quantitatively. For instance, it has the correct behavior—linear in k—at the $k = 0$ band edge.

5. – Details and comments on the calculation.

(4.5) completes the set of formulas necessary to do the simple type of calculation we have set ourselves. Let us summarize the set of basic equations. The density of states is to be calculated from (4.1)

$$(5.1) \qquad \varrho(E) = \frac{1}{\pi} \sum_l (2l+1) \, \text{Im } a_l \left[\frac{\mathrm{d}\delta_l(\varkappa, E)}{\mathrm{d}E} - R_s^2 \mathscr{J}_l^2 \frac{\partial \gamma_l(\varkappa, R_s)}{\partial \varkappa} \right].$$

The bracketed terms may be calculated directly if we have γ_l at any radius $R \leqslant R_s$ and $\geqslant R_m$ tabulated as a function of E: δ_l from (3.6):

$$(5.2) \qquad \text{ctg } \delta_l(k, E)_R = \frac{k n'_l(kR) - \gamma_l n_l(kR)}{k j'_l(kR) - \gamma_l j_l(kR)},$$

(with $k = \varkappa$ this is independent of R) and $\gamma_l(\varkappa)$ by continuing from R to R_s and using

$$(5.3) \qquad \gamma_l(\varkappa, R_s) = \varkappa \left(\frac{\cos \delta_l \, j'_l(\varkappa R) - \sin \delta_l \, n'_l(\varkappa R)}{\cos \delta_l \, j_l - \sin \delta_l \, n_l} \right),$$

a_l is to be computed from (4.5), where $\mathscr{J}_l$ and $\mathscr{N}_l$ are defined by (2.1) and (2.14)

$$(5.4) \qquad \mathscr{J}_l(\varkappa r) = \cos \delta_l(\varkappa, E) j_l(\varkappa r) - \sin \delta_l(\varkappa, E) n_l(kr),$$

$$(5.5) \qquad \mathscr{N}_l(\varkappa, r) = \cos \delta_l(\varkappa, E) n_l(\varkappa, r) + \sin \delta_l(k, E) j_l(\varkappa r)$$

and k is determined by (4.2) and (4.3).

The calculations were carried out on an IBM 7090. A full run took less than a minute, using the set of γ_l $0 \leqslant l \leqslant 10$ computed by L. F. Mattheiss and kindly lent us by him. (As mentioned earlier, these refer to practically the same potential and muffin-tin height used by him for his band-theoretic calculations. These did not disagree within our accuracy with those of Wood, which we took for comparison.) To determine k we used T_l only for $l = 0, 1, 2$ since the contributions of higher l were small, uncertain and varied little. A subroutine based on the recurrence relations was used for all the spherical Hankel functions, complex as well as real. Contributions to (4.1) up to $l = 5$ were computed, but $l > 3$ never contributed significantly.

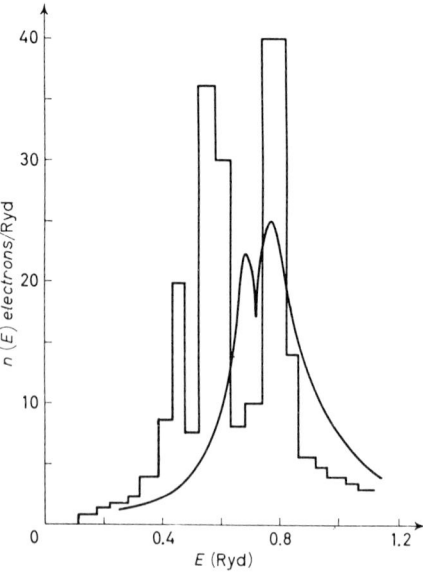

Fig. 5. – Results for $\varrho(E)$ from theory of this paper, as compared with a somewhat smoothed version of Wood's density of states as computed by MATTHEISS [18] for b.c.c. Fe: hystogram b.c.c. (WOOD); solid curve: « liquid Fe » theory.

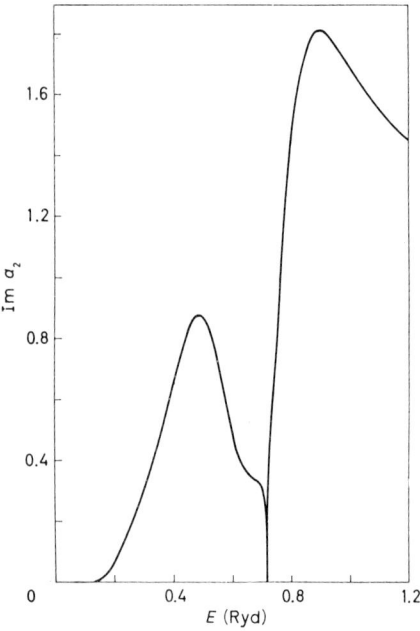

Fig. 6. – Im a_2, demonstrating sharp minimum at center of resonance and side peaks.

The results are shown in Fig. 5, along with the computed density of states for the crystalline solid. We see that even in the completely amorphous, uncorrelated liquid the density of states has begun to show a two-peaked structure and to have roughly the correct width and position of the d band relative to the Fermi level, indicating considerable support for our original notion that the primary features of the electronic structure of d-band metals, like those of ordinary polyvalent metals, are not strongly dependent upon the crystal structure. To a great extent they result from the interactions of the atom and the electron gas. In the polyvalent metals, the interaction is relatively weak in general and the nearly-free-electron approximation is valid. Here, the interaction is strong—a resonant scattering—and the properties of the electron gas as a medium are strongly affected by it; but the detailed structure is not enormously important.

The result for Im a_2 shown in Fig. 6 is perhaps even more interesting than the density of states. Here we see the large correction to the simple Friedel result which we anticipated from eq. (4.7b): a « hole » in the middle of the resonance where $k \to 0$ and peaks, actually on both sides, coming from regions where R is negative and there is a true resonance.

It is noticeable that the d band is more severely split in crystalline Fe than in our calculation, and that there are actually 3 peaks of the spectrum and a pronounced, fairly isotropic dip in the lowest energy branch in most crystal directions near the zone boundary. We suspect that this feature *is* a crystal structure effect, which however appears to be roughly isotropic in k space. We think this might well be reproduced, and the structure of the d band in the solid even more nearly duplicated, by using in the liquid structure factor as Edwards, for example, has suggested [19]. He has discussed the effect of the correlated structure of the liquid on the E vs. k curves, and shown that inclusion of the structure factor can lead to band-gap-like phenomena in the multiple-scattering theory. An approximate way to include such effects would be to include 2-atom correlations to second order in the coherent potential by means of some such formula as

$$(5.6) \qquad V_c(k) = n \left\{ T_{kk} + \sum_{k'} T_{kk'} \frac{1}{E(k') - E(k)} T_{k'k}(c_2(k - k') - 1) \right\},$$

where c_2 is the two-body correlation function in k-space. This is the direct analogue using scattering matrices T of Edwards formula (2.13), $(c_2 - 1)$ entering because the T matrices include the second-order terms for uncorrelated atoms. A calculation using something like (5.6) for a reasonable c_2 is clearly called for as the next step of our program.

Meanwhile, we suggest that our density of states is likely to be a fair approximation to that for the actual uncorrelated « liquid » we have treated, but not as good a one to that of true liquid transition metals—or of crystalline alloys, for that matter. One difficulty with (5.6) is that it may lead to two solutions for k as a function of E, and we do not yet have a scheme for matching into a birefringent medium in order to calculate a_l.

6. – Results, conclusions and remarks.

We started this work with the idea that the information contained in the modern A.P.W. calculations of electronic bands in metals could also be used to carry the problem of the electronic structure of disordered metals to the stage of practical computation. While an extensive purely theoretical literature on the disordered lattice problem exists, little or no serious attention has been paid to actual calculations of the density of states with realistic atomic potentials. We have certainly shown how the simplest version of multiple scattering theory can be immediately applied, with surprising success, even to d-band metals such as iron. As noted by Ziman independently, the basic concept is that the d level is a virtual resonance coming in the midst of the free electron levels and severely perturbing them.

The big surprise of the work was the discovery that while multiple scattering theory gives excellent results for the propagation of free-electronlike waves through the lattice, it fails completely, even in principle, in giving us a technique for calculating the density of states in the d-band resonance. This failure is quite fundamental, in our opinion, in the sense that the claim that the multiple scattering theory is an expansion in the density of scatterers, or at least that it is a *useful* expansion, is not correct. We have developed a method for combining a proper treatment of the density of states with the multiple scattering theory, and applied it to the artificial problem of uncorrelated « liquid » metals, mostly just to demonstrate its computational feasibility.

Actually, we might almost *a priori* have guessed that a state-counting problem must come up in passing through the d band. Consider the question of the Fermi surface of liquid Ni, Cu, or Cu-Ni alloys. We should ask ourselves: at what point does the Fermi surface—if there is a reasonably well defined one, as is envisaged in any multiple scattering theory—stop enclosing a volume large enough for all the d states? Must that transition not take place continuously? We believe so, and in consequence we suggest also that at least in liquids and disordered lattices the theorem that the Fermi surface of an element contains an exact integral number of electron states may fail—of course, the problem of definition of a Fermi surface is a serious one.

The implications of our theory for d-band metals are fairly extensive. Our original interest had been in the problem of magnetism: if we could treat a disordered lattice of resonant states, we could study spin disorder in a magnetic metal and its effect on the electronic energy of the band—*i.e.*, we could work toward an understanding of the physics behind the parametrization of the magnetism problem by MORIYA [20]. We can deal with magnetic as well as nonmagnetic alloys of d band metals on a realistic basis, perhaps eventually develop methods of treating at least the average coupling of d electrons to lattice vibrations, etc.

An even more distant and less obvious prospect has also suggested itself to us. The central feature of our method is that we can treat the wave function on an individual atom correctly on a *local* basis, no matter how complicated and disorderly the surroundings may be. The other atoms enter only as a boundary condition which determines the weight to be given to a particular atomic state. In particular, the surroundings need not be solid or liquid: they could be part of a strictly finite, more or less complicated, molecule. We have here a *local* technique which might be persuaded to help demonstrate why the properties of chemical bonds are insensitive of the global nature of their surroundings, without giving up the relative rigor which comes from using the molecular orbital—*i.e.*, band theoretical—point of view. To do this problem we may have to give up the simplification of sperical symmetry, but that is not very hard to do.

* * *

We acknowledge helpful conversations with J. M. ZIMAN, M. LAX and P. SOVEN, and the loans mentioned in the text by L. F. MATTHEISS of his working material from several A.P.W. calculations. Our interest in the problem started from discussions witt T. MORIYA.

REFERENCES

[1] M. I. GOLDBERGER and K. M. WATSON: *Collision Theory* (New York, 1964). As discussed in that book (expecially Chapter VIII) similar properties of the phase shift were used as early as 1936 by E. BETH and G. E. UHLENBECK: (*Physica* **3**, 727 (1936)); but the earliest discussions which are directly relevant here were those in ref. [2], [4] and [5].
[2] J. FRIEDEL: *Adv. in Phys.*, **3**, 446 (1954).
[3] See for example J. R. KLAUDER: *Ann. of Phys.*, **14**, 43 (1961); S. F. EDWARDS: *Proc. Roy. Soc.*, A **267**, 518 (1962); J. L. BEEBY and S. F. EDWARDS: *Proc. Roy. Soc.*, A **274**, 395 (1963). The simple form of multiple-scattering theory mentioned here is merely a predecessor of the sophisticated techniques of these authors but no serious differences would be entailed by including more complication, since we do not wish to discuss atomic correlation effects in detail. The whole scheme goes back to LAX (ref. [12]).
[4] N. FUKUDA and R. G. NEWTON: *Phys. Rev.*, **106**, 1558 (1956).
[5] B. S. DE WITT: *Phys. Rev.*, **103**, 1565 (1956).
[6] W. B. RIESENFELD and K. M. WATSON: *Phys. Rev.*, **104**, 492 (1956).
[7] J. M. ZIMAN: *Proc. Phys. Soc.*, **86**, 337 (1965).
[8] L. HODGES and H. EHRENREICH: *Phys. Lett.*, **16**, 203 (1965); F. H. MUELLER: to be published, *Phys. Rev.*
[9] J. H. WOOD: *Phys. Rev.*, **126**, 517 (1962).

[10] Our calculations were actually based on the potential for iron used by L. F. MATTHEISS: *Phys. Rev.*, **134**, A 970 (1964), which agrees with (9) to well within our accuracy; this was used because Dr. MATTHEISS was kind enough to lend us his tables of computed logarithmic derivative values.
[11] F. S. HAM: in *The Fermi Surface*, W. HARRISON and M. W. WEBB, ed. (New York, 1960).
[12] M. LAX: *Rev. Mod. Phys.*, **23**, 287 (1951). The basic philosophy is more fully discussed in ref. [1].
[13] J. KORRINGA: *Physica*, **13**, 392 (1947).
[14] W. KOHN and N. ROSTOCKER: *Phys. Rev.*, **94**, 1111 (1954).
[15] P. PHARISEAU and J. M. ZIMAN: *Phil. Mag.*, **8**, 1487 (1963).
[16] J. L. BEEBY: *Proc. Roy. Soc.*, A **279**, 82 (1964); only this of these two references actually presents a formula for $\varrho(E)$ and no proof of the weighting factor is given.
[17] P. MORSE and H. FESHBACH: *Methods of Theoretical Physics* (New York, 1053), p. 1694.
[18] L. F. MATTHEISS: *Phys. Rev.*, **136**, A 1893 (1965).
[19] S. F. EDWARDS: *Proc. Roy. Soc.*, A **267**, 518 (1962).
[20] T. MORIYA: *Solid State Comm.*, **2**, 239 (1964).
[21] F. M. MUELLER: *Phys. Rev.*, **153**, 659 (1967).
[22] J. C. PHILLIPS: *Phys. Rev.*, **153**, 669 (1967) and this volume.
[23] V. HEINE: *Phys. Rev.*, **153**, 673 (1967).
[24] P. W. ANDERSON: *Phys. Rev.*, **124**, 41 (1961).
[25] P. W. ANDERSON: *Proc. Intern. Conf. on Magnetism 1964*, Institute of Physics, London (1965), p. 17.

Kondo Effect I
Comments on Solid State Physics **1**, No. 2, 31 (1968–69)

Kondo Effect II
Comments on Solid State Physics **1**, No. 6, 190 (1968–69)

The Kondo Effect III
Comments on Solid State Physics **3**, No. 6, 153 (1971)

The Kondo Effect IV
Comments on Solid State Physics **5**, No. 3, 73 (1973)

This series is virtually unavailable and I know of no comparable review which summarizes in a reasonably brief compass what the issues which puzzled us were, and how they gradually became resolved.

The Kondo Effect. I

I title this so only in order to have a starting point, and the Kondo effect has certainly been the starting point for the kind of anomalous scattering effect at the Fermi surface I want to talk about, and the focus of the theoretical activity in the field.

"The Kondo Effect": rare and delightful indeed it is to have an effect named after the theorist who explained it, not the experimentalist who stumbled on it. de Haas, de Boer, and van den Berg,[1] who may have been the first to see it, called it the "Resistance Minimum" and the name stuck for about 30 years. To give it the quantitative description first given by van der Leeden,[2] we may define it as the appearance of a term

$$\rho \propto -Ac \ln T \tag{1}$$

in the resistivity of a dilute alloy at low temperatures. Since the phonon resistivity goes as T, and the usual residual resistivity due to other impurities approaches a constant as $T \to 0$, this leads to a minimum in the resistivity at a temperature

$$T_{min} \propto c^{\frac{1}{5}}, \tag{2}$$

the empirical law evinced by Kondo[3] as evidence for his theoretical expression (1).

Kondo postulated that the phenomenon occurred only for magnetic impurities. In the early days, while the classic work showed that a minimum was associated with magnetic impurities in many cases, neither metallurgy nor an understanding of when impurities are magnetic were up to establishing this exactly; but now papers are to be found which casually assume that observation of a $\ln T$ term means the impurity is magnetic—I believe rightly. (Note: I discuss for the time being only the term linear in c. The resistance maximum and other concentration-dependent effects are a whole separate story to themselves.)

Kondo's model for a magnetic impurity is simply a site in the lattice

31

possessing a spin-vector S which interacts with the free electrons of the otherwise nonmagnetic metal via an exchange Hamiltonian

$$H_K = J\, \mathbf{S} \cdot \mathbf{s}(\mathbf{r})$$
$$= J \sum_{\sigma,\sigma'} \sum_{k,k',\alpha} S^\alpha c^\dagger_{k\sigma} s^\alpha_{\sigma\sigma'} c_{k'\sigma'}, \qquad (3)$$

the second equality defining the local spin-density s(r) in second-quantized notation. If the sign of (1) is to be correct, J must be antiferromagnetic: positive as defined.

The canonical case everyone discussed at that time was Mn very dilutely dissolved in Cu, on which much experimental work had been spent to demonstrate that such a model approximately explained its magnetic and resonance properties; the Hamiltonian (3), though called for Kondo-effect purposes the "Kondo Hamiltonian", was invented by these early investigators, particularly Kittel and Yosida.[4] In a later discussion of the source of (3), we will see this was not a particularly fortunate choice, since J here has its minimum possible value, 0.5-1 eV.

From (3), whence (1)? Much earlier Yosida,[4] after clearing up the problem of the spin polarization caused by such a magnetic impurity, had calculated the resistivity due to spin disorder scattering to second order in J. This simple calculation is instructive, so let us describe it.

In the absence of a magnetic field one third of the scattering comes from the S^z term in (3) which acts like an ordinary potential JM scattering $k\sigma \to k'\sigma'$. The matrix element is JM and the transition probability by the golden rule is

$$J^2 \langle M^2 \rangle \rho(E) = \tfrac{1}{3} J^2 S(S+1) \rho(E).$$

It is irrelevant whether or not we take the exclusion principle into account in the final state because the rate of the transition $k \to k'$ is $f_k(1 - f_{k'})$ and that for $k' \to k$ is $f_{k'}(1 - f_k)$, so the direct and reverse transitions are necessarily *both blocked or neither*. Thus we might as well take the transition probability as proportional to f_k. The Fermi surface has no effect.

The situation is quite different for spin flips S^+, S^- even though the result at first appears the same; each contributes another $\tfrac{1}{3}S(S+1)$. The probability for $M \to M - 1$, $c_{k-} \to c_{k'+}$ is

$$J\,|\langle M - 1 | S^- | M \rangle|^2 \rho f_{k-}(1 - f_{k'+}) W(M),$$

while the probability for the reverse process is proportional to $W(M - 1)$. Thus the $f \cdot f$ terms cancel only if $W_M = W_{M-1}$, i.e. only if $H = 0$, and more important, in principle the Fermi surface—i.e. the occupation $f_{k'+}$—does enter the picture. In fact when H is turned on, a hole in the transition probability does appear at the Fermi surface, with edges as sharp as $f(\epsilon/kT)$, because the exoergic scattering process is less probable than the reverse one. This is the effect which is responsible for the giant negative magnetoresistance which often accompanies the Kondo effect; and it is interesting to realise that this already is a Fermi-surface effect, showing that spin phenomena are not susceptible to the old theorem that the Fermi surface does not affect scattering probabilities.

What of that theorem? Why can it be wrong? The basic reason for the theorem is the idea that the relationship between incoming and outgoing states, before and after a single scattering, is a unitary transformation. A unitary transformation on the underlying Hilbert space of any many-body fermion wave function leaves the exclusion principle automatically satisfied.

But the S matrix is only unitary for the states of *target and projectile*. When the *target* changes state there is no guarantee that the projectile states *alone* undergo a unitary transformation. Two electrons from different initial states may perfectly well try to be scattered into the same final state; thus the exclusion principle can play a role.

Kondo's calculation[3] shows very directly how it does. Consider the scattering amplitude for $k+ \to k'-$ in second order in J. This may occur two ways: $k+ \to k''+ \to k'-$ with amplitude

$$\frac{J^2(M+1 \mid S^z \mid M+1)(M+1 \mid S^+ \mid M)}{\epsilon_{k''} - \epsilon_k}(1 - f_{k''})$$

or $k''+ \to k'-$, then $k+ \to k''+$, with amplitude

$$-\frac{J^2(M+1 \mid S^+ \mid M)(M \mid S^z \mid M)}{\epsilon_{k'} - \epsilon_{k''}} f_{k''}.$$

From this we find for the amplitude

$$\frac{J^2(M+1 \mid S^zS^+ - S^+S^z \mid M)}{\epsilon_{k''} - \epsilon_k}(1 - 2f_{k''})$$

$$+ J^2(S^zS^+ + S^+S^z).$$

Integrated over the intermediate-state energy, this leads to a term

$$J^2 S^+ \int \frac{[1 - 2f(\epsilon'')]\, d\epsilon''}{\epsilon'' - \epsilon}, \tag{4}$$

which behaves like

$$J^2 \begin{cases} \ln \epsilon, & \epsilon \gg kT, \\ \ln kT, & kT \gg \epsilon. \end{cases} \tag{5}$$

This logarithmic behaviour is indeed the Kondo effect. When the scattering amplitude is squared to give the resistivity, this J^2 term multiplies the ordinary J term giving a $J^3 \ln T$ resistivity.

Notes:

(a) It depends on the noncommutativity (i.e. dynamics) of the local spin: it requires the "target" to have internal degrees of freedom.

(b) The sign depends on the sign of J because of the interference. This is easily remembered as follows: when the exchange is *antiferromagnetic*, the spin pulls electrons of opposite spin closer to itself, so that they can then spin-flip scatter more easily; thus ρ is larger; and vice versa for ferromagnetic J.

Thus the effect is basically a "vertex correction" of diverging magnitude. Almost all cases are antiferromagnetic, a fact which we will explain next time.

Another effect which was almost immediately explained by this calculation was the so-called "giant thermoelectric power" which accompanies the resistance minimum; the sharp variation of scattering cross sections with energy is the essential point here.[5]

Let me depart from historical order to close this first contribution, which is confined to the striking successes of lowest order Kondo theory, with the phenomenon of spin-flip tunneling. In 1964 Adrian Wyatt[6] observed anomalies in the tunneling current through certain oxide layers between metals (notably tantalum–tantalum oxide layers, but the phenomenon is quite widespread) and I noticed that the anomalies were logarithmic and suggested some kind of Kondo-effect origin for them. Rowell[7] demonstrated that they depended both on the oxide and on the metal, which pointed towards a coupling with magnetic impurities in the oxide.

Appelbaum[8] finally fitted this phenomenon correctly into the Kondo-effect framework. His contribution was to observe that there could be *both* an exchange coupling J of the form (3) between a magnetic impurity and one of the metals *and* an *exchange tunneling* amplitude T_J which coupled the spin with the tunneling amplitude:

$$H' = J\mathbf{S}\cdot\mathbf{s}_A(\mathbf{r}) + T_J \mathbf{S}\cdot \sum_{k,k'} c^{A\dagger}_{k\sigma} \mathbf{s}_{\sigma\sigma'} c^{B}_{k'\sigma'}, \tag{6}$$

where $c^{A,B}_{k'\sigma'}$ acts in metal A or B, respectively. There is no particular reason why T_J should be enormously smaller than ordinary tunneling amplitudes T, while J can easily be of the usual order for metallic impurities, since the magnetic center may be right at the surface of the metal (e.g. a surface state). Kondo's argument now goes through with no formal difference at all, giving for the tunneling amplitude:

34

$$T_J \to T_J + T_J J\rho \int \frac{1 - 2f(\epsilon'/kT)}{\epsilon' - \epsilon} d\epsilon', \tag{7}$$

where this is the tunneling amplitude for a particle of energy ϵ from the Fermi surface. The tunneling probability then contains a logarithmic term proportional to $T_J^2 J$, which can easily be of the order of the anomalies observed. The tunneling version of the Kondo effect has the nice feature that it is possible to check both the energy (ϵ) and T dependence of (7), as well as the dependence upon H and the existence of the second-order "hole" in the tunneling probability as a function of ϵ for finite H. This check has been carried out by Shen and Appelbaum in work to appear shortly. On the other hand, the tunneling case almost intrinsically leaves us in the dark about the nature of the spin centers.

Let me close with a comment about the sign. In tunneling, the current measures essentially a "scattering probability"—the probability for scattering of an electron through the oxide layer. The resistance of a bulk sample measures also the scattering probability. Thus tunneling conductivity is the quantity corresponding to the resistance; each in fact normally increases logarithmically as $T \to 0$, corresponding to *antiferromagnetic* sign of J.

All readers will have noted that $T \to 0$ leads to an impossible divergence of the scattering probability. Before that happens, the perturbation series implied by Kondo's calculation for the scattering matrix K:

$$K = J + \alpha J^2 \rho \ln T + \beta J^3 \rho^2 (\ln T)^2 + \cdots$$

will certainly have diverged. The early authors felt $J\rho$ would always be so small this "Kondo temperature" defined by $J\rho \ln T \sim 1$, $T_K \propto e^{-1/\alpha J\rho}$ would be unphysically low.

Later articles will examine this incorrect assumption and the peculiar and still mysterious phenomena occurring in this region. But before that we will discuss the sign and magnitude of J.

P. W. ANDERSON

References

1. W. J. de Haas, J. de Boer, and G. J. van den Berg, Physica **1**, 1115 (1933/34).
   A full review of the older experimental data is given by G. J. van den Berg, *Progress in Low Temperature Physics IV*, edited by C. J. Gorter (North-Holland Publishing Company, Amsterdam, 1964), pp. 194–259.
   The most conclusive experiments were those of A. N. Gerritsen and J. O. Linde, Physica **17**, 537 (1951).
2. P. van der Leeden, Thesis, Leiden, 1940.
3. J. Kondo, Progr. Theoret. Phys. (Kyoto) **32**, 37 (1964).
4. M. A. Ruderman and C. Kittel, Phys. Rev. 96, 99 (1954).
   K. Yosida, Phys. Rev. **106**, 893 (1957); **107**, 396 (1957).
5. J. Kondo, Progr. Theoret. Phys. (Kyoto) 34, 372 (1965).
6. A. F. G. Wyatt, Phys. Rev. Letters 13, 401 (1964).
7. J. M. Rowell and L. Y. L. Shen, Phys. Rev. Letters **17**, 15 (1966).
8. J. Appelbaum, Phys. Rev. Letters **17**, 91 (1966).

The Kondo Effect. II

As I closed in the first article of this series, I remarked that Suhl,[1] who seems to have been the first to emphasize the mathematically intractable nature of the Kondo singularity in the scattering amplitude, was followed by most early writers in assuming that the effective free-d exchange integral was of the order of or less than an atomic s–d exchange integral, i.e. a volt, so that ρJ tended to be ~0.2 or so and the temperature of the "Kondo singularity" defined by

$$1 \cong -\rho J \ln\left(\frac{kT_K}{E_F}\right)$$

was supposed to be unphysically low. In fact, only for Mn of the d-level ions and for the rare earths is this definitely the case. The physics of the parameter J is basically fairly well understood as the result of a series of developments extending from about 1958 through '68 and in fact into the present, although here too there appeared an intractable many-body problem about which a literature of considerable complication and confusion has grown up: the so-called Anderson (or Wolff) model (to a great extent due to Friedel, true to nomenclatural tradition in this field).

In the fall of '59 a delightful little discussion meeting on magnetism in metals was held in Brasenose College, Oxford. Aside from hearing the first public exposure of spin density waves, this meeting also, in my mind, was noteworthy in that the two basic components of the correct electronic theory of local moments made a remarkably elastic collision. Blandin presented the idea of virtual states and I the conceptual basis for antiferromagnetic s–d exchange, without any understanding, at least on my part, that the two ideas belonged together. The only immediate positive scientific result of the meeting was that I won a wager on the sign of the Fe hyperfine field on the basis of these ideas.

The idea of the virtual state has come to play the major role in the one-electron theory of d electrons in metals even in the pure case. [Note: what has come to be called a virtual state in solid state theory

190

is more properly called a resonance or resonant state in the nomenclature of scattering theory as developed by particle theorists.] If the d electron is to participate at all in metallic properties, its energy level must be above the mean effective potential for free plane waves, but because of the centrifugal potential barrier $l(l+1)/r^2$ its communication with the free electron states is weak and it forms a more or less sharp resonant state—characteristically, in the $3d$ series, 2–3 volts wide and a few to 10 volts up in energy from the zero. That is, its position in energy is as in Fig. 1 and its density of states as in Fig. 2.

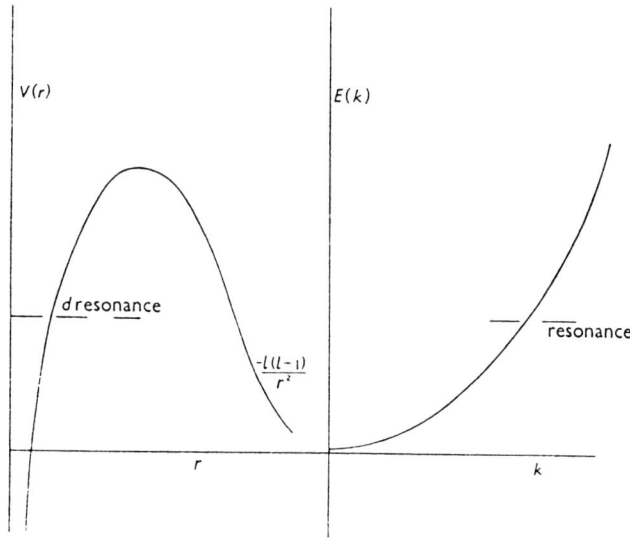

FIGURE 1.

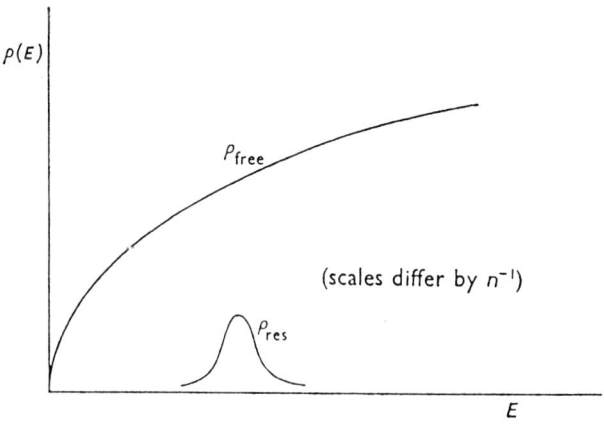

FIGURE 2.

In an almost-free electron metal like Cu, Mg, Zn, or Al, the only valid description of such a state is as a scattering resonance. It may be formally parameterized in terms of a variation of scattering phase shift with energy like

$$\tan \delta_l(E) = \frac{\Delta}{E - E_d}, \tag{1}$$

where Δ and E_d are slowly energy-dependent quantities representing the resonance width and position. In Friedel's work,[2] these parameters are taken arbitrarily to fit the experimental behavior. Much progress has been made recently in evaluating them from first principles, particularly in the milestone paper of Heine.[3] The interest there was in pure metals, where the resonance is merely the first stage of a calculation of the band structure via the Korringa–Kohn–Rostocker method or any of a number of others, but his methods are equally applicable to the isolated impurity. By these methods, or by actual direct calculation as carried out by McMillan and myself,[4] one finds a resonance $\sim 1\text{–}2\,\text{eV}$ wide, depending on whether it is lower or higher in the band; for $4d$ or $5d$ elements quite a bit wider.

In my work[5] I found it more convenient to assume a "resonant state orbital" ϕ_d coupled by matrix elements V_{dk} to the free electron states in the band. The resulting model Hamiltonian is

$$H_A = \sum_{k,r} \epsilon_k n_{k\sigma} + \sum_{k,\sigma} V_{dk}(c^\dagger_{k\sigma} d_\sigma + \text{c.c.}) + U n_{d\uparrow} n_{d\downarrow}. \tag{2}$$

This leads to a phase shift behavior like (1). In two papers[4,6] I discussed the essential equivalence of this model with Friedel's, but a more satisfactory treatment is Dworin's recent one[7] using Weinberg scattering theory. With this latter theory it is actually possible to define the ϕ_d eigenstate formally, as the eigenstate of an operator, $(E - H_0)^{-1}V$, a concept towards which I had been fumbling myself.

Finally a very popular model has been the so-called "Wolff model" where the psychologically more satisfying assumption is made that the local state is merely one of the Wannier functions of the band under consideration, perturbed by a local change in mean potential.[8] This model, while simple conceptually, has serious difficulties; it is completely misleading for the impurity in a free electron metal, where there are no valid Wannier functions and where the necessary antiresonance to make up the total count of states simply doesn't exist. Lately it has been made clearer that it may be very roughly valid in a d-band metal

192

where the modern Ehrenreich–Mueller model Hamiltonian[9] does allow one a partially tight-binding type of band. But these Hamiltonians in turn rely on the treatment of the entire d-band in terms of resonances for their validity.

The most important concept of the one-electron theory of resonances is the "Friedel theorem" evolved from Friedel's theory of screening of charges in metals and his "sum rule".[10] The Friedel theorem states that a scattering phase shift of δ_l at E ($l = 2$ in all interesting cases) implies that there are $[(2l + 1)/\pi]\delta_l$ extra electrons in the states of angular momentum l and given spin in the region immediately surrounding the scatterer. This theorem is immediately accepted if one just thinks in terms of the node-counting theorem, that the ordinal number of a state is its number of nodes, and $\delta_l = \pi$ implies one extra node; but its detailed proof is rather subtle. This enormously important theorem is the basic reason for the unsatisfactory nature of the Wolff model; since in going through an $l = 2$ resonance we draw in ten (five of each spin) extra electrons, no less than five bands must be considered in order to find the necessary extra states. It also plays a role, together with the sum rule on δ which is enforced by electrical neutrality, by telling us that an impurity with a charge of Z relative to the background must have the sum of its spin up and spin down δ's adding up to $\pi Z/2(2l + 1)$. Langer and Ambegaokar[11] have essayed a proof that the Friedel theorem is valid even in the presence of electron interactions, but whether this proof really extends to magnetic situations is only a conjecture which, however, is generally accepted. The scattering theory also sets very important upper limits on the scattering cross sections, namely that (neglecting other δ_l's)

$$J_{\max} = (2l + 1)\lambda^2 \sin^2 \delta_l \leqslant (2l + 1)\lambda^2,$$

which we will call the "unitarity limit", and which plays a great role in the Kondo effect.

The second ingredient of the theory of local moments is the idea of a strong exchange-like interaction confined to the locality of the resonant state. Friedel speaks vaguely of "exchange" but it seems to have been the introduction of the analogy between antiferromagnetic superexchange in insulators and the exchange process here which clarified the actual process.[5]

The Coulomb interaction of two antiparallel electrons in the same d or f state is of the order of a rydberg, the exchange interaction J between two states in the same shell about 1/10–1/5 of that. With these large energies available one finds that the unrestricted Hartree–Fock

193

solution of (3) with the term in U_{eff} included, for a partially filled resonant state, can split it into at least two levels of opposite spin, of which one will tend to be below the Fermi level and one above (in order to maintain neutrality). (See Fig. 3.) The condition that this splitting should occur is roughly (in my representation)

$$\frac{\pi^2 \rho \langle V_{dk} \rangle^2}{U_{\text{eff}}} = \frac{\pi \Delta}{U_{\text{eff}}} < 1, \tag{3}$$

where V_{dk} is the effective d–k matrix element, Δ the half-width of the resonance, ρ the density of states, and U_{eff} the net effective exchange.

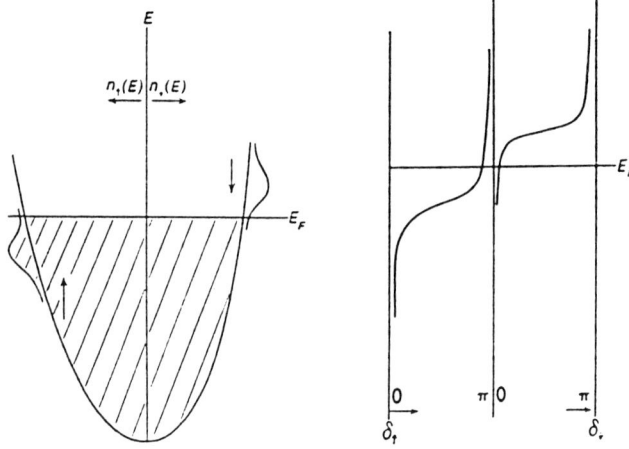

FIGURE 3.

As I emphasized (in Ref. 5), this is not necessarily an easy condition to satisfy and in normal, non-transition polyvalent metals where $E_F \sim 1\,\text{Ry}$ it is now known that it normally is not for d-levels, though virtually always for f's. Only in noble metals with one free electron or in d-band metals do magnetic states normally occur.

If (3) is satisfied and the virtual level splits, several quite important conclusions may be drawn.

(i) Depending on the strength of the intra-atomic integrals, the d-level may retain its orbital angular momentum (as for most f atoms) or lose it (as for all known d's). The former happens only if Eq. (3) is satisfied for

$$U_{\text{eff}} = U - \text{interatomic integrals } J, F$$

as well as for

$$U_{\text{eff}} = U + 4J;$$

194

that is, intra-atomic exchange *helps* the quenched state but *hurts* the unquenched state. Moriya[12] has made a more detailed analysis of the true 5-fold degenerate d-level, and Caroli and Blandin[13] have corrected a number of errors of detail in my and in Moriya's analyses and given a proper formalism for dealing with orbital degeneracy.

(ii) As pointed out by Clogston and myself,[14] using perturbation theory, before the full localised state theory appeared, and reiterated in that theory,[5] the mechanism leads directly to an antiferromagnetic effective exchange between free electrons and the localized state. The nonintegral effective polarization due to this exchange is intrinsic to the model, being represented by the tails of the virtual levels on the "wrong" side of the Fermi surface (see Fig. 3). So is the observation that the scattering phase shifts for the two spins are opposite in sign, because the antiparallel resonance is necessarily *above* the Fermi surface (*attractive* phase shift $\delta < \pi/2$) and the parallel resonance below (*repulsive* phase shift $\delta > \pi/2$). This exchange mechanism was analysed in more detail by Schrieffer and Wolff,[15] who extended the Clogston–Anderson perturbation theory into a complete perturbative canonical transformation to an effective Hamiltonian containing a local spin, and $S_x s_x$ and $S_y s_y$ as well as $S_z s_z$ effective exchange terms. This process again has been studied more completely by Caroli and Blandin.[13]

One can ask whether it is possible to avoid perturbation theory in arriving at a model Hamiltonian with an effective local spin and exchange interaction. Oddly enough, the answer appears to be "no"; the attempt to produce a theory equivalent in generality to the unrestricted Hartree–Fock theory already involves us in one of the many ramifications of the generalized Kondo problem. One very direct way of seeing this is as follows. It was already well known that the unrestricted Hartree–Fock theory leads to states which are not eigenstates of total spin—basically because the orbital states for spin-up and spin-down electrons are not the same, so that a spin rotation modifies the orbitals too. By far the simplest and best procedure for completing the unrestricted Hartree–Fock state to make it a suitable eigenstate is that of Heine,[16] but it is an easy matter to see that Heine's procedure (or any correct one) applied to the Friedel–Anderson localized magnetic state is not convergent: the additions which are small in an atomic problem are logarithmically infinite in this case. Thus the early hopes that the localized state model Hamiltonian could become one of the classical model problems of many-body theory have been only too well satisfied, in that as yet the problem appears even more intractable than, if related to, the Kondo problem.

195

(iii) Leaving out of account these difficulties it is interesting to ask how localized magnetic moments of this type interact. This question was first studied by Alexander and myself[17] but was covered in very much greater generality and more effectively by Moriya.[12] The interesting result here is that all the qualitative behaviors of magnetically interacting d-band atoms can be very clearly understood with a reasonable set of parameters for the width and the effective U's of the appropriate resonances. Although this theory is very simple and provides, to my mind, the clearest, and almost the only effective approach to the magnetic metals, it is not directly related to my subject matter here and I will not go into it further.

(iv) The last comment which is relevant to the Kondo effect is about the magnitude of the effective exchange integral: again a point which is implicit in much of the literature but was made best by Caroli and Blandin,[13] and later in my own work.[18] According to the Friedel theorem, only when the d states are about half full can the effective free-d exchange integral be small: that is, only for Mn in a nonmagnetic background like Cu, Ag, Au. This is because the d resonance contains five electronic states, and if fewer than five are to be accommodated (unless the orbital degeneracy is split as in the f shell) that means that the phase shift must be of order $n\pi/5$ where $n = 1, 2, 3, 4$. This is not small and can correspond to a very large exchange integral, especially in such cases as Cr or V in Au. In such cases there is no reason to expect T_K to be small, or for that matter no reason to expect the Kondo-model Hamiltonian to represent a reasonable approximation to the real one. Another way of putting this is to say that as a resonance approaches and crosses the Fermi level we can look at the phenomena in two ways: either as an increase in T_K to above normal temperatures or as the disappearance of magnetism. Present opinion has it that these are two aspects of the same phenomenon, which of course puts the solution of the Kondo problem on a new and much more generally important footing. In the later articles of this series we will examine the striking anomalies that are observed on the edge of the magnetic region.

P. W. ANDERSON

References

General: most of the material discussed herein is more completely treated in my article "Localized Moments", in *Many-Body Problem*, 1967 Les Houches Summer School, edited by C. de Witt and R. Balian (Gordon and Breach, New York, 1968). This reference I abbreviate "Les Houches". Also a number of articles in another summer school proceedings are repeatedly referred to: *Theory of Magnetism in Transition Metals*, Proceedings of the 1966 Varenna School, edited by W. Marshall (Academic Press, New York, 1967). I will refer to this book below as "Varenna".

1. H. Suhl, Phys. Rev. **138**, A515 (1965) and Varenna, p. 116.
2. J. Friedel, Nuovo Cimento **7**, 287 (1958); J. Phys. Radium **23**, 501 (1962); also in *Metallic Solid Solutions*, edited by Friedel and Guinier (W. A. Benjamin, New York, 1963), p. 19.
   A. Blandin, Varenna, p. 393.
3. V. Heine, Phys. Rev. **153**, 673 (1967).
   See also J. C. Phillips, Varenna, p. 22.
4. P. W. Anderson and W. L. McMillan, Varenna, p. 50.
5. P. W. Anderson, Phys. Rev. **124**, 41 (1961).
6. P. W. Anderson, Proceedings of the International Conference on Magnetism, 1964 (Institute of Physics, London, 1965), p. 17.
   N. Rivier, thesis, Cambridge University, 1968, comes to similar conclusions.
7. L. Dworin, Phys. Rev. (to be published).
8. P. A. Wolff, Phys. Rev. **124**, 1030 (1961).
9. L. Hodges and H. Ehrenreich, Phys. Letters **16**, 203 (1965).
   L. Hodges, H. Ehrenreich, and N. D. Lang, Phys. Rev. **152**, 505 (1967).
   F. M. Mueller, Phys. Rev. **153**, 659 (1967).
   J. C. Phillips, Varenna, p. 72.
10. J. Friedel, Advan. Phys. **3**, 446 (1954).
11. J. S. Langer and V. Ambegaokar, Phys. Rev. **121**, 1091 (1961).
12. T. Moriya, Progr. Theoret. Phys. **34**, 329 (1965); Varenna, p. 206.
13. B. Caroli and A. Blandin, J. Phys. Chem. Solids **27**, 503 (1966).
14. P. W. Anderson and A. M. Clogston, Bull. Am. Phys. Soc. **6**, 124 (1961).
15. J. R. Schrieffer and P. A. Wolff, Phys. Rev. **149**, 491 (1966).
16. V. Heine, Phys. Rev. **107**, 1002 (1957); Czech. J. Phys. **13**, 619 (1963).
17. S. Alexander and P. W. Anderson, Phys. Rev. **133**, A1594 (1964).
18. P. W. Anderson, various unpublished talks.
    J. R. Schrieffer, J. Appl. Phys. **38**, 1143 (1967), takes the same point of view.

Kondo Effect III: The Wilderness—Mainly Theoretical[†]

In the first article of this series, I set up the "Kondo Problem" of the interaction of a single local spin S, via an exchange integral J, with a Fermi gas of free electrons, and pointed out that the second term in a high temperature Born series for the scattering amplitude diverges logarithmically as $T \to 0$. This is called the "Kondo Effect" and leads to the "Resistance Minimum." In the second article I showed where the Kondo problem comes from physically—the "Anderson Model" (not physically different from the "Wolff Model")—and that this too is a model many-body problem with important mathematical difficulties.

To bring this problem up to the same point as I did the Kondo problem in Part I, I should note that Scalapino[1] worked out that the Anderson model leads to a Curie-Weiss law, and Hamann[2] that its behaviour also shows the Kondo singularity. These two papers represent the fourth-order and sixth-order terms in a systematic development of the Green's function in powers of V_{kd}, and correspond reasonably exactly to one's intuitive expectations from the Schrieffer–Wolff transformation. We will see much later, in Part IV, that in fact no substantive distinction need be drawn between the Anderson and Kondo models in their low-temperature behaviour.

I also pointed out that J (or V_{kd}) can often be quite large enough to make the divergences occur in accessible temperature regions.

In 1965 Suhl[3] conjectured that attempts to renormalise away the Kondo divergence by summing up the series of divergent terms were going to make it worse rather than better. Rough estimates suggested that the series would be geometric for the scattering matrix, giving

$$T \simeq J + J^2 \rho \ln T/T_F + J^3 \rho^2 \ln^2 + \cdots$$
$$\simeq J/(1 - J\rho \ln T/T_F),$$

which has a pole, for antiferromagnetic sign of J, at $T_K \simeq J_F e - 1/J\rho$, the so-called Kondo temperature (I have called it the "Kondo–Suhl" tempera-

[†] Work at the Cavendish Laboratory supported in part by the Air Force Office of Scientific Research Office of Aerospace Research, U.S. Air Force, under Grant Number 1052-69.

153

ture). This observation provoked a literature of extraordinary magnitude and mathematical complexity, involving the application of a remarkable number of techniques, some of them new to many-body theory. Some of the qualitative aspects of a correct solution emerged very rapidly; but other features of every solution were manifestly wrong. At the same time, elusive anomalies kept showing up in the experimental data. This "wilderness period" lasted for almost exactly 5 years, and I hope in this article to find a reasonably safe path through it, slighting, I am afraid, much of the very exciting work which was done during it. There is no question, however, that many of the concepts and methods developed during this period are of real permanent value, and that they are an important background for modern developments. In the past two years the end of the wilderness has quite suddenly begun to appear; these beginnings will be the subject of my next article, and the experimental results which have disposed of many of the anomalies will be touched upon in (I hope) a final one.

I have elsewhere[10] suggested that the best classification of methods is into two contrasting types: (a) High-temperature methods extrapolated down to and past T_K; (b) "Bound state" assumptions as to the ground state at $T = 0$, sometimes carried up to $T > T_K$. A third class (c) starts from the Anderson model itself and also extrapolates perturbation series: the "localised spin fluctuation" idea. But in this case the perturbation series is a different one, as we shall see.

It turned out in the end that all the valid methods under (a) gave equivalent sets of equations. In chronological order these were Suhl's method,[3] which was noteworthy also as the first really sophisticated use of dispersion theory in the many-body problem; Nagaoka's method[4] based on the equations of motion and double-time Green's functions; and Abrikosov's method[5] of replacing the local spin by a pseudo-Fermion and summing "Parquet" diagrams (see also Doniach's "drone-Fermion" method).[5] But at the beginning they all appeared to be very different.

Suhl[3] observed that a divergence of the scattering amplitude is not really possible. Since the Kondo Hamiltonian as normally written allows only s-wave scattering, the scattering matrix T must be less than the "unitarity limit":

$$|T| \leq \frac{2l+1}{\pi\rho(E)} \left(= \frac{1}{\pi\rho} \quad \text{if} \quad l = 0 \right). \tag{1}$$

The way one deals with divergences in scattering amplitudes in particle theory is to use the analytic properties of the scattering amplitudes—"dispersion theory"—and then any approximation at least leaves things finite. Suhl set up the equations appropriate for applying dispersion theory

154

to this problem, expressing the unitarity of the scattering amplitude (the "optical theorem")

$$(\text{Im } T)_{ii} = \pi\rho(T\tilde{T})_{ii} = \pi \sum_{\substack{\text{all real} \\ \text{scattering} \\ \text{processes } j}} T_{ij} T_{ji} \qquad (2)$$

and the analyticity requirement or dispersion relation

$$T(E) = \int \frac{\text{Im } T(E') dE'}{E - E'} + \text{entire functions}. \qquad (3)$$

Into these relations he inserted two assumptions. (1) That the scattering amplitude has the same formal expression at all temperatures as the perturbing Hamiltonian V:

$$V = V_0 + JS \cdot \sigma(r) \rightarrow T = T_0 + T_1 S \cdot \sigma \qquad (4)$$

and that the intermediate real states j (not virtual states. The advantage of the dispersion method is just that indefinitely complicated virtual transitions are taken into account) are strictly single scattered particle states. These two related assumptions, powerful as they are, unfortunately can very precisely be falsified by the actual behaviour at absolute zero, as we now know: as $T \rightarrow 0$ a pure singlet "bound" state appears, in which S has no meaning and (4) becomes incorrect, because the symmetry itself of the local spin becomes modified by the interaction. In addition, assumption (1) fails because of the now famous "infra-red divergences."[6]

In the first work on Suhl's theory a singularity did indeed appear at a certain temperature of order T_K; however, it soon appeared[7] that this was a mathematical artefact of the method, as one could guess from the general physical expectation (now verified by Hepp's exact theorem)[8] that a single scattering centre could have no mathematically singular behaviour at finite T.

Almost simultaneously with Suhl's work appeared Abrikosov's diagrammatic theory[5] and Nagaoka's double-time Green's function[4] one. It was not until much later that it was realised, primarily due to the efforts of Silverstein,[9] that the three are all mathematically identical. Prior to that it was unfortunately assumed that Nagaoka's rather rough approximate solution of his Green's function equations indicated, again, two regions: a bound state of antiparallel free electron spins compensating the local one, at low temperature, separated from a true magnetic state at higher T. This was a fortunate mistake for Nagaoka in that his name has been attached to the bound state idea, which, as a matter of fact, Suhl put forward earlier.

Nagaoka's equations of motion for the double-time Green's functions are truncated at the very first nontrivial order, and it is not difficult to show, as I did,[10] that this is an identical approximation to Suhl's. Silverstein[9] makes a

155

great point of the fact that the Abrikosov theory, which clearly "sums the most divergent diagrams," does not unequivocally lead to Suhl; it is my feeling that the additional analyticity requirements of the Suhl–Nagaoka theories which make them unique are simply evidence of their superiority, in some measures, over "most divergent diagrams." Suhl and Wong, Hamann,[11] Fowler,[11] and Bloomfield[12] and, in the greatest details, Zittartz,[13] have worked out mathematically "exact" solutions of the Suhl–Nagaoka–Abrikosov theory. Unfortunately, these solutions are often rather confusingly billed as "exact" solutions of the Kondo problem, which they have no claim to be, and in fact are not in any sense. I believe that they are, as yet, the best approximations available over the entire temperature range, but their asymptotic properties as $T \to 0$ are obviously incorrect. Zittartz[13] shows that the ground state has a fractional entropy and a tiny but finite magnetic moment, even for $S = \frac{1}{2}$, and as of now experiment and theory are agreed that there can be no moment at least in this case.

It is rather hard to trace the first hypothecation of a bound state at absolute zero: I believe Suhl and Nagaoka both mentioned it, but it may have been Takano and Ogawa[14] who first tried to make a formal theory for the Kondo Hamiltonian (in the Anderson one, early work of Alexander[15] was my first contact with this idea). But Takano and Ogawa postulate an unphysical anomalous amplitude, and the first reasonably carefully worked out bound state theory was written down by Kondo.[16] This theory, extended to finite temperatures and improved by Kondo and Appelbaum[17] working together, gave singular behaviour in the $T \to 0$ limit, though weak: a resistivity behaving like $(T \ln T)^2$ and a susceptibility like $\ln T$. This, most of its predecessors, and a rather complicated attempt by myself[18] which is noteworthy only for some new mathematical results on determinants of scattered wavefunctions, all had the weakness of giving the wrong exponent in T_K and in the binding energy, an error which gets very bad ($\sim e^{+1/J}$) in the weak coupling limit $J \to 0$. The exponent is certainly given *correctly* by the "most divergent diagram" aspect of the type (a) theories; the difficulty of type (b) theories is that the bound state energy tends to incorporate an unknown fraction of the much larger perturbation theory energy which is also present at higher temperatures. The series in J for this is now realised to be only asymptotic, with an error of the same order as T_K (see my next article), so it may not be possible *in principle* to do a correct variational theory.

Yet another line, which seems to give results of much higher quality, is the technique of Yosida, Okiji and co-workers.[19] These workers started by systematically minimising the ground state energy using, first, mixtures of (one bound electron) and (one bound electron + one excited pair), generalising to first 5 and then 7 excitations, and then extrapolating their results somewhat intuitively to give an integral equation for the wavefunction and energy. The

156

bound state energy and some of its properties seem to be roughly correct; but the difficulty of the method and its rather intuitive approach have led to its not being widely accepted as *the* solution. In particular, the integral equation contains a series which is arbitrarily *assumed* to be geometric. I would not, however, be surprised if in the end it turns out to be equivalent to one of the newer methods.

A final line of investigation, which in some ways was more successful than the other two, was instituted by Lederer and Mills,[20] Rivier,[21] and Suhl.[22] The idea of this approach was to begin with the Anderson model or equivalent and to use absolutely orthodox many-body perturbation theory starting with the assumption that the ground state is a *nonmagnetic* one. This is called the "localised spin fluctuation" theory, from an analogy with the currently fashionable theory of nearly magnetic metals. Now in cases in which the impurity is indeed *not* very magnetic—Mn in Al, for instance[23]—such a theory is very successful. One gets a non-singular, slightly decreasing susceptibility, and a resistivity which may increase or decrease as T^2. Again, since the localised state *is* indeed nonmagnetic, such a theory works, with renormalised parameters, over any limited temperature range. But how the renormalisation works in such a way as to give one the Kondo temperature, and the proper relationship between low and high temperature regimes, are questions which had to wait for the more modern methods of Hamann, Schrieffer, etc. which we will cover next time. Suhl's original attempt to do this part of the problem was suggestive but certainly wrong.

I should, finally, mention the experimental work which also left us in considerable suspense during most of this period. Two types of systems took a central place: the Cu–Fe, Cu–Cr and similar systems, in which T_K was of order 10–20°K so that one could run fairly easily from high to moderately low temperatures, and get a fairly clear idea of the T_K region; but it was difficult to get precise results for asymptotic behaviour as $T \to 0$.[24] On the other hand there were systems like Rh–Fe[25] and Au–V,[26] where T_K appeared to be very high, but one hoped to be measuring asymptotic properties quite accurately.

Two conclusions came out of the rather massive, very patient attempts by the experimentalists to measure with precision the properties of the increasingly dilute alloys the theorists demanded (in order that the theory of "isolated" scatterers should be correct). The first was a conviction that the bound state and T_K were a reality, fitted quite well qualitatively by the Suhl–Nagaoka type of theory. The second was that throughout the period theory *and* experiment joined in appearing to give singular behaviour as $T \to 0$: the $T^2 \ln^2 T$ of the Kondo–Appelbaum theory or a linear T term in R, and singular susceptibilities behaving like weird inverse powers or logs. The feeling persisted that we were confronted with a new set of "critical

157

exponents" as in the critical point problem. We had, in other words, no really adequate theory of the ground state of our system, which seemed to be separated by an impenetrable barrier from the ordinary behaviour which was not hard at all to observe. The nature of this barrier will become clear in my next article.

P. W. ANDERSON

References
1. D. J. Scalapino, Phys. Rev. Letters **16**, 937 (1966).
2. D. R. Hamann, Phys. Rev. Letters **17**, 145 (1966); Phys. Rev. **154**, 596 (1967).
3. H. Suhl, Phys. Rev. **A138**, 515 (1965).
4. Y. Nagaoka, Phys. Rev. **A138**, 1112 (1965).
5. A. A. Abrikosov, Physics **2**, 5 (1965); S. Doniach, Phys. Rev. **144**, 382 (1966).
6. P. W. Anderson, Phys. Rev. Letters **18**, 1049 (1967); P. Nozieres and C. T. de Dominicis, Phys. Rev. **178**, 1097 (1969).
7. H. Suhl and D. Wong, Physics **3**, 17 (1967); H. Suhl, Physics **2**, 39 (1965); a good review is in *Theory of Magnetism in Transition Metals*, W. Marshall, ed., p. 116 (Academic Press, New York, 1967).
8. W. Hepp, preprint.
9. S. D. Silverstein and C. B. Duke, Phys. Rev. **161**, 456 (1967).
10. P. W. Anderson, in *Many-Body Problem*, C. de Witt and R. Balian, eds. (Gordon and Breach, New York, 1968).
11. D. R. Hamann, Phys. Rev. **158**, 570 (1967); an essentially equivalent result was found by D. S. Falk and M. Fowler, Phys. Rev. **158**, 567 (1967).
12. P. E. Bloomfield and D. R. Hamann, Phys. Rev. **164**, 856 (1967).
13. J. Zittartz and E. Muller-Hartmann, Z. Physik **212**, 380 (1968).
14. F. Takano and J. Ogawa, Prog. Theoret. Phys. **35**, 343 (1966).
15. S. Alexander, private communication.
16. J. Kondo, Prog. Theoret. Phys. **36**, 429 (1966).
17. J. Kondo and J. W. Appelbaum, Phys. Rev. Letters **19**, 906 (1967); Phys. Rev. **170**, 542 (1968).
18. P. W. Anderson, Phys. Rev. **164**, 352 (1967).
19. K. Yosida, Phys. Rev. **147**, 223 (1966); **164**, 879 (1967); Prog. Theoret. Phys. **36**, 875 (1966); A. Okiji, Prog. Theoret. Phys. **36**, 714 (1966); A. Yoshimori, Phys. Rev. **168**, 493 (1968); A. Yoshimori and K. Yosida, Prog. Theoret. Phys. **39**, 1413 (1968); **A42**, 753 (1969).
20. P. Lederer and D. L. Mills, Solid State Commun. **5**, 131 (1967).
21. N. Rivier and M. Zuckermann, Phys. Rev. Letters **21**, 904 (1968); N. Rivier, thesis, Cambridge, 1968.
22. H. Suhl, Phys. Rev. Letters **19**, 442 (1967); Phys. Rev. **171**, 567 (1968).
23. A. D. Caplin and C. Rizutto, Phys. Rev. Letters **21**, 746 (1968).
24. See M. D. Daybell and W. A. Steyert, Rev. Mod. Phys. **40**, 380 (1968) for a review.
25. B. R. Coles, Phys. Letters **8**, 243 (1964); B. R. Coles, J. H. Waszinki, and J. Loram, Proc. Intl. Conf. Nottingham, p. 165 (1964).
26. K. Kume, J. Phys. Soc. Japan **22**, 1116, 1309 (1967); **23**, 1226 (1967).

158

Kondo Effect IV: Out of the Wilderness

I have delayed over two years since the last of these articles because I had hoped that Comment IV in this series on Kondo effect could represent the solution, in full detail. Unfortunately, while a quantum jump, in sophistication and power of the methodology and in our qualitative understanding of the Kondo problem, has taken place, the last few steps in achieving and experimentally confirming an actual numerical solution, rather than a solution in principle, have yet to be completed; so what I have to give you here is yet another progress report.

First let me explain what I will *not* talk about. To Comment V, I relegate the full discussion of the experiments and of most of the theories which go by the name of "localized spin fluctuations", which are to a great extent just a verbalization of the experimental results. This unfortunately excludes direct confrontation with experiment. Few theorists so far seem willing to try to cope with the true multichannel, coupled Coulomb-exchange problem for d electrons in real metals. I have, however, a strong intuition, but only an intuition, that the $S = \frac{1}{2}$ Kondo problem will turn out to be closely related to the experiments. The only mathematically sophisticated theories of "localized spin fluctuations" are by Hamann[1] and Schreiffer, Evenson and Wang,[2] who have developed rigorous path integral methods for dealing with the $S = \frac{1}{2}$ s-state symmetrical Anderson model. The result is simply to render absolutely rigorous the Schreiffer–Wolff equivalence of Anderson and Kondo models for that case. Both of these followed the original path-integral work of Anderson and Yuval,[3] which is the first stage of the work reported here.

What I *will* discuss here is the appearance of a set of new methods which conclusively solve "in principle" (if you believe one group, of which I am a member), or at least enormously improve, our understanding of the $S = \frac{1}{2}$ Kondo problem, as well as conclusively disproving the whole spectrum of "wilderness" methods which I discussed in Comment III [Comm. Sol. State Phys. **3**, 153 (1970)].

What these methods have in common is that they do not try to solve the whole problem all at once; instead, the main technique is to transform away a little of the problem at a time, and thus to prove the equivalence of whole

73

classes of different problems. The result is a "scaling equation" for converting different problems into each other. The chief advantage of this "scaling equation" is that it makes it extremely clear what the nature of the mathematical difficulty of the Kondo problem is.

There are two ways of getting at the scaling equations. The simplest for this particular result is the so-called "renormalization group" method of Fowler–Zawadowski[4] and Abrikosov–Migdal,[5] which is very near to the "poor-man's method" of Anderson.[6] (An excellent paper on these methods by Fowler will appear shortly[7] and some of my discussion will simply be a paraphrase of this.)

The problem of the Kondo Hamiltonian

$$\sum_{k\sigma} \varepsilon_k n_{k\sigma} + J_z S_z s_s + \frac{J_\pm}{2}(S_+ s_- + S_- s_+) \qquad (1)$$

(we write it in this generalized anisotropic form for later purposes) is convergent only if there is an upper cutoff E_c on ε_k (measured from E_F). Thus the Kondo problem includes this cutoff as a parameter. But, if J is small, the effects of a small number of states in an energy range $E_c > |\varepsilon_k| > E_c - dE_c$ near this upper cutoff will be very small and can be very accurately estimated by perturbation theory. So long as $kT \ll E_c$ and $J \ll E_c$, the low-energy behavior could be calculated either from the original Hamiltonian or from a new Hamiltonian in which we have eliminated these few states so that the sum extends only to $E_c - dE_c$. But then the Js have to be modified (and also a change in total energy made). This scaling procedure can, if the new J remains small, be repeated *ad infinitum*. The result is a set of differential equations of which the lowest-order terms are

$$\frac{d(J_\pm \rho)}{(J_\pm \rho)(J_z \rho)} = -\frac{dE_c}{E_c} = \frac{d(J_z \rho)}{(J_z \rho)^2} \qquad (2)$$

$$dE_g = -\tfrac{1}{2} J_\pm^2 \rho \frac{dE_c}{E_c}. \qquad (3)$$

These equations tell us the Js to use for a given cutoff E_c to obtain the same low-energy results.

Equations like these are not new to the Kondo problem; they arise in the "parquet graph" techniques of Abrikosov[8] and in Suhl's scattering theoretic methods. What is new is their interpretation as a set of relations rigorously connecting different problems, not equations which are used in the course of solving a fixed type of approximation.

That they lead to the standard Kondo effect is clear. When the cutoff has been reduced to kT, one is left with an uncomplicated problem with no further singularities, and with an effective coupling constant given by (in the isotropic case)

$$(J\rho)_{\text{eff}} = \frac{J\rho}{1+J\rho \ln(kT/E_c^\circ)}, \quad (4)$$

which has the standard Kondo behavior so long as

$$kT > kT_k \simeq E_c^\circ \exp(-1/J\rho). \quad (5)$$

But one is not directed to mindlessly allow $(J\rho)_{\text{eff}}$ to blow up according to Eq. (4) as $T \to T_k$. Rather, one recognizes that in this regime the whole problem has become an entirely different one, in which completely new methods and ideas are going to be necessary. But the first thing one can do is to stop and ask: how much information can we get from scaling alone?

The answer is, a very great deal: we can essentially solve the whole question of analytic behavior as a function of J, since the singularities which are associated with the Kondo effect must come at $J\rho \to 0$, where the scaling laws are as accurate as one likes. The need to express this singular behavior clearly is the basic reason for introducing anisotropic Js; the consequences are most clearly shown in the $J_z\rho$, $J_\pm\rho$ plane as in Fig. 1. (On the abscissa we have plotted actually the Anderson–Yuval variable $\varepsilon \simeq 2J_z\rho + \frac{1}{2}J_z^2\rho^2$ rather than $J_z\rho$.)

The hyperbolas shown ($J_\pm^2 - J_z^2 = \text{const}$) are the curves connecting equivalent problems, and the arrows leading to the right (larger $J_z\rho$) are the direction in which E_c decreases. Now in this diagram one can get continuously from the ferromagnetic ($J_z = -|J_\pm|$) to the antiferromagnetic ($J_z = +|J_\pm|$) Kondo problems, and we see that these two cases lie on the asymptotes of the scaling curves. To the left of the ferromagnetic case, all curves scale into

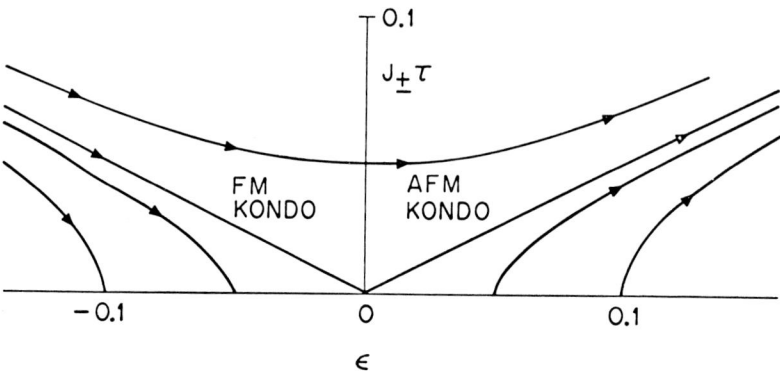

FIGURE 1.

the axis $J_\pm = 0$, which is a soluble problem because it represents simply a static, Ising model type spin perturbing the free electron gas. Thus the ferromagnetic case is, for all practical purposes, soluble. The antiferromagnetic case, unfortunately, scales into the region of large $J\rho$, along with all cases in which $J_z > -|J_\pm|$. We will discuss later whether a soluble case may be found in this region. But we emphasize that the behavior *as a function of $J\rho$* is completely summarized in this diagram. It is interesting, for example, to write down the equation connecting the cutoff with the Js:

$$\frac{E_c^\circ}{E_c(\varepsilon)} = \left\{ \left|\frac{\varepsilon+\varepsilon_0}{\varepsilon_i+\varepsilon_0}\right| \left|\frac{\varepsilon_i-\varepsilon_0}{\varepsilon-\varepsilon_0}\right| \right\}^{1/|\varepsilon_0|}, \quad (6)$$

where
$$\varepsilon_i = (\varepsilon)_{\text{initial}} \simeq 2J_z^\circ \rho$$

and ε_0 is the intercept on the $J_\pm = 0$ axis defined by

$$\varepsilon_0^2 = \varepsilon_i^2 - (2J_\pm^\circ \rho)^2 \quad (7)$$

(ε_0 indeed becomes imaginary above the asymptotes).

The expression (6) has extraordinarily nasty mathematical properties at the Kondo singular point, the full complexity of which I do not wholly understand. It is clear that one encounters an essential singularity as one approaches the physical cases, which have $\varepsilon_0 = 0$. This appears to be the main difficulty with the rather widely publicized method of Emery and Luther[10]: that their method is valid on the $J_\pm = 0$ axis everywhere except near $\varepsilon = 0$, but does not treat this very queer singularity correctly. (Incidently, the motivation for their work was a homogeneity relationship which appears to be exact, and is indeed satisfied by Eq. (6). Luther and Emery seem not to have noticed this agreement.)

A second point which can be firmly solved by the scaling equations is the Kondo temperature T_k. Oddly enough, as Armytage and I have recently noticed, the conventional expression for T_k is incorrect. T_k is not a sharp temperature but may be adequately defined by letting $kT_k = E_c$ (ε of order unity), i.e., T_k is the temperature scale at which the equivalent problem becomes a strong coupling one. To get a precise estimate, we go back to the scaling equation (2) and keep higher order terms:

$$-\frac{dE_c}{E_c} = d(J_z\rho)\left(\frac{1}{(J_z\rho)^2} + \frac{a}{J_z\rho} + b + \ldots\right).$$

The coefficients a and b are difficult, but not impossible, to compute. We note that b and all higher terms are not very meaningful, because they modify kT_k only by factors of order unity, but the same is not true of a, which gives

$$\frac{kT_k}{\varepsilon_c^\circ} = (J\rho)^a \exp[-(1/J\rho)]. \quad (8)$$

Armytage calculates $a = 3/2$, which gives a rather large correction. This correction is *not* available from the "wilderness" methods and represents a definite advance for the new ideas.

That is, however, as far as we can go with the scaling equations alone. Those equations tell us very clearly the nature of the Kondo problem: that, no matter how small the exchange integral J, there is a correspondingly low scale of energy [given by Eq. (7)] at which the problem of the exchange coupling with free electrons will become a strong-coupling problem. We can describe the scaling towards this strong-coupling problem accurately, but we must now solve it.

Unfortunately, there are two opinions about the nature of the solution. Abrikosov–Migdal, and Nozières,[5] among others, believe that there is a second singular point at a finite value of $J\rho$, $(J\rho)_0$, to which the scaling must inevitably lead and at which there are singularities in the behavior as a function of the physical variables T, H, etc.

Fowler, and I and my coworkers,[11] on the other hand, argue that the final strong-coupling problem is a highly nonsingular one of a very uninteresting nature, much like the naive spin fluctuation theory gives. Our arguments are of two types. Fowler feels that it is likely that there is no singularity in the "renormalization group" equations short of $J\rho \to \infty$, so that eventually all problems become equivalent to infinitely strong coupling. But infinitely strong coupling is no coupling at all: the Kondo spin will simply couple itself indissolubly to $J = 0$ with a spin made up from the free electrons, and aside from that there will remain no magnetic scattering at all, because there is no net spin. Thus, at low enough temperatures, the spin will simply look like a more or less sharp resonance in the ordinary scattering amplitude.

Our argument makes use of the rather more complicated route to the scaling equations by which we initially found them.[11] We first transformed the Kondo problem into a one-dimensional statistical mechanical problem by means of a path integral argument. The principle of this transformation was to observe that a typical time-history of the local spin could consist of a sequence of flips from spin-up to spin-down and back. We showed that the effect of the free electrons was to weight the probability amplitude for a given sequence of flips by a logarithmic interaction between flips t_i and t_j,

$$V_{\text{flip}} = \pm(2-\varepsilon)\ln\left|(t_i - t_j)\right|.$$

Then the path integral in "time" becomes formally the same as the configuration integral of a certain statistical problem in a one-dimensional space with V_{ij} as its interaction potential.

We could also show that a second problem has the same type of path integral expression as its solution: the soluble problem of an *ordinary* resonant

77

scatterer. But in this problem, as pointed out by Toulouse,[12] the coefficient $\varepsilon = 1$, invariably. Different resonance widths, etc., lead only to different flip amplitudes and cutoff behaviors. But once we had the scaling equations, this was enough: we can scale to $\varepsilon = 1$ and then all problems are scaled into soluble ones.

A second argument against the Abrikosov–Migdal singularity is simply the nature of the one-dimensional statistical problem which, almost self-evidently, has no further singularities.

Fitting together the nonsingular "Toulouse limit" type of solution with the correct branch of the scaling equation is turning out to be a fairly difficult problem, but that is clearly the next order of business. Meanwhile, we are in the frustrating position of feeling sure we have a solution in principle, but not being able to display any single numerical prediction for verification, except for remarks so general they appear trivial:

1) The scattering phase shift at $T = 0$ must be $\pi/2$ (either from the $J\rho \to \infty$ or the Toulouse limit).

2) There is a finite susceptibility of order $1/T_k$ (again either picture agrees).

3) The resistivity and other quantities vary nonsingularly (i.e., as const $+ aT^2$) as $T \to 0$.

In conclusion, then, the status is this: we understand very clearly the physical nature of the Kondo phenomenon, which is beautifully expressed in terms of Fowler's picture of scaling: electrons of high enough energy interact with the weak, bare interaction and the bare Kondo spin, but as we lower the energy the effects of other electrons gradually strengthen the effective interaction until finally, at energies near T_k, the effective interaction starts to get so large that we must allow the local spin to bind a compensating spin to itself, and the Kondo spin effectively disappears, being replaced by a large resonant nonmagnetic scattering effect. My own opinion is that the low temperature behavior is totally nonsingular, the Kondo impurity looking simply like a localized spin fluctuation site, but others believe that there may remain a trace of singular behavior. On the other hand, for clear, verifiable predictions for experimentalists, I am afraid we must wait for Comments V, VI, and possibly VII.

P. W. ANDERSON
Bell Laboratories,
Murray Hill, N.J.
and Cavendish Laboratory,
Cambridge, England

References

1. D. R. Hamann, Phys. Rev. B **2,** 1373 (1970).
2. W. E. Euenson, S. Q. Wang, and J. R. Schreiffer, J. App. Phys. **41,** 1199 (1970).
3. P. W. Anderson and G. Yuval, Phys. Rev. B **1,** 1522 (1970).
4. M. Fowler and A. Zawadowski, Sol. State Comm. **9,** 471 (1971).
5. A. A. Abrikosov and A. A. Migdal, J. Low Temp. Phys. **3,** 519 (1970); P. Nozières, private communication.
6. P. W. Anderson, J. Phys. C **3,** 2346 (1970).
7. M. Fowler, submitted to Phys. Rev.
8. A. A. Abrikosov, Physics **2,** 21 (1965).
9. P. W. Anderson and G. Yuval, J. Phys. C **4,** 607 (1971).
10. V. J. Emery and A. Luther, Phys. Rev. Lett. **26,** 1547 (1971).
11. P. W. Anderson, G. Yuval and D. R. Hamann, Phys. Rev. B **1,** 4464 (1970).
12. G. Toulouse, Compt. Rend. **268,** 1200 (1969).

Superconductivity in the Past and the Future

Superconductivity, ed. R. Parks (Marcel Dekker, New York, 1968), p. 1343, Vol. 2, Vol. 1, p. 47, epilogue

This is an elaboration on and postscript to paper 9.

23

SUPERCONDUCTIVITY IN THE PAST AND THE FUTURE

P. W. Anderson

BELL TELEPHONE LABORATORIES, INCORPORATED
MURRAY HILL, NEW JERSEY

The learned authors of the other chapters of this book have explained, even more clearly and completely than I expected when I accepted the commission to write a critical summary chapter, the present status of the field of superconductivity. In many cases the chapter authors have also done my work of summarizing and integrating the material as well as of presenting the basic information, and perhaps one of the most valuable services I can do the reader is to call attention to a number of places where this is to be found. (Of course, almost all the chapters contain some good material of this kind, and I mention specific sections here only to save myself labor by pointing out the most complete examples.)

The first part of Chapter 14, for instance, not only summarizes the ideas involved in type II superconductivity, but reviews the concepts behind flux quantization and the Landau–Ginzburg theory very ably. Chapter 18 in its early part summarizes nicely one of the unifying principles which run through superconductivity from BCS to the latest wrinkle, the vital role of time-reversal invariant pairing and the contrast between pair-breaking and pair-conserving perturbations.

I found Chapter 20, rather as expected, an excellent treatment of the whole question of the coherent quantum field aspect of superconductivity as well as of some other areas. Chapter 19 leaves me little more to say about the resistive state; Chapter 7 is in its entirety an able summary of the history of the gauge-invariance squabble. Chapter 1 brings in well one of the less appreciated threads in the historical development, that associated with Pippard's nonlocal electromagnetism. Chapters 10 and 11 both contain able reviews of the history and present status of the microscopic theory.

Fortunately for me, even though the present is very well covered by the rest of the book, I am left with almost two-thirds of the field: most of the past, and all of the future. Superconductivity has had a remarkably exciting past, and I hope it will have an equally exciting future. Since so much of my allotted job has been done for me I would like to try to convey some of this excitement to the reader, and I may be forgiven I hope if I take the liberty of presenting this history from a personal point of view.

My own involvement even with the periphery of the field dates from an afternoon in the spring of 1950, watching Bardeen and Fröhlich discussing, in a sunny windowseat of Charles Kittel's house in New Vernon, N.J., their soon to be published independent theories of superconductivity (*1,2*) involving electron interaction via phonons. As Blatt says in his book (*3*), the modern history of the field started then: that was the point, with Serin's and Maxwell's discovery (*4,5*) of the isotope effect and the resulting focus on the electron–electron interaction via phonons, at which people began to feel the problem *could* be solved. It wasn't–in Japan in 1953, Fröhlich was already convinced the exchange selfenergy was not the answer–but it is true that from that work came a series of developments which were the seeds of the correct solution: Fröhlich's introduction of field theory methods for the electron–phonon Hamiltonian, Bardeen and Pines's (*6*) discussion of the electron–electron interaction and, far in the future, the vital work of Migdal (*7*) on the self-energy in the normal state.

As an aside which may help us put some of the modern developments in perspective, let us go into these preliminary developments a little further. Both theories were based on the suggestion that the phonon exchange self-energy of the individual electron could become larger than its band energy, causing an instability of the normal Fermi surface and some kind of major rearrangement of the occupation numbers. The diagram we would now use to describe this self-energy is shown in Fig. 1; it is strongly energy-dependent because of the Fermi factor for occupation of state $k + q$, when state k is within an average phonon frequency of the Fermi surface. Nowadays we would express the

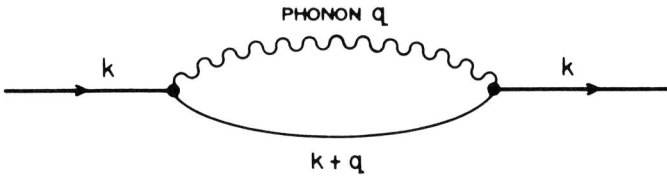

Fig. 1.

Fröhlich and Bardeen theories by writing an effective quasi-particle energy as

$$(E_k)_{\text{eff}} \simeq \epsilon_k + \Sigma(k, E_k)$$

Σ is the self-energy due to the exchange process in Fig. 1. In those days one assumed Σ was just a function of k:

$$\Sigma_k = \Sigma(k, \epsilon_k)$$

and the instability is obvious if

$$\partial \Sigma_k / \partial \epsilon_k < -1$$

(see Fig. 2).

Kohn and Vachaspati (8), Wentzel (9), and others objected that the self-energy correction of the phonons caused by electron pair exchange (see Fig. 3)

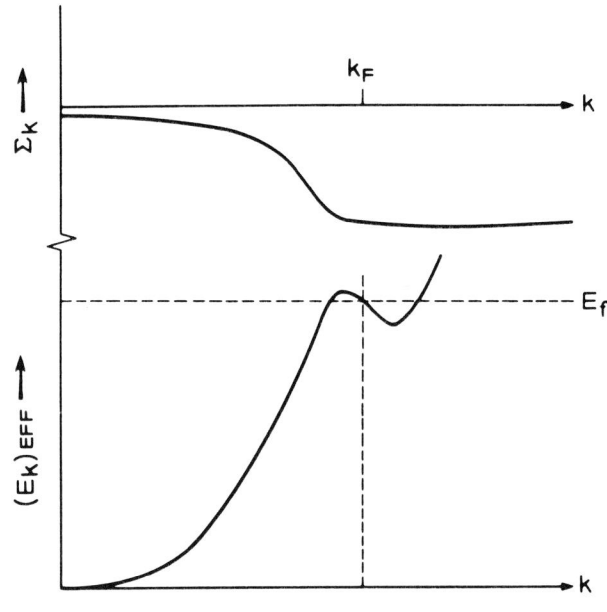

Fig. 2.

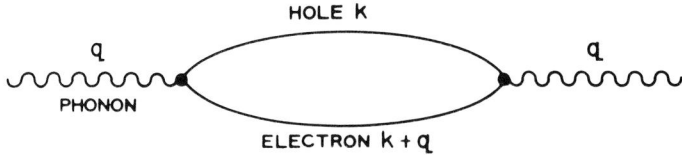

Fig. 3.

was actually more likely to cause an instability by exceeding the original phonon frequency, thus causing an instability in the phonon system before one could occur for the electrons. Such an instability would show up as a modification of the crystal structure, not as superconductivity. This consideration alone places an upper limit on the coupling strength which excludes the Fröhlich–Bardeen theories. But in fact Migdal resolved the problem even more completely by showing that Σ is a much more sensitive function of ω than of k—i.e., it is local in space, retarded in time, thus depends on ω much more sharply—so that the large correction to $(E_k)_{\text{eff}}$ comes from $\partial\Sigma/\partial\omega$:

$$(E_k)_{\text{eff}} = \epsilon_k + E_k(d\Sigma/dE_k) \qquad E_k = Z\epsilon_k$$

and

$$Z = \frac{1}{1 - (\partial\Sigma/\partial\omega)}$$

is always positive: the Fröhlich–Bardeen instability simply does not occur. The complete treatment shows that phonon instability and the first singularity of Z actually occur precisely at the same coupling strength. The fact that the coupling strength is limited by phonon stability can also, as Migdal explained, be used to show that the whole Fröhlich Hamiltonian could be treated essentially exactly within perturbation theory. This ingenious reversal of the old argument was little appreciated at the time, but now plays a great role in the whole field of superconductivity and phonons and electrons in metals; Schrieffer in his book (*10*) and Scalapino in his article here (Chapter 10), describe how this apparent "proof" of the impossibility of superconductivity is now used to give us a rigorous theory of it, the last of many instances in which impossibility theorems have played a useful role in clarifying our ideas of superconductivity.

It was clear from the isotope effect that superconductivity must have to do with electron–phonon interaction; if any confirmation was needed the strong correlation both Fröhlich and Bardeen found of transition temperature and electron–phonon coupling as estimated from high-temperature resistivity was it. This early dream of correlating resistance and T_c is now a reality, but in a way neither would have predicted: superconductivity is now understood to give us a much more reliable estimate of the total effect of electron–phonon coupling (telling us immediately the renormalization constant and thus the value of $\partial\Sigma/\partial\omega$, for instance) than is resistance, and we now have to use, in addition, the detailed knowledge of coupling constants we get from superconducting tunneling to predict, with any accuracy, the electrical resistance of the superconducting elements. This story is told in the chapters of McMillan and Rowell (Chapter 11) and Scalapino (Chapter 10).

What convinced both Fröhlich and Bardeen so quickly that the theory wasn't right? As I understand it, the two most serious problems were that it was quite impossible, first to understand the transition temperature, which should have been $\sim \Theta_D$, the Debye temperature, and second to understand the electrodynamics even crudely. No simple modification of the Fermi surface gives superconductivity.

1950 was in the midst of the McCarthy era, a time of which one of the silliest manifestations was the banning of Russian scientific publications in the United States. Some were even dumped into the harbors. The JETP containing the paper of Ginzburg and Landau (*11*) which, more even than the Western developments, marks the beginning of the modern era, was one of these. Partly because for many years it was available here only in a poor translation or on microfilm, this staggering achievement of understanding and intuition was not quick to be appreciated in the West. It even contained the first published remark on the idea now called "off-diagonal long-range order" that was shortly to be published by Penrose (*12*), and which represents the formalization of the possibility of coherent quantum particle fields, which is utterly basic to the dynamics of superconductivity and superfluidity.

There followed a tragicomic period of over a decade which should be fascinating to the historians of science and to those concerned with the relationships between science and society, during which the interaction between Russia and the West in the subjects of superfluidity and superconductivity resembled a comic opera duet of two characters at cross purposes rather than a dialogue. To what extent this was a result of cold war difficulties, either outright censorship such as the Ginzburg–Landau paper had, or the lack of face-to-face communication, and to what extent it may have been a fundamental difference in styles and points of view, it is hard to say. Whatever it was, it is hard to avoid the conclusion that if the groups on either side of the iron curtain had fully appreciated what was going on on the other, the science of superconductivity would have been five years further ahead by 1960 or so. (Much the same thing was going on in the case of helium, but fortunately that's out of my area here.)

On the Russian side, Ginzburg–Landau was a guiding principle for a much deeper understanding of intermediate state phenomena, the internal states of superconductors and their phenomenology, than was then realized in the West. Experiments on domain phenomena were interpreted by Ginzburg in 1955 (*13*) in terms of specific numerical values of the penetration depth and coherence length parameters and their ratio κ, and in 1956 Abrikosov extended the theory, in his magnificent, classic paper (*14*) on type II superconductors, to the point where little of importance has ever had to be added. The West failed almost completely to appreciate this paper as well as the data of Schubnikov (*15*) which it explained, which together founded and almost completed the science of type II superconductivity. There was a strong feeling that such phenomena were basically "dirt effects" which was assuaged by the Mendelssohn sponge theory (*16*) and supported by the merely fortuitous fact that almost all type II systems are alloys. High critical fields continued to be interpreted in terms of fine particles until finally Goodman broke the dam of misunderstanding in 1961 (*17*). This particular lack of communication was of great practical importance: It was left to Western empiricism to discover that superconducting magnets were practical, five years after Abrikosov had pointed out that high critical fields are a stable, intrinsic property of very many superconducting materials. The Russians seem to have been as uninterested in the applications of the theory as we were unaware of its real implications.

To add a footnote: Only one really fundamental idea was left to be added in this field—that flux quanta not only penetrate type II superconductors, but that they move in response to the Lorentz force $J \times B/c$, and provide a dissipation controlled by the Josephson relation $V = (h/2e)(dn/dt)$. That they do so was shown by Kim and myself in 1962 (*18,19*); the precise rules of the game belong to the future [although the work of Bardeen and Stephen (*20*) probably gives a qualitative, and Caroli and Maki recently perhaps a quantitative (*21*), understanding of this phenomenon at its simplest].

More fundamental was the Russian understanding that Ginzburg–Landau was close enough for all practical purposes to the real phenomenology of superconductivity: that it was the task of a microscopic theory simply to explain the Ginzburg–Landau type of phenomenology, not the wide range of disparate —and often, of course, interesting and significant as well—microscopic data which we were focusing on at the time. Now one understands that this *is* the phenomenon of superconductivity where any other characteristic may be missing —the energy gap, for instance. Without this understanding on any of our parts it was possible in 1959 for Blatt, Fröhlich, and others at the Cambridge meeting to complain that BCS had not explained persistent currents, in the face of the Russian announcement that Gor'kov had derived Ginzburg–Landau from BCS (*22*), (which was greeted with no excitement in the West).

Yet one must not ascribe all wisdom and understanding to the Russian side. For instance, after the original hint in the first Ginzburg–Landau paper, Landau seems to have rejected or at least deemphasized the off-diagonal long-range order point of view, particularly in the case of helium, but also for superconductivity. In particular, the flux quantization ideas which accompanied London's (*23*) and Onsager's much less sophisticated discussions of superconductivity were ignored in the Abrikosov paper, and Abrikosov, who gave there a full theory of a single flux quantum, referred vaguely to Feynman and never once remarked that the flux was related to hc/e. So the flux quantum was again left to be discovered, or at least rediscovered, through Western empiricism (*24*).

Thus in the absence of the revealed truth of Ginzburg–Landau, we in the

West were struggling along step by step with the various heuristic hints and half-formed ideas which *did* in the end lead to the solution of these problems.

One area of importance was the study of the electron–phonon interaction which Bardeen and Fröhlich revived after their 1950 attempt. As we have remarked, as far as electron–phonon effects from a fundamental many-body theoretic point of view were concerned, the Migdal 1957 paper settled most questions; but the less formally satisfactory but physically simpler paper of Bardeen and Pines, and the interest in the actual computation for real metals which led to the dielectric approach of Bohm and Staver (25) (going back to Bardeen, of course), Pines, Nozières, etc. [and the more recent work of the pseudo-potential school (26)] led to a much more solid intuitive feeling for the reality of the interaction and the relevant magnitudes. To this date almost all Russian papers ignore the physical nature of the interaction, or at best work with the unphysical "Fröhlich Hamiltonian." Only last year several of us had the frustrating experience of being unable to convince one of the Russian leaders in the field that we really *could* do the microscopic theory quantitatively in simple metals.

A second area of very great historical importance was the question of the nonlocal electromagnetism of Pippard and of the energy gap, for which it provided the first evidence (27). As Bardeen argued (28), if there were an electronic energy parameter of the order of T_c it would explain the experimental fact of nonlocal electromagnetism; and if it were a gap it might explain the London equation as well. (Previous authors had rejected this idea on the basis of gauge-invariance arguments, rightly of course, in a sense, since the BCS ground state "breaks" local gauge symmetry.) By then—1956—several experimental lines also converged on an energy gap: specific heat and microwave measurements, and soon optical ones.

The third important idea which was traveling (literally) around the U.S. was electron pairs. The most dramatic moment I remember having to do with superconductivity came at the end of Feynman's talk at the Seattle theoretical physics conference in September 1956, when he announced that, although he had solved the problem of superfluidity, he had spent many months computing on the problem of superconductivity and had failed utterly. Some radical new idea was needed to solve the problem, he said. At this point John Blatt leaped up on the stage and announced, "We *have* that idea, and we *have* solved the problem." The idea was pairs, indeed, but the Australian group was very far from a formal solution of the problem (later they were to show that it was feasible with their methods, although staggeringly difficult).‡ It is amusing also that Blatt worked with Matsubara on superconductivity about then, but it was the Russians

---

‡ References to all of this work are exhaustively given in (29).

who, at this very time, were perfecting the Matsubara temperature Green's function techniques to the point where they would become the best method for working with superconductivity. It is an amusing speculation that if they had also picked up the pair idea the Gor'kov theory might have appeared independently of BCS.

The difficulties Blatt and co-workers had were two-fold but related: that they avoided the apparatus of quantum field theory, even second quantization, and that they founded their ideas too closely on the then existing theories of Bose–Einstein condensation in helium. Helium is, in a sense, too easy: it can be done without introducing particle fields, macroscopic quantization, or off-diagonal long-range order, especially if one concentrates on the condensation and ground state rather than the dynamics; whereas in the end it was only this aspect which was to turn out to be common to the two superfluids.

The same summer—although he seems not to have crossed paths with Blatt (or Schafroth, contrary to a statement in Blatt's book occasioned by a confusion of names)—Leon Cooper was traveling around with a calculation suggesting that the ground state of the electron gas was unstable to the formation of bound pairs of electrons, using essentially the Tamm–Dancoff method. While it is true that the electron gas is apparently unstable to any kind of perturbation if you use the Tamm–Dancoff method, this idea did, as we know, lead Cooper, together with Bardeen and Schrieffer, to the correct result.

There has been a tendency for the Landau group to ascribe the really crucial step to Cooper. Aside from the fact that Tamm–Dancoff calculations such as his are often wrong for the electron gas, this is unfair not only to Blatt and his co-workers, who probably invented bound electron pairs first (leaving out of account a speculation of Ogg's), but to the achievement of the BCS group. (Perhaps there is a certain, partially justifiable, pique with the failure to appreciate the Ginzburg–Landau paper.) Anyone who listened, as I did, to Blatt and Cooper during that summer, had to realize that there was just no conception of what the state of the superconductor was, nor of how bound pairs could explain quantitatively any single experimental aspect of superconductivity (except perhaps, in Cooper's gase, T_c). Cooper's or Blatt's pairs were a speculation, while BCS is a theory which fit immediately so many of the experimental facts that no plausible alternative was presented from that point on.

Most of the subsequent history has been covered in detail in this book, so I will pause to take in only some of the more exciting or more acrimonious subsequent developments. It seems, looking back as a participant, as though everything was happening at once during the succeeding few years.

It is amazing how much actually was in BCS (although the authors had been kind enough to do the formalism in one of the least satisfactory of possible ways, so that many of us amused ourselves in the beginning by making formal improvements, of which Bogoliubov's, Gor'kov's, or Nambu's, and Gor'kov + Nambu = Eliashberg–Schrieffer are now the most used.‡ The idea that $c_k c_{-k} = b_k$, the pair field, has a finite mean value is there, although oddly enough the statement that Bose–Einstein condensation of the pairs is irrelevant is also there. The reason is that at that time Bose–Einstein condensation was thought of as, of necessity, a condensation of preexisting particles, not as the appearance of a macroscopic coherent particle field, as we now should understand it.

The idea that the response to a time-reversal invariant perturbation is dramatically opposite to the response to a time-reversal noninvariant (pair-breaking or magnetic) one is very centrally there. The response to the pair-breaking electromagnetic field or spin (nuclear or electronic) fluctuation is sharply reduced in its real part—which is the essence of the Meissner effect—but large in imaginary part—which causes the increase in spin relaxation rate. The pair nonbreaking response—such as to phonons—has a smaller imaginary part (the famous coherence factors) and is essentially unchanged as to real part: Phonons don't shift, the dielectric properties don't change.

It was said as late as 1960 that BCS was a theory which predicted nothing which was not a manifestation of the energy gap; actually "this coherence-factor" effect is both distinct from the energy gap and is at the heart of a majority of the experimental results about superconductors. It seemed to me as soon as I realized (30) that the time-reversed pairing explained also the facts about the effects of magnetic impurities in reducing the transition temperature, and of nonmagnetic ones in not doing so [Casimir's "mile of dirty lead wire" (31)] that the crucial experimental proof of the theory had been found (if not sooner, of course).

The energy gap was in the theory—too prominently, probably, because it represented one of the major sources of the ideas. But experimentally it was magnificently confirmed, in exactly the form of the BCS prediction, by optical measurements and then, considerably later (1960) by the beautiful new technique of tunneling. Unfortunately it was a while—although shown formally in 1959 by Abrikosov and Gor'kov—before we realized that a true gap is not essential, but rather, again, the $\psi^*\psi^*$ pair field parameter. This is responsible for the energy gap in the uniform nonmagnetic case, but doesn't in general imply one.

The statistics were there (Bogoliubov is often given credit for them for some reason) in a form which gives entirely satisfactory agreement with the best data. And a rough but workable estimate of the transition temperature as related to electron–phonon coupling was there. Nonetheless a number of objections had to be answered. The loudest—the Sum Rule Shout—came from people who had rejected the gap idea in the past, because it seemed to violate gauge

‡ A thorough discussion of Bogoliubov is to be found in (29), and of the others in Schrieffer's book (10).

invariance. This complicated hassle had the effect primarily of clarifying our ideas about condensation and broken symmetry, leading to the complex of ideas often called the Goldstone theorem, because they are not a theorem and the ideas came from Nambu and others (see Chapter 7). As far as superconductivity is concerned, one had merely to show that the sum rules were valid and the longitudinal collective modes unchanged even in the presence of condensation, and a number of authors accomplished this. The most salutary effect of all this is that no one now objects on theoretical grounds to a coherent particle field being present.

The Great Knight Shift Knockabout was another confusing but rather irrelevant mess. One of the vital results of the pair-breaking argument is that the Pauli spin susceptibility must vanish at absolute zero, while in fact experimentally the Knight shift does not. Since the latter is, in the simple theory, caused by the former, this is a paradox. The relevant point is that in the presence of Elliott spin-orbit scattering the susceptibility is only fractionally Pauli, the rest being high-frequency or Van Vleck–like (32). Strenuous experimental effort has finally proved the many theories which provided more complicated explanations wrong, and this simple idea right. (In most cases; other contributions are essential in most transition metals.) The magnetic reduction of T_c and many other indirect measures of the true spin susceptibility had actually always outweighed, in many people's minds, the difficult if deceptively more direct measurement of the Knight shift.

Finally, there was the Isotope Effect Imbroglio. The early emphasis on the isotope effect had left people with the impression that all theories using the electron–phonon interaction stood or fell on the isotope effect $T_c \propto M^{-1/2}$. Thus when experiments showed Ir and Rh had $T_c \sim M^{0 \pm .1}$ (33), a great to-do was made about alternative interactions. This discussion had at least a positive effect on the science of superconductivity: it stimulated the first serious quantitative work on the transition temperature and the interaction.

Preliminary work by Swihart (34), like a more serious attempt by Morel and myself (35), simply harked back quantitatively to a formula of Bogoliubov's which stated that very weak coupling superconductors, with $T_c \ll \Theta_D$, should indeed have isotope effects considerably below $\frac{1}{2}$, because of the sharp competition between phonon interactions, which *do* scale with $M^{-1/2}$, and the Coulomb interactions and electronic screening, which don't. Present-day data agree much better with this idea than did the data then—the history of the isotope effect is a sorry one for the experimentalists—but it may be, as Garland suggested, that, in addition, because of the narrower bands the reduction is even stronger in transition metal superconductors (36). Certainly no outstanding paradoxes beyond the often very poor experimental accuracies exist at present.

More important, this problem stimulated us to look seriously at the real gap equation, which had been formally expressed by Eliashberg in 1959 (see Chapter 11), combining the Migdal treatment of electron–phonon interactions and the Gor'kov–Nambu formalism for superconductivity. Surprisingly, Eliashberg's formalism leaves one with numerically manageable equations which Morel was able to approximate reasonably well for given assumptions about the interaction; the basic simplification being the point, due to Migdal, that the interactions are short-range in space, even if long in time. Then one day when I was giving a talk in Birmingham about this work, Prof. Peierls asked me if the structure we predicted in the energy gap parameter as a function of energy might not be observable.

It was a thought which I am ashamed to say had not struck me before, but I mumbled something about optical spectroscopy or tunneling, not realizing that two weeks later I would hear from the Bell Labs group that spectacularly detailed structure was being seen in Pb in tunneling. (Giaever had seen and published some structure at ω_D, but it was Rowell who was responsible for establishing this structure as a spectroscopic measurement.) I was almost more depressed than excited at this news; I remember remarking that this meant the end of superconductivity as a wide-open, speculative field and the beginning of a thoroughly quantitative and exact era; how right I was is detailed in Chapters 10 and 11, by Scalapino and McMillan and Rowell. There is no longer any room for uncertainty, even at the 10% level, about the mechanisms of electron–electron and electron–phonon interaction for any metal for which tunneling is feasible.

This, then, I think, brings us right up to the present and on to the future. One bit of the future of superconductivity is clear: as a tool for the quantitative study of electrons and phonons in metals, the tunneling technique in all its variations combined with superconductivity has no competitor, and will remain an important technique in metal science for many years. Innovations continue —the Tomasch effect, anisotropy of the gap—and will make this a steadily more valuable tool.

One of the most exciting moments I remember I have left until now, because I think it too belongs basically to the future: the day in '62 a slight, reserved Cambridge student came into my office and said that while studying Cohen, Falicov, and Phillips' work (37) on the theory of tunneling, he had found an extra and very peculiar term in the tunneling current between two superconductors. I already knew Josephson well enough not to seriously question its existence, but it was not until after Rowell and I had actually found the current that we worked out a presentation of the effect which really satisfied us (38). Nonetheless, we both were immediately fascinated by the fact that the current was phase-dependent and gave a measurable physical reality to the phase of the pair wave function. While in principle there is little in the Josephson effect that is not foreshadowed in Ginzburg–Landau, in practice the former and its relatives make many kinds of interference effects measurable which were inaccessible before. One can hardly doubt that the whole field of quantum interference opened up by Josephson and elaborated by Rowell, Mercereau, Shapiro, and others (see Chapters 8 and 9) will have an exciting threefold future: as basic science, as the source of an increasing family of measuring instruments and techniques—already we have the precision measurement of h/e and the micro-microvoltmeter of Clarke—and as a technological breakthrough which could go so far as to become the heart of a whole new form of computer, and will certainly produce many devices of value. Among these will also be unique methods for detecting—and possibly also generating—ac signals in new frequency ranges.

As for superconducting magnets and type II, the problems of the motion of flux lines and of domains, and, the sources of dissipation, remain the biggest theoretical puzzle of the present, tantalizingly close to solution but not solved (see Chapter 19). Thus the whole subject of transport in superconductors remains in limbo with the central part unsolved, although much of the periphery is chewed away. Collective motions of line structures and many other complications remain ahead of us. Giaever's dc transformer shows us that the lines move—but how is still a question.

One of the authorities in the field tells me he believes around 800,000 G is the present extreme limit of what one might hope to build in a superconducting magnet. That involves making only reasonable extrapolations from materials available at present: the major component of such an extrapolation being the increase in the "Clogston limit" one might hope to get by maximizing spin–orbit scattering from heavy impurities. Of course, there are many other facets of the question, but mostly they boil down to the semiempirical, semitheoretical fact that materials so far have a maximum transition temperature of around $20\,°K$, and thus energy gap of around $50\,°K$. The upper critical field H_{c2} is qualitatively

$$H_{c2} \simeq \Phi_0(\Delta/\hbar D)$$

where D is the diffusion constant, and with values of D appropriate to metals, even assuming $l \sim 1$ atom radius, this cannot be raised above about 10^6.

This brings us to the ticklish question of what might be expected from radically different materials. What T_c might be achieved, and if so—and this question is almost never asked—how useful might a higher T_c material be?

In the first place, it seems definite that ordinary metals and the orthodox phonon mechanism cannot carry us much farther. Work of McMillan (not included in his chapter in this book) following the main lines of the present theory of metals strengthens somewhat the conclusion many of us reached quite a few years ago: that the phonon mechanism gives a definite upper limit to T_c, not very far above the values already achieved. As McMillan points out, the energy gap (which is about $3.5\,kT_c$) must be a certain amount less than the energy of the modes of vibration to which the electrons are strongest coupled, to take advantage of the attractive region of the coupling as well as to avoid severe reduction of T_c due to scattering of the electrons reducing the pair lifetime: "lifetime effects." A second important limitation is that the electron coupling reduces the frequency of the relevant modes: that frequency is controlled by the dielectric response function of the electron gas to the ion motion, which involves the same coupling which leads to superconductivity. Thus for a given system there is an optimum, not very large, coupling constant, and most reasonably high-temperature superconductors are not very far from their optima. This theory is not accurate enough to predict a maximum T_c to 10%, but it makes transition temperatures higher by a factor of 2 seem most unlikely.

One might hope that some other interactions acting in a metal might lead to higher T_c's—for instance, the often-suggested but mysterious effect of an open f-shell in La. Unfortunately, one has always the phonons, whether one is relying on them for the interaction or not, and at the phonon frequency the dielectric function—which controls the electron–electron interaction—behaves singularly. Thus other interactions cannot be discussed as independent entities: Phonon lifetimes will reduce T_c, and phonon resonances undoubtedly change the sign of the interaction over some range of energy, and probably—although of course not certainly—it will be virtually impossible to raise Δ above the dominant phonon frequencies (which tend to be $\sim 100\,°K$).

Most systems one can think of, such as conjugated organic materials, appear to be subject to the same difficulties except insofar as the relevant electrons may represent only small fractions of the total electron gas, as in semimetals and semiconductors. Perhaps the most realistic discussion of bizarre mechanisms for superconductivity has been the studies of M. L. Cohen, partially reported in this book, on the semiconductor and semimetal problem. He finds that, while most computations lead to quite low T_c's for these systems, there could in principle be semiconductor systems with quite large pair binding energies between electron polarons. When the optical mode coupling is enhanced by polarity, and its frequency raised, the coupling of the electrons to transverse or acoustic modes may become so weak that it doesn't necessarily break up the pairs. Thus perhaps high T_c's are not necessarily excluded in special polar systems.

In all of this work, as in the many less well-founded speculations on inhomogeneous systems, etc., the implicit assumption that the pair binding energy gives T_c is of course made. But that is not true, as we know from liquid helium: There the fermions are bound with energies of rydbergs, but $T_c \sim 2\,°K$. Once bound the pairs must also move: they must couple their phases over macroscopic distances with an energy adequate to give us the phenomenon of superconductivity. We can perhaps estimate that in a dilute system such as conjugated molecules or a polar semiconductor T_c is the lower of the pair binding energy and some kind of Fermi energy of the electrons, which may be quite low. In any case, in such systems pair binding will not necessarily lead to phenomena we readily identify as superconductivity—as it took many years for the energy gap and anomalous moments of inertia of nuclei to be recognized as a form of superconductivity.

While we're on the subject of bizarre mechanisms, it is well to mention the hope of many of us that simple time-reverse pairing will not always remain the only possibility. In very pure magnetic metals at low temperatures or under unusual circumstances none of us can predict, one can hope that p- or d-wave pairing can become a reality. If so, we would have a phase of matter with extremely strange properties (*39,40*).

Let me perhaps, then, summarize my ideas on the future of superconductivity and try to restate clearly to what extent firm predictions can be made. In the first place, the nature of the phenomenon in ordinary metals and its microscopic explanation are remarkably well understood, the biggest hole visible at present being our lack of a really adequate quantitative theory of transition metals. The future here is most promising as a research tool of the science of metals.

The whole field implied by macroscopic quantum coherence: flux line motion, interference phenomena, etc., seems to me to have an incalculably rich technological and perhaps even scientific future about which any one should hesitate to speculate much further.

Finally, of the possibilities for radically new types of superconductors and interactions, the only safe prediction is that conventional extrapolations of any type—including what types of materials to look for or what "superconductivity" itself may mean—are completely unsafe. Room-temperature magnets or superconducting skis? Probably not. For one thing, so many diverse types of materials exist in nature or have been made in the laboratory that large, obvious effects would probably already have been seen. But room-temperature interferometers? Who knows?

As the reader who has reached this point already knows, in the above I have expressed very much my own personal point of view and opinions, even to the extent of taking the forbidden liberty of inserting personal recollections and actual incidents of the history of the field. I have done this partly to liven up the book and partly to give the reader a sense of what we've felt we were up to this past decade. Perhaps the future historian of science will find some of this interesting. In any case, I hope the reader will recognize that much of what I have said is to be taken in the sense of personal remarks and opinions rather than scientific truth.

ACKNOWLEDGMENT

I would like to acknowledge help in writing this chapter from so many people that individual names, except for my wife's, can hardly be mentioned.

REFERENCES

1. H. Fröhlich, *Phys. Rev.* **79**, 845 (1950).
2. J. Bardeen, *Phys. Rev.* **80**, 567 (1950).
3. J. M. Blatt, *Theory of Superconductivity*, Academic Press, New York, 1964.
4. E. Maxwell, *Phys. Rev.* **78**, 477 (1950).
5. C. A. Reynolds, B. Serin, W. H. Wright, and L. B. Nesbitt, *Phys. Rev.* **78**, 487 (1950).
6. J. Bardeen and D. Pines, *Phys. Rev.* **99**, 1140 (1955).
7. A. B. Migdal, *Zh. Eksperim. i Teor. Fiz.* **34**, 1438 (1958); *Soviet Phys. JETP* **7**, 996 (1958).
8. W. Kohn and Vachaspati, *Phys. Rev.* **83**, 462 (1951).
9. G. Wentzel, *Phys. Rev.* **83**, 168 (1951).
10. J. R. Schrieffer, *Theory of Superconductivity*, Benjamin, New York, 1965.
11. V. L. Ginzburg and L. D. Landau, *Zh. Eksperim. i Teor. Fiz.* **20**, 1064 (1950).
12. O. Penrose, *Phil. Mag.* **42**, 1373 (1951).
13. V. L. Ginzburg, *Nuovo Cimento* **2**, 1234 (1955).
14. A. A. Abrikosov, *Zh. Eksperim. i Teor. Fiz.* **32**, 1442 (1957); *Soviet Phys. JETP* **5**, 1174 (1957).
15. L. V. Shubnikov, W. J. Chotkevich, Yu. D. Shepelev, and Yu. N. Riabinin, *Physik Z. Sowjet.* **10**, 165 (1936).
16. K. Mendelssohn and J. R. Moore, *Proc. Roy. Soc. (London)* **A151**, 334 (1935); **A152**, 34 (1935).
17. B. B. Goodman, *IBM J. Res. Develop.* **6**, 63 (1962).
18. Y. B. Kim, C. F. Hempstead, and A. R. Strnad, *Phys. Rev. Letters* **9**, 306 (1962); P. W. Anderson, *Phys. Rev. Letters* **9**, 309 (1962).
19. P. W. Anderson and Y. B. Kim, *Rev. Mod. Phys.* **36**, 39 (1964).
20. J. Bardeen and M. Stephen, *Phys. Rev.* **140**, A1197 (1965).
21. C. Caroli and K. Maki, *Phys. Rev.*, to be published.
22. L. P. Gor'kov, *Zh. Eksperim. i Teor. Fiz.* **37**, 833 (1959); *Soviet Phys. JETP* **10**, 593 (1960).
23. F. London, *Superfluids*, Vol. I, Wiley, New York, 1950.
24. R. Doll and M. Näbauer, *Phys. Rev. Letters* **7**, 51 (1961); B. S. Deaver and W. M. Fairbank, *Phys. Rev. Letters* **7**, 43 (1961).
25. D. Bohm and T. Staver, *Phys. Rev.* **84**, 836 (1952).
26. W. A. Harrison, *Pseudopotentials*, Benjamin, New York, 1967.
27. T. E. Faber and A. B. Pippard, *Proc. Roy. Soc. (London)* **A231**, 336 (1955).
28. J. Bardeen, *Handbuch der Physik*, Vol. 15 (S. Flügge, ed.), Springer, Berlin, 1956, p. 274.
29. J. M. Blatt, *Theory of Superconductivity*, Academic Press, New York, 1964.
30. P. W. Anderson, *J. Phys. Chem. Solids* **11**, 26 (1959).
31. H. B. G. Casimir, *N. Bohr and the Development of Physics* (W. Pauli, ed.), Pergamon Press, New York, 1955, p. 118.
32. R. A. Ferrell, *Phys. Rev. Letters* **3**, 262 (1959); P. W. Anderson, *Phys. Rev. Letters* **3**, 325; M. Tinkham, private communication, 1959.

33. T. H. Geballe, B. T. Matthias, E. Corenzwit, and G. W. Hull, Jr., *Phys. Rev. Letters* **6**, 275 (1961); **8**, 313 (1962).
34. J. C. Swihart, *IBM J. Res. Develop.* **6**, 14 (1962).
35. P. Morel and P. W. Anderson, *Phys. Rev.* **125**, 1263 (1962).
36. J. W. Garland, Jr., *Phys. Rev. Letters* **11**, 111, 114 (1963).
37. M. H. Cohen, L. M. Falicov, and J. C. Phillips, *Phys. Rev. Letters* **8**, 316 (1962).
38. P. W. Anderson, *The Many-Body Problem*, Vol. II (Ravello notes) (E. R. Caianello, ed.), Academic Press, New York, 1964.
39. K. A. Brueckner, T. Soda, P. W. Anderson, and P. Morel, *Phys. Rev.* **118**, 1442 (1960).
40. R. Balian, *The Many-Body Problem*, Vol. II (E. R. Caianello, ed.), Academic Press, New York, 1964.

Macroscopic Coherence and Superfluidity

Contemporary Physics (International Atomic Energy Agency, Vienna, 1969)
Vol. 1, p. 47

Trieste, 1968; Salam's last convening of the whole of theoretical physics. This meeting followed on from Kyoto 1953 and Seattle 1956, both of which I attended but gave no major talk. (At Seattle I spoke briefly about my first blundering steps towards localization.) The meeting came in June '68; I remember hearing the news of Robert Kennedy's assassination as I drove across Northern Italy from a NATO meeting in the French Alps which was dominated by discussions of politics. But at Trieste the only fuss was made by various of the Nobelists, who objected to being sequestered in Duino Castle. There was an excellent pre-meeting program on molecular biology by Crick, Brenner and Klug which to me was the most exciting part. I was dissuaded from talking on the Infrared Catastrophe; by request, instead, I repeated much of papers 10 and 11.

MACROSCOPIC COHERENCE AND SUPERFLUIDITY

P. W. ANDERSON[*]
Cavendish Laboratory,
University of Cambridge,
Cambridge, United Kingdom

Abstract

MACROSCOPIC COHERENCE AND SUPERFLUIDITY. The problem of the coherence in a superfluid, namely a substance which possesses ODLRO in the Yang sense, is treated, with a particular account of the problem of dissipation in the superfluid. The relation between dissipation and quantized vortices is also explained, both for superfluids and superconductors.

My feelings at this point are best expressed by describing a cartoon which appeared in the New Yorker a number of years ago. It showed a father whose little boy, dressed in pyjamas, was sitting on his knee, looking up at him trustingly, and saying "Daddy, tell me again the story of how jazz came up the river from New Orleans". I have been asked to retell on this occasion our success story in the theory of superfluidity, a story which I believe is now mostly a source of fruitful ideas for other fields. I should like, in addition, to touch some of the boundary points of our knowledge, especially where the problems are of some real theoretical interest.

It seems most likely that almost all non-magnetic systems (at least) which have enough zero-point energy to remain fluid in their ground-states will, in the end, be found to be superfluids, in the sense that they spontaneously transform from the so-called normal Bose or Fermi liquid state into a state which has a certain definite type of correlation. It can certainly not be so proven, because we cannot imagine all the other types of possible ground-states for many-body systems which might have yet lower energy. The motivation for this ordering can be expressed in terms of what I shall say shortly, but a much deeper microscopic discussion, at least for metals, is given by Schrieffer in these Proceedings.

We thus just define a superfluid as a substance which has the type of correlation exhibited by liquid helium (postulated by Penrose and Onsager) and superconductors (BCS and Gor'kov) and named ODLRO by Yang:

$$\langle \psi^\dagger(r)\, \psi(r') \rangle = F^\dagger(r)\, F(r') + \text{short-range terms}$$

In the case of a metal we must define $\psi = \psi_{e\sigma}\, \psi_{e-\sigma}$, a field operator for pairs of electrons, not single particles as in helium.

This definition is compatible with a number of prejudices many-body theorists have since discarded, especially that the phase of the field operator is not a physical quantity and that coherence cannot exist between states of different particle numbers, since this mean value conserves N. However, this formulation is most inconvenient, if barely usable, for discussing the

---

[*] Present address: Bell Telephone Laboratory, Murray Hill, N.J., United States of America.

coherence phenomena of superfluidity. In studying coherence phenomena for light, we use slits, half-silvered mirrors, etc. — all devices for separating out different pieces of light whose relative phase we want to test. In the case of superconductivity and superfluidity the great advance came when we invented some corresponding entities: Josephson junctions and their relatives, and the orifice, which can serve as switches connecting or disconnecting different samples of our particle field at will. Imagine, now, two buckets of superfluid connected by an orifice that can be opened or closed (Fig. 1). If this system has ODLRO, it is easy to see that, if r is on one side of the orifice and r' on the other, one must have coherence between two states

$$\psi_{NN'} : N_1 = N_1, N_2 = N', \text{ and } \psi_{N+1, N'-1}$$

because these are the states which are connected by $\psi^\dagger(r)\psi(r')$. Thus when we close the door and, at least in principle, can then handle the two parts independently, each part has an internal coherence which cannot be described by ODLRO alone, but which does involve coherence of states of different N and a real physical meaning for the phase. For instance, an experiment which can be done in principle at least, is to subject the two sides to different gravitational potentials and re-open the door, comparing phases by the resulting instantaneous current.

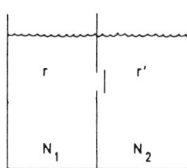

FIG. 1. Two buckets of superfluids connected by an orifice.

We can also get a glimpse of the motivation for the ODLRO from this picture. We can easily understand that when the orifice is open, there is a kinetic energy term in the Hamiltonian which transfers particles from one side to the other

$$\mathcal{H}_K = -T(C_1^\dagger C_2 + C_2^\dagger C_1)$$

and this term can be caused to have a negative mean value by having coherence. This coupling energy is a very central concept in superfluidity.

Although it is possible, by treating one's system as a part of a larger one, to avoid grasping this particular nettle, it seems far more convenient to do so and to satisfy the ODLRO condition by the radical assumption

$$\langle \psi^\dagger(r_1) \rangle = F^\dagger(r_1, t_1) \quad \langle \psi(r_2) \rangle = F(r_2, t_2)$$

i.e. that the phase and amplitude of the quantum particle fields have macroscopic mean values in the superfluid.

What kind of state has this property of macroscopic quantum coherence? Clearly, it has a wave-packet state made up from different N-values

$$\Psi = \sum_N a_N e^{iN\varphi} \Psi_N.$$

where by actual calculations we find

$$\frac{1}{V}\int d\tau \langle \psi(r) \rangle = \frac{1}{V}\left(\sum_N a_{N-1} a_N (\psi_{N-1}, \psi\psi_N)\right) e^{i\varphi}$$

For a perfect Bose gas, if the a_N's spread over a wide enough range $\Delta N \gg 1$, $\langle \psi \rangle \sim (N/V)^{\frac{1}{2}}$; for actual superfluids, the value ranges from ~ 0.1 for He to $\sim 10^{-4}$ for superconductors.

We notice that the transformation coefficient from states with definite N to definite φ is $e^{iN\varphi}$. Except for minor corrections for small N (which do not alter the rest of this paper), it is precisely valid to treat N and φ as conjugate dynamical variables, the latter being treated as a true macroscopic statistical parameter of the system in the same sense as P, V, S, T (and no less abstract than the latter pair). The Hamilton's equations expressing the motion of these two variables are the basic equations of superfluidity

$$\hbar \frac{d(\varphi)}{dt} = \langle \frac{\partial \mathcal{H}}{\partial N} \rangle = \mu$$

$$\hbar \frac{dN}{dt} = -\langle \frac{\partial \mathcal{H}}{\partial \varphi} \rangle$$

The second equation is the general equation ensuring the presence of supercurrents. If, for instance, there is indeed a coupling force across our orifice holding the phases of the two buckets together, a free energy

$$F \propto \frac{U_0 (\varphi_1 - \varphi_2)^2}{2}$$

there is then the possibility of a current between them

$$\frac{dN_1}{dt} = -\frac{dN_2}{dt} = -U_0 (\varphi_1 - \varphi_2)$$

It seems clear that supercurrents exist if and only if one has phase coupling: the phase is the only variable conjugate to N. This is the nearest thing there is to a rigorous characterization of superfluidity.

For electrons, the charge 2eN is coupled to the electromagnetic field and the phase alone is not a gauge-invariant quantity but $\nabla\varphi - 2eA/\hbar c$. Thus supercurrents can flow in response to magnetic fields as well as to phase gradients. Incidentally, by a definition which can be checked at T = 0 and for homogeneous samples against Galilean invariance, we may define, if we like, $v_s = (\hbar/m)\nabla\varphi$ and $n_s = \partial^2 U/\partial(\nabla\varphi)^2$ and then $j = n_s v_s$. Quantization of flux and of circulation are merely two manifestations of the elementary idea that $\langle \psi \rangle$ is single-valued and thus $\oint \nabla\varphi \cdot dS = 2n\pi$.

It is the other of the superfluid equations which may be more interesting here.

In one form, it is the oldest existing description of superfluidity, i.e. London's acceleration equation

$$m \frac{dv_s}{dt} = m \frac{\hbar}{m} \frac{d}{dt} \nabla \varphi = \nabla \mu = \vec{F}$$

This is the basic characteristic of superfluidity: a superfluid cannot sustain a true force — in the sense of a gradient of the thermodynamic potential — without undergoing acceleration. This clearly means that in the absence of vorticity we must have flow without force. Conversely, if we have a difference in chemical potential $\mu_1 - \mu_2$ between two reservoirs of superfluid, their relation phases must be changing at precisely the rate

$$\frac{d\varphi}{dt} = \frac{\mu_1 - \mu_2}{\hbar}$$

Note that, by at least two arguments, we can see that there are no intrinsic corrections to this equation. First, the derivation draws only on the meaning of φ and the actual definition of μ as dE/dn. Second, gauge invariance tells us that $\hbar \, \partial \varphi / \partial t - 2 \, eV$ is the only way in which φ can occur physically. Of course, we can make mistakes in measuring either μ (temperature or contact potential differences, for instance) or $d\varphi/dt$ (we can count wrong), but that is an experimental problem.

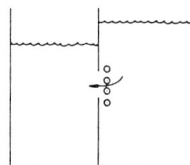

FIG. 2. Phase slippage due to vorticity.

There are two very similar ways the phase change can manifest itself. Most simple is a Josephson device, a connection between superconductors which exerts a coupling energy $U \propto -\cos(\varphi_1 - \varphi_2)$ and thus a <u>current</u> $\propto \sin(\varphi_1 - \varphi_2)$. More generally interesting is the possibility of phase slippage due to vorticity. A quantized vortex is a line singularity around which $\oint \nabla \varphi \cdot ds = 2\pi$. One can see immediately that every time a vortex passes between two points, it changes their relative phase by 2π: so we can sustain a $\overline{\mu_1 - \mu_2}$ if at the same time we have a rate of motion of vortices of $dN/dt = \overline{(\mu_1 - \mu_2)}/h$ (Fig. 2).

This phenomenon is behind all manifestation of dissipation in superfluids and superconductors. Incidentally, after this equation had been applied to superfluids, we found that it is actually a perfectly respectable — if undiscovered — theorem in classical perfect fluid hydrodynamics. φ is simply a quantized version of a velocity potential, and going through precisely the same arguments for potential flow one finds

(mean chemical potential difference) = mean transverse flow of vorticity

Whether this has any hydrodynamic or magnetohydrodynamic applications, I don't know. Incidentally, the superconducting version also has a classical equivalent: $d\Phi/dt$ = E.M.F. (expressed in quantized flux units). As in the hydrodynamic case, it is important to realize that the relationship to $dN/dt = \Delta\mu$ is tricky. This relationship of electromagnetism and hydrodynamics has a fascinating 19th-century ring to it.

The original and most elegant way of demonstrating and measuring the Josephson a.c. equation was suggested by Josephson in 1962 and carried out a year later for superconductors, two years later for superfluids; it uses the phenomena of <u>entrainment of frequency</u>. I want to mention a fact out of a book on non-linear mechanics about this. It appears that entrainment was discovered by Huyghens, who noticed that two clocks hanging on the same wall ran in precise synchronism. More mundanely, anyone familiar with TV knows that a strong a.c. signal applied to a non-linear oscillator has the capability of entraining the frequency of the non-linear oscillator in rather precise synchronism with the external signal, and this is in fact the most accurate way to measure $\hbar/2e$ for electron pairs, or $\hbar/m$ for helium atoms, using the Josephson equation. One can induce steps in an I-V characteristic at multiples of $V = h\nu/2e$ (Fig. 3). Unfortunately, one can also induce them at subharmonics and on occasion at absolutely arbitrary frequencies in ways which will only be understood when we have the actual non-linear mechanics under more precise control. But it is fairly easy to set up conditions where the intrinsic accuracy of this phenomenon — which is in principle a mechanism, if you like, rather than a measurement in the usual sense, and thus almost infinitely accurate — is far better than other uncertainties such as in absolute measurement of voltage.

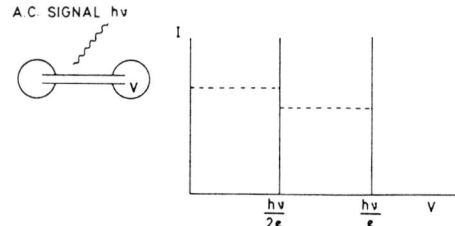

FIG. 3.   Entrainment of frequency.

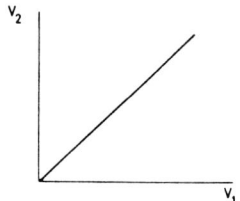

FIG. 4.   Experiment of Ivar Giaevar.

Let me return now to the statement that dissipation in superfluid flow is always the consequence of the motion of quantized vortices. A very beautiful demonstration of this fact was given recently by Ivar Giaevar in one of those experiments which convince on the spot. He superposed two superconducting films in a magnetic field, separating them by an oxide layer which was quite thick enough to provide good insulation, but thinner than the distance $(B/\Phi_0)^{-\frac{1}{2}}$ between quantized matrices. When a current flowed through one (the primary), he observed a voltage in the second practically equal to that in the first: essentially a perfect "d.c. transformer" (Fig. 4).

This is not to imply that these dissipative effects are understood perfectly. In both superfluid helium and superconductors, one of the major theoretical problems remaining is the question of the detailed explanation of dissipation. In the case of helium, our worst problem is to understand how the vortices get there in the first place, a question which has been emphasized by Vinen. The energy of a quantized vortex per unit length is enormous, i.e. $\sim(\hbar^2/me^3)\ln(R/a)$, which for a vortex ring of only barely macroscopic size (~ 100 Å) is already tenths of an electron volt (as we verify from the Reif experiment). Thermal fluctuations of this magnitude are hopelessly rare and thus the nucleation of vorticity must be terribly difficult. One can only hope that the experimentalists have so far always been so clumsy as to introduce microscopically rough places in their apparatus or the like at which the nucleation takes place.

Incidentally, I definitely do not wish to imply that all problems of the microscopic theory and the detailed behaviour of liquid helium are solved — quite the contrary. There is a fairly respectable microscopic theory to which the number of contributors is simply too great to mention any of them here, which seems to give ~ 20-50% accurate answers and nothing seriously wrong in principle, but many details remain to be worked out; I only mention vortex nucleation because it is the biggest puzzle at the moment. One hopeful sign for the microscopic theory is the experimental possibility of an equivalent to superconducting tunnelling — the velocity distribution for evaporation of particles from the superfluid surface, an experiment which John King and students at MIT are carrying out.

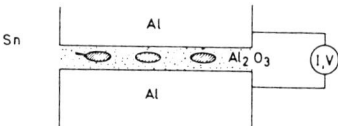

FIG. 5.   Experiments on tin particles.

In the case of superconductors, since the work of Bardeen and Stephen on the flow of flux lines, and mine on their creep, the orders of magnitudes of the viscosity and other properties of flux line motion are not unreasonable. But both the qualitative two-fluid picture of Bardeen and Stephen and the beautiful perturbation calculation of Caroli and Maki seem to be hung up quantitatively at least on a rather old dilemma in the field — to what extent do the flux lines flow downstream with the electron gas? This (using our canonical equation saying that potentials are due to transverse vorticity flows) determines the Hall effect, and while it is now apparent both experimentally and theoretically that the Hall effect is of the same order of magnitude as the normal one, the fact is that Bardeen and Stephen give the wrong answer quantitatively, while Caroli and Maki do not answer the question at all.

One of the most interesting limiting cases for the phenomenon of superfluidity is that of very small particles. For nine or ten years we have had one example of superconductivity in very small particles, namely the pairing phenomenon in nuclear matter. However, nuclei are not easy things to work with; we can apply the equivalent to H-fields by rotating them, but we cannot do thermal or coherence experiments on them, nor can we go to the macroscopic limit of nuclear matter. Thus it is nice that Giaevar and a co-worker have come up with a good series of experiments on tin particles in the interesting range $\sim < 10^2$ Å. Giaevar evaporates very tiny amounts of tin on to an oxidized Al substrate, which then forms separate globules of fairly sharply defined sizes ~ 15 to 150 Å. He then oxidizes again, covers all with Al, and measures the tunnelling characteristic (see Fig. 5).

The question is, of course, what sign is left of the superconductivity of Sn. There are basically two new energy parameters introduced by the smallness of the particles: the electrostatic energy per electron

$$\frac{e^2}{C} = \frac{e^2 d}{\epsilon A}$$

$\sim 10/R^2$(Å) eV ~ 1 MeV for 100 Å ($10°$K), and the granularity of the energy levels, $E_F a^3/R^3 \sim 100/R^3$ eV.

The first effect of the electrostatic energy is to destroy the coupling between particles — the Josephson current — since the electrostatic energy difference between states of different N will cause $\Delta N \rightarrow 0$, $\Delta\varphi \rightarrow \infty$ when it is greater than the rather weak phase coupling.

One might wonder if electrostatics will prevent internal phase coupling. The idea that the internal coupling could also be broken up appears in the literature, but it is wrong. What is essential is the effective quasi-particle interaction, which includes the ionic and electronic charge clouds — the actual renormalized charge of a quasi-particle is zero, the charge all appearing at the particle surface. Thus indeed the tunnel characteristic shows evidence of superconductivity down to ~ 50 Å. How can it, when electrostatics alone introduces a gap in the spectrum of order $10/R^2$? Because the Fermi levels of the neutral particles have a random distribution, and some few match the Al matrix perfectly and cancel the gap. Granularity should begin to be very effective at ~ 50 Å. This is rather a complicated effect and when it is really bad, of course, the superconducting gap becomes meaningless, since there is already a larger gap; but in determining an effective T_c, it does indeed weaken the tendency for pairs to break up, so T_c increases more or less as $\sqrt{(\Delta^2 + \Delta_{gram}^2)}$ before disappearing (Parmenter). This is indeed observed.

I would like to end by making a few remarks on the generally interesting subject of "blue-sky" superconductors. It seems to be fashionable to write hopeful papers about the possibility of really high-temperature superconductors, preferably made from complicated organic molecules. I have no pipeline to the Deity so cannot pronounce on the existence of such things. I can, however, remark on the three most important points which have not been discussed in these papers.

(1) Organic materials are generally rather loose vibrationally and have large observed electron-vibration couplings. The incoherence introduced by such couplings and their influence on the effective quasi-particle interactions cannot be ignored.

(2) Any pairing strong enough to overcome this as well as the intrinsic "granularity" energy gaps of the order of at least $0.1 - 1$ eV will have very large chemical effects. In particular, the structural chemistry of a substance with such pairing must be peculiar, and the delicate adjustment of bond lengths and angles which led to the Double Helix structure, for instance,

could not be useful for such a substance: to find an organic superconductor look for a structure which Pauling, Crick, et al. <u>cannot</u> solve.

(3) Large pairing energies may or may not mean superconductivity. The energy gap in He, for instance, is 20 eV, but T_c is 2°K. Both pairing and the zero-point kinetic energy which propagates the pairs must be present, and T_c is given by the smaller of the two.

The Fermi Glass: Theory and Experiment
Comments on Solid State Physics **2**, 193 (1970)

I left localization in 1958–9 for superconductivity, not because I had any doubts about the result, since the paper was practically a proof and Feher's ENDOR method couldn't work without it, but because I was bemused by the strong interactions between real electrons. Mott deserves all the credit for nursing my dream into the real world, and here is the first evidence that even I had become a believer. I saw at last that Mott and Anderson insulators were complementary and indispensible to each other. That is, repulsive interactions will necessarily make even fewer final states available for coherent hopping processes at the Fermi level than are available for non-interacting electrons. Elihu Abrahams and Nevill Mott had, between them, built a whole transport theory on my ideas, which Mott proceeded to correlate with experiments. This paper marks the prodigal's return to the subject.

The Fermi Glass: Theory and Experiment

About a dozen years ago I wrote an obscure paper, now widely quoted but perhaps not as widely read, on "The Absence of Diffusion in Certain Random Lattices".[1] Special interest has attached to it recently because of its being used in discussions[2,3] of the underlying physics of conduction in semiconducting glasses, which are of interest partially because of their use in Pearson–Dewald–Ovshinsky devices.[4]

The central idea of my paper was to suggest that under some circumstances the exact eigenstates in a macroscopic random lattice might be localized rather than extended throughout the lattice, and to present a mathematical argument for localization (including a definition of that concept) for one model.

Mott, particularly, has used this concept as the basis for an important series of discussions of electronic conduction in impurity bands and in amorphous or glassy semiconductors.[5,6] One point I would like to make here is that whatever one's opinions on the theoretical issues may be, the experimental question is overwhelmingly settled in favour of localization. First of all, the work on impurity conduction by Fritzsche and Lark-Horowitz[7] and subsequent work by Fritzsche and others (for references see Ref. 7), and the interpretation by Twose[8] and Miller and Abrahams[9] admits of no other hypothesis. A striking extension is Mott's prediction that at low temperatures $\ln \sigma$ should vary as $-\alpha/T^{1/4}$, which in amorphous Si and Ge is satisfied over ten decades of conductivity. The prediction of Pollak and Geballe[10] that for frequency ω the conductivity behaves like $\omega^{0.8}$ is often confirmed, and depends on the same hypothesis. Mott,[11] following earlier experimental and theoretical work by Tanaka *et al.*,[12] predicts optical absorption such that $\sigma(\omega)$ varies as ω^2, and there is some evidence that this is observed in amorphous materials.[13,14]

And finally there seems no doubt that for many amorphous substances the conductivity tends to zero exponentially with $1/T$, while the density of states does not vanish in the band gap. My point is that

193

the discussion of all these phenomena depends on the assumption of localization.

The model that I used, a fair enough approximation for the relevant cases, contained "sites" (impurities in the impurity-band case, dangling bonds in glasses etc.) with local energies E_i distributed randomly with density $\rho(E_i) = 1/W$, and with hopping matrix elements V_{ij} connecting them, which might or might not be stochastic. I proved that if W/V_{ij} exceeded a certain value depending on the coordination number, there was no diffusion. The actual eigenfunctions are local, and therefore, an electron initially localized on one site, or in any larger volume, would not diffuse away.

In studying the motion of electrons in random systems, one first instinctively applies the Golden Rule. The Golden Rule predicts that an electron placed initially at site i will decay at a rate

$$\frac{2}{\hbar} \langle V_{ij}^2 \rangle_{\text{av}} \rho(E_i).$$

With any reasonable assignment of parameters this formula leads to a conductivity practically independent of temperature and frequency, many orders of magnitude larger than that observed. This naive Golden Rule answer was that proposed (quite sensibly: in the problem he was studying, that of spin diffusion, it sometimes works, for reasons not yet understood) by Portis[15] in the paper I was answering at the time, and in the past few years a number of papers have appeared[16] in which the same result is obtained and on the basis of it my paper is called into question. My second purpose here is to point out that the difficulty in these papers is that they do not seem to accept the central argument of my work even though Ziman's paper, which is the predecessor of the most recent group of critics, seems to restate my arguments most ably.[17]

This central remark is that if one is to study localization, one must take account of the fact that in a random array no site is like any other, in any other place in the crystal or in any member of the ensemble. Thus it is wrong to study averages, but instead all quantities must be characterized by their probability distribution functions.

This is true in all random lattice problems but especially so in case the question is one of localization, since then ergodicity or homogeneity approximations prejudge the answer. What quantity is best studied in the appropriate probabilistic way? The quantity I chose (re-expressed in modern terms) was the Green's function

$$G_{ij}(\omega) = \left(\frac{1}{\omega - \mathcal{H}} \right)_{ij}$$

194

for slightly complex $\omega = E + is$. G is referred to site wave functions ψ_i and ψ_j, and we study particularly the diagonal matrix element

$$G_{ii}(\omega) = \sum_{\alpha} |(i|\alpha)|^2 \frac{1}{\omega - E_\alpha} \tag{1}$$

(by definition) where the exact wave functions ψ_α are

$$\psi_\alpha = \sum_i (\alpha|i)^*\psi_i. \tag{2}$$

We see from (1) that the poles of G_{ii} on the real axis are the eigenstates related to site i and since ψ_i is normalized the sum of the residues at these poles, which is the sum of the probabilities of all states, is unity. As the system becomes very big, there are two ways in which we are used to seeing the distribution of poles behave:

(1) in otherwise empty regions of energy, true bound states may appear: isolated poles;

(2) in others, a continuum limit of poles evenly spaced along the axis with rapidly diminishing spaces between them. Then the function G acquires a cut, so that

$$\operatorname{Im} G(\omega + is) = -\operatorname{Im} G(\omega - is) = \text{density of poles}.$$

This is the usual free-particle case in which transport exists.

What actually happens in the localized case is a third option, rather surprising to most theorists, but perfectly possible mathematically and I believe my paper showed it to be true quite unequivocally in some cases:

(3) G has a discrete denumerable distribution of poles which, while randomly arranged in energy, in amplitude fall off like an absolutely convergent series. That is, if we want to pick up all but ϵ of the total amplitude, we can always find a finite number $M(\epsilon)$ of states (poles) which do so. There are, indeed, an infinite number of poles, but their amplitudes decrease so rapidly that most of them are utterly irrelevant; the nth pole, characteristically, has a residue $\propto e^{-\alpha n^{1/3}}$. At the risk of repeating myself let me reiterate that this peculiar analytic structure is just equivalent to the statement that wave functions are localized: each pole α which contributes to i is a wave function localized near i.

The technique of showing this simply involved, first, showing that it is always true if one takes the first perturbation theoretic term in the self-energy Σ:

$$\Sigma_{ii} = \sum_{j \neq i} (V_{ij})^2 \frac{1}{\omega - E_j} + \sum_{j,k \neq i} \frac{V_{ij} V_{jk} V_{ki}}{(\omega - E_j)(\omega - E_k)} + \cdots,$$

195

which enters G_{ii} as

$$G_{ii} = \frac{1}{\omega - E_i - \Sigma_{ii}(\omega)}.$$

Then, with more difficulty, I showed that the perturbation series for Σ_{ii} converges in a probability sense if $\langle V_{ij} \rangle$ is small enough [of order 1/(number of neighbours)] relative to the breadth W of the band of energies E_j, so that, with suitable renormalizations, there are many cases where one may simply assume the form of the first term is exact. Note that if the density of states is variable in energy the series may converge at one value of ω and diverge at another, which is the idea behind Mott's *mobility edge*[2] and critical energy E_c separating localized from nonlocalized states. That is, G_{ii} can perfectly well have a "discrete-random" structure of poles at one energy and a truly continuous one at another.

With the first term of Σ on display we can see the fallacy of the van Hove type of averaging used by Lloyd et al.[16,17] Clearly there are often very large values of the terms of Σ where ω is near E_j, and particularly of the imaginary part

$$\text{Im} \, \Sigma \cong \sum_j \frac{s}{s^2 + (\omega - E_j)^2} (V_{ij})^2.$$

In fact, in the limit $s \to 0$ the *average* of $\text{Im} \, \Sigma$ remains finite. But we must not consider the average: looking instead at the probability distribution of $\text{Im} \, \Sigma$, for different values of E and i, we find that $\text{Im} \, \Sigma \to 0$ with probability 1. *Averaging Σ leads to nonsensical results*, as we discussed earlier. Only physical quantities like σ or $n(E)$ may be averaged.‡

A few final remarks on this subject. First, it turns out that this first term already gives Mott's ω^2 conductivity in principle, using standard Green's function conductivity theory.

Second, Ziman[18] has expressed a rather more sophisticated doubt about the theory. If finite temperatures, or the introduction of phonons, lead to the individual states E_j having an intrinsic breadth, won't it become necessary again to use the average, not the probability value, for $\text{Im} \, \Sigma$? As far as I can see, introducing a breadth actually improves

‡ This point seems to be the difficulty in Lloyd's paper, which otherwise is, as far as I can see, perfectly valid: he states that only ImG may really be averaged, which is correct—it is the physical quantity "density of states"—but then he assigns a meaning in terms of an invalid definition of localization to other averages. In fact, if ansatz (3) is used in the definition of a physical decay constant or of the conductivity, their averages do indeed converge to zero.

196

the convergence and has a small and calculable effect at the near neighbors which seems to indicate that just the opposite of such an amplifying effect happens. Experimentally also it clearly does not happen.

The third and final major point is to justify the title of this essay. As far as I know this point has not been made before explicitly, so perhaps mentioning it is a bit out of bounds here, but it is implicit in all of Mott's work on the subject. It has become well known that the electrons in a metal, the nucleons in nuclear matter, or the atoms in He^3, may all be treated as independent particles, moving in an average field, for sufficiently low temperatures T or energies of excitation E. This "Fermi liquid" theorem is a rigorous consequence of the exclusion principle: it happens because the phase space available for real interactions decreases so rapidly (as E^2 or T^2).

The theorem is equally true for the localized case: at sufficiently low temperatures or frequencies the noninteracting theory must be correct, even though the interactions are not particularly small or short-range: thus the noninteracting theory is physically correct: the electrons can form a "Fermi glass".

P. W. ANDERSON

References

1. P. W. Anderson, Phys. Rev. **109**, 1492 (1958).
2. N. F. Mott, Festkoerperprobleme **9**, 22 (1969), is the most complete discussion of this aspect.
3. M. H. Cohen, Proceedings of the 1969 Cambridge Conference on Amorphous Semiconductors, to be published in J. Amorphous Materials.
4. A. D. Pearson, W. R. Northover, J. F. Dewald, and W. F. Peck, *Technical Papers, Sixth International Congress on Glass, Washington, D.C., 1962* (Plenum Press, New York, 1962), pp. 357–65.
    S. R. Ovshinsky, Phys. Rev. Letters **21**, 1450 (1968).
5. N. F. Mott and W. D. Twose, Advan. Phys. **10**, 107 (1961).
6. N. F. Mott, Phil. Mag. **13**, 989 (1966); Advan. Phys. **16**, 49 (1967); Phil. Mag. **17**, 150 (1968).
    N. F. Mott and E. A. Davis, Phil Mag. **17**, 1269 (1968).
7. H. Fritzsche and K. Lark-Horowitz, Phys. Rev. **99**, 400 (1955).
8. W. D. Twose, Ph.D. thesis, Cambridge University, 1955; see also Ref. 5.
9. A. Miller and E. Abrahams, Phys. Rev. **120**, 745 (1960).
10. M. Pollak and T. H. Geballe, Phys. Rev. **122**, 1742 (1961).
11. N. F. Mott, Phil. Mag. **19**, 835 (1969).
12. S. Tanaka, M. Koboyashi, E. Hanamura, and K. Uchinokura, Phys. Rev. **124**, A256 (1964).
13. A. E. Owen, Glass Industry **48**, 695 (1967).
14. R. F. Shaw, thesis, Cambridge University, 1969.
15. A. M. Portis, Phys. Rev. **104**, 584 (1956).
16. V. L. Bonch-Bruevich, Phys. Letters **18**, 260 (1965).
    P. Lloyd, J. Phys. C **2**, 1717 (1969).
    F. Brouers, J. Non-Crystalline Solids (in press).
17. J. M. Ziman, J. Phys. C **2**, 1230 (1969).
18. J. M. Ziman, private communication.

Space-Time and Scaling Techniques in the Kondo Problem

Proc. of 12th Int. Conf. on Low Temp. Physics, ed. Eizo Kanda
(Academic Press of Japan, 1971), pp. 657–660.

ILTP Kyoto, 1970. My "Infrared Catastrophe" paper which, after Mahan's work, instituted the subject of Fermi surface singularities, led on through Nozières-de Domenicis' ideas to the long-range interaction between spin flips in the time domain which is the secret of the Kondo effect. Then to solve the resulting statistical problem we had to invent a version of the renormalization group. It is not the same as Wilson's but it is very useful and it was earlier. And we did solve the Kondo problem, although Ken Wilson has never given us credit for our priority. This is a very brief review of these developments, which were very much furthered by Gideon Yuval, with John Hopfield as a hidden collaborator and Don Hamann contributing an important generalization.

We were fortunate that John Hopfield was in Cambridge that year. Setting out the full and very complex history of these developments will have to await a stronger person than I. Repercussions and relationships were (1) via Thouless, to the Kosterlitz-Thouless theory, where the ubiquitous "K" diagram for the renormalization trajectories reappears; (2) via Ursula and Dieter Schotte, to bosonization; (3) via Solyom and Zawadowski, to field-theoretic methods and the field theorists' RNG. These and other repercussions continue to intrigue theorists in a variety of fields.

I say we "solved" the Kondo problem. In essence, this solution consisted in showing that the fixed point was a non-interacting one. The numerical solution, which was later done by my student Armytage, was only accurate to about 10%, where Wilson's figure was < 1%; but we obtained the famous "Wilson factor" to be 2 exactly, while his "2" was a numerical guess.

Also at Kyoto I stood in for Brain Josephson who received the London Prize. As a talk I presented a little history of the Josephson effect which necessarily (given the occasion) underplayed my own role. This was not the history I would have told a historian of science, and for that reason is not included here. No one ever asked me about my unbiased version.

SPACE-TIME AND SCALING TECHNIQUES IN THE KONDO PROBLEM

P. W. Anderson

> The essential difficulties of the problem of the ground state of the Kondo Hamiltonian have been cleared up by: (1) showing the equivalence of $<\exp(-\beta\mathcal{H})>$ to the partition function of a classical one-dimensional statistical problem with long-range interactions. (In one version, an Ising "$n = 2$" problem). The two parameters of the anisotropic Kondo problem, $J_\pm \rho$ and $J_z \rho$, go into two parameters of the classical problem, T and the ratio of short to long-range interaction. (2) Then a set of "scaling laws" are derived which connect different parameter sets to each other (this may also be done by "cutoff renormalization" or "renormalization group" techniques in the original Kondo problem). All ferromagnetic Kondo problems are equivalent, and are equivalent to the classical Ising problem at its critical point; the nature of the singularity which prevents solution of the Kondo problem by perturbation methods is thus revealed. The antiferromagnetic Kondo problem, on the other hand, scales into the Ising problem at high temperatures, which is nonsingular and rather trivial. The equivalent Kondo problem, however, is in its strong coupling limit, so that again perturbation theory in J is not relevant. The low temperature behavior of the isolated Kondo system is highly *nonsingular* and thus it appears that many of the experiments are dominated by interaction effects.

In the past two years a completely new line of approach to the Kondo problem has begun to give us both an appreciation of why it has been so difficult, and an understanding, not necessarily of the precise solution in numerical terms, but of its physical nature and behavior. I cannot hope to do anything in the time I have here but summarize the nature of this approach and the results of the dozen or so rather long and difficult papers which have been written about it; for details I can only suggest you read the papers.[1,11]

In the first place, what is the Kondo problem, what has it got to do with physics, and what would we like to know about it? The Kondo problem in simplest terms consists of a certain deceptively simple Hamiltonian,

$$\mathcal{H} = \sum_{k,\sigma} \epsilon_k n_{k\sigma} + J \sum_{kk',\sigma\sigma'} c^+_{k\sigma} c_{k'\sigma'} (S \cdot s_{\sigma\sigma'}) \qquad (1)$$

and in my opinion it is physically uninteresting to allow $S > 1/2$. This Hamiltonian, which is basically due to Yosida and Kittel, is meant to describe a single impurity with spin S interacting via exchange with the free electron gas in the metal; as you all know, it was Kondo who pointed out that if J is antiferromagnetic $\mathcal{H}$ leads to a logarithmic divergence in T in the scattering at the Fermi surface, and shortly thereafter Nagaoka[12] and Suhl[13] posed the Kondo problem proper: the question of what really does happen at $T \rightarrow 0$, and remarked that probably the spin binds an antiparallel spin to itself leading to a net singlet.

The question arises as to when and whether the Kondo Hamiltonian is physically relevant. Hamann[9] and Schrieffer[10] have shown that the same space-time technique we apply to the Kondo Hamiltonian will, with care, reduce the symmetric, weak coupling regime of the Anderson Hamiltonian to the same problem with J given by the Schrieffer-Wolff transformation. Some other more naive and direct attempts do not, I believe, give correct answers because of the incorrect use of asymptotic expressions. But, I believe, the extremely strong-coupling, asymmetric cases like Fe-Ir are far better treated by the methods of "localized spin-fluctuation" theory[14] which may be discussed elsewhere at this conference. In any case, the problem of rigorously taking into account the orbital degeneracy of the real $l = 2$ levels must be faced before meaningful numerical comparisons with experiments can be undertaken.

There is, however, an important kind of experimental comparison which must be considered. Most of the approximate theories which have been presented over the past five years, especially those which have looked reasonably good over the whole temperature and J range, have suggested a singular behavior of the exchange scattering or of the susceptibility as $T \rightarrow 0$. They give behaviors like (scattering amplitude) $\sim \frac{1}{\ln T}$ (Kondo-Appelbaum)[15] or $\sim \frac{1}{T^{1/2}}$ (my work)[16] or at worst $T \ln T$. Such behaviors violate certain important elementary ideas about the cross-section being dominated by the phasespace for scattering—simply, that the lower the energy of the particle from the Fermi surface, the fewer states it has available to be scattered into because of energy conservation—and while no rigorous proofs were available, these ideas seem very plausible a priori, and on the other hand are absolutely essential to the whole "localized spin fluctuation" scheme. *If phasespace was violated in the Kondo problem, there is every reason to*

– 657 –

believe that no LSF scheme, no matter how clever, will ever work. Fortunately, our results show that all these singularities violating phase-space were in fact artifacts of the methods. Their experimental appearance has always been open to question because of interaction effects.[17]

With this survey of what we really need to know, we see that the physically interesting Kondo problem is usually only the case of relatively small antiferromagnetic J. Nonetheless, the very first thing we shall do is to introduce a whole host of completely *unphysical* Kondo problems, and try to solve them all together: that is, the first step of our solution is to allow J to be axially *anisotropic* with a different value for J_z, as opposed to $J_x = J_y = J_\pm$:

$$H_{\text{int}} = J_z S_z s_z + \frac{J_\pm}{2}(S_+ s_- + S_- s_+) \quad (2)$$

(where $S_+ = S_x + i S_y$ etc.)

and the Fermion operators are understood. It turns out that this expansion of the parameter space gives us a very much clearer insight into the relationships. Why?

In the first place, we note that only the sign of J_z determines whether (2) is ferromagnetic. A reversal of sign of $J_\pm$ is equivalent to a reversal of sign of S_x and S_y, which is simply a proper rotation and does not change S's commutation relations: thus just a relabeling. So by keeping $J_\pm$ fixed and taking J_z continuously to $-J_z$ we obtain the ferromagnetic case from the antiferromagnetic.

The form of the Kondo energy: $\propto e^{-1/J\tau}$, tells us that $J = 0$ is an essential singularity, so we must be able to maneuver past it to understand the ferromagnetic-antiferromagnetic relationships. It is, in addition, just this relationship which is the sticking point for all *simple* bound state theories: even though perturbation theory and most other indications suggest that the ferromagnetic case has no bound state, it is extremely hard to avoid getting one.

We now have two parameters, J_z and $J_\pm$: is that all? There are actually two other important numbers aside from kT, and one of these must be understood quite clearly. A simple one is the density of states ρ: this may be scaled at will by use of the normalization volume. But a very important nontrival parameter remains: the high-energy cutoff $E_c \sim \hbar/\tau$. Clearly the band and the interaction energy J do not in fact extend to infinitely high energy and must be cut off. Why is this so important? Because there are also divergences: "ultraviolet divergences" which occur if $E_c \to \infty$: most simply, if we just calculate the second order perturbation of the ground state energy it diverges as $E_c \to \infty$ linearly. This is a standard type of divergence and we simply have to live with it: but we must keep the cutoff and we find that it does interact in a sense with the "infrared" divergence which is the Kondo effect. But we make one all-important assumption: that if J is small enough, the precise nature of the cutoff cannot possibly make any difference: the single parameter, E_c or τ, is all that the low-frequency behavior of such things as the magnetization can possibly depend on. By appropriate adjustments, in fact, we can show that everything is a function only of the three dimensionless parameters

$\frac{J_\pm \tau}{\hbar}, \frac{J_z \tau}{\hbar}, \beta\hbar/\tau = \frac{\hbar}{kT \cdot \tau}$; for instance, (setting $\hbar = 1$)

$$\text{free energy } F = \frac{\hbar}{\tau} F_{\text{dim}} \cdot (J_\pm \tau, J_z \tau, \frac{\beta}{\tau}) \quad (3)$$

the ground state energy $E_g = \frac{1}{\tau} E_{g\,\text{dim}} (J_\pm \tau, J_z \tau)$

We have then, if, as from now on, we are interested in only the ground state, only a two-parameter space to worry about, and (see Fig. 1) the two cases of physical interest are the straight lines $J_z = \pm J_\pm$. In Fig. 1 we use instead of $J_z \tau$ the parameter $\epsilon = 2 J_z \tau +$ (higher order).

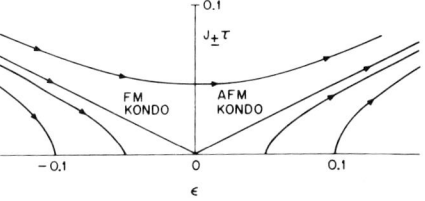

Fig. 1.

From this point two main lines diverge. The simpler technique is related to the "renormalization group" method of quantum field theory, and has been carried out by us[4] and, with more sophistication, Fowler and Zawadowski.[18] The essential trick is to eliminate the highest energies: make a small charge in $E_c \to E_c - \Delta E$, and do perturbation theory to first order in ΔE to get all the new quantities. (see Fig. 2.) Then we get simple differential equations like

$$d(J_\pm \tau) = (J_\pm \tau)(J_z \tau) \frac{d\tau}{\tau}$$

$$d(E_g \tau) = \left(\frac{J_\pm \tau}{2}\right)^2 \frac{d\tau}{\tau^2} \text{ etc.}$$

for the changes in the different parameters, and the assumptions (3) lead to a set of "scaling laws": the results for one set of parameters are related to those for another by a definite set of rules, so that one is back to a one parameter set of solutions. The scaling curves defined by these scaling laws are shown in Fig. 1; they are just the hyperbolas

$$(J_\pm \tau)^2 - (J_z \tau)^2 = \text{const.} \quad (4)$$
$$= \epsilon_0^2$$

Unfortunately, these curves can only be usefully followed in one direction. This direction is to the *right*: This is the direction in which E_c gets smaller, τ gets larger:

$$d(\ln \tau) = \frac{d(J_z \tau)}{(J_\pm \tau)^2} = \frac{d(J_z \tau)}{(J_z \tau)^2 - \epsilon_0^2}$$

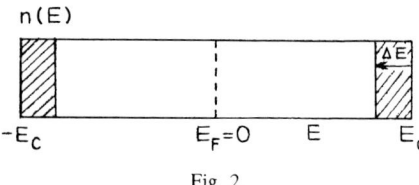

Fig. 2.

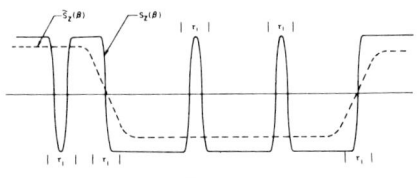

Fig. 3.

Looking at the chart, we see that this rule means that the ferromagnetic isotropic cases all scale into the point $J \equiv 0$, i.e., zero coupling: this is a trival point where there is no interaction and a magnetic moment exists, pointing in its original direction forever. Cases in which the ferromagnetic J_z is even greater also scale to $J_\pm \tau \equiv 0$, where again no spin-flips take place. But all the cases to the right of the ferromagnetic line scale upwards to the right, towards large $J\tau$.

This unfortunately includes the antiferromagnetic, physical case. In this case one reaches values of $J\tau \sim 1$ at a time τ given by the inverse of the Kondo energy $\exp\left(-\frac{1}{J\tau}\right)$. Thus the meaning of this energy is that it is at this point that the problem becomes a "strong coupling" one.

Unfortunately, the perturbation theory does not work satisfactorily once $J\tau$ is ~ 1. This is the difficulty with the renormalization group technique, and means that it cannot be used to solve the real Kondo problem rigorously. One can give a heuristic argument—that as $J\tau$ becomes larger, one should shift to a perturbation theory in $\left(\frac{1}{J\tau}\right)$—and I even believe that is correct—but I have not found a really acceptable way of dealing with this problem.

For this reason we must go back now to the original approach with which Yuval and I began two years ago.[1-3] This is the space-time approach patterned after the Feynman space-time approach to field theory and the polaron.[18]

Feynman would suggest that we calculate not E_g directly, but some physical amplitude like
$$<0 | e^{-\beta \mathcal{H}} | 0>$$
where $<0|$ is the "unperturbed" ground state. Feynman usually used $\beta = it$, t a time, but we find it most convenient to use a real negative β because then in the limit of large β $<0|e^{-\beta \mathcal{H}}|0> \to e^{-\beta E_g}$, and actually a rather simple trick[2] also allows us to do finite temperatures: β is basically $1/kT$. Feynman's theory now allows us to express the amplitude as the sum over all the possible "paths" for the motion, of the relevant amplitude for each path.

$$<0 | e^{-\beta \mathcal{H}} | 0> = Z = \int d(\text{paths}) \, e^{-\int S(\text{path})}$$

Now if $\mathcal{H}$ involved only S, paths would be very simple objects: $S_z(\beta) = \pm 1/2$, with an indefinite number of flips: a path would look like Fig. 3, simply a step function with flips at $\beta_1, \beta_2..\beta_n$. What it turns out, rather surprisingly, to be possible to do for the Kondo problem is to *eliminate* the free electron gas entirely, using a technique related to that used in the so-called x-ray problem by myself[19] or by Nozieres-de-Dominicis,[20] and indeed to write the path integral solely as an integral over possible paths of the local spin. I will write this equivalence down in a formula:

$$Z = \sum_{n=0,2..}^{\infty} \int_0^\beta d\beta_1 \int_{\beta_1+\tau}^\beta d\beta_2 \int_{\beta_2+\tau}^\beta \cdots \int_{\beta_{n-1}+\tau}^\beta d\beta_n$$
$$\times \left(\frac{J_\pm}{2}\right)^n \exp(2-\epsilon) \sum_{i>j} (-1)^{i-j} \ln \frac{\beta_i - \beta_j}{\tau} \quad (6)$$

where $\epsilon \simeq 2J_z\tau = \frac{4}{\pi}(\delta\uparrow\downarrow - \delta\uparrow\uparrow) - \frac{2}{\pi}(\delta\uparrow\downarrow - \delta\uparrow\uparrow)^2$ is related to $J_z\tau$ and is, in fact, the given function of the scattering phase shifts.

This frightening equation giving the amplitude for the paths merely expresses the obvious fact that Z, the path integral, could equally well be described as the partition function of a certain classical problem in the statistical mechanics of one-dimensional chains. In this form, it is a statistical mechanics of hard rods of length τ, interacting via a logarithmic potential, but there are a number of different classical problems each of which takes this form.

Now when we did this we were very happy because we thought we would just look in the "one-dimensional" book by Lieb and Mattis[21] and find the answer; but it turned out that in two separate forms, this classical problem had not quite been solved by Dyson.[22] The most interesting form is the Ising problem with long-range interactions falling off as the square of the distance, which has been a great bone of contention in the one-dimensional world. The relationship between the Kondo and Ising problems is given by the equations:

$$\mathcal{H}_{\text{Ising}} = \frac{1}{2} \sum_i J_{NN}(S_i S_{i+1} - 1)$$
$$+ \frac{1}{2} \sum_{i>j} J_{LR}(i-j)^{-2}(S_i S_j - 1)$$

and

$$\frac{J_\pm \tau}{2} = \exp\left(-(J_{NN} + J_{LR}(1+c))/T\right)$$

$$2 - \epsilon \simeq 2 - 2J_z\tau = \frac{2J_{LR}}{T}$$

which are plotted in Fig. 4, as the dashed curves: the Ising model temperature T increases to the right.

We note that these curves cross the ferromagnetic line

— 659 —

$J_z = -J_\pm$; we guess, and it turns out to be correct, that this must give the transition temperature for the Ising model. In a recent paper[11] we have worked out the other relevant results for the Ising model.

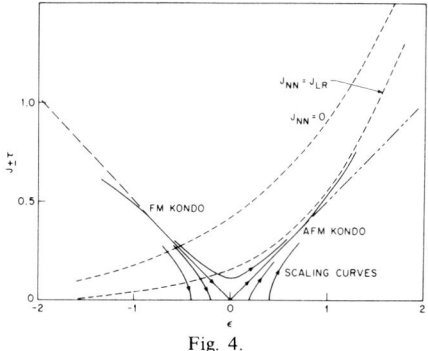

Fig. 4.

The scaling laws, which we got by a perturbation technique of reducing E_c the other way, are here gotten by increasing τ. The trick is terribly simple but long-winded; we eliminate close pairs of spin-flips by averaging over them, which is easiest to visualize from Fig. 3. A close pair of spin-flips is a brief period of reversed spin, so will decrease the effective spin to be used to compute the interaction in the next stage, etc. I will not repeat the scaling equations here. Really the only thing this method has to add is that certain rigorous inequalities can be worked out which limit the range of the correlations once we have scaled the problem far enough to the right. These inequalities are such as to show that the final low-temperature limiting behavior is just such as to agree with localized spin fluctuation theory.

This has been a long and rather mathematical talk, so I suspect it is important for me to summarize the basic result. This is that the essential solution of the Kondo problem is that it changes its nature depending on the energy scale (or time scale) at which you look at it. Initially, in terms of the whole energy band, or of atomic times, it looks like a weak coupling problem; but as one looks for longer and longer times the coupling becomes stronger and stronger until at the time corresponding to the Kondo temperature ($\tau_k = \frac{\hbar}{kT_k}$) the problem has become a strong coupling one, and changed character entirely. At this point one must simply shift gears and solve it in one or another completely different way. The best form for this solution has yet to be found, but its nature is clear at this point. All this is the antiferromagnetic case; for the ferromagnetic case, the corresponding effect is that the coupling gets continually weaker, and one can carry the time scale to ∞ in a continuous fashion. But unfortunately the mathematically more intriguing ferromagnetic case never occurs in nature. Meanwhile, to my mind the main concern about the Kondo problem—that it presented new difficulties for many-body theory in principle, and that one really could not reconcile its behavior with standard concepts —has been essentially settled. There still remains the entire question of the realistic description of experimental cases; but I confess that I personally have not the strength to continue with that.

References

[1] P.W. Anderson and G. Yuval, Phys. Rev. Letters, **23** (1969) 89.
[2] G. Yuval and P.W. Anderson, Phys. Rev., **B1** (1970) 1522.
[3] P.W. Anderson, G. Yuval and D.R. Hamann, Phys. Rev., **B1** (1970) 4464.
[4] P.W. Anderson, J. Phys. (C), to be published.
[5] G. Toulouse, Thesis, Orsay (1968) unpublished; Comptes Rendus, **288** (1969) 1200, 1257.
[6] K.D. Schotte and U. Schotte, Phys. Rev., **182** (1969) 429.
[7] M. Blume, V.J. Emery and A. Luther, preprint.
[8] M. Fowler and A. Zawadowski, preprint.
[9] D.R. Hamann, Phys. Rev. Letters, **23** (1969) 95, Phys. Rev., **B2** (1970) 1373.
[10] S.Q. Wang, W.E. Evenson and J.R. Schrieffer, Phys. Rev. Letters, **23** (1969) 92.
[11] P.W. Anderson and G. Yuval, preprint.
[12] Y. Nagaoka, Prog. Theor. Phys., **37** (1967) 13.
[13] H. Suhl, Phys. Rev., **138** (1965) A515.
[14] P. Lederer and D.L. Mills, Phys. Rev., **165** (1968) 837; N. Rivier and M.J. Zuckerman, Phys. Letters, **23** (1969) 95; A.B. Kaiser and S. Doniach, preprint.
[15] J. Appelbaum and J. Kondo, Phys. Rev., **170** (1968) 542.
[16] P.W. Anderson, Phys. Rev., **164** (1967) 352.
[17] W.M. Star, Proc. 11th Int'l. L.T. Physics Conf., St. Andrews (1968), pp 1250, 1280.
[18] R.P. Feynman, Phys. Rev., **97** (1955) 660.
[19] P.W. Anderson, Phys. Rev. Letters, **18** (1967) 1049.
[20] P. Nozieres and C.T. De Dominicis, Phys. Rev., **178** (1969) 1097.
[21] E. Lieb and D.C. Mattis, "Mathematical Physics in One Dimension", Academic Press, (1966).
[22] F.J. Dyson, J. Math. Phys., **3** (1962) 140, 156, 166, 1191; Commun. Math. Phys., **12** (1969) 91.

Comments on the Maximum Superconducting Transition Temperature
(with M.L. Cohen)
AIP Conf. Proc., p. 17, 1971

Comment on "Model for an Exciton Mechanism of Superconductivity"
(with J.C. Inkson)
Phys. Rev. B **8**, 4429 (1973)

This paper has an odd history. Marvin gave it and wrote it up when I was on the other side of the Atlantic, on the basis of brief conversations and a bit of correspondence; it was kind of him to include me as a co-author. But as a result, the full force of the arguments was not adequately covered in the paper, and the best reasons for believing in a pretty strong upper limit to superconducting T_c remained in my notebooks for many years. I didn't think they would be as important as they turned out to me. My full argument is contained in paper (30), which however also contains an unsuccessful attempt to produce a theory for the "bad actor" superconductors; it is also summarized in some of Ginzburg's papers. I believe the result of (18) or, if you like (30) is still as correct as when it was written, and much more relevant! Not just the cuprates, but fullerenes, barium bismuthates, and probably the chevrels, at least, are all outside the simple BCS-Eliashberg theory, and involve electronic as well as phononic mechanisms.

COMMENTS ON THE MAXIMUM SUPERCONDUCTING TRANSITION TEMPERATURE

Marvin L. Cohen*
University of California, Berkeley, Cal. 94720

P.W. Anderson[+]
Cavendish Laboratory, Cambridge, England
Bell Laboratories, Murray Hill, New Jersey, 07974

ABSTRACT

Using current superconductivity theory and some assumptions about the normal state properties of solids, estimates of the maximum superconducting transition temperature are made. The optimum resonant frequency for an attractive interaction, the role of umklapp scattering, and the appearance of lattice instabilities are discussed.

INTRODUCTION

At the present stage of our understanding of solids, it is possible to make reasonable estimates of superconducting transition temperatures for many materials. Some problems do arise when one attempts to study transition metals and complicated solids but these do not appear to be associated with the foundations of the theory of superconductivity of Bardeen, Cooper and Schrieffer,[1] but rather with our knowledge of the properties of the normal state of these solids. In fact, using the BCS theory and more recent modifications[2-6], McMillan[7] was able to analyze the properties of a large number of metals and arrive at an equation which gives estimates for the superconducting transition temperature, T_c, in terms of a few parameters. McMillan also expressed these parameters in terms of quantities which could be measured (like the Debye temperature) or calculated (like λ the electron-phonon coupling constant). More recently, pseudopotential calculations of λ[8] have yielded values for the parameter which are in good agreement with values obtained using the McMillan equation for a known T_c. In addition, since λ also gives a measure of the effective mass arising from electron-phonon coupling, estimates of λ can be obtained from measurements of the heat capacity. These values are consistent with the calculated values and those determined from T_c.

Considering these results, which indicate that theoretical estimates of the strength of the electron-phonon interaction are con-

---

* Supported by National Science Foundation Grant GP 13632.

[+] Work at the Cavendish Laboratory supported in part by the Air Force Office of Scientific Research Office of Aerospace Research, U.S. Air Force, under Grant Number 1052-69.

sistent with experiment, and considering the success of the Eliashberg[2] form for the frequency dependence of the phonon induced interaction in explaining superconducting tunneling, one has reason to have confidence in the statement that it is the knowledge of the normal state which is limiting[10] us in calculating T_c. Accepting this limitation we can still say something about the maximum superconducting transition temperature, assuming that the systems we consider are ideal in some sense. The properties we assume will be described in detail later. So the aim of this paper is to use the present theory of superconductivity and some general properties of solids to estimate T_c^{max}.

INTERACTIONS, T_c and T_c^{max}

The two conventional interactions involved in the pairing of electrons in the theory we will be using are the repulsive Coulomb interaction and the attractive phonon-induced interaction. The matrix element of the Coulomb interaction for electrons scattered from states k to k+q with energy transfer ω can be represented (assuming unit volume) by

$$V^c(q,\omega) = \frac{4\pi e^2}{q^2 \epsilon_e(q,\omega)} \qquad (1)$$

where $\epsilon_e(q,\omega)$ is the electronic dielectric function. The quantity necessary for a solution of the BCS integral equation has the form $N(0)V^c \equiv \mu$ where V^c is an average of the Coulomb interaction over the Fermi surface and $N(0)$ is the density of states at the Fermi energy. At first we will ignore the frequency dependence of μ. If a Random Phase Approximation is made for $\epsilon(q,\omega)$ and the appropriate Fermi surface averages are made, μ is fairly constant up to energies of the order of the plasma energy. So the most important property of the ω dependence of μ to be considered is the cut off frequency. Since all interactions between electrons are basically electrostatic, V^c is unavoidable and the other interactions may be thought of as merely screening effects acting on V^c.

For estimates of the phonon induced interaction, one can choose a simple Bardeen-Pines[11] or the more correct Eliashberg[2] interaction. In both cases resonant terms appear and the strength of the interaction is governed by the square of the electron-phonon matrix element. We can choose the λ parameter to give the strength of the interaction; then λ is the zero frequency value of the phonon kernel of the integral equation, i.e. $K^{ph}(0)$ or $N(0) V^{ph}(0)$. For our purposes the frequency dependence of $K^{ph}(0)$ can be represented by a Bardeen-Pines type interaction of the form

$$K^{ph} = \frac{\lambda \omega_{ph}^2}{\omega^2 - \omega_{ph}^2} \qquad (2)$$

where ω_{ph} is a characteristic phonon frequency.

Now that we have the form of the interactions our next job is to solve for T_c and T_c^{max}. Let us start with the simplest model for solving the BCS integral equation, i.e. the original BCS model where

$$N(0)V = \lambda - \mu = \text{constant} \tag{3}$$

for electron energies less than ω_{ph} and zero otherwise. The results are (we ignore the 1.14 prefactor throughout)

$$T_c(BCS) = \omega_{ph} e^{-1/N(0)V} . \tag{4}$$

This famous expression for T_c has been the source of much conflict between experimentalists and theorists. Since ω_{ph} and $N(0)$ are measurable, the experimentalist tends to assume V is given by God and to complain when it turns out to depend on the other two. (4) is formally correct but a tactical error; neither V nor $\lambda-\mu$ is independent of ω_{ph} and $N(0)$, and attempts to verify (4) are therefore meaningless. However, since as yet we have placed no limits on λ and μ, it appears that for T_c^{max} we would want $\mu \to 0$, $\lambda \to \infty$ and

$$T_c^{max}(BCS) = \omega_{ph} . \tag{5}$$

To get the largest T_c we want the largest ω_{ph} possible.

If λ gets large we will need the modifications of strong-coupling theory.[7] Using the framework of the more recent formulation[2-7] of the BCS theory which includes renormalization the parameter $\lambda \to \lambda^*$ where $\lambda^* = (1+\lambda)/\lambda$. If we again let $\mu \to 0$ then

$$T_c(\text{Renorm}) = \omega_{ph} e^{-(1+\lambda)/\lambda} , \text{ and} \tag{6}$$

$$T_c^{max}(\text{Renorm}) = 0.37\omega_{ph} \tag{7}$$

assuming λ can be as large as we want the situation is unchanged and we want as large an ω_{ph} as possible.

We have not as yet included the fact that $\mu \neq 0$. Before doing this it is important to include in our theory some of the effects arising from the frequency dependence of μ. The simplest first step is to use a different frequency cut off for μ than for λ. Appropriate cut offs are: $\mu \neq 0$ for $-E_F \leq \omega \leq \omega_{plasma}$. But it is more convenient and still a good approximation to take: $\mu \neq 0$ for $-E_F \leq \omega \leq E_F$. If we take the same form of K_{ph} as before: $K_{ph} \neq 0$ for $-\omega_{ph} \leq \omega \leq \omega_{ph}$, the result is a two square well model (2SW)[12,5,10] for the kernel of the BCS integral equation. The results for T_c are

$$T_c(2SW) = \omega_{ph} e^{-1/(\lambda-\mu^*)} \tag{8}$$

where

$$\mu^* = \frac{\mu}{1 + \mu \log E_F/\omega_{ph}}$$

The net effect is to reduce μ and to introduce an ω_{ph} dependence in μ^* which is important for calculating the isotope effect parameter. If we assume $T_c \sim M^{-\xi}$, then

$$\xi = \tfrac{1}{2}[1 - (\mu^*/(\lambda - \mu^*))^2] \quad . \tag{9}$$

The 2SW model gives values for ξ which are consistent with experiment and ξ can be used to estimate μ^*. An interesting aside is that ξ can be negative. This was interpreted by some researchers to signal an absence of the phonon mechanism. This attitude seems strange since $\xi > 0$ still gives an M dependence to T_c. Also Eq. (9) says that ξ is never greater than 1/2. To our knowledge there are no experimental contradictions; i.e. $\xi \not> \tfrac{1}{2}$.

What about T_c^{max}(2SW)? Once again if we have the freedom to choose any λ or μ we want, the 2SW form is uninteresting as once again we let $\lambda \to \infty$ and μ or $\mu^* \to 0$ giving the same limits as before. This just gives $T_c^{max} \propto \omega_{ph}$ and again we would want the largest ω_{ph} possible. However, we can use the 2SW model another way. Since the argument of the exponential now depends on ω_{ph} there is an $\omega_{ph} = \omega_{ph}^m$ which maximizes T_c for a <u>fixed</u> λ and μ which is given by

$$\omega_{ph}^m = e^{-(2/\lambda - 1/\mu)} \tag{10}$$

and

$$T_c^{max}(2SW) = E_F e^{-(4/\lambda - 1/\mu)} \tag{11}$$

for a given choice of λ and μ (to account for renormalization, λ can be replaced by λ^*). This tells us that the optimum ω_{ph} (Eq. 10) depends on the size of λ and μ. It is <u>not</u> necessarily the largest frequency attainable as it sometimes appears to be in theories or models[13,14] attempting to achieve high T_c's.

Before estimating λ and μ, we compare our results with the McMillan[7] equation to see if all the essential features of his successful equation have been included. McMillan gets

$$T_c(McM) = \frac{\theta}{1.45} \exp - \frac{1.04(1 + \lambda)}{\lambda - \mu^*(1 + 0.62\lambda)} \tag{12}$$

where θ is the Debye temperature. The McMillan expression fits experimentally measured T_c's quite well, but for our purposes it introduces no new features not contained in Eq. (8) when we let $\lambda \to \lambda^*$.

So we are left with T_c^{max} calculations which depend on λ and μ. Even though we know the most favorable ω_{ph} to use if we had the freedom to change it, in our model ω_{ph}^m now depends on λ and μ, and we therefore need estimates of λ^{ph} and μ.

LIMITS ON λ AND μ

The most important limit on λ and μ arises from a stability requirement for the crystal. Ignoring umklapps, at $\omega = 0$, the total dielectric function $\epsilon(q,\omega=0) \geq 0$. If this were not the case, e.g. if $\epsilon(q_0,0) < 0$, then the energy which is ϵ times the square of the electric field would become negative for $q = q_0$ and the crystal would be unstable to q_0 deformations.

If we now consider the total pairing interaction at $\omega = 0$ it can be expressed as

$$V(q,\omega=0) = \frac{4\pi^2}{q^2 \epsilon(q,\omega=0)} \tag{13}$$

where the total dielectric function contains both electronic and ionic polarizabilities. So we have the condition

$$V(q,0) \geq 0 . \tag{14}$$

For the maximum choice of λ, $V(q,0) = 0$ and $\lambda = \mu$. This imposes restrictions on T_c^{max}.

Before using this value for λ we can make things more general by assuming that the attractive interaction between pairs is caused by the exchange of some boson (not necessarily a phonon) with a resonance frequency ω_0. This implies that an interaction of the form given in Eq. (2) with $\omega_{ph} \to \omega_0$ is appropriate. Now using the Coulomb repulsion for the repulsive interaction we can use the 2SW model to estimate T_c^{max}. Solving for the optimum ω_0^m we can use Eq. (10) with $\lambda = \mu$

$$\omega_0^m = E_F e^{-1/\lambda} \tag{15}$$

and

$$T_c^{max}(\lambda=\mu) = E_F e^{-3/\lambda} . \tag{16}$$

If we include renormalization $\lambda \to \lambda^*$ and we have

$$\omega_0^m = E_F e^{-(2 + \frac{1}{\lambda})} \tag{17}$$

$$T_c^{max}(\lambda^*,\lambda=\mu) = E_F e^{-(4 + \frac{3}{\lambda})} . \tag{18}$$

The expressions for the optimum ω_0 and for T_c^{max} contain λ (or

μ). If we now use our model (Eqs. 17 and 18) and realise that we have little freedom to modify μ because it comes from the straightforward Coulomb interaction (the largest possible μ is about $\frac{1}{2}$) then for $E_F \sim 7$ eV, $\omega_0 \sim .1$ eV and $T_c \sim 10^\circ$K. This shows that the typical phonon frequencies which one normally gets are not too far from the optimum frequency and that an exotic mechanism with a high ω_0 isn't enough to ensure high T_c's. All the exotic mechanisms suggested so far[13,14] are such as to exclude umklapps and thus they cannot lead to high T_c's, no matter what ω_0 is chosen.

Now a maximum of 10°K is low, but we have not included the effects of umklapp scattering in λ. For short wavelengths with umklapps, the phonons and electrons can see different interactions and the causality and stability statements we've insisted upon don't necessarily apply[15]. Another way to express it is that the stability of the lattice is determined by the <u>local field at the ions</u>, while the effective field the electrons see can be larger.

If we consider umklapp scattering and let the momentum transfer we've been using $\underset{\sim}{q} \to \underset{\sim}{q} + \underset{\sim}{G} = \underset{\sim}{Q}$ where $\underset{\sim}{G}$ is a reciprocal lattice vector, then

$$V \propto \frac{4\pi e^2}{Q^2 \epsilon(Q)} \left[1 - \frac{\Omega_p^2}{\epsilon(Q) \omega_{ph}^2(q)} \right] . \qquad (19)$$

(19) is easy to derive from the following ideas and simplifications: (1) Nearly free-electron theory: the scattering through $Q=q+G$ on the unperturbed spherical Fermi surface is by the effective potential at wave number Q. (2) Approximately point ion potentials, screened by $\epsilon(q,\omega)$. (3) The ion cannot experience its own self field so the effective local field on an ion is $E - E_{self}^i = E + LP_i$ where L is the local field constant $= 4\pi/3$ in a cubic lattice. The role of the local field is merely to allow $\lambda > \mu$ because the ions and electrons see <u>different</u> fields.

We calculate the motion of the ions and their resulting field at the wave vector $Q = q+G$ (q in 1st B.Z.), when a static ($\omega=0$) electronic charge density wave at Q is present.

The basic equation expresses the fact that the ions must move until the field they see is zero, since $\omega=0$.

$$E^{ext} + E_{ions}^{eff} = 0 \qquad (20)$$

But from the equation for the phonon frequencies we can evaluate E_{ions}^{eff} as a function of the ionic polarization P_q:

$$M_i \ddot{x}_q = M\omega_q^2 x_q = Ze\, E_q^{eff}$$

or

$$P_q = ex_q = \frac{Ze^2}{M\omega_q^2} E_q^{eff}$$

$$= \frac{1}{4\pi} \frac{\Omega_p^2}{\omega_q^2} E_q^{eff}$$

On the other hand, E_{ext} is caused by the assumed external charge distribution ρ_Q

$$\nabla \cdot E_{ext} = -\frac{4\pi \rho_Q}{\epsilon_Q}$$

and thus we have, inserting (20)

$$Q \cdot P_q = \frac{\Omega_p^2}{\omega_q^2} \frac{\rho_Q}{\epsilon_Q}$$

Finally, the field seen by the electrons, at $Q = q+G$, is the total, not the effective, field due to the ions,

$$E_{tot} = E_{ext} - \frac{4\pi (P_q \cdot Q)Q}{\epsilon(Q)Q^2}$$

$$Q \cdot E_{tot} = -\frac{4\pi \rho_Q}{\epsilon_Q}\left(1 - \frac{\Omega_p^2}{\omega_q^2 \epsilon(Q)}\right)$$

This gives (19) for the dielectric constant. We have simplified the expression by ignoring polarization (setting all ω_q's equal) but it is clear that different phonons of polarization ϵ_{iq} enter in proportion to $\epsilon_{iq} \cdot Q$.

This model is probably reasonably accurate for good metals. As yet we have not found a correspondingly simple treatment for tight-binding bands, but we suspect $\lambda \sim 1/\omega^2$ may be a general result. As McMillan has already observed, this leads to a severe limitation on T_c.

We see that $\lambda \propto \frac{\Omega_p^2}{\omega_{ph}^2 \epsilon}$ where Ω_p^2 is the ionic plasma frequency and ϵ is the dielectric constant at short wavelengths. This is the same dependence that McMillan[7] found. Now λ can be larger than 0.5, the umklapp contributions are important, and symmetric structures are favored, e.g. assuming simple cubic and fcc structures, values of λ with umklapp are large.

The umklapps imply short wavelength distrubances, and any instabilities associated with them will be charge fluctuations of size quite small compared to the unit cell, associated with motions of the atoms. Such an object is a fair description of a covalent bond.

When one of our ω_{ph}^2 passes through zero we have a second-order transition to a more covalent structure; but of course most phase-transitions are first order, and what we expect is that when these interactions get too strong we will find our structure unstable relative to covalent insulating or semimetallic structures, as is nicely illustrated in the 5th and 6th columns of the periodic table.

Covalency is stronger for light elements; hence light elements with high Ω_p^2 do not necessarily have high T_c. It's as if the strong pairing is a dynamic attempt towards bond formation.

At this point it is easy to include the dependence of λ on ω_{ph} (for $\mu^* \to 0$.) in a calculation of T_c^{max}. McMillan[7] has also made this estimate. One chooses $\lambda \propto \Omega_p^2/\omega_{ph}^2$ and the λ which gives T_c^{max} is $\lambda(T_c^{max})=2$. This is quite large and one begins to worry about instabilities because of the strong λ coming from the umklapps. The role of instabilities can be seen using simple models. An example is given in the next section.

SIMPLE MODEL FOR INSTABILITIES

One of the nice aspects of studying superconductivity in degenerate semiconductors is the fact that the starting point is an undoped crystal whose properties are presumably known. When the doping is added, parameters of the theory such as phonon frequencies can be renormalized in an appropriate way. Metals have their electrons at the beginning, and it's difficult to remove them in an effort to start over. In this section we will discuss a model appropriate to a semiconductor but it can be thought of in some sense as applying to metals, e.g. the undoped semiconductor can correspond to a primordial metal with phonon frequencies at Ω_p.

We begin by finding the total dielectric function ϵ in terms of phonon and electron dielectric functions ϵ_p and ϵ_e. If we couple for example[10] to one ionic phonon mode ω_ℓ then

$$\epsilon_p = \epsilon_\infty \frac{\omega^2 - \omega_\ell^2}{\omega^2 - \omega_t^2} \quad \text{and} \quad \epsilon = \epsilon_e + \epsilon_p - \epsilon_\infty. \text{ This gives}$$

$$\frac{1}{\epsilon} = \frac{1}{\epsilon_e} + \frac{(1/\epsilon_\infty - 1/\epsilon_0)\omega_\ell^2}{\left(\frac{\epsilon_e}{\epsilon_\infty}\right)^2 (\omega^2 - \tilde{\omega}^2)} \qquad (21)$$

where the renormalized $\tilde{\omega}$ is

$$\frac{\tilde{\omega}^2}{\omega_t^2} = \frac{\epsilon_\ell + \epsilon_0 - \epsilon_\infty}{\epsilon_\ell} \qquad (22)$$

which looks like a modified Lyddane-Sachs-Teller relation. The total interaction is now

$$\frac{V^c}{\epsilon} = \frac{4\pi e^2}{q^2 \epsilon}.$$

The first term is the Coulomb interaction while the second is the Bardeen-Pines interaction with renormalized phonons and a screened electron-phonon matrix element $\overline{M}$ where

$$M = i\left[\frac{-2\pi e^2 \omega_\ell}{q^2}\left(\frac{1}{\epsilon_\infty} - \frac{1}{\epsilon_0}\right)\right]^{\frac{1}{2}} \qquad (23)$$

which is the usual M used in polaron coupling. Using Eq. (21) with a bare Coulomb interaction and Eq. (22) one can solve[10] for T_c and $\tilde{\omega}^2$. As the coupling constant increases it is possible to explore the dependence of T_c and $\tilde{\omega}$ on M.

It is convenient to write Eq. (22) in the form

$$\tilde{\omega}^2 = \omega_\ell^2 - \frac{2M^2 \omega_\ell}{V_c/\epsilon_\infty}\left(1 - \frac{\epsilon_\infty}{\epsilon_e}\right), \qquad (24)$$

and it becomes obvious that M^2 cannot be made arbitrarily large i.e. $\tilde{\omega} \to 0$ when $M^2 \to M_1^2 = V_c/\epsilon_\infty \, \omega_\ell/2 (1 - \epsilon_\infty/\epsilon_e)^{-1}$. This imposes a restriction on λ. In this model

$$\lambda = N(0) \frac{2\overline{M}^2 \omega_\ell}{\tilde{\omega}^2} = \frac{\alpha M^2}{\tilde{\omega}^2} \qquad (25)$$

which is similar to the McMillan form.

Setting $\mu^* = 0$ we can calculate T_c^{max} as McMillan did and again find similar results, i.e. $\omega^m = \sqrt{\alpha/2}\, M$, however now we can also explore the dependence on the bare coupling constant M^2. The result is

$$M_{max}^2 = \frac{\omega_\ell^2}{\alpha}\left[\sqrt{1 + \frac{2\alpha M_1^2}{\omega_\ell^2}} \mp 1\right]; \qquad (27)$$

assuming the $-$ sign,

$$T_c^{max} = \omega_\ell (1-y)^{\frac{1}{2}} e^{-(1+y/2)} \qquad (28)$$

where $y = M_{max}^2/M_1^2$. So T_c will increase with M^2 until $M^2 = M_{max}^2$ and then decreases to zero with larger coupling. If $\omega_\ell^2 \to \Omega_p^2$ for a metal we draw similar conclusions.

For the positive sign $M_{max}^2 > M_1^2$ and the system deforms before reaching the maximum transition temperature. The above model can be put in a more general form,[16] and the observations on the stability of the model have been made independently by C.S. Koonce.[17]

SUMMARY

So our comments on the maximum superconducting transition temperature are:
(1) Reasonable theoretical estimates of T_c can be made.
(2) Oversimplified models imply that resonant boson frequencies, ω_0, should be as large as possible and this is not true.
(3) Two square-well (2SW) models (like McMillan's model) should give reasonable estimates of T_c and T_c^{max}.
(4) Assuming no umklapps and using a 2SW model and a stability condition we find, ω_0 (optimum) $\sim e^{-4}E_F$ and for $E_F \sim 7$ eV, $T_c \sim 10°K$. Phonon frequencies are not too far from the optimum value.
(5) Umklapps lead to large electron-phonon couplings, hence we want symmetric structures. There is a tendency to form bonds and to have instabilities.
(6) We present a simple model which exhibits the dependence of λ on ω_{ph} and a lattice instability for large coupling constant.
(7) The only system which might have a high T_c would involve high ω_0 and umklapp scattering. We can see no physical method of realizing this.

ACKNOWLEDGEMENTS

One of us (MLC) would like to thank Carmen Varea de Alvarez, Philip Allen and Calvin Koonce for discussions. PWA is indebted to E.I. Blount and P.C. Martin for constructive criticism which led to the present form of the paper.

REFERENCES

1. J. Bardeen, L.N. Cooper, and J.R. Schrieffer, Phys. Rev. 108, 1175 (1957).
2. G.M. Eliashberg, JETP 11, 696 (1960).
3. Y. Nambu, Phys. Rev. 117, 648 (1960).
4. L.P. Gorkov, JETP 7, 505 (1958).
5. P. Morel and P.W. Anderson, Phys. Rev. 125, 1263 (1962).
6. J.R. Schrieffer, "Theory of Superconductivity" (W.A. Benjamin, Inc. New York, 1964).
7. W.L. McMillan, Phys. Rev. 167, 331 (1968).
8. P.B. Allen and M.L. Cohen, Phys. Rev. 187, 525 (1969).
9. W.L. McMillan and J.M. Rowell in "Superconductivity", ed. R.D. Parks (Marcel Dekker Inc., New York, 1969), p. 561.
10. It has been possible to make some reasonable estimates of T_c for degenerate semiconductors where the normal state properties are fairly well known, c.f. M.L. Cohen in "Superconductivity", ed. R.D. Parks (Marcel Dekker, Inc., New York, 1969), p. 615.
11. J. Bardeen and D. Pines, Phys. Rev. 99, 1140 (1955).
12. N.N. Bogoliubov, V.V. Tolmachev, and D.V. Shirkov, "A New Method in the Theory of Superconductivity", Academy of Sciences, Moscow, 1958 (Consultants Bureau, New York, 1959).

13. W.A. Little, Phys. Rev. **134A**, 1416 (1964).
14. V.L. Ginsberg, Uspekhi **13**, 335 (1970).
15. P.C. Martin, Phys. Rev. **161**, 143 (1967).
16. C.S. Koonce and M.L. Cohen, Phys. Rev. **177**, 707 (1969).
17. C.S. Koonce, private communication.

Comment on "Model for an Exciton Mechanism of Superconductivity"

J. C. Inkson
Cavendish Laboratory, Cambridge CB2 3RQ, United Kingdom

P. W. Anderson*
Cavendish Laboratory, Cambridge CB2 3RQ, United Kingdom
Bell Laboratories, Murray Hill, New Jersey 07974
(Received 2 March 1973)

The excitonic mechanism of superconductivity in a metal-semiconductor system is studied from the point of view of the complete electron-electron interaction. It is shown that recent calculations of an enhancement of the superconducting transition temperature by way of virtual excitons involves a double counting of these processes. Once this is taken into account the enhancement disappears. The local-field effects in the semiconductor are discussed but it is shown that the off-diagonal elements of the interaction are no help in recovering the enhanced superconductivity.

There has been a recent flurry of interest in what might be called the superconducting Schottky barrier.[1] In this system, enhancement of the superconducting critical temperature of a metal is achieved by placing it in intimate contact with a semiconductor. The electrons in the metal interact by way of virtual excitons in the semiconductor so that an extra attractive interaction is obtained. This system has been aired previously[2] but a recent detailed analysis by Allender, Bray, and Bardeen[1] (hereafter referred to as ABB) suggests that a real effect does exist and should be observable. They derived an expression for the exciton coupling constant to be included in the BCS equation in which the average band gap of the semiconductor (ω_g) appeared thus:

$$\lambda_{ex} \propto \omega_p^2/\omega_g^2 , \qquad (1)$$

ω_p being the plasmon energy. This equation would suggest that as the average band gap decreased and our semiconductor tended towards a free-electron metal, the coupling constant would shoot off to large values and astronomically high superconducting transition temperatures would be achieved; quite the opposite of what we would expect to happen. This is of course unfair in that the approximations of AAB's method would break down long before this. We believe in fact, however, that a large amount of double counting has taken place in deriving the above expression. When this is taken into account the anomaly disappears as does, unfortunately, the possibility of real enhancement of the metal superconducting transition temperature.

The model we are concerned with is this[1]: On placing a metal in contact with the semiconductor, metal electrons near the Fermi level tunnel into the semiconductor band gap, where they interact with each other by way of virtual excitons. The depth of penetration of the electrons is, at most, a few angstroms,[3] so that we are in a region where the electron-electron interaction is highly nonlocal.[4] We will ignore this, as ABB have done, and assume that when in the semiconductor the electrons interact by way of the Coulomb interaction reduced by the semiconductor dielectric function $\epsilon_s(\vec{q}, \omega)$. In the metal we have

$$V_c(\vec{q}, \omega) = 4\pi e^2/q^2 \epsilon_m(\vec{q}, \omega) , \qquad (2)$$

which leads, by averaging over momentum transfers over the Fermi sphere, to the required parameter $\mu(\omega) = N(0)\langle V_c \rangle$. The symbols have their usual meaning, and of course $\mu(\omega)$ is then approximated by a square-well interaction in energy for the purposes of the BCS equation.[5] In a similar way the electron-phonon interaction is characterized by the parameter λ_{ph}. The total, in favorable circumstances, is an attractive interaction, and superconductivity occurs. We assume in the present model that λ_{ph} is the same for semiconductor and metal. Now we have to include the exciton effects, and here we differ from ABB. It is not enough just to add them onto the μ and λ_{ph} for the metal. The excitons are included in the semiconductor dielectric function, so what we shall look at first is

$$V_{cs}(\vec{q}, \omega) = 4\pi e^2/q^2 \epsilon_s(\vec{q}, \omega) . \qquad (3)$$

If we average this in the normal way over the Fermi surface, we get an awkward answer for defining an "effective" μ. Because $\epsilon_s(0, 0)$ is finite we get a logarithmic divergence in $\langle V_{cs} \rangle$ for low energies reflecting the residual long-range repulsion in semiconductors. This of course would be disastrous for any form of superconductivity. This logarithmic divergence will be cut off by screening by the metal layer and may not be particularly strong. On the other hand, we find, upon detailed analysis, that enough remains so that the net effect of the semiconductor layer is to

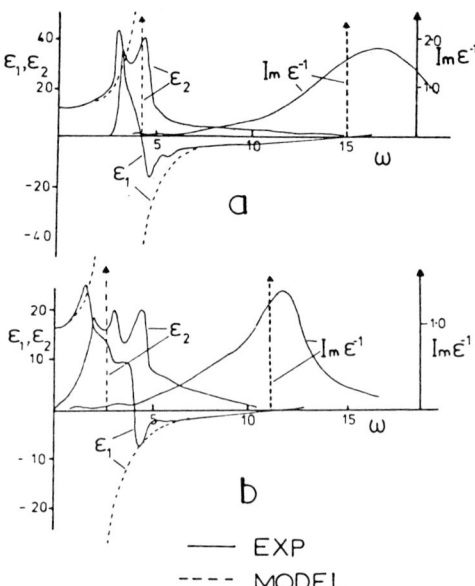

FIG. 1. Comparison of the long-wavelength model dielectric function with the experimental results of Ehrenreich and Phillipp: (a) silicon; (b) indium antimonide.

increase the repulsion, not the attraction. There is a simple form of semiconductor dielectric function which illustrates this[4];

$$\epsilon_s(\vec{q}, \omega) = 1 + \frac{\epsilon_0 - 1}{1 + (q^2/k^2)\epsilon_0 - (\omega^2/\omega_R^2)\epsilon_0} , \quad (4)$$

where

$$\gamma^2 = k^2 \frac{\epsilon_0}{\epsilon_0 - 1}, \quad \omega_R^2 = \omega_p^2 \frac{\epsilon_0}{\epsilon_0 - 1} ,$$

and k^{-1} and ω_p are the screening length and plasmon energy of the equivalent electron density metal. This dielectric function satisfies the sum rules; it has the plasmon peak in $\text{Im}[1/\epsilon(\vec{q}, \omega)]$ and an "exciton" peak in $\text{Im}\,\epsilon(\vec{q}, \omega)$ and also contains the essentials of the structure of the experimental[6] and theoretical[7] semiconductor dielectric functions that we have (Fig. 1). We feel that Eq. (4) is entirely accurate enough for qualitative purposes. Inserting this into Eq. (4), we find that

$$V_{cs}(\vec{q}, \omega) = \frac{4\pi e^2}{q^2} \frac{1}{\epsilon_0} + \frac{\epsilon_0 - 1}{\epsilon_0} \left(\frac{4\pi e^2}{q^2}\right) \frac{1}{\epsilon_M(\vec{q}, \omega)} , \quad (5)$$

where $\epsilon_M(\vec{q}, \omega)$ is the metallic dielectric function within the same approximation.[4]

One essential point here is that, as we already remarked, the "excitons" and other excitations are poles of the dielectric function, but these are zeros of the interaction because that involves $1/\epsilon$. The main pole of $1/\epsilon$ is at the plasma frequency,

just as in a metal.

If we averaged Eq. (5) we would have the normal μ factor reduced by $(\epsilon_0 - 1)/\epsilon_0$, but what we are left with to add on is the long-range Coulomb repulsion, which only disappears as we go to the metallic limit $\epsilon_0 \to \infty$. This leaves us with the conclusion that the effect of the presence of the semiconductor will be to suppress rather than enhance the pair attraction.

What about the excitonic effect of ABB? Let us look at it in more detail. The mechanism is illustrated in Fig. 2(a). We have interactions involving the emission and absorption of virtual excitons. This is second order in perturbation theory and, so, of course, is attractive. By various approximations ABB were available to evaluate the matrix elements and obtain the result of Eq. (1). Consider now the similar process which occurs in the metal—the interaction by way of virtual excitation of electron-hole pairs [Fig. 2(b)]. Obviously there is no essential difference between the two processes. In diagrammatic form the second process corresponds to that shown in Fig. 3(a), and in Fig. 3(b) we show how this process fits into the total Coulomb interaction by a simple resumming of the RPA diagrams for the interaction. We see that if we take out this process we leave the *bare* Coulomb interaction (modified by the third term in the series). By adding it onto the screened interaction, we would have serious double counting of this process. Of course in the metal case this is not done, but we see that this is just what has happened in the semiconductor.

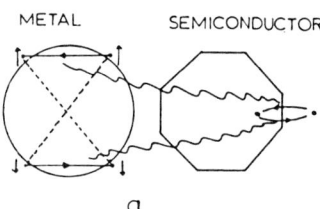

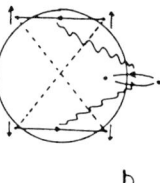

FIG. 2. Schematic representation of (a) Allender, Bray, and Bardeen's (Ref. 1) second-order virtual-exciton process; (b) the equivalent process which occurs in the metal.

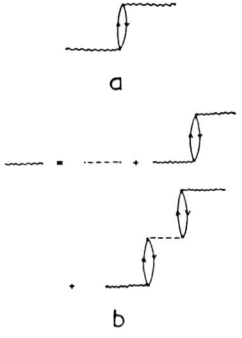

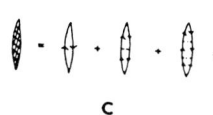

FIG. 3. Diagrammatic representation of (a) the interaction by way of a virtual "exciton"; (b) the splitting of the total Coulomb interaction; (c) the summation for the "exciton" bubble.

If we are to calculate the second-order processes in the manner of ABB, we must use essentially the unscreened interaction in calculating the μ. We now see the way the anomaly of Eq. (1) disappears: As we go towards the metal-metal situation the virtual "exciton" merges into the electron-hole excitation of the metal case, and any effect (be it enhancement or depression) will disappear.

We shall say a few words about the exciton and its contribution to the dielectric function. The excitons to which we have been referring are really electron-hole pairs with a minimum possible energy equal to about the band-gap energy. The changes owing to their being in fact bounded by the Coulomb interaction have been ignored. We would justify this by appeal to the experimental fact that in the type of semiconductors to which we are referring, i.e., those with static dielectric functions larger than about 10, true excitonic effects are very small.[8-10] In energy-loss experiments, for instance, there is no evidence for a significant exciton peak[6]: It is the plasmon-loss peak which dominates, as it is in the reflectance data of Ehrenreich et al.[7] (Fig. 1). The coupling between electrons and excitons is small. Even if excitons effects were large, however, our main conclusion, which is that you can not separate off the excitonic effects from the interaction without grave danger of double counting, still holds; we would just have to reinterpret the bubbles of Figs. 3(a) and 3(b) as the "excitonic" bubble obtained by summing the series of laddered bubbles as in Fig. 3(c). As the screening becomes more complete ($\epsilon_0 \to \infty$), this will tend towards the polarization bubble for the metal, and we are back where we started.

There remains one last mechanism by which the interaction is different for metals and semiconductors; that is the relative size of off-diagonal elements in the electronic dielectric function. There is a general sum rule obtained from lattice-stability requirements[11] by which the diagonal and off-diagonal elements of the static dielectric function are connected, but it does not give us actual estimates of the quantities involved. The off-diagonal elements are associated with local-field effects and give, in our case, umklapp processes involving the excitons. For this it is the relative strength of the poles in $1/\epsilon$ which is important, not their static values. We can actually calculate the off-diagonal elements by introducing nearly-free-electron wave functions in the expression for the polarization P[12]:

$$P(\vec{q}, \vec{q}+\vec{G}, \omega) = -\sum_{\vec{k}, \vec{Q}, \vec{Q}'} \frac{f_0(E_{\vec{Q}'}(\vec{k}+\vec{q})) - f_0(E_{\vec{Q}}(\vec{k}))}{\omega - [E_{\vec{Q}'}(\vec{k}+\vec{q}) - E_{\vec{Q}}(\vec{k})]}$$
$$\times \langle \vec{k}+\vec{q}, \vec{Q}' | e^{i \vec{q} \cdot \vec{r}} | \vec{k}, \vec{Q} \rangle$$
$$\times \langle \vec{k}, \vec{Q} | e^{i(\vec{q}+\vec{G}) \cdot \vec{r}} | \vec{k}+\vec{q}, \vec{Q}' \rangle, \quad (6)$$

where $\vec{G}, \vec{Q}, \vec{Q}'$ are reciprocal-lattice vectors and $|\vec{k}\vec{Q}\rangle$ is a wave function of reduced wave vector k ($\vec{k}+\vec{Q}$ in extended zone) and energy $E_{\vec{Q}}(\vec{k})$. For small $\vec{q}$ we get the approximate forms, after some algebra,

$$\epsilon(\vec{q}, \vec{q}+\vec{G}, \omega) \approx -\frac{\vec{q} \cdot \vec{G}}{q^2} \frac{1}{\omega^2 - E_\ell^2} A, \quad (7a)$$

$$\epsilon(\vec{q}+\vec{G}, \vec{q}, \omega) \approx -\frac{\vec{q} \cdot \vec{G}}{(\vec{q}+\vec{G})^2} \frac{1}{\omega^2 - E_\ell^2} A, \quad (7b)$$

A is a constant that includes band effects in the integration over $\vec{k}$ in the polarization. The off-diagonal element $\epsilon(\vec{q}, \vec{q}+\vec{G}, \omega)$ is *not* small compared with the diagonal element $\epsilon(\vec{q}, \vec{q}, \omega)$ for small $\vec{q}$ (it goes as $1/q$), but in the inversion of the matrix $\{(\epsilon)_{\vec{G}, \vec{G}'}\}$ it is the product of factors like $\epsilon(\vec{q}, \vec{q}+\vec{G}_1, \omega) \epsilon(\vec{q}+\vec{G}_2, \vec{q}, \omega)$, which matters; this *is* small compared to the terms involving the diagonal element for the value of A we have estimated. A good approximation to the inverted matrix elements are

$$\epsilon^{-1}(\vec{q}, \vec{q}+\vec{G}, \omega) = \frac{\epsilon(\vec{q}, \vec{q}+\vec{G}, \omega)}{\epsilon(\vec{q}+\vec{G}, \vec{q}+\vec{G}, \omega)\epsilon(\vec{q}, \vec{q}, \omega)}, \quad (8a)$$

$$\epsilon^{-1}(\vec{q}+\vec{G}, \vec{q}, \omega) = \frac{\epsilon(\vec{q}+\vec{G}, \vec{q}, \omega)}{\epsilon(\vec{q}+\vec{G}, \vec{q}+\vec{G}, \omega)\epsilon(\vec{q}, \vec{q}, \omega)}. \quad (8b)$$

As far as the excitons are concerned, we again get a zero of the interaction [we have one factor $1/(\omega^2 - E_\ell^2)$ on top and two below], and what we said before still holds. As we tend towards the metallic situation, the importance of these off-diagonal elements will decrease still further.[13]

In conclusion, then, we find no evidence for possible excitonic enhancement of the superconducting transition temperature for metal-semiconductor systems; rather the lack of complete screening in the semiconductor would tend to increase the pair repulsion in that region. The mechanism suggested by ABB has been shown to be due to including the effect of virtual-"exciton" transitions twice while ignoring the incomplete screening in the semiconductor. We would suggest that if excitonic effects are to be found, it would be in a medium which offers excitons *plus* complete screening.

The authors would like to thank Dr. E. Tosatti for stimulating the discussions during the course of this work.

---

*Work at the Cavendish Laboratory supported in part by the Air Force Office of Scientific Research (AFSC), U. S. Air Force under Grant No. AFOSR 73-2449.

[1] D. Allender, J. Bray, and J. Bardeen, Phys. Rev. B **7**, 1020 (1973).

[2] V. L. Ginzberg, Usp. Fiz. Nauk **101**, 185 (1970) [Sov. Phys.-Usp. **13**, 335 (1970)].

[3] V. Heine, Phys. Rev. **138**, A1689 (1965).

[4] J. C. Inkson, J. Phys. C **5**, 2599 (1972).

[5] J. Bardeen, L. N. Cooper, and J. R. Schrieffer, Phys. Rev. **106**, 162 (1957); Phys. Rev. **108**, 1175 (1957).

[6] J. P. Walter and M. L. Cohen, Phys. Rev. B **2**, 1821 (1970).

[7] H. Ehrenreich and H. R. Phillipp, Phys. Rev. **128**, 1622 (1962).

[8] C. J. Powell, Proc. Phys. Soc. Lond. **76**, 593 (1960).

[9] E. Tosatti (private communication).

[10] J. C. Phillips, Solid State Phys. **18**, 55 (1966).

[11] L. M. Pick, M. H. Cohen, and R. M. Martin, Phys. Rev. B **1**, 910 (1970).

[12] C. M. Bertoni, V. Bertoloni, C. Calandra, and E. Tosatti, Phys. Rev. Lett. **28**, 1578 (1972). (Our method of the evaluation of the expression for P differs in detail somewhat from theirs as we are interested in the energy dependence.)

[13] J. C. Phillips, Phys. Rev. **123**, 420 (1961); and Phys. Rev. **128**, 2093 (1962).

Conference Summary

Proc. of Nobel Symposium at Goteborg, Sweden, June 13, 1973
Collective Properties of Physical Systems, ed. B. & S. Lundqvist
(Academic Press, New York, 1974), p. 266.

A summary of the Nobel symposium at Aspenasgarden near Goteborg, June 1973. A light treatment with some still useful thoughts about the renormalization group. Like many such summaries, it was composed hastily to say the least; however, unlike many, it had a great deal to report on: the Kondo effect, the phase transition work of Kadanoff, Wilson, Fisher, the first serious progress on superfluid He3, and the x-ray edge work were all very recent. It was the second great period of condensed matter physics.

Conference Summary

If past experience at this conference is any guide at all, I will still be talking to you at least an hour from now. And what's more after I sit down somebody will ask a question and I will suggest that I have about three more preprints or reprints to describe; therefore, so that I will bore you on my own time and not anybody else's, I think I'd like to start by acting as a second representative of the visitors here to express our appreciation for what, for many of us I'm sure, is the most magnificent scientific meeting we've ever attended. And I think we all owe a very hearty vote of thanks to our great leaders Stig and Bengt (Lundqvist)[1]. Everything as far as I could see was done perfectly. I think I should mention particularly that I've never seen a projection setup that worked as consistently well as this one. Harry Suhl referred to many other things last night, but of course Harry could not comment on last night's banquet because he hadn't had it yet. I think I must say of the banquet that at just about the time when I said to Gerry Mahan "they've laid on just about everything but dancing girls," in came the dancing girls.

Of course we have also to give a hand to the other Swedish members of the committee as well, to Professors (Lamek) Hulthén and (Lars) Hedin and to all the young people whom we've come to rely on so heavily for practically everything.

And then I think we do have to say something about the power behind the throne, i.e. John Wilkins. Actually he was sometimes to be found well in front of it with his neck stuck out. But I think it's obvious to all of us how valuable John's help has been, not least in insisting in a break after every major talk. I don't know when I've met in more frequently changed air and that's been surprisingly helpful. I have to mention that John's effect (and that of several of the others I have mentioned), on the meeting has not been entirely positive, because while certain aspects of the late night hospitality have been very pleasant, at the moment I'm suffering from at least some of them and perhaps some of the rest of you are. I was still reading my notes at 2:30 last night.

So if this seems a bit incoherent that's the reason why.

I would also like to thank John for always being the man to ask the question I wish I'd thought of to make things clear in my mind, though I'm afraid there were times when things weren't clear nonetheless.

Now, at this time I think I'd like to get about the business of awarding a number of prizes. Considering the title of the Symposium one can hardly object if these become Nobel prizes in the minds of their holders in the future. I think the first prize is the prize for Top Banana that should have gone to John Wilkins, but he gave it away to Bill Brinkman so we'll have to make Bill Brinkman the Top Banana.[2] I was going to give a prize for the most clearly, concisely and beautifully presented talk to Seb Doniach but unfortunately he just disqualified himself.[3] It was a clearly and beautifully presented talk but so were most of the others. Therefore that prize has got to be shared between John Hubbard and I. Dzyaloshinsky. And I suspect there were various other people who copped out (or were pushed out) whose names never have formally appeared on the program and I express our appreciation for the forbearance of all of them. A memorable remark on this subject not on the list (which you will soon see) was by some chairman or organizer who said "Well, we've put that headache off for awhile". I won't remind you which paper that was except to say it was a good one.

The prize for the most artistic doodler unquestionably goes to L. P. Gorkov. I don't know if all of you have seen any of his productions. Last night Michael Fischer and I were scrambling about among the papers up here looking for an example to show you, but unfortunately either the students have stolen them or he's taken them home with him. Michael came in second.

Next I wanted to award prizes for a few of the

---

[1] Editors remarks.
[2] John Wilkins presented an object billed as a Triplet Banana to Bill before his repeatedly postponed talk.
[3] Doniach's talk had been cancelled but time was found for it in the last session.

Nobel 24 (1973) Collective properties of physical systems

memorable things people said during the meeting. I noted a few of them myself, and others I collected last night wandering around asking what had struck people one way or another, as amusing but also with some thematic application to what we have been hearing. I'll present the list first:

Quotes—sometimes out of context
"In this calculation I have everything going for me." (K.G.W.)

"RNG is as general and vigorous as the partition function approach to statistical mechanics was." (M.E.F.)

"Right for partly the wrong reason." (P.W.A.)

"When Einstein speaks, I listen." (J.R.S.)

"Like South America, the Kondo effect will always have a great future." (H.S.)

"What did Smetana do in Göteborg?" "He suffered." (a Lundqvist)[1]

"Five years in superconductivity and five years in liquid crystals—this field was *made* for me!" (P. G. de Gennes)

The first one struck a lot of people as funny; I suppose because of the overtones of God being on the side of the big battalions. I was sorry I didn't get more votes for Ken's (Wilson)[2] remark about his role being to replace a 30% by a 10% theory.

The second one was said with such conviction and caused such discussion that it just had to be put down here.

The third—I do remember in the early days of the BCS theory and a bit before there was a very eminent theorist, who will be nameless, who was invariably wrong for absolutely the right reason. I have been accused of preferring the other way around.

The next two are self-explanatory. Harry Suhl provided a problem of selection—we will mention at least one other of his.

The next one—I should hasten to explain my reference here.[1] I have a very poor attendance record at most meetings—I believe I average about 25% of the sessions—but here I've been

at absolutely every session, suffering from the responsibility for this talk. But I didn't suffer actually all that much because there certainly was no single session that I didn't feel had been very much worth my going to.

Again, the last one is self-explanatory.

Now, finally, I'd like to run very informally through the program and say a little about what we actually did hear, try to put some flesh on the bones of Harry Suhl's remarks last night that amounted to essentially saying that Many-Body Theory is alive and well and living right here in Lerum.

It was an incredibly rich meeting from a scientific point of view. There were four developments of absolutely major significance. I hardly need to tell you what they were: the renormalization group on the first day, the ^{3}He experimental data and all its consequences on the second day, the solution of the Kondo problem on the third day, and the announcement of the results of Heeger on the third day as well. These were all so exciting that a lot of things may have slipped by that were really pretty major, without people noticing their importance and weight. Of course the monumental work that Walter Kohn described is an example, but there were other things that were just described in seconds.

Here are some examples picked at random: in most meetings the announcement of the first reasonably practical calculations of LEED intensities would have been of major importance. De Gennes' discussion remark on smectic B is a very important, exciting development. I thought Gorkov's very exciting new ideas on V_3Si and on one-dimensional systems are very much worth our serious consideration, though perhaps later I'll say things about what remains unsatisfying about them.

Now to take the program day by day, more or less systematically:

The first day, we were of course very grateful for the road show that blew in from Temple in the States, (a very similar sequence of talks had been given at a full three-day meeting in Philadelphia two weeks before) and we also realized it may have been a little dull for them because they were saying, one presumes, many of the same things they'd just been saying two weeks before. But it was not dull, by any means, for any of us who have not followed the field all that carefully.

---

[1] Those who attended the Symposium will remember that we spent quite a bit of time milling about the Smetana room in the Town Hall of Gothenburg on Wednesday night.
[2] Editors remarks.

Nobel 24 (1973) Collective properties of physical systems

What that day really was about was the application of the renormalization group techniques, the new renormalization group as opposed to the old renormalization group, of course, to the problem for which they were invented, and especially the epsilon expansion and the $1/n$ expansion.

I guess I forgot to put down a particularly memorable thing Harry Suhl said in his talk which was the remark that if a theory satisfies the fundamental principles of conservation of momentum and energy and is mathematically correct, it always seems to be worth doing. I would have, a few years ago, felt that the study of critical points was somewhere near that category. It should be done of course, but only because it satisfies momentum and energy and is correct. But now I think Harry's remark has come home to us very seriously in this case.

I think that we may or may not have something that is as rigorous as equilibrium statistical mechanics. I think we have to keep in mind that these fixed points are at least the limit points of lines of essential singularities, namely the droplet singularities; and God knows what other singularities of peculiar Thouless-Kosterlitz type or something like that might often be associated with them. But it certainly is a classification, a way of thinking, and a quantitative approach that can be thought of in terms of expansions of various kinds, i.e. it's got an enormous potential in its own field. But I think what one will see more and more of is an extension of these ideas into other fields, an expansion of them into doing different kinds of things of the general sort that we saw in the Kondo problem. There are very wide possibilities.

Bob Schrieffer said something privately to me (I don't know if he wants to have it quoted) which opened my eyes to some of these possibilities. He said that this is the kind of qualitative idea that may be of value in very distant fields—information handling, sociology, psychology. It's the kind of thing that can apply to a very much wider range of situations—the basic ideas of scale change, irrelevant variables, limit points, etc. are very general.

Now another application: renormalization group is one of these things like speaking prose we've been doing all our life. One special example, for instance, is the Landau Fermi liquid theory which is an effective Hamiltonian valid in a certain range. It's an effective Hamiltonian which is invariant to the form and behavior of the starting parameters and always gives you the same final Hamiltonian with perhaps different values of the coefficients. So one thinks that the Landau Fermi liquid theory must be the equivalent, in the electron gas, of the Gaussian fixed point, the ordinary fixed point that everything goes to in the end. Then there are obviously at least two other fixed points. I have made a little diagram of Fermi Hamiltonian space with its singular points. There's the Landau Fermi liquid theory fixed point, and as you see everything runs through Fermi sea Hamiltonian space and eventually arrives at that. But one can very well tell if in the course of its trajectory through Fermi sea space it has passed near what one might call the Doniach-Schrieffer or paramagnon fixed point. There's undoubtedly at least one other point, the Brinkman-Rice antiferromagnon point you might call it, and I wonder if perhaps there isn't a line of singularities connecting these two. But unquestionably that is something like the structure of the Hamiltonian space as you cut off renormalized Fermi sea theories. That's the full three-dimensional version of the same diagrams that Ken Wilson talked about on the third day, where, as someone was saying this morning at breakfast, the localized state Hamiltonian is the single atom version of the Hubbard Hamiltonian.

To see how things might go with the renormalization of the strong-coupling Fermi sea, let's rephrase the Kondo solution. The Kondo renormalization starts from a localized, strong coupled

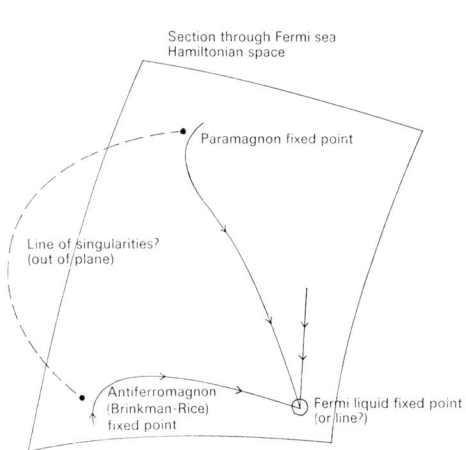

Nobel 24 (1973) Collective properties of physical systems

spin fluctuation, renormalizes it down, and eventually gets, at the low temperature end, to a weak coupling localized spin fluctuation which is nonmagnetic and apparently uninteresting, kind of the equivalent of the Landau Fermi liquid theory, but by no means in all cases has that Kondo center forgotten that it went very close to a singularity, to an unstable fixed point. I don't think I'll take time to speculate further except to remark that this program is in fact perhaps underway already in the case of the paramagnon, that what Hertz and Edwards have done looks very much like a beginning of the full renormalization group approach to the paramagnon problem.

This perhaps then suggests what I hoped would be another theme of my talk, which is universality and breadth of techniques. The fact is that the techniques which were developed for this at first apparently very specialized problem of a rather restricted class of special phase transitions and their behavior in a restricted region are turning out to be something which is likely to spread over not just the whole of physics but the whole of science. A second example can perhaps be introduced by talking about the ^{3}He work. I guess we all felt that was a very exciting day. I found the most exciting moment was when I finally began to realize that (a) I understood Tony (Leggett's) longitudinal resonance and (b) that he maybe understood what I was saying and then (c) that we weren't in any real disagreement. That's the kind of thing that meetings like this are for. There have been many other discussions but that perhaps was the most fruitful that actually took place on the meeting floor.

Well, the point about ^{3}He is the very exciting experimental results. I was asked by (Robert)[1] Richardson to announce one which just arrived in a telegram from Wheatley. The fact that the B phase, Wheatley's B phase, is the same as the Cornell group's B phase is now experimentally confirmed at least on the basis that what Wheatley has been calling the AB transition has a first order magnetization change. So, yet more experimental data are coming in.

It seems very likely that this incredibly unlikely phase that's been postulated is the correct description. There are quite a number of pretty solid experimental vs theoretical checks, and several further indications that one can see qualitatively are strongly in the right direction, and I know of nothing which is seriously contradictory at this point. So, I think, perhaps, this will have been the first conference at which such a wide spectrum of experiment and theory have seemed to come into agreement that one begins to be able to suspend disbelief in the superfluid phase.

Lars Hedin asked me, "Well, all right, so it's superfluid. What are you going to do with it?" And the answer is, "Absolutely nothing." I don't think that that's irrelevant nonetheless. As Harry says: "It satisfies energy and momentum, it must have some use." One remembers that back in 1957, after Bob (Schrieffer)[1], Leon (Cooper)[1] and John Bardeen made their advances, the theoretical developments in understanding superconductivity eventually leaked over into high energy physics; the most exciting modern developments in high energy physics are based on one or another of those concepts that we had then: The concept of broken symmetry, the concept of Goldstone bosons, the concept of what we now call Higgs bosons—the plasmons, in fact. ^{3}He is a nice example of a field with a broken internal symmetry, of non-Abelian type, which is just exactly what people are doing with the renormalization group in high energy physics, at this point. One must assume that having an actual physical model is going to serve again as a guide for that field of physics.

The Kondo effect: I think one must associate Ken Wilson's magnificent solution ... to me it's an incredible intellectual feat ... with the other uses of the renormalization group and again emphasize the point I think he's been making that while, yes, it is very important, in doing things with the renormalization group, that one understands the classification of singular points, in the end what one will want to be doing it for is actually getting from a bare Hamiltonian to an effective Hamiltonian, and knowing really how to get down from A to B and then calculate some numbers and do some things with them. We really basically understood the Kondo problem from a classification point of view. It hadn't been proved by any means, but more than speculations certainly existed, and yet this new solution is infinitely more valuable.

Then Heeger's new data and all the new questions about what kind of one-dimensional fluctua-

---
[1] Editors remarks.

Nobel 24 (1973) Collective properties of physical systems

ting states can have this conductivity which appears to be of a superconducting type. What was most exciting to me, because I guess I'd been a bit familiar with these data, was the new data on the thermoelectric power which are extremely striking, and I think Alan (Luther)[1] is quite right to claim those are something which one really has to contend with. They show ... this is another Harry Suhl quote (or is it C. Kittel?) they show that Forces are at Work.

I would like to run through the days of the meeting and remind you, aside from those four very major things, of some of the things that went on.

On the first day I think I have one more prize that I want to award and that is for maximum overload on my information handling capacity. I award that to Gerard Toulouse. The main problem of course was that he had an enormous amount to say, and I think that this anatomy of the n–d plane is a delightful and fascinating game, and more than a game. Again it satisfies energy conservation and momentum conservation (although I'm not sure that negative dimensionality satisfies anything). I do wish that one of the questions John Wilkins had asked was, what has zero dimensionality got to do with percolation theory and self-avoiding walks. I had it explained to me once but it's completely gone.

Nonetheless, it was a very good, very interesting day, not just the three major talks, really, but all the remarks that were made in addition. I was glad Dzyaloshinsky reminded some of these phase transition buffs that there are n's greater than 3, lots and lots of them.

^{3}He: I guess I really have covered that pretty much. Perhaps I should mention the feeling of intellectual excitement that Pierre de Gennes gave us all in his "anatomy of the liquid ^{3}He horse". That's perhaps why I have quoted that last sentence from his talk. Then the Kondo problem. We've talked about that, but can add that Fred Zawadowski, of course, has pointed out that that particularly simple logarithmic renormalization group is not confined to that one problem at all. There are two simpler problems which it does completely: the so-called $n = 2$ one-dimensional Ising model and the ordinary X-ray problem. So there are quite a few problems to which this can be applied, even though it doesn't quite in its simplest form successfully do the Kondo problem.

There were the few-dimension sessions and various things ... another of those things that was whipped past very fast is this very strange puzzle of the $(H_c)_2$ of layered superconductors which Alan Luther mentioned, which suggests again that Forces are at Work.

I did mention the Gorkov talk which I thought was very important, and of course we all were very excited by the discussion of the Heeger session.

In surfaces there was the monumental paper and sequence of work that Walter Kohn has assembled on this subject. I personally had not realized how closely one could get agreement with experiment. It seems the basic equilibrium processes that happen on surfaces are beginning to be very straightforwardly understood. There do of course remain problems.

I think the discussion session will be valuable on the question of the plasmon contribution.

Yet another of these problems that passed over our heads rather rapidly, in an extremely exciting field of physics, is the question of superstructures that André Blandin talked about. I know of at least four groups that are making progress in the understanding of various kinds of surface superstructures. It's a very spectacular phenomenon from time to time. I have seen a diffraction pattern, and nothing seemed wrong about it to me, that claimed to be a 43 by 43 superstructure on a triangular layer. There are certainly 13 by 13 superstructures on TaS_2 and, of course, on germanium and silicon there are 7 by 7s and 8 by 8s that have been known for many years. These are very interesting phenomena. They may well end up being understood as excitonic transitions. One may hope so.

And then, finally, of course, there are the threshold effects. Perhaps that session seemed a little tame in comparison with the rest of the conference, not very tame I must say, because that was the problem we solved last year instead of this year, and all the rest of the conference has been about the problems we either solved this year or are only about to solve. And there people have done some pretty magnificent things, but they seem to have a lot of it very much under control. One mustn't forget, however, that they are important and exciting many body effects.

---

[1] Editors remarks.

Nobel 24 (1973) Collective properties of physical systems

Now I'd like to say some things in summary and conclusion about open questions and especially about the biology talks. We are fortunate in that the "cat" has by no means lapped up all the "cream" in our field, there is more and more coming along. One can think of problem after problem. In Harry Suhl's talk the peculiar problem of the compensation of chemical activations was mentioned. John Hopfield has also worked on that—an incredible puzzle. One may pick out many others. Certainly the one-dimensional structures are full of puzzles that one has no particular hope of understanding. For instance, I would certainly like to suspend my disbelief in Gorkov's theory of the Nb_3Sn, V_3Si transitions, but I find myself very unhappy with the fact that one uses a purely one-dimensional approach. I know that in the layer metals an actual band calculation gives one interlayer hopping integrals of the order of 1 eV which is two orders of magnitude higher than 50 K, which was being assumed I think in this calculation, and those layers are very far apart and are separated by van der Waals' bonds, whereas Nb_3Sn is a very, very rigid tough metal with obvious three-dimensional character. And what's more, one is going out of one dimension in order to get the right relevant states, which I am sure Gorkov has got. So, can one suggest that perhaps it renormalizes towards a one-dimensional fixed point of some sort? Perhaps if one starts with two pieces of Fermi surface which are parallel within 1 eV and also a strongly interacting system, when one renormalizes they become more parallel. That's a wild suggestion, but it's the only thing that I might understand. There is evidence after evidence that these crystals are one-dimensional. That one of them could be is perfectly reasonable. That all of them could be, I will believe when I see a band structure calculation that I absolutely believe. But they could be renormalized one-dimensional systems... The other thing about them of course is that the transitions do take place and real one-dimensional transitions don't take place. To what extent is one justified in eating one's cake, and having it too, using one-dimensionality at one point, and not using it at another? So there are some very interesting problems ... I mentioned that at such length because this is not by any means an isolated problem. The existence of anomalous quasi-one-dimensional and quasi-two-dimensional fluctuations is a very widespread phenomenon that occurs both in insulators and in metals, in ferroelectrics and in soft mode systems: and remains one of our major puzzles. So we have a great future ahead of us, like South America.

We also have a future ahead of us travelling outside our fields as Harry Suhl and Leon Cooper and John Hopfield showed us so ably. All of these talks I think illustrate why it is that many body theorists can and should invade wider fields of science, why they can do so well in it. I think we are uniquely those physicists who are completely eclectic in our methodology, and eclectic in our point of view. We are not unwilling to take methods and ideas from the most abstract mathematical level, as a couple of these papers make quite obvious. (I certainly was lost at some points.) We're also many of us not by any means unwilling to get our hands dirty and to get down and really look at the phenomena, really look not just at the phenomenology, but at the electrons and atoms as they move around, and with one technique or another, dive in and really understand it. And if it's a matter of just doing kinetics right as opposed to the incredible hash that the literature was in before, that's something we can at least do ... if we're as good as John Hopfield ... and something which was not done until many-body theorists moved into that particular field.

So, we have a good field to work in. We also have an excellent opportunity, particularly for those of us who are somewhat younger than I am, to move out of the field and look at a wider spectrum of things. I think we must consider ourselves very fortunate.

P. W. Anderson

Nobel 24 (1973) Collective properties of physical systems

Asymptotically Exact Methods in the Kondo Problem

(with G. Yuval)

Magnetism, Vol. V (Academic Press, New York, 1973), ed. H. Suhl, p. 217

Another version of paper 17: more complete, perhaps less careful. Yuval wrote it and he is a bit slapdash.

7. Asymptotically Exact Methods in the Kondo Problem

P. W. Anderson

*Bell Telephone Laboratories, Inc.
Murray Hill, New Jersey and
Cambridge University
Cambridge, England*

G. Yuval[†]

*Joseph Henry Laboratories
Princeton University
Princeton, New Jersey*

I.	Introduction	217
II.	The Kondo Problem: A Discrete Path-Integral Approach	219
III.	Eliminating the Fermi Gas	221
IV.	The Method of Schotte and Schotte	224
V.	Summing Up over the Paths	225
VI.	Finite Temperatures	226
VII.	Numerical Results	227
VIII.	The Scaling Method	230
IX.	The Kondo Temperature	232
X.	Physical Implications	233
XI.	"Renormalization Group" Methods in the Kondo Problem	233
	References	235

I. Introduction

In 1964, Kondo [1] discussed the Born series for the scattering of the electrons in a Fermi gas by a single localized spin. At absolute zero the second-order term turned out to be infinite; at a finite temperature T, this term behaves like $\log T$. Thus, below a certain temperature—the Kondo temperature—the second-order term is larger than the first-order one, and the convergence of the Born series is in doubt. The behavior of the system when the interaction J approaches zero is thus

[†] Present address: Racah Institute of Physics, The Hebrew University, Jerusalem, Israel.

not a small variation of the behavior when $J = 0$. The trouble seemed to be that the system's behavior was not *understood*, and not that any great complication appeared in the mathematics.

One might have thought that these difficulties arose because the local spin was just postulated to exist, and its origin in the repulsion between electrons in a tunneling resonance [2] was ignored. However, the convergence difficulties due to Kondo appear also in the Anderson model, which is not open to that criticism. The approach to that problem by methods similar to those we discuss here is given in Chapter 8 by Hamann, and leads to essentially the same basic problem.

A great many *approximate* methods have been applied to the Kondo problem, several of which are described in this volume. While many of these give an adequate numerical account of the behavior in the region near and above the Kondo temperature, none seems to have led to an adequate qualitative understanding of the problem, specifically on two points:

1. Many authors (not all) agree that the state at $T = 0$ involves a bound compensating spin in the antiferromagnetic case, but the nature of this state and its connection to the finite T behavior were unclear; specifically whether $T = 0$ was a singular line was unknown.

2. Mathematically, $J \to 0^{+}$ is certainly a singular point, as can be seen most simply by noting that $J < 0$ (antiferromagnetic) and $J > 0$ (ferromagnetic) behavior are qualitatively different; the heart of the "Kondo problem" is the nature of this singularity (almost all methods specifically neglect terms of relative order $J\rho$ compared to the leading ones).

The methods discussed in this chapter have in common that, although they are only beginning to produce numerical results for comparison with experiment, the qualitative statements they make are not approximate in any real sense. That is, these methods do not modify the nature of the mathematical singularities, but do deal approximately to some extent with their coefficients. They are, then, to a real extent asymptotically exact formalisms in the $J\rho \to 0$ limit.

The most important and most powerful of these methods are the path-integral ones, which are the subject of the first ten sections of the chapter. These lead first to exact equivalences between the Kondo problem and certain classical one-dimensional statistical ones. These methods have been exploited to furnish computer results for the susceptibility in good agreement with experimental behavior. Secondly, the

---

† See Section II for definitions.

classical problem can be approached in certain regions to give a precise answer for the ferromagnetic case, and to demonstrate nonsingularity of the $T \to 0$ limit, as well as elucidating the real nature of the Kondo problem. Some other results, such as the mathematical nature of the ground-state energy, are available.

In Section XI we discuss the set of methods using the idea of the "renormalization group." Precise equivalences between different sets of parameters of the Kondo problem are established which can reproduce some of the results of the path-integral method.

II. The Kondo Problem: A Discrete Path-Integral Approach[†]

With the wisdom of hindsight, we can say that the Kondo problem [1] is a textbook exercise in Feynman's path-integral methods.[‡]

In the Hamiltonian of this problem, we have a noninteracting Fermi gas

$$H_0 = \sum_{k\sigma} \varepsilon_k a_{k\sigma}^+ a_{k\sigma}, \qquad (2.1)$$

a localized spin $\mathbf{S}[S^2 = \frac{1}{2}(\frac{1}{2} + 1) = \frac{3}{4}]$, and an interaction between them

$$H_1 = J_z S_z \cdot s_z(0) = (J_z/2) S_z \cdot \sum_{k,k',\sigma} \langle \sigma | S_z | \sigma \rangle C_{k\sigma}^* C_{k'\sigma} \qquad (2.2)$$

$$H_2 = (J_\pm/2)[S_+ \cdot s_-(0) + S_- \cdot s_+(0)]$$

$$= (J_\pm/2) \left[S_+ \sum_{k\sigma;k'\sigma'}{}' \langle \sigma | S_- | \sigma' \rangle C_{k\sigma}^+ C_{k'\sigma'} + hc \right]. \qquad (2.3)$$

The total Hamiltonian is $H_0 + H_1 + H_2$. For an isotropic system (to which we shall keep until Section V), $J_z = J_\pm$; we shall then call this number J. If it were not for H_2, the problem would be trivial—the spin-up electron would be scattered by a *scalar* potential due to S_z, the sign

---

[†] The Kondo problem's chief attraction for theorists has been the difficulty in understanding it, that is, in describing the system's behavior by a simple picture. Therefore we shall concentrate here on the physical ideas we introduce, referring the reader to the original articles for details of the mathematics.

[‡] Most of the path-integral techniques used for the Kondo problem have been developed from very similar, but simpler, methods for Mahan's x-ray edge problem [3]. In this problem, we have an infinitely heavy hole at the origin (with creation and annihilation operators b^+, b) and $H = E_h b^+ b + \sum \epsilon_k a_k^+ a_k + b^+ b \cdot n(0)$, where $n(0)$ is the number of Fermions at the origin. Spin is ignored, and E_h is assumed to be very large.

Whenever a method is introduced for the Kondo problem, we shall, in a footnote, refer to the similar method for the x-ray problem (XRP).

depending on that of S_z, and so would the spin-down electron, but the sign would be opposite. The same result holds, of course, if H_1 were equal to $JS_x \cdot s_x(0)$ or $JS_y \cdot s_y(0)$. The sum of these last two terms is H_2 and we see that any attempt to solve the problem in a manifestly spherically symmetric manner will make the $H_0 + H_1$ problem look as difficult as the $H_0 + H_1 + H_2$ problem—the Kondo problem—which it is not. For this reason, keeping H_1 and H_2 separate seems preferable.

In classical mechanics, it is possible to find the behavior of a system by considering all its possible paths $X_i = X_i(t)$ (where X_i are the system's coordinates) between an initial configuration and a final one, calculating for each path a functional—the action—and choosing the path for which this functional is a minimum.

The quantum mechanical analog of this is Feynman's path-integral approach: We take all possible histories (paths) of the system, and give each an amplitude (which depends on the action integral along it). Instead of a wave function depending on the particles' position (and spins) at all times, integrals over the paths give us various properties of the system in a way similar to that in which integrals over the coordinates do this in ordinary quantum mechanics.

If we perform this path-integral calculation for the Kondo Hamiltonian, we can then consider all the paths of the system in which the paths $S_z = S_z(t)$ of the localized spin are identical, whatever the history of anything else may be. This partial set of paths describes the motion of the noninteracting electrons under the influence of a time-dependent single-particle Hamiltonian, due to their scattering by exchange interactions with the time-dependent spin $S_z(t)$. For this new problem, we no longer have to use path integrals; if we want to sum over the Fermi gas paths,[†] we only have to take its ground state, let the time-dependent scatterer due to S_z act on it, and project back to the ground state. The amplitude $\langle 0 | \exp[i \int H(t) \, dt] | 0 \rangle$ of this projection is then the amplitude for the relevant path $S_z = S_z(t)$ of the localized spin.[‡] Since all states of a spin are linear combinations of its spin-up ($S_z = \frac{1}{2}$) and spin-down ($S_z = -\frac{1}{2}$) states, the only possible paths for the local spin consist of alternate flips of S_z up and down at times t_i (where i has one parity when S_z flips up, and another when S_z flips down). This leaves us with two problems: (a) how to calculate the amplitude for each path of S_z; and (b) what to do with these amplitudes, once we get them.

Feynman [5] has shown how to eliminate harmonic oscillators (i.e., Bosons) from a problem, leaving only an effective interaction within the

---

[†] We assume that, in the initial and final configuration, $H_0 + H_1$ is in its ground state $|0\rangle$ with $S_z = +\frac{1}{2}$.

[‡] See Noziéres and de Dominicis [4] for this approach in the XRP.

remainder of the physical system[†]—in the Kondo problem, this would mean a (noninstantaneous) self-interaction of S_z with itself. If the resulting self-interaction of S_z were instantaneous, the problem would be trivial. It is the interaction's noninstantaneous nature that forces us to work with path integrals rather than wave functions.

III. Eliminating the Fermi Gas[‡]

Because of the conservation of spin, an lectron in the Fermi sea must flip its spin up whenever the local spin flips down, and vice versa. Thus, if the $S_z s_z$ interaction term $H_1 = J_z S_z \sum_{k\sigma}$ were not there, we would only have to follow the behavior of the two Fermi gases—one for $s_z = \frac{1}{2}$ and one for $s_z = -\frac{1}{2}$, when creation and annihilation operators[§] at the origin are applied to them whenever S_z flips. If we start with the ground state, and project upon the ground state at the end, we get the product of two many-electron Green's functions; this product must be multiplied by $(J_{\pm/2})^{2n}$ (for $2n$ spin flips) because each S term in H_2 has the coefficient $(J_{\pm/2})$ in front of it.

That is, the amplitude for this path is

$$\langle 0 \mid H_2(t_1) H_2(t_2) H_2(t_3) \cdots H_2(t_m) \mid 0 \rangle$$
$$= (J_{\pm/2})^{2n} \langle 0 \mid \psi_\uparrow^+(t_1, r = 0) \psi_\uparrow(t_2, r = 0) \psi_\uparrow^+(t_3, r = 0) \cdots \mid 0 \rangle$$
$$\times \langle 0 \mid \psi_\downarrow(t_1, 0) \psi_\downarrow^+(t_2, 0) \cdots \mid 0 \rangle. \tag{3.1}$$

Since the particles are free, each of these Green's functions reduces (for a path with $2n$ spin flips) to a sum of $n!$ products of one-electron Green's functions

$$\langle 0 \mid \psi_\sigma^+(t', 0) \psi_\sigma(t'', 0) \mid 0 \rangle.$$

This sum of $n!$ products of n Green's function turns out [7] to be a determinant.

[†] The most familiar example is eliminating the photons from quantum electrodynamics, leaving a delayed (or advanced) interaction between the charges. This method is also used for the polaron problem.

[‡] This approach was used in the XRP by Noziéres and de Dominicis [4]. It was the first [6] path-integral method applied to the Kondo problem. After seeing these results, Hamann and Schrieffer and Evenson applied similar methods to the Anderson model; these are the subject of Chapter 8 of this volume.

[§] Flipping an electron's spin up is equivalent to creating a spin-up electron and annihilating a spin-down one.

In order to proceed, we need the form of the one-electron Green's function. As we only need this function when the creation and annihilation operators in it are at the origin, we shall ignore its behavior in the rest of (three-dimensional) space. Thus $r = 0$ will be assumed from now on. The real physical exchange interaction has a form factor, but we shall limit the bandwidth rather than introduce a $J(r)$.

If we assume an infinite structureless band, the behavior of $G_0 (E, r = 0)$ is clear:

$$\operatorname{Im} G = \int \delta(E - E_k) \cdot \operatorname{sgn}(E_k - E_F) \cdot n(E_k) \, dE_k \qquad (3.2)$$

and hence it has one value for $E > E_F$, and the opposite value for $E < E_F$. If we put $E_F = 0$, we have $\operatorname{Im} G_0(E) \pm \operatorname{sgn}(E)$. $\operatorname{Re} G$ is indeterminate but may be set equal to zero. Fourier transforming, we find up to a scale factor

$$G_0(t) = 1/it. \qquad (3.3)$$

In order to keep expressions real, we shall here and henceforward use imaginary times, $t = -i\tau$. Then

$$G_0(\tau) = 1/\tau,$$

and the determinant of n Green's functions is

$$\left| \frac{1}{\tau_{2i} - \tau_{2j-1}} \right|.$$

For a finite band, G_0 will have a short-time cutoff due to the bandwidth. For instance, for a rectangular band of width $2/t_0$, symmetric about the Fermi surface, we find

$$G_0(\tau) = \frac{1 - \exp(-|\tau/t_0|)}{\tau}.$$

This short-time cutoff is essential to keep the expression finite, and plays an important role in later developments. The determinant of free Green's functions equals [8]:

$$D = \frac{\prod_{i<j}(\tau_{2i} - \tau_{2j})(\tau_{2i-1} - \tau_{2j-1})}{\prod_{i,j}(\tau_{2i} - \tau_{2j-1})}. \qquad (3.4)$$

The effect of a short-time cutoff in G on D is to impose such a cutoff on D too† (ref. [7]). There are two such determinants: one for the spin-up electrons, and one for the spin-down ones. Then, if it were not for the time-dependent potential scattering due to S_z, D^2 would be the amplitude of the path in which S_z flips at the times t_i.

Because of the potential scattering due to S_z, the Green's function in the electron gas differs from the free electron function. However, it is still possible to find a closed-form expression for $G(\tau, \tau')$.

The Green's function obeys Dyson's equation

$$G(\tau, \tau') = G_0(\tau - \tau') + J \int_{-\infty}^{\infty} G_0(\tau - \tau'') S_z(\tau') G(\tau'', \tau') d\tau''$$

$$\cong \frac{1}{\tau - \tau'} + J \int_{-\infty}^{\infty} \frac{1}{\tau - \tau''} S_z(\tau'') G(\tau'', \tau') d\tau''. \tag{3.5}$$

This is a singular integral equation of the type studied by Muskhelishvili [9]; for the case when S_z flips between two values, we find [7]

$$G(\tau, \tau') = \frac{1}{\tau - \tau'} \prod_i \left(\frac{\tau_i - \tau}{\tau_i - \tau'}\right)^{\pm \delta/\pi}, \tag{3.6}$$

where the $\pm$ sign depends on whether S_z flips up or down at time t, and δ is the scattering phase shift due to J_z.

Therefore, the effect of the time-dependent $S_z(t)$ term H_2 is to multiply each of the $n!$ products of n Green's functions by

$$\left[\frac{\prod_{i<j}(\tau_{2i} - \tau_{2j})(\tau_{2i-1} - \tau_{2j-1})}{\prod_{i,j}(\tau_{2i} - \tau_{2j-1})}\right]^{2(\delta/\pi)} = D^{2(\delta/\pi)}. \tag{3.7}$$

Thus D is multiplied by $D^{2(\delta/\pi)}$.

There is one other effect due to S_z. It perturbs the ground-state Fermi sea. Indeed, the two many-body ground states of $H_0 + H_1$ with $S_z = +\frac{1}{2}$ and with $S_z = -\frac{1}{2}$ are orthogonal [10] for any value of J; thus a sufficiently long interval in which S_z is reversed will cause the overlap

$$\langle 0 | \exp\left[i \int H(t) dt\right] | 0 \rangle$$

to be very small.

---

† It is shown in ref. [7] that, although the cutoff modifies the Cauchy nature of the determinant when times are too close together, (3.4) holds for all widely separated times; it is asymptotically exact in the sense of the introduction. That is, as $J_\pm \to 0$, the important paths have fewer and fewer t's, and thus $t_i - t_{i-1} \to \infty$. But it will turn out that nearest neighbor times are not very important anyhow.

One should note that this effect appears without any reference to the creation and annihilation operators. This effect can be included [4], [7] by summing up over the closed-loop diagrams (Fig. 1). This

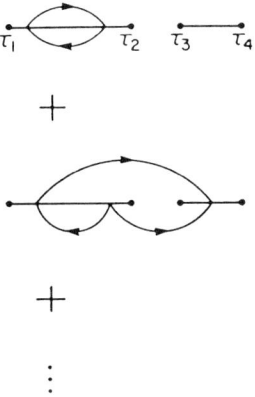

FIG. 1. The closed-loop diagrams.

sum can be obtained from $G(\tau, \tau)$, as done by Yuval and Anderson [7], and the result is to multiply D further by $D^{(\delta/\pi)^2}$. Therefore, the amplitude corresponding to the path is

$$D^{2[(\pi+\delta)/\pi]^2} = \left[\frac{\prod_{i<j} (\tau_{2i} - \tau_{2j})(\tau_{2i-1} - \tau_{2j-1})}{\prod_{i,j} (\tau_{2i} - \tau_{2j-1})} \right]^{2[(\pi+\delta)/\pi]^2}. \quad (3.8)$$

Since $J_i 2$ appears with each $S_\pm s(0)_\pm$ operator, we have to multiply this last expression by $(J_i 2)^{2n}$.

IV. The Method of Schotte and Schotte

Schotte and Schotte [11] transformed the Fermi gas into a Bose gas, using the method due to Tomonaga [12].

Since the interaction terms (H_1 and H_2) act on the Fermi gas at the origin, we are only interested in s electrons; those with higher orbital angular momentum do not interact with S. The s electrons form a one-dimensional gas; in the neighborhood of the Fermi surface, the energy E is a linear function of the momentum k. This is where we again make the asymptotic assumption that J is small, and that we need treat only electrons near the Fermi surface accurately.

7. ASYMPTOTICALLY EXACT METHODS IN THE KONDO PROBLEM

If we consider density wave operators in this one-dimensional gas

$$\rho_q = \sum_{\substack{E_k < E_F \\ E_{k+q} > E_F}} a^\dagger_{k+q} a_k$$

we find

$$[H, \rho_q] = vq\rho_q, \tag{4.1}$$

where v is the Fermi velocity. The ρ_q obey Bose statistics so long as $q \ll k_F$ and the Fermi gas is only weakly excited.

Having transformed the Fermi gas into a Bose gas, we have to transform the electron creation and annihilation operators (due to flips of S_z) into Boson operators. Since a change of π in the scattering phase shift causes an extra bound state to appear (or disappear), without otherwise affecting the scatterer, we expect such a change in the scattering phase shift at the origin to imitate the creation and annihilation operators fairly well. This does indeed happen [11], and the effect of the electron creation and annihilation operators is to replace the weak scatterer $\pm J$, giving a phase shift $\pm \delta$, by a much stronger scatterer with a phase shift of $\pm(\pi \pm \delta)$.

Now that we have a time-dependent potential exciting an array of harmonic oscillators, we can use Feynman's original expression [5] for the resulting amplitude due to each oscillator, and find the amplitude for any path $S_z(t)$ by multiplying them together. Since these amplitudes are each the exponential of a bilinear expression in the perturbing potential, we will find $(\pi \pm \delta)^2$ appearing as a part of an exponent in the expression for the amplitude. Summing up over the oscillators, Schotte and Schotte find an amplitude

$$(J_\pm/2)^{2n} \left[\frac{\prod_{i<j}(\tau_{2i} - \tau_{2j})(\tau_{2i-k} - \tau_{2j-1})}{\prod_{i,j}(\tau_{2i} - \tau_{2j-1})} \right]^{2[(\pi+\delta)/\pi]^2} \tag{4.2}$$

as in (2.3) [$(J_\pm/2)^{2n}$ being again due to the $(J_\pm/2)S_\pm \cdot s_\mp(0)$ term in H_2].

V. Summing Up over the Paths

We now have an amplitude for each path of S_z (over imaginary time). These amplitudes are all real and positive.† Moreover, the amplitudes are products of terms

$$(\tau_i - \tau_j)^{\pm 2[(\pi+\delta)/\pi]},$$

† This only holds if we work on the imaginary-time line, which is why we work there.

involving two times each. Therefore we associate an "energy" E with each path, so as to make its amplitude equal to[†] $e^{-\beta E}$, and we find that E is a sum of two-body logarithmic interactions between the flips. That is, we may write

$$E = \sum_{i,j} (-)^{i-j} \cdot (2 - \varepsilon) \ln(\tau_i - \tau_j). \tag{5.1}$$

If we now regard the amplitudes as if they were probabilities, we find the Kondo problem equivalent to the statistical mechanics of a gas of *classical* particles (the flips at times t_i) on a straight line, interacting via a logarithmic potential, and with a chemical potential $\ln(J_{\pm/2})$ (since a path with $2n$ flips has a factor $J_{\pm}^{2n}$ in its amplitude).

Integrating by parts twice, the interaction "energy"

$$\iint_{|\tau-\tau'|>\tau_0} d\tau \, d\tau' \ln(\tau - \tau') \frac{dS_z}{d\tau} \frac{dS_z}{d\tau'} \tag{5.2}$$

between flips is equivalent, so long as no two flips occur within a time τ_0 of each other, to an interaction

$$\iint_{|\tau-\tau'|>\tau_0} d\tau \, d\tau' \cdot \frac{1}{(\tau - \tau')^2} \cdot S_z(\tau) \cdot S_z(\tau') \tag{5.3}$$

between the spins S_z at various (imaginary) times. Thus instead of a logarithmic interaction between the spin flips, we have an inverse-square interaction between the spins $S_z(t)$ at any two times. If we had let τ_i, τ_j approach each other arbitrarily close (corresponding to a really infinite band), we might have had trouble about convergence, since the integral $\iint d\tau \, d\tau'[-(2 + \varepsilon) \ln(\tau - \tau')]$ has a divergence for $\tau - \tau' \ll 1$ if $\varepsilon > 0$. If we impose a cutoff, say that $|\tau_i - \tau_j|$ must always be greater than a minimum t_0 (which is approximately the inverse bandwidth, because of the uncertainty relationship $\Delta t \, \Delta E > h$), these divergences disappear, and we have a finite integral; instead of diverging, it gives us a contribution that behaves (in the region of interest) like $\ln t_0$. We recall that the Kondo problem arose because of difficulties in perturbation theory for arbitrarily small J, that is, in the region where flips are rare, and where $\varepsilon = \delta/\pi + 2(\delta^2/\pi^2)$ is very small, so that the logarithmic divergences at long range dominate the behavior of the classical gas.

Logarithmically divergent integrals depend only very weakly on the detailed shape of the cutoff, and we therefore expect the thermodynamics

[†] β^{-1} is a "temperature" which we shall choose to be unity.

of the system to remain essentially unaltered if, instead of the τ_i being continuous variables subject to the condition $|\tau_i - \tau_j| > t_0$, we make all the τ_i different integer multiples of t_0. Let us consider then a one-dimensional Ising chain specified by

$$E = \sum_{i,j} \frac{J_{LR}}{(i-j)^2} S_i S_j + \sum_i J_{NN} S_i S_{i+1}. \tag{5.4}$$

It is shown [13] that this is mathematically equivalent, as far as the long-range interactions between spin reversals along the chain are concerned, to the hard-rod model above, and thus also the spin $\frac{1}{2}$ Kondo problem. The formulas giving the correspondence are

$$\left| \frac{J_\pm t_0}{2} \right| = \exp(-\beta[J_{NN} + (1+C) J_{LR}])$$
$$2\beta J_{LR} = 2 - \varepsilon \simeq 2 - 2J_z t_0, \tag{5.5}$$

where C is Euler's constant.

Unfortunately, this Ising problem remains the subject of controversy in the literature, although a solution based on our work on the Kondo problem is now beginning to be accepted.

VI. Finite Temperatures

The methods used in the last two sections to study the ground-state behavior of the Kondo problem can also be used for the finite-temperature case. The approach is the same with both methods: We impose the requirement that the (imaginary-time) behavior be periodic modulo $i\beta = i/kT$. Another way of formulating it is to limit the spin flips, or the Ising spins, to a line of length $i\beta$, and have the interaction behave as

$$\log\left(\frac{\beta}{\pi} \sin \pi \frac{\tau_i - \tau_j}{\beta}\right)$$

instead of $\log(\tau_i - \tau_j)$ between the flips, or as

$$\frac{\beta^2/\pi^2}{\sin^2(\pi/\beta)(\tau_i - \tau_j)}$$

instead of $1/(\tau_i - \tau_j)^2$ (between the Ising spins). We can summarize the equivalence between the Kondo system and the two classical systems as shown in Table I. Since no finite system can exhibit a rigorous phase

TABLE I

COMPARISON OF THE KONDO SYSTEM WITH THE TWO CLASSICAL SYSTEMS

Kondo system	Charged-rod model of the flips	Ising model
β	System is on circle of circumference β	System is on the circle of circumference β
J_z	Logarithmic interaction proportional to $1 + J_z t_0$	Inverse-square interaction proportional to $1 + J_z t_0$
$J_\pm$	Chemical potential proportional to $\ln \mid J_\pm \mid$	Nearest-neighbor interaction proportional to $\ln \mid J_\pm \mid$
Scaling factor in the interactions	Temperature	Temperature

transition, we only expect nonanalytic dependence on J_z, $J_\pm$ in the properties of the ground state (corresponding to an infinite system), and not at any finite temperature (corresponding to a finite system).

The finite-temperature behavior of the Kondo problem is thus equivalent to the thermodynamics of a *finite* classical system, as against the infinite system, found for the ground-state behavior. Since it takes a system of a certain size to tell whether we are above or below the phase transition "temperature," we expect to find a sharp difference between the antiferromagnetic Kondo system (for which we shall show the classical system to be uncondensed) and the ferromagnetic system (for which the classical system is condensed) only at $T = 0$. One way of defining the Kondo temperature is as the temperature at which large differences between the two systems show up.

VII. Numerical Results

Using Monte Carlo or other computational methods, it is possible to go directly from the classical models to properties of the Kondo system. One such calculation has been made by K. D. Schotte [14], who has been kind enough to let us present his results. The magnetic susceptibility at any temperature may be directly related to the $\langle S_z(0) S_z(\tau) \rangle$ correlation function, which is calculable in terms of charged rods on a ring. One approximation is made: a canonical rather than grand canonical distribution of the number of spin flips; but this is increasingly unimportant as the temperature is lowered. The basic limitation on computation time is that as $\beta \to \infty$ ($T \to 0$) the ring becomes infinitely large; in practice no more than 50 spin flips were treated.

Three separate values of $J\tau_0$ were treated ranging from 0.175 $[T_K = \exp(-1/J\tau_0) = 3.3 \times 10^{-3}]$ to $0.225 (T_K = 12.4 \times 10^{-3})$. Since

7. ASYMPTOTICALLY EXACT METHODS IN THE KONDO PROBLEM

these do not differ greatly, we have placed them on the same graph Fig. 2) along with results of Ting [15] from the Suhl equations, for contrast between the exact and the approximate methods. The two agree very well at T_K. Above T_K the deviation comes about because Ting's formulas neglect $J\rho \simeq J\tau_0$ in the sense that terms of relative order $J\tau_0 \ln(T/T_K)$ are explicitly neglected, whereas no essential approximation occurs in the Schotte results.[†]

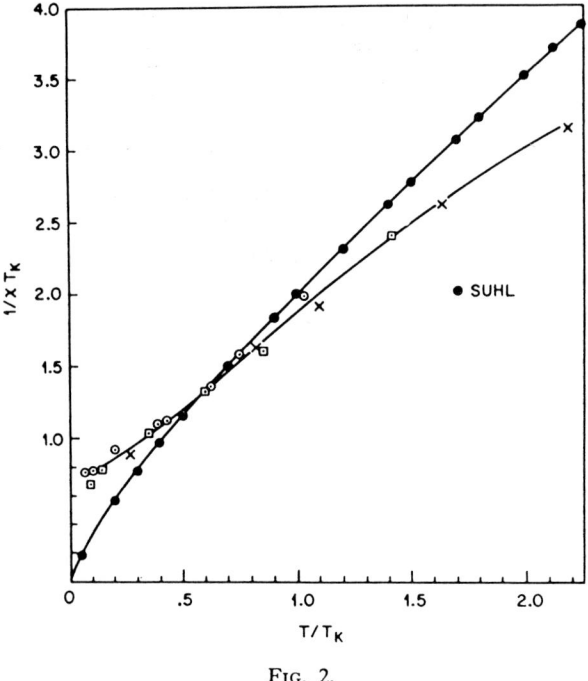

FIG. 2.

Below T_K the Suhl results, and those of almost all other methods as well which are valid at high T, lead to a divergent susceptibility, and thus $1/\chi \to 0$, usually with vertical slope. Schotte's results curve upward instead, as they must if a bound state is to be reached and Nernst's theorem is valid (as it need not necessarily be!). It is noticeable that the region of interest near T_K is almost linear, in agreement with the majority of experimental results, which all agree that

$$\chi \simeq \frac{C}{T + ``T_K"}. \qquad (7.1)$$

[†] Mr. John Armytage has found that actually, Schotte's results are not accurate at higher T because of the microcanonical assumption, so that only the lower T differences are germane.

This is a consequence of the inflection point which must separate negative curvature at high T/T_K from positive at low T/T_K.

VIII. The Scaling Method

From previous work on the Kondo problem (in particular Mattis [16]) it seems clear that the behavior of the system is entirely different in the ferromagnetic case ($J > 0$) and in the antiferromagnetic case ($J < 0$). For such a dramatic change to occur in the thermodynamics of the classical gases discussed in Section V, we have to have a phase transition separating the $J > 0$ and $J < 0$ regions. The Kondo problem, which is really a problem of very low J values, is thus closely related to the critical-point behavior of the equivalent classical system.

In order to look for such a phase transition, we want to normalize away the fluctuations due to pairs of flips near each other (Fig. 3). This renormalization process turns out to be the crucial point of the solution[†]: We use the model in which the τ_i are continuous variables with $|\tau_i - \tau_j| < t_0$, and consider the pairs of flips (we shall call them *close pairs*) with $t_0 < |\tau_i - \tau_j| < t_0 + dt$. These close pairs are very rare (since dt is infinitesimal), and thus they do not interact with each other. Since they are small, and the other spins are usually far away from them[‡], it is a good approximation to assume that they appear uniformly along the t line. Therefore, we can ignore (i.e., renormalize away) all these pairs, so long as we adjust the interaction of the flips that remain; this adjustment is necessary, because a region that looks (neglecting the close

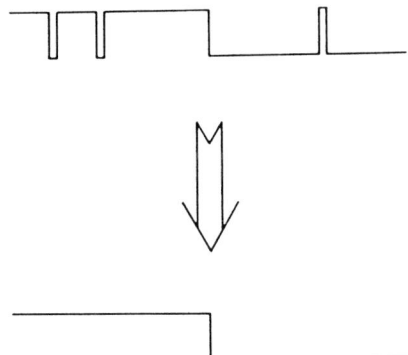

FIG. 3. Renormalizing the fluctuations away.

[†] See Anderson *et al.* [17] for details of the mathematics.

[‡] The interaction between a close pair and any other flips is weak, because the two flips in the pair give terms that nearly cancel each other out.

pairs) as if $S_z = \frac{1}{2}$ all along it has a finite probability of having in it an $I_z = -\frac{1}{2}$ region of length t_0; thus all interactions between the flips are multiplied by a factor.

In this way, we can renormalize t_0 upward in infinitesimal steps,[†] and in the process change the values of the various interactions in the classical system. To follow the behavior of the classical gas under the renormalization any further, it is easier to consider isotropic and anisotropic Kondo models at the same time; J appears twice in the Hamiltonian, in $H_1 = J_z S_z \cdot \sum \langle \ \rangle$ and in $H_2 = J_{\pm/2} \cdot (S_+ \sum \langle \ \rangle + \text{h.c.})$. If $J_\pm \neq J_z$, we have two parameters to replace J: In the classical gas, J_z gives the strength of the interaction between the spin flips, while $J_\pm$ gives their chemical potential.

When we change t_0 continuously in the renormalization process, $|J_\pm|^{\ddagger}$ and J_z move over a two-dimensional diagram as in Fig. 4

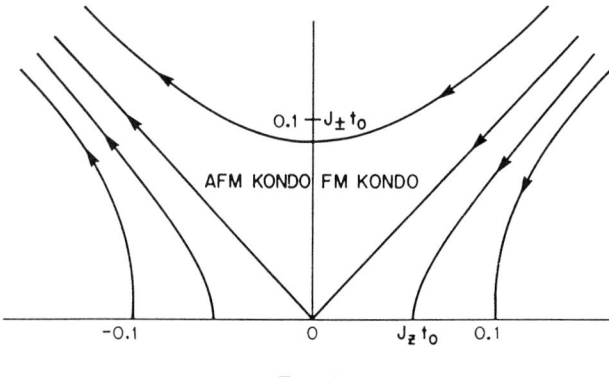

FIG. 4.

(this figure is reversed with regard to the figures of Anderson et al. [17]).

Thus, if we start in the region $J_z > |J_\pm|$, the renormalization process carries the system into one on the X axis, on which $|J_\pm| = 0$. Such a system cannot have any spin flips, and corresponds to a completely condensed gas. For $J_z < |J_\pm|$, renormalization will ultimately bring us to a region where $|J_\pm|$ grows larger and larger; therefore, in that region the flips become more and more frequent (on the scale of the renormal-

---

[†] Since t_0 is approximately the inverse bandwidth, this renormalization is equivalent to a downward renormalization of the bandwidth. The latter can be performed without reference to the path-integral method; see Section XI.

[‡] The sign of $J_\pm$ has no effect on the mathematics.

ized t_0), and the system is clearly uncondensed.† The phase boundary is on the isotropic ferromagnetic line $J_z = |J_\pm|$, but states on this line behave as if they were below the transition.

The phase transition in this model (and in the Ising model of Section V) has a discontinuity in the order parameter (i.e., the magnetization). Thouless [18] has already shown that such a discontinuity must occur here if there is a phase transition at all; we find that the order parameter just below the transition has precisely the minimum value it can have according to Thouless. Dyson [19] has rigorously shown such a phase transition for a special case of his hierarchical model. Several other properties of the phase transition are also anomalous. In our previous work [13, 17] we have discussed the thermal properties of the infinite Ising model, which correspond to the ground-state energy of the Kondo problem, and demonstrated that its analytic behavior agrees with that calculated from perturbation theory by Kondo.

IX. The Kondo Temperature

We have previously mentioned that the Kondo effect has a characteristic temperature scale—the Kondo temperature. It should be emphasized that the Kondo temperature is *not* the temperature of a phase transition; it is only a characteristic temperature for a gradual change in the properties of the system. Using the scaling method, this characteristic temperature can be given a precise meaning.

When we renormalize τ_0 upward, β/τ_0 scales downward. As we have seen in [17], there is a precise equivalence between two Kondo systems with different values of β/τ_0, if J is scaled accordingly.‡ Using this equivalence in the opposite direction, two Kondo systems with different values of J (but the same sign) are equivalent to each other if we consider them at two *different* temperatures, the ratio between which depends on the J values. There is thus a natural unit of temperature, depending on the system (a Kondo temperature), such that in terms of this unit, all ferromagnetic Kondo systems have the same behavior, and so do all antiferromagnetic systems. This Kondo temperature is found to be

$$T_K = \exp(-1/J\tau_0).$$

† Indeed, if we continue the scaling sufficiently far into the left-hand side of the diagram, we reach a point where the interaction between flips gives a $(\tau_i - \tau_j)^{\pm 1}$ term in the amplitude [rather than $(\tau_i - \tau_j)^{\pm 2}$ in the $J \to 0$ limit]. As Toulouse pointed out, this point corresponds to the path-integral problem for the spin-up electron tunneling in and out of a resonant state, without any spin-down electrons, which is exactly soluble.

‡ A change of both β and τ by the same factor has no effect on the mathematics, if $J\tau$ is kept constant.

X. Physical Implications

We have found out that, in the path-integral formalism, the ferromagnetic and antiferromagnetic Kondo systems correspond to the statistical mechanics of a classical system just above and just below its phase transition, respectively. To see the implication of this for the original quantum mechanical problem, we consider the long-range order.

In the ferromagnetic case, there is infinite-range order, and if $S_z > 0$ at one point, the expectation value of S_z will remain positive infinitely far from this point. Since the space dimension in the classical system is the time line, this means that S_z has a memory for arbitrarily long time intervals. This agrees with a triplet ground state: If the total angular momentum points in the positive z direction, S_z is always more likely than not to be positive.

In the antiferromagnetic case, S_z has no infinite-time memory; this agrees with a singlet ground state, which is nondegenerate, and has no preferred direction in space. Since, for small J, the system is just above its transition, we have correlations (i.e., spin memory) extending over extremely long regions (i.e., time intervals), and above the temperature corresponding to the correlation range (which is again the Kondo temperature), we can ignore the singlet nature of the ground state.

Far enough below the temperature, the system to which the physical one is equivalent is one with a *large* $J\tau_0$ and a singlet ground state. Near zero temperature it behaves, then, simply like a localized spin fluctuation (LSF), but so far the types of theories which have been applied to that problem are incapable of handling the essential stage of connecting together the LSF and magnetic regimes, as we can.

XI. "Renormalization Group" Methods in the Kondo Problem

Another group of methods have been applied to the Kondo problem which also have the (as yet incompletely exploited) possibility of retaining intact the analytical nature of the problem, and which lead at first to a very similar scaling procedure to that already derived. Since the results do not seriously differ (insofar as they remain exact) from those already discussed, this section is rather brief and merely serves to call attention to these other approaches.

The papers we discuss here are those of Anderson [20] and Fowler and Zawadowski [21]. These papers have different emphasis and points of view, although there is a close initial similarity. Anderson's is very much simpler mathematically and easier to follow; Fowler and Zawadowski use the full panoply of the renormalization group method and try to achieve a complete solution, but make certain approximations which Anderson avoids in principle if not in practice.

Let us first sketch the simple Anderson method. The idea here was to carry out a *projection* of the problem with an upper cutoff energy D onto the set of states appropriate to a problem with lower cutoff energy $D - \varepsilon$. When the problem is reexpressed in terms of the exact resolvent operator $G = (E - H)^{-1}$, this projection can be done by summing over perturbation theory diagrams involving the high-energy states between D and $D - \varepsilon$, which will be of order $(\varepsilon^n/D^n)J^n$ for a diagram of nth order. Clearly only the lowest order is relevant since we choose ε at will.

In a general problem such a procedure will lead to a small change in the effective interaction $V \to V + VP_\varepsilon V/D$, which is not useful because it is more complicated than the original V. Here, however, we find that to a very good approximation the new interaction is almost exactly like the old one. This is what makes it possible again to write down scaling differential equations like those obtained by the fluctuation renormalization method:

$$d(J_\pm \tau) = (J_\pm \tau)(J_z \tau) \frac{d\tau}{\tau},$$

$$d(Et\tau) = \left(\frac{J_\pm \tau}{2}\right)^2 \frac{d\tau}{\tau^2},$$

(11.1)

and so on (where $d\tau = \varepsilon$, $\tau = D^{-1}$).

Unfortunately, this procedure is not quite exact because the change $d(J\tau)$ depends to some extent on the energy of the initial and final states of the scattering process. This energy dependence becomes severe when the effective exchange integral itself becomes large. In fact, if we follow the equations (11.1) to their logical conclusion, eventually $J\tau$ increases without limit: but only for electrons right at the Fermi level.

Anderson's solution [20] to this problem was to suggest that this continuous scaling be stopped at a definite point where $(J\tau)$ has become reasonably large, and that this resulting strong coupling problem be treated by entirely separate methods. This is the problem discussed by Mattis [16], for instance, of a spin strongly coupled to the nearest Wannier function, with relatively small transfer matrix elements (given by the cutoff band-width $1/\tau = D$). Perturbation theory in D/J should give correct answers for this problem. But Anderson did not show that $J\tau$ could be scaled to values large enough to make this perturbation theory converge rapidly, nor did he carry out that calculation. Nonetheless, this method is a very useful simple visualization of the essentially more exact results of the space-time methods.

Fowler and Zawadowski [21], on the other hand, specifically emphasized the frequency dependence of the effective interactions. They worked from the start with Abrikosov's [22] pseudo-Fermion technique and conventional Green's functions. Studying the dependence of the renormalized Green's functions on the physical input parameters J, D, and the frequency variable ω in the scattering process, they observed that different values of input parameters could give the same renormalized scattering amplitudes. One then looks for the group of transformations on the inputs which gives the same physical results: This is the "renormalization group."

There are two regions of the parameters where the renormalizations are very small. First, if we set the input J nearly equal to the scattering matrix T itself (which in this theory is called the "invariant coupling") and the frequency and cutoff energies are very small, we have very little renormalization. Second, in the case where D is large but ω is at some sufficiently large value also, then all perturbation denominators are very large (there is a regularization procedure here introduced by Bogoliubov), and for the physical J also renormalization is small. These two regimes are equivalent under the renormalization group, and can be scaled into each other by use of a differential equation involving only the latter regime. Here, however, it becomes clear that ω dependence plays a crucial role, so that we improve the accuracy by taking into account higher diagrams. In lowest order, one gets the conventional Abrikosov results; in the next order, the very accurate approximation of Noziéres [23].

By the nature of the method it is still only a very sophisticated way of generating approximations. But again it has the great advantage of bringing out clearly the fact that the antiferromagnetic Kondo problem scales into a strong coupling problem for arbitrarily small values of J, and that the factor giving the scaling is the conventional "Kondo temperature" energy.

ACKNOWLEDGMENT

We would like to acknowledge the help of Mr. John Armytage, especially in preparing Fig. 3 and in the preparation of Section XI.

References

1. J. Kondo, *Progr. Theor. Phys.* **32**, 37 (1964).
2. P. W. Anderson, *Phys. Rev.* **124**, 41 (1961).
3. G. D. Mahan, *Phys. Rev.* **163**, 612 (1967).

4. P. Nozièrcs and C. de Dominicis, *Phys. Rev.* **178**, 1097 (1969).
5. R. P. Feynman, Ph. D. Thesis, Princeton, Univ. Princeton, New Jersey, 1942.
6. P. W. Anderson and G. Yuval, *Phys. Rev. Lett.* **23**, 89 (1969).
7. G. Yuval and P. W. Anderson, *Phys. Rev. B* **1**, 1522 (1970).
8. G. Polya and G. Szegö, "Aufgabe Und Lehrsätze aus der Analyse." Dover, New York, 1945.
9. N. I. Muskhelishvili and D. A. Kveselava, *Tr. Tbilis. Mat. Inst.* **11**, 141 (1942).
10. P. W. Anderson, *Phys. Rev.* **164**, 352 (1967).
11. K. D. Schotte and U. Schotte, *Phys. Rev.* **182**, 479 (1969).
12. S. Tomonaga, *Prog. Theor. Phys.* **5**, 544 (1950).
13. P. W. Anderson and G. Yuval, *J. Phys. C* **4**, 607 (1971).
14. K. D. Schotte and U. Schotte, *Phys. Rev. B* **4**, 2228 (1971).
15. C. S. Ting, *J. Phys. Chem. Solids* **31**, 777 (1970).
16. D. C. Mattis, *Phys. Rev. Lett.* **19**, 1478 (1967).
17. P W. Anderson, G. Yuval, and P. R. Hamann, *Phys. Rev. B* **1**, 4464 (1970); *Solid State Commun.* **8**, 1033 (1970).
18. D. Thouless, *Phys. Rev.* **187**, 732 (1969).
19. F. J. Dyson, *Commun. Math. Phys.* **21**, 269 (1971).
20. P. W. Anderson, *J. Phys. C* **3**, 2436 (1970).
21. M. Fowler and A. Zawadowski, *Solid State Commun.* **9**, 471 (1971).
22. A. A. Abrikosov, *Physics (Long Island City, N.Y.)* **2**, 5 (1965).
23. P. Nozières, unpublished, 1970; also A. A. Abrikosov and A. A. Migdal, *J. Low Temp. Phys.* **3**, 519 (1970).

Many-Body Effects at Surfaces
Elementary Excitation in Solids, Molecules, and Atoms
(Plenum, New York, 1974), pp. 1–29

This is a summary of the physics in Inkson's thesis which explains fairly clearly why local self-energies (the LDA) don't get energy gaps right. Inkson and I felt that these considerations were widely ignored, and then rediscovered, without much acknowledgement, by those more in the mainstream. The methods resemble those of the Swedish school. Inkson was for various reasons unable to publicize our work very much, while this rather obscure place is the only time I found to talk about it, so we are at least partially culpable if there is any blame to be assigned for its neglect.

MANY-BODY EFFECTS AT SURFACES

P.W. ANDERSON

Bell Laboratories, Murray Hill, New Jersey

and

Cavendish Laboratory, Cambridge, England

The lectures I will give here will be based in the first instance on the thesis of one of my Cambridge students, Dr. John C. Inkson [1], which has been the source of a number of articles in the literature [2]. John's work in turn leans heavily as far as formalism is concerned on some papers by D.M. Newns [3], and on Denis Newns' advice and help during the period when he was at Cambridge. All of his work is concerned with many-electron effects at interfaces; the main intention was to study the metal-semiconductor interface with a view to understanding the so-called surface state phenomenon, but in the end we found ourselves concerned with all kinds of surfaces and with the whole range of electronic excitations which may occur at surfaces, whether metal-vacuum, metal-insulator or metal-semiconductor. I will also draw on work on surface states by various people at Bell and Cambridge: Heine, Pendry, Appelbaum and Hamann.

In fact, the ideas go beyond surfaces to the very fundamentals of the electronic theory of metals and insulators, which is the main reason I chose it as a topic to talk about here. There is in fact so much physics to talk about that I will continually dodge the formal mathematics of the problem, which in some cases is rather formidable. This formal mathematics is to be found in Inkson's papers, which may be looked up in the literature or obtained by writing him.

The genesis of this problem which concerned me lies far back, in the ideas of Bardeen [4] and Shockley [5] which played an important role in the invention of the transistor. What they were trying to do was to create what is now known as the 'field effect transistor': to apply sufficient voltage to the surface

1

of a piece of semiconductor to move the Fermi level relative to the
energy bands and thus to control the flow of current inside it

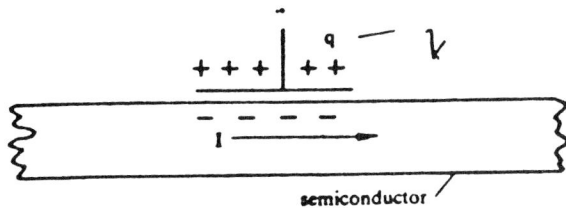

Why should have this device worked as, incidentally, they
now do — it is amusing, and I think very instructive in understanding the nature of science, to realize that the work which
earned Bardeen his first Nobel Prize was essentially completely
out of date by the time he got his second. (This absolutely is
not to say that he didn't deserve it, any more than Einstein detracts from Newton).

To see this we have to understand the nature of rectification
and the Schottky barrier layer on a semiconductor. In general,
it *is* true that the Fermi level at the surface of a semiconductor is not the same relative to the valence and conduction
bands as it is in the bulk. In the bulk it is controlled by the
nature and density of impurity levels within the band gap (P in
Si, for example),

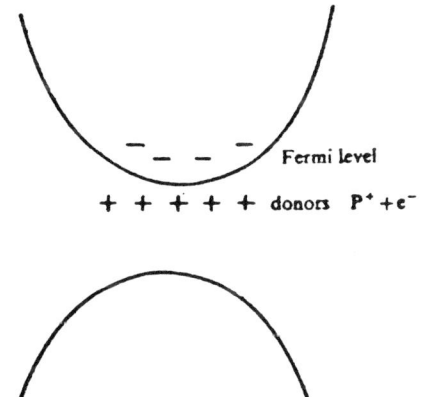

but at the surface it is determined by surface properties, by
any metal we may put on, etc.

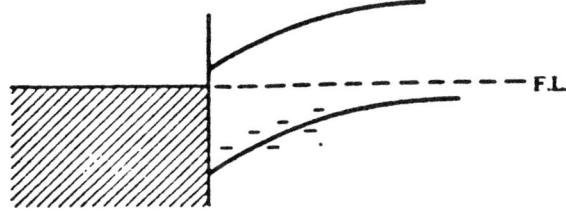

The solution to this problem is the Schottky barrier: the bands bend according to $\nabla^2 V = 4\pi\rho$, the charge being provided by charged impurities: in the above case, extra electrons captured by the acceptors; B^- bends the bands down, and we get a barrier

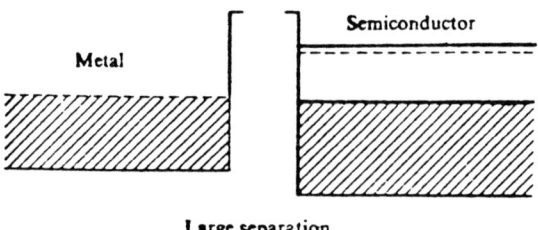

Large separation

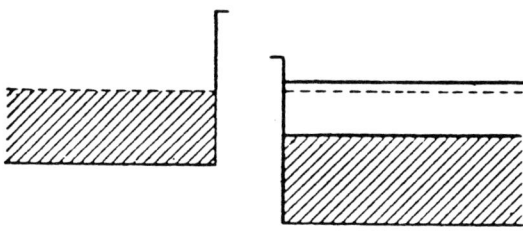

Fermi levels equalised

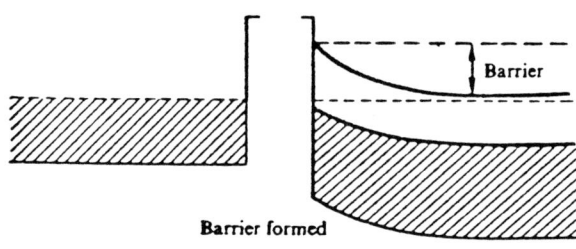

Barrier formed

Classical Schottky barrier formation

Figure 1

region of space charge. (See also figure 1)

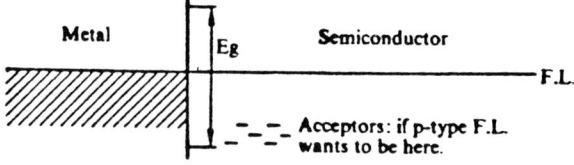

The thickness of this barrier region is quite large if the material is pure (screening length = $(kT/4\pi ne^2)^{1/2}$) and it therefore rectifies current, for reasons easily available in solid state physics texts.

What was expected was that extra charge applied to the surface, say by applying an electric field to a metal electrode or by changing the work function of the metal, would change the barrier very considerably. Since this extra charge would have to be compensated by just that many more charged impurities.

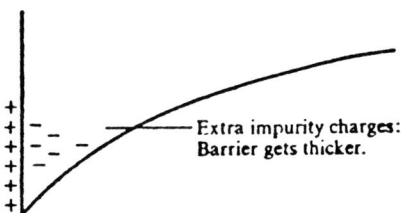

But they found essentially that this effect never works, at least on the semiconductors (Si and Ge) they had available then.

Bardeen's answer to this problem was to invoke a relatively old idea due to Tamm [6]: surface states. In general, at any point where the periodic structure of a solid is interrupted, one has the possibility of bound states appearing in the forbidden energy gap. It was Tamm who first pointed out that a surface is just such a point, and that the mere interruption of the periodic potential of the crystal at the surface would, in general, lead to a band of 'surface states' in the forbidden energy gap.

The old-fashioned way of describing this is to observe that the allowed band is distinguished from the forbidden one by whether or not the propagation constant k in $\Psi = u(r)e^{i\vec{k}\cdot\vec{r}}$ is real or imaginary. If it is complex no normalizable states in the bulk are possible, but at the surface the wave-function may decay into the bulk. The oldest surface state is probably the Rayleigh surface wave on an elastic continuum.

Shockley in particular gave good arguments for the idea that there would always be surface states in the gap of the usual covalent semiconductor — arguments which seem to be correct. Therefore, they argued, there will be a very high density of the equivalent of impurity states at the semiconductor surface. A band of surface states will have 2 states/surface atom, or $\sim 10^{15}\text{-}10^{16}/\text{cm}^2$, in the 1-2 eV energy gap, which is capable of absorbing any amount of charge without allowing any appreciable excursion of the Fermi level. Any great excursion of the Fermi level will charge up the surface states and it is thus pinned at a point in the gap. If I have time, I would like to talk to you about surface states on clean semiconductors as they have been calculated by Appelbaum and Hamann, and very probably measured by Rowe at Bell; the old ideas of Shockley are confirmed and extended by their results, so all of this probably *is* actually true of a clean semiconductor surface in perfect vacuum. However, the experiments until recently have all referred to an entirely different physical situation. Twenty-five years ago it is likely that all surfaces were heavily oxidized or otherwise chemically contaminated; and it is now known that an oxide layer has an enormous effect. But even more important, most of the interesting experiments are not at semiconductor-vacuum interfaces at all, but at semiconductor-metal contacts. What seems not to have bothered people at all until Volker Heine wrote his 1965 paper [7] is that at a semiconductor-metal contact the original Tamm and Shockley arguments for surface states are wholly meaningless, since whatever the electronic state behaves like on the *semiconductor* side of the barrier, on the metal side we must have a propagating state with a real k value, so that there can be no question of true bound states in the energy gap. What Heine suggested, and made a start at discussing, is that whatever electronic states exist in the semiconductor are the decaying *tails* of the wave functions from the metallic side, and that the really correct point of view is to ask what the metal electrons see when they get to the semiconductor surface, and whether they can be expected to have short or long experimental tails. This is what will determine the amount of charge on the semiconductor surface, and above its Fermi level.

Since Heine's paper, it has become clear that there are actually two distinct types of semiconductor-metal interface. This is the result of a series of investigations by Mead, Kurtin et al., [8] which are summarized in figure 2. The experiment Mead et al. do is simply to study the type of surface barrier which

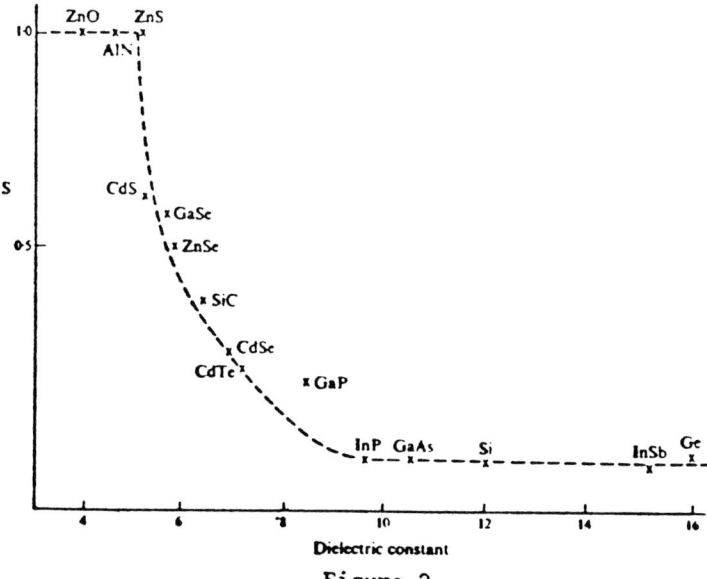

Figure 2

results when metals with a wide variety of work-functions are put on the surface of a given semiconductor. It turns out that there is rather a sharp distinction. The 'classical' covalent semiconductors Si, Ge and many others all behave as though the surface were covered with a dense layer of Bardeen surface states — Phillips has called this a 'Bardeen' Barrier. That is, independently of where the Fermi level in the metal is, an amount of charge adequate to bring the Fermi level at the surface of the semiconductor to a fixed level about 1/3 up in the gap always forms:

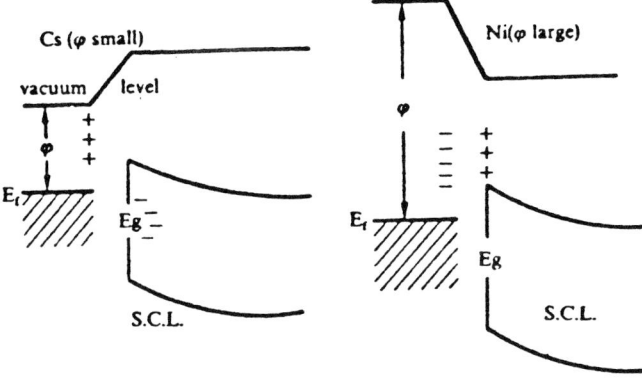

On the other hand, great numbers of other semiconductors, especially the ionic ones, with wide band gaps, behave in essentially the classic, Schottky-barrier way: E_F at the surface just equalizes to that in the metal, with no large surface charge effects.

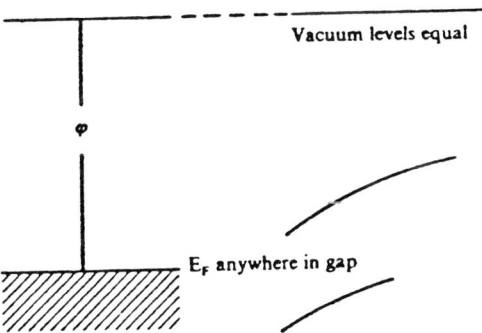

It was this dichotomy in behavior that I called to Inkson's attention, and that we feel we found an explanation for. Our explanation envisages a qualitative difference between the two types of semiconductors near a metal: the ones which form 'Bardeen' barriers essentially become metals themselves in the first layers next to a metal, losing their energy gaps entirely, while the others remain insulators, with a well-expressed gap at the metal surface (see figure 3).

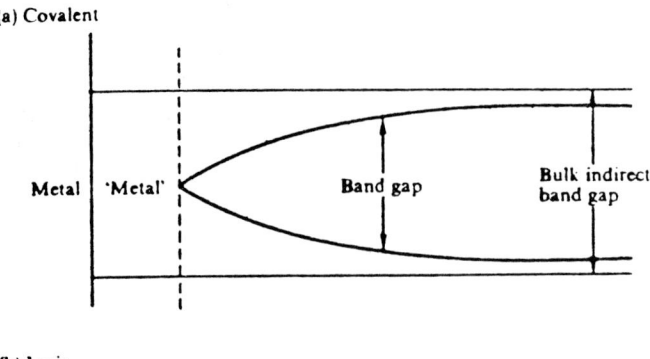

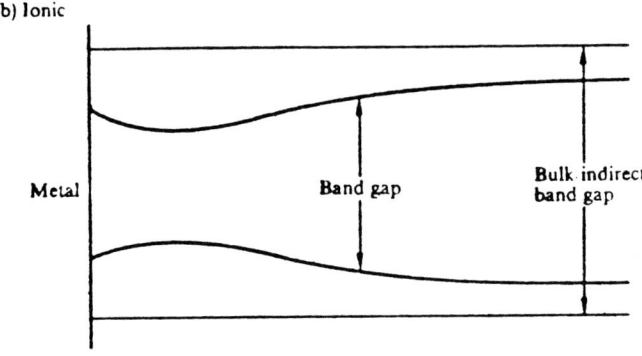

Indirect band gap distortion (schematic)
Figure 3

Like one of the seven blind men confronted with an elephant, then, I see what interests me most when confronted with nature: a many-body effect of the nature of a *surface* metal-to-insulator transition. I should emphasize that I have now changed the distance scale and on these figures the distances are not tens or hundreds of angstroms as they are for Schottky barriers, but one or two atom layers — very short indeed. But, of course, if the gap is absent even in one atom layer the metal states will spill right into the semiconductor, and charge neutrality of the semiconductor surface will require that the Fermi level be pinned very strongly at a position near the middle of the energy gap — actually we even find that it should be a bit below the middle. In other words, no matter what the relationship between the Fermi level and the vacuum potential in the metal, a dipole layer must build up at the surface which sets the semiconductor Fermi level somewhere near the middle of its gap, the electrostatic potential level floating to the necessary height. In the other case, the Fermi level may be almost anywhere in the semiconductor gap, since the charge in the semiconductor is rather independent of where the Fermi level is, and it is then controlled just by the electrostatic potential barrier at the metal surface, just as though the two were separated by a vacuum.

It is probably equally valid to look at this elephant from a chemical point of view and to say that the one group of semiconductors forms a *metallic bond* between the last atomic layer and the metal, while the second does not form any strong chemical bonds, or only electrostatic ones. One of our points is to be that one can predict whether or not this essentially chemical process takes place by studying the dielectric properties of the two substances and their effect on the electrons.

In any semiconductor, as I have shown on the figure, at relatively large distances from the metal the band gap will decrease. This is the result of the electrostatic image force for an electron in the metal surface. The image potential is of course given by $-e^2/2\varepsilon a$, where ε is the macroscopic dielectric constant of the semiconductor.

On the other hand, the image force effect on a hole in the valence band of the semiconductor is also an attractive potential: that is, removing one electron from the valence band leaves behind a *positive* charge, which attracts an equal *negative* charge to the surface of the metal, giving

$$V_{hole} = - \frac{e^2}{2\varepsilon a}.$$

But an *attraction* for holes is equivalent to a *repulsive* force for electrons: in effect, the valence band *rises* as we approach the metal, and the gap is given by

$$(E_g)_{a \gg a_0} = E_g(\infty) - \frac{e^2}{\varepsilon a}.$$

With the dielectric constant of Si, 12, the image force term reaches the energy gap of about 1 eV at 1.2 Å from the surface, and the gap is reduced by half at 2.4 Å, which is certainly a distance at which macroscopic electrostatics should be valid.

Now you can see the reason why we went off into many-body theory. This very peculiar effect which is *opposite* in sign for valence and conduction bands does not fit into any one-electron theory — it simply cannot be handled as an ordinary potential, or even as an exchange effect, since the electron 2.5 Å from the metal certainly is not exchanging with any of the metal electrons. So in order to keep this simple-sounding and very big effect in the problem we have got to go to a full many-body treatment of the interactions of the electrons.

In order to keep in the many-body effects, we have to hope that it is not too much of a simplification to use relatively schematic models for the metal and the semiconductor. It is probably quite accurate to treat the metal as a free electron gas with a plane surface — as far as low-energy and relatively long-range effects are concerned, that should be a good approximation. We shall be considering metals with screening lengths of order $\frac{1}{2}$ Å, and that should be some kind of indication of the thickness of the metal surface.

For the semiconductor we shall also take a homogeneous model — the so-called Penn model [9], actually due to Phillips and Cohen, which is a good approximation to covalent semiconductors and a rough one to most others. The idea of the Penn model is to ignore the detailed crystal structure and treat the Jones zone at which the energy gap appears as a sphere with a uniform gap across it. The gap is caused not by simply displacing the free electron states by a fixed amount but by mixing the states k and $k_F - k$ by a potential V_g, just as though it were a true gap due to scattering by a periodic potential. We assume that this gappy free electron gas also has a plane surface at the interface. I will shortly give reasons why this ignoring of the periodic structure of the two substances is a pretty good approximation.

What we now wish to do is to study the propagation of the electron in the presence of these two substances and this interface. To do so we must calculate the *self-energy* of the electron,

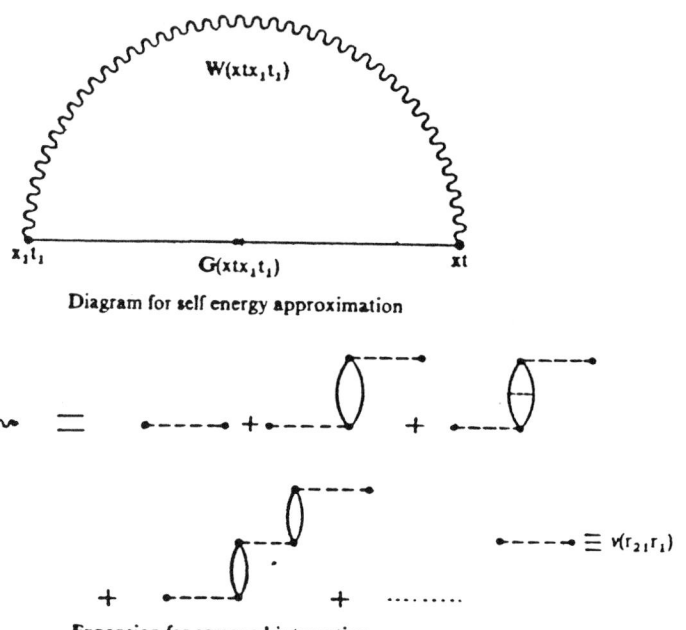

Diagram for self energy approximation

Expansion for screened interaction

Figure 4

i.e. we must learn to deal with the exchange and correlation energy. It has already been pointed out by Bennett and Duke [10] that a great fraction of the work function of most metals is caused by the attraction of the electron to its exchange-correlation hole, of which, again, the image potential is the longest-range piece. Every electron in a metal is accompanied — because of the exclusion principle — by a positive hole of exactly the same charge, called the exchange-correlation hole. When the electron is outside the metal, its hole is the image charge: as it merges into the metal, the image charge becomes its exchange hole, and the attraction to this hole is the sum of its exchange and correlation energies. Our problem is to see how this process goes when the vacuum is replaced by a semiconductor.

What we cannot at first do is to draw the kind of diagram we have been drawing so far, in which we simply draw out the band structure as a function of position. That is because in many-body theory the energy of a particle is not a local operator, as it is in one-electron quantum mechanics. In one-electron theory, we write

$$\mathcal{H}\Psi = \left[\frac{p^2}{2m} + V(r,t)\right]\Psi = i\hbar \frac{\partial \Psi}{\partial t}$$
$$= E\Psi,$$

but as you no doubt know, in Hartree-Fock theory the exchange part of the potential becomes nonlocal:

$$V(\vec{r})\Psi(\vec{r}) \rightarrow \Psi(r)\int \rho(\vec{r}')V(\vec{r} - \vec{r}')d^3r'$$

$$- \int d^3r' \sum_{\substack{occupied \\ states\ n}} \phi_n^*(r)\phi_n(r')V(\vec{r} - \vec{r}')\Psi(r'),$$

the last term being a nonlocal exchange operator which relates two different points $\vec{r}$ and $\vec{r}'$.

The full generalization of this concept is the *self-energy* operator, which summarizes all of the complicated things which can happen to an electron. In general, the self-energy is nonlocal in time as well as space. It represents the sum of all the possible scattering processes which can take an electron from point r,t to point r',t': in diagrams (which you do not need to understand thoroughly in order to follow what I am going to say),

where ———→ is an electron propagator $G_0(\vec{r},\vec{r}',t - t') = i\langle\Psi(\vec{r}t)\Psi\dagger(\vec{r}'t')\rangle_0$ and - - - - is the interaction $V(\vec{r} - \vec{r}')$. Σ is the correction to the energy due to the interactions; we express this by doing the diagram sum $G = G_0 + G_0\Sigma G_0 + G_0\Sigma G_0\Sigma G_0 + \ldots$ which represents the 'Dyson equation' where G_0 is the propagator in the absence of any interactions

$$\left(E - \frac{p^2}{2m}\right) G_0 = 1,$$

$$G_0 = (E - \mathcal{H}_0)^{-1} \quad \text{and} \quad G = (E - \mathcal{H}_0 - \Sigma)^{-1}.$$

We will not by any means use the full complication of this kind of formalism; what we will use is called, in the case of uniform systems, the *screened interaction approximation* of Quinn and Ferrell [11], and Hedin and Lundqvist [12]. This approximation may be described diagrammatically as *neglecting all 'vertex corrections'*. It is basically just a Hartree-Fock theory in which we use not the Coulomb interaction $V = e^2/r$ but the fully dynamically screened interaction W, which in a uniform system would be the Fourier transform of $V(q,\omega)/\varepsilon(q,\omega) = 4\pi e^2/q^2 \varepsilon(q,\omega)$. The errors made in this kind of theory are harmless for our purposes; the short-range parts of the correlation energy are slightly overestimated. What is neglected are repeated strong scatterings: paramagnons, hard cores, and the like. The RPA, as well as Migdal's theory of electron-phonon interactions, are included within this approximation. The content of the approximation is to set

$$\Sigma = G \cdot W,$$

where W is the fully screened interaction. Thus once we understand how our semiconductor-metal junction acts to screen the Coulomb interaction, we will easily be able to work out the self-energy of an electron in the vicinity of the junction.

Again following rather similarly to what Hedin and Lundqvist chose to do, Inkson decided that any possible simplification for such a complicated problem was more than justified. Thus he chose to use certain analytic approximations for the two dielectric constants $\varepsilon_{metal}(q,\omega)$ and $\varepsilon_{sc}(q,\omega)$ which determine the screened interaction. What he uses for the metal already has a distinguished history; it was used by Lundqvist as an approximation to the Lindhard dielectric function of the free energy gas.

$$\varepsilon_M = 1 + \frac{k_s^2}{q^2 - \frac{\omega^2}{\omega_p^2} k_s^2},$$

where

$$k_s^2 = \left(\frac{6\pi n e^2}{E_F}\right) \qquad \omega_p^2 = \frac{4\pi n e^2}{m}$$

are the Thomas-Fermi screening length and plasma frequency. As you see, it behaves correctly both for q and $\omega \to 0$:

$$\varepsilon_M(q = 0) = 1 - \frac{\omega_p^2}{\omega^2},$$

$$\varepsilon_M(\omega = 0) = 1 + \frac{k_s^2}{k^2},$$

and is chosen on this basis plus two simplifying conditions: (1) the only pole of $1/\varepsilon$ is at $\omega_p^2(q)$ where

$$\omega_p(q) = \omega_p \left(1 + \frac{q^2}{k_s^2}\right)^{\frac{1}{2}},$$

and (2) this pole satisfies the sum rule.

It is interesting to contrast the behavior of ε — which determines optical properties, for instance — with that of $1/\varepsilon$, which determines the interaction and the self-energy. The former has its poles at a median free-electron excitation energy $\omega_e = q v_F/\sqrt{3}$, the latter has these poles completely suppressed and one sees only the plasma frequency, as in the electron energy losses. We will see the same phenomenon in the semiconductor, and it is a very important effect.

This dielectric constant is not new, but the one John invented to fit the semiconductor is, and I think has a lot of very useful and transparent properties. He uses,

$$\varepsilon_s(q,\omega) = 1 + \frac{\varepsilon_0 - 1}{1 + \frac{q^2}{\gamma^2}\varepsilon_0 - \left(\frac{\omega^2}{\omega_R^2}\right)\varepsilon_0}.$$

Here ε_0 is the observed electronic dielectric constant, given in terms of the 'average (or isotropic) gap' ω_g of the Penn model by

$$\varepsilon_0 = 1 + \frac{\omega_p^2}{\omega_g^2}.$$

The other two parameters are

$$\omega_R^2 = \omega_p^2 \left(\frac{\varepsilon_0}{\varepsilon_0 - 1}\right) = \omega_g^2 + \omega_p^2,$$

$$\left(\omega_p^2 = \frac{4\pi n e^2}{m}, \quad \text{again}\right),$$

and

$$\gamma^2 = \left(\frac{\epsilon_0}{\epsilon_0 - 1}\right) k_s^2.$$

Again we have a dichotomy between the poles of ϵ, which occur at the 'excitonic' frequency

$$\omega_e^2 = \omega_g^2 + \frac{q^2 v_F^2}{3},$$

and the plasmon poles of $1/\epsilon$, at

$$\omega_p^2(q) = \omega_R^2 + \frac{q^2 v_F^2}{3}.$$

ω_R^2 in fact, in the case of Si, is a slightly better fit to the observed bulk plasma frequency than is ω_p^2.

I have a couple of figures showing how remarkably good this Ansatz for the dielectric constant is (see figures 5,6), both in terms of $\epsilon(0,\omega)$, which can be measured optically, and of $\epsilon(q,0)$ which can be calculated from the Penn model. I think this dielectric constant is one of the nice achievements of Inkson's thesis; it makes a lot of results in semiconductor physics rather transparent - such as the plasma frequency. He has recently been extending this to study the nonuniform parts of the dielectric constant — that is, the off-diagonal matrix elements in reciprocal space, $\epsilon_{q \to q+G}$ where G is a reciprocal lattice vector. These tell us how nonuniform the dielectric response is because of the pileup of electronic charge in the covalent bonds. It is part of the Phillips mythology of dielectric effects that the 'bond charge' is given by the missing screening of the atomic charge, i.e.

$$\text{bond charge} = \frac{4}{\epsilon_0} \div 2 \text{ bonds/atom} = \frac{2}{\epsilon_0}.$$

This myth is in fact supported by some arguments due to Pick, Cohen and Martin [13] based on sum rules connecting the diagonal and off-diagonal parts of the dielectric constant; it appears that $1/\epsilon_0$ is in fact a good measure of the relative importance of nonuniform terms. Thus in the good covalent semiconductors like Si and Ge with ϵ_0 of 12 or 16 these terms are indeed less than 10% of the uniform ones, which justifies, to some extent, our rough continuum model.

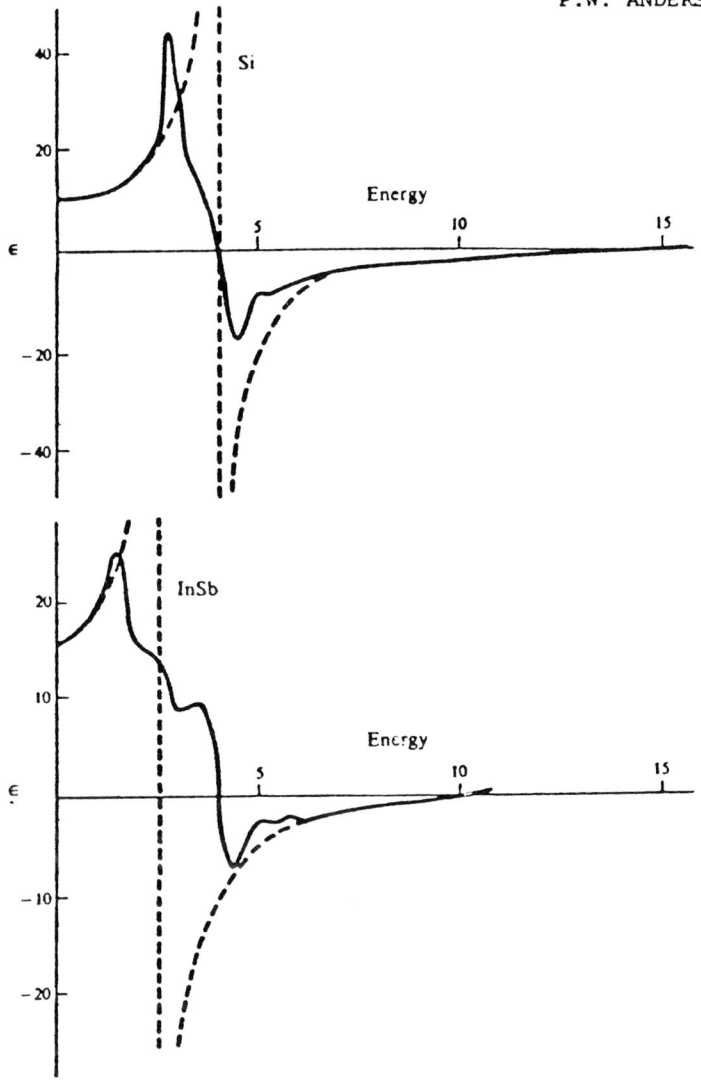

Comparison of model and experimental dielectric functions

Figure 5

It is these simple, and yet at the same time very realistic, approximations to the dielectric functions which make it possible to get anywhere with the problem at all. Already with these dielectric functions it is possible to get good estimates of something which was not previously available in the literature — surface plasmons for semiconductor surfaces.

The long-wavelength, observable surface plasmons follow directly from a simple argument due to Stern [14]. We want to know at what frequency a charge fluctuation of infinite wavelength at the surface will give an infinite E-field, i.e. we can

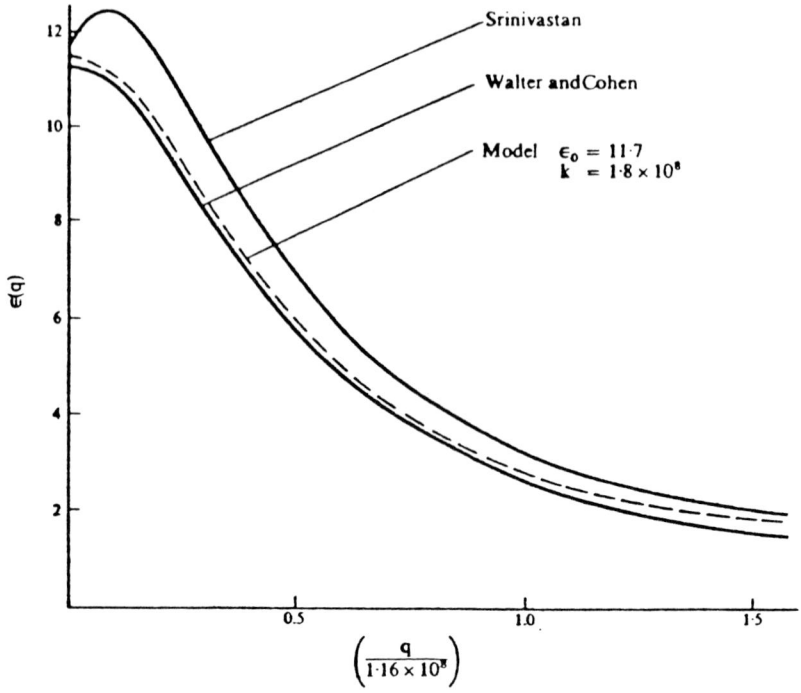

Theoretical static dielectric function
Figure 6

have a spontaneous oscillation of the E-field. We have two media of dielectric constants ϵ_1 and ϵ_2,

$$D_1 = \epsilon_1 E_1, \qquad D_2 = \epsilon_2 E_2.$$

From Poisson's equation

$$\nabla \cdot D = 4\pi\rho$$

we get

$$\epsilon_1 E_1 - \epsilon_2 E_2 = 4\pi\rho_{surface}$$

and thus if $E_1 = -E_2 = E$ (no external sources for the fields, or, equally, V constant at ∞),

$$(\epsilon_1 + \epsilon_2)E = 4\pi\rho_{surface}.$$

Thus

$$\epsilon_1 + \epsilon_2 = 0$$

gives us a finite E for zero ρ.

This gives us the famous $\sqrt{2}$ factor for a metal-vacuum surface:

$$1 - \frac{\omega_p^2}{\omega_s^2} + 1 = 0,$$

$$\omega_s = \frac{\omega_p}{\sqrt{2}}.$$

For the semiconductor-vacuum interface, using the Inkson dielectric constant we get

$$\omega_s = \omega_R \left(\frac{\epsilon_0 + 1}{2\epsilon_0}\right).$$

This does not fit Rowe's observations any better than a simple factor $\sqrt{2}$, but cannot be distinguished from it.

In the metal-semiconductor interface a new and interesting phenomenon occurs: there are *two* surface plasmons, not one; one gets pushed below the gap frequency ω_g, the other up above the metal and semiconductor plasmon frequencies. The physics behind these is not complicated — the lower one can just be estimated from the formula

$$\omega_{M-s} = \frac{\omega_p}{(1 + \epsilon)^{\frac{1}{2}}},$$

where ϵ, the semiconductor dielectric constant, is somewhat greater than ϵ_0 in this range. Since ω_R and ω_p are about the same, and $\epsilon_0 = \omega_p^2/\omega_g^2$, ω_{M-s} is about ½ to 2/3 the gap at long wavelengths.

To do the dispersion of the surface plasmons it is not adequate to apply the Stern-Ferrell relationship, since we cannot guess that the field is uniform in the z direction, and in fact it presumably decays exponentially. This result, however, is one which falls naturally out of our central problem; the calculation of the potential of interaction of electrons in the surface region - i.e. we put a charge at a distance a from the surface and ask for the potential $\Psi(\rho,z)$ at a point z from the surface with the radial coordinate ρ. (We will always work in frequency space rather than time; V and Ψ are Fourier components with frequency ω).

Since this calculation is central to all the further work, I will do it out, at least in the notes, in a bit of detail. Of course it is important to the self-energy problem, because our approximation is just to take the self-energy as that of the appropriate charge distribution in the given medium - basically, the straightforward generalization of the image potential problem. So what we need most to know is $\Psi(\rho \to 0, z \to a)$ which gives us the self-potential. Second, the *poles* of this interaction in frequency space are the plasma frequencies — they give a potential response in the absence of charge. In general, they will

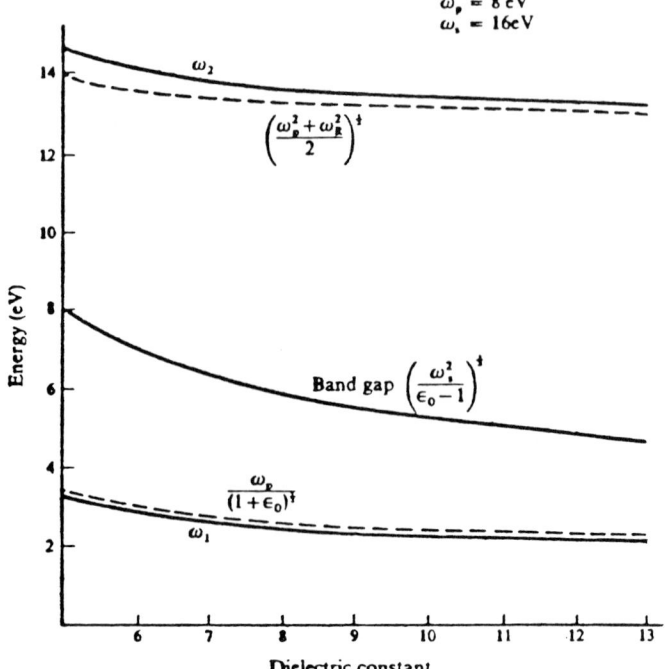

Long wavelength surface plasmon energies

Figure 7

be complex — decaying — but usually only slightly so. These poles will form a continuum for a point charge, but if we Fourier transform in the transverse direction there will be one or two discrete ones for each transverse momentum m. It is convenient to use circular coordinates in this space and thus we transform using Bessel rather than Fourier series,

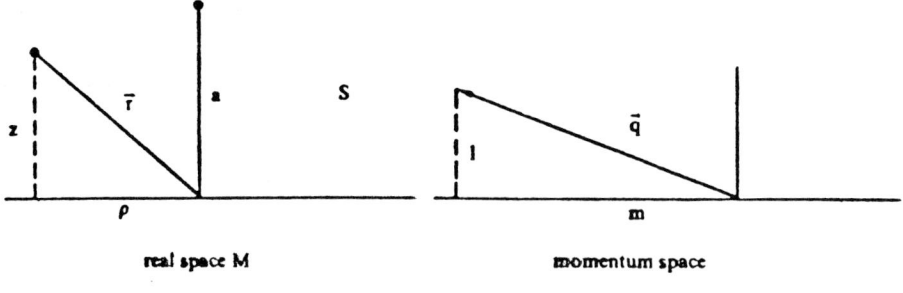

real space M   momentum space

$$V(\rho,z) = \frac{1}{(2\pi)^2} \int m\,dm\; V(m,z) J_0(m\rho)$$

and the problem separates in the transverse momentum m (there is no angular dependence, of course).

The solution is carried out using the continuity conditions at the surface,

$$\Psi_1(\rho, z = 0) = \Psi_2(\rho, z = 0)$$

and

$$\frac{\partial \Psi_1}{\partial z} = \frac{\partial \Psi_2}{\partial z},$$

and a general method of images. Within either substance the potential obeys

$$\nabla^2 \Psi(q,\omega) = \frac{4\pi\rho}{\epsilon(q,\omega)}.$$

The only charge which is *inside* a region is the charge source $a, 0$, and so a part of the potential obeys

$$V = \frac{4\pi e}{(2\pi)^2} \int \frac{m d\ell dm}{\ell^2 + m^2 \epsilon_1((\ell^2 + m^2)^{\frac{1}{2}}, \omega)} e^{-i(z-a)\ell} J_0(m,\rho).$$

To satisfy the boundary conditions we may add to this the potential of a charge distribution which is wholly outside region 1, but computed as though region 1 filled all space; this might as well be on the surface

$$\Psi = V + \Phi,$$

$$\Phi = \frac{4\pi e}{(2\pi)^2} \int \frac{m d\ell dm}{(\ell^2 + m^2)\epsilon_1} f_1(m,\omega) e^{-i\ell z} J_0(m,\rho),$$

where f_1 is the charge distribution. Correspondingly, the potential in region 2 can be taken to be the potential of some arbitrary surface charge distribution f_2. f_1 and f_2 are functions of m which are related by the two continuity conditions, which thus determine them completely. We get two equations of the form

$$\alpha + \beta f_1 = \gamma f_2,$$

$$\Delta - \nu f_1 = \mu f_2,$$

where α, β, γ, etc. are integrals of the type

$$\int \frac{d\ell e^{ia\ell}}{\epsilon_{1,2}(\ell^2 + m^2)} \quad \text{or} \quad \int \frac{\ell d\ell e^{ia\ell}}{\epsilon_{1,2}(\ell^2 + m^2)}.$$

These integrals would be hopeless with real dielectric functions

MANY-BODY EFFECTS AT SURFACES   19

but with the simple analytic approximation we have chosen they
are not very difficult by contour integration; there are two
contributions, one coming from the pole at $\ell = im$ and the other
from the zero of the dielectric constant at

$$m^2 + \ell^2 = q^2 = k_s^2 \left[\frac{\omega^2}{\omega_p^2} - 1 \right].$$

But of course the resulting equations for f are complicated and
while one can indeed make a great deal of sense of the result
physically, it is so complicated as not to be worth our writing
it down.

One thing which *is* worth noting is that in general f_1 and f_2
represent the response of the surface to the charge. Thus the
point at which they blow up is clearly a surface plasma, and it
is to be obtained by the singular case for solution of the linear equations for f_1 and f_2:

surface
plasmon:
$$\begin{vmatrix} \beta(m,\omega) & -\gamma \\ \nu & \mu \end{vmatrix} = 0.$$

The integrals are in this case quite simple and John has worked
out the two cases. In the metal-vacuum case,

$$\omega(m) = \frac{\omega_p}{2k_s} (m + (m^2 + 2k_s^2)^{\frac{1}{2}}),$$

which is a bit surprising in that it gives a linear dispersion
law, but perfectly permissible.

For the metal-semiconductor, the solutions had to be obtained numerically and are given in the figures (see figure 8).
We note that the dispersion is quite complicated. One noteworthy feature which we have *not* analyzed, but would like to
think further about, is whether there is any appreciable electron-electron resonant attractive coupling via the low-frequency
surface plasmon. The strength of this pole in the interaction
is rather low, and as you can see its dispersion is steeply
varying, but one suspects it is far more strongly coupled to the
electrons than any so-called 'excitonic' excitation.

So this completes one of the results which is my excuse for
entering a school on 'elementary excitations' — surface plasmons.
The second result is what we will now take up — how elementary
electron and hole excitations behave near the surface. Here we
keep in mind the surface state problem; although actually several other applications of these ideas have been made or suggested — one very important one, for instance, being the calculation of potentials and energy losses for medium-energy electrons
impinging on a metal or semiconductor surface, which is a key
question for LEED.

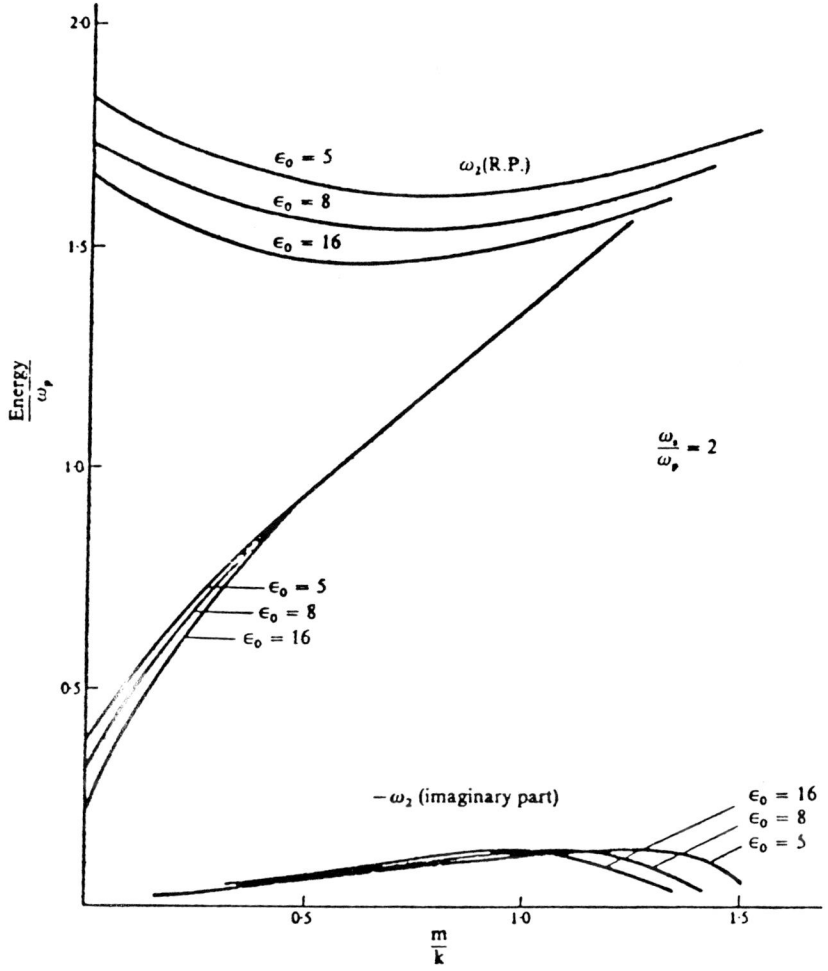

Surface plasmon dispersion relation (dielectric constant varying)

Figure 8

Next I would like to bring out a very over-simplified approximation which nonetheless does bring out a great deal of the physics we are after. This is what Inkson calls a 'Thomas-Fermi' approach but it is actually even oversimplified relative to that: we could call it linearized Thomas-Fermi. The basic approximation is to neglect the frequency-dependence of ϵ entirely, and just to treat the two media as dispersive classical state dielectrics, and the electron and hole as static point charges. In other words, what if we generalize the image potential idea by taking the short-range dielectric behavior of metal and semiconductor correctly into account?

This had already been done by Newns [3] for the metal-vacuum interface, where one can use the static metal dielectric constant

$$\epsilon = 1 + \frac{k_s^2}{q^2},$$

or equally the equivalent linearized Thomas-Fermi equation

$$\left(\nabla^2 - k_s^2\right)\phi(r) = 0,$$

can be integrated to give the potential in the metal. The same thing can be done in the semiconductor for our dielectric constant

$$\epsilon = 1 + \frac{\epsilon_0 - 1}{1 + \frac{q^2 \epsilon_0}{\gamma^2}}.$$

If we introduce the vacuum potential U which satisfies $\nabla^2 U = 4\pi\rho$, the real potential obeys

$$(\nabla^2 - \gamma^2)\phi = -\frac{\gamma^2}{\epsilon_0} U(r).$$

By using this differential equation, or just returning to our image techniques of before, one may arrive at the image potential for a classical point charge approaching the metal, which is shown in the figure 9.
One sees that it is very heavily dependent on the ratio of the screening constant γ in the semiconductor to that (k_s) in the metal. When the *short-range* screening in the semiconductor is better, the image potential reverses near to the metal and can even become positive. That is, where the electron density — at least where the electrons are — is higher in the semiconductor, it will in fact have a higher exchange-correlation energy and thus the image attraction will be reversed near the metal. In general, this *is* the direction of the trend which is seen; and while this is not the whole story — in particular (a) the potential is nonlocal, in general, where this is local; (b) the hole is not a classical point charge, by any means — I think this reversal gives one a clear understanding of why so many semiconductors do *not* lose their band gaps near a metal.

While this picture, if oversimplified, gives a good account, it is enlightening to go on and try to achieve some degree of realism in the same calculation. This is unfortunately the point at which one must really leave the world of the living for the complex plane, but I think one can discuss the physics, if not the math, with some realism.

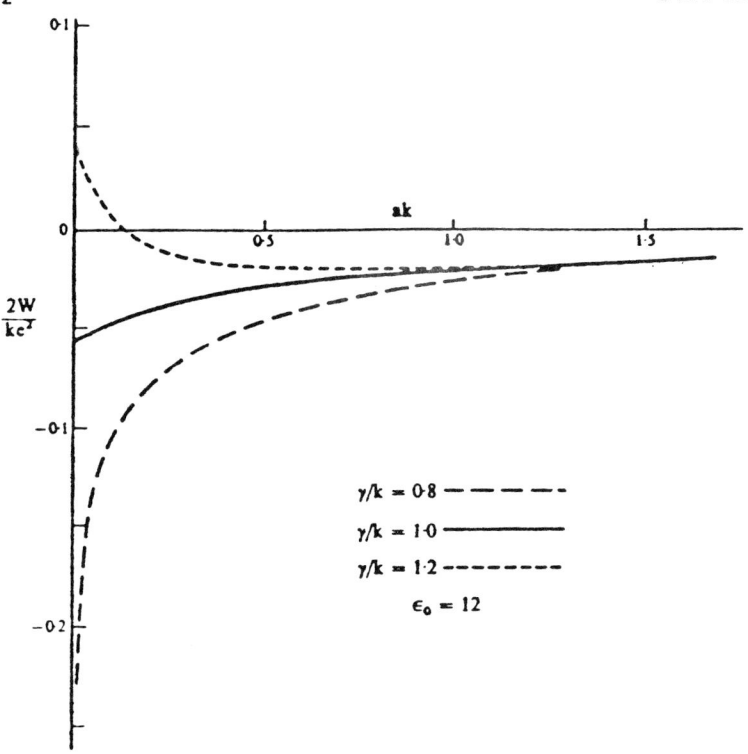

Variation of image potential with distance

Figure 9

The self-energy in space-time terms is the product

$$\Sigma(x,x',t-t') = G(x,x',t-t')W(x,x',t-t')$$

of G, the amplitude for the electron to get from x to x', and W, the interaction. In Fourier space then it is a convolution integral

$$\Sigma(x,k,\omega) = \frac{i}{(2\pi)^4}\int W(x,q,\omega)G(x,k-q,\omega-\omega')e^{-i\delta\omega'}dqd\omega',$$

where we have Fourier transformed in time and in $x - x'$ space. One has left over the x-dependence, which is necessary here because we do not have a homogeneous problem. This divides naturally into two parts, which we can describe as the 'correlation' and the 'screened exchange'. If W were an ordinary time-dependent potential, the exchange term would be the only contribution to the ω' contour integration, and would arise from the poles of G at $\omega - \omega' = E_n$, the occupied state energies; so the screened exchange term is taken to be just that term from the poles of G. There is also a 'correlation' term, which comes from the poles

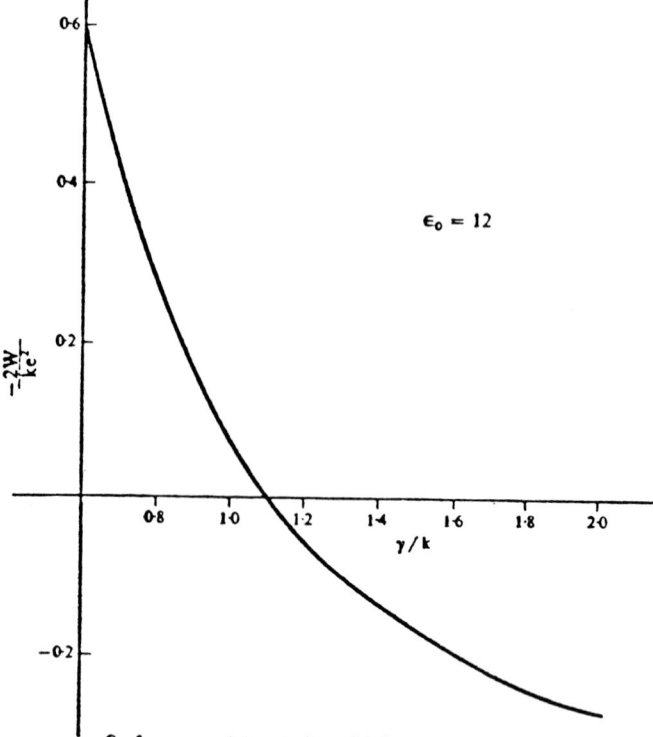

Surface potential variation with Thomas-Fermi wavevector

Figure 10

of W: the various bulk and surface plasmons we have been discussing. Of course there is correlation energy, as normally defined, in each term.

The model for G which is used is the kind of two-band model we have already described — incidentally, one of the surest approximations we can make is to ignore the effect of the boundary on G, and use the unperturbed semiconductor G;

$$G(x,x',\omega) = \sum_n \frac{\phi_n(x)\phi_n^*(x')}{\omega - E_n + i\delta\,\mathrm{sgn}(E_n - \mu)}$$

is its definiton. In the two-band model

$$\phi_{\pm k}(x) = \frac{1}{(1 + \alpha^2(k))^{\frac{1}{2}}} (e^{ik\cdot x} + \alpha e^{i(k-G)x}),$$

$$E_\pm(k) = \tfrac{1}{2}[E^0(k) + E^0(k - G) \pm (E_g^2 + (\Delta E^0)^2)^{\frac{1}{2}}],$$

$$\alpha^\pm = \tfrac{1}{2} \frac{E_g}{E^\pm - E^0} .$$

This gives

$$G(x,p,\omega) = \frac{1}{(1 + \alpha^2)^{\frac{1}{2}}} \left\{ \frac{\phi_{-p}(x)}{\omega - E_-(p) + i\delta} - \frac{\alpha\dagger(p)\phi_{-p}(x)}{\omega - E_+(p) - i\delta} \right\}.$$

Only the first term has a pole on the right side of the axis — corresponding to occupied states — to contribute to the integral in Σ.

Given G, then, and given W, we can calculate the self-energy of the electron near the surface. Before doing a final realistic calculation it was useful to go to the long-range limit $a, z \gg 1$. This was, for quite a while, a stumbling block for us because we were not able to find the physical image potential effect which we knew must be there. There was no way to calculate the *correlation* term, as we understood it, which did not give almost exactly the same answer for valence and conduction bands. For these terms the main difference between valence and conduction bands is the value of ω — positive for one, negative for the other. Only for the low-frequency surface plasmon term could this possibly make a difference, and for that term there is no mixing of the two terms involving k and $k - G$ and therefore no effect. So the 'correlation' term is practically identical for the two bands, and in fact follows our main 'Thomas-Fermi' calculation to a good approximation. Thus at long distances it goes simply to $- e^2/2a\varepsilon_0$: the classical image potential *for an electron* — no dichotomy of the two bands.

The screened exchange term, on the other hand, behaves quite differently. At long range this term is actually zero for the conduction band, and goes as $+ e^2/a\varepsilon_0$ for the valence band, thus giving us back our original semiclassical result. There is a good physical reason for this. The conduction band wave-function is orthogonal to all of the occupied states, so insofar as $W(x,x') = $ constant, when $x \to \infty$,

$$\int dx' W(x,x') \sum_{occ} \phi_n(x')\phi_{cond}(x') \approx 0.$$

On the other hand, the valence band wave functions do *not* have this orthogonality property and will see a full exchange term.

There is a very perspicuous way of seeing that this difference in screened exchange exists, which also brings us back to a rather fundamental point of semiconductor physics. First we note the fundamental Wannier theorem on full bands: that the wave functions of a filled valence band with a true band gap may always be written as linear combinations of exponentially localized Wannier functions:

$$\phi_k(x) = \sum_n W_n(x) e^{ik \cdot R_n}.$$

For a multiple set of bands as in a covalent semiconductor there may be several W_n's per atom — in that case actually the best set are of the symmetry of sp^3 hybrid bond orbitals.

The second remark is that the exchange energy of an arbitrary wave-function is the sum of the *Coulomb self-energies of its overlap charges with all occupied* wave-functions. That is,

$$\sum_{ex}(\phi) = \sum_{\substack{n \\ occ}} \int dx \int dx' \phi(x)\phi_n^*(x)V(x - x')\dot{\phi}^*(x')\phi_n(x').$$

So we can define the *overlap charge*

$$\rho(\phi,\phi_n) = \phi(x)\phi_n(x).$$

The fact that one can always describe a filled band in terms of a Wannier representation means that it is equally true (for static V, at least) that the exchange self-energy is the sum of the self-energies of the overlap charges with the Wannier functions. For a bulk crystal all *these* terms are the same and we get

$$V_{ex}(\phi) = W \int dx \int dx' \phi(x)W_n(x)V(x - x')\phi(x')W_n(x').$$

Here we see a sharp dichotomy between conduction and valence bonds: namely

$$\int \phi(x')W_n(x')dx' = \frac{1}{\sqrt{N}} \quad \text{(valence band)}$$

$$= 0 \quad \text{(conduction band)}$$

i.e. the conduction band wave functions have no *net* overlap charge. What our difference in image effect is, then, becomes obvious: the exchange charge in the valence band induces an image and is therefore *screened* by the presence of the metal, leading to a net reduction of the exchange energy, which overcompensates the ordinary image effect. In the conduction band, on the other hand, there is no net overlap (or exchange) charge and the metal screening has no great effect.

This point relates at a number of places to physical facts. Firstly, it tells us a limitation on the effectiveness of the band-gap closure idea. It is a very reasonable one: simply, we cannot treat the image force on a more local basis than the size of a Wannier function. For average semiconductors this is not a severe limitation; while the exponential tails may extend out to neighboring atoms and bands, the bulk of the Wannier function can be rather sharply localized on its bond site. Thus I cannot see at all why this exchange screening difference cannot be big enough to close the gap entirely, and in fact as shown in the figures 11-14 actual calculations with the two-band model do show band closure in the suitable cases.

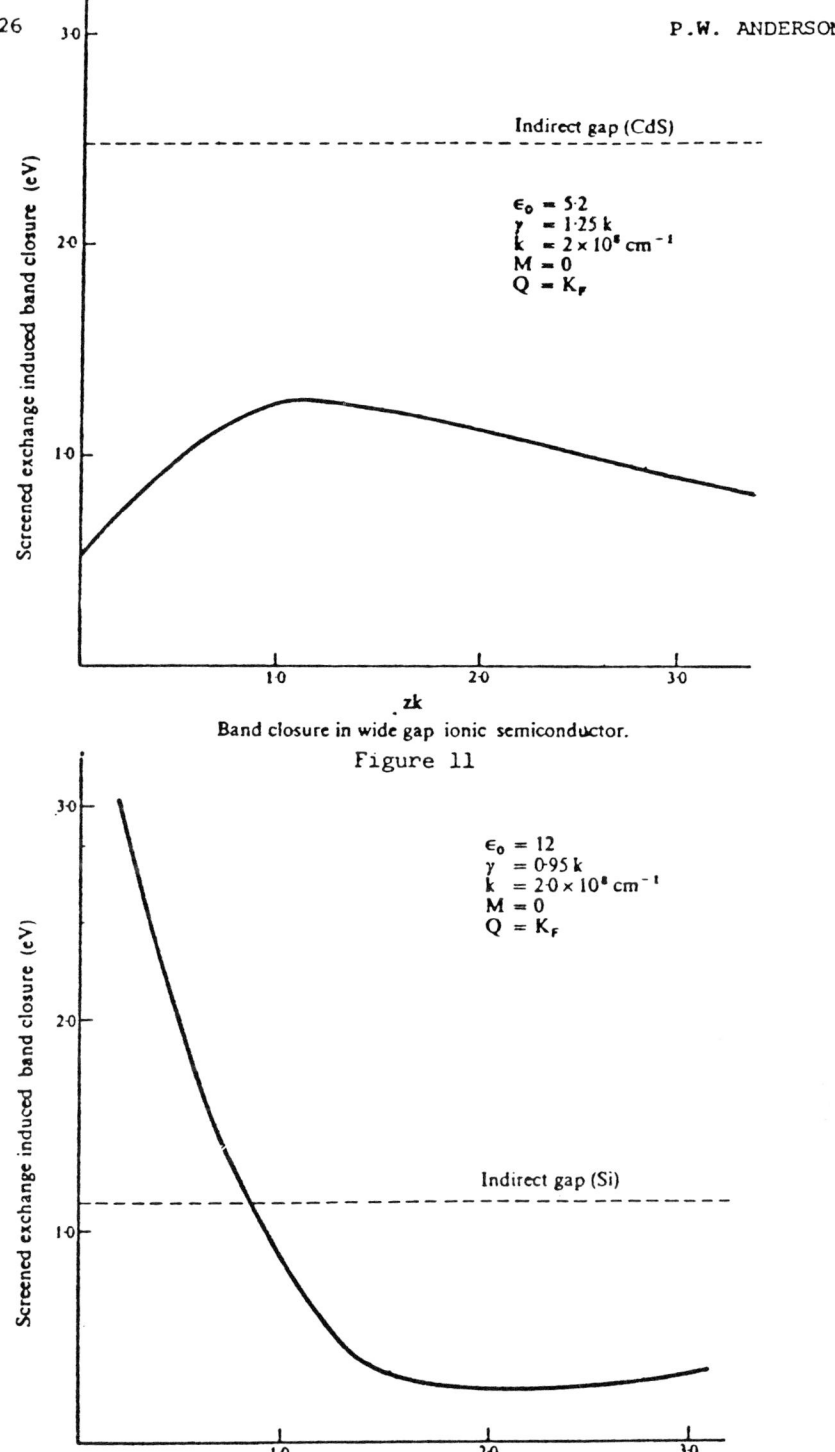

Band closure in wide gap ionic semiconductor.
Figure 11

Band closure in narrow gap covalent semiconductor
Figure 12

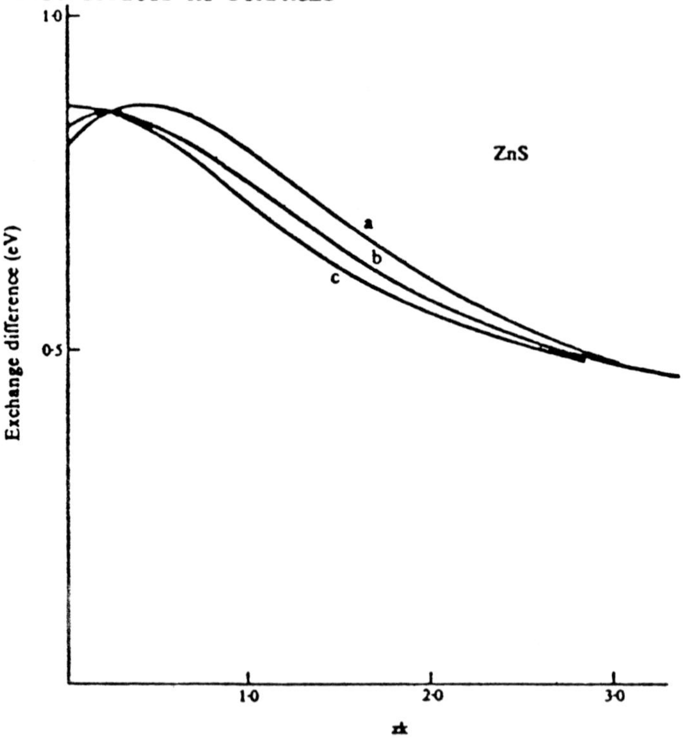

MANY-BODY EFFECTS AT SURFACES

Figure 13

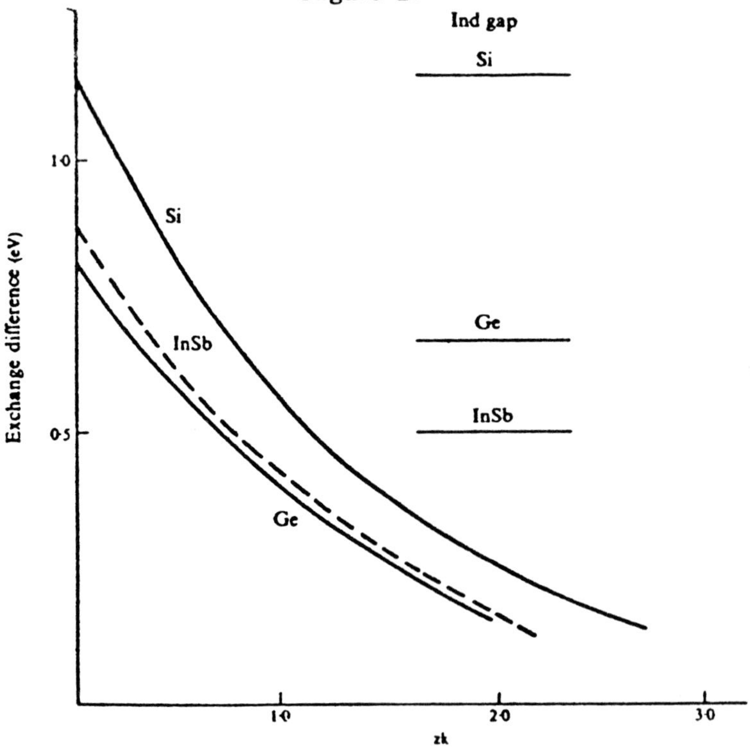

Figure 14

Second is a tie-in to a rather old but still inadequately understood bit of physics: the Mott or Mott-Hubbard gap and its relation to the energy gap in true semiconductors. One is taught — at least in reasonably good solid-state physics courses — that there are two kinds of insulators: simple band theory ones, where the energy gap is entirely caused by the strong pseudo-potential leading to large gaps at the zone boundaries; and the Mott type, where it is caused by the Coulomb correlation effect in a relatively dilute material. This simple distinction is here seen to break down — in fact, it was to my knowledge E.O. Kane who first pointed out that the existence of a true gap in most semiconductors is a consequence of large exchange differences. In fact, the difference in exchange and correlation energy between valence and conduction bands *is* just the Mott gap — he described it as the extra energy necessary to separate electron and hole against their Coulomb attraction, which is equal to the Coulomb self-energy of a valence band Wannier function, i.e. $e^2/\varepsilon a_0$ where a_0 is the Wannier function mean radius.

Thus, while in general the total or 'average' gap V_g in semiconductors is, mostly, a true potential effect, contributions of order 1-2 volts (as calculated by Inkson) are due to the exchange effect; in most cases these are at least as big as the indirect or minimum gap, and are thus responsible for it.

To summarize this last bit of argument: the energy gap in covalent semiconductors has a contribution of order 1-2 eV from the screened exchange term, which may be interpreted as the Mott gap, i.e. the energy necessary to remove an electron from a Wannier function to ∞. This contribution is screened by the presence of a metal surface, and thus the gap may close near the surface. This will have the effect of pinning the Fermi surface to the center of the band gap, which will simulate the effect of surface states. In the less covalent semiconductors, on the other hand, the Mott contribution to the gap is less than the minimum gap, and screening cannot close the gap near the surface, leading to a 'Schottky' type behavior.

If time permits, a brief qualitative description of Hamann and Appelbaum's theory of true surface states on a (111) surface will be given as well.

REFERENCES

1. Inkson, J.C. (1971). (Thesis), (Cambridge).
2. Inkson, J.C. (1971). *Surf. Sci.*, 28, 69; (1972). (Surface Plasmon), *J. Phys.*, C5, 2599; (1973). (Band Gap Reduction), *J. Phys.*, C.
3. Newns, D.M. (1970). *Phys. Rev.*, B1, 3304.
4. Bardeen, J. (1947). *Phys. Rev.*, 71, 717.
5. Shockley, W. (1939). *Phys. Rev.* 56, 317.
6. Tamm, I. (1932). *Phys. Z. Sowjetu.*, 1, 732.
7. Heine, V. (1965). *Phys. Rev.*, A138, 1689.
8. Kurtin. S., McGill, T.C. and Mead, C.A. (1969). *Phys. Rev. Lett.*, 22, 1433.

Uses of Solid-State Analogies in Elementary Particle Theory
Proc. of Conf. on Gauge Theories and Modern Field Theory,
ed. R. Arnowitt and P. Nath (MIT Press, Cambridge, MA, 1976), p. 311

This is a retrospective view of my slant on the origins of broken symmetry in particle and condensed matter theory; staking my claim to have first seen the necessity for the Goldstone boson as well as the Higgs.

USES OF SOLID STATE ANALOGIES IN ELEMENTARY PARTICLE THEORY

Philip W. Anderson*

The solid state background of some of the modern ideas of field theory is reviewed, and additional examples of model situations in solid state or many-body theory which may have relevance to fundamental theories of elementary particles are adduced.

*Bell Laboratories, Murray Hill, N.J. 07974.

As I have been trying during the past weeks to catch up with what you all have been doing in the past decade or more since last I took a detailed look, I find that more and more ideas from _my_ field are showing up in _your_ field. I thought it might perhaps be interesting for you to hear some discussion of the actual many-body theoretic context of these ideas, not only for purely historical reasons but in order to stimulate the drawing of yet more analogies in your field to the strange phenomena we solid staters now and then encounter in ours.

Another thing that I have found on the relatively rare occasions when I _do_ discuss these things with E. P. theorists is that very often ideas which seem to them to be absolutely trivial and obvious, and to go without saying, are to me the most esoteric and difficult results; but also of course almost everything I say to them which seems very easy and simple to me they find abstruse, and difficult or impossible to understand. This is by way of pointing out that you and I have a very difficult problem of communication here, but I do think we have things to say to each other, so I hope you will bear with me.

Uses of Solid State Analogies

I will first indulge myself in some ancient history about the many-body theorems which correspond to Goldstone and Higgs, trying to place these theorems which are so familiar to you into their original context. Next I'd like to continue in the same vein with some ideas about broken symmetry which are only just now appearing to be relevant to the E. P. field: phase transitions, and order parameter singularities. Finally, I'd like to use any time remaining to me to say a few things about how modern asymptotically free quark-gluon theories of strong interactions look to a many-body theorist, and why I think they behave like the Kondo effect, which may serve as rather a model of how such theories can behave.

Let me make the basic nature of the quantum many-body to E. P. analogy clear, if I can. So far as I know Nambu deserves the credit for this very fruitful idea. The ground state of a many-body quantum system is compared to the vacuum state of a field theory. The fact that many-body Hamiltonians very often have ground states which do not have the same symmetry as the Hamiltonian itself is the broken symmetry phenomenon; and many, if not most of the field theories you discuss here have this same feature in common. In general, you break the symmetry by introducing a field - usually scalar - which is taken not to have zero expectation value: this is the analog of what we call an "order parameter": a physical variable, defined locally, which describes the distortion of our system from its unperturbed ground state. In our theories symmetry-breaking is always "dynamical", at least in some sense. Such order parameters are usually, but not always, **not** constants of the motion - i.e. do not commute with the Hamiltonian; this is the typical milieu of the Goldstone-Higgs phenomenon.

The first conscious use of a Goldstone boson was in the problem of the ground state of an antiferromagnet in 1952.[1] Of course, the idea of phonons in solids as well as Landau's and Bogolyubov's theories of phonons in liquid helium already existed, but there was no concept of the role of these excitations in a consistent description of the ground state and of the quantum theory of broken symmetry systems.

The problem in the case of antiferromagnetism was the incompatibility of a theorem of Hulthen's,[2] that the ground state of the antiferromagnetic Heisenberg Hamiltonian

$$\sum_{ij} J_{ij} S_i \cdot S_j$$

had to be a singlet, with the obviously non-singlet character of the accepted Neel hypothesis for the ground state

$$= \pi_{A \text{ subl.}} \chi_\uparrow(i \text{ on } A) \quad \pi_{B \text{ subl.}} \chi_\downarrow(i \text{ on } B).$$

The solution of this problem of incompatibility was first suggested, at least to me, by Conyers Herring; that the true ground state had to be in some sense a symmetric linear

Uses of Solid State Analogies

combination of all the apparently orientationally degenerate Neel states pointing in different directions. But that this symmetry argument had dynamical consequences both for the energy level spectrum and for the long-wavelength fluctuations of the order parameter - that is, the sublattice magnetization - was, I believe, first discussed in ref. 1. The fact is that the broken symmetry requires a Goldstone boson, the antiferromagnetic spin wave, which in turn has a zero-point fluctuation which diverges in the long-wavelength limit in just such a way that the exact ground state is rotationally invariant. But infinitely close to the exact ground state, in the $N \to \infty$ limit, is an infinitely degenerate manifold of states which can be recombined to give the broken symmetry ground state. In '58, I thought the same phenomenon might be responsible for the gauge invariance difficulties of the BCS theory which preoccupied people at the time. But in the superconducting case, a new feature turned out to be there.(3) The long-range nature of the electromagnetic forces destroys the argument from continuity in Q which requires the existence of the Goldstone boson which would be there in a neutral B.C.S. system - in fact has now been demonstrated as 4th sound in the B.C.S. anisotropic superfluid He_3.

In fact, the Goldstone boson is raised to a finite frequency and joins the two components of the photon in a triad of excitations which, at least at $Q = 0$, mimic the 3 components of a massive vector boson. The three components are required by rotational symmetry: in the many-body system, where the order parameter and the underlying many-body systems carry conserved quantum numbers such as baryon number, mass and charge, the ground state is not even Galilean invariant, much less relativistic, and analogies to field-theoretic ideas of broken symmetry must be handled with great care. Nonetheless the Higgs mechanism is very nicely demonstrated, in the example of the charged Bose gas, or the equivalent charged B.C.S. pair condensation, where one makes up the charged scalar field which breaks the symmetry out of a Fermion pair field. All of these phenomena were described in a paper in 1962[4] which is usually not the one of mine quoted in this context, for reasons I don't understand.

In the course recently, of writing on the idea of broken symmetry, I had occasion to try to put together in my own mind all of the existing general ideas and rules about broken symmetry in the many-body sense, and it was of course then an obvious thing to do to ask which of these things might have significance in the field-theoretic context. It happens that two of them have actually appeared in the literature within the past year or two.

Long before any dynamical ideas about broken symmetry existed, Landau seems to have been the first to emphasize the importance of symmetry classification of phases of macroscopic systems.[5] Landau, in fact, is responsible for the first meaningful - if trivial - theorem of broken symmetry: that a symmetry element can never be discarded gradually, so that a symmetry change <u>always</u> requires a sharp thermodynamic phase transition. Usually - but not always - the higher symmetry phase allows more fluctuations and has therefore more entropy, and occurs at higher temperature, while the low symmetry phase occurs at low temperature. Realizing then that modern unified theories involve broken symmetry in the vacuum - ground-state, one is immediately tempted to ask whether in the original cosmic fireball of the big bang theory, the temperature could have been so high as to restore full symmetry and equate, for instance, the weak and electromagnetic interactions in a single global gauge symmetry. Kirshnitz and Linde,[5] and later and in more detail Weinberg[7] as well as Jackiw[8] have answered this question in the affirmative.

A much more recent development in the discussion of broken symmetry in the many-body and solid-state context is the idea of classification of singularities of the broken symmetry, which to my knowledge was initiated by deGennes only a few years ago. To describe what I mean by a singularity it is probably best simply to give examples, which I do in Table 1. In general, when a solid state type of system condenses into a broken symmetry state it has to satisfy a number of external constraints - it will always have boundaries, for instance, where there will be some constraint on the order parameter (symmetry-breaking field,

Table 1

Order Parameter Singularities in Broken Symmetry

"Dim" of Order Parameter	Broken Group	Example	Types of Singularities
1 (real scalar)	point (finite)	"Ising Model" anisotropic ferromagnet	domain boundary
2 complex scalar, 2-dim vector	continuous Abelian	superfluid, supercond., x-y model	" + line (vortex) $\oint d\varphi = 2\pi$
3 or more	non-Abelian continuous	crystal liquid xtal He$_3$	" + point : vacancy, Brinkman point in He$_3$ $\hat{n}$

't Hooft opole?

what else?

Uses of Solid State Analogies

that is) or it will simply initiate its broken symmetry differently in different places, which are incompatible when they meet; or weaker forces such as dipole forces in a ferromagnet will be left over and enforce certain conditions. As a result in general the order parameter is not uniform in space. If Weinberg's phase transition occurs such a variation may be expected. But this non-uniformity is not continuous but achieved in such a way that ψ has its usual bulk value almost everywhere, but the order is seriously disturbed in restricted regions of a certain fixed geometry: planes, lines or points. Examples are given in Table 1. The key intuition here is geometrical: one must ask how or whether it is possible to get <u>continuously</u> from one stable value of the order parameter to another. If the possible routes can be placed in one-to-one correspondence with one-dimensional paths (an Abelian group) lines are necessary - essentially the "strings" of usual monopole theories. But in the non-Abelian theories, the possible routes can correspond to a 3-dimensional continuum: point singularities are possible. Such a singularity is a vacancy in a crystal; others are certain point singularities of liquid crystals or helium 3 - the latter are sometimes called "Brinkman spheres". In each case there is a quantization condition equivalent to the idea that a vector order parameter around a point must span 4π of solid angle, or that a vacancy in a crystal must be a whole vacancy, not a piece of one. Often the larger singularity as well as the smaller remains: dislocations in crystals, vortex lines in He_3; but I suspect boundaries are not possible in the Weinberg type of theory. Such extended singularities would be very peculiar astrophysical objects.

There is a precise analogy of the point singularity to the 't Hooft magnetic monopole[9] - the "'t Hooft opole" - in a unified gauge theory. I hope no one will construe this comment as an endorsement of any experimental result, but it would be delightful if such objects existed.

In ordinary many-body theory "condensates" - the matter exhibiting the order parameter - often carry conserved currents: baryon number or mass for helium, charge for a superconductor, mass for a crystal, etc. This is the appropriate milieu for what I have called "generalized rigidity" - such behavior as rigidity, superfluidity, etc. - which clearly must _not_ be the case in a field theory context: your condensates must essentially carry only the quantum numbers of the vacuum.

In my final few minutes I'd like to make some abbreviated remarks about what one might call the "Theory of _Un_broken Symmetry". This is a type of theory which we have only just in the past few years been realizing exists in ordinary low-energy examples, and which I suspect is very much a model for the currently favored color gauge theories of strong interactions. One point of analogy is that the kind

Uses of Solid State Analogies

of logarithmic approach to asymptotic freedom which is characteristic of their renormalization group behavior is fairly common in ours: it occurs, among other cases, in the B.C.S. theory, in the Kondo effect, and in one-dimensional quantum systems. It signifies two things: first, that the high-energy behavior is going to be harmless; and second, that something spectacular will happen at the infrared end of things. I don't think elementary particle physicists have quite appreciated the strength of the first statement: that an asymptotically free theory is not only renormalizable, it can be instantly renormalized without further nonsense by the simple process of inventing a "model" - like the B.C.S. or Kondo "models", both of which are terribly simple Hamiltonians which in fact are merely models which correctly describe the infrared behavior of complex systems. The idea is to introduce a real physical cutoff and coupling constant, which are to a great extent arbitrary and can be chosen more or less at will, so long as they are satisfactorily into the asymptotically free regime. Asymptotic freedom from this point of view is merely the statement that nothing above such a cutoff has any relevance to low energy physics.

One way of understanding this point is to realize that there are actually two renormalization group schemes, and it seems to be a source of confusion to the quantum field theoretic community that these two schemes go by the same name, and to a great extent satisfy the same equations.

The "classical" RNG of Goldberger, etc. is a useful sequence of statements about renormalizable field theories which follow from the arbitrariness of our choice of scale in such theories. An entirely different scheme is the one which we low-energy people, especially Ken Wilson, and myself on a smaller scale, have been applying to our various phase transition problems. This is the kind of thing Wilson already mentioned - the Block Spin technique. In this system, the scaling equations are not statements about mathematical solutions of the same problem, they are descriptions of equivalence of the results of measurements on different physical systems. The Callan-Szymancik equation,

$$\frac{d_g}{(d(\ln\Lambda)} = Cg^3 + ..$$

in this description can be understood as connecting different physical problems with different high-energy cutoffs Λ, which need not necessarily be handled in a relativistically invariant way, but all of which correctly describe the same physics. It may not even need to have perfect gauge symmetry.

Asymptotic freedom in this context of course implies infrared troubles: necessarily, the perturbative scaling equation breaks down at low temperatures and the nature of the new physical problem is in doubt. In fact, we have in low-energy physics two types of exact solutions of such problems, each of which is interesting in the field-

Uses of Solid State Analogies

theoretic context. The first is the broken symmetry case of B.C.S. or many other phase transitions: the system has an infrared-<u>unstable</u> fixed point below which its behavior changes radically, and in particular it breaks its original symmetries: one makes the kind of sudden transition to a wholly different symmetry I talked about earlier. In this case the high energy RNG has little or no relevance to what really happens and is only a convenience: it may be much simpler, as in the B.C.S. case, to deal with the cutoff "model" problem than with the real one.

A second type of behavior has been found lately in two other types of systems.[10,11] In these - in both of which the precise character of the solutions is known through solution of a soluble case - there is no phase transition and no change from the aboriginal symmetry, which remains unbroken. In these systems it seems to be possible to squeeze down the high-energy cutoff to ordinary physical energies without changing the essential nature of the model - although it develops new parameters which affect numerical results. The essential nature of the system is that as $\Lambda \to 0$ $g \to \infty$: the scaling equations carry right through to a fixed point with infinite coupling constant.

In the Appendix I shall try to take you through an infinitely rapid course on this process in the case of the Kondo effect, where, to me, the formal analogy to what may be the true behavior of color gauge theories is very striking. I have in fact written a note suggesting that the $g \to \infty$ case can be converted to a reasonable sort of lattice model for elementary particles. What is particularly striking about such models is that in every case the actual low-energy spectrum, while perfectly exhibiting the symmetry of the original theory, has <u>changed</u> <u>qualitatively</u> in that certain types of excitation are missing. Thus this is a type of model in which quark confinement without recourse to broken symmetry seems perfectly possible.

I have one last point to make about this kind of system: that there is one way to avoid the infrared divergences which can also be useful. This lies in the suggestion by Collins and Perry[12] that a very high density of baryons will see only very weak forces - essentially the quarks will become asymptotically free if their Fermi energy is driven up into the asymptotic region. I, for one, believe this is so - and I also believe that one of the most promising ways to study how the quarks and gluons confine themselves and make real baryons is to do so in a series in 1/density or in $\frac{1}{E_F}$: the reason being that I am quite sure that all infrared divergences are eliminated in such a system by Fermi exclusion factors and by screening.

I thus conclude not with a many-body analogy but with a many-body problem which I think may be a very important task of elementary particle physics: the real theory of high density nuclear matter.

Uses of Solid State Analogies

I hope I have at least said enough here to indicate to you that it is high time our two fields reestablished regular contact.

APPENDIX: The Kondo Problem

The Kondo problem is a very much over-simplified model of a magnetic impurity in a metal. The magnetic impurity is represented by a local spin S = 1/2 interacting via an exchange integral J with a Fermi sea of free, otherwise non-interacting electrons. The appropriate Hamiltonian is

$$H = \sum_{k\sigma}^{\varepsilon_k < E_c} \varepsilon_k n_{k\sigma} + J \mathbf{S} \cdot \mathbf{s}(0)$$

(s(0) being the spin density at the origin of space coordinates). The J term can cause spin-flip scattering of the free electrons, which is enhanced by the non-spin flip attractive elastic scattering. The result of the interaction of these two effects is a characteristic infrared divergence of the spin-flip scattering cross section as calculated in second Born approximation, $\sim \ln T/E_c$.

The density of states $\rho(\varepsilon_k)$ and the energy J are dependent on the normalization volume. The actual physical parameters of H are (a) the cutoff energy E_c (dimension energy) (b) a dimensionless coupling constant $j = J/E_c$, which is $\ll 1$ in the canonical Kondo Problem.

Recently[10] it has been realized that the renormalization group leads to an understanding of the Kondo H. Progressive reduction in the cutoff E_c leads to modification of j according to

Uses of Solid State Analogies

$$dj\left(\frac{1}{j^2} - \frac{1}{2j} + \ldots\right) = -d\ln E_c.$$

Below T_K, one has a behavior first guessed at in detail by Mattis (although qualitatively suggested by Kondo and many others): the coupling constant becomes so large as to bind one "free" electron to S with opposite spin, with a <u>constant</u> binding energy T_K, and in all ways the resulting system scatters like a <u>non</u>-magnetic impurity; essentially, in a very real sense the spin S is "confined". A beautifully precise computer calculation by Wilson verifies the essentials of this result, for instance that the multiplicity of the levels in a finite model of the system is opposite to that required by S = 1/2.

As Wilson and Nozieres have shown, the result is equivalent to a renormalized normal non-magnetic impurity. Luther and Emery[11] have demonstrated a similar "confinement" phenomenon in the triplet excitations of a one-dimensional electron gas, where the essential mathematics has strong similarities to the Kondo problem.

REFERENCES

1. P. W. Anderson, Phys. Rev. $\underline{86}$, 694 (1952).
2. L. Hulthen, Thesis, Uppsala, 1938.
3. P. W. Anderson, Phys. Rev. $\underline{110}$, 837, 985 (1958).
4. P. W. Anderson, Phys. Rev. $\underline{130}$, 439 (1963).
5. L. D. Landau and E. M. Lifshitz, Statistical Physics, #79, 134, Pergamon, London, 1958.
6. D. A. Kirshnitz and A. D. Linde, Phys. Lett. $\underline{42B}$, 471 (1972).
7. S. Weinberg, Phys. Rev. $\underline{9D}$, 3357 (1974).
8. L. Dolan and R. Jackiw, Phys. Rev. $\underline{9D}$, 3320 (1974).
9. G. 't Hooft, Nucl. Phys. $\underline{B79}$, 276 (1974).
10. Kondo theory: P. W. Anderson, G. Yuval, D. R. Hamann, Phys. Rev. $\underline{B1}$, 4464 (1970); K. G. Wilson, Revs. Mod. Phys. $\underline{47}$, 773 (1975).
11. One-dimensional Electron Systems: A. Luther and V. J. Emery, Phys. Rev. Lett. $\underline{33}$, 589 (1974); A. Luther and L. Peschel, Phys. Rev. $\underline{B9}$, 2911 (1975); S. T. Chui and P. A. Lee, Phys. Rev. Lett. $\underline{35}$, 315 (1975).
12. J. C. Collins and M. J. Perry, Phys. Rev. Lett. $\underline{34}$, 1353 (1975).

Uses of Solid State Analogies

DISCUSSION

B. ZUMINO (CERN)

You spoke about the "Goldstone" and "Higgs" mechanisms. In relativistic field theory one now has examples of Goldstone spinors (which are fermions) and of Higgs-like phenomena in which Goldstone spinor of spin ½ disappears to give mass to a previously massless gauge field of spin 3/2. Is anything like this known in many body theories? Also, in relativistic local field theory there cannot be Goldstone particles of spin 1 or higher (in four dimensional space time). Are there Goldstone particles of spin larger than zero in non relativistic many body physics?

ANDERSON

I tried to think about fermions whether there are any fermion many body "Goldstone-on" and I don't think there are any. There are none that I know of. As far as the other theorem is concerned I think Bruno is trying to get me to mention one thing that I mentioned to him namely that probably there's at least one tensor "Goldstone-on" or "Higg-on" in many body physics. Namely, if you have a large enough solid body its elastic modes interact with relativistic modes and the resulting mess is at least of tensor symmetry if not more complicated.

S. WEINBERG (Harvard)

This is a statement but the question is whether you agree with it. I was trying to understand in the last few months why - although there are these two uses of the renormalization group that you've described - nevertheless it is possible to reproduce a lot of the second use of the renormalization group, that is the going from one effective Hamiltonian to another effective Hamiltonian, by methods which were really designed for the first application which was to discuss the scale invariance of Green's functions at very large momentum. And as far as I can understand it, the reason is that (I'm thinking particularly of the work of Brezin and his group) if you define a renormalization point and define coupling constants by giving the values of Green's functions of that renormalization point, then whether the theory is renormalizable or non-renormalizable - of course if it's nonrenormalizable you'll have to do it an infinite number of times - then all the integrals are effectively cut off at a momentum of the order of the renormalization point. When the momentum in the Feynman graph gets large compared to the value of the renormalization point, you see that the integral begins very rapidly to converge. So that in effect by introducing a variable renormalization point even though it looks like what you're doing is eliminating ultra-violet divergences, at the same time you're providing a method of having a floating cut off just like Wilson, and Kadanoff, and you do in applying the renormalization group to the study of effective Hamiltonians.

ANDERSON

I think I will agree with that, yes. Well in fact I will agree with that in the sense that I've really never thought that hard about it but it sounds right. Besides Steve Weinberg said it.

S. BLUDMAN (Penn)

Can you spell out the necessity or virtue of 't Hooft monopoles in removing cosmological order parameter singularities?

ANDERSON

I'm not saying that they're all that necessary. You know they might have all annihilated against each other if they came in equal numbers or perhaps the cosmos was small enough so that it all went the same way. So it is not an essential thing that some singularity be there, but it certainly is a possible thing that there could have been singularities. What's nice from my point of view is that one doesn't have to have such horrible things as boundaries. Whether one has to have lines I think depends on the groups and the broken symmetries and what has actually happened in

Uses of Solid State Analogies 331

that global decrease in the size of the areas that Weinberg put on that last slide of his [e.g. in the group symmetries remaining after successive symmetry breakdowns].

S. BLUDMAN   (Penn)

Well I understand in the abelian case why you have to have Dirac strings. What I would like you to spell out is the role of the point singularities in the nonabelian case.

ANDERSON

I can only give you the many body example that I know which is if one had a tube of liquid He_3, it happens to have a vector order parameter called $\vec{n}$ and occasionally, and in the case of a tube, the boundaries I think make $\vec{n}$ want to be parallel to the boundaries. But one is stuck with the situation that - on the top it may have condensed with $\vec{n}$ pointing up the tube and on the bottom with $\vec{n}$ point down. The best way to solve that problem is obviously to stick a monopole in between. Supposing now one had that on some kind of cosmic scale, with a big boundary in between, and the boundary can be condensed perhaps into a sequence of these monopoles and the monopole density would give you the total amount of order parameter. I guess I really don't know enough about gauge theories to know whether this analogy is a solid one. This kind of thing incidently occurs in crystals. In crystals you have grain boundaries. In fact if you look at a grain boundary, you often find that it has condensed itself into a sequence of dislocations. There is a theorem that a grain boundary at long range is equivalent to a sufficient density of dislocations which are the line singularities of that theory.

K. WILSON (Cornell)

I know that I don't like the symptom that you read letters in the newspaper and they are always answering other letters, but I think I must defend myself against Steve Weinberg. I would like to stress the importance of the idea of the model Hamiltonian (or a completely new Hamiltonian) for describing the infra-red behaviour of these systems with infra-red catastrophies, which cannot be done by the Callan-Symansyk equations. The Callan-Symansyk equations can tell you how the strength of that interaction changes. But what they cannot tell you is if you have to have a whole new Hamiltonian to describe what's going on. It's not quite that bad in the Kondo problem, I mean you can start with the same Hamiltonian just with $j = $ infinity, but if you're really going to get a complete description of its low energy behavior, you have to throw in some extra scattering terms and you only get that from a non perturbative renormalization group. The situation is clearly worse in the quark case where the spectrum itself has to change or presumable has to change from being free quarks to confined quarks and I can't conceive of the Callan-Symansyk equations telling you what the effective Hamiltonian for confined quarks is.

ANDERSON

Before Steve answers that if he wants to, I think I would have a couple of comments about that. Yes, I think I agree with you that the great thing about the Wilson renormalization group was the statement that the Hamiltonian wasn't necessarily qualitatively the same and - where universality is sufficiently adequate - one doesn't necessarily have to introduce new terms but it contains the general possibility of introducing new terms. The Callan-Symansyk equation on the other hand I think is useful so long as the coupling constant is small, it gets you to something with a very small coupling constant to a moderately small coupling constant and can be used so long as perturbation theory holds.

S. WEINBERG   (Harvard)

I'm clearly out of my depth but I'll try anyway. In many cases where the Callan-Symansyk equation has been used to study critical phenomena, I presume that it is being used because the model has not qualitatively changed and it is not a case like the ones you've been talking about today Ken. But even in the case where it is and you're really talking about a model Hamiltonian, it doesn't seem to me that one is restricted, even though it is a model Hamiltonian, from imposing a cut off instead of by a lattice by the method of introducing a variable renormalization point.

K. WILSON   (Cornell)

Now the problem is how do you know what terms are *in* the effective Hamiltonian?

S. WEINBERG   (Harvard)

Well how do *you* know what terms there are in the effective Hamiltonian?

K. WILSON   (Cornell)

Well that's my trade secret. That's what I get out of the computer.

VOICE FROM THE AUDIENCE

Could you tell me how difficult it is to go from something like the BCS Hamiltonian which involves self-interacting fermions to a phenomenological theory involving just the order parameter, and in particular can you use the BCS theory to calculate things like the scattering in collective excitations and such? Is that a well understood problem in solid state?

ANDERSON

Yes, fairly. It's essentially the problem of finding the appropriate generalized Ginzberg-Landau functional for the system and in many cases that doesn't describe all of the collective excitations of the system. One has a serious problem of time scales. One has this zero frequency but finite space scale problem, and then as one goes to higher frequencies one often has to introduce new time behaviors. One will have a collisionless regime, and a non collisionless regime depending on the time scale relative to the space scale.

VOICE FROM THE AUDIENCE

Do you use the BCS Hamiltonian to calculate the parameters in the phenomenological theory?

ANDERSON

One can yes. Well, first one can take the phonons and the electrons. That can then in many cases be transcribed into some effective BCS Hamiltonian and that finally in turn into a Landau-Ginzberg functional which tells you how the order parameter varies. Each of these are fairly complicated stages.

B. WARD   (Purdue)

In your introduction, you mentioned both the Kondo problem and the results of Luther. You chose to describe to us only the Kondo problem, presumably because of the lack of time. Could you please describe how the result of Luther would have fitted into your discussion if you had had more time?

ANDERSON

Well what Luther essentially does is to take an asymptotically free case for instance and show that the renormalization group equations will carry you to a certain case with a finite coupling constant and then that case is mathematically soluble, and it is again a theory in which there are no phase transitions, no changes in symmetry. But in this case there is a gap in the triplet excitation spectrum, even though in the original Hamiltonian there appeared to be no gap in the excitation spectrum for triplet excitations as opposed to singlet excitations. So then a gap is appearing out of nowhere without symmetry change. Ken Wilson emphasized that there are a lot of caveats here but there are also a lot of opportunities. The standard dictionary says energy gap equals mass. That isn't necessarily always true in going from a many body analogy to a field theory analogy. Energy gap may be instead of that a certain distance off the mass shell rather than the mass. I think it may be in the quark gluon case that is the way that it goes, that when an energy gap appears in the theory one might interpret that as saying that there isn't any stable particle of that sort. I think that's not an impossibility.

Possible Consequences of Negative U Centers in Amorphous Materials
J. de Physique Colloque, No. 4, pp. C4-339 (1976)

Somewhat speculative but I still believe that the negative U center plays a very big and interesting role in the electrical properties of amorphous semiconductors, photosynthetic centers, etc. I remember walking a hair-raising trail 1500 meters above Grenoble with a big group including Friedel, which had consumed great quantities of wine with lunch, as part of the Au Trans meeting at which this was given. The late Michael Schluter, as well as Gene Baraff and Don Hamann, actually showed later that such centers are possible by explicit calculation of the Si vacancy center. But the *mobile* bipolaron which many people have proposed for superconducting phases is undoubtedly a figment.

INTERRELATION OF CORRELATION AND DISORDER. EXPERIMENT AND THEORY.

POSSIBLE CONSEQUENCES OF NEGATIVE U CENTERS IN AMORPHOUS MATERIALS*

P. W. ANDERSON

Bell Laboratories Murray Hill, New Jersey 07974, U. S. A.
and Princeton University

Abstract. — I have proposed, and Mott and others elaborated, a model of amorphous semiconductors in which there is a fairly high density of localized centers near E_F which have effectively negative U and can hence accommodate zero or two electrons, in conjunction with a mobility gap which is of order $|U|$. Two new aspects of these centers will be mentioned here : (1) Varma and Pandey (private communication) have proposed that such centers form at metal-insulator contacts and there may be direct experimental evidence for them. They could be responsible for the well-known Fermi level pinning effect ascribed by Bardeen to surface states (2). Several arguments suggest that the one-electron gap itself will be a function of the pair state occupation. If so, and the relaxation rates of these states are very slow and cover a broad range as expected, they may be seen to lead to : (1) $1/f$ resistance noise ; (2) Long-period photo-electric phenomena ; (3) Switching.

1. Introduction. — Two general approaches to the problem of understanding the physical properties of amorphous materials can be identified. Our understanding of ordinary solids is very heavily based on proceeding from detailed structural information and deducing spectra from it (band structure, Fermiology, phonon dispersion), proceeding thence to understanding the macroscopic properties. Structural studies of amorphous materials are thus the natural starting point for many of us, but in the necessary absence of exact information we may, for example, tend to hypothecate structures and speculate about the resulting properties without realizing that the electronic and atomic energies determine the structure and not vice versa.

The second possible starting point is to abandon detailed structural ideas at the outset, the structure perhaps to be understood as a consequence rather than as a postulate. In this approach the idea is to search for *structure-independent generalizations*. Examples are the ideas of localization and the mobility edge which are common to most random structures ; the idea of tunneling centers ; and even the idea of long-wavelength phonons in random structures, which follows from conservation laws and homogeneity.

The latest such structure-independent generalization is the postulation of *negative U* or *two-electron* centers as a general property of amorphous materials [1], especially in equilibrium or quasi-equilibrium situations such as glasses. This idea follows most naturally from conceiving of the amorphous material as a random collection of pair-bonds and other possible electron pair states such as lone pairs and filled ionic rare gas shells. The overwhelming majority of substances have their electrons paired up in such states, as is evinced by the prevalence of diamagnetic materials in nature, and amorphous semiconducting and insulating glasses are no exception. The general reason is that if a site can accept one electron, in general the atoms will move in response to that electron's presence in order to lower its energy and make it favorable for a *second* opposite-spin electron also to occupy the site. Often the binding energy follows from this tendency, as for instance for the H_2 molecule. It is only relatively rarely that, as in for example donor states in semiconductors or transition metal salts such as MnO, the interelectronic Coulomb repulsion outweighs the coupling of the electrons to atomic motions and *repels* a second electron with the positive Van Vleck-Hubbard repulsion U, leading to unpaired electron spins.

We imagine, then, the glass as full of electrons in pair bonds, lone pairs, etc. Without prejudice we will often refer to these states as *bonds*. In a regular material the energies of the possible states are all the same or regularly distributed, so that there can be, and often is, a gap between occupied and empty states even if we ignore atomic motion. In the irregular structure, however, we do not expect an absolute gap, although since the structure will be stabler, the lower the one-electron energies, there will tend to be a relatively low density of states near the Fermi energy, even an order of magnitude or more less than average. But there should always be a finite density of pair states in the sense that there will be a *last* pair state occupied as we add more electrons : a bond, for instance, which is the weakest and would be the first to break if we removed electrons.

(*) Work at Princeton partially supported under NSF grant DMR-76-00886.

The two-electron center model describes this density of randomly situated states by a Hamiltonian

$$\mathcal{H}_1 = \sum_{i\sigma} E_i n_{i\sigma} \qquad (1)$$

where E_i is the one-electron energy in roughly the equilibrium configuration, and the density of states $\rho(E_i)$ may be imagined to look as in figure 1, with low but finite value near the Fermi energy.

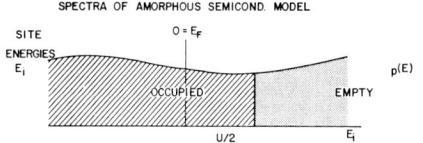

FIG. 1. — Spectra of amorphous semiconductor model : site energies.

To this we add the net effect of the Coulomb repulsion and the coupling of atomic motion to the electron density at site i, which leads by our postulate to a *negative effective U* for thermal equilibrium processes :

$$\mathcal{H}_2 = -\sum_i U_i^{\text{eff}} n_{i\uparrow} n_{i\downarrow} . \qquad (2)$$

Two things should be said about $\mathcal{H}_2$: First, as I have indicated, U_i^{eff} is not a fixed constant but also a random distribution like E_i and we will see that (1) and (2) do not necessarily lead to a sharp gap.

Second, U_i^{eff} is the consequence of relatively large amplitude atomic motions, and for rapid motions (2) is not valid. Among other things, there will be a distribution of relaxation times for thermal readjustment of the occupation numbers n_i which will probably extend from very slow to very fast, having a roughly uniform distribution of activation energies V_i

$$\tau_i^{-1} = \omega_D\, e^{-V_i/kT} \qquad (3)$$

$$\rho(V_i) \sim \text{const}.$$

(1) and (2) imply that at $T = 0$ the sites i will, to a first approximation, be occupied ($n_i = n_{i\uparrow} + n_{i\downarrow} = 2$) up to $E_i = U/2$ and empty above that, measuring the energy from a Fermi energy $E_F = 0$. This is because the energy necessary to add a pair of electrons is, per electron

$$E_{2-e} = \frac{2 E_i - U}{2}. \qquad (4)$$

The spectra of unperturbed energy E_i and of E_{2-e} are contrasted in figures 2a and 1.

There are several other spectra which are important to the problem. As we pointed out before, the spectrum even of localized *one-electron* states has a gap if U_i is finite for all i, even though neither the 2-electron nor the *unperturbed* one-electron one does. This comes about because it costs at least $U/2$ to create an electron in an empty state or a hole in a filled one, because of the interaction energy-U in the latter case. This is shown in figure 2b.

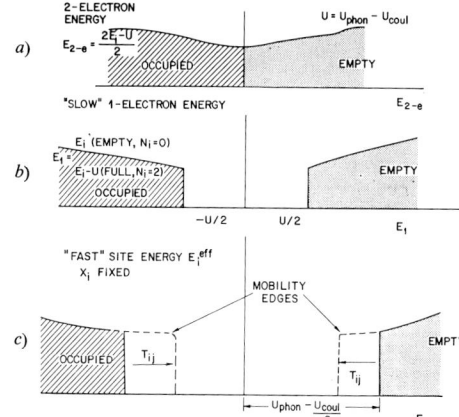

FIG. 2a. — Spectra of amorphous semiconductor model : 2-electron energy.

FIG. 2b. — Spectra of amorphous semiconductor model : slow 1-electron energy.

FIG. 2c. — Spectra of amorphous semiconductor model : fast site energy E_i^{eff}.

Also relevant is the spectrum of mobile or extended one-electron states. To understand this we must recognize again that (2) is not an instantaneous interaction ; U^i is the resultant $U^i = U_{\text{phon}} - U_{\text{coul}}$ of a true coulomb interaction U_{coul}^i and a phonon displacement effect

$$U_{\text{phon}} = \frac{\lambda_i^2}{m\omega_i^2}.$$

The latter comes about indirectly from the electron-lattice coupling at the site i :

$$\mathcal{H}^{e-l} = -\lambda x_i(n_{i\uparrow} + n_{i\downarrow}) \qquad (5)$$
$$+ \frac{m\omega_i^2 x_i^2}{2} = -\lambda \eta_i x_i + \frac{m\omega_i^2 x_i^2}{2}$$

and when we choose x_i by

$$\frac{\partial E}{\partial x_i} = 0 \quad \text{or} \quad x_i = \frac{\lambda}{m\omega_i^2} n_i$$

we get the attractive phonon terms. For mobile electrons also we must include a *hopping* Hamiltonian

$$\sum_{ij\sigma} T_{ij} C_{i\sigma}^+ C_{j\sigma} \qquad (6)$$

and if T_{ij} is of normal size x_i cannot follow the motion of an electron in an extended state and must be taken as fixed.

The net effect of this is that the effective Hamiltonian for mobile electrons becomes

$$\mathcal{H}^{\text{mobile}} = \sum_i E_i^{\text{eff}} n_{i\sigma} + \sum T_{ij} C_{i\sigma}^+ C_{j\sigma} \qquad (7)$$

where E_i^{eff} is given by

$$E_i^{\text{eff}}(\text{empty}) = E_i$$

$$E_i^{\text{eff}}(\text{occupied}) = E_i + U_{\text{coul}} - 2\,U_{\text{phon}}; \qquad (8)$$

that is, the gap in the *site* energy spectrum for mobile states is even bigger than that for localized ones (see Fig. 2c). But this gap gets filled in by the effect of the bandwidth due to T_{ij} and the mobility edge may be above or below the gap for one-electron *localized* states. I argued that in many glasses the experimental evidence suggests it is below that gap, or at least that there are few localized one-electron states. Figure 3 shows the final outcome of all this discussion, suitably smeared for random variations of U^i. Please note that while the gap for localized states need not be sharp, and presumably seldom is, the mobility edge is by definition a sharp phenomenon. This, to my mind, is one of the stronger experimental confirmations of the general picture given here : that amorphous systems if anything show a sharper, steeper gap edge than regular ones. The only vertical edges in nature that I know of are the Fermi level and the mobility edge.

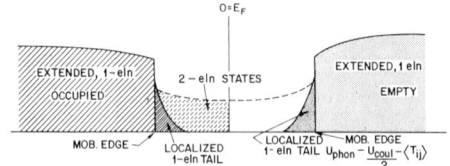

FIG. 3. — Spectra of amorphous semiconductor model : final state density with random smearing.

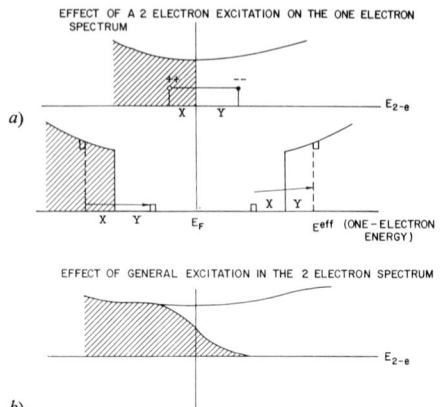

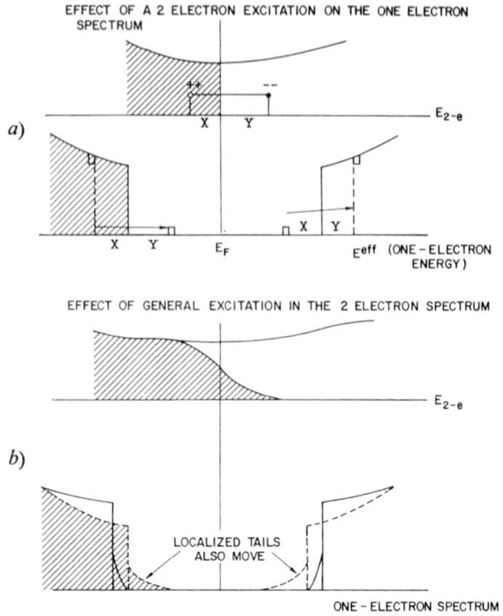

FIG. 4a. — Effect of a 2 electron excitation on the one electron spectrum.
FIG. 4b. — Effect of general excitation on the 2 electron spectrum.

All of the above is exactly what would follow in detail from the rather cryptic discussion of my letter of a year ago. Now I would like to point out some new general directions in which consequences might flow from this. The first general area has to do with properties of bulk amorphous semiconductors, and is work I am doing in collaboration with Ki Ma at Princeton. The crucial observation here is another consequence which follows from the above model : *the energy gaps for one-electron states depend on the occupation of the two-electron states*.

That is (see Fig. 4) if we take two electrons from a state with

$$E_{2-e} < 0, \qquad E_i = \frac{U}{2} - X$$

to another nearby site where

$$(E)_{2e} > 0, \qquad E_i = \frac{U}{2} + Y,$$

we also modify the basic densities of one-electron states. Specifically, we shift the effective one-electron spectrum as shown in the figure 4 : There is now an *empty* state *below* $U/2$ by X, and a full one *above* $-U/2$ by Y. Thus both the localized and extended one-electron spectrum have their gaps reduced in proportion to the number of such 2-electron excitations and to their energy shifts. This effect is non-negligible both because of the very high density of such states and because the resistivity depends exponentially on E_g/T, so that a small shift in E_g can have a large effect.

We propose that several well-known effects may follow from this.

(1) *Switching* : A high field will tend to redistribute carriers among the 2-electron centers essentially randomly. The distribution (3) of relaxation times means that slowly relaxing centers will remain out of equilibrium, at least in the presence of a continued current, and as a result the gap will remain smaller. That is, a high field can effectively redistribute the bonding sites, resulting in a decreased gap and much lower resistance. Once the resistance is low the presence of a current will maintain the disequilibrium in the 2-electron states. Experimentally, one might try to measure the gap in switched material to confirm this.

The *memory* switch can occur if some of the relaxation times are very long, or, more likely, if the new, relatively unstable switched phase has a tendency to recrystallize or otherwise macroscopically modify itself chemically or physically. We emphasize that the above hypothesis has much in common with other ideas about switching. We are postulating that there is an effective increase in the temperature of the 2-electron states without prejudice as to whether or not carrier and phonon temperatures track with it or not, so it is a quasi-thermal hypothesis. We also postulate that real chemical changes take place in the sense of a redistribution of the bonds between atoms, nor do I exclude actual physical migration of atoms over short distances. For the *memory* switch our hypothesis is very similar to others as well.

(2) Long period photoelectric effects. Again, we rely on the idea that excited carriers will not recombine at first into the most stable possible chemical bonds but can become trapped in less stable ones, resulting in a lowered mobility gap. Thus high photo-excitation can lead to a temporarily unstable state with much smaller gap and higher conductivity.

(3) Resistance noise. In thermal equilibrium the bonds will form and break at a rate given by the spectrum of relaxation times τ_i of (3). In addition to other electrical effects, there will be a corresponding fluctuation in the mobility gap locally, and thus in the number of mobile carriers. This resistance noise is one of the few direct effects which can allow us a *spectroscopy* of the bond states, their field screening effect being perhaps the major other one. If the distribution of the activation energies V_i is uniform in energy, the spectrum will be of the canonical $1/f$ type. Although $1/f$ noise was discovered originally in amorphous carbon, we have been unable to find many definitive measurements of it in characteristic amorphous systems, and we would urge noise studies as an important approach to the fundamental phenomena in these materials.

As a final topic I would like to move out of the normal range of amorphous physics to the surfaces of semiconductors. Specifically, these are ideas stemming from a suggestion by K. C. Pandey and C. M. Varma [3], who remarked that except under the most rigorous conditions we may surely expect an amorphous, or at least highly disordered, situation at any metal-semiconductor contact. The metal-semiconductor bonds will in general not all have the same character or strength, and again we have every reason to expect that many of them will be weaker than the internal bonds of either component specifically those in the semiconductor. They will therefore lead to 2-electron states in the semiconductor gap, which could have the effect of pinning the Fermi level near the middle of the gap just as they do in an amorphous semiconductor, and with a high density of such states $\sim 10^{15}/cm^2$ or more, there would be no difficulty in explaining the well-known pinning phenomenon ascribed by Bardeen to surface states. D. Haldane, my student at Princeton, has been studying the properties of 2-electron resonances with negative U (a *negative U anderson model*) in contact with the metallic Fermi sea and finds that the same $n = 0 \leftrightarrow n = 2$ transition takes place as for localized states, so that Fermi level pinning can take place. Interestingly, recent PS and other data of Rowe and Margaritondo reveal very amorphous-semiconductor like sharp edges to the gap as the first few layers of metal are laid down on a semiconductor (see Fig. 5). It is noteworthy that semiconductor surfaces and contacts are well-known to be strong sources of $1/f$ noise.

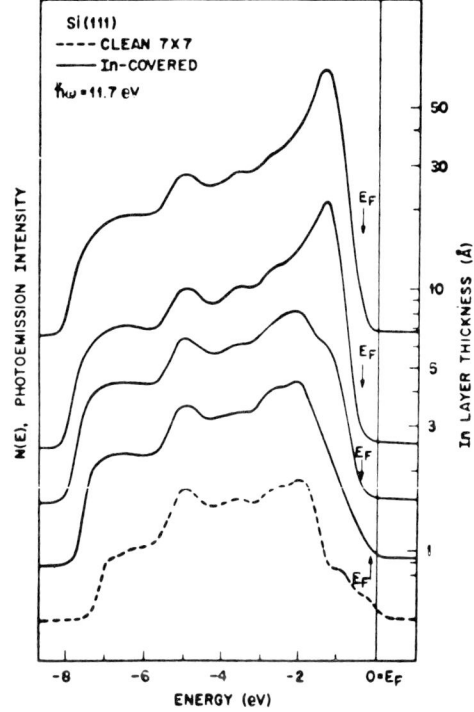

FIG. 5. — XPS spectra of Si surface as atomically thin layers of Ga are added. Note disappearance of 1-electron spectrum in gap, appearance of sharper gap edge. After J. Rowe.

In conclusion, the postulate of negative U bond states seems to lead to natural explanations of most of the outstanding mysteries of amorphous semiconductor physics, and promises to extend to surface studies as well via the negative U Anderson model.

Acknowledgements. — I would like to acknowledge the use of unpublished remarks from C. M. Varma, F. D. M. Haldane and Ki Ma as stated in text. Discussions with R. G. Palmer, J. Rowe, C. M. Varma and N. F. Mott have been most useful.

References

[1] ANDERSON, P. W., *Phys. Rev. Lett.* **34** (1975) 953. The main results of that paper are restated and elaborated in a special case by
ADLER, D. and YOFFA, E., *Phys. Rev. Lett.* **36** (1976) 1197.

[2] STREET, R. A. and MOTT, N. F., P. R. L. **35**, 1293 have specialized the arguments of ref. [1] to a specific electronic model, as have Adler, Kasner and Fritsche in a preprint. I confess to being somewhat disturbed by these papers' referring to the two electron centers as *defects*, since one of the strongest features of ref. [1] is that it is consistent with the fact that a glass properly has no defects per se : it has only a statistical distribution of weaker or stronger two- (or sometimes one-) electron bonds. It is, of course, obvious that atoms with lone pair orbitals are more prone to have weakly bound pairs, as both of these papers emphasize.

[3] PANDEY, K. C. and VARMA, C. M., to be submitted.

Survey of Theories of Spin Glass
Amorphous Magnetism II, ed. R.A. Levy and R. Hasegawa
(Plenum Press, New York, 1977), pp. 1–16

While others meet at famous tourist spots, devotees of amorphous magnetism go to post-riot Detroit or Troy, N.Y. — an interim view during the growth years of spin-glass theory.

SURVEY OF THEORIES OF SPIN GLASS

P. W. Anderson*

Bell Laboratories

Murray Hill, New Jersey 07974

    I hope the organizers of this Conference are as confused about what a "keynote" address should contain as I am. Certainly they have given me no firm instructions, while the politicians have not provided me with the right kind of examples. It would take someone more presumptuous than I to tell you what you <u>should</u> say in the rest of the meeting, and Uri Geller to tell you what you <u>will</u> say. My own response to this charge is, first, to express our pleasure and gratitude at being here to talk about a most fascinating part of physics, and second, to get about doing so as immediately as possible.

    Solid state physics has been enormously successful in dealing with many aspects of regular crystalline solids - so successful that many of us feel called upon to branch out and try to find ways of managing systems which don't have simple periodicity properties. The fascination, for me at least, of the general problem of random and amorphous systems is the necessity of developing a whole new conceptual structure not based on conventional band structure and translational symmetry, and requiring newer ideas such as percolation, localization and other forms of the general type of phenomena one might call non-ergodicity, as well as various other new techniques and concepts. The problem of the spin glass is one of the most complex and interesting, and most characteristic, in this new field. Thus it is a very suitable subject for such an opening talk.

---

* Also at Department of Physics, Princeton University, Princeton, N.J. 08540. Work at Princeton was supported in part by the National Science Foundation Grant #DMR76-00886.

Under the title "Theories of Spin Glass" I can very severely restrict myself, partly because as one of the co-inventors of the term I can use it to mean what I like. My first criterion is that I will discuss only theories which take seriously the possibility that there is something like a genuine thermodynamic phase transition in systems with random exchange interactions, and that this transition is a property of the right kind of randomness and not of some mysterious ad hoc anisotropy field or "clustering" phenomenon. Since no one seems to have expected a phase transition before the experimental results of Canella and Mydosh[1] which appeared, as did the term "spin glass",[2] for practical purposes, at the first of these meetings 3 or 4 years ago in Detroit, this limits me to very recent history. I will also ignore a number of artificial models which have been invented to reduce the spin glass to some trivial result; if there is anything we surely know it is that the spin glass problem is not trivial, and it is easy in each case to see how, in these models, the baby has gone out with the bath water, in that the models do not contain the spin glass physics.

This leaves me with the following outline for my talk:

I. Brief discussion of experiments
  (a) Wide variety of materials, semidilute noble - transition alloys, "ferromagnetic superconductors", glass, CrO_2-VO_2[3] some metallic glasses (see this conference) etc.
  (b) Cusps in X at T_c; frozen in random moments below.
  (c) "Micto" magnetic phenomena starting at T_c.[4]
  (d) No Csp anomaly at T_c, linear at $T \to 0$.[5]
  (e) Computer experiments mimic (b,c,d) with random sign J_{ij}. (Binder,[6] Walker,[7] etc.)
II. Theories type 1: $n \to 0$, "replica" theories: Edwards-Anderson[8], Sherrington-Kirkpatrick,[9] Fischer, etc.
III. Type 2: Thouless, Anderson, Palmer: a mean field theory exact as $Z \to \infty$.
IV. Type 3: attempts at critical behavior: Lubensky and Harris[12] q^3, is "micto" magnetism = marginal dimensionality and/or line of critical points?

Although the characteristic spin glass alloys are the solutions of type Cu-Mn .1-10% or Au-Fe, a great variety of random magnetic materials seems to show the characteristic cusps and mictomagnetic behavior, and I have listed some of these above. I do not place great emphasis on the importance of RKKY oscillatory interactions; what I do think is vital is that the interactions be of random sign, or at least that the different interactions via different paths between two given spins be competing: this competition is the essence of the spin glass phenomenon and is the missing feature of most artificial models. I need hardly show you the characteristic cusp-like susceptibility which you will have seen so often, and which defines experimentally the spin glass phenomenon. One example is shown in Fig. 1. (After Canella, Mydosh and Budnick[1]) At this same T_c at which the cusp appears it is clear from several experimental data such as Mossbauer and μ-meson spectroscopy[3] that an essentially D.C. frozen in moment appears at each site. There is a freezing phenomenon, but all of our experiments as well as computations show that there is no <u>regular</u> structure. Along with the cusp and the frozen moment goes a very high and anomalous degree of field sensitivity of the susceptibility near T_c, as well as the kind of time and history dependence below T_c which we call micto-magnetism.[4]

A few weeks ago I received a letter from Ralph Hudson of the NBS objecting to this term, on the basis that he thought that the only other word in the English language using the same root was "micturation" and the root was Latin for "urine". I think myself that the term is very descriptive: back in the Middle West we used to refer to something as "p---poor" if it was not worth anything more substantial, and that is a good description of this kind of magnetism. The remanance is very small and very sluggish, there are peculiar training phenomena, and the susceptibility is often very history dependent. Unfortunately, I am assured by Collin Hurd and by the OED that Hudson is incorrect and that "micto" is a legitimate Greek root meaning "mixed".

To my mind there are two clear results which must be accounted for by any theory: (1) It is clear from every kind of data that no real ferromagnetism is involved: the frozen-in local moment is many orders of magnitude smaller than in ferromagnetism, and a quite different shape as a function of temperature: it just makes no sense to ascribe the remanance to isolated clusters. (2) There is clearly a definite T_c at which all of the phenomena begin, which is the same T_c as for the susceptibility: again we are confronted with a sharp transition.

My final point about experiments has to do with the other branch of experimental physics, computer experiments. The computations, particularly the exhaustive ones of Binder,[6] very realistically mimic the actual behavior of spin glasses, based on a short-range, random sign J_{ij} (in his case an Ising interaction, for others a more realistic Heisenberg one). These computations, I believe, confirm that the model we are using is correct: what is necessary is to solve it. So on to the theory.

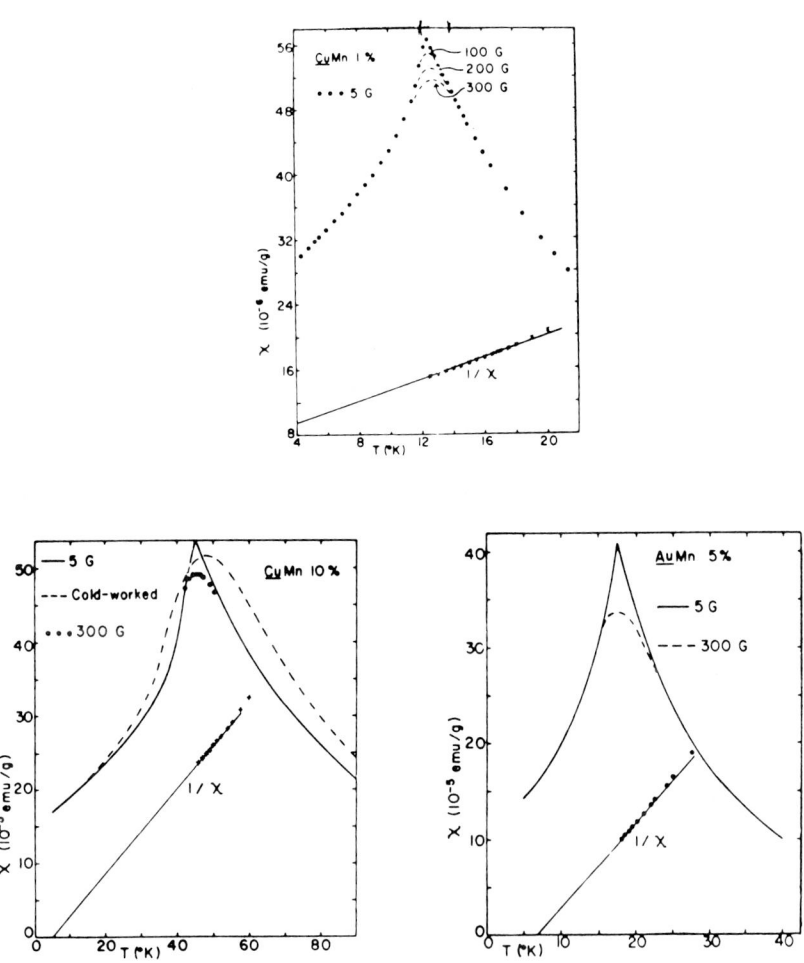

Figure 1 - Typical examples of susceptibility cusps.

THEORIES OF SPIN GLASS

The first attempts to understand how such a transition might occur were by Sam Edwards and myself[8] and much the same line has been followed on by Sherrington[9] and associates - Sothern, Kirkpatrick, and by Fischer.[10] Our basic initial contribution was to identify something which might serve as an order parameter for such a transition: namely we decided that since we believed there was no long-range order in space, we had to introduce a long-range order in time:

$$q = \lim_{t' \to \infty} \langle\langle S_i(t=0) S_i(t=t') \rangle\rangle$$

and we assumed that q is zero at high temperatures and that at a sharp transition point T_c it suddenly assumes finite values. (Note that in a finite magnetic field q is always finite so there is no sharp transition!)(See Fig. 2) In all ordinary phase transitions there is time long-range order, i.e. a change to non-ergodic behavior, but there is also an ordinary spacial order parameter. What is fascinating here is the absence of the ordinary kind, and the really exciting question for the theorist is whether that is possible and whether it represents a true phase transition. My own feeling is that it is and does, but that is by no means proven conclusively yet. What is clear to me already is that if the transition occurs, its properties are very markedly different from those at ordinary phase transitions, and my best guess is that those great differences are responsible for the complicated mictomagnetic phenomena which are observed.

The actual formal technique which Edwards brought to the problem is so simple I can show it to you. If you want to calculate the free energy of a system you use

$$e^{-\beta F(T)} = \text{Tr} e^{-\beta H} = Z$$

It is also obviously true that

$$e^{-\beta n F} = (\text{Tr} e^{-\beta H})^n = Z^n$$

for any n.

Now it turns out to be easy to average expressions like $e^{-\beta H}$ over a random distribution of J_{ij}'s; this is true in general but it's particularly true of a Gaussian.

$$P(J_{ij}) = e^{-J_{ij}^2 / 2\overline{J_{ij}^2}}$$

so if for some integer n we write

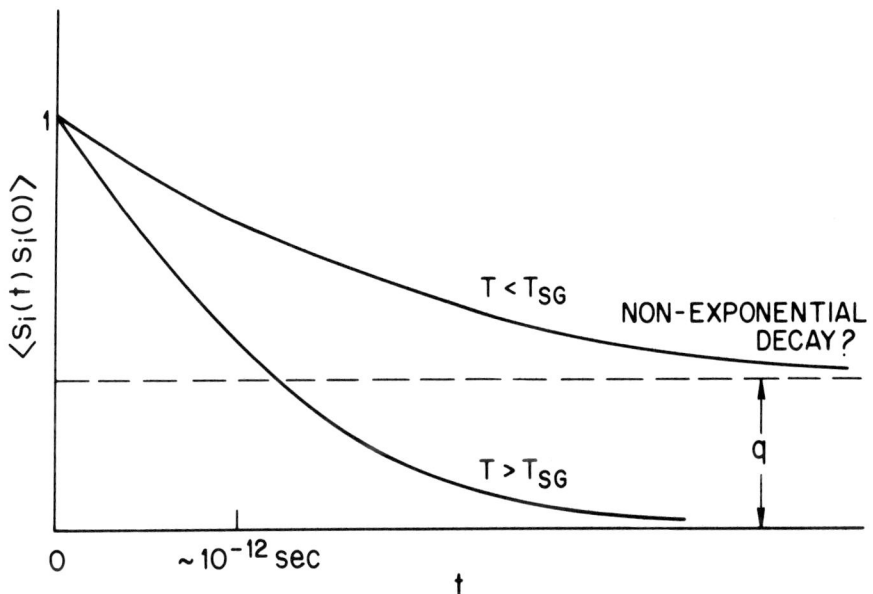

Figure 2 - Schematic diagram of $\langle S_i(t)S_i(0)\rangle$ vs t.

THEORIES OF SPIN GLASS

$$(\text{Tr} e^{-\beta H})^n = \text{Tr}_{S_i(1)} \cdots \text{Tr}_{(S_i(\alpha))} e^{-\beta \sum_{ij} J_{ij} \sum_\alpha^n S_i(\alpha) S_j(\alpha)}$$

and take the average

$$\langle \quad \rangle_{P(J_{ij})} \sim \text{Tr}_{S_i(\alpha)} \cdots \text{Tr} \, e^{\sum_{ij} \frac{\overline{\beta^2 J_{ij}^2}}{2} (\sum_\alpha^n S_i(\alpha) S_j(\alpha))^2}$$

$$= \text{Tr}_{S_i^\alpha} e^{-\beta \sum_{\alpha\beta} H_{\alpha\beta}}$$

$$H_{\alpha\beta} = \sum_{ij} \overline{\beta J_{ij}^2} \, S_i(\alpha) S_j(\alpha) S_j(\beta) S_i(\beta)$$

It looks as though we have just reduced the average free energy problem to a trivial one in that there is no longer any random interaction J_{ij}, but only the mean $\overline{J_{ij}^2}$ which is translationally invariant. But it's not that trivial, since it is not safe to go to the large N, thermodynamic limit of an expression like $e^{-\beta H}$ because H is proportional to N and its fluctuations grow with N: if we did it straightforwardly we'd end up with the free energy of some special favorable configuration of J_{ij}'s, and this is indeed shown by the fact that F isn't independent of n. The only limit in which this average is safe is the limit $n \to 0$, where it becomes $\langle \log Z \rangle$, so the trick is to do the problem for all integer n to get an expression for F(n) and take the limit as $n \to 0$ when finished.

The formal structure which we get, as you see, involves a coupling between the different "replicas" α, and for reasons which are obscure to everyone but Sam Edwards, this replica-replica correlation

$$q_{\alpha\beta} = \langle S_i^\alpha S_i^\beta \rangle$$

seems to behave in exactly the same way as the long-time correlation

$$\langle S_i(0) S_i(\infty) \rangle \, .$$

Most of the work done by this method treats $H_{\alpha\beta}$ strictly within the simplest kind of mean field theory. The most ambitious treatment was that of Sherrington and Kirkpatrick,[9] who realized that mean field theory is traditionally supposed to be exactly correct in the limit as the number of neighbors $Z \to \infty$. Their treatment seems at first sight to have this property. For simplicity they do Ising spins and allow $N = Z \ggg 1$, and they solve exactly for all integer n and extrapolate to $n \to 0$, achieving results which look almost exactly like those of Edwards and Anderson: a cusp both in X and Csp, a linear specific heat as $T \to 0$, etc. The only trouble is a very serious one: it is quite obvious internally that the low temperature results are wrong: the entropy is negative, the free energy is a maximum, not a minimum, and numerical computations have since shown that the ground state energy is too low. Since the <u>only</u> approximate step in the calculation is the $n \to 0$ limit, this has the unfortunate result of showing that that limiting process is not reliable. In fact, Richard Palmer[14] has shown that the crucial step is that they have taken the limit $N \to \infty$ <u>before</u> the limit $n \to 0$, which is not allowable.

In this disastrous situation Thouless, Palmer and I decided that the first essential was to understand this one example thoroughly, and to see if it couldn't be solved by a completely independent method. Rather to our surprise we found that it could, and that the results, while significantly different from those of SK at low temperatures, <u>confirm</u> the transition which SK found, and give the same physical properties at that point. This is no place to go into mathematical detail as to how we did this. The main idea was that we exploited the usual high-temperature diagram expansion in which you expand

$$\ln \text{Tr} e^{-\beta H} = \ln \text{Tr} \left[1 - \beta H + \frac{\beta^2 H^2}{2} + \cdots \right]$$

which is a power series in $1/T$, describing each term as a diagram in which lines are interactions J_{ij} and points are products of spins S_{ij}. In the large Z limit the mean square interaction must fall off as $1/Z$ to give a finite T_c:

$$\overline{J_{ij}^2} = \frac{\tilde{J}^2}{Z}$$

so that any diagram that has too many interaction lines in it drops out in this limit. Thus we can classify diagrams according to powers of $1/Z$. The biggest are the simple chains (Fig. 3) which give the S-K high temperature result:

$$F_{HT} = -N\tilde{J}^2/4T \ .$$

The next biggest, of order N/Z and hence apparently negligible, are

THEORIES OF SPIN GLASS

"CHAINS":

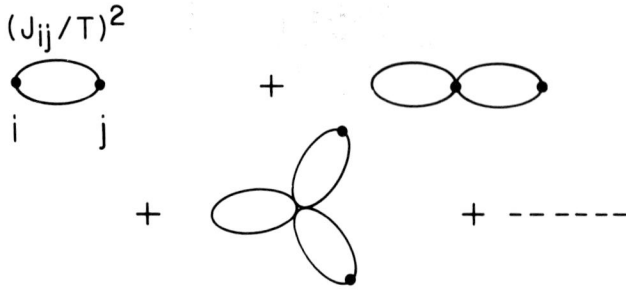

Figure 3 - Chain diagrams summed by the S-K method at high T.

all the simple rings (Fig. 4). These are indeed negligible if they converge, but we find that at $T_c = \mathcal{J}$ they <u>diverge,</u> signalling the transition found by S-K. That these diagrams are necessary to find the transition is heartening, since they are the simplest ones which involve competition among different paths.

Below T_c we found that the only way to make our diagram series converge was to introduce a random mean spin m_i at every site satisfying the peculiar mean field equation

$$m_i = \tanh h_i/T$$

$$h_i = \sum_i J_{ij} m_j - \frac{m_i}{T} \sum_j J_{ij}^2 (1-m_j^2)$$

which as you see is not the same as naive mean field theory, which has only the first term, and in fact T_c is exactly 1/2 that given by ordinary mean field theory (but equal to SK, which is $n \to 0$ mean field theory). This equation is extremely complicated to solve - note that it involves the random matrix J_{ij} - but with the help of the computer we believe we did find solutions and some of our results, contrasted with SK's, are shown in Figs. 5 and 6. One

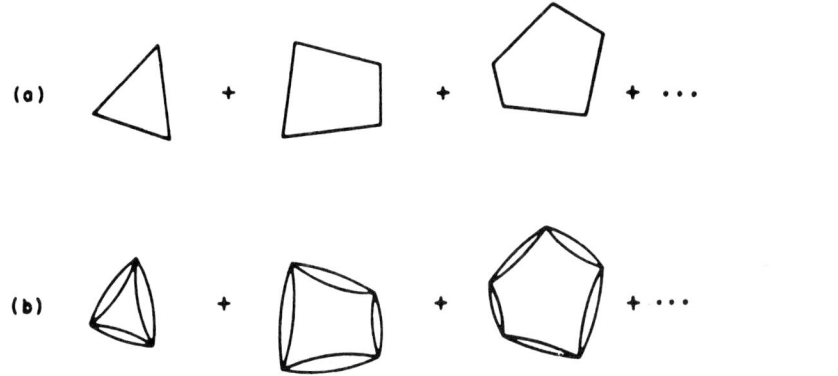

Figure 4 - Single (a) and multiple (b) ring diagrams whose divergence causes the transition.

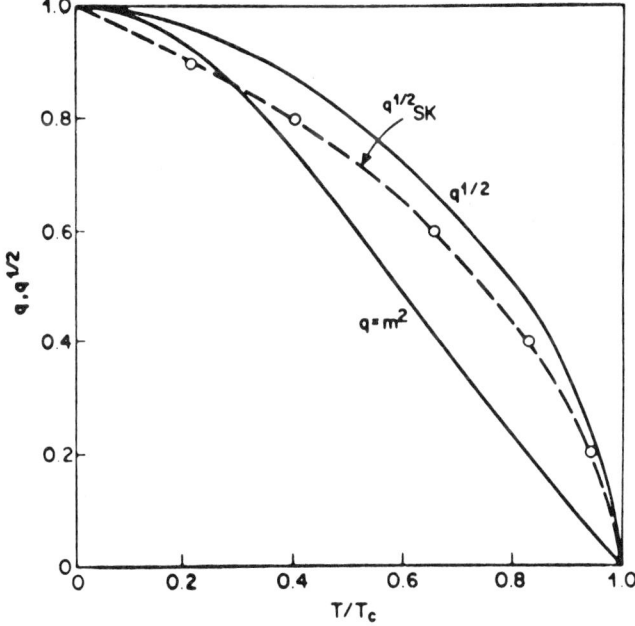

Figure 5 - Experimental results (SK and exact) for q(T).

THEORIES OF SPIN GLASS

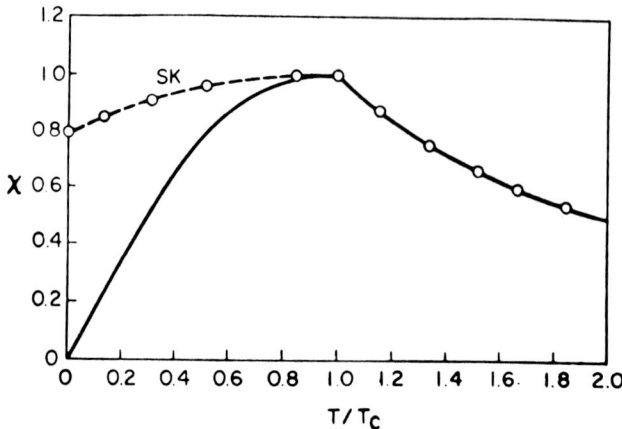

Figure 6 - Same for $X(T)$.

further result is shown in Fig. 7: we found from our numerical computation that as $T \to 0$ the probability distribution $P(h_i)$ of the mean local field becomes linear in h_i: this means the famous linear specific heat would not even result in this Ising model case! Thus the linear specific heat of the EA-SK theories is, as we all suspected, an artifact.

The really spectacular result of our theory, however, is a conjecture which seems to be borne out by our computer solution, and represents what may be a very strange and important property of spin glasses. It is possible for us also to write down an expression for the free energy as a function of the mean magnetization:

$$F_{MF} = - \sum_{ij} J_{ij} m_i m_j$$

$$- \frac{1}{2T} \sum_{ij} (1-m_i^2)(1-m_j^2) J_{ij}^2$$

$$+ T/2 \sum_i \left[(1+m_i) \ln \left(\frac{1+m_i}{2}\right) + \left(\frac{1-m_i}{2}\right) \ln \left(\frac{1-m_i}{2}\right) \right]$$

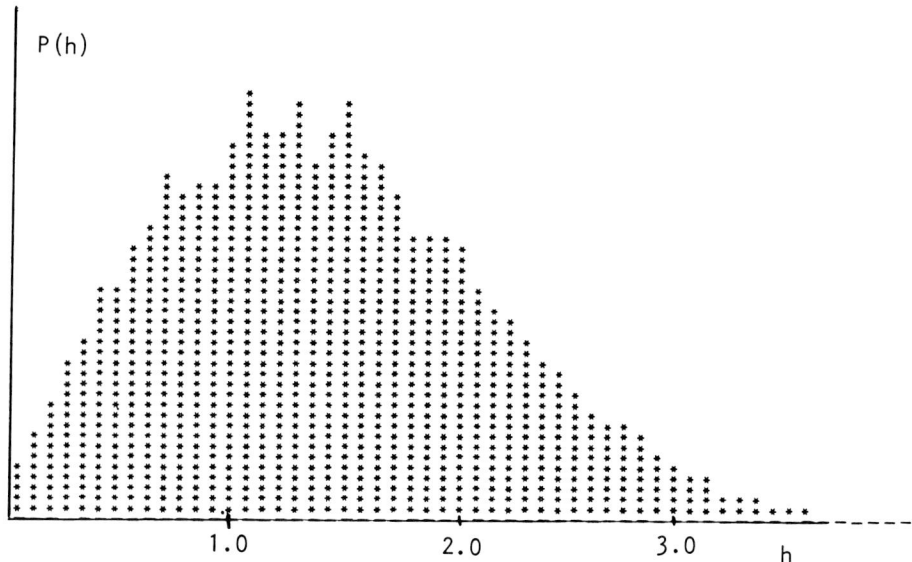

Figure 7 - Histogram of a numerical calculation of P(h).

but this free energy is meaningless unless the ring diagram series converges. We believe in fact that states with m_i^2 less than the largest solution of the mean field equation are forbidden by this ring sum condition: they have anomalously high free energies. What we find, when we look at F in the region near T_c and near $T = 0$ where we can solve the problem, is that it has the peculiar structure shown in Fig. 8: it has a horizontal inflection point at the mean field solution, i.e. $F \sim [(m_i)-(m_i(0))]^3$ <u>at</u> <u>all</u> temperatures. Now obviously one must identify our converging mean field m_i^2 (actually m_i, but that makes no difference to these considerations) with the order parameter q of the other types of theory, and what this says is that <u>at</u> <u>all</u> $T < T_c$

$$F(q) \sim |q-q_0|^3$$

THEORIES OF SPIN GLASS

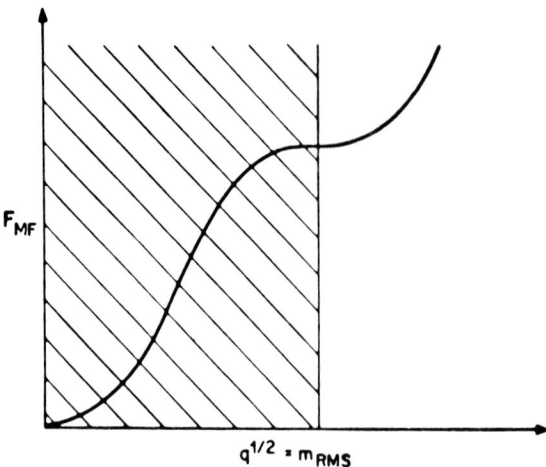

Figure 8 - Schematic of $F(m_i^2)^{1/2}$ below T_c, showing exclusion region.

not, as in usual phase transitions, $F \sim (q-q_0)^2$. If q is to be taken seriously as the order parameter, this means that not just the transition temperature, but all $T < T_c$, has some of the fluctuation properties of a critical point: the critical fluctuations of q persist to all $T < T_c$.

This is the point at which I must go on to the last, and most speculative, part of my talk. So far, we have been discussing results obtained with a very crude model in which, in order to make it exactly soluble, we have assumed infinitely long range interactions - or to put it another way, we are discussing the mean field theory only and not allowing any effects of fluctuations such as nowadays are handled by renormalization group methods. A very important first step towards a real theory was taken by Lubensky and Harris[12] when they started to speculate what kind of renormalization group theory could be based on a mean field theory in which the order parameter is a variable like q. The first thing they found is that such a theory would be very different from the conventional critical point theories because of the nature of the variable q. q, being a correlation function, has the property of being essentially positive. This property can be expressed in terms of the replica version of q by

$$\text{Tr} q^3 = \sum_{\alpha \beta \gamma} S_{i\alpha}^2 S_{i\beta}^2 S_{i\gamma}^2$$

$$= n(n-1)(n-2)$$

One effect of this is that in the mean field $F(q)$ odd power terms such as q^3 can appear, where for an ordinary magnet one may not have M^3, only M^4. This, as Lubensky points out, changes the upper critical dimensionality at which mean field theory is correct from 4 to 6, so that critical fluctuations are much worse. Our TAP mean field theory, incidentally, has exactly this property, <u>not only at</u> T_c, but rather surprisingly a similar behavior at all $T < T_c$ as well. What this means is that even in mean field theory the whole temperature axis below T_c is a line of critical points: a phenomenon which occurs in a few other situations (like the 2-D xy model) but never in mean field theory. Since it is my belief that TAP will in the end prove to serve as a mean field "Ginzberg-Landau" free energy about which fluctuations must be taken into account, fluctuation phenomena will be very large at <u>all</u> $T < T_c$, not just near T_c.

One of the possible consequences of this increase in the strength of fluctuations is an increase in the "lower critical dimensionality" or "marginal dimensionality" below which no ordering is possible. In ordinary systems this is 1 for Ising and 2 for Heisenberg models. I believe I can show that this is 2 for Ising spin glasses and 3 for Heisenberg ones. Thus in physical spin glass systems we are in a very difficult and marginal situation in which the existence of a q is only barely possible if at all, and in which in any case very long range and slow critical fluctuations are occurring at T_c, and even possibly far below it. Under these circumstances I believe it is not at all surprising that the very complicated large-cluster phenomena, which we call Mictomagnetism in spite of Ralph Hudson, occur and dominate the behavior. It may be a long time before they or such things as C_{sp} and field sensitivity, can be understood in detail, but I certainly do not despair of doing so.

In closing let me caution you that a great deal of the last part of my talk is preliminary and speculative. This does not apply to the TAP mean field results, but to all implications I may draw from them as to real systems. But every experimental and theoretical indication leads me to believe that these delightful and unexpectedly elegant mathematical complications may really have a great deal to do with the most inelegantly named phenomena of mictomagnetism.

REFERENCES

1. V. Canella, J. A. Mydosh, and J. I. Budnick, J. App. Phys. <u>42</u>, 1689 (1971); V. Canella, In <u>Amorphous Magnetism</u> Hooper and

de Graaf, eds., (1973) p. 195; V. Canella and J. A. Mydosh, Phys. B $\underline{6}$, 4220 (1972).

2. P. W. Anderson, <u>Amorphous Magnetism</u>, op cit., p. 1.

3. C. Schlenker, B. K. Chakraverty, J. de Physique, to be published (Autrans Conference Proceedings).

4. J. S. Kouvel, J. Phys. Chem. Solids $\underline{24}$, 795 (1963); P. A. Beck, Met. Trans. $\underline{2}$, 2015 (1971).

5. L. E. Wenger and P. H. Keesom, Phys. Rev. B $\underline{11}$, 3497 (1975).

6. K. Binder and K. Schroder, Solid State Communications $\underline{18}$, 1361 (1976); Binder, to be published.

7. L. R. Walker and R. E. Walstedt, International Conference on Magnetism, Amsterdam, 1976 (to be published).

8. S. F. Edwards and P. W. Anderson, J. Phys. F $\underline{5}$, 965 (1975).

9. D. Sherrington and S. Kirkpatrick, Phys. Rev. Lett. $\underline{35}$, 1792 (1975).

10. K. Fischer, Phys. Rev. Lett. $\underline{34}$, 1438 (1975).

11. D. J. Thouless, P. W. Anderson, R. G. Palmer, Phil. Mag., to be published.

12. A. B. Harris, T. C. Lubensky, J.-H. Chen, Phys. Rev. Letter $\underline{36}$, 415 (1976).

13. C. E. Violet and R. J. Borg, Phys. Rev. $\underline{149}$, 540(1966) and other work; A. T. Fiory, this conference.

14. R. G. Palmer, to be submitted.

DISCUSSION

R. Tournier: Do you consider the remanent magnetization an intrinsic property of a true spin glass?

P. W. Anderson: The one thing I know is that these phenomena can almost be infinitely complicated. It is very likely that we have here a question of an enormous range of time scales so that something like the remanence could be basically the same whether you measure it on a time scale of 100 sec or 10^{-5} sec and yet if you measure at infinity it might be zero.

N. Rivier: I would like to make sure I understand correctly what you said when you mentioned in your theory that the slow time dependence of the order parameter was dependent on the fact that the free energy was a saddle point.

P. W. Anderson: Let us put it that that will cause a slow time dependence but that there may be other causes as well.

T. Mizoguchi: Can you comment a little on why the spin glass system appears to be field dependent at the transition temperature?

P. W. Anderson: I think I understand why but I have not proven why the system is so very field dependent. Incidentally, in the Thouless, Anderson and Palmer model we do not get an anomalous field dependence. To get that I think one has to include critical behavior. However, everything one believes says that if you are this close to the lower marginal dimensionality you have what is called an infinite order phase transition in which the correlation length is growing exponentially with $(T - T_c)$ which implies that the field dependence will be logarithmic. The specific heat already tells us that you have something like an infinite order phase transition because that exponent is essentially one over the order of the phase transition so that both the specific heat and the field dependence would be explained if you had something close to an infinite order phase transition.

T. Kaneyoshi: In dealing with the theories of spin glass do you think it is necessary to use a gaussian distribution function?

P. W. Anderson: Not at all, for this infinite range thing it would not matter if you use a gaussian or not. The distribution function would become gaussian because the law of large numbers hold. However, in any case I do not think it is necessary.

Disorder: A Frontier of Theoretical Physics
Ziran Zazhi **4**, No. 2, 83 (1981) (*Nature Journal*, Shanghai, China)

A somewhat popularized view of developments as of that time in spin glass, localization, etc. I visited China to give lectures at Qing Tao Institute of Technology, and was fascinated to lecture in the actual buildings where the "Hundred Day War" which inaugurated the Great Cultural Revolution was fought.

DISORDER: A FRONTIER OF THEORETICAL PHYSICS

P. W. Anderson

Bell Laboratories
Murray Hill, New Jersey 07974

and

Princeton University*
Princeton, New Jersey 08544

Physics is a subject of infinite variety, a discipline which allows one to approach Nature in many aspects and from many points of view. Some physicists focus on the artificial creation of energetic particle beams which blow apart the most elementary building blocks of Nature and reveal the hidden inner symmetries of the fundamental particles. Others focus on the creation of unique conditions of pressure and temperature in order to mimic reactions otherwise present only inside stars, and to generate energy in new and unnatural ways; yet others turn their eyes to the cosmos, searching for the ultimate beginning in the big bang, or probing the endless reaches of space for new phenomena and unique objects.

These explorations of outer space, or other space, can be contrasted with another kind of physics; namely, the attempt to fully understand the everyday world of the here and now. This kind of exploration of the real nature of everyday objects has an intellectual fascination which, while perhaps a bit more subtle, is no less gripping than that of the other kind of physics, and to those of us engaged in this pursuit, these other areas often seem unsatisfying in being less demanding, and less constrained by reality itself than what we do.

Our métier, then, is to proceed from the rather simple laws of atomic and electronic physics and try to understand in a fundamental and honest way all of the complexity of the world about us. The fascination of this exercise lies in a number of directions, some obvious and some perhaps not so obvious. One is that the intellectual puzzles we encounter are simply in themselves uniquely difficult, mainly because the everyday world is so *large* compared to the atomic level at which our physics begins that a great deal of complexity intervenes: 10^{23} atoms can do very many things which a single atom cannot. What is more, when we think we have solved a problem, we always can confront our hopeful solution directly with experiment: we theorists of condensed matter are at our best when closest to experiment.

A second subtle point is one of the key conceptual facts of the physics of real matter: the way in which the complexity of the everyday world emerges from the simplicity of atomic physics. We have the simplest and most important laboratory for the study of what has been called emergent properties: those properties of objects, including ourselves, which are not built into our microscopic descriptions; properties such as rigidity, superfluidity, ferromagnetism, life, consciousness... there perhaps we make contact not only with deep philosophical questions but also with the world of elementary particle physics where all kinds of hidden entities and hidden symmetries, not manifest in the real world, are today being postulated to exist on the microscopic level.

Finally, I have to mention the one aspect of our work which is intellectually perhaps not its most stimulating, but which is the source of what public enthusiasm one occasionally encounters: the fact that in the end we strive for understanding of the practical objects of everyday life, and again and again this understanding translates itself into useful technology. To emphasize this useful aspect and promise a new energy source for each tiny advance is not honest, and undervalues and misleads people as to our motivations and means. But the

---

* The work at Princeton University was supported in part by the National Science Foundation Grant No. DMR 78-03015, and in part by the U.S. Office of Naval Research Grant No. N00014-77-C-0711.

possibility - even probability - of new technology does exist. We also are of great practical use not only to technology but also to other branches of science, such as, for instance, in understanding the properties of condensed objects such as neutron stars.

This has been a long preamble about aims, but one which I think very much needs saying. For you to understand where I am aiming is far more important than to know how I get there.

In the general field of the physics of matter, the problems of gases, plasmas and other relatively rarified fluids exhibit no surprises on the microscopic level, although their macroscopic behavior remains an active frontier. It is the physics of *condensed* matter: by which is meant solids, and dense fluids, particularly superfluids, where real microscopic problems arise. Very early after the discovery of the quantum laws of atomic physics, the first key steps in understanding most solid matter were taken, when Bloch, Wigner and others recognized the importance of taking the crystalline and other symmetries of the solid into account properly; and solid state physics has been living on these key ideas for half a century, adding to them a number of other concepts contributed by such key theorists as Landau and Onsager.

When confronted with a substance we immediately perceive whether it is rigid or not, whether it is opaque or transparent, and a few simple measurements tell us if it conducts electricity well like a metal or poorly and is an insulator. More sophisticated measurements - but not much more - will tell us if it ferro- or antiferromagnetic, superconducting or superfluid, ferroelectric, or has any of the other gross properties which solid matter can have. Clearly our first chore is a qualitative understanding of these gross properties, and for regular solids we understand them reasonably well by now - although from time to time some pretty spectacular new phenomena do show up, even now. For an example of such a gross property, rigidity is caused by the presence of the crystal lattice (Fig. 1) which melts and disorders in the transition to the liquid, metallic conduction occurs if free quasi-electrons occur at the fermi surface, as we will discuss, while transparency is a result of a large energy gap between bands; the magnetic phases involve a regular structure of the atomic magnetic moments, etc., etc.

After this work, there remained a no-man's land of materials and problems involving condensed phases in which the lack of crystalline regularity which is, in fact, always present, plays an essential role. Characteristically, an isolated imperfection in a crystal is in principle as easy to deal with as a free atom (Fig. 2) while a low density of imperfections scatter the various kinds of waves characteristic of the perfect crystal without changing their character radically, (Fig. 3). But what if our concern is only with the gas of imperfections - as in the so-called impurity banding phenomenon in semiconductors (Fig. 4) or if the crystal contains no "perfect" regions at all, but is entirely an irregular structure like glass? (Fig. 5) The conventional theory of solids gives no answers, or quite incorrect ones, to even the gross questions about such systems. Until recently, these problems had been left almost exclusively to the more empirical chemical approach, which while often very succesful in developing new materials and devices, does not make much contact with underlying theory. It was only about 20 years ago that a first real basis for dealing honestly with such problems began to be laid, which was my contribution. I would like here to give you an overview of a few kinds of problems which are encountered in such systems, some of which are still mostly or partly unsolved; then concentrate on the most fundamental and most nearly solvable, the problem of electron localization and metallic conduction in disordered systems. (Fig. 6)

As an example of a problem which is in my opinion still unsolved, in fact perhaps the most puzzling and important unsolved problem in condensed matter physics, I give you the glass transition. We believe we have at least a qualitative understanding of what happens when a liquid freezes into a conventional solid such as ice or lead or rock. The atoms of the liquid, which were interchanging freely with each other, presumably because they occupy fluctuating, random positions, suddenly sit down in a regular lattice and, as a consequence, relative motions are now impeded. The energy is quite clearly dependent on the amount of distortion of the lattice which leads to elastic rigidity. (Fig. 7) It is not because any one bond in the crystal can't break but because they all work together to form a rigid structure, that we observe rigidity in crystals.

Now let us look at a liquid-glass transition as we supercool a glass-forming liquid below its true melting point. (Fig. 8a) In this case, a microscopic examination shows no perceptible distinction between liquid and glass. What is more, instead of happening as a *sudden,* discrete thermodynamic phase transition - which, as pointed out by Landau, melting must be, because it is a change in symmetry (or *broken* symmetry) - all that happens is a *rapid* but *continuous* increase in viscosity when one reaches about 2/3 of the melting temperature of the corresponding solid. (See Fig. 8) This increase in viscosity obeys a law discovered over 52 years ago by Vogel:

$$\eta = \eta_0 \, e^{A/T - t_o} \qquad (1)$$

with T_o less than T_g, the glass-forming temperature, by 10-20%. (See Fig. 9) As far as I know, Eq. (1) is still not plausibly explained; and since it is the key fact of the glass transition, we are at a complete loss to understand - as opposed to describe - the glass-forming phenomenon.

I will pass over the next most mysterious item in my agenda - the mysterious amorphous magnetic phase known as the spin glass - and go on to the question of conduction in random systems, where the record is by now *almost* a success story.

Let me first explain to you the conventional solid state textbook theory of metallic conduction and insulation. This depends on a terribly simple argument first given in the early days of quantum mechanics about solutions of Schrödinger's equation in a periodic array of atomic potentials, such as one has in a regular crystalline solid, using the symmetry properties of this solid. Starting from Schrödinger's equation

$$(T+V)\psi = H\psi = E\psi$$

in such a system with $V(r+a) = V(r)$, we observe that

$$H(r+a) \, \psi(r+a) = E\psi(r+a)$$

or

$$H(r) \, \psi(r+a) = E\psi(r+a)$$

so $\psi(r+a)$ is also a solution

$$\psi(r+a) = e^{ka}\psi(r)$$

k can be either real - in which case ψ blows up exponentially when one steps off a large number n of steps in this way:

$$\psi(r+na) = e^{kn} \, \psi(r)$$

and this is a forbidden energy. On the other hand, where k is pure imaginary

$$\psi(r+na) = e^{ikn}\psi(r)$$

and the wave function behaves very much like that of a free electron plane wave - it is a so-called *Bloch wave.*

Thus there are *bands* of allowed energy states, where the electrons are quasi-free, and bands where no states are allowed. (See Fig. 10) These empty regions are called "band gaps" and are a purely quantum effect. When all the states in a band are full, it will cost a lot of energy to excite an electron to the next band and as a result, the material is transparent and cannot conduct electricity. This is a so-called *insulator*. If, on the other hand, after the Pauli exclusion principle is satisfied, we nonetheless find that the last energy state is in the middle of a band, the electrons conduct electricity freely and a photon from outside can be absorbed at any energy we like, so that the material is opaque and shiny - we have a metal. This clear distinction, corresponding very much to our everyday experience of dichotomy between the two states of matter, was the earliest triumph of the quantum theory of solids.

But it is clear that this explanation leans heavily on the periodic structure of the material. All real materials are impure and irregular to some extent, so it is always relevant to ask what is done to this picture as we make the substance more irregular.

Two things happen when we do so. One is that the free electron-like states are scattered by the impurities, so that they do not travel indefinitely but have a correlation in velocity $\vec{v}$ or in momentum $\vec{k}$ only over a certain distance, the mean free path l. It is easy to get the conductivity by a simple kinetic theory argument. Applying an electric field E, the electron acquires a velocity $\delta v = \dfrac{eEl}{mv_F}$ before it is scattered again. Here v_F is the "Fermi velocity" enforced by the Pauli exclusion principle which pushes the electrons up to states with a finite velocity. For a simple band, we have that $mv_F = p_F = \hbar k_F$ with $k_F^d \approx n$, the total number of electrons (d is the dimensionality).

This mean velocity leads to a current J and a conductivity $\sigma = \dfrac{J}{E} = \dfrac{ne^2 l}{mv_F} = \dfrac{e^2}{\hbar} k_F^{d-2} (k_F l)$. (All of this if the system is a metal and has electrons at a Fermi surface in the middle of a band.) Thus irregularity limits the otherwise infinite conductivity but at first it doesn't change things radically. But as we make the material more irregular, we run up against a fundamental quantum limitation pointed out by Yoffe: the mean free path l can't be much less than the de Broglie wavelength $\lambda_F = \dfrac{2\pi}{k_F}$ so that $k_F l > 1$. Thus there is a *minimum* metallic conductivity

$$\sigma_{min} \gtrsim \dfrac{e^2}{\hbar} k_F^{d-2}$$

What happens then? Mott proposed that at this point something very radical happens: conductivity (except for thermally activated motions) stops altogether !

To see how you could come to this conclusion, go on to the second effect which irregularity has. In the irregular material, the potential energy must fluctuate from place to place, so that some parts of the material are especially favorable and others unfavorable. (See Fig. 11) Thus the band edges, which are *sharp* in the regular material, smear out and in particular there will locally be attractive places in the material where there will be bound electronic states which fall off exponentially from the attractive centers. Thus there will not really be sharp occupied bands and sharp empty ones: The Fermi level will always sit in a region where there is a finite density of electronic states.

If one knows a little quantum mechanics, but not too much, one comes to the conclusion that these exponentially localized bound states always can carry current because the electrons can tunnel, by the quantum mechanical tunneling effect, from one to another and really they just form another version of an energy band: the so-called impurity band. In fact, when there are high densities of impurities, we *do* observe so-called "impurity conduction" which is like metallic conduction in being temperature-independent, but for λ_F we have to substitute a, the distance between these impurities, in our metallic formula. (See Fig. 12)

What I discovered two decades ago, however, is that taking a little more care with our quantum mechanics, leaves us with the possibility that the tunneling cannot take place in this regime: that the exponentially localized wave functions can overlap and still not delocalize, in that they simply may not be able to conserve energy - the wave function may have no others of the same energy close enough to tunnel to. This is indeed what happens at Mott's minimum metallic conductivity: everything stops and no further conduction takes place except by activated motion. I show another figure (Fig. 13) of the hundreds or thousands of experimental data which now exist showing that such a real transition takes place. In particular, it is striking that data already existed in 1914 which showed that Bi films *never* exhibited metallic resistance per square higher than 30,000 ohms, which happens to be approximately $\dfrac{h}{e^2}$.

My paper was written in 1958, and until 1968 only Mott showed any interest. In 1968 the rest of the world began to rediscover our work, a process which completed itself in 1977, a decade later, with the Nobel Prize. It may be timely that 20 years later we are in process, we think, of taking yet another major step in the understanding of the limitations on metallic conductivity.

Like so much of modern solid state physics, this is a story of dimensionality: the realization that physics can change radically in flatland or in four dimensions, and that dimensionality can even be treated as a continuous variable. One of the very first results added to my ideas by Mott was the observation that localization is especially true in one dimension. That is, any irregular potential in one dimension causes localization. David Thouless, using his important insight that $\hbar/e^2$ is a universal number of ohms, expressed this very succinctly by saying that *any* linear metal system with a resistance higher than about 10,000 Ω will show localization. This at first seems an absurdity: we are all familiar with fine wires - slide-wires, wire-wound resistors, which are good metals and have high resistance. But David did mean it: he showed that the key question was whether the electrons were scattered inelastically by phonons, or by electron or nuclear spins *before* they reached their limiting localization length, or not. This then meant that it was necessary to fabricate extremely fine wires and take them to extremely low temperatures to show the failure of metallic conductivity. This has been done by Dolan and Osheroff, at Bell, and by Giordano of Yale: I show a figure (Fig. 14) of Dolan and Osheroff's data for a one-dimensional wire. The nonlinearity of current vs voltage is because the voltage *heats* the electrons, increasing the rate of inelastic scattering which helps the electrons delocalize. These wires are ≈ 1000Å wide, 40 thick, and 1-10 mm long; the temperatures $\approx .020°$K. This gives us a remarkably clean way of demonstrating the presence of localization and probing its laws, but it also means that one must operate with *extremely* sensitive and noise-free apparatus. We have a striking confirmation of David's remark.

More exciting, really, is the fact that in the course of rethinking the theory and of searching for this effect, we have suddenly come upon the fact that localization is universal in flatland, too ! *There are no 2-dimensional metals at the absolute zero of temperature.*
It is only in the ordinary 3-dimensional world that real metals exist. It was already guessed that two dimensions is what is called a "critical dimensionality": in $2 + \epsilon$ dimensions, there are metals, while $2 - \epsilon$ behaves like 1d in being localized. But only now do we realize that the dividing line is at $2+\epsilon$, and in 2 dimensions we have a whole new phenomenon of *logarithmic localization:* conductivity falls off with the logarithm of the size of the film, or of the inelastic scattering length and hence of V and T. This behavior experimentally demonstrated is shown in the last figure, #15.

This work on the conductivity of impure metals typifies the kind of result which makes condensed matter theory so fascinating: a careful and thoughtful examination of apparently trivial concepts dealing with the everyday world of metal films and wires suddenly opens in front of us a chasm of ignorance and a dazzling new result. The world around us seems an inexhaustible reservoir of subtlety and complexity.

Some General Thoughts About Broken Symmetry
Symmetries and Broken Symmetries in Condensed Matter Physics,
ed. N. Boccara, IDSET (Paris, 1981), pp. 11–20

This is my most complete summary of the theory of broken symmetry in condensed matter systems.

Some general thoughts about broken symmetry

P. W. Anderson

Bell Laboratories, Murray Hill, New Jersey 07974, USA

and

Princeton University *, Princeton, New Jersey 08544, USA

I. Introduction

It is appropriate to commemorate the discovery of piezoelectricity, one of the most useful of broken symmetry phenomena, with a general discussion of broken symmetry as it manifests itself in condensed matter systems. Arguments from broken symmetry, however, go back even further in French science : Louis Pasteur's deduction that fermentation was a spontaneous life process on the basis of the optical activity of fermentation products is to me one of the most miraculously early and deep insights in the history of science. It is striking that so much of that history has taken place within a few hundred meters of where we now stand.

The more theoretical physicists penetrate the ultimate secrets of the microscopic nature of the universe, the more the grand design seems to be ultimate symmetry and ultimate simplicity. But all of the interesting parts of the universe, at least to us, are, like the earth itself as well as our own bodies, markedly complex and markedly *unsymmetric*. In the most elementary sense, then, we are surrounded by « broken symmetry », the result undoubtedly of some sequence of catastrophes. What I want to do here is to discuss the general rules which govern this process of the development of complexity and the breaking of symmetry in particular kinds of cases.

In particular, I want to point out that there is a complete, rather satisfactory theoretical structure describing one particular type of broken symmetry, namely that which occurs in equilibrium condensed matter systems, such as the crystals which exhibit piezoelectricity. It is this kind of broken symmetry object which allows us to build structures, communicate, make measurements, calculate, locate ourselves — in essence, carry on all the everyday business of life.

It is clear that an equally important kind of broken symmetry occurs in many dissipative systems when they are driven far from equilibrium. These systems exhibit nonlinear instabilities at which new structures appear, which have often been called [1] « dissipative structures » and discussed on a parallel basis with the equilibrium structures of broken symmetry systems. Surely there are obvious parallels between these two types of systems — for instance, they both come under the general umbrella of « catastrophe theory » [2], and the second kind often — but not always — changes the symmetry of the system in which it occurs, if only in some trivial fashion.

What I want to do here is to pin down the properties of condensed matter systems exhibiting broken symmetry in some detail. After all, the existence of broken sym-

* The work at Princeton University was supported in part by the National Foundation Grant No. DMR 78-03015, and in part by the U.S. Office of Naval Research Grant No. N00014-77-C-0711.

metry in itself may not be of any use or significance : for instance, fully developed turbulence can be thought of as a broken symmetry state, yet seems to be totally chaotic and without definable structure. When one speaks of « dissipative structures » and makes the analogy to equilibrium phases, there is an implied analogy to their structural properties. Thus it is relevant to exhibit these and their underlying sources.

The second part of the talk will consist almost entirely of questions rather than answers. To my mind there exists neither a theoretical nor an experimental basis for deciding whether (or not) dissipative systems have structural properties analogous to equilibrium ones, and I am essentially presenting a program of questions one may ask in this field.

II. Equilibrium broken symmetry and the concept of rigidity

Landau was the first person to emphasize the important role which symmetry plays in the phase transitions of equilibrium condensed matter systems [3]. It is this, rather than Ehrenfest's concept of 1st vs. nth order, which gives us our most fundamental classification scheme of phases and the most basic theorems of solid state physics.

In modern terms, one envisages a basic symmetry group of the underlying particles and of space, G_0, containing not only such elements as rotations (isotropy of space) and translations (homogeneity of space) but time-reversal invariance, in some cases spin-rotation invariance, etc. Landau observed the very important fact that condensed states often exhibit lower symmetry than G_0; for instance, where a molecular liquid is homogeneous and isotropic and exhibits the group G_0, a nematic liquid crystal is anisotropic, a smectic one inhomogeneous. The new group of, for instance, the nematic, G_1, has only rotation symmetry about the director D.

FIG. 1a. — Nematic liquid crystal in the disordered state. The line segments represent the rodlike molecules of the nematic. Averaging molecular orientations over macroscopic distances yields zero.

FIG. 1b. — For a suitable choice of thermodynamic parameters, the nematic enters the ordered state, with the appearance of a macroscopic order parameter (the director **D**). The system is no longer isotropic, but has chosen a special direction : rotational symmetry has been broken.

Thus phase transitions often involve a change of symmetry. We classify phase transitions into 3 basic classes :

1) Same symmetry in the two phases : as, e.g., liquid-gas, the Mott metal-insulator transition [4]. In this case the transition may be first-order, and the line of first-order transitions can, and often does, end in a critical point where the free energies of the two phases coincide and the transition simply disappears. Too often the formal similarity of this case to certain aspects of case 2) is allowed to obscure the physical difference in principle.

2) G_1 is a subgroup of G_0 : the symmetry of G_0 is « broken ». In this case there can never be a disappearance of the transition line, which may be first or second order depending on details. *Symmetry cannot change continuously* : what I have called the First Theorem of condensed matter physics.

Since the two phases differ in symmetry, they may have the same free energy and the same physical parameters given by the derivatives of the free energy such as $T = \partial F/\partial S$, $P = \partial F/\partial V$, and yet be distinct : a higher-order transition is permitted, even along a line of critical points, which line may continue into a line of first-order transitions.

3) The uninteresting case is unrelated symmetries G_0 and G_1 ; in this case only a first-order transition is possible, since only by impossible coincidence could all parameters coincide.

In order to do this Landau added one concept which is vitally important to all of these questions : that of the « order parameter » η. Unfortunately, the concept of the order parameter still remains somewhat mysterious, although in most simple cases its choice is obvious. This concept is only of interest in the « broken symmetry » case 2). In case 1), any of the relevant thermodynamic variables will serve to distinguish the degree of difference of the two phases, as for instance in the Mott transition the number of free carriers, etc.

In the broken symmetry case where a loss of symmetry is present, Landau introduces an « order parameter » η indicating the degree of broken symmetry which, as he points out, is usually also the degree to which the system has « ordered » — for instance, in the liquid crystal, the degree to which the molecules are no longer arbitrarily oriented but have now oriented themselves along a specific direction.

In this example, one might use $\langle \cos^2 \theta \rangle - 1/3$ as the order parameter, where θ is the angle of a molecule with the director D. This « order parameter » is very important in many ways, not the least of which is that it is a new thermodynamic variable, and often either is or contains a dynamic one as well.

As far as I know there is no complete characterization of the « order parameter » but it must certainly have the following important properties (and I have the impression that the term is, unfortunately, often used without understanding these restrictions) :

1) It must be a variable which is operated on by the group generators of $G_0/G_1 = H$, the group representing the amount of lost symmetry. This is in fact often called the « group of the order parameter » and defines « order parameter space » [5, 6].

2) Since these group generators — at least for all continuous groups G_0 — are

necessarily *dynamical* as well as *thermodynamic* variables, since they are operators in the Hilbert space of the underlying quantum mechanics, the order parameter must *contain*, at least, a dynamical variable or variables conjugate to these.

3) Finally, the order parameter must be a useful quantitative measure of the degree to which G_0 has been broken. It has then a *thermodynamic* as well as *dynamic* significance. In such cases as order-disorder transitions (described by the Ising model) where the broken symmetry is discrete, this may be its only role : the order parameter can be merely the difference in mean population of the two or more relevant states on a given site.

In the more interesting continuous situation, there are two cases. First, there is the case of spins, where the algebra is totally defined by the spin components, which are themselves the generators of spin rotations. In this case the group generators, S_x, S_y, S_z, are selfcontained, and their mean value $\langle S \rangle$ can be taken as the order parameter. Since the average of any of these group generators is, by symmetry, a constant of the motion, the ferromagnetic case represents a very rare example — a case in which the order parameter is also a constant of the motion, which has important consequences for the nature of the relevant Goldstone bosons. It is too bad that in illustrating the concept of broken symmetry, the ferromagnetic example is often used : it is extremely atypical.

More typical in its complication is the antiferromagnetic case. Here an additional but discrete symmetry is broken, that of the lattice translation, and the order parameter is not the mean magnetization but the sublattice magnetization, or more generally a Fourier component of the magnetization. While this is not a constant of the motion, it still can be related to generators of the continuous part of the symmetry group.

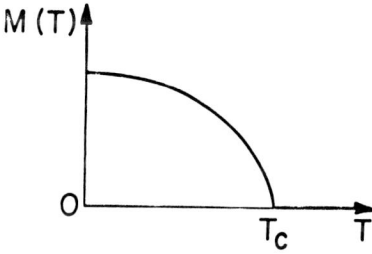

FIG. 2. — Variation of magnetization M with temperature T in a simple ferromagnet. This is a typical second-order phase transition, in which the order parameter grows continuously from zero as T is lowered below a critical temperature T_c.

The cases of density waves — solid lattice or smectic liquid crystal — and superfluidity are more typical. Let us first discuss the simple one, superfluidity [7]. Here the broken symmetry is the gauge symmetry of interactions which locally conserve particle number, and the group generator is then the number operator, which may be written $i\dfrac{\partial}{\delta \varphi}$, φ the phase or gauge variable. But the phase itself is not a suitable order parameter because it is periodic and its origin is meaningless. Hence it is natural to use $\langle e^{i\varphi(r)} \rangle = \psi(r)$ as an order parameter.

This exhibits the most straightforward type of order parameter. The *phase* of ψ is the relevant dynamical variable, independently of the magnitude which is purely statistical in nature. This is the general rule: the order parameter contains a phase-angle like quantity which is both a *dynamical variable* and *reflects the original symmetry* — in the sense that the free energy cannot depend on the *value* of φ, only on relative values in different parts of the sample. φ may move about freely in the space defined by the group H (in this case the group of gauge symmetry $U(1)$). In the isotropic ferro- and antiferromagnetic cases, « angle » is the spin direction, a dynamical variable free to rotate on the sphere. In real (as opposed to model) anisotropic ferromagnets, the spin retains its dynamical character in such phenomena as domain wall motion and spin waves. The nematic liquid crystal behaves in much the same way.

But the density waves bring in added complications. The phase angle variable in this case is position in space, and it is reasonable to use the Fourier components of the density as order parameters:

$$\rho_G = \langle e^{i\mathbf{G}\cdot\mathbf{r}} \rangle.$$

Clearly, this contains the displacement **u** as a phase parameter:

$$\rho_G(u) = \langle e^{i\mathbf{G}\cdot(\mathbf{r}+\mathbf{u})} \rangle$$

is a density wave displaced by a distance **u**. Thus the strain **u** is the dynamical variable and $|\rho_G|$ the statistical one for this case.

In the crystal lattice case, where the number of independent G's is equal to the dimensionality of space, **u** is the only independent angle variable in a certain sense. This is despite the fact that **G**, of course, also can rotate freely in space. However, as Halperin and Nelson have emphasized [8] (and as was already evident in the old Shockley-Read construction of a grain boundary as a dislocation array) one must introduce a large, finite density of defects in the strain, essentially destroying *positional* order, before *orientational* readjustment of parts of the crystal independently is permitted. Thus in the true lattice case, only the strain is a relevant dynamical object, although the crystal overall may rotate as well.

This is not the case in the cholesteric and smectic, which are perhaps the most difficult to characterize in principle of all ordered states. Here strain **u** *and* **G** are both free angle variables, but there is what Volovik and Mineev [9] have called an « integral constraint » restricting the angular variation of G: namely,

$$\oint \mathbf{G}\cdot d\mathbf{s} = 0.$$

That is, the layers may bend but the distance between layers must be constant. The nature of singularities, at least, in these two cases is not understood as yet.

As important and little-appreciated caution must be added on the dynamics of the density wave situation. The « strain » or « position » variable « u » is that of the density wave itself, which only *usually*, not always, means the position of the whole substance. As Overhauser first pointed out for spin density waves [10], and Leggett for solids [11], the mass being carried as the density wave moves by the density wave fluctuations or « phasons », *may* or *may not* be the whole mass density. This question is a subtle and complex one related to the question of Mott or Wigner metal-insulator transitions. If there is a true energy gap for excitation of particles in the self-consistent lattice potential — as there is for all real solids, but not for smectics or electron density waves in some cases — the lattice carries all the mass.

From the order parameter and its dynamic nature flow many of the useful and important properties of condensed broken symmetry systems: the Goldstone and Higgs boson excitations, the long-range elastic-like forces (such as Suhl-Nakamura interactions in magnets) but most important of all the property I call *generalized rigidity*. Important examples of generalized rigidity are true rigidity, superconductivity, superfluidity, and hysteresis in magnets.

The order parameter, which is invariably a thermal average of some local quantity, can be defined locally, at least if its variation is sufficiently slow. The magnetization direction of an antiferromagnet, or the phase of a superfluid, can vary from place to place in a given sample. It is only reasonable to suppose that the extra free energy caused by such variation grows only slowly as the rate of variation increases: i.e.,

$$F = F_0(|\eta|, T, --) + F_1(|\varphi|, T) : (\nabla\varphi)^2 + \cdots + \alpha(\nabla|\eta|)^2 + \cdots.$$

We write the gradient terms only schematically, as far as their tensor character goes.

This free energy determines the degree of fluctuation of η and hence φ by conventional Gibbs theory (treating η as a conventional thermodynamic variable whose local average value we can constrain at will for purposes of calculating F). F_0 does not depend on φ at all, by our original symmetry. The $(\nabla\varphi)^2$ term is necessary if the broken symmetry ordered state is to be stable. This equation implies the rigidity property. First, it is clear that with even a very small force applied only locally we can move φ about at will in the whole sample. This is because the dependence of F_0 on the origin of φ vanishes by symmetry, so that φ can move about at will; while the $\nabla\varphi$ term will enforce uniform φ. Equally, without breaking down ordering, i.e., increasing F_0 at least locally, a force applied on φ at one end of the sample will be transmitted to the other: this is *rigidity*, and I consider it one of the key consequences of broken symmetry in condensed systems, since only with rigidity can structures be formed of these systems, or information or energy be transmitted through them.

III. Generalization of the order parameter and broken symmetry concept, especially « dissipative structures »

In several special cases one has attempted to define an « order parameter » which does not have the full properties, for instance, that of being a dynamical variable — the present version of the order parameter in the spin glass is one example [12]. It is not at all clear, at least theoretically and probably experimentally, that true rigidity exists in this system. My own preference is to leave such aberrant cases to one side and recognize that the analogy is a dangerous one.

It is for this reason that I am disturbed by the common uses of the terms « broken symmetry », « order parameter », and « dissipative structure » in the theory of nonlinear instabilities of driven systems. The attempt is to draw the analogy with equilibrium structures of broken-symmetry systems. It is proposed that if such pro-

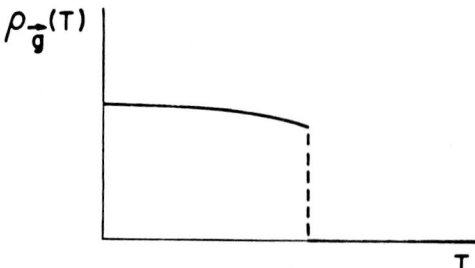

FIG. 3. — In a first-order transition, such as the liquid to solid crystal transition shown here, the order parameter will exhibit a discontinuous jump at the transition with an associated release (or absorption) of latent heat.

perties existed, they would have important consequences in our understanding of the self-organization of living systems [13, 14].

One can argue endlessly about words rather than meanings in this area so I would like to make a very clear distinction between what one might hope to be a *useful* « dissipative structure » as opposed to something which, while it has broken symmetry per se, and contains visible structure and something which might be described as an order parameter, is nonetheless an artefact which does not have the properties which might be useful for purposes of self-organization.

The two properties which I would consider essential for self-organization are :

a) Autonomy.

b) Rigidity.

Both of these are properties of condensed broken symmetry systems, and we may ask if they are exhibited in any well-understood dissipative systems.

FIG. 4. — Illustration (somewhat schematic) of generalized rigidity. An external force (the crank) couples to the order parameter at one end of the system, represented as a gear. A change in the order parameter at any point in the ordered system is transmitted to all other parts of the system (first gear turns the second gear). The second gear turns the second crank : a force has been transmitted from one end of the system to the other *via* the order parameter.

By autonomy of a structure I mean that its space or time structure should not be predetermined in terms of the scale of the external boundary conditions (as opposed to the microscopic scale of atoms or molecules, for instance). An example of a dissipative system which *does* have autonomy of scale is a dye laser or any laser where the precise mode of oscillation is not predetermined by careful mode selection techniques : the wavelength of light is an autonomous microscopic scale irrelevant to the scale of the apparatus. This is not the case in such classical systems as Bénard or Couette convection cells, which are controlled in size by one of the apparatus dimensions. Autonomy is necessary if one is to speak of *self*-organization as opposed to predetermined organization.

The second property, that of rigidity, it seems to me is also essential. If the structure is to be stable, carry out actions, and, above all, to serve as a *substrate for information*, it is essential that it be rigid : that is, that it have the two properties of (1) having internal degrees of freedom which are not predetermined ; (2) having a freedom which may be stably manipulated and which can exert action at a distance.

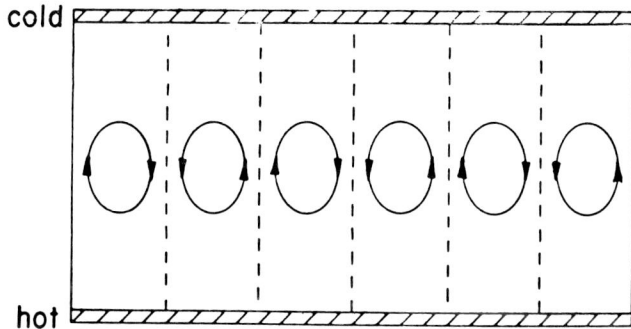

FIG. 5. — The Bénard instability in rectangular geometry. A layer of fluid between two horizontal rectangular plates is heated from below. When a sufficient thermal gradient is reached between top and bottom plates, convection arises in the form of rolls. In this cutaway edge-on view, the arrows represent the fluid velocity.

I am unaware of any work in the literature which demonstrates these properties in any well-understood dissipative system. Most of the conventional hydrodynamic systems which show regular roll patterns are not really autonomous. In these systems one often defines an « order parameter » which is the inhomogeneous component of velocity or flow, and under sufficiently restrictive conditions a kind of Gibbs free energy functional of the order parameter can be derived which gives the equations of motion near the instability by differentiation. But the assumptions which go into this derivation seem to preclude its general use, or the derivation of rigidity properties from it ; nor is it clear that this « order parameter » has the real properties of the condensed systems order parameters, such as freedom to move within an order parameter space, locality, etc.

In lasers and in turbulent systems, as well as in some chemical oscillation systems, autonomy seems to be available but not order or rigidity ; in general, these systems

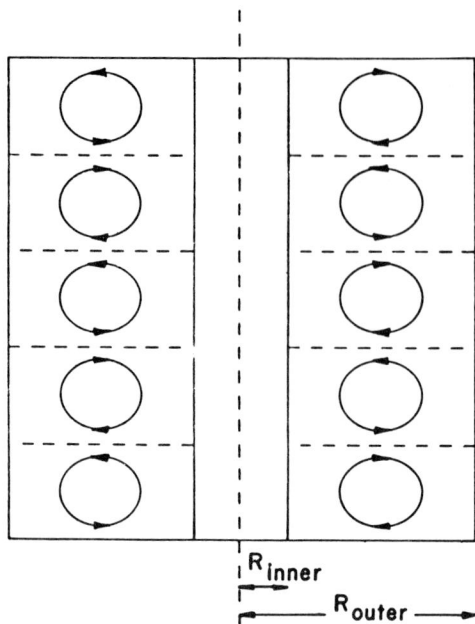

FIG. 6. — Couette flow : A fluid is placed between two cylinders with different rotational velocities about their axes. When the velocity gradient exceeds a critical value, rolls of vortices form. In this view the cylinder is cut along its length.

appear to be very chaotic. Whether this is a general state of affairs needs to be studied more carefully.

If one enquires how living systems seem to go about building autonomous, rigid structures, one finds a fascinating mixture of dissipative and condensation processes at work. One seems to see dissipative stages initiating condensation, such as the formation of membranes, for example, and the condensed systems in turn controlling dissipative stages. Haken has emphasized this structure. It is not clear that the present idea of « dissipative structures » in the theory of nonlinear systems is at all relevant to this process.

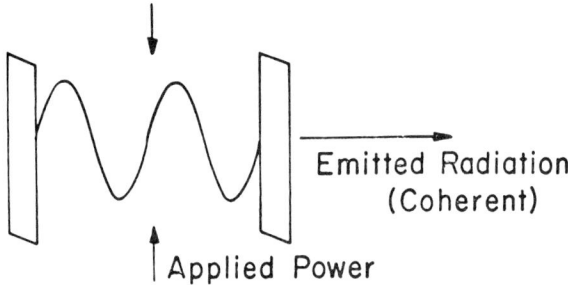

FIG. 7. — In a laser, a standing wave of excitation density is set up between two end plates, or mirrors, resulting in emission of a beam of coherent radiation.

Acknowledgments

The above ideas have evolved over many years and months, and in discussions with many people. Their present form is much influenced by recent discussions with D. Stein, L. Sneddon, A. Iberall and H. Soodak. Some of the ideas about broken symmetry evolved in a lecture course given at Quing Hua University, and I am grateful for their hospitality and the assistance of Charlies Xie.

References

[1] For example, H. Haken, *Synergetics* (Springer-Verlag, Berlin) 1977.
[2] R. Thom, *Structural Stability and Morphogenesis* (W. A. Benjamin, New York) 1975.
[3] L. D. Landau and E. M. Lifshitz, *Statistical Physics* (Addison-Wesley) 1958.
[4] N. F. Mott, Metal Insulator Transition (Taylor and Francis, London, 1974).
[5] L. Michel, *Rev. Mod. Phys.* **52** (1980) 617.
[6] N. D. Mermin, *Rev. Mod. Phys.* **51** (1979) 591.
[7] P. W. Anderson, *Rev. Mod. Phys.* **38** (1966) 298.
[8] See B. I. Halperin, in N. Boccara, *Symmetries and Broken Symmetries in Condensed Matter Physics* (IDSET, Paris) 1981, p. 183.
[9] G. E. Volovik and V. P. Mineev, *Zh. Eksp. Theor. Fiz.* **72** 2256; **73** (1977) 767.
[10] A. W. Overhauser, see for instance *Phys. Rev.* **167** (1968) 691.
[11] A. J. Leggett, *Phys. Rev. Lett.* **25** (1970) 1543.
[12] S. F. Edwards and P. W. Anderson, *J. Phys. F* **5** (1975) 965.
[13] H. Haken, to be published (Proc. Dubrovnik Conference on Self-Organization, E. Yates, ed.).
[14] See G. Nicolis, and J. Prigogine, in N. Boccara, *Symmetries and Broken Symmetries in Condensed Matter Physics* (IDSET, Paris) 1981, p. 35.

The Rheology of Neutron Stars: Vortex Line Pinning in the Crust Superfluid
(with M.A. Alpar, D. Pines and J. Shaham)
Philos. Mag. **A45**, 227 (1982)

The "Glitch" phenomenon is nearly an unequivocal proof that neutron stars are at least partially solid. It is very hard to obtain two time scales differing by $\sim 10^{13}$, with no intermediate scales, without a breakdown of rigidity. Ruderman, Pines and Shaham proposed actual "starquakes" caused by the strains generated by spindowns, but soon accepted the mechanism Itoh and I proposed in a brief letter in *Nature*, of "vorticity jumps", equivalent to "flux jumps" in a superconducting magnet. Our letter was given an award by the *Observer* magazine for "the most incomprehensible title of the year". Pines and Shaham are responsible for introducing me to the Aspen Center for Physics, which played a great role in my life on my return to the US in 1975 and for 15 years thereafter. We wrote too many papers on too few data, of which this is a typical one, but this is perhaps natural in view of the generally apathetic response to our work in the astrophysics community.

The rheology of neutron stars
Vortex-line pinning in the crust superfluid

By P. W. Anderson

Bell Laboratories, Murray Hill, New Jersey 07974, U.S.A.
and Princeton University‡, Princeton, New Jersey 08544, U.S.A.

M. A. Alpar†

Bogazici University§, Istanbul, Turkey
and Princeton University‡, Princeton, New Jersey 08544, U.S.A.

D. Pines†

University of Illinois, Champaign-Urbana, Illinois 61801, U.S.A.

and J. Shaham†

Racah Institute of Physics, Hebrew University, Jerusalem, Israel

[Received 10 April 1981 and accepted 15 July 1981]

Abstract

After a discussion of the general physics of neutron stars, we give a brief discussion of the 'glitch' phenomenon and its relation to superfluidity, and finally a rather detailed study of the physics of vortex-line pinning in the crust lattice.

§ 1. Introduction

Very soon after the discovery of pulsars and the convincing evidence that these were Landau's 'neutron stars', rotating at frequencies of about a hertz, observation began to show very-large-period irregularities in their rotations. The famous Crab pulsar, 900 years old and still rotating about 30 times per second, showed noisy irregularities including one, or sometimes a few, large events (which came to be called 'glitches' rather than noise) where the rotation frequency changed suddenly by $\Delta\Omega/\Omega \sim 10^{-9}$–$10^{-8}$. The Vela pulsar is somewhat quieter on average, but has up to now shown four giant glitches where the period changes by $\sim 2 \times 10^{-6}$. These are enormous events in terms of total energy ($\sim 10^{43}$ erg), although unaccompanied by visible radiations. Two other,

---

† Work supported in part by the National Science Foundation, through Grant NSF PHY78-04404 and by NASA Grant NSG-7653.
‡ The work at Princeton University was supported in part by the National Science Foundation Grant No. DMR78-03015 and in part by the U.S. Office of Naval Research Grant No. N00014-77-C-0711.
§ The work at Bogazici University was supported in part by the Space Sciences Research Unit of the Scientific and Technical Research Foundation of Turkey.

© 1982 Bell Telephone Laboratories, Incorporated

older and slower, pulsars have exhibited giant Vela-style glitches, as well as, possibly, the X-ray stars Her X-1 and Vela X-1.

Ruderman (1969) had already, before the first of these observations, pointed out that the neutron star is covered by a solid crust of nuclei in a crystalline lattice, embedded in a Fermi gas of neutrons and electrons, the former gas increasing sharply in density with depth. A solid layer naturally provides the rigidity necessary for catastrophic breakdowns, such as 'starquakes', and a starquake mechanism was immediately proposed (Ruderman 1969, Baym and Pines 1971, Pines, Shaham and Ruderman 1973).

Apart from the planets, these glitches are the only evidence of rigid matter anywhere in the universe. Everything else is either gaseous or at best dusty, as far as we can ascertain. Even the white dwarf stars are likely to be too hot to solidify, and in any case show us no evidence of rigidity. It is only the coincidence of having the remarkable rotating 'searchlight' of pulse emission from the neutron stars (as yet by no means completely understood) which allows us to time their rotations with spectacular accuracy, and hence to follow the internal dynamics of these stars. It is now accepted that the energy source which fuels the continued emissions of these stars is the flywheel energy of rotation, of the order of 10^{46} erg for a conventional 1 s period pulsar, and $\sim 10^{49}$ erg for the fast rotators Vela and Crab. This rotation, left over from the angular momentum of the parent supernova, has stored much of the gravitational energy of the original star. It carries around the highly compressed magnetic field of the star ($\sim 10^{12}$ G) and, by mechanisms which are plausible but not precisely known, this leads to pulses of radio emission at each rotation of the magnetic pole. The energy radiated matches the regular rate of slowdown which is observed. Most pulsars are slowing down at a rate indicating an age of only $\sim 10^6$ years; the Crab is only 930 years old (by Chinese observation of the supernova) and the Vela about 10^4 years. Apparently most pulsars 'turn off' for unknown reasons at a few million years of age; the older ones shine only when they have a binary companion to shower down accreting matter upon them, as in Her X-1.

If these fascinating objects are our only solid companions in the universe, it may behove us to try to understand their modified version of solid-state physics and, in particular, its most striking manifestation, the glitch. Mechanisms not involving solidity have been proposed, notably encounters with planetary objects or hydrodynamic instabilities, but neither of these can agree either with the remarkable between-glitch stability of the period, nor the sudden, spectacular events which take place (in a day or so at most—no glitch has yet been followed through its actual occurrence).

Thus attention must focus on the solid-state aspects of the star, on the 'rheology' of the neutron star matter in some general sense: what kinds of catastrophic 'fracture' or breakdown events can occur, and what kinds of frictional phenomena may follow upon these events?

It is in this context that a paper dedicated to Nicolás Cabrera can deal with the apparently exotic astrophysical subject of neutron star physics. It was in the dislocation theory of the strength of solids that many of the concepts used in these considerations first became clear: the vital importance of the topology of elementary defects in condensed phases, and of the interaction of defects with each other and with the lattice.

§ 2. Hypotheses to explain glitches

As we pointed out, the first hypothesis immediately suggested is a starquake. As the star slows down, its equilibrium shape becomes less oblate and a rigid elastic layer must eventually reach its elastic limit and break. This may still be reasonable for the very recent Crab pulsar.

As for the Vela pulsar, however, it is not possible to store and release 10^{-6} of the star's rotational energy every two years as elastic energy in the solid crust, since that is roughly 1% of the total energy loss of the entire star in that period. The total amount of energy due to elastic rigidity of figure can be at most $(\Delta I/I)^2$ times the total star energy, and the centrifugal acceleration $a = \Omega^2 R$ in Vela is only about 10^{-3} of the gravitational acceleration g, hence $(\Delta I/I)^2 \sim 10^{-6}$. The elastic energy is only a fraction of this, perhaps 1%, because of elastic relaxation. Hence not even the entire elastic energy amounts to 10^{-6}, much less the amount accumulated in a few years due to slowdown.

In fact, the possible mechanisms of the glitch phenomenon are very severely restricted by the unthinkably large magnitude $\Delta\Omega/\Omega = 2 \times 10^{-6}$. A corresponding event on Earth would be an earthquake of magnitude 15 or 16 on the logarithmic Richter scale! Alternatively, an asteroid 100 miles on a side would have to fall in every 2 or 3 years without tidal disintegration in a gravitational field 10^8 times our own.

We have estimated (Anderson et al. 1980) from the number of Vela-type events observed in older ($\sim 10^6$ years) pulsars—namely two—and the total duration of timing observations on such stars (of the order of 200 star-years) that approximately 1% of all pulsar slowdown (for quiescent pulsars, i.e. other than the Crab) takes place in connection with these catastrophic events. Since all sources of energy other than the rotation of the stars are known to be negligible in these neutron stars (nuclear processes are dead, gravitational collapse has reached its endpoint, and heat is just much too small, as is the energy of the stellar magnetic field and electric current system), this fundamental rotation must be the source powering the glitches as well.

We were led unequivocally to a single type of event. To understand what might be taking place, we have to understand the structure of the star in more detail. Rather accurate estimates of the nature of the neutron-rich matter which makes up these stars are possible, since in most of the star its density is not very different from that in ordinary nuclei, and in much of the rest it is rather low-density and may be calculated from known scattering data.

Ruderman's early estimates were refined by nuclear physicists, notably Negele and Vautherin (1975) and Baym and Pethick (1975). It is generally accepted that the star consists of 'crust' and 'core' regions (fig. 1). There is, first, a core composed of a mixture of neutron and proton superfluids with a relativistic electron Fermi gas, about 10 km in radius, at a density of about 10^{14} g cm^{-3}. The bulk of the mass is a BCS superfluid of paired neutrons. The evidence for superfluidity of both neutrons and protons in the terrestrial nucleus is overwhelming; hence we are quite sure it occurs at the same density in the neutron star. Once the density of neutrons has decreased with radius to somewhat below nuclear density, it turns out that because of the attractive interactions of neutrons and protons, it is energetically favourable for some of the neutrons and all of the protons to clump together into moderate-sized nuclei containing perhaps 50 protons and about three times as many neutrons.

Fig. 1

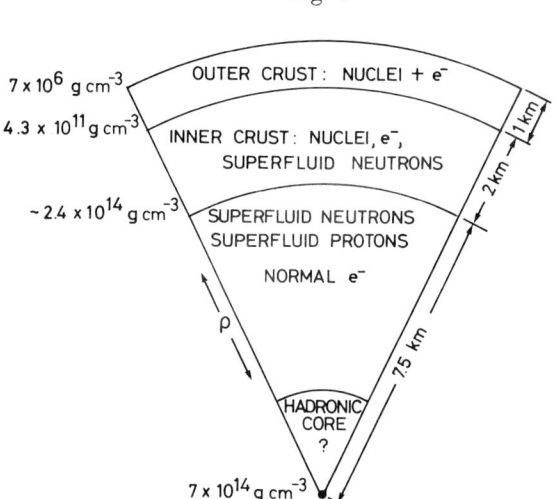

Sketch of the structure of a neutron star. The magnitudes of the various radii and densities can vary by ~50% depending on estimates of the parameters in the equation of state.

The situation is made very clear in fig. 2, from Negele and Vautherin (1975). The nuclei must remain about this size, otherwise the protons' electrostatic repulsion will destroy them. The electrons which neutralize the charge of the protons are very uniform in density because they are relativistic in energy with an E_F of ~200 MeV and a very long screening distance; hence the nuclei see quite large electrostatic fields from each other (about 30 or 40 nuclei are within a screening radius of any given one). Thus the forces which maintain the nuclei in a rigid lattice are exactly calculable: the nuclei form a pure electrostatic 'Wigner' lattice, all of whose elastic properties are fairly easily determined.

The background neutron fluid between nuclei continues to be a superfluid, interpenetrating the nuclei and electrons. In fact, the energy gap and T_c of this superfluid (shown in fig. 3) (Hoffberg, Glassgold, Richardson and Ruderman 1970, Takatsuka 1972) are expected to rise as the density decreases, at first, and then to fall once the density falls below about 10^{13} g cm^{-3}. The nuclei have about the same superfluid energy gap as the neutrons in the core. Thus this crustal matter is very interesting: it is a lattice as well as two superfluids, one in the nuclei and one outside them. This is not at all impossible: precisely the same description is true of terrestrial superconductors, in which the electrons are superconducting but the nuclei form a rigid lattice.

The crust is a few kilometres thick; near the surface the density falls below 10^{11} g cm^{-3}, where the neutron fluid disappears and one has 'ordinary' white dwarf matter. This outer region has negligible dynamical influence because it is so light.

The 'core superfluid' played an important role in early theories. It was observed that immediately after a glitch (which always represented a sudden increase in rotation frequency) the rate of slowdown would increase by an

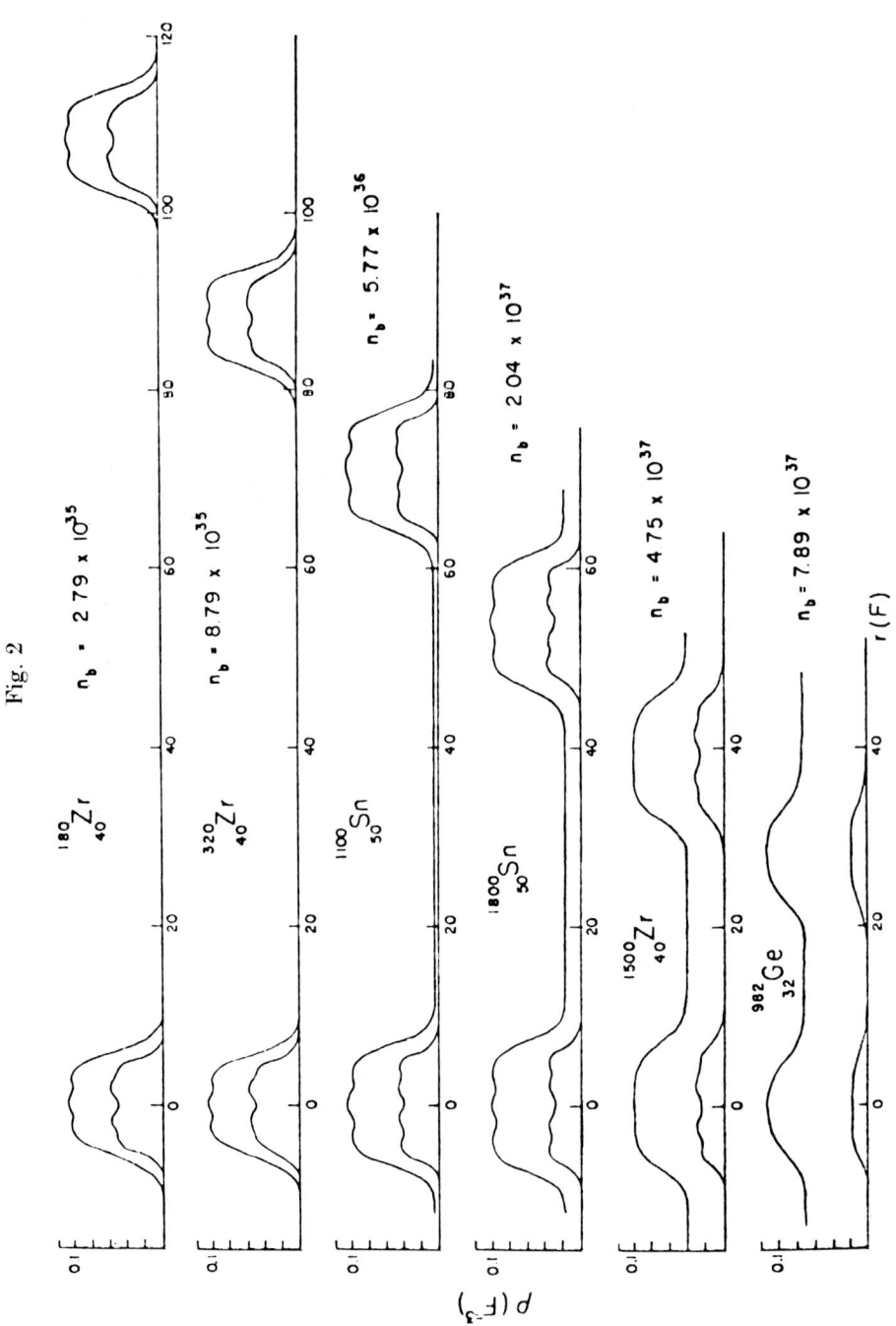

Fig. 2 Typical density variation in the crust region. The horizontal axes denote proton and neutron density distributions occurring along a line joining the centres of two adjacent unit cells. (After Negele and Vautherin 1975.)

amount which is fractionally large compared to $\Delta\Omega/\Omega$: $\Delta(\dot{\Omega})/\dot{\Omega} \simeq 10^{-2}$ in Vela (fig. 4, from Downs (1981) shows a typical event (see also Reichley and Downs 1971)). The observed glitch, however caused, was in this theory presumed not to have affected the neutron superfluid because that fluid was dynamically independent of the crust and of the charged components which rotate with the crust nuclei because of the enormous strength of the electromagnetic coupling via the magnetic field. The spin-down after a glitch was then caused by the slow spin-up process of this core superfluid.

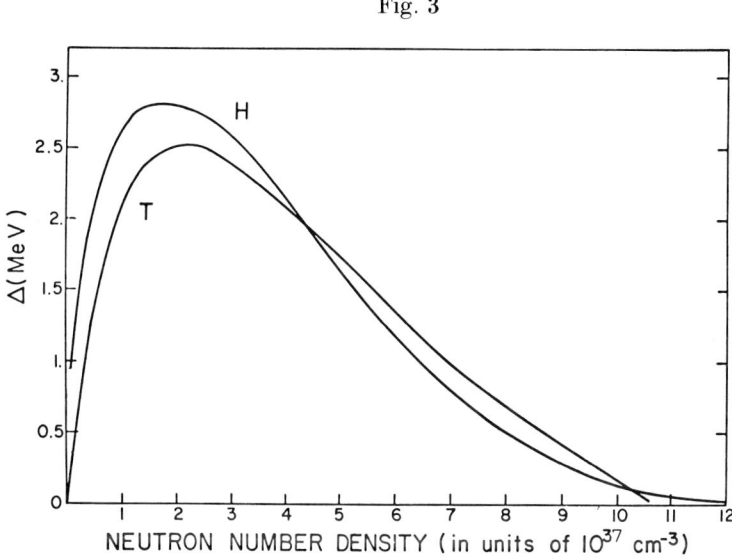

Fig. 3

Energy gap of superfluid neutrons as a function of density. The curve labelled H is from Hoffberg *et al.* (1970); T is an independent calculation for comparison.

More recent theories (Alpar 1977, Anderson and Itoh 1975, Anderson, Alpar, Pines and Shaham 1980, Alpar, Anderson, Pines and Shaham 1981, Pines, Shaham, Alpar and Anderson 1981) have had to come to terms with the idea that coupling processes can be found which couple in the core superfluid, and that the entire dynamics of the glitch process is related to the superfluid component of the crust.

As we have discussed in detail elsewhere (Alpar 1977), we believe that there is a considerable layer of the star's crust where the quantized vorticity of the neutron superfluid is pinned to the lattice. (The coherence length in these stars, unlike that in ordinary superconductors, is smaller than the lattice constant and therefore vorticity can pin to the lattice itself, not just to imperfections.) Thus the angular velocity Ω of the superfluid can only change by vortex creep or by vorticity jumps, the latter causing the glitches. The detailed scenario which we propose has been described elsewhere (Alpar *et al.* 1981). We now concentrate on the actual physical processes related to vorticity pinning.

Fig. 4

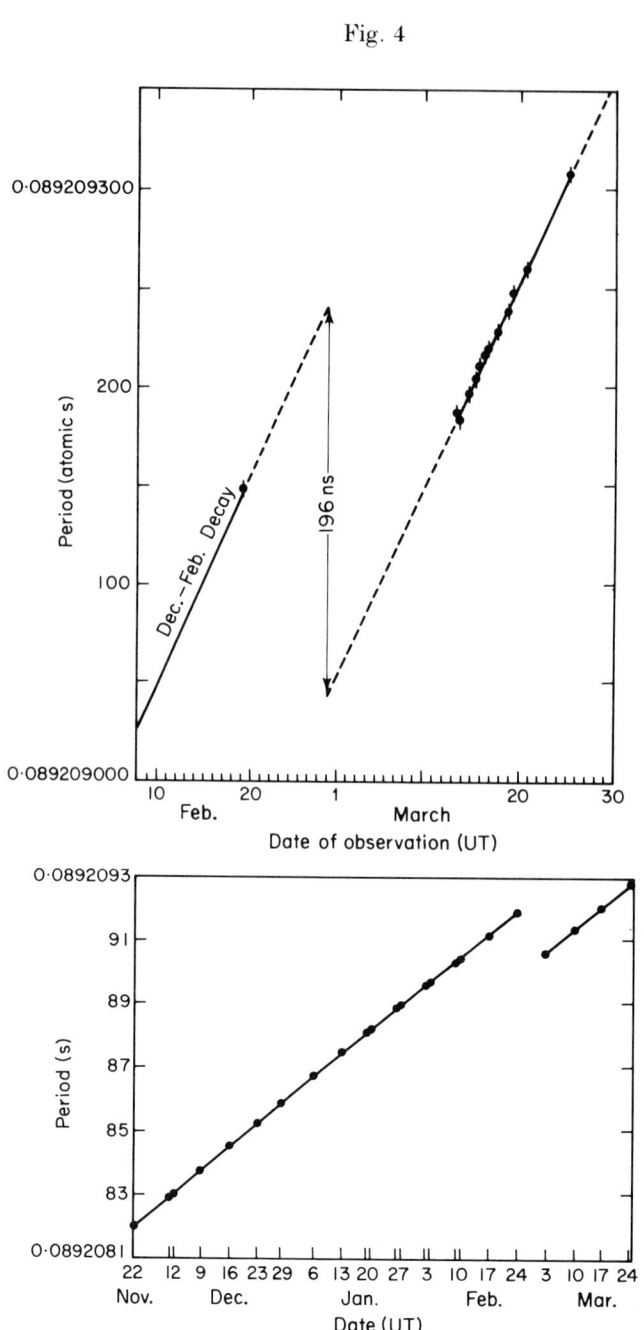

A typical Vela glitch event. The change in slope is barely perceptible on this scale, but is easily measurable.

§ 3. Weak and strong pinning of vortex lines in the neutron-star crust superfluids

Most of the excess neutrons occupy Bloch states of the lattice. These neutrons have different local densities inside and outside the potential wells set up by the nuclei at lattice sites. The difference in the corresponding values of the superfluid gap $\Delta(\rho)$ (Hoffberg et al. 1970, Takatsuka 1972) results in a difference between the energy cost of a vortex line going through a nucleus and one that threads through the interstitial regions. This leads to forces that pin vortex lines to nuclei or repel them from nuclei, depending on which situation is energetically favourable (Alpar 1977). Since the rotational properties of a superfluid are determined by the density and distribution of quantized vortices in it, pinning has important consequences for pulsar dynamics. In particular, we have proposed the sudden release of pinned vorticity as the cause of pulsar glitches, and the coupling of the pinned superfluid as the cause of the subsequent slowdown of pulsars, through a vortex creep process (Anderson et al. 1980, Pines et al. 1981) very similar to creep of dislocations in solids.

In this paper we describe the behaviour of pinning forces as a function of density in the neutron star crust. The pinning energy per nucleus is:

$$E_\mathrm{p} = \tfrac{3}{8} \left[\rho_1 \frac{\Delta^2(\rho_1)}{E_\mathrm{F}(\rho_1)} - \rho_0 \frac{\Delta^2(\rho_0)}{E_\mathrm{F}(\rho_0)} \right] V, \qquad (1)$$

where ρ_1 and ρ_0 are the superfluid neutron densities inside and outside the nuclei, and V is the volume of the nucleus. The pinning force F_p is obtained by dividing E_p by the larger of the coherence lengths ξ, and R_N is the nuclear radius. ($\xi = \hbar v_\mathrm{F}/\Delta$.)

Pinned vortices move together with the crust lattice, at the angular velocity Ω_c of the neutron-star crust. However, the angular velocity of the superfluid, $\Omega_\mathrm{s}(r)$, is determined by the number of vortices enclosed in a circular contour at radius r from the star's rotation axis. Thus $\Omega_\mathrm{s}(r)$ is greater than Ω_c because of pinning. A vortex will move with the ambient fluid unless a force is exerted on it. The pinning force keeps the vortices moving at $\Omega_\mathrm{c} < \Omega_\mathrm{s}$. The equation of motion is

$$\mathbf{f}_\mathrm{p} = \rho \mathbf{K} \wedge (\mathbf{V}_\mathrm{s} - \mathbf{V}_\mathrm{c}) = \rho \mathbf{K} \wedge [(\Omega_\mathrm{s} - \Omega_\mathrm{c}) \wedge \mathbf{r}], \qquad (2)$$

where f_p is the force per unit length of vortex line, ρ the superfluid density and K the vorticity of the line. For vortex lines in the neutron BCS superfluid, $K = h/2M_\mathrm{N}$, where h is Planck's constant and M_N the neutron mass. The right-hand side of eqn. (2) is called the Magnus force. Unpinning will start when $\Omega_\mathrm{s} - \Omega_\mathrm{c}$ increases, through the continuous slowdown $\dot{\Omega}_\mathrm{c}$ of the pulsar, to a critical value which the available pinning force cannot sustain. We obtain a scale for the critical angular velocity difference for unpinning by balancing f_p, the pinning force per unit vortex length, against the Magnus force:

$$\delta \Omega_\mathrm{cr} \equiv (\Omega_\mathrm{s} - \Omega_\mathrm{c})_\mathrm{cr} = \frac{F_\mathrm{p}}{a \rho K r} = \frac{E_\mathrm{p}}{x a \rho K r} \qquad (3)$$

where x is the larger of ξ and R_N, and a the distance between pinning nuclei along the vortex line. In the spherical geometry of the neutron star, for regions of the crust close to the rotational axis, r is much smaller than the actual radius of the star, $r_* \sim 10^6$ cm. However, the equatorial regions, with $r \sim r_*$, can already have $\delta\Omega_{cr}$ much stronger than the critical difference $\dot{\Omega}_c t_g$ that builds up between successive glitches in some layers, while in other layers in the equatorial plane pinning can be weak enough to give $\delta\Omega_{cr} \lesssim \dot{\Omega} t_g \simeq 10^{-2}$ in the Vela pulsar. We therefore investigate a cylindrical model of the star, i.e. we take the equatorial plane, where $\delta\Omega_{cr}$ is not strengthened by the r^{-1} factor, and investigate its variation as a function of density through the crust. $\delta\Omega_{cr}$ can be calculated as a function of the density ρ in the neutron star crust, using lattice structure and gap values appropriate to each density value (Alpar 1977). From eqn. (3) we see that $\delta\Omega_{cr}$ is largest when a is the nearest-neighbour spacing in the crust lattice. This can happen if the lattice is oriented so that a line of nuclei coincides with each vortex line, that is, the lattice is oriented parallel to Ω, the current angular velocity of the neutron star. We do not expect this to be the usual case, since conditions when the lattice froze would have been entirely different and undoubtedly cracking and flow of the crust have occurred on many occasions. A vortex line is likely to find the lattice oriented at some angle to its preferred direction $\hat{\Omega}$. The coupling between vortices in the vortex array of spacing $l_v = (K/2e)^{1/2}$ limits the amplitudes and wavelengths of distortions on individual vortex lines to less than l_v; but since pinning encounters occur on length scales of the order of the lattice, the contribution of the vortex array rigidity is negligible—the energy involved, $(\rho K^2/2\pi)\frac{1}{2}(a/l_v)^2$ per unit length of vortex line displaced a length a, is much smaller than the energies of pinning and vortex line bending we discuss below. We therefore neglect the other vortices and ask if a vortex line can optimize its pinning by displacing the nuclei from their equilibrium sites in the lattice. The pinning force is to be compared with the force needed to displace the charged nuclei from their lattice equilibrium sites by a distance comparable to the size of each nucleus, the length scale of the pinning force. This lattice displacement force is

$$F_L \simeq \frac{z^2 e^2}{b^3} R_N. \tag{4}$$

Only in the layers of the crust where the pinning energy is larger than about 3 MeV will the vortices be able to displace nuclei so as to achieve optional pinning along much of the length of the vortex, and under these conditions eqn. (3) can be used with a a few times the lattice spacing. In fig. 5 we plot $\delta\Omega_{cr}$ as a function of the density in the neutron star crust, and indicate the transition between the strong ($F_p > F_L$) and weak ($F_p < F_L$) pinning layers. At still stronger values of the pinning force, there has been a proposal that the pinning forces may lead to a breaking of the lattice (Ruderman 1976). This will happen when F_p reaches the critical value F_c, the force per nucleus that will break the lattice:

$$\mu\phi_c = \frac{L}{a} F_c n_v. \tag{5}$$

where μ is the shear modulus $z^2 e^2/a^4$, ϕ_c is the critical strain angle of the lattice, L the length of a vortex line going through the pinning lattice ($L \sim 10^4$–10^5 cm, the thickness of the crust pinning layer) and n_v the area density of vortices,

Fig. 5

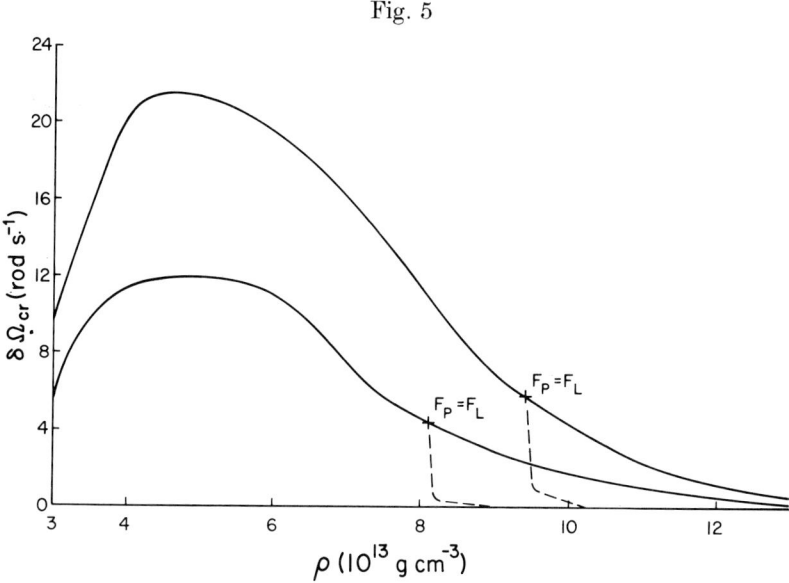

$\delta\Omega_{cr}$ resulting from pinning as a function of density in neutron star crust. The upper curve was calculated using $\Delta(\rho)$ given by Hoffberg et al. (1970). The lower curve is the result of reducing $\Delta(\rho)$ by a factor of three-quarters. The transition at $F_L = F_p$ is indicated for each curve.

$n_v = 2\Omega/K$. Ruderman (1976) has used $\phi_c \sim 10^{-4}$–10^{-5} to argue that F_c is smaller than both F_L and F_p, so that the lattice breaks before individual nuclei can be displaced or vortex lines can unpin from the nuclei.

We feel that this argument misconstrues the role of the rigidity of the crust. The crustal matter as a whole is almost in gravitational equilibrium (except for a small degree of shape distortion due to elastic rigidity). Only crustal fractures which are either neutral to, or helpful with respect to, gravitational equilibrium are possible. An outward flow of crustal matter is unthinkable; yet such a flow is the only way to respond dynamically to the pinning forces. Hence the overall, global pinning force is not contained by elasticity but by the much stronger gravitational shape forces, to which the force is conveyed by the energies of compositional and gravitational equilibrium. We feel that the crust-breaking mechanism is not dynamically plausible. However, we do feel that it is easy—in fact likely—for vorticity motion to trigger true elastic quakes of the type originally proposed (Ruderman 1969).

Figure 5 shows $\delta\Omega_{cr}$ calculated using the gap values given by Hoffberg et al. (1970). The transition to strong pinning as discussed above is indicated. Clearly, the $\delta\Omega_{cr}$ values from eqn. (2) are too high (except in the inner and outer boundaries of the crust) to have $\delta\Omega_{cr} = \dot{\Omega} t_g = 10^{-2}$ (for the Vela glitches) which we require to build up $(\Omega_s - \Omega_c)$ to $\delta\Omega_{cr}$ to start the next glitch by unpinning. However, $\delta\Omega_{cr}$ will be reduced from the values given by eqn. (2) for two different reasons. First, an uncertainty in $\Delta(\rho)$ by a factor of two means that $\delta\Omega_{cr}$ (which is proportional to Δ^2 or Δ^3, depending on whether the coherence length exceeds R_N) could everywhere be smaller than calculated using the $\Delta(\rho)$

given by Hoffberg et al. (1970). The $\delta\Omega_{cr}$ resulting from two model $\Delta(\rho)$ runs are shown in fig. 5. Such general reduction of $\delta\Omega_{cr}$, however, does not lead to sufficiently small values of $\delta\Omega_{cr}$.

A possibly more significant reduction will occur because of the disruption of the lattice near a vortex line. The vortex line may unpin simply by carrying its accompanying nuclei along with it through the lattice, as originally suggested by Anderson and Itoh (1975). This means that the effective F_p is limited by F_L, and is probably considerably smaller than F_L. In fact, all errors are clearly in such a direction as to reduce F_p, and thus the excessive size of $\delta\Omega_{cr}$ is not important here.

On the other hand, where $F_p < F_L$, the vortex lines cannot pin optimally by pulling individual nuclei out of their equilibrium sites. They also cannot, for practical purposes, bend, because their energy per unit length is $E_F k_F$, which is greater than 100 MeV/Fermi. Instead, the lattice will deform more smoothly, on a length scale L which we can estimate by equating the pinning energy to the energy of distortion per site, $\sim E_L b/L$, where E_L is a typical lattice energy $E_L = F_L b = z^2 e^2 R_N/b^2$. The length scale L will be the distance between effective pinning sites, and will go as

$$\frac{b}{L} \sim \frac{E_p}{E_L}$$

so that, finally, the effective pinning energy is

$$E'_p \simeq E_p^2/E_L \quad \text{or} \quad F'_p \simeq \frac{F_p^2}{F_L}. \tag{6}$$

The pinning force falls off very much more rapidly than F_p as F_p gets smaller; this rapid fall-off is shown in fig. 5 as the dashed line. Eventually L will become so long that the lattice can be assumed to be rigid and the line encounters nuclei as it would if it were a random straight line, i.e. at

$$L = \frac{b^3}{\xi^2},$$

i.e. about one nucleus every 10^{-2}–10^{-3} lattice constants. At this point the pinning force must be estimated statistically, and we will also begin to have bending at about this scale. In any case a reduction as given approximately by eqn. (6) is not a bad estimate.

Thus, as we proceed to higher densities from the strong pinning layer boundary $F_p = F_L$, we have a rather sudden transition to random pinning and a weak pinning layer, given by eqn. (6). For all higher densities up to the inner boundary of the crust, we have weak pinning with $\delta\Omega_{cr}$ as low as or lower than $\dot{\Omega} t_g$.

§ 4. Conclusion

The vortex–lattice interaction in neutron stars provides an interesting example of a transition from strong to weak random pinning, depending on the relative strength of binding forces in the lattice and pinning forces. Reduced effective values of pinning forces, which result from vortex line energy as well as lattice geometry and randomness, lead to important dynamical effects in pulsars.

In particular, we have ascribed (Alpar *et al.* 1981) the 'glitches' to a sudden release of the pinned vorticity in a volume containing about 10^{-2} of the moment of inertia of the star. This is triggered by the high density of vorticity built up in the boundary layer between weak and strong pinning regions. This hypothesis seems to be the only one which explains all the data, especially when such a large fraction of the slowing-down process is involved in the glitches.

The more detailed consequences for astrophysics of these rheological considerations are not appropriately discussed here, and have been the subject of a previous paper. We felt that in a volume dedicated to Nicolás Cabrera, it might be very appropriate to discuss the 'farthest out' consequences of the kinds of detailed understanding of defect structures and motion in solids in which he was a very early and important pioneer.

REFERENCES

ALPAR, M. A., 1977, *Astrophys. J.*, **213**, 527.
ALPAR, M. A., ANDERSON, P. W., PINES, D., and SHAHAM, J., 1981, *Astrophys. J. Lett.*, **249**, L29.
ANDERSON, P. W., ALPAR, M. A., PINES, D., and SHAHAM, J., 1980, *Proceedings of the Symposium on Pulsars*, IAU Symp. No. 95 (International Astronomical Union).
ANDERSON, P. W., and ITOH, N., 1975, *Nature, Lond.*, **256**, 25.
BAYM, G., and PETHICK, C. J., 1975, *A. Rev. nucl. Sci.*, **25**, 27.
BAYM, G., and PINES, D., 1971, *Ann. Phys.*, **66**, 816.
DOWNS, G. S., 1981, *Astrophys. J.*, **249**, 687.
HOFFBERG, M., GLASSGOLD, A. E., RICHARDSON, R. W., and RUDERMAN, M. A., 1970, *Phys. Rev. Lett.*, **24**, 775.
NEGELE, J. W., and VAUTHERIN, D., 1975, *Nucl. Phys.* A, **207**, 298.
PINES, D., SHAHAM, J., ALPAR, M. A., and ANDERSON, P. W., 1981, *Prog. theor. Phys.*, Supplement No. 69, p. 376.
PINES, D., SHAHAM, J., and RUDERMAN, M. A., 1973, *Physics of Dense Matter*, edited by C. J. Hansen, IAU Symp. No. 53 (Dordrecht : Riedel), p. 189.
REICHLEY, P., and DOWNS, G. S., 1971, *Nature, Phys. Sci.*, **234**, 48.
RUDERMAN, M. A., 1969, *Nature, Lond.*, **223**, 597 ; 1976, *Astrophys. J.*, **203**, 213.
TAKATSUKA, I., 1972, *Prog. theor. Phys.*, **48**, 517.

Localization Redux
Physica **117B** and **118B**, 30–36 (1983)

This is a final summary of my work in the second "weak localization" period of localization theory, which was inaugurated by the "Gang of 4" papers in 1978. During this period we all (Abrahams, Fukuyama, Vollhardt and Wölfle, Ramakrishnan, Lee, Thouless and occasionally some Russians) met regularly at Aspen in the summers. This cooperative stress-free atmosphere seems to have been a casualty of high-T_c, funding and job shortages, the cutbacks at Bell and IBM, or a combination of these: it doesn't exist any more.

LOCALIZATION REDUX

P. W. Anderson

Bell Laboratories
Murray Hill, New Jersey 07974, U.S.A.

and

Joseph Henry Laboratories
Princeton University, Princeton, N. J. 08544, U.S.A.

I survey recent progress in the field of localization. Theory and experiment are in excellent agreement in two-dimensional systems such as inversion layers and thin metal films. The metal insulator transition in three-dimensional systems remains an interesting problem as exemplified by highly accurate data on doped semiconductors, but on the metallic side perturbational theory has had a number of successes.

Localization,[1] a very old concept and a continuing controversy, keeps bringing itself back to the center of the stage, year after year, which is why I use the word "Redux".

The source of the latest reintroduction is a series of developments, both experimental and theoretical, which have begun to bring scientific precision into what used to be a highly speculative and controversial field. One of the characteristics of the field has been an almost worldwide, semicollaborative effort on the theoretical side, and parallel evolution at different places on the experimental side.

The theoretical cooperation was much aided by two timely workshops, one in Russia at Yerevan, one at Aspen last summer, so that in naming important workers in the field I have to pinpoint groups from at least six countries. My own collaborators[2-4] have included especially Ramakrishnan, Licciardello, Lee, Abrahams, Fisher and Shapiro from the Bell Labs-Princeton-Rutgers group, as well as Thouless[5] and Fukuyama;[6] particularly useful work which found its way into the workshops has been done by Wegner,[7] Götze,[8] Vollhardt and Wölfle[9] in Germany; Altshuler, Aronov[10] and others[11] in Russia; Kawabata,[12] Hikami and others[13] in Japan; McMillan[14] in the U.S.; McKane and Stone[15] in England; and many, many others.[16] In experiment, the Bell group,[17] especially Dynes and collaborators, and the Cambridge group with Pepper and others[18] have run neck and neck on two-dimensional work, while there is important work by a Belgian group with Deutscher;[19] on three dimensions the only precision work as yet published is by Thomas, Rosenbaum and coworkers[20] of Bell-Princeton. Notable work at Illinois,[21] IBM[22] and elsewhere[23] on various aspects is very worth mentioning.

1. INTRODUCTION TO QUANTUM TRANSPORT

Let us briefly introduce the set of concepts which have brought us to where we are. The basic problem is that of the fundamental limitations of quantum transport, of which the most characteristic example is metallic conduction at low temperature, where, as we all know, there is maximum conductivity at $T \to 0$ which decreases due to thermal agitation, in contrast to ordinary transport processes which only occur due to thermal agitation. Realizing that it is sure to fail in the end, we borrow from simple Boltzmann transport theory the expression for the conductivity

$$\sigma = \frac{ne^2\tau}{m} = \frac{ne^2}{mv_F}\ell$$

$$= \frac{e^2}{h}\frac{2k_F}{3\pi}(k_F\ell) \quad \text{(in 3 dimensions)}$$

$$= \frac{e^2}{h}(k_F\ell) \quad \text{(in 2 dimensions)}$$

Note the appearance of the characteristic quantum unit of conductance

$$\frac{e^2}{h} = \frac{1}{25,000} \text{ (ohms)}^{-1}.$$

Using the quantum Einstein relation between conductivity and diffusion, we relate these Boltzmann transport theory results to a second aspect of quantum transport, diffusion by tunneling from site to site:

$$e^2 \frac{dn}{dE} D = \sigma$$

$$\text{i.e.} \quad D \sim \frac{h}{m}(k_F\ell)$$

where h/m is the fundamental quantum unit of diffusivity and $k_F\ell = \frac{2\pi\ell}{\lambda}$ (λ the de Broglie wavelength) appears in this view as a coherence factor for tunneling at different sites. (We note that, for a tight-binding band of energy width B,

$$\frac{h}{m} \sim a^2 \left(\frac{B}{h}\right) \sim a^2/(\text{time for a quantum jump}).$$

The concept of localization[1] evolved from my observation that D couldn't be reduced below the $k_F\ell \sim 1$ limit because of a failure to break up coherence and prevent reverse tunneling, which Mott equated to the Ioffe limit $k_F\ell > 1$ on the ratio of mean free path to wavelength. Over the years, Mott developed the concept[24] of a "minimum metallic conductivity" of about $e^2/\hbar a$, which many data seemed to support, at which the conductivity dropped to zero.

2. SCALING THEORY: GANG OF IV AND SCATTERING THEORY

By combining diagrammatic transport theory with a scaling technique, the Gang of Four (myself, Abrahams, Ramakrishnan, Licciardello[2]) gave a probably correct account of the localization regime proper. A basic component is the Thouless conjecture[5] that (a) conductance can be defined on a <u>microscopic scale</u>, for a sample of only a few atoms, even; (b) the conductance g in units of e^2/h is the only dimensionless coupling constant characterizing the wave functions and energy states, or at least characterizing the diffusion of wave packets.

The mean conductance, then, of a sample of side 2L made up from a group of smaller samples of size L chosen from the appropriate statistical universe must be a function only of the mean conductances of the smaller samples of size L:

$$\bar{g}(2L) = f\left[\bar{g}(L)\right]$$

which may be written

$$\frac{d \ln \bar{g}}{d \ln L} = \beta(\bar{g})$$

To this we add the macroscopic result that $g \to \infty$ is the classical Ohm's law limit, for which $g = \sigma L^{d-2}$ (d the dimensionality) and $\beta = d-2$, as well as the localization result for the limit $g \to 0$, i.e., $g \sim e^{-\alpha L}$ or $\beta \sim \ell n g$, so that the curves of β vs. g must look roughly as in Fig. 1.

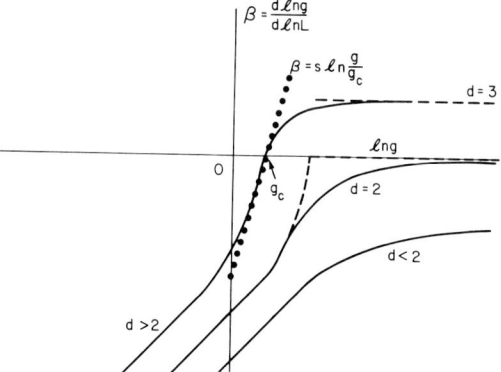

Fig. 1: Schematic diagram of scaling function data vs. log g from G_{IV} paper.

Finally, to get precise results we took advantage of an old perturbation theory scheme of Langer and Neal.[25] The ordinary theory is the lowest-order diagrams involving scattering of an electron-hole pair by the impurity potential:

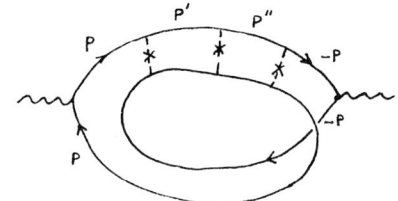

and the first corrections to β, which are of order $1/g$, come from summing all of what we called the "maximally crossed diagrams" like:

which closely resemble the "Cooper pair" diagrams of superconductivity theory,[26] and are now called "cooperons".

Since g is proportional to ℓ which is proportional to $1/v^2$, v being the impurity scattering potential, a series in scattering is a series in $1/g$ and it is natural for this series to start with $1/g$; but oddly enough all higher terms have been shown to vanish, and the actual behavior for large g is dominated by the one term.

In the case of two dimensions, since d-2 vanishes the $1/g$ term is the whole story and leads to the famous logarithmic correction to the conductivity: a correction which is verified in thin films by Dolan and Osheroff[17] and others[19,22,23] and in MOSFETs by Bishop et al[17] and Uren et al.[18] It can be demonstrated unequivocally by its strong, anisotropic magnetic field dependence: the "cooperon" diagrams are, as in superconductivity, very field dependent. The actual formula is:

$$\sigma(T,H) = \sigma_0 - p \frac{e^2}{2\pi^2 \hbar} \ln T/T_0$$

$$+ \frac{e^2}{2\pi^2 \hbar} \{\psi(a+1/2) - \psi(a)\}$$

$$a = \frac{\hbar c}{2eH\ell_e \ell_i}$$

ℓ_i = inelastic length $v_F \tau_i$
ℓ_e = elastic length

Note this allows determination of the inelastic scattering time τ_i, which we shall see later is very important.

In three dimensions the same calculation predicts not a mobility edge but a critical point, but apparently all behavior near this point is modified by interactions. That there really is a critical point and not a minimum metallic conductivity is beautifully demonstrated by the experimental data of Paalanen et al[20] shown in Fig. 2. It is likely that Mott[24] has severely underestimated the critical range of σ, and that all the measurements shown here are within the critical range. One-dimensional experimental results are also as yet in flux,[27] although Skocpol et al[28] at Bell Labs Holmdel have promising data, as well as Tinkham's group[29] and Sacharoff et al[30] at Harvard.

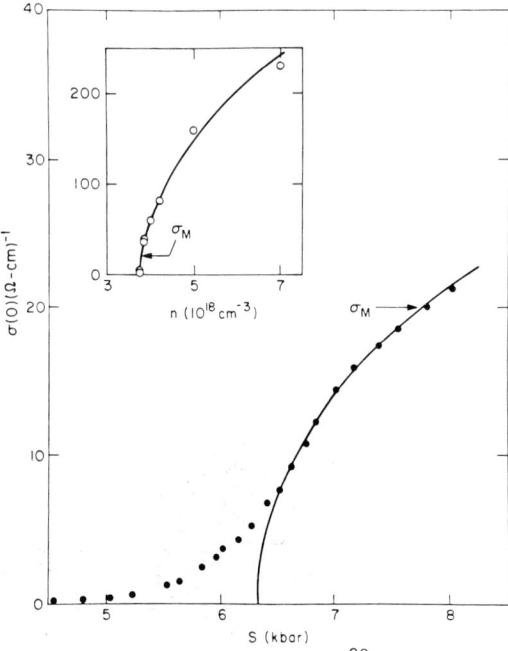

Fig. 2: Results of Paalanen et al[20] on conductivity in the impurity band near the metal insulator transition. In the lower curve stress is used as a vernier parameter.

The mathematical assumptions of the scaling theory have been very completely investigated in the one-dimensional case by myself and co-workers,[2-4] Az'bel[31] and more recently others. The result is to place the basic Thouless and G_{IV} conjectures on a very sound footing, in addition to teaching us some very subtle bits of the mathematics of scaling theory.

3. INTERACTIONS: THE DIFFUSON CONCEPT

Unfortunately, this very beautiful story is extremely far from complete: there is a second, and indeed often dominant, source of corrections of similar type to conduction in poor metals, namely the electron-electron interactions. These corrections were pointed out years ago for the special case of electron-phonon interactions by Schmid,[32] and some of the theory for three dimensions given by Altshuler.[10] But it wasn't until the Yerevan workshop that the two-dimensional case was done by Altshuler, Aronov and Lee, and the experimental observability[17-20] of the effects began to be obvious.

The basic discovery here is that it is *not* really possible to treat random impurity scattering and the effects of interactions completely independently, as we have always done: essentially, letting the interactions renormalize the particles and make them into quasi-particles with new numbers but not new properties. For instance, there are scattering corrections to quantities as simple as the exchange self-energy (which determines the effective mass) and which make the density of states change in a singular manner at the Fermi surface. These are called "diffuson" corrections and originally were put in as a "vertex correction" at each interaction due to scattering. The diagram for self-energy is as shown:

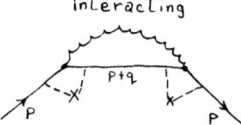

The physical meaning is that after emitting a phonon, for instance, the electron does not propagate away as fast, so interacts more strongly with it. Lee, Abrahams, Ramakrishnan and I[5] showed how to introduce diffuson corrections using the exact scattered wave functions and much simplify and generalize the calculation. But essentially it always involves introducing a "diffuson propagator"

$$\frac{1}{i\omega + Dq^2}$$, and doing integrals

$$\int \frac{(dq)^d}{i\omega + Dq^2}$$

gives us

$$\ln \omega \quad (2d)$$
$$\omega^{-1/2} \quad (1d)$$
$$\omega^{1/2} \quad (3d).$$

Where in many of the corrections $\omega \to \frac{kT}{\hbar}$, or alternatively $\frac{\mu_B H}{\hbar}$ under a magnetic field. The

$\omega^{1/2}$ corrections for 3d have been nicely demonstrated in tunneling density of states by Bob Dynes:[17] in fact, they are a very old and puzzling result we saw 15 years ago and did not explain. Some data are shown in Fig. 3.

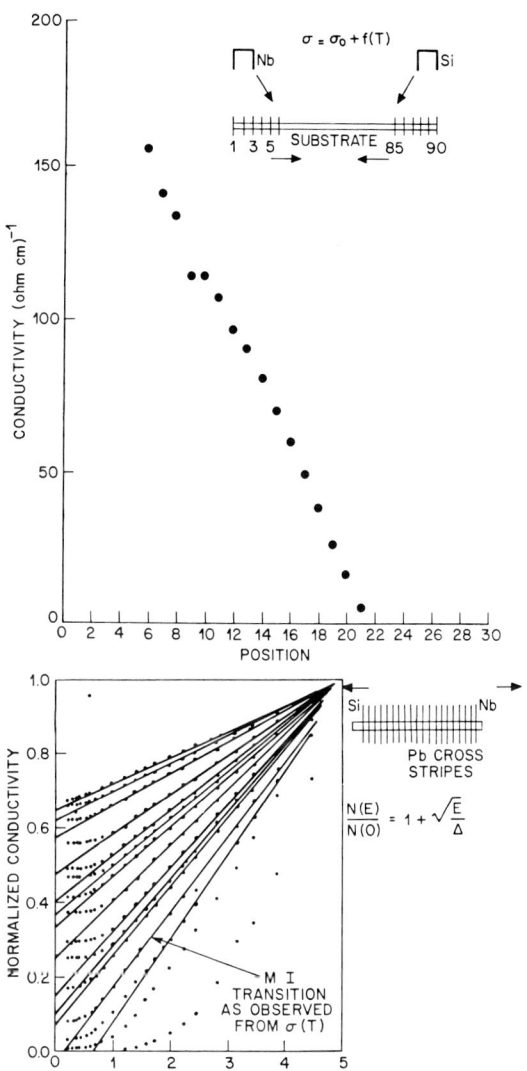

Fig. 3: Data from Hertel et al[33] snowing tunneling conductivity from amorphous Si-Nb. Inset shows experimental set-up with varying Nb content. Tunneling conductance measures density of states at Fermi level.

There are also effects on the conductivity, which <u>add</u> to those of localization. For a few months we had a good strong controversy going as to which was important, but we now know <u>both</u> are. The fortunate fact is that the magnetic field dependence is entirely different from that of localization,[20] so the three groups of experimentalists[17-19] have very neatly been able to identify separately the contributions to the logarithmic behavior of localization and of interactions. I give for the record the extra resistivity in its dependence on magnetic field, and assure you that Uren et al,[18] Bishop et al[17] and Deutscher et al[19] have verified this formula in detail:

$$(\Delta\sigma)_{int} = + \frac{e^2}{4\pi^2\hbar}(2-F)\ln\frac{T}{T_0} - \frac{e^2}{2\pi^2\hbar}\frac{F}{2}g(h)$$

$$h = \frac{\mu_B H}{kT}, \quad 0 \leq F \leq 1 \quad (F = \text{screening factor})$$

$$g(h) = \int_0^\infty d\omega \frac{d^2}{d\omega^2}\left(\frac{\omega}{e^\omega-1}\right)\ln\left|1 - \frac{h^2}{\omega^2}\right|$$

The verification of the localization contribution has left us with detailed measurements of the inelastic scattering time τ_i. In all of the experimental results, this scattering time has been found to be surprisingly short and very slowly varying with temperature. This was a source of much of the early confusion: using classic methods, we supposed that this would go as T^{-p} with $p \sim 3$ or greater. There are, however, very striking diffuson effects, especially on the inelastic scattering by phonons. The most recent work by Lee, Altshuler and Aronov[10] has somewhat revised our earlier work but still gives a large value for τ^{-1}, linear in T, in 2d, and $T^{2/3}$ in 1 dimension (a result which implies the failure of Fermi liquid theory in that case!). The large value and its agreement with theory is shown in Bishop et al's figure[35] (4), which he has loaned me. It appears that the large $T^{2/3}$ result is not at all incompatible with the most recent 1d data.[28-30]

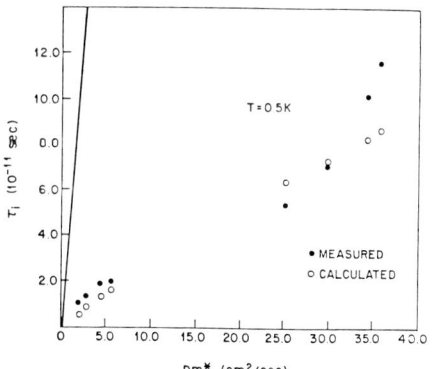

Fig. 4: Data from Bishop et al.[34] Inelastic scattering time τ_i in inversion layers as a function of rate of scattering to illustrate arge diffuson effects.

Let us conclude with a remark and a problem still unsolved for future theorists. The remark is that G_{IV} implies that σ, and therefore D, is a function of L - effectively, of q itself. The true "diffusion kernel" near the mobility edge is not

$$\frac{1}{i\omega + Dq^2} \quad \text{but} \quad \frac{1}{i\omega + D_0 q^d},$$

a much more singular function in 3 dimensions, for instance, implying that at the mobility edge very major changes take place. Muttalib, Ramakrishnan and I[35] have proposed a theory of impurity effects on superconductors using this, for instance.

The problem is that even using this, the best available theories as yet do not explain the observations in 3d. I showed you a figure of the Paalanen et al[20] data already (Fig. 2), showing that apparently σ drops off as $(n-n_c)^\eta$ with $\eta = 1/2$.

Theory and experiment are in agreement that (again in contradiction to Mott's assertions over many years) there is an Efros-Shklovskii gap[36] or pseudogap at the mobility edge, but its nature and shape are not understood. I should reemphasize that agreement of weak localization and interaction experiment and theory is in pretty good shape, even to the extent of seeing the predicted negative sign $T^{1/2}$ singularities[20] and the reverse-sign spin orbit effect (beautiful data of Bergmann[37]). It is very interesting that the exponent $\eta = 1/2$ does not satisfy the Harris-Mott criteria[24,38] for stability against inhomogeneity, and indeed the Paalanen et al data[20] show an unexplained and unexpected exponential tail. I leave you with a last intriguing puzzle: the close resemblance of Paalanen's σ vs. n and optical absorption α vs. ω for the band tails in amorphous semiconductors.[39]

REFERENCES

1. P. W. Anderson, Phys. Rev. 109, 1492 (1958); A. F. Ioffe and A. R. Regel, Prog. in Semiconductors 4, 237 (1960); N. F. Mott, Metal-Insulator Transitions (Taylor and Francis, London, 1974).
2. E. Abrahams, P. W. Anderson, D. C. Licciardello and T. V. Ramakrishnan, Phys. Rev. Lett. 42, 673 (1979); ibid 43, 718 (1979).
3. E. Abrahams, P. W. Anderson, P. A. Lee and T. V. Ramakrishnan, Phys. Rev. B 24, 6783 (1981).
4. P. A. Lee, Phys. Rev. Lett. 42, 1492 (1979); P. A. Lee, J. Non-Cryst. Sol. 35, 21 (1980); P. W. Anderson, D. J. Thouless, E. Abrahams and D. S. Fisher, Phys. Rev. B 22, 3519 (1980); E. Abrahams and T. V. Ramakrishnan, J. Non-Cryst. Sol. 35, 15 (1980); B. Shapiro and E. Abrahams, Phys. Rev. B 24, 4889 (1981).
5. D. C. Licciardello and D. J. Thouless, J. Phys. C 8, 4157 (1975); ibid C 11, 925 (1978); D. J. Thouless, Phys. Rev. Lett. 39, 1167 (1977); D. J. Thouless, Solid St. Comm. 34, 683 (1980).
6. H. Fukuyama, J. Phys. Soc. Jpn. 48, 2169 (1980); ibid 49, 644 (1980); ibid 50, 3407 (1981); Y. Ono, D. Yoshioka, H. Fukuyama, J. Phys. Soc. Jpn. 50, 2143 (1981); D. Yoshioka, Y. Ono, H. Fukuyama, J. Phys. Soc. Jpn. 50, 3419 (1981).
7. F. J. Wegner, Z. Phys. B 35, 207 (1979).
8. W. Götze, Solid State Comm. 27, 1393 (1978); W. Götze, J. Phys. C 12, 12/9 (1979); W. Götze, P. Prelovsek, P. Wölfle, Solid St. Comm. 30, 369 (1979); W. Götze, Philos. Mag. 43, 219 (1981); A. Gold, W. Götze, J. Phys. C 14, 4049 (1981); D. Belitz, W. Götze, Philos. Mag. B 43, 517 (1981); D. Belitz, A. Gold, W. Götze, Z. Phys. B 44, 273 (1981).
9. D. Vollhardt, P. Wölfle, Phys. Rev. Lett. 45, 842 (1980); Phys. Rev. B 22, 4666 (1980).
10. B. L. Altshuler and A. G. Aronov, Zh. Eksp. Teor. Fiz. 77, 2028 (1979) (Soviet Phys. JETP 50, 968 (1979)); B. L. Altshuler, A. G. Aronov and P. A. Lee, Phys. Rev. Lett. 44, 1288 (1980); B. L. Altshuler, D. Khmelnitskii, A. I. Larkin and P. A. Lee, Phys. Rev. B 22, 5142 (1980); B. L. Altshuler, A. G. Aronov, D. E. Khmelnitskii and A. I. Larkin, Zh. Eksp. Teor. Fiz 81, 768 (1981) (Soviet Phys. JETP 54, 411 (1981)).
11. L. P. Gor'kov, A. I. Larkin and D. E. Khmelnitskii, JETP (Soviet) Lett. 30, 228 (1979); A. I. Larkin, JETP Lett. 31, 219 (1980); K. B. Efetov, A. I. Larkin, D. E. Khmelnitskii, JETP (Soviet) 52, 568 (1980).
12. A. Kawabata, Solid St. Comm. 34, 431 (1980); J. Phys. Soc. Jpn. 49, Suppl. A., 375 (1980); Solid St. Comm. 38, 823 (1981).
13. S. Hikami, A. I. Larkin, Y. Nagaoka, Progr. Theor. Phys. 63, 707 (1980); S. Hikami, Phys. Rev. B 24, 2671 (1981).
14. W. L. McMillan, Phys. Rev. B 24, 2739 (1981).
15. A. J. McKane, M. Stone, Ann. Phys. 131, 131 (1981).
16. F. J. Wegner, Z. Phys. B 25, 327 (1976); F. J. Wegner, Phys. Rev. 67, 15 (1980) and references therein; B. Shapiro, Phys. Rev. Lett. 48, 823 (1982); B. Shapiro, Phys. Rev. B 25, 4266 (1982); M. Kaveh and N. F. Mott, J. Phys. C 14, L177 (1981).

17. D. J. Bishop, D. C. Tsui and R. C. Dynes, Phys. Rev. Lett. 44, 1153 (1980); ibid 46, 360 (1981); R. C. Dynes and J. P. Garno, Phys. Rev. Lett. 46, 137 (1981); G. J. Dolan and D. D. Osheroff, Phys. Rev. Lett. 43, 721 (1979).
18. M. J. Uren, R. A. Davies and M. Pepper, J. Phys. C 13, L985 (1980).
19. G. Deutscher, O. Entin-Wolman and Y. Shapira, Phys. Rev. B 22, 4264 (1980); T. Chui, G. Deutscher, P. Lindenfeld and W. L. McLean, Phys. Rev. B 23, 6172 (1981).
20. M. A. Paalanen, T. F. Rosenbaum, G. A. Thomas and R. N. Bhatt, Phys. Rev. Lett. 48, 1284 (1982); T. F. Rosenbaum, K. Andres, G. A. Thomas, P. A. Lee, ibid 46, 568 (1981); T. F. Rosenbaum, R. F. Milligan, G. A. Thomas, P. A. Lee, T. V. Ramakrishnan, R. N. Bhatt, K. DeConde, H. Hess and T. Perry, ibid 47, 1758 (1981); T. F. Rosenbaum, K. Andres, G. A. Thomas and R. N. Bhatt, ibid 45, 1723 (1980); G. A. Thomas, T. F. Rosenbaum and R. N. Bhatt, ibid 46, 1435 (1981); G. A. Thomas, A. Kawabata, Y. Ootuka, S. Katsumoto, S. Kobayashi and W. Saski, Phys. Rev. B 24, 4886 (1981); H. A. Hess, K. DeConde, T. F. Rosenbaum and G. A. Thomas, ibid B 25, 5585 (1982); G. A. Thomas, Y. Ootuka, S. Katsumoto, S. Kobayashi and W. Sasaki, ibid B 25, 4288 (1982); R. N. Bhatt, ibid B 24, 3630 (1981 and ibid, in press.
21. W. L. McMillan and J. Mochel, Phys. Rev. Lett. 46, 556 (1981); B. W. Dodson, W. L. McMillan, J. M. Mochel and R. C. Dynes, Phys. Rev. Lett. 46, 46 (1981).
22. P. Chandhari and H.-U. Habermeier, Phys. Rev. Lett. 44, 40 (1980), and Solid St. Comm. 34, 687 (1980); A. B. Fowler, A. Hartstein and R. A. Webb, Phys. Rev. Lett. 48, 196 (1982).
23. Y. Kawaguchi and S. Kawaji, J. Phys. Soc. Jpn. 48, 699 (1980); K. Morizaki, Phil. Mag. B 42, 979 (1980); R. G. Wheeler, Phys. Rev. B 24, 6783 (1981); Z. Ovadyaku and Y. Imry, Phys. Rev. B 24, 7439 (1981); S. Kobayashi, F. Komori, Y. Ootuka and W. Sasaki, J. Phys. Soc. Jpn. 49, 1635 (1980); ibid 50, 1051 (1981).
24. N. F. Mott, Phil. Mag. 26, 1015 (1972); ibid, B 44, 265 (1981).
25. J. S. Langer and T. Neal, Phys. Rev. Lett. 16, 984 (1966).
26. J. Bardeen, L. M. Cooper, J. R. Schrieffer, Phys. Rev. 106, 162 (1957); ibid 108, 1175 (1957).
27. N. Giordano, Phys. Rev. B 22, 5635 (1980); J. T. Masden and N. Giordano, Physica 107B, 3 (1981); N. Giordano, W. Gilson and D. E. Proler, Phys. Rev. Lett. 43, 725 (1979).
28. W. J. Skocpol, L. D. Jackel, E. L. Hu, R. E. Howard and L. A. Fetter, unpublished.
29. A. E. White, M. Tinkham, W. J. Skocpol and D. C. Flanders, Phys. Rev. Lett. 48, 1752 (1982).
30. A. C. Sacharoff, R. M. Westervelt and J. Bevk. unpublished.
31. M. Az'bel, J. Phys. C 13, L797 (1980); Phys. Rev. Lett. 46, 675 (1981); ibid 47, 1015 (1981); Phys. Rev. 25, 849 (1982) and references therein.
32. A. Schmid, Z. Physik 259, 421 (1973); ibid 271, 251 (1974).
33. G. Hertel, D. J. Bishop, R. C. Dynes, E. Spencer and J. M. Rowell, unpublished.
34. D. J. Bishop, R. C. Dynes and D. C. Tsui, Phys. Rev. B 26 (1982).
35. P. W. Anderson, K. A. Muttalib and T. V. Ramakrishnan, unpublished.
36. A. L. Efros and B. I. Shklovskii, J. Phys. C 8, L49 (1975).
37. G. Bergmann, Phys. Rev. Lett. 49, 162 (1982); ibid 48, 1046 (1982); ibid 43, 1357 (1979); and Phys. Rev. B 25, 2937 (1982).
38. A. B. Harris, J. Phys. C 7, 1671 (1974).
39. For example, J. Tauc in The Optical Properties of Solids, ed. F. Abeles (North Holland, Amsterdam, 1970) p. 277.

Chemical Pseudopotentials
(Symposium to mark the 70th Birthday of G.H. Wannier, Eugene, Oregon, 1983)
Physics Reports **110**, #5–6, 311 (1984)

This is my best review of the "chemical pseudopotential" theory, a methodology which is meant to clarify many of the empirical facts of chemistry, such as locality of bonding, the meaning of ions in molecules and solids, the meaning of directed bonds and bond-angle forces, etc. It seems to be wholly out of synch with modern trends in theoretical chemistry; chemists do not ask the "why?" questions but only calculate particular cases.

Chemical Pseudopotentials

P.W. ANDERSON

Bell Laboratories, Murray Hill, N.J. 07974, U.S.A.

and

Joseph Henry Laboratories, Princeton University, Princeton, N.J. 08544, U.S.A.

The subject of so-called chemical or localized pseudopotentials will be reviewed. These pseudopotentials are based on the concept of Wannier functions and one can derive a self-consistent wave equation which such localized functions satisfy. As pointed out by Adams and by Gilbert, the optimum such functions are not orthonormalized, as was remarked by Wannier himself many years ago. These functions have been the subject of many successful chemical calculations by D.J. Bullett, which will be reviewed.

My early career in physics was furthered by a number of very lucky breaks, and one of these was the happenstance that I arrived at Bell Labs in 1949 on the same day as Gregory Wannier. Both of us had been shoe-horned in over the quota or "nosecount", and as a result there was no office space, and we were assigned to sit together in a small conference room for the time being.

Gregory very quickly decided that a physics department had to have afternoon tea, and that conference room – whose successor still exists – remained the tea room even after we were found offices. But in the meantime I, who had dropped out of the solid state physics course at Harvard, had had time to take an apprenticeship from Gregory, and in particular to learn the magic of Wannier functions, and the Wannier transformation, at the hands of the originator. The Wannier functions are still one of the most useful but underutilized methodologies of solid state physics, and in particular it is in the language of Wannier functions that I feel the *chemical* implications of band theory are most effectively expressed.

Of course the reason is that most of the concepts of chemistry are *local* concepts, such as bonds, ions, complexes, etc., while band theory is a *global* structure, in which the wave functions permeate the entire system and the eigenenergies depend on the position of every atom everywhere. There is no a priori reason why band theory should lead to such a chemically intuitive result as that the carbon–carbon single bond should have roughly the same energy and bond length whenever it appears, the O^{--} ion should have a constant radius and negative electron affinity, etc.

This same weakness is shared by band theory's chemical equivalent, the molecular orbital theory of Hund and Mulliken. From the very first, there was a vain attempt to restore locality by the use of atomic states and the valence-bond idea, very much advocated by Pauling, but only Pauling's great ingenuity in applying the vague concept of "resonance" and his enormous prestige kept this scheme afloat as long as it has been: it is just not a valid way of doing quantum mechanics, and fails completely in the case of metals and of the organic chemical equivalent of metals, namely aromatic compounds and graphite, and is not very useful elsewhere except in the hands of a master empiricist such as Pauling.

Nonetheless many – by no means all – chemical properties seem to be roughly understandable on a local basis: how is this compatible with quantum mechanics? This is the enigma to which Wannier functions give us a very precise and clear answer. Not only that, but with a bit of ingenuity it is possible

0 370-1573/84/$2.70 © Elsevier Science Publishers B.V. (North-Holland Physics Publishing Division)

to modify the local functions in such a way as to give one a simple, accurate and serviceable method for quantum chemical calculations. The participants in this work have been Gregory himself, W.H. Adams Jr., Tom Gilbert, myself (also with one paper jointly with John Weeks and Alan Davidson), and most particularly David Bullett, who has been almost solely responsible for the chemical and solid state applications.

Let us then recall some of the basic facts about Wannier functions. Bloch's general theorem requires that the eigenstates of a periodic potential $V_\tau(r)$ with

$$V_\tau(r + \tau) = V_\tau(r) \tag{1}$$

be expressible as

$$\phi_k(r) = e^{ik \cdot r} u_k(r) \tag{2}$$

where u_k is periodic with period τ. The corresponding energy ε_k must also be periodic with period $G = 2\pi(\tau)^{-1}$ since $\phi_{k+G}(r)$ is also a solution equivalent to ϕ_k.

The Wannier function is defined as [1]

$$W(r - R) = \frac{1}{\sqrt{\Omega}} \sum_k e^{-ik \cdot R} \phi_k(r) \tag{3}$$

(the sum is over a Brillouin zone of k's). Using (2)

$$W = \frac{1}{\sqrt{\Omega}} \sum_k e^{ik \cdot (r-R)} u_k(r).$$

The form of this function depends upon R, but if $R \to R + \tau$ it becomes the *same* function of $(r - (R + \tau))$. W is not an eigenstate: instead the energy matrix elements are the Fourier transform of the eigenenergy ε_k:

$$(W(r - R)|H|W(r - [R + \tau])) = \frac{1}{\Omega} \left(\sum_k e^{+ik \cdot R} \phi_k(r), \varepsilon_k\, e^{-ik \cdot (R+\tau)} \phi_k(r) \right) \tag{4}$$

$$= \sum_k e^{ik \cdot \tau} \varepsilon_k$$

and in particular $(W|H|W) = \bar{\varepsilon}_k$, the average eigenenergy in the band.

It is equally possible to reverse the Fourier transform and get back to ϕ_k by

$$\phi_k = \sum_\tau e^{ik \cdot (R+\tau)} W(r - R + \tau). \tag{5}$$

A first approximation to a simple tight-binding band is (5) with $W \to \psi(r - R_i)$ with ψ the atomic function centered around R_i, but the ψ's are not orthogonal. To get to the orthonormalized W from the non-orthogonal ψ we can use the Wannier–Bloch transformation, normalize the resulting pseudo-Bloch function to get the Bloch functions and return; or use the Löwdin technique

$$W = (1 + S)^{-1/2} \psi$$

where S is the overlap matrix.

Some important properties of W: clearly W is not unique; one may insert in (3) an arbitrary phase or "gauge" function $e^{i\eta(k)}$ and obtain an equivalent W. Blount [2] has shown that minimizing $\langle r^2 \rangle$ gives a unique most localized W with certain simple properties. In particular, for a band separated from all others by *gaps*, the best W is *exponentially localized*:

$$W(r - R) \sim e^{-\kappa|r-R|}$$

(a theorem due to Kohn) [3] where κ may be shown to be related to the widths of the band gaps. (To be precise, κ as a function of energy moves into the complex plane at the band gaps, and κ is the maximum of $|\text{Im} k|$ in the band gap.)

Finally, as Wannier showed [4] W satisfies a pseudopotential-type wave equation:

$$(H - H_{ps}) W(r - R) = \bar{\varepsilon}_k W \tag{6}$$

where

$$H_{ps} W = \left[\int d^3 r' \sum_\tau W(r - (R + \tau)) W(r' - (R + \tau)) H W(r' - R) \right] \tag{7}$$

is an integral operator $\int d^3 r' F(r', r) W(r' - R)$. The kernel depends on the Wannier function itself, but in principle, (7) could be solved self-consistently for W without solving the Bloch equation at all; but in practice, the non-uniqueness of W in particular makes this exercise almost impossible.

Nonetheless, already at this stage one can see that for all substances where there is an appreciable energy gap at the Fermi surface the chemical – that is, the energetic – properties are determined by the Wannier functions of the occupied bands, which are local and satisfy the local – if self-consistent – wave equation (6). This is one of the most important and least regarded theorems of solid state physics. It also extends to the electronic response functions (by slight generalization) so it also means that in these cases essentially local force systems can explain vibration spectra and other atomic properties. But the key point is that the Wannier function energy is the *mean* energy of the whole band and, with all bands full or empty, the detailed band structure is energetically irrelevant.

Some generalizations are necessary before we see the full value of this idea. The most important is to "symmetry orbitals" [5]: it may often be that a group of bands is separated from the others by gaps, even though the individual bands are interleaved in a complex way. The most famous example is the s and p bands in the diamond structure, where in general a group of four valence bands are intimately connected and separated by gaps from others. When this is the case one can always find a set of Wannier functions representing the whole multiple set of bands, symmetry – related to each other with more than one per unit cell. In the diamond case, there is one per covalent bond, and they are in essence bonding linear combinations of sp^3 valence orbitals.

It is only necessary to run through a series of examples to see that invariably the Wannier functions of the occupied bands mirror the chemical reality. As contrasted with the bond orbitals of covalent crystals, in ionic crystal structures the valence band Wannier functions are the p orbitals of the negative halogen or chalcogen ions. In complexed structures such as SO_4^- or SO_3^- they are the bonding orbitals of the negative ion complex, etc.

Long after Wannier's work Adams and Gilbert made the breakthrough which allowed these ideas to take practical form: the abandonment of orthogonality as a criterion for Wannier functions of filled bands. These authors pointed out the unexpected result that a complete, linearly independent but *non*-orthogonal set of orbitals can be found which are more localized, and often *much* more localized, than the Wannier functions, and can in general be determined *uniquely*.

To see that this is certainly so, let us calculate, for instance, $\overline{r^2}$ for a function in which we subtract off a fraction ε of the neighboring Wannier functions:

$$\phi(r-R) = N\left[W(r-R) - \varepsilon \sum_\tau W(r-(R+\tau))\right]$$

$$(\phi|\phi) = N^2(1+\varepsilon^2) = 1$$

$$(\phi|r^2|\phi) = N^2\left[\overline{r_W^2} - 2\varepsilon \sum_\tau (W_R|r^2|W_{R+\tau}) + O(\varepsilon^2)\right].$$

Clearly N and all corrections are of order ε^2; the off-diagonal term is of order ε and may be chosen to decrease $\overline{r^2}$, so W *cannot* have the minimum $\overline{r^2}$. Both Adams and Gilbert [6,7] gave much more exhaustive discussions showing that a very appreciable improvement could be made.

Interestingly, Gregory suggested this to me many years before, in terms of a remark of Wigner's: that actually, the original atomic functions were sure to be more satisfactory in some way than his orthogonal ones. In fact, as we will see the "ultralocalized" functions often closely resemble atomic ones. These authors, especially Gilbert, pointed out that such functions can be defined by satisfying self-consistent equations like (6) and (7) which there were strong arguments to believe gave more localized solutions. But certain prejudices and limitations left their work purely on a theoretical level.

Next came my own contribution [8], which was to point out that there are sets of ultralocalized or "ultra-Wannier" functions which obey a far simpler self-consistent wave equation than (6), so simple in fact that it can be used as the basis for sophisticated band or molecular calculations. In other words, the idea was to combine the conceptual advances of the pseudopotential theory, so valuable in band calculations, with those of the ultralocalized pseudo-atomic Wannier functions, to make a practical calculational scheme based on local, atomic concepts and not on band theory. The strong locality of the wave functions means that all energetic quantities are much more local than Wannier theory suggests.

The simplest version of this theory is presented in my Physical Review Letter and in a Physical Review paper [9] on the use of these methods as a basis for the famous "Hückel" theory of aromatic organic compounds. The Hückel system of π-electrons on the C atoms of an organic compound is perhaps the neatest possible illustration of the method.

To a good level of approximation, we may schematize the potential energy as a sum of atomic potentials, centered on the individual atoms,

$$H = T + \sum_n V_n(r). \tag{8}$$

We define the atomic "ultra-Wannier" function $\phi_n(r)$ by a self-consistent wave equation

$$H^n \phi_n(r) = (T + V_n(r))\phi_n(r) + \sum_{m \neq n} \left[V_m(r) \phi_n(r) - \left(\int \phi_m(r') V_m \phi_n(r') \right) \phi_m(r) \right]$$
$$= E_n \phi_n(r). \tag{9}$$

In fact, it is unnecessary to be as inaccurate as (8); one may use as "V_m" the difference between the true effective potential (as obtained by density functional methods or, as in most of our work, by the "Wigner trick" of locating the correlation hole on the local atom, which is for chemical problems often more accurate) and the single-atom potential V_n. The method allows remarkable flexibility.

In general, the pseudopotential operator

$$H^{ps}(r, r') = \sum_{m \neq n} V_m(r)(\delta(r - r') - \phi_m(r)\phi_m(r')) \tag{10}$$

is (a) relatively small, (b) repulsive: the occupied or valence band states usually oversaturate the potential V_m (as one can easily show using completeness of the full set of wave functions), hence the wave function ϕ_n is either close to the unperturbed atomic function or even more localized. Thus a remarkably accurate approximation is usually $(T + V_n)\phi_n \cong E_n \phi_n$: the atomic wave functions themselves. This is the same near-miracle which happens in conventional pseudopotential theory: the naive approach works remarkably well.

A second, and even more surprising, miracle occurs when we examine the secular equation which determines the band energies. The form of the pseudopotential equation allows us to expand the actual "Bloch" wave functions ϕ_α as linear combinations of the ultra-Wannier functions ϕ_n (note that there is absolutely no need to orthogonalize at any point): completeness alone allows us to write

$$\phi_\alpha = \sum_n (\alpha|n)\phi_n \tag{11}$$

and we can write

$$E_\alpha \phi_\alpha = H\phi_\alpha = \sum_n (\alpha|n) H\phi_n$$
$$= \sum_n (\alpha|n)(E_n \phi_n + \sum_m (n|V_m|m)\phi_m).$$

Picking out the coefficient of ϕ_n, we get

$$(E_\alpha - E_n)(\alpha|n) = \sum_m (n|V_m|m)(\alpha|m) \tag{12}$$

so that the secular equation is

$$\|(E_\alpha - E_n)\delta_{nm} + V^m_{nm}\| = 0.$$

The secular equation has three remarkable qualities: (1) the off-diagonal matrix elements are simple *potential* matrix elements – hence very short-range, and devoid of the messy three-center integrals of so-called "a priori" methods; (2) overlap terms are totally absent. This is the well-known character of the "Hückel" secular equation, and a very accurate value of "β" was determined in the original paper; (3) note that the total energy of the band is simply $\Sigma_n E_n$: only the energy eigenvalue of the local pseudo-wave function enters the total energy. This means that this scheme is particularly adapted to chemical bonding problems where it is the mean energy of the whole band which is relevant. For instance, Bullett has shown that, instead of atomic wave functions, in covalent systems *bonding* ultra-Wannier functions may be derived which give an excellent account of bonding in tetrahedral semiconductors for instance [10].

Following on this lead, these same methods have been taken up by Dave Bullett especially, in Cambridge and later at Bath [11]. I don't want to talk at too great length about work which is not my own, but I want to brag a little bit about the extraordinarily complex quantum chemical problems Dave has been able to demonstrate their usefulness on. Fig. 1 [12] is an example of his work on complex borides, systems where even the best of other techniques have been unavailing, whereas he is able to put together the full quantitative quantum chemical picture for these phases. Let me emphasize once

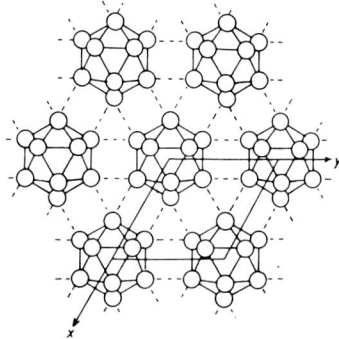

Fig. 1a [12]. Schematic section through the α-rhombohedral boron structure perpendicular to the three-fold axis. Broken lines symbolize the three-center bonds linking adjacent icosahedra in the basal plane; the other atoms are linked to icosahedra in neighboring layers by regular two-center bonds.

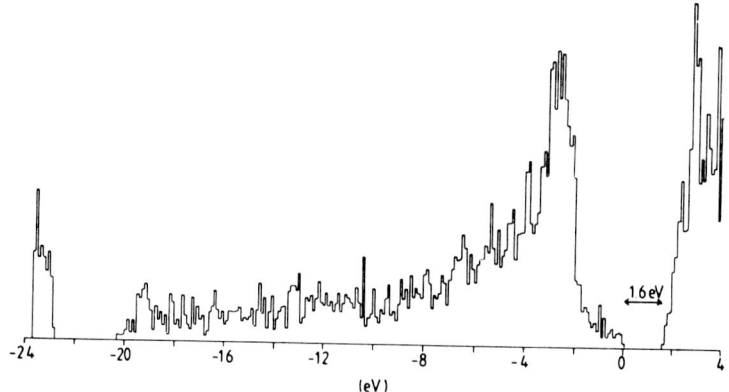

Fig. 1b [12]. Calculated density of states histogram for electrons in α-rhombohedral boron. Energies are measured in eV relative to the valence band maximum.

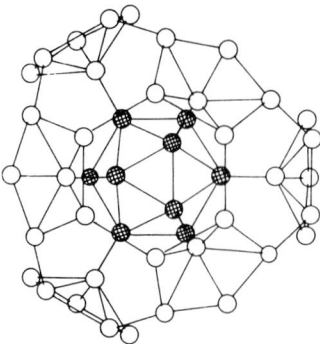

Fig. 1c [12]. Part of the 84-atom unit that forms the basic element of the β-rhombohedral boron structure. Each vertex of the central (shaded) icosahedron is attached to a half-icosahedron; these link together along the rhombohedral cell edges.

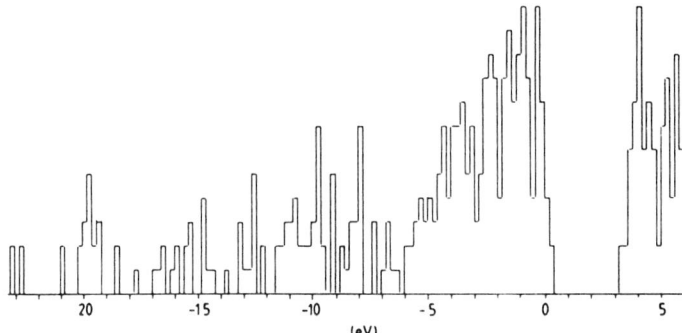

Fig. 1d [12]. Calculated density-of-states histogram for the "ideal" β-rhombohedral boron structure. Energies are measured in eV relative to the Fermi level; the full valence band would accommodate 320 electrons per unit cell, compared with the $105 \times 3 = 315$ electrons available.

again that there is really nothing "semi-empirical" or "pseudo" or phony about these techniques: they are as capable as any others of starting strictly from atoms or even earlier and putting together an entire complex structure. Much of their relative obscurity has been the feeling of physicists and quantum chemists that anything so easy must be fake in some way: but this is no more true of these than of other pseudopotential methods, and perhaps less; and they have the added blessing that one deals with physically significant atomic wave functions throughout.

What I want to do last is to talk about some of the original dreams I had for these methods that have not yet been fully realized. The method is a particularly clean way to understand the overlap repulsion (often miscalled "exchange repulsion") effect which is the fundamental repulsive force which keeps molecules apart (among other things: I would argue that it is also responsible for bond angle forces, for instance). This force results not from mysterious many-body "exchange forces" but simply from the Pauli principle limitation which prevents two electrons from occupying the same piece of Hilbert space together. We recognize that in our formalism this force must appear in the mean effect of the pseudopotential on the energy eigenvalues E_n. We can derive its value directly and *rigorously* (except of course for the usual premise that we must subtract out half of the interelectronic repulsive energy so as not to double count) from the eigenvalue equation. We multiply through by $\phi_n(r)$ and integrate:

$$E_n \times (\phi_n, \phi_n) = (\phi_n|T + V_n|\phi_n) + \sum_m (\phi_n|V_m|\phi_n) - \sum_m (\phi_n|\phi_m)(V_{nm}^m)$$

so the repulsive force is just the mean value of the pseudopotential:

$$(\delta E_n) = (\phi_n|V_m|\phi_n) - S_{nm} V_{nm}^m.$$

It is interesting to understand why this is indeed repulsive. In particular, this depends very much on the nature of ϕ_n, and it is rather a subtle argument which convinces one that the attractive effect of the mean potential $(\phi_n|V_m|\phi_n)$ is overcompensated by the overlap term.

Where ϕ_n is relatively a very smooth, flat wave function, in fact like a plane wave, the net interaction is very small, as we know in the case of ordinary pseudopotentials: the effect of orthogonalization to the core wave functions is almost negligible, and it is only the outer, mean part of the potential which affects the energies. Why don't we get then near compensation here?

Physically, one can draw a very simple picture which makes this clear. It is a fact that eigenfunctions like ϕ_m are more extended in space than the potentials V_m which they satisfy. Their shape looks like fig. 2. Now when we calculate the overlap matrix element $\int \phi_m \phi_n$, most of its value will come from the large volume where V_m is negligible, while the contribution V_{nm} will be concentrated in the overlap region. A way of estimating this is to point out that $\phi_n^2(r) V_m(r)$ will be proportional to $\phi_m^2(R_m) \cdot \bar{V}_m$ while S_{nm} will be essentially

$$\phi_n^2(r_m) \bar{V}_m \times \frac{\text{volume in which } \phi_n \phi_m \text{ is roughly const.}}{\text{volume in which } V_m \text{ is const.}}.$$

In general, this volume ratio is quite large if ϕ_n and ϕ_m are similar wave functions, and this accounts nicely for the normal repulsive forces between molecules. On the other hand, if ϕ_n is a weakly bound wave function (like a lone pair orbital on O^{--}) and ϕ_m is tightly bound (like a σ-bond between C and H) the repulsion is quite weak, and the compensation of the two terms nearly complete. This, I am convinced, is the true story behind the mysterious H-bond: the *absence* of the conventional overlap repulsion effect, rather than the presence of a peculiar kind of bond. This then brings us back to

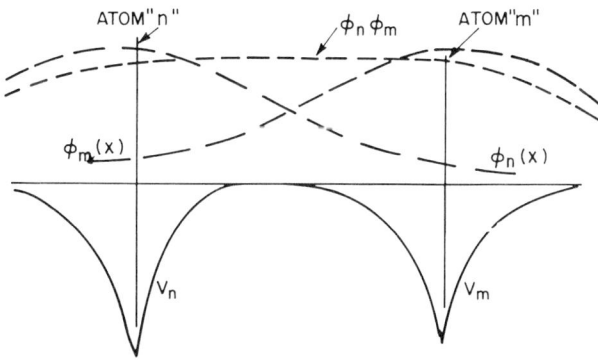

Fig. 2. Repulsion due to overlap of orbitals which extend outside the region of their respective potentials.

Gregory Wannier; among all the many problems we discussed together back in those days at Bell Labs, the interesting questions he first called my attention to, perhaps the H-bond remains the one most nearly unsolved.

References

[1] G.H. Wannier, Phys. Rev. 52 (1937) 191.
[2] The best review giving many results on Wannier functions is
E.I. Blount, in: Solid State Physics, Advances in Research and Applications, eds. F. Seitz, D. Turnbull, and H. Ehrenreich (Academic Press, NY) 13 (1962) 305.
[3] W. Kohn, Phys. Rev. 115 (1959) 809.
[4] See ref. [2].
[5] J. Lennard-Jones, Proc. Roy. Soc. Lond. A198, No. 1 (1949) 14.
[6] W.H. Adams Jr., J. Chem. Phys. 34 (1961) 89; 37 (1962) 2009; Chem. Phys. Lett. 11 (1971) 71, 441.
[7] T.L. Gilbert, Molecular Orbitals, a Tribute to R.S. Milliken, eds. P.O. Löwdin and B. Pullman (Academic Press, New York) p. 405.
[8] P.W. Anderson, Phys. Rev. Lett. 21 (1968) 13.
[9] P.W. Anderson, Phys. Rev. 181 (1969) 25.
[10] D.W. Bullett, in: Solid State Physics, Advances in Research and Applications, eds. H. Ehrenreich, F. Seitz, and D. Turnbull (Academic Press, NY) 35 (1980) 129.
[11] D.W. Bullett, sequence of papers: J. Phys. C 8 (1975) 2695, 2707, 3108; Sol. St. Comm. 17 (1975) 843; Phil. Mag. (8) 32 (1975) 1063; J. Phys. C, many papers 1976–83.
[12] D.W. Bullett, J. Phys. C: Solid State Phys. 15 (1982) 415.

Some Remarks on Strong Electron-Phonon Coupling Metals
(with C. C. Yu)
Proceedings of the Int'l School of Physics, "Enrico Fermi", July 1983
ed. F. Bassani, F. Fumi and M. Tosi, Varenna, Italy, July 1983

I anticipated this in the remarks on paper 18. For years Rowell, Dynes and I had occasionally held what we called the "Little Dirty Seminar" where we discussed unsolved problems of the complex metallic compounds, especially the transition metal compounds. Among these problems were the "martensitic" transitions, the peculiar resistivity behavior (saturation with temperature and impurity content), of course the occasionally high superconducting T_c, insensitivity to magnetic rare earth components as shown by the Chevrels, etc. The first part of this set of lectures describes this dilemma well; the second is an unsuccessful attempt by Clare Yu and me to find a mechanism.

Some Remarks on Strong Electron-Phonon Coupling Metals.

P. W. ANDERSON

Bell Laboratories - Murray Hill, NJ 07974
Joseph Henry Laboratories, Princeton University - Princeton, NJ 08544

C. C. YU (*)

Bell Laboratories - Murray Hill, NJ 07974

1. – Introduction.

These lectures are amusing to give here at this school where many historic figures in our science of solid-state physics are or will be gathered, because they have a great deal to do with the history of solid-state physics. They do so in two ways. One is that I will be discussing—far from solving, I assure you from the start—one of the historic problems of the quantum physics of solids, the problem of the strong-coupling superconductors—the «bad actors» which do not fit into the conventional scheme of BCS superconductivity and simple electron-phonon interaction. These materials have been around for a long time—since 1954, to be precise, when HULM and MATTHIAS almost simultaneously discovered the two «classics» V_3Si and Nb_3Sn (T_c's 17.1 and 18.2 K, respectively). It is very likely that NbC, discovered about 1930, and NbN, in the 40's, are also members of the class, but I do not know of any measurements of the normal metal properties which would certify them. Clearly several other A15's are, including the present record-holder Nb_3Ge (23 K). More recently, ROWELL and DYNES, in particular, have emphasized that at least two other classes of superconducting cluster compounds exhibit the kind of anomalies we will be discussing, the chevrels $X(Mo_6S_8)$ where X = Cu, Pb, Gd, Ho, etc.—which have, in addition to high T_c's, extraordinarily high critical magnetic fields H_{c2}—and the rhodium borides like $Er(RhB)_4$ (**).

(*) Permanent address: Department of Physics, Princeton University, Princeton, NJ 08544.
(**) I have often thought, in agreement with C. M. VARMA (private communication), that the CDW dichalcogenides like $NbSe_2$ may also belong in this group.

The second historically intriguing aspect of this subject is that it is almost a textbook case of the discrepancies between the ideal, classical version of the scientific method which we all think we learned in school, and practice every day, and the actual reality of how science is done. Some of the realities of the scientific process were revealed by T. S. KUHN some years ago; but even KUHN never studied a subject confused by the imprint of such personalities as Bernd MATTHIAS of much regretted memory, to pick unfairly on the one of us who cannot answer back.

KUHN speaks of « normal science » and « anomalies »: normal science is said to go chugging along, following accepted theories and using so-called « paradigms » which have been successful in the past, while « anomalies » confront normal science with results which are incompatible with its present ideas. The idealistic model of science is that the scientist immediately abandons his previous theories when confronted with such anomalies, and that successful theories have been repeatedly tested and never found wanting. In actual fact, the most common response we all exhibit to such anomalies is to ignore them. Very often this is quite correct: the great majority of anomalous results turn out to be simply mistakes, and the scientist who believes every unexpected result he hears is completely incapacitated for reasonable work. (Murray GELL-MANN calls this « ambulance chasing ».) But there are also theories and hypotheses which survive in the textbooks in spite of the fact that the great majority of *all* experimental results contradict them: two good examples are the Debye-Eyring theory of dielectric relaxation (which is a very rare case) and Matthiessen's rule for the additivity of phonon and impurity resistance, which within the past few years has been discovered to be false in principle.

The BCS theory, of superconducting transition temperatures and electron-phonon coupling, as modified by MOREL and myself, ELIASHBERG, SCHRIEFFER and MCMILLAN is much better supported than that: it works beautifully and quantitatively for all such ordinary elemental superconductors as Pb, Sn and Hg, and to an extent even for Nb and Ta, and one may produce a really basic microscopic theoretical structure behind it. It works quite well qualitatively for ordinary transition metals; but with the « bad actors » one has not just a quantitative discrepancy but a rather severe qualitative one: the conventional theory of electrons and phonons in metals simply fails in a qualitative sense, and there is *no* way around that fact.

2. – Resistivity data: the puzzle; other anomalies of the « bad actors ».

I would like first to show you a brief sample of the kind of data which convinces me of this fact, comparing it with crude but realistic theory, just to make it clear that there is a real problem. Then I will indulge myself in trying to point out how irrelevant most theoretical work in this area is. Then

I will come back and re-examine the conventional electron-phonon theory in some detail before describing the extremely unconventional ideas which I have developed in collaboration with Clare YU.

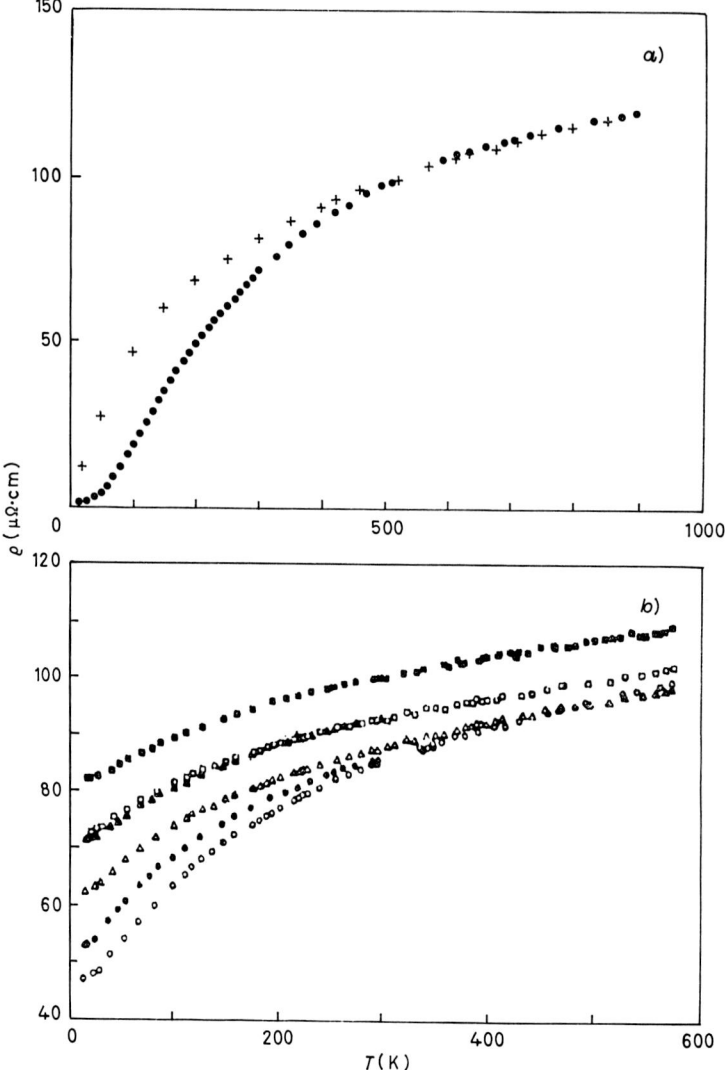

Fig. 1. – a) Resistivity data for • Nb_3Sb and + Nb_3Sn [1]. b) Resistivity data for Nb_3Ge. Increasing residual resistivity is correlated with decreasing T_c [2].

Figure 1 shows the resistivity of samples of one of the canonical A15 metals, say Nb_3Sn, progressively damaged by irradiation with α-particles. What we immediately notice is the very striking deviations from Matthiessen's rule. This is a well-known phenomenon called « resistivity saturation » and has been extensively discussed from an experimental point of view by GURVITCH [3], DYNES [4], MOOIJ [5] and others and from a theoretical one by ALLEN [6] and many others. It appears certain that this saturation occurs where the resistivity approaches the «Mott limit» where the mean free path becomes of atomic size. When the impurity resistivity reaches this magnitude, it is felt that phonons assist conductivity rather than the other way around—the electrons are close to the localization limit and are *delocalized* by thermal agitation. Thus the general trend of reaching a resistivity where ϱ vs. T flattens out is roughly understood; on the other hand, the actual theory of this region is absolutely not complete.

Let us estimate the Mott limit. The conductivity of a Fermi gas is

$$\frac{ne^2\tau}{m} = \frac{ne^2 l}{mv_F};$$

inserting $mv_F = \hbar k_F$ and $k_F^3/3\pi^2 = n$, we get

$$\sigma = \frac{e^2}{3\pi^2 \hbar} k_F(k_F l).$$

The limit occurs at a value of $k_F l \sim \pi$; some more careful numerology gives

$$\sigma_{\text{Mott}} \simeq \frac{e^2}{2h} k_F$$

and k_F is, for most metals, of order 10^8 cm^{-1}, while e^2/h is the famous quantum conductance $1/25\,000$ Ω^{-1}, so we get

$$\sigma_{\text{Mott}} \simeq \frac{10^8}{50\,000} \simeq \frac{1}{500} \, (\text{m}\Omega \text{ cm})^{-1}.$$

This agrees rather well with typical saturation values. One may object that this does not take into account all kinds of complicated band structure and many-body effects, but, in fact, we note that σ_{Mott} is basically a *geometric* quantity: dynamic effects drop out, special properties of the Fermi surface are irrelevant, and it is only the *mean free path* and the *area* of the *Fermi surface* which enter.

A mere glance at the figure reveals that the phonon—*i.e.* temperature-dependent—resistivity of the pure material at $\theta_D \sim (250 \div 300)°$ is a major fraction of σ_{Mott}: a *very strange* and *important fact*: $kl_{\text{phonon}}(\theta_D) \sim 1$ for the *bad actor* superconductors.

Why is this strange? It turns out that such a tiny value of l is *completely incompatible* with conventional electron-phonon theory. We see by a very simple argument that this is a gigantic rate of scattering: what it says is that the random potential due to the presence of ~ 1 phonon/atom—*i.e.* essentially 1 vibrational quantum $\hbar \omega_D$ for each atom—is of the order of the entire electronic band width E_F—*i.e.* the scattering potential on each atom must be as strong as the kinetic energy itself. This is a very strange result, when we recognize how small the average amplitude of atomic motion is:

$$M\omega^2 \overline{x^2} = \hbar\omega,$$

$$\overline{x^2} = \frac{\hbar}{M\omega} \sim \frac{10^{-27}}{10^{-22} \cdot 2\pi \cdot 3 \cdot 10^{13}} \simeq \frac{1}{2} \cdot 10^{-18} \text{ cm}^2,$$

i.e. the atomic motion is $\sim 10^{-9}$ cm $\sim 1/20 \div 1/50$ of typical interatomic distances. The phonon energy $M\omega^2 x^2$ itself comes from the effective potential of interaction between electrons and ions, and we would expect that the potential fluctuation should be of order $\hbar \omega_D$, not E_F. Any reasonable measure for E_F in Nb_3Sn or V_3Si must be $\sim (2 \div 5)$ eV; this contrasts with $\hbar\omega \sim 0.02$ eV.

In fact, it is easy to see that the dimensionless electron-phonon coupling constant $\lambda_M = N(0)V$ defined in the early days of superconductivity by MOREL and myself, SCHRIEFFER, MCMILLAN, etc. is essentially the ratio

$$\lambda \simeq \frac{\langle V_{ph}^2 \rangle}{\hbar\omega} N(0),$$

where $\langle V_{ph}^2 \rangle$ is the mean square potential amplitude due to the presence of a single phonon. (This can be derived in a number of ways—most simply by classical response theory: In the presence of a charge density wave $\varrho(q)$ the phonon potential energy is

$$\Delta E = -\varrho(q)V(q) + \frac{M\omega_q^2 x_q^2}{2}.$$

The potential $V(q)$ is proportional to x_q by a constant K: $V_q = Kx_q$. Minimizing ΔE, we get

$$x_q = \frac{K\varrho(q)}{M\omega_q^2} \quad \text{and} \quad \Delta E = -\frac{1}{2}\frac{K^2}{M\omega^2}\varrho_q^2.$$

To evaluate K:

$$V_{ph}^2 = K^2 \cdot (\overline{x_q^2})_{1 \text{ phonon}} = K^2 \cdot \frac{\hbar}{M\omega},$$

so $\Delta E = -\frac{1}{2}(V_{ph}^2/\hbar\omega_q)|\varrho_q|^2$. The same result—even more briefly—comes from second-order perturbation theory and the golden rule.)

If we now insert the fact that, in the case of the A15 materials, $\langle V_{ph}^2 \rangle^{\frac{1}{2}} \sim E_F$, which in turn is of the order of 1 eV, we recognize that

$$\lambda \sim \frac{((1 \div 2)\ \text{eV})^2}{300\ \text{K}} \frac{1}{(1 \div 2)\ \text{eV}} \simeq 100.$$

This is a ridiculously strong electron-phonon coupling and is completely incompatible with the values of λ which have been derived phenomenologically from various superconductivity data such as tunneling spectroscopy. Most simply, people have used the McMillan formula

$$T_c \sim \frac{\omega_D}{1.2} \exp\left[-\frac{1.04(1+\lambda)}{\lambda - \mu^* - 0.62\lambda\mu^*}\right]$$

with the observed ω_D's, or observed phonon spectra. All estimates [7] agree on a set of λ values of which 1.4 to 1.5 for Nb$_3$Ge is possibly the largest. Figure 2

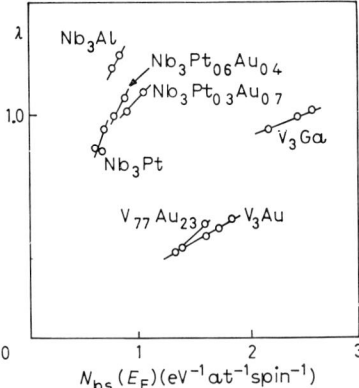

Fig. 2. – Typical values for λ [8].

shows a typical group of values. In fact, the usual estimate of the Coulomb pseudopotential $\mu^* \sim 0.13$ or so is probably not okay for A15's, but any reasonable value only adds a little to λ. An additional peculiarity in having such enormous values of λ is that the stability of phonons is also supposed to be indicated by the magnitude of the electron-phonon coupling: this is perhaps most simply expressed by asking: how big is the shift in phonon frequency caused by the response of the electron gas? (The appropriate diagram is shown in fig. 3.) The value of this self-energy is

$$\Sigma_{\text{ph}} \sim V_q^2 \cdot \chi_{\text{electronic}} \sim V_q^2 N(E)$$

and $\Sigma_{\text{ph}}/\omega \simeq \lambda$: so the renormalization of the phonon frequency should be much bigger than that frequency itself! This suggests the possibility of phonon instability and, therefore, of instability of the lattice structure, which is (see [9]) the supposedly most fundamental limitation on the magnitude of superconducting T_c's. Incidentally, at this level even the broadening of phonons due to

Fig. 3. – Phonon self-energy.

electron scattering is nonnegligible. At room temperature $\mathrm{Im}\,\Sigma$ can be of order $\lesssim \omega_D$, which, in fact, it probably is (see fig. 4). The kinds of temperature changes seen here are quite unusual.

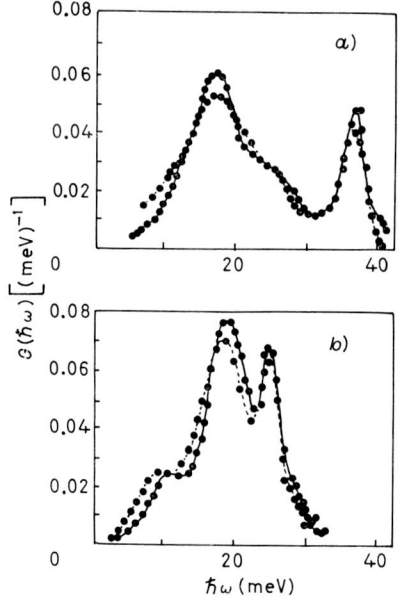

Fig. 4. – $G(\omega) \sim$ phonon density of states for a) Nb_3Al and b) Nb_3Sn [10].

This is also a very important component of the fundamental microscopic theory of superconductivity. The mythology has always been that $\lambda \lesssim 1$ (which was enforced by the stability condition) is a fundamental limitation on the size of electron-phonon effects, and, in particular, this restriction is the basis of an idea called « Migdal's theorem » which states that all complicated perturbation effects can be neglected. This whole mythology now fails at least for the A15's, and the standard methods of superconductivity theory simply must not be used.

This is the very large anomaly which has been accessible to us for over 20 years and has not been seriously considered previously. I have yet to find any mention of this problem in the literature, in spite of the fact that I have lectured on it before quite sophisticated audiences on several occasions. I really feel that there may be something special in the sociology of this particular community: perhaps an excessive self-confidence, a habit of being protected from the necessity of thinking really clearly by the complexity of the subject and by one's own erudition. In any case, in the reams of literature which one can read on this problem, including several excellent books [11] (Parks series, the three d- and f-band conferences, the Lake Geneva ternary superconductor conference), perhaps only one or two papers have ever focused on the strangely bad resistivity of the high-T_c superconductors.

Let me dispose of at least one red herring immediately. (Maybe two.) The resistivity in fig. 1 clearly has a characteristic temperature near the accepted θ_D of these materials, which are well studied as far as neutron spectroscopy and specific heat are concerned (by FISK among others). Thus electron-electron scattering is not the primary mechanism, although older papers make much of the fact that some small fraction of ϱ often goes like T^2 and *may* be interband scattering (not *must* under present understanding of diffuson effects). The bulk is clearly phonon-related. GURVITCH has emphasized the obvious point that the small T-dependence for very dirty samples is illusory. This is the cancellation of what must be very *large* effects. The pure material is the appropriate one to study.

One paper in particular which really has focused on these problems was a talk given at a Gordon Conference a couple of years ago by John ROWELL called « Bad Metals, Good Superconductors ». There he particularly emphasized that several kinds of high-T_c or otherwise anomalous superconductors (high H_c, magnetic, etc.) all show many symptoms of the anomalies and, in particular, have very high phonon resistivities. Particularly striking cases are $ErRh_4B_4$ and some of the chevrel phases.

Table I is from an old talk of my own and summarizes most of the anomalies of these compounds. Do not worry if you do not recognize all the physics in this table—I will run it down briefly here, but I will return later to everything that is of importance. I would emphasize as most important 1) resistivity, which I have just discussed; 2) the very large and anomalous effects of defects and damage, both on ϱ and on T_c; most normal superconductors simply do not even care if they are stoichiometric or not; 3) the strange insensitivity to magnetic effects; Mo with 10^{-6} Fe is not even superconducting at all, while Ho, Er, Gd, Mo_6S_8 are all high-T_c superconductors with stoichiometric magnetic ions; 4) the many diverse indications of strongly anharmonic atomic motions; 5) the T-dependent susceptibility of the same sign in all cases. This we will discuss in relation to our theory.

TABLE I.

Property	«Conventional»	«Anomalous»
T_c	Usually $< 10°$, agrees well with phonon λ as measured from γ in C_{sp}.	Often $> 15°$, peculiar systematics, limited above.
H_{c2}	Fits theory with $\xi^2 = \xi_0 l$ evidence of Clogston limit.	Extraordinarily high in chevrels, some A15's no Clogston limit?
Magnetic impurities	Strong pair breaking effect on T_c. Spin glasses; RKKY in normal state.	Not clearly different from ordinary ones. Coexistence of ordered magnetism + superconductivity.
Nonmagnetic impurities	Weak effect fits Langreth-Kadanoff-Anderson theory.	Effect can be strong, a major determinant of T_c, for instance.
χ_{spin}	T-independent, fits reasonable $n(E)$.	Strong but smooth T-dependent at all T. Apparent sharp $n(E)$ structure.
γ	T-independent, fits reasonable $n(E)$. Also see T_c.	Strong but smooth T-dependent at all T. Apparent sharp $n(E)$ structure. γ often large but not extraordinarily so.
σ	Conventional phonon-impurity theory, $\gg$ Mott limit; constant as $T \ll T_D$.	T-dependent at all T, $\sim$ Mott limit. R_{300}/R_0 correlates with T_c.
Defects	Stoichiometry or stable, homogeneous solid solution with Hume-Rothery rule okay.	Defect structure common.
Polymorphism	Few long-wavelength structures.	Density waves and B-B «martensitic» transitions occur.
Elasticity and phonons	Nearly T-independent fluctuations normal.	Elasticity T-dependent. Strong anharmonicity. Large fluctuations.

3. – Previous theories: why $N(E)$ spikes do not work.

Now, partly in order to introduce some of the physics, I would like to convince you that most theoretical approaches have been ineffectual—primarily, I would say, because they all ignore the basic implications of fig. 1 and concentrate on one or another of the other aspects which are more manageable. Among these approaches are:

1) Most common: ignore fig. 1 and treat as though they were conventional superconductors with slight differences. This is the majority view and has several subsets:

Bell Labs—central: use McMillan phenomenology, calculations of phonons, etc. Try to squeeze down into conventional BCS shape (DYNES, VARMA; an offshoot at Stony Brook under ALLEN). I do not know what the Italian term is but in central Illinois we would have called this a « blivet » theory, a blivet being two kilos of something unpleasant in a 1 kilo container. The Stony Brook group at least emphasizes the saturation phenomenon, but tries to calculate it in simple perturbation theory. Nonetheless, the key problem here —aside from the overall structure of table I, which says that this is really a total phenomenon—is the fact that $T_c \ll \omega_D$ yet λ from resistivity ~ 100.

Let me not be too harsh. VARMA and WEBER [12], in particular, will appear in my story as among the first to recognize that the phonon properties of transition metals are really different. Also much useful phenomenology has been done here.

2) Many of the same group, plus generations of band theorists, have ascribed the phenomena to fortuitously sharp peaks in the density of electronic states at the Fermi level (fig. 5). Some of this aspect will be discussed under the one-dimensional men of 3); but the key point here is somewhat different. It is true that some—but not all—of the highest-T_c materials have a quite high specific heat γ. Here are some numbers: According to theory,

$$\gamma = \frac{\pi^2 R}{2} \frac{Z}{T_F},$$

where R is the gas constant (2 cal/mol), $kT_F = E_F$ for free electrons, which is a good version of how narrow the relevant pieces of band structure are. Some γ values and appropriate Z's and E_F's appear in table II. The γ value for Nb_3Sn is, of course, quite large, but it is clear—at least to me—that the relatively large γ's of many of these materials are strongly enhanced at low temperatures

TABLE II. – *Specific heats: is there a « peak »?*

Element	γ	Z	E_F (eV)	Renormalized for a reasonable λ (eV)
Pb	7.5 cal/mol degree^2	4	5	10
Ti	8	4	5	6÷7
V	15	5	3	6
Nb	21	5	2	5
Nb$_3$Sn	45/atom degree^2	4.75	1	2÷3
La	16	3	5	6÷8

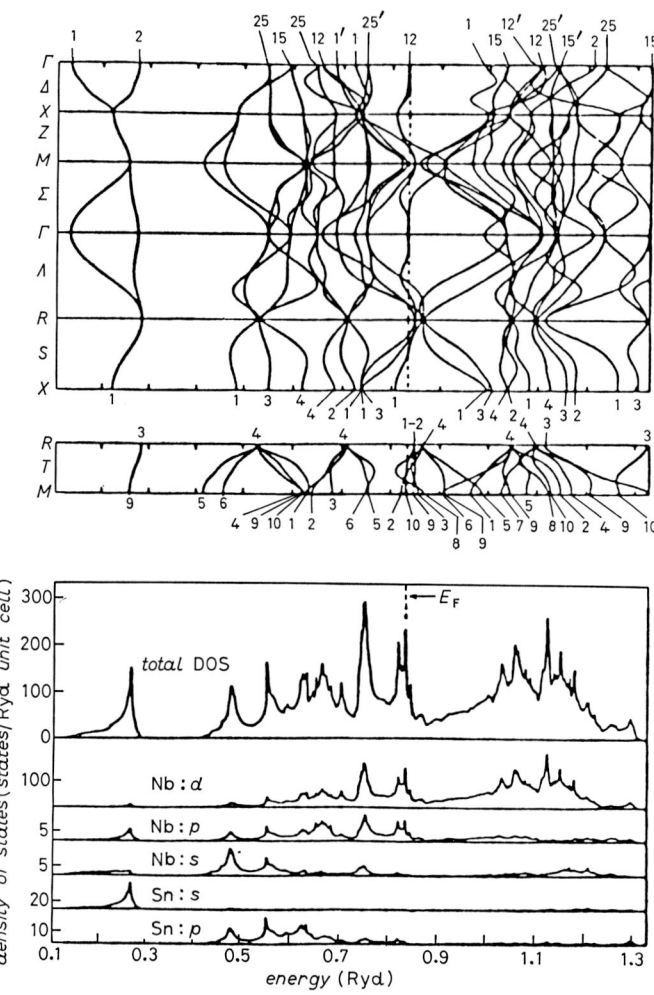

Fig. 5. – Band structure and densities of states for Nb_3Sn [13].

by phonon effects, which, even if conventional theory were reasonable, would leave us with conventional-size band widths, perhaps twice as narrow as typical d metals.

In the calculated band structures sharp peaks often appear, with breadths more nearly comparable to $k\theta_D$ (which would be necessary to explain the strong T-dependence of so many effects) (fig. 5). People often do not realize the true spikiness of things: if you have a *very* complicated band structure, as you do here, its most spiky features are sure to be much spikier than the average. I believe, however, that these spikes, far from being (as many authors, including many of the most respected experimentalists, seem to believe) the key to the system, are almost irrelevant. Consider the following:

a) The strangest of all the phenomena, I keep repeating, is ϱ. σ is the conductivity of all bands in parallel, not just of the heaviest electrons, and nobody is so rash as to claim that there are not several bands crossing the Fermi surface of which only one can be spiky at once. Conductivity depends on Fermi-surface area, not density of states, so the high γ-high mass bands lose back their advantage by losing Fermi velocity at the same time, and all bands contribute equally.

b) All phenomena are rather insensitive to stoichiometry. The calculations are done in idealized structures; but stoichiometry changes will move the Fermi surface and the bands around arbitrarily so any coincidental spikes will move off. Yet—as we will see—such data as χ almost always act as though the density-of-states peak sat right at the Fermi surface (see fig. 6).

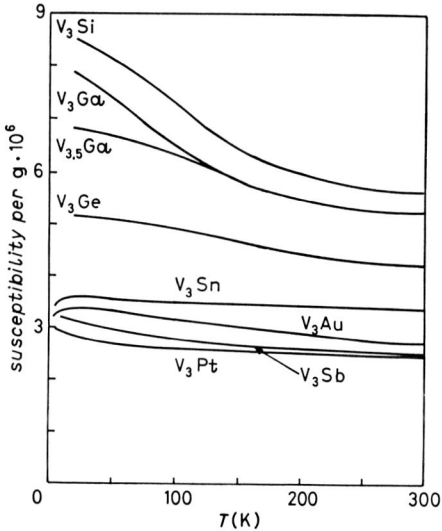

Fig. 6. – Susceptibility data for some V_3X compounds [14].

c) Incidentally, in conventional theory phonon effects cannot influence χ, so the T-dependence of χ is normally ascribed to these band spikes and requires θ_D level structure in the underlying bands. Since we will discard Migdal's theorem, we will not be quite as disturbed by $\chi(T)$ as previous theories.

d) The phonon resistivity itself implies that the mean free path for electronic excitations of order 300 K ~ 0.03 eV is of order 10 Å. Thus k cannot be defined on a scale sharp enough to really show the detail which the fictitiously perfect crystals of band structure calculations provide. This is why I have really been so casual in tracing the figure: none of the detail has any meaning. Incidentally, one way in which this lack of meaning is never discussed is the presence of zero-point corrections to the actual band structure calculation itself. That is, if 300 K worth of phonons can cause $\sim \frac{1}{2}$ to 1 eV energy uncertainty at finite T, the zero-point motion of $\hbar\omega_D/2$ causes exactly half as much change, at the very least, so the potentials used in these calculations cannot be very accurate.

e) Finally, let me counter the naive tendency to equate complex crystal structure with weak coupling of atoms and narrow band widths, especially in the A15 structures (see fig. 7, which rewards some study, for the nature of

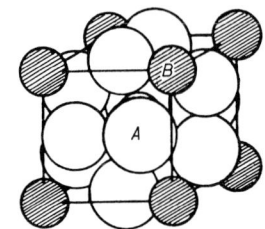

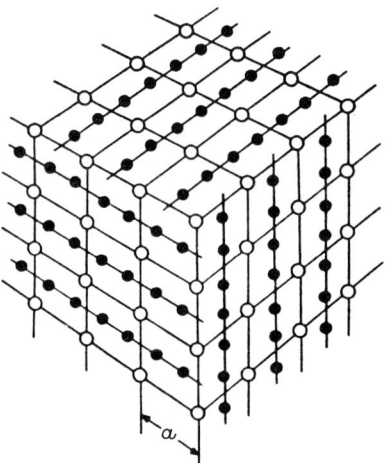

Fig. 7. – A15 structure [15].

this structure.) The Nb atoms are actually closer than in Nb metal—the chains are very strongly bound covalently by a « d_σ » band of d levels pointing along the chain. The Nb-Si distances are only 12% greater and the Si atoms—which share 6 neighbors from 3 *separate* chains—are strongly involved in d_π bonding via the $m_d = 1$ orbitals of the chains. The relevant overlap integrals are in the one case > 5 eV and in the other 3 or 4, as big as in a typical organic carbon compound for typical σ and π bands. Hence it is ridiculous to isolate any of the atoms or the electrons from each other, and all or almost all of the electronic states participate in bands with a full set of overlaps > 2 eV.

3) A final group of hypotheses are closely related to the former ones, namely the one-dimensional advocates. These include GOR'KOV, WEGER, FRIEDEL and their followers. Essentially all of the difficulties of the previous category and a few more are encountered by this kind of theory.

The basic idea is to take advantage of the complexities of one-dimensional physics by focusing on the dominant crystal structure feature, those strong, prominent Nb chains. Somehow wishing away the Si as well as the direct overlaps between Nb d-functions of various sorts (the nearest Nb-Nb interchain distance is only 22 % bigger than that within the chains) it is attempted to make up the material of quite independent one-dimensional bands, which leads to all the complexities for which one-dimensional chains are famous. In the first place, even if this theory were correct, the giant electron-phonon coupling would lead to Peierls transitions not at the freezing temperature of 20 K as observed, but at $T \sim 10^4$ K. (We know this actually occurs in most real one-dimensional structures, as Schrieffer's lectures will emphasize.) The real problem is that the intrinsic energy scale of the band structure is *far too big* if one-dimensional effects are relevant. Interchain interactions are clearly already larger than the temperature at which the real action takes place. We add, of course, the usual question of what about the effect of the other bands on the conductivity?; and why is not the density wave dominant as it is in most other one-dimensional systems? But basically I do not think any realistic energy band calculation will support a fundamentally one-dimensional electronic structure.

I should emphasize that, critical as I have been, I really give all of the above one-dimensional authors a positive grade relative to the majority of those who have discussed this subject; they have, in general, been willing to accept that there is a problem and have only been driven to take the rational point of view of proposing an implausible solution to an impossible set of data. I had better not criticize *too* deeply since I will propose quite implausible hypotheses myself. In fact, I am driven to re-examine a number of aspects of conventional theory, and in order to do that I will start by teaching you some of that.

4. – Conventional theory: screening, Migdal and all that.

A very simple treatment of conventional theory is found in Cohen-Anderson in the first d- and f-band superconductor book, and I follow that.

If the electron-electron interaction is linear—perhaps the soundest assumption made in the various theories—, we can surely assume that it follows from the response $V(q)$ of the substance to a charge density wave

$$\varrho_q = \sum_{k\sigma} C^\dagger_{k+q\sigma} C_{k\sigma}$$

which will in turn interact with ϱ as $-\varrho V(q)$. The response is simply the potential of the charge screened by the total dielectric constant ε:

$$V(q, \omega) = \frac{4\pi\varrho^2}{q^2 \varepsilon_{tot}(q, \omega)}.$$

For superconductivity ω is small so it is important to study $\omega \to 0$. We assumed that $\varepsilon(\omega = 0) > 0$, for stability (otherwise a potential would grow without limit); this is not quite exactly right as argued by the Russians, but it turns out that in fact it is never violated seriously. Hence $V(q, 0) \geqslant 0$ and there can be no attractive interactions! How can superconductivity really occur?

The answer to this appears in two easy stages. The first makes essentially all real metals superconducting, but at a rather low temperature. This is that $V(q, \omega)$ has a dynamic structure even in a featureless « jellium » electron gas:

$$V(q, \omega) \simeq \frac{4\pi e^2}{q^2 \varepsilon_{electronic}(q)} \left(1 - \frac{\Omega^2}{\omega_{ph}^2 - \omega^2}\right),$$

where, to keep $V(0) > 0$, $\Omega^2 < \omega_{ph}^2$ and Ω is the jellium frequency. The basic thing is that there are two frequency scales, and the phonon frequency scale is much lower than the electron one. This means that the electrons have a much bigger range of frequencies to use in dodging the repulsive part of the interaction, and they do so very effectively, leading to a solution of the BCS equation first guessed by BOGOLIUBOV and used by MOREL and myself,

$$T_c \simeq \omega_{ph} \exp\left[-\frac{1}{\lambda - \mu^*}\right],$$

$$\mu = \left\langle \frac{N(0) \cdot 4\pi e^2}{q^2 \varepsilon(q)} \right\rangle \simeq \frac{1}{2} \text{ in most metals},$$

$$\mu^* = \frac{\mu}{1 + \mu \ln(\omega_p, E_F/\omega_{ph})},$$

$$\lambda = \mu \left\langle \frac{\Omega^2}{\omega_{ph}^2} \right\rangle.$$

If we go further and realize that the phonons (as shown by BOHM and TREES) are basically ion plasmons screened by ε as well, we find that

$$V = \frac{4\pi e^2}{\varepsilon(q)q^2}\left(1 - \frac{\Omega_{\mathrm{pi}}^2}{\varepsilon(\omega_{\mathrm{ph}}^2 - \omega^2)}\right)$$

with $\omega_{\mathrm{ph}}^2 \simeq \Omega_{\mathrm{pi}}^2/\varepsilon$, so $\lambda = \mu \simeq \frac{1}{2}\Omega_{\mathrm{pi}}$ is the ion plasma frequency. (Note that at low frequencies it is a good estimate to write

$$\frac{4\pi e^2}{\varepsilon q^2} = \frac{4\pi e^2}{K_{\mathrm{s}}^2 + q^2}, \quad \text{so} \quad \varepsilon = 1 + \frac{K_{\mathrm{s}}^2}{q^2}, \quad K_{\mathrm{s}}^2 \propto N(0), \quad \omega_{\mathrm{ph}}^2 \simeq \frac{\Omega_{\mathrm{pi}}^2}{K_{\mathrm{s}}^2}q^2.)$$

This leads to a universal T_c in the neighborhood of 3 or 4 K for almost anything. It also puts a permanent kibosh on any excitonic superconductivity mechanisms, since the logarithmic factors are actually optimized for $\omega_{\mathrm{ph}} \simeq 10^3$ K, $T_\mathrm{c} \simeq 10$ K.

How then does anything get above 5 K? The answer may be variously described as local fields or umklapps, which amounts to the same thing, essentially the statement that a solid really has ions and electrons which have a very nonuniform density, so the actual contents of the unit cell matter. In particular, the electrons and ions are not in the same place and do not see the same local field or dielectric constant. The electrons in a polyvalent metal like lead occupy an «extended zone» Fermi surface which involves scatterings $k \to k' = k + Q$ with Q's much larger than the size of the Brillouin zone which limits the q's of the phonons. The phonon $q = Q - G$, where G can be some reciprocal-lattice vector. We did a simple local-field analysis in that paper, which I reproduce in total here:

If we consider umklapp scattering and let the momentum transfer we have been using $\boldsymbol{q} \to \boldsymbol{q} + \boldsymbol{G} = \boldsymbol{Q}$, where $\boldsymbol{G}$ is a reciprocal-lattice vector, then

(19) $$V(\omega = 0) = \frac{4\pi e^2}{Q^2 \varepsilon(Q)}\left[1 - \frac{\Omega_{\mathrm{pi}}^2}{\varepsilon(Q)\omega_{\mathrm{ph}}^2(q)}\right].$$

The above is easy to derive from the following ideas and simplifications: 1) Nearly free-electron theory: the scattering through $Q = q + G$ on the unperturbed spherical Fermi surface is by the effective potential at wave number Q. 2) Approximately point ion potentials, screened by $\varepsilon(q, \omega)$. 3) The ion cannot experience its own self-field, so the effective local field on an ion is $E - E_{\mathrm{self}}^i = E + LP_i$, where L is the local-field constant $= 4\pi/3$ in a cubic lattice. The role of the local field is merely to allow $\lambda > \mu$ because the ions and electrons see *different* fields.

We calculate the motion of the ions and their resulting field at the wave vector $Q = q + G$ (q in 1st B.Z.), when a static ($\omega = 0$) electronic charge density wave at Q is present.

The basic equation expresses the fact that the ions must move until the field they see is zero, since $\omega = 0$,

(20) $$E^{\text{ext}} + E^{\text{eff}}_{\text{ions}} = 0 \, .$$

But from the equation for the phonon frequencies we can evaluate $E^{\text{eff}}_{\text{ions}}$ as a function of the ionic polarization P_q:

$$M_i \ddot{x}_q = M\omega_q^2 x_q = Ze E^{\text{eff}}_q$$

or

$$P_q = ex_q = \frac{Ze^2}{M\omega_q^2} E^{\text{eff}}_q = \frac{1}{4\pi} \frac{\Omega_{\text{pi}}^2}{\omega_q^2} E^{\text{eff}}_q \, .$$

On the other hand, E_{ext} is caused by the assumed external charge distribution ϱ_Q

$$\nabla \cdot E_{\text{ext}} = -\frac{4\pi \varrho_Q}{\varepsilon_Q}$$

and thus we have, inserting (20),

$$Q \cdot P_q = \frac{\Omega_{\text{pi}}^2}{\omega_q^2} \frac{\varrho_Q}{\varepsilon_Q} \, .$$

Finally, the field seen by the electrons, at $Q = q + G$, is the total, not the effective, field due to the ions,

$$E_{\text{tot}} = E_{\text{ext}} - \frac{4\pi (P_q \cdot Q) Q}{\varepsilon(Q) Q^2} \, ,$$

$$Q \cdot E_{\text{tot}} = -\frac{4\pi \varrho_Q}{\varepsilon(Q)} \left(1 - \frac{\Omega_{\text{pi}}^2}{\omega_q^2 \varepsilon(Q)} \right) .$$

This gives (19) for the inverse dielectric constant at $\omega = 0$. We have simplified the expression by ignoring polarization (setting all ω_q's equal), but it is clear that different phonons of polarization ε_{iq} enter in proportion to $\varepsilon_{iq} \cdot Q$.

This model is probably reasonably accurate for good metals. In the next section we will discuss tight-binding bands, which do not obey this theory. We suspect $\lambda \sim 1/\omega^2$ may be a general result. As McMillan has already observed, this leads to a severe limitation on T_c.

Repeating the final result, eq. (19) of that paper,

$$V = \frac{4\pi e^2}{Q^2 \varepsilon(Q)} \left[1 - \frac{\Omega_{\text{pi}}^2}{\varepsilon(Q) \omega_{\text{ph}}^2(q)} \right] .$$

This is at $\omega = 0$; at finite ω, $\omega_{\text{ph}}^2 \to \omega_{\text{ph}}^2(q) - \omega^2$.

For polyvalent metals $\varepsilon(Q) \sim \text{const} \sim 1$ or 2, so we get

$$\lambda \lesssim \frac{1}{2} \left\langle \frac{\Omega_{\text{pi}}^2}{\omega_{\text{ph}}^2(q)} \right\rangle$$

and λ can be of the order of $1 \div 2$ but the phase space of really small ω_{ph}^2 is seldom large enough to make it much bigger. This phase-space requirement also is a very serious difficulty for all esoteric superconductivity mechanisms.

Having reached this point, it is rather disturbing to realize that the same arguments with little or no modification also limit the magnitude of electron-phonon scattering in resistivity. Writing the coupling constant as $N(0) \cdot (\text{M.E.})^2 / \hbar \omega_{\text{ph}}$, we see that $(\text{M.E.}) \sim \sqrt{\Omega_{\text{pi}} / \omega_{\text{ph}}}$ is the only concession to giant coupling constants which we can wrest from Nature. (For resistivity, one of the two ω_{ph} disappears and is replaced by kT.)

5. – Electron-phonon coupling in transition metals.

This is actually particularly disturbing, as I discovered in preparing these lectures, in that the « bad actors » are by no means alone in having rather large phonon resistivities. When I started to insert a curve into fig. 8 showing how much larger the resistivities of the bad actors were, I found that, while large, they by no means completely outclassed those of other transition metals. Nb and Ta, for instance, show resistivities only 5 or 6 times smaller, and a good 10 times bigger than Ag or Au. A bit of this comes from umklapp, *i.e.* the factor of $2 \div 4$ gained by the increase in λ^2, and one would expect at most, say, 5, but the remarkably large resistivity of Y, Zr, Lu and the like is striking and unexpected. This is by no means reflected in the superconducting T_c's, in most cases, and one may only suppose that there are fundamental difficulties in transition metal superconductivity theory even subtler than what I will talk about here.

In thinking about this I began to feel that I was arriving at a somewhat more straightforward physical insight on the large electron-phonon coupling of these metals, which I wanted to insert as a sidelight to these lectures. This insight comes from following the tight-binding approach to these metals which FRIEDEL has so long advocated, and which in its relevance to phonon theory has been to some extent followed up in the work of Varma and Weber. But what I want to do here is to go beyond Varma and Weber and other more or less conventional methods to describe the « chemical pseudopotential » methods which I advocated a number of years back and which have been refined and exploited in many systems of materials by Dave BULLETT.

What these methods do is to give the classical LCAO tight-binding methods and assumptions a much more precise rationale and a quite precise, and at

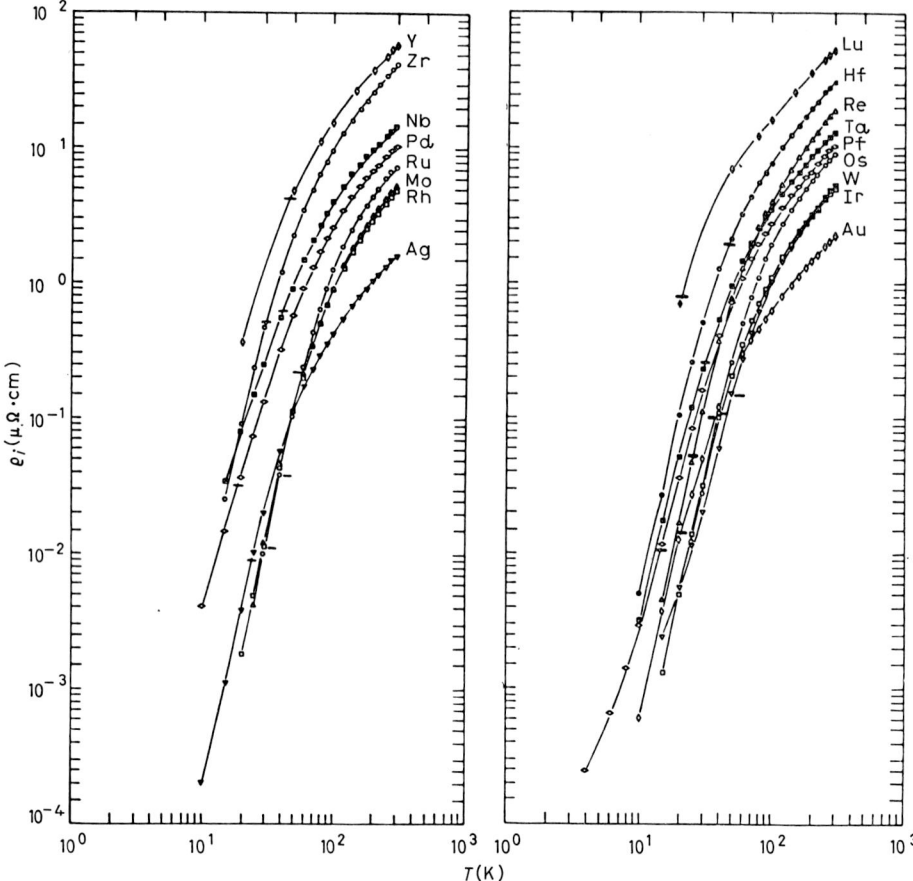

Fig. 8. – Resistivity data for transition metals [16].

the same time simple, set of prescriptions for calculating the relevant matrix elements from first principles.

Let us suppose that we have a band or set of bands which may be described by a set of atomic orbitals. It is not difficult, incidentally, to use a mixed free-d representation with techniques described a number of years ago by WEEKS and myself [17], similar to those of Phillips, Hubbard and Heine, but most of the physics is contained in the tight-binding part.

What this method does first is to provide an effective pseudopotential wave equation for determining the best possible atomic function.

The scheme goes as follows. First we divide the potential around a given atom (or site, in some applications) into a « home » part and a « foreign » part. In the simplest case we just write it as the sum of atomic parts for each atom

$$H = T + \sum_i V_i \,.$$

We approach the determination of the wave functions self-consistently, so we assume that we have a good estimate of all the wave functions φ_j which do not belong to the atoms of interest φ_i. The technique is to subtract away the best linear combination of all those atoms to minimize their effects on φ_i, which gives us

$$E_i \varphi_i = (T + V_i)\varphi_i + \sum_j (V_j \varphi_i) - \sum_j (i|V_j|j)\varphi_j,$$

$$(i|V_j|j) = \int \varphi_i(r) V_j(r) \varphi_j(r) \, \mathrm{d}^3 r$$

(and, as I said, the φ's then are determined self-consistently). This gives us two things aside from φ_i: the energy E_i of the function φ_i and the matrix element $(i|V|j)$ for hopping between neighboring atoms. Oddly enough, it is easy to prove that these are not *estimates* of the appropriate values, but the actual numbers themselves: they are mathematically identical to the energies of the corresponding properly orthogonalized Wannier (symmetry) orbitals representing the relevant bands. To prove this, we simply write an exact wave function as a linear combination of the *nonorthogonal* atomic orbitals φ_i

$$\psi = \sum_i a_i \varphi_i$$

and then we find

$$H\psi = E \sum_i a_i \varphi_i = \sum_i E_i a_i \varphi_i + \sum_{i,j} a_i(i|V_j|j)\varphi_j.$$

Assuming only linear independence, we extract from this the coefficient of φ_i which is the equation

$$\sum_i [(j|V_i|i) + E_i \delta_{ij}] a_j = E a_i,$$

so the secular equation involves only E_i and V_{ij}.

In order to think about a possible tight-binding treatment of phonons, let us think of the simple case in which we have two atoms close together, with potentials V_1 and V_2 and wave functions φ_1 and φ_2. If we change the distance between them, φ_1 and φ_2 will not change very much because of the cancellation effect, but $E_{1,2}$ and V_{12} will change a lot. We may calculate E_1 by using its defining equation and multiplying through by φ_1:

$$\varphi_1 E_1 \varphi_1 = \varphi_1(T + V_1)\varphi_1 + \left(\varphi_1 V_2 \varphi_1 - \varphi_1 \varphi_2 \int \varphi_2 V_2 \varphi_1\right)$$

and integrating (assuming $(T + V_1)\varphi_1 = E_1^0 \varphi_1$)

$$\Delta E_1 \simeq \int \mathrm{d}^3 r \, \varphi_1^2(r) V_2(r) - S_{12} V_{12}.$$

This leads to a repulsive potential *not* because of the first term, which is actually attractive and represents the penetration of φ_1 into the deep attractive core of V_2, but because of the second, which represents the Hilbert-space restriction due to overlap of φ_1 and φ_2. (Note incidentally that the long-range part of the first term simply cancels the Coulomb repulsion of the cores and both it and the core-core interaction play no role.) Because the S overlap $(S_{12} = = \int \varphi_1 \varphi_2 \mathrm{d}^3 r)$ comes from the large volume between the atoms where the potential is not strong, and V_2 is basically a deep, localized potential, S_{12} is anomalously large and this causes the repulsion.

If both orbitals are filled, as in the repulsion of rare-gas atoms or closed-shell molecules, that is it. But, if we have an open shell, V_{12} causes covalent mixing of φ_1 and φ_2, or band formation, and there is an attractive component about equal to V_{12}. For two atoms alone there is no electronic redistribution due to V_{12}, one simply has a covalent bond. But in a band situation, the motion of a single atom to the right enhances that V_{12} and decreases V_{13} to the other side (see fig. 9).

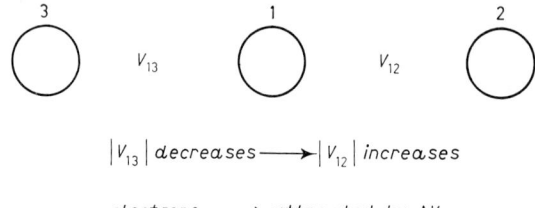

Fig. 9. – An atom vibrating between two neighbours: electron-phonon coupling in the tight-binding case.

The slopes, of course, balance if we were in equilibrium, but the phonon frequency is determined by the curvature

$$M\omega^2 = \frac{\delta^2 E_{tot}}{\delta x^2}.$$

The force balance gives us essentially that

$$\frac{\delta V_{12}}{\delta x} = \frac{\delta}{\delta x}(V_{12} S_{12} - (V_2)_{11})$$

and the overlap terms on the right depend on *two* wave function overlaps $\exp[-2\alpha r_{12}]$, so will be much steeper relative to their magnitude than V_{12}: hence, the magnitude of attractive covalent bonding dominates the repulsion at equilibrium. We expect this relative steepness also to be the case for the curvature, in fact the repulsive term would dominate the restoring force if the electrons did not have the ability to respond to the change ΔV_{12} in the matrix elements of the hopping integrals.

When the atom moves to one side, ΔV_{12} is opposite in sign to ΔV_{13}, and the electron gas sees a perturbing potential to which it can respond by drawing electrons into the stronger potential: thus there will be a compensating potential

(A) $$(\Delta E)_{\text{band response}} = -\frac{(\Delta V)^2 \chi}{2} \simeq \frac{(\Delta V)^2}{W}.$$

This, too, involves the second derivative of overlaps and is capable of being as large as the first term. ΔV is, of course, the electron-phonon coupling and (A) is the response to it. W is the band width and χ is the compressibility of the electron gas.

Now we might expect that, as the electrons move in response to ΔV, they will to some extent act to compensate the potential to which they are responding and screen it out. This is what happens in the conventional jellium metal: in proportion as the electrons respond and the phonons get softer, the coupling constant λ itself gets screened away and its maximum possible value is limited by the screening effect. The point of instability is reached at a *finite* coupling constant λ.

The same is not nearly as true here. Indeed, the electron response does act in some measure to reduce the change in matrix element ΔV, $V_2(r)$ itself is reduced. But the overlap charge $\varphi_1(r)\varphi_2(r)$ is primarily in the exterior region between the two atoms, while the potential V_2 which is being screened is predominantly interior to the atom, so that the screening is quite ineffective and λ can become quite large without necessarily leading to instability of the phonons.

This is really not completely distinct from the « effective field » argument which we have just used for strong-coupling s-p band cases. The question again has to do with the fact that the electron response to phonon motion has fewer high-momentum components, so it is not effective in screening out the steep potentials of the ionic motion. It is by no means a one-to-one correspondence—for instance, the unperturbed large phonon frequency in the absence of electronic response is in the one case purely electrostatic, in the other comes from the overlap forces; but the mechanisms are both related to the ions and electrons not seeing the same local fields.

The upshot of all of this is that again, once we detach the electrons and the phonons, the electron-phonon coupling constant becomes essentially proportional to

$$\lambda \propto \frac{(\Omega^2)_{\text{intrinsic}}}{(\omega_{\text{ph}}^2)_{\text{observed}}},$$

where $(\Omega^2)_{\text{intrinsic}}$ can be thought to be a material constant having to do with the intrinsic stiffness of the wave functions. In particular, this kind of argument tells us one *very important fact*: why it is that the strong-coupling superconductors more often than not come from the class of «cluster compounds» defined by GEBALLE as those in which transition metal atoms have low coordination and are very close to each other in clusters. These materials seem to have net stiffnesses and Debye frequencies comparable to those of others, but it must be that their intrinsic stiffness is considerably larger than that of other materials. Thus they have big λ's not particularly because ω_{ph}^2 is soft, but because $(\Omega^2)_{intrinsic}$ is big.

But along with this solution of an old problem, we are left with a very serious puzzle: why is there so little correspondence between the λ_ϱ^2 deduced from resistivity and the λ_{sc}^2 deduced from T_c (and in some cases from tunneling data)? This discepancy is very bad for the bad actors but is clearly present in such simple, innocuous-looking cases as Ti, Lu, Hf and even slightly wrong for Nb and Ta. (I hesitate to put in numbers, but clearly there is a considerable difficulty here.) The possibility of a major renormalization of λ is, in fact, what I am about to discuss with regard to the A15's, but the mechanism we propose is rather bizarre and clearly not sufficiently general. I can only conclude that there is some slightly more general way of violating the Migdal argument, that there is no renormalization of electron-phonon effects, or possibly that there has been a tendency to underemphasize electron-electron effects in some of these cases, so that μ^* can be not quite so small as has been supposed in some transition metals.

6. – A new and rather speculative theory for A15 and other bad actor superconductors.

The following is based on the Ph. D. thesis which is being written by Clare C. YU at Princeton. (A draft version of that thesis is available here at Varenna.) Clare has done a remarkable amount of work on this and deserves to be co-author of this part. She acknowledges help from a number of people but in our early discussions of this work we were stimulated particularly by J. P. SETHNA and A. Z. ZAWADOWSKI, while my own early thoughts were much influenced by F. D. M. HALDANE.

After the full introduction I have given you I hardly need to emphasize that whatever is to be proposed for these materials must have at least some seriously novel aspects. The first idea that attracted me for a number of years was that in these cases the electron-phonon coupling had become so strong as, in some sense, to bind electron pairs together prior to the appearance of their Bose condensation into a superconducting state. Such a concept, however, runs up against both experimental and theoretical difficulties. Experimentally, one would expect a *decrease*, not an increase, in the Pauli susceptibility.

Theoretically, the worst problem is self-trapping and a corresponding enormous increase in the pair mass, so that they would act purely as heavy classical particles, and it is likely that instead the system would form some new stable, covalently bonded crystalline phase.

Yet again, experimentally we have the hard fact of the giant resistivity to confront, without any decrease in other metallic properties or any unconscionable increase in γ, which belies any self-trapping or binding mechanism.

My attention then was caught by the equally bizarre properties of the so-called « mixed-valence » or, better, « fluctuating valence » metals like SmB_6, SmS, $CeAl_3$, etc. These materials are now accepted to have quantum fluctuations between two different values of valence, e.g. Sm^{++} and Sm^{+++}, and among their many properties they also show very large temperature-dependent resistivity, approaching the Mott limit at high temperatures; and yet, like the « bad actors », they settle down and become metals, if somewhat peculiar ones, at low temperatures; one or two even become superconducting ($CeCu_2Si_2$, for instance).

The presently accepted model for these metals is a lattice of « Anderson models »: a lattice of f-orbitals into which « free » electrons—actually mixtures mostly of s and d orbitals—can tunnel. Overall free-energy conditions require a mixture of two different occupancies of the f-orbital (an « asymmetric » Anderson model), but the quantum tunneling of the electron prevents any individual atom from settling down permanently in one or the other valence. This suggested to me a version of the Anderson model coupled to phonons which had at one time been discussed by my then student Duncan HALDANE.

With these vague suggestions in mind, as well as with rather a lot of evidence about the peculiar anharmonicity effects which were often seen in the A15 materials, along with Miss YU I evolved the following general outlines of a model. In the first place, it is essential, it seemed to me, that quantum fluctuation must play a large role. If we are to have such massive electron-phonon coupling as would, in a classical mean-field theory, lead to deep instability of the phonons, we can only be rescued from that by quantum fluctuations of the type which occur in Kondo-like systems such as the mixed-valence metals. If these fluctuations are large, let us start with a model which in some sense takes electron-phonon coupling as H_0 and neglects other couplings—in particular, the coupling between local lattice oscillators which converts the Einstein phonon model into a Debye one. In fact, in the electron-phonon problem the acoustic branch, for which propagation of phonons is important, has a relatively negligible influence on resistivity or superconductivity in any case. So we replace the phonons by a lattice of *local* oscillators, uncoupled one to another: we neglect *phonon* hopping matrix elements, while, of course, allowing electrons to be essentially free particles. This is the key step in the physics and I make no excuses for it. It is a high-temperature approximation focusing on the resistivity problem, not superconductivity. Even that is not simple enough to begin with.

For reasons we are only beginning to understand, in the Kondo and mixed-valence problems the properties are often well approximated by independent local systems, and that is what we do here, at first: try to solve the problem of a *single* lattice oscillator, over-coupled to the electron gas and permitted to be as anharmonic as we like. Our paper [18] gives the relevant Hamiltonian:

$$H = H_{ph} + H_{el} + H_{e\text{-}p},$$

$$H_{ph} = \frac{p^2}{2M} + \frac{1}{2} M\Omega^2 Q^2,$$

$$H_{e\text{-}p} = -\sum_{kk'} (C^\dagger_{k's} C^\dagger_{k'p}) \begin{pmatrix} 0 & \lambda Q \\ \lambda Q & 0 \end{pmatrix} \begin{pmatrix} C_{ks} \\ C_{kp} \end{pmatrix},$$

$$H_{el} = \sum_k \varepsilon_{ks} C^\dagger_{ks} C_{ks} + \varepsilon_{kp} C^\dagger_{kp} C_{kp}.$$

The three components involve:

H_{ph} is an harmonic local phonon. $H_{e\text{-}p}$ and H_{el} express the fact that the lattice has necessarily a parity symmetry $Q \to -Q$, $\varphi_s \to \varphi_s$, $\varphi_p \to -\varphi_p$. Thus any instability necessarily leads to a symmetric double well. This cannot be done without at least two electronic channels «s» and «p» which are merely meant to indicate odd and even representations of the point group. In the first pages of our paper [18] is given a formal theory of a purely linear electron-phonon system, replacing the electrons by Tomonaga bosons, which simply goes unstable at the relevant point. However, this theory fails not only because of the intrinsic phonon anharmonicity but also because of the limitations of the Tomonaga theory itself: Tomonaga is only valid so long as the relevant perturbations on the electrons are small, *i.e.* for small λ and small Q. But when we express the Hamiltonian in the equivalent form of a coupling to a local *electron* state (as in eq. (13) of the paper; we have diagonalized the electron-phonon part by a linear transformation), we get

$$H_{el} = \sum_k \varepsilon_k n_{k\sigma} + \sum_{\sigma\pm} E_0 \psi^\dagger_{0\sigma} \psi_{0\sigma} + \sum_{k\sigma} V_{k\sigma} C^\dagger_{k\sigma} \psi_{0\sigma} + \text{h.c.},$$

$$H_{e\text{-}ph} = -(\psi^\dagger_{0+} \psi^\dagger_{0-}) \begin{pmatrix} \lambda Q & 0 \\ 0 & -\lambda Q \end{pmatrix} \begin{pmatrix} \psi_{0+} \\ \psi_{0-} \end{pmatrix}$$

and we see that, in fact, the phonons are coupled to a *single* pair of electronic states, which by the exclusion principle can only be occupied once, and $H_{e\text{-}ph}$ is *bounded* as a function of Q by $|H_{e\text{-}ph}| < \lambda Q$. This provides an intrinsic non-linearity when λ is large enough, which had not been previously exploited. In this form we can see the resemblance between the local electron-phonon part of our problem and the « dynamic Jahn-Teller effect ». (This resemblance was pointed out to us by TOSATTI.) The local state is unstable at the symmetric point no matter how stiff the phonon may be because of the intrinsic degeneracy of the electronic states $\psi_{0\pm} = \psi_{0s} \pm \psi_{0p}$ and the linear coupling to the oscillator Q.

What we now do is to go to what VAN VLECK would have called « our mathematical slave » (in this case, Don HAMANN) to eliminate the electronic degrees of freedom from this system and to express the solution in terms of a so-called « path integral » or « space-time » formulation: to get an effective « action » which is a functional of the time dependence of the local phonon $Q(t)$. This is the characteristic trick of the space-time method for these problems, to invert Feynman's polaron trick and eliminate fermions rather than bosons. Then all probability amplitudes can be expressed as

$$\int d[\text{paths}\,(Q(t))]\exp[-S(t)].$$

For example, the partition function is

$$Z = Z_0 x \int d(\text{paths}) \exp\left[-\int_0^\beta dt\,(S(\tau))\right].$$

In this case $S(\tau)$ contains an « adiabatic potential » which represents the result for a classical, stationary Q, and a hopping part which depends on Q as a function of τ. The adiabatic potential is

(B)
$$\begin{cases} V(\tau) = \dfrac{1}{2}M\Omega^2 Q^2 - \dfrac{1}{\pi N(0)}\ln|1+\gamma^2(\tau)|,\\[6pt] \gamma(\tau) = \pi N(0) E(Q(\tau)) = -\pi N(0)\lambda Q(\tau). \end{cases}$$

Note the intrinsic anharmonicity of $V(\tau)$. The time-dependent part is

(C) $\quad T(\tau) = \dfrac{1}{2}M\left(\dfrac{dQ}{d\tau}\right)^2 - \dfrac{1}{\pi^2}\int_0^\beta d\tau'\ln|\tau-\tau'|\cdot$

$$\cdot\frac{d\gamma(\tau')}{d\tau'}\frac{d}{d\tau}\left\{\frac{\gamma(\tau)}{\gamma^2(\tau)-\gamma^2(\tau')}\ln\left|\frac{1+\gamma^2(\tau)}{1+\gamma^2(\tau')}\right|\right\} \simeq$$

$$\simeq \frac{1}{2}M\left(\frac{dQ}{d\tau}\right)^2 - \frac{1}{\pi^2}\int_0^\beta d\tau'\ln|\tau-\tau'|\left(\frac{d}{d\tau'}\gamma(\tau')\frac{d}{d\tau}\gamma(\tau)\right)$$

and this determines the degree to which Q is allowed to fluctuate. The first term represents the intrinsic massiveness of the phonon and limits the rate of motion of Q, while the second is the attractive interaction between « hops » of Q of opposite sign which is responsible for the « Kondo-like » renormalization in problems of this type.

In solving for physical results from these equations we have been driven to a rather crude approximation in which $V(Q(t))$ is taken to be a rather deep double-well potential. This is probably unrealistic in that, if it were, the M term in (B) would reduce the fugacity of genuine flips of Q from one well to another to unacceptably low values. Nonetheless, I believe the double-well picture to be an excellent crude guide to the physics.

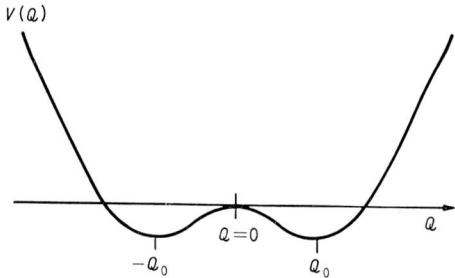

Fig. 10. – Double-well shape of $V(Q)$.

In such a double-well picture in which we look upon $V(Q)$ as having the shape sketched in fig. 10, the interesting paths in Q are the Kondo-like hopping paths shown in fig. 11.

Because of the logarithmic interaction term in $T(\tau)$, there will be a rather large number of close pairs of hops, which will in the course of scaling gradually diminish the effective size of Q_0 and height of the barrier until, at a low enough scale of energy kT_K, the renormalized Q_0 will vanish and the barrier disappear.

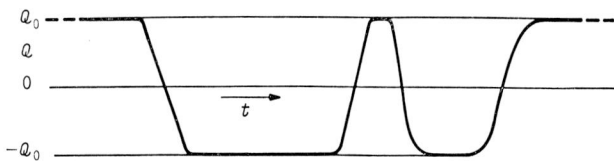

Fig. 11. – Kondo-like paths for Q.

It is our hypothesis that this scaling process takes place over an energy range from E_F to somewhat below room temperature, and that in the interesting materials T_K is in the ~ 100 K range. In our paper reasonable numerical fits to such high-temperature properties as resistivity and magnetic susceptibility are shown using this theory. (An example is given in fig. 12.)

Conceptually, then, the relationship among energy scales which we envisage is as follows. At very high temperatures, $E_F > kT > \hbar\omega_{ph}$, the oscillators are independent and anharmonic, gradually scaling towards less and less anharmonic behavior, and smaller and smaller average excursions, as $T \to \hbar\omega_\text{D}$. But in the range above room temperature, where very high resistivity is observed, the mean excursion is still $\overline{x^2} > \hbar/M\omega$ and the mean fluctuation in phase shift η is comparable with unity. At lower temperatures two things happen. First, at T_K the oscillators tend to become effectively harmonic; and also in this

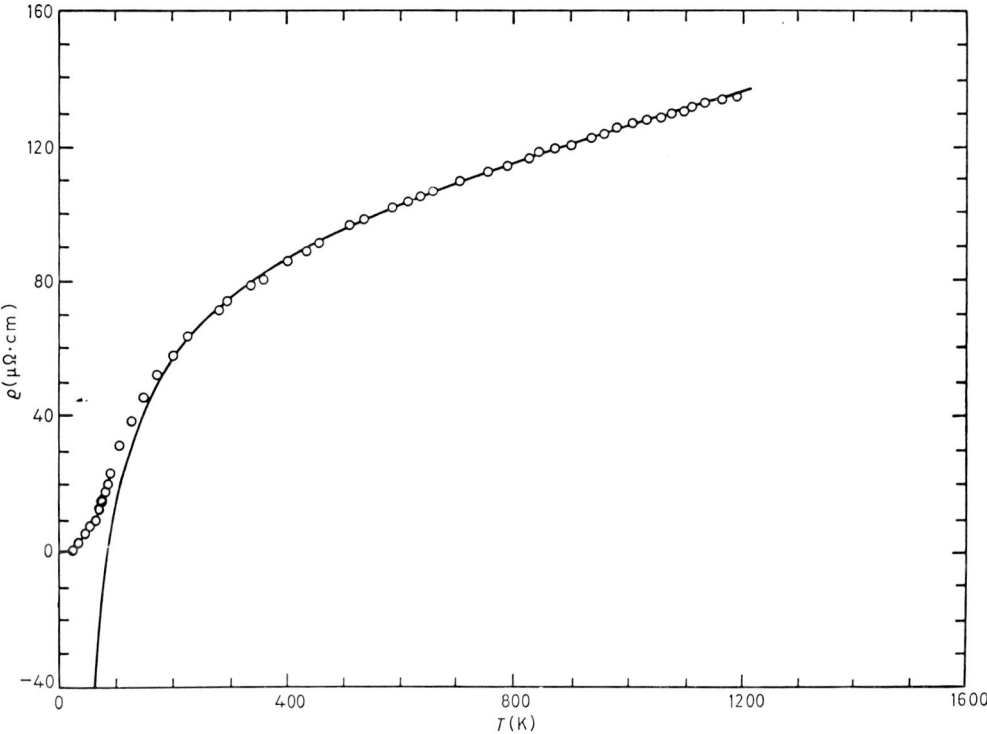

Fig. 12. – Fit to resistivity data of Marchenko [18, 19], V_3Si, $\varrho(T) = A' - B'/T + \varrho_1 T$, $A' = 88\ \mu\Omega\cdot\text{cm}$, $B' = 8.1\cdot 10^3\ \mu\Omega\cdot\text{cm}\cdot\text{K}$, $\varrho_1 = 4.6\cdot 10^{-2}\ \mu\Omega\cdot\text{cm/K}$.

range the interoscillator coupling which has been neglected reasserts itself and a true set of renormalized harmonic phonons can be defined. The only observable remnant of the anomalous fluctuations on a shorter time scale would be anomalous amplitudes of Debye-Waller factors and of atomic excursions as measured by such experiments as channeling [20]. It would also be true that throughout the scaling region the effective harmonic stiffness as measured by elastic constants would be quite temperature dependent. (We must realize that the anharmonic localized oscillators are only one or a few of the $\sim 10 \div 20$ degrees of freedom in the unit cell and that we would not expect greater than $\sim 10 \%$ effects on the elastic constants.) (See fig. 13.)

An experimental result which is very directly explained is the large effect of damage on the resistivity. An unsymmetric double-well potential does not renormalize to an effective symmetric well, but instead the phonon becomes permanently biased to one side or another. Relatively small local strains

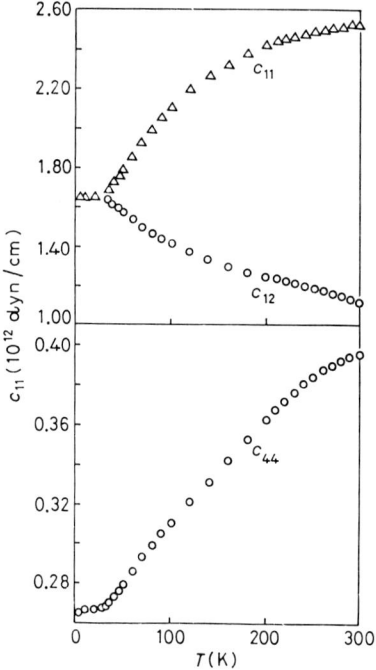

Fig. 13. – Elastic moduli for Nb_3Sn [21].

should be capable, according to our calculations, of leaving the oscillator « stuck » on one side, which would then lead to an average scattering phase shift of the order $\pi \lambda Q_0 \sim 1$ rather than zero. In effect, in a random strain field such as could be caused by damage the high-frequency fluctuations become « frozen in » at low temperatures.

As for the superconducting properties, we have so far made no attempt to carry our theory past the phonon recoupling point, as would be necessary to have a coherent theory of low-temperature properties. It is also probably true that we shall have to include spins and a Hubbard interaction in our theory, which is, up to now, spinless and does not make any attempt to estimate the modifications of μ which follow from the large, fluctuating scattering. These are no doubt large (as I have already suggested in collaboration with MUTTALIB and RAMAKRISHNAN [22]) and their net effect may account for a large fraction of the discrepancy between $(\lambda)_\varrho$ and $(\lambda)_{T_c}$.

***

Many discussions have been acknowledged in the text. A full list of our indebtedness must await the publication of our paper, but we should especially mention the hospitality of the Aspen Center for Physics.

One special remark should be made. We have recently learned that C. M. VARMA, from quite a different point of view, has developed some ideas about large anharmonicities, short coherence lengths and violations of Migdal's theorem in similar systems [23]. We should emphasize that, while the two methods are quite independent, they are expressing almost the same physical realities and should not be seen as competitive except as to mathematical methodology.

The work done at Princeton University was supported in part by NSF grant DMR 8020263.

REFERENCES

[1] Z. FISK and G. W. WEBB: *Phys. Rev. Lett.*, **36**, 1084 (1976).
[2] H. LUTZ, H. WEISSMANN, O. F. KAMMERER and M. STRONGIN: *Phys. Rev. Lett.*, **36**, 1576 (1976).
[3] M. GURVITCH: in *Superconductivity in d- and f-Band Metals*, edited by H. SUHL and M. B. MAPLE (New York, N.Y., 1980), p. 317.
[4] J. M. ROWELL, R. C. DYNES and P. H. SCHMIDT: in *Superconductivity in d- and f-Band Metals*, edited by H. SUHL and M. B. MAPLE (New York, N.Y., 1980), p. 409.
[5] J. H. MOOIJ: *Phys. Status Solidi A*, **17**, 521 (1973).
[6] P. B. ALLEN: in *Superconductivity in d- and f-Band Metals*, edited by H. SUHL and M. B. MAPLE (New York, N.Y., 1980), p. 291; *Phys. Rev. Lett.*, **37**, 1638 (1976).
[7] A good source is G. GLADSTONE, M. A. JENSEN and J. R. SCHRIEFFER: in *Superconductivity*, Vol. **2**, edited by R. PARKS (New York, N.Y., 1969), p. 665.
[8] J. MULLER: *Rep. Prog. Phys.*, **43**, 641 (1980).
[9] M. L. COHEN and P. W. ANDERSON: in *Superconductivity in d- and f-Band Metals*, edited by D. H. DOUGLASS (New York, N.Y., 1972), p. 17.
[10] B. P. SCHWEISS, B. RENKER, E. SCHNEIDER and W. REICHWARDT: in *Superconductivity in d- and f-Band Metals*, edited by D. H. DOUGLASS (New York, N.Y., 1976), p. 189.
[11] *Superconductivity*, Vol. **1** and **2**, edited by R. PARKS (New York, N.Y., 1969); *Superconductivity in d- and f-Band Metals*, edited by D. H. DOUGLASS (New York, N.Y., 1972); *Superconductivity in d- and f-Band Metals*, edited by D. H. DOUGLASS (New York, N.Y., 1976); *Superconductivity in d- and f-Band Metals*, edited by H. SUHL and M. B. MAPLE (New York, N.Y., 1980); *Ternary Superconductors*, edited by G. K. SHENOY, B. D. DUNLAP and F. Y. FRADIN (New York, N.Y., 1981).
[12] C. M. VARMA and W. WEBER: *Phys. Rev. Lett.*, **39**, 1094 (1977); *Phys. Rev. B*, **19**, 6142 (1979).
[13] B. M. KLEIN, L. L. BOYER, D. A. PAPACONSTANTOPOULOS and L. F. MATTHEISS: *Phys. Rev. B*, **18**, 6411 (1978).

[14] H. J. WILLIAMS and R. C. SHERWOOD: *Bull. Am. Phys. Soc.*, **5**, 430 (1960).
[15] L. R. TESTARDI: *Physical Acoustics*, Vol. **10** (New York, N.Y., 1973), p. 193.
[16] G. GLADSTONE, M. A. JENSEN and J. R. SCHRIEFFER: in *Superconductivity*, Vol. **2**, edited by R. PARKS (New York, N.Y., 1969), p. 665.
[17] P. W. ANDERSON and J. D. WEEKS: in *Computational Methods for Large Molecules and Localized States in Solids*, edited by F. HERMAN, A. D. MCLEAN and R. K. NESBET (New York, N.Y., 1972), p. 251; also P. W. ANDERSON: *Phys. Rev.*, **181**, 25 (1969).
[18] C. C. YU and P. W. ANDERSON: to be published.
[19] V. A. MARCHENKO: *Sov. Phys. Solid State*, **15**, 1261 (1973).
[20] L. R. TESTARDI, J. M. POATE, W. WEBER and W. M. AUGUSTYNIAK: *Phys. Rev. Lett.*, **39**, 716 (1977); J. L. STAUDENMANN and L. R. TESTARDI: *Phys. Rev. Lett.*, **43**, 40 (1979).
[21] K. R. KELLER and J. J. HANAK: *Phys. Rev.*, **154**, 628 (1967).
[22] P. W. ANDERSON, K. A. MUTTALIB and T. V. RAMAKRISHNAN: *Phys, Rev. B*, **28**, 117 (1983).
[23] C. M. VARMA and A. L. SIMONS: *Phys. Rev. Lett.*, **51**, 138 (1983); C. M. VARMA: to be published.

Spin Glass Hamiltonians: A Bridge Between Biology, Statistical Mechanics and Computer Science
Emerging Syntheses in Science, ed. D. Pines, Santa Fe Institute
(Addison-Wesley, 1984)

My contribution to SFI's founding Workshop. The evolutionary biology line has been much exploited by Kaufmann, Weisbuch, Fontana and others associated with SFI.

P. W. ANDERSON
Joseph Henry Laboratories of Physics, Princeton University, Princeton, NJ 08544

Spin Glass Hamiltonians: A Bridge Between Biology, Statistical Mechanics and Computer Science

A remarkable number of fields of science have recently felt the impact of a development in statistical mechanics which began about a decade ago[1] in response to some strange observations on a variety of magnetic alloys of little or no technical importance but of long-recognized scientific interest.[2] These fields are:

1. Statistical mechanics itself, both equilibrium and non-equilibrium;
2. Computer science, both special algorithms and general theory of complexity;
3. Evolutionary biology;
4. Neuroscience, especially brain modelling;
5. Finally, there are speculations about possible applications to protein structure and function and even to the immune system.

What do these fields have in common? The answer is that in each case the behavior of a system is controlled by a random function of a very large number of variables, i.e. a function in a space of which the dimensionality is one of the large, "thermodynamic limit" variables: $D \to \infty$. The first such function of which the properties came to be understood was the model Hamiltonian

$$H = \sum_{ij}^{N} J_{ij} \, S_i S_j \qquad (1)$$

(J_{ij} is a random variable, $P(J_{ij}) \alpha e^{-J_{ij}^2/2(\bar{J}_{ij})^2}$, S_i a spin variable attached to site i) introduced[1] for the spin glass problem. This Hamiltonian has the property of "frustration" named by G. Toulouse[3] after a remark of mine, which roughly speaking indicates the presence of a wide variety of conflicting goals. A general definition suitable for a limited class of applications has been proposed:[4] imagine that the "sites" i on which the state variables reside constitute the nodes of a graph representing the interactions between them—simply a line for every J_{ij} in the case of (1), for instance. Let us make a cut through this graph, which will have a certain area $A (\propto N^{d-1}$ in case the graph is in a metric space of dimension d). Set each of the two halves in a minimum of its own H, normalized so that $H \propto N$. Then, reunite the halves and note the change in energy ΔH. If the fluctuations in ΔH are of order $\sqrt{A}$, H is "frustrated"; if they can be of order A—as in (1), they will be if the J are all of the same sign—it is "unfrustrated." The dependence on $\sqrt{A}$ means that when the interactions within a block of the system are relatively satisfied, those with the outside world are random in sign; hence, we cannot satisfy all interaction simultaneously.

A decade of experience with the spin glass case has demonstrated a number of surprising properties of such functions as H. As the Hamiltonian of a statistical mechanical system, for most dimensionalities it has a sharp phase transition in the $N \to \infty$ $limit^1$. At this transition it becomes non-ergodic in that different regions of phase space become irretrievably separated by energy barriers which appear to be of order N^p where p is a power less than unity.[5] As the temperature is lowered these regions proliferate, exhibiting an ultrametric multifurcation.[6] It is suspected that the number of such regions at or near the minimum (ground state) of H has no entropy (not of order e^N), but may be exponential in some power of N. Many unusual properties of the response functions, and some strange statistical mechanical and hysteretic behaviors, have been explored at length. Recent work has generalized the Hamiltonian (1) and also shown that even first-order phase transitions may occur for some models.

In computer science, there are a number of classic optimization problems which have been studied both as objects for heuristic algorithms and as examples for complexity theory. These include the spin glass itself (sometimes under other names), the graph partition problem (which can be transformed into a spin glass), graph coloring (close to a Potts model spin glass), the Chinese postman (in some cases equivalent to a spin glass) and the famous travelling salesman (Design a tour through N cities given distances d_{ij}, of the minimum length $L = \sum d_{ij}$). As I indicated, several of these are spin glasses—there even exists a very inefficient transformation due to Hopfield[7] $TS \iff$ spinglass—and all are well-known NP-complete—i.e. hard—problems, of which it is speculated that no algorithm will solve the general case faster than $O\left(e^N\right)$.

On the level of heuristic algorithms, Kirkpatrick[8] has suggested that the procedure of annealing using a Mitropolis–Teller Monte Carlo statistical mechanics algorithm may be more efficient for some of these problems than the conventional heuristics. In any case, the knowledge that a "freezing" phase transition exists and that, for values of the function below freezing, one may be stuck forever in an unfavorable region of phase space, is of great important to the understanding of the structure of such problems. To my knowledge, the computer science community only knew of freezing as a bit of folklore, and has not yet absorbed its fundamental importance to the whole area—which includes great swatches of problem-solving and AI. Incidentally, a workshop at BTL came up with the limited but interesting conclusions that (a) simulated annealing works; (b) sometimes—not always—it beats previously known heuristics.[9]

Equally important should be the knowledge that there is a general theory of *average* properties of such problems, not limited to the mathematicians' type of worry over worst cases, but able to make statements which are overwhelmingly probable. For instance, we also know analytically the actual minimum energy to order N for "almost all" cases of several kinds of spin glass. A student (Fu) and I have an excellent analytic estimate for the partition problem on a random graph, etc. We also can hope to achieve a real connection between algorithmic solution and non-equilibrium statistical mechanics: after all, the dynamic orbits of a system are, in some sense, the collection of all paths toward minimum energy, and, hence, of all algorithms of a certain type (D. Stein is working on this).

In evolutionary biology, we can consider the fitness—the "adaptive landscape" as a function of genome to be just the kind of random function we have been talking about. The genome is a one-dimensional set of sites i with 4 valued spins (bases) attached to the sites, and the interaction between the different sites in a gene is surely a very complicated random affair. In the work of Stein, Rokhsar and myself,[10] we have applied this analogy to the prebiotic problem, showing that it helps in giving stability and diversity to the random outcomes of a model for the initial start-up process. Here we see the randomness as due to the tertiary folding of the RNA molecule itself.

G. Weisbuch has used a spin glass-like model for the evolutionary landscape[11] to suggest a description of speciation and of the sudden changes in species known as "punctuated equilibrium." Most population biology focuses on the near neighborhood of a particular species and does not discuss the implications of the existence of a wide variety of metastable fixed points not far from a given point in the "landscape."

In neuroscience, Hopfield[12] has used the spin-glass type of function, along with some assumed hardware and algorithms, to produce a simple model of associative memory and possibly other brain functions. His algorithm is a simple spin-glass anneal to the nearest local minimum or pseudo-"ground state." His hardware modifies the J_{ij}'s appropriately according to past history of the S_i's, in such a way that past configurations $\vec{S} = \{S_i\}$ are made into local minima. Thus, the configuration $\{S_i\}$ can be "remembered" and recalled by an imperfect specification of some of its information.

Finally, for our speculations for the future. One of these concerns is biologically active, large molecules such as proteins. Hans Frauenfelder and his collaborators have shown that certain proteins, such as myoglobin and hemoglobin, may exist in a large number of metastable conformational substates about a certain tertiary structure.[13] At low temperatures, such a protein will be effectively frozen into one of its many possible conformations which in turn affects its kinetics of recombination with CO following flash photolysis. X-ray and Mössbauer studies offer further evidence that gradual freezing of the protein into one of its conformational ground states does occur. Stein has proposed a spin glass Hamiltonian to describe the distribution of conformational energies of these proteins about a fixed tertiary structure as a first step toward making the analogy between proteins and spin glasses (or possibly glasses) explicit. In any case, this field presents another motivation for the detailed study of complicated random functions and optimization problems connected with them. Yet another such area is the problem of the immune system and its ability to respond effectively to such a wide variety of essentially random signals with a mechanism which itself seems almost random in structure.

REFERENCES

1. S. F. Edwards and P. W. Anderson, *J. Phys.*, F. **5**, 965 (1975).
2. Original observation: Kittel Group, Berkeley, e.g. Owen, W. Browne, W. D. Knight, C. Kittle, *PR* **102**, 1501-7 (1956); first theoretical attempts, W. Marshall, *PR* **118**, 1519-23 (1960); M. W. Klein and R. Brout, *PR* **132**, 2412-26 (1964); Paul Beck, e.g. "Micto Magnetism," *J. Less-Common Metals* **28**, 193-9 (1972) and J. S. Kouvel, e.g. *J. App. Phys.* **31**, 1425-1475 (1960); E. C. Hirshoff, O. O. Symko, and J. C. Wheatley, "Sharpness of Transition," *JLT Phys* **5**, 155 (1971); B. T. Matthias et al. on superconducting alloys, e.g. B. T. Matthias, H. Suhl, E. Corenzwit, *PRL* **1**, 92, 449 (1958); V. Canella, J. A. Mydosh and J. I. Budnick, *J.A.P.* **42**, 1688-90 (1971).
3. G. Toulouse, *Comm. in Physics* **2**, 115 (1977).
4. P. W. Anderson, *J. Less-Common Metals* **62**, 291 (1978).
5. N. D. Mackenzie and A. P. Young, *PRL* **49**, 301 (1982); H. Sompolinksy, *PRL* **47**, 935 (1981).
6. M. Mézard, G. Parisi, N. Sourlas, G. Toulouse, M. Virasoro, *PRL* **52**, 1156 (1984).
7. J. J. Hopfield and D. Tank, Preprint.
8. S. Kirkpatrick, C. D. Gelatt and M. P. Vecchi, *Science* **220**, 671-80 (1983).
9. S. Johnson, private communication.
10. P. W. Anderson, *PNAS* **80**, 3386-90 (1983). D. L. Stein and P. W. Anderson, *PNAS* **81**, 1751-3 (1984). D. Rokhsar, P. W. Anderson and D. L. Stein, Phil. Trans. Roy. Soc., to be published.
11. G. Weisbuch, *C. R. Acad. Sci. III* **298** (14), 375-378 (1984).
12. J. J. Hopfield, *PNAS* **79**, 2554-2558; also Ref. 7.
13. H. Frauenfelder, *Helv. Phys. Acta* (1984) in press; in *Structure, etc.*

Measurement in Quantum Theory and the Problem of Complex Systems
Talk given at the "Niels Bohr Centenary Symposium", Oct. 1985 (Copenhagen),
The Lessons of Quantum Theory, ed. J. de Boer,
E. Dal and O. Ulfbeck (Elsevier Science Publishers, 1986)

This is the most precise (and concise) statement I can make of my attitude towards the fundamentals of quantum theory. It says nothing which does not seem self-evident to me, yet Leggett disagrees violently.

The Lesson of Quantum Theory, edited by J. de Boer, E. Dal and O. Ulfbeck
© Elsevier Science Publishers B.V., 1986

Measurement in Quantum Theory and the Problem of Complex Systems

Philip W. Anderson
Princeton University
Princeton, New Jersey, USA

Contents

1. An approach to the Bohr–Einstein debate 23
2. Canonical Stern–Gerlach experiment 25
3. Phase variable of superfluid helium. Broken symmetry 29
4. Conclusions 32
References 32
Discussion 33

1. An approach to the Bohr–Einstein debate

Every physicist who has studied the quantum theory is familiar with the great debate on the uncertainty principle between Bohr and Einstein which began at the 1927 Solvay Conference but which then simmered on for many years. Perhaps the commonest view is that Bohr simply won hands down; in Einstein's somewhat mistranslated phrase, God *does* "play dice with the world" and the outcome of a quantum measurement is truly uncertain. In a way, this view corresponds to the truth—at least to the pragmatic truth that now, after 60 years, there is no doubt in any quantum physicist's mind that the whole quantum theory, including the original quantum prescriptions of measurement theory, are correct in describing the results of experiment. Why, then, re-open the question?

It may be slightly provocative, but surely in a spirit Bohr himself would have enjoyed, to suggest that many of those who have thought beyond the textbook level have come to the conclusion that the really correct answer to which one was right is —neither! It is not true that quantum mechanics is a purely probabilistic theory, any more so than statistical mechanics, for instance. That it is, seems to have been Bohr's point of view, in so far as we can follow his statements about complementarity as a general philosophical principle. In fact, generations of philosophers have been misled, probably against Bohr's intent, to give a deep meaning to this "uncertainty". On the other hand, so far as I can see, one cannot avoid the probabilistic results of measurements on atomic scale systems, as Einstein wished. Einstein's reasons are clear enough: he believed firmly that general relativity was the appropriate model for future theories, with its fundamental emphasis on the geometry of space-time, and it disturbed him deeply to have that geometry indeterminate and fluctuating—in particular, how can an indeterminate geometry properly maintain causality? The difficulties of quantizing general relativity vindicate Einstein's misgivings to some extent. Whether or not Einstein would have felt comfortable with the present liberties being taken with the structure of space-time is hard to know; one would have thought he would have felt a proprietary interest in four dimensions. It is true that we are beginning to return to his geometrical view and that at some ultimate scale everything may yet change. But really that is a bit of a diversion.

I want to talk here about simple non-relativistic quantum theory, although I believe everything I say is equally valid in relativistic field theory at any scale short of the Planck mass, e.g. during most of the events which took place during the "big bang" including the symmetry-breaking phase transitions.

What I consider to be the correct approach to the Einstein–Bohr debate began very much with one of my own personal idols, Fritz London. Interestingly enough, London trained originally as a philosopher, and one sees, in following his career, a certain philosophical theme: how can we deal with the quantum theory at the level of macroscopic objects? He is most justly famous for his ideas on the two macroscopic quantum phenomena which most severely test quantum mechanics in this regime, superfluidity and superconductivity; but I refer here to a seminal paper in the theory of measurement by London and Bauer (1939), who I believe were the first to take the point of view I advocate here, that the central problem of measurement theory is not the quantum mechanics of atoms, which is simple and easy, but the fact that macroscopic everyday objects are very difficult indeed for the quantum theory to deal with properly. To mis-quote another famous Dane: "The fault, dear Horatio, is not in our atoms but in ourselves."

The sticking point is twofold. The first problem is that the quantum theory is in principle *not* probabilistic but deterministic. The equation for the time variation is a simple deterministic one:

$$i\frac{\partial \psi}{\partial t} = H\Psi, \tag{1}$$

where H is a given—in principle—function of a maximal set of arguments of Ψ at the present time—or in relativistic theory, at a space-like surface. Of course, there is a mathematical equivalence between eq. (1) and a sum over classical trajectories, which resembles a stochastic process, but every attempt to replace quantum mechanics with a truly stochastic theory ends in failure. Thus in principle, given an initial $\Psi(t=0)$, we know precisely what Ψ is. What is more, if quantum mechanics is a complete theory, $\Psi(t)$ should be a complete specification of the physics including measurements, though not necessarily vice versa.

What London points out to be the second prong of the problem is that the Ψ involved must be taken to be the wave function describing the measuring apparatus and the experimenter as well as the experiment. It is philosophically repugnant to suppose that the prescriptions of quantum mechanics are different depending on whether we include the apparatus and the observer in the system or not. I follow Everett (1957) in his important, if perhaps somewhat incomplete paper, on "relative states" sometimes called "many worlds". Everett gives the following statement of the question: "The wave function must be taken to be the basic physical entity... interpretation comes only *after* an investigation of the logical structure of the theory", hence also only after an investigation of the quantum mechanics of large or rigid or sentient objects. A clear necessity is that Ψ must tell us what it may know about itself, not vice versa—there is no necessity that Ψ be a measurable thing in general, only that it describes what we actually observe.

The large objects which are involved in a conventional measurement behave in essentially non-quantum ways in at least three respects, each of which, at one time or another, has been conjectured to be the source of the anomalies of measurements.

(1) They must be or appear irreversible, so that at least the result of the measurement is recorded after it has taken place, not predicted before; measurement is of the essence of an irreversible process in actual experimental fact.

(2) They must be rigid; in order to make the appropriate kinds of measurements (we will see that "rigidity" can be defined in a much generalized way) they must have the properties of pointers.

(3) They must be sentient, to communicate it. Now (3) involves us in enormous difficulties with the different levels of computational complexity and with unknown aspects of the brain; fortunately, as London and others have pointed out, measurements can be made and recorded by purely automatic machinery, and one hardly sees physics depending on the final process of peeking at the instrument. I believe both (1) and (2), on the other hand, play an important role.

2. Canonical Stern–Gerlach experiment

It is important to think a little about the actual measurement process at this time. Let us start for instance with the canonical Stern–Gerlach experiment, which has been described as using the position variables of the atomic beam as an instrument to measure the spin of the atoms (fig. 1). But we must realize that just separating the two beams is not a measurement per se; a measurement must in some sense, as everyone agrees, destroy the possibility of interference between the two beams.

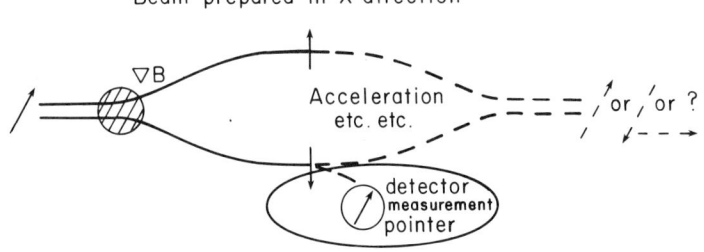

Fig. 1. Stern–Gerlach experiment.

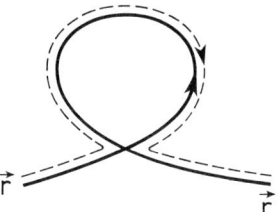

Fig. 2. A self-intersecting Feynman path for an electron to propagate from r to r'. Propagation along solid and dashed paths can interfere.

A measurement, according to the old viewpoint which agrees with experiment, definitely sets the relevant quantum number equal to that appropriate to the one beam alone, but interference with the other modifies the values of the quantum number and destroys the property of being in the one eigenstate, so if such an interference experiment remains possible, no true measurement in the canonical sense has taken place. This will be an important observation to the remainder of the chapter. In fact, in a Stern–Gerlach apparatus we very clearly can, even after accelerating the beams and passing them through all kinds of lenses etc., return them to coincidence and observe interference effects. This kind of experiment has been successfully carried out to demonstrate the effects of potentials on neutrons, for instance. (Macroscopic quantum coherence is *not* relevant here. That is enforced by generalized rigidity and minimization of free energy and is irrelevant to measurement theory.) The existence of an interference effect after subjecting beams of particles to all kinds of vicissitudes is the basis of the phenomena of "weak localization" in metals, where we demonstrate interference between electrons travelling in a state and its time—reversed, i.e. precisely backscattered, version. Strong localization, in fact, occurs when this backscattering prevents irreversible loss of coherence entirely (fig. 2). A most striking experiment was carried out first by Sharvin and Sharvin (1981) at the suggestion of Altshuler et al. (1981). In this experiment the interfering pairs of states are carried around an array of holes in the metal, or along an array of wires in a magnetic field, leading to Bohm–Aharanov interference phenomena at half-flux quanta periods between paths for these so-called "cooperons" which take opposite direction around the holes. An excellent review which describes these phenomena is by Lee and Ramakrishnan (1985), from which fig. 2 is borrowed, showing schematically the identical paths of the particle and the backscattered hole which interfere in such a way as to affect the response function which gives the resistivity due to elastic scattering. This response function is closely related to the electron–hole pair propagator. In fig. 2 we see two possible routes for this pair around a hole containing magnetic flux, which can interfere constructively or destructively depending on the flux. Figure 3 shows an example of the delicate microlithographic methods employed to fabricate a network containing many such tiny holes, and an example of such a network (from Bishop et al. 1985); fig. 4 shows some examples of the resulting interference pattern. The amplitude of this pattern measures the degree of coherence and vanishes if the path around the hole is larger

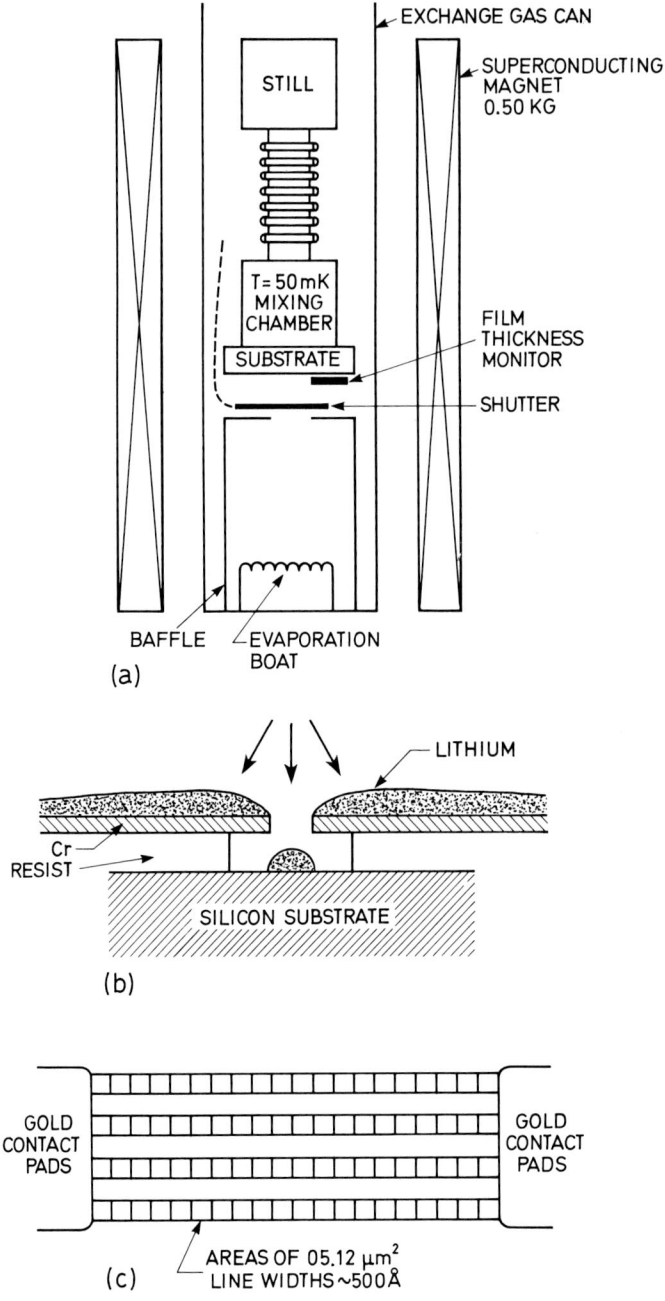

Fig. 3. (a) Apparatus shown schematically; (b) the overhanging photoresist and (c) the grids for our normal metal flux quantisation experiment.

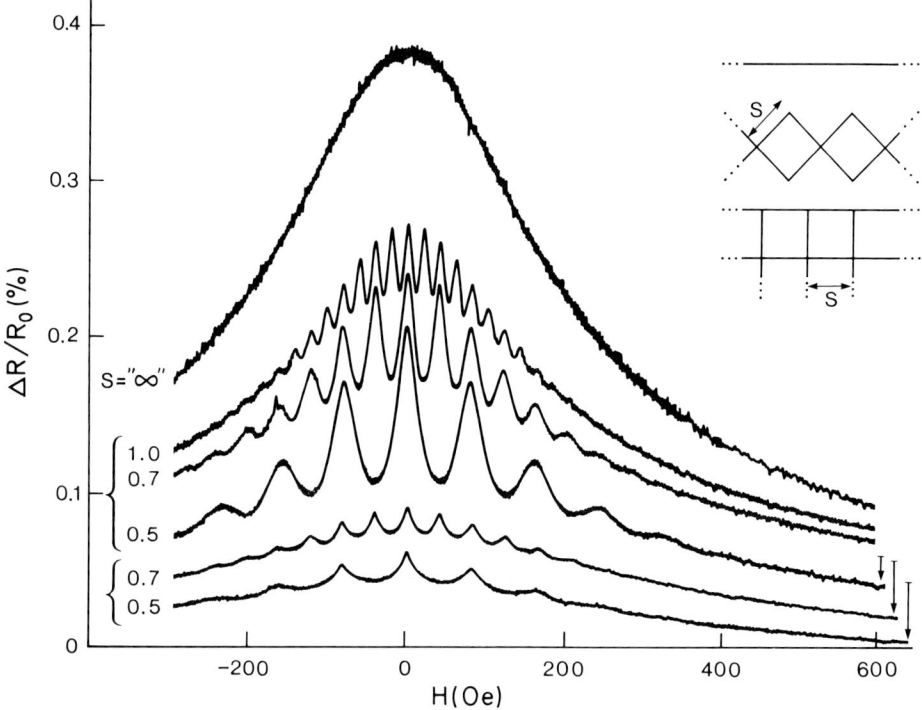

Fig. 4. Relative change of resistance showing interference between particle and backscattered hole as a function of magnetic flux enclosed by the path.

than the "Thouless length" $\sqrt{l_e l_i}$ [l_e and l_i are the elastic and inelastic (coherence-destroying) mean free paths]. This is a particularly clean system for studying the effects of inelastic scattering of the particles on the interference phenomenon. As suggested initially by Thouless (1977), in these experiments it is possible to measure quantitatively the rate of loss of coherence. Fukuyama and Abrahams (1983), Altshuler, Aronov and Khmelnitski (1982) and Fukuyama (1984) have been studying — with some controversy among themselves, but in essential agreement with experiment — the problem of how to quantify the definition of an inelastic collision event which destroys coherence between the electron and its backscattered partner, as opposed to carrying away a given amount of energy or momentum. This is most significant in that, to my knowledge, it is the first example of measurement and prediction of coherence - destroying inelasticity rates, in the true sense. A second interest is that the Russian group has emphasized that the relevent scattering can be thought of as coming from the Nyquist noise, generated by the density fluctuations of the electrons themselves. Because it deals with a completely isolated many-particle system, this work gives the feeling of having finally gotten down to the actual nuts and bolts of the phenomenon of irreversibility itself.

Once inelastic scattering has taken place, we would have to recohere the additional excitations in order to reconstruct the beam, and then each of our

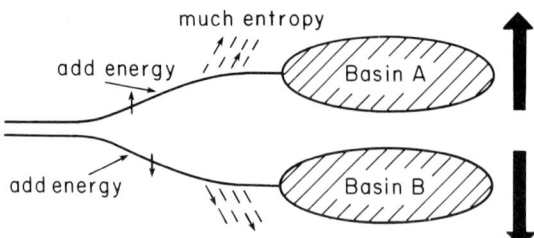

Fig. 5. Two stable basins of attraction.

excitations itself scatters inelastically, etc. Perhaps, in principle, a perfect reversible quantum computer (Bennett 1973) could keep track of every scattering, but in effect the two halves of the beam are by now following different, untraceable paths through wave function space and we must treat the outcome in terms of probabilities and not probability amplitudes.

So far, however, no "measurement" has taken place. For instance, no "particle" has been registered as having gone through a channel. In a general measurement process, we add energy—e.g. accelerate the beams, or cause them to impinge on metastable grains of silver salt, or some other similar scheme. This allows us to enhance the rate of dissipation—i.e. production of entropy—so that the trajectory of each piece of the wave function in Hilbert space can satisfy Liouville's theorem by spreading out in irrelevant variables, but contracting strongly in certain relevant ones which have been coupled to the variables to be measured (see fig. 5). The relevant variables, which are normally the coordinates of some macroscopic, rigid object (a "pointer"), become trapped in a "basin of attraction" which is different for the different values of the original quantum number. By virtue of registering that fact in his neurons, which are themselves switching devices with multiply stable basins of attraction, the observer too becomes entrapped in beam A or beam B, since in essence the Hilbert spaces of the macroscopic objects involved are essentially disparate and no physical operator acting on a few coordinates at a time, expressing the only possible physical processes internal to the system, can connect states in A to states in B. (Of course, an external force can move a rigid pointer by the application of an overall force coherently to every atom of the pointer simultaneously, but this merely redefines "A" and "B".)

3. *Phase variable of superfluid helium. Broken symmetry*

These properties of macroscopic systems are much more clearly shown if we look at an example. One example of measurement, and of quasi - degeneracy and of rigidity properties, which has always intrigued me, is that of the phase variable of superfluid helium. This is particularly attractive because this is a broken symmetry variable whose nature is not confused by the presence of many other objects with the same kind of broken symmetry. At this point let me insert a brief primer on broken symmetry in condensed matter physics. [see Anderson (1984) for more details.] This takes the form of five points.

(1) The initial Hamiltonian of our system has a large group G (e.g. local homogeneity and isotropy for a liquid).

(2) The lowest-energy state, or mean field fixed point, has a lower symmetry H (as a crystal, with discrete rotation and translation symmetries).

(3) Hence there is an "order parameter" describing the state, which has "phase angles" free to move in a space isomorphic with the factor group G/H.

(4) Interactions enforce generalized rigidity of these phase angles: there is a (free) energy $\propto (\nabla\phi)^2$. This allows dissipationless "action at a distance" via rigidity, supercurrents, etc.

(5) Hence all atoms are correlated, in highly non-quantum but very familiar behavior—*we* are broken symmetry objects with quasi-degeneracy and rigidity, after all, so we have no trouble in dealing with these concepts except when we try to make them compatible with quantum theory.

Now, in order to explore the relationship of these peculiar—if you are a quantum person—or ordinary—if you are macroscopic—properties to measurement problems, let us do the following sequence of "Gedanken" experiments. [With recent advances in technique perhaps these need not all remain "Gedanken" experiments forever (see Avenel and Veroquaux 1985).]

As a first stage, cool down isothermally a bucket of liquid helium from above T_λ to near $T = 0$, starting, if one likes, from a microcanonical ensemble so that the resulting state has a fixed number N of particles. The bucket will subside into its lowest state, which can be written as

$$\Psi = \int d\phi \, e^{iN\phi} \, \Psi(\phi),$$

where Ψ is a quasi-classical coherent state in which the order parameter is taken to be

$$\langle \Psi \rangle = |\Psi_0| e^{i\phi}.$$

The phase ϕ will be uniform everywhere because of the "generalized rigidity" term in the free energy

$$\tfrac{1}{2}\rho_s v_s^2,$$

where

$$v_s = \frac{\hbar}{m} \nabla\phi.$$

I would assert that at this point, already, the different components of the wave function have ceased to interfere and that our bucket is already in the "many worlds" representation; within very fine, calculable limits enforced by the N-ϕ uncertainty principle, ϕ has become a classical variable, and while no experiment can determine what the actual overall value of ϕ is—since it is gauge-

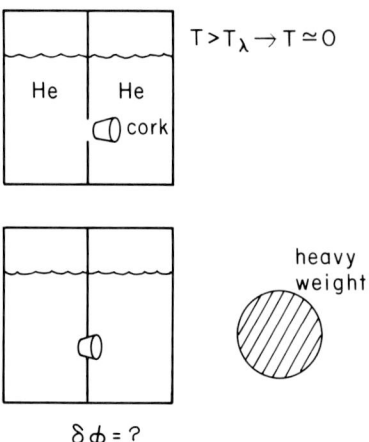

Fig. 6. Change in the condensate phase due to gravitational potential.

dependent—any future experiment will be interpretable as though ϕ was fixed. For instance, as a second stage of our experiment, let us suppose our original bucket was nearly divided into two, and at a certain time we close off the septum between the two halves of the bucket, and thereafter subject the two halves to different gravitational histories—e.g. by putting a large object like a locomotive near one half and not near the other half. If the difference in gravitational potential is ΔV, we expect to observe a change in phase of $\Delta\phi = \hbar^{-1}\int\Delta V \, dt$ (see fig. 6).

Each half of the bucket, we assert, is now in a state in which it has a wide variation of N, ΔN, and in which there is coherence between the wave functions belonging to different values of iN:

$$\langle N_{1+1} | \Psi^* | N_1 \rangle = \Psi_0 \, e^{i\phi_1}$$

and similarly for the other half of the bucket. When we reunite them, a measurable interference current will initially flow depending on their relative gravitational histories. This interference current, of course, violates Wigner's "superselection rule" but that is not our problem. What is relevant to measurement theory is that when the two are reunited the experimentalist does not see any interference between different values of the overall phase. This is not quite exactly the case; the phase in each bucket does diffuse slightly, because compressibility forces make the states of different N slightly inequivalent. The time scale, for that, however, is on the order of 10^8 years.

A third experiment illustrates this point. I would suppose that if the experimenter now cools down two entirely different, non-communicating buckets, each of which initially had a fixed N and no preferred phase, that he, upon opening an orifice between the two, would see initially with equal probability any *fixed* value of the phase difference, and thereafter no experiment he tried could recover the components of the wave-function which started out with different relative phases. He would not see *zero* interference current, which would be the result if he was to average over all of his many worlds.

That is, we have done a macroscopic Stern–Gerlach experiment, dividing the states of a macroscopic object into beams with different quantum numbers.

The above experiments essentially represent the equivalent of a "reference system" for phase. It is rather like the development in the early universe of the first gravitational inhomogeneities—prior to that time, "location" would not have had a measurable meaning.

In another sense this experiment resembles the original big bang. During the earliest moments, present theory suggests that one or several symmetry-breaking phase transitions take place leaving us with a present set of particle interaction symmetries which are much reduced, as well as with values of certain "Higgs" fields which have, in principle, arbitrary phase angles—otherwise, of course, they would have broken no symmetry. Coincidences which, to me, suggest that the whole story is not yet in, leave us with no experimental handles on that phase-angle and no corresponding rigidities; nonetheless we can be quite sure that physics averaged over all possible values of that phase-angle is not what we live with, even though in the initial big bang—corresponding to our helium bucket above T_λ—the symmetry-breaking field averaged to identically zero and its phase was meaningless, so that below T_λ we must be in a linear superposition of all possible quasi-degenerate worlds with different values of that phase-angle. (This is of course, quite independent of the existence of monopoles and the singularities of the symmetry-breaking fields, which are disturbances of the relative, not absolute, values of the symmetry-breaking field.)

4. Conclusions

It is symptomatic of the depth of Bohr's thinking that his simple resolution of this deep problem still stands up as the practical way to deal with these hard questions. What I have tried to show is that nonetheless they remain a fascinating subject for subtle experiment and intriguing theory.

Let me close with a quotation from Niels Bohr remarking on the early days of quantum mechanics: "Although the spectroscopic successes of the quantum theory were more spectacular, the explanations of the macroscopic properties of matter were more satisfying and more fundamental". This has always been my view.

References

Altshuler B.L., A.G. Aronov and B.Z. Spivak, 1981, JETP Lett. **33**, 94.
Altshuler, B.L., A.G. Aronov and D.E. Khmelnitski, 1982, J. Phys. **C15**, 7367.
Anderson, P.W., 1984, Basic Notions in Condensed Matter Physics (Benjamin–Cummings, New York).
Avenel, O., and E. Veroquaux, 1985, Phys. Rev. Lett. (to be published).
Bennett Jr, C.H., 1973, IBM J. Res. Dev. **17**, 525.
Bishop, D.J., J.C. Licini and G.J. Dolan, 1985, Appl. Phys. Lett. **46**, 1000.

Everett III, H., 1957, Rev. Mod. Phys. **29**, 454–462.
Fukuyama, H., 1984, J. Phys. Soc. Jpn. **53**, 3299.
Fukuyama, H., and E. Abrahams, 1983, Phys. Rev. **B27**, 5976.
Lee, P.A., and T.V. Ramakrishnan, 1985, Rev. Mod. Phys. **57**, 287.
London, F., and E. Bauer, 1939, in: Actualités Scientifiques et Industrielles: Exposé de Physique Géneral, no. 775. ed. P. Langevin (Hermann, Paris) reprinted in: J.A. Wheeler and W.H. Zuerek, 1982, Quantum Theory of Measurement (Princeton Univ. Press, Princeton, NJ).
Sharvin, D.Yu., and Yu.V. Sharvin, 1981, Pis'ma v Zh. Eksp. Teor. Fiz. **34**, 285 [JETP Lett. **34**, 272].
Thouless, D.J., 1977, Phys. Rev. Lett. **39**, 1167.

Discussion, session chairman W. Kohn

Thirring: Don't you think that some axioms of the orthodox interpretations of quantum mechanics should be modified, namely that the measuring device has to be described in terms of classical variables only. These variables should certainly not be limited to the p's and q's alone because, as you pointed out, the phase of liquid He-II can be used for measurements.

Anderson: I turn your question around to a statement and say: I agree with it.

Weisskopf: You have connected your remarks with Everett's theory of many worlds, and I am not quite clear how you interpret this. For example: In your experiment with the two buckets one could say that there was an equal probability for any phase difference. But the one observed is the one that happens to be realized in this case. I personally cannot see why one has to assume that all others are also present, but your reference to Everett makes me think that you made this assumption.

Anderson: I am afraid I do mean that. I agree with John Wheeler who once said that that is much too much philosophical baggage to carry around, but I can't see how to avoid carrying that baggage.

Weisskopf: Why are you forced to carry it along in this case?

Anderson: Let's say I carry it along in the philosophical, and therefore not very serious, part of my mind.

Ambegaokar: While I thoroughly agree that, for the problems treated by most physicists, many paradoxes are avoided by treating subsystem *and* environment as a single quantum-mechanical system, is it not a rather large extrapolation to apply quantum mechanics to the universe as a whole? Is there some *evidence* that this is a justified or prudent assumption?

Anderson: No, there is no evidence. So far there is, in fact, a disturbing absence of evidence in that people have tried to predict a number of singularities in the hypothetical symmetry-breaking field left around: for instance, transition monopoles and things like that, which have not been seen. It is hard to know whether the phenomenon of inflation is to be considered as a confirmation or a further puzzle and, as will be discussed later, there is also the strange question of the cosmological constant. Note, incidentally, that all of this takes place only in the *relative* phases and says nothing about all the other universes.

Casimir: I am afraid I don't quite understand the remarks by Thirring and Anderson about a measurement not always having to be reduced to classical, macroscopic observations. I accept that one can use the phase in liquid helium for measurements. But how do you determine that phase? How do you read that instrument? You have got to arrive at something classical. As an intermediate step, you can certainly use things which are not classical, not macroscopic. But I like to maintain that you can always, in the last instance, reduce the measurements to classical phenomena.

Anderson: You are, of course, right. I guess I would make two caveats: One is at what point the measurement actually takes place; and the other is that someday we may make a computer entirely out of Josephson junctions, which is itself capable of making the relevant observations.

Kohn: You have made extensive reference to macroscopic quantum phenomena. In the last two or three years we have seen this remarkable development of the electron-tunnelling microscope where we see quantum phenomena, namely tunnelling, literally on a one-atom scale. This has been an exciting experimental development. A lot of work has been done since its invention and all kinds of applications have emerged. But I wonder what your thoughts are on the future relevance of that direction of experimental work for measurement theory?

Anderson: I haven't really thought about it. The only thing I have thought about a little bit is that it would be amusing to use some of the electronics in the Josephson mode. I don't think that anyone has done it yet. There could be some things that you might be able to do with adjustable junctions of that sort, in terms of turning Josephson currents on and off.

Ginzburg: If I understand correctly you have in mind to measure a difference (or gradient) of the phases between two buckets with He-II. But this means that you measure the velocity or the mass flux of helium. These, however, are macroscopic quantities, measured in a macroscopic way. So I do not understand what is the difference here with the usual interpretation of quantum-mechanical measurements using macroscopic devices.

Anderson: see answer to Casimir above.

Peierls: At the very beginning you stated that quantum mechanics is deterministic because the wave function satisfies a causal equation. Does this not imply that one regards the wave function as a physical object? This leads to all kinds of difficulties.

Anderson: I guess my answer to that is: If it's not a physical object—what else have we got? Quite seriously, I am saying that it contains all the physics but some parts of it are hard or impossible to observe because we are in it, and if you like, we don't observe alternative versions of ourselves.

It's Not Over Till the Fat Lady Sings
Talk given at *History of Superconductivity*, APS meeting, March 1987

This was mostly paraphrased by Lilian Hoddeson in her book on the history of solid-state physics; this is the original, unpublished version.

IT'S NOT OVER TILL THE FAT LADY SINGS
History of Superconductivity — APS, March 1987

P.W. ANDERSON

Department of Physics: Joseph Henry Laboratories

Jadwin Hall

Princeton University, Princeton, N.J. 08544

For 46 years prior to the BCS paper in 1957, superconductivity had baffled the best minds in theoretical physics. For example, Bob Schrieffer has described Feynman's self-confessed bafflement in 1956 very graphically. After BCS, it took remarkably little time, as such things go, for most experimentalists, especially those not previously in the field, to accept the basic tests given in the first paper and in its immediate aftermath; it was much longer before the theorists' skepticism was quelled. A few very senior theorists remain skeptical to this day. I would like to describe the processes by which most of that skepticism eventually was dispelled, and the healthy additions and changes to the BCS theory which were added in the course of this struggle. In fact, the changes were so great as to inspire my title (borrowed from the Philadelphia Flyers): the game was not over in 1957 with the BCS paper; perhaps the fat lady sang "America" sometime in 1963; or perhaps — we learn this year — it is still not really over. By being over in 1963, I mean that at that point any <u>rational</u> objection could be answered to the satisfaction of the answerer at least, if not the objector.

Most theorists' objections to BCS focused initially on gauge invariance, quite properly, because many previous theories — for instance Bardeen's early one — had foundered on that. The question is very straightforward: London's equation as it comes forth from perturbation theory with an energy gap reads not

$$\nabla \times J = -\frac{1}{\lambda^2} H$$

but

$$J = -\frac{1}{\lambda^2} A$$

which is not gauge invariant; and this form is easy to derive incorrectly from a variety of non-superconducting states. These points were made especially by

Wentzel and by Blatt and Schafroth, following Buckingham's lead; but the question was raised again and again in public presentations — as Nambu recalls doing at Schrieffer's talk in Chicago.

The question is in fact a non-trivial one. The first response chronologically, and possibly the only fully correct one, was probably my own first paper, which I discussed with Bardeen before BCS actually appeared (it is acknowledged in a somewhat backhanded way in the BCS paper itself). A similar argument appeared independently in Bogoliubov's very rapid series of papers already appearing as a book in late 1958. Bogoliubov, as well as Nambu in a more complete (and technically very useful) paper in 1959, argued that the system would have a branch of longitudinal acoustic-like excitations (we would now call them zero sound) which would cancel out the longitudinal currents but not the transverse ones, and satisfy the relevant sum rules. From the start I accepted that zero sound waves would appear and be relevant in a neutral BCS gas but that what we now call the "Higgs mechanism" would operate to eliminate them in favor of plasma modes for the charged gas. Both explanations relied on what we now call the "broken symmetry" of the BCS state. This broken symmetry aspect was simply not available in the original BCS theory — as can easily be seen from the Ginsburg-Landau formula for the current,

$$J \propto \psi^* \left(i \nabla - \frac{e^* A}{\hbar c} \right) \psi$$

where the phase of the order parameter, which is absent in BCS, plays a key role. The derivation of this formula depends vitally both on correct treatments of collective excitations <u>and</u> on allowing fluctuations in the order parameter in these collective excitations.

In fact, with surprising rapidity this explanation of the gauge problem was accepted. The controversy had, however, an important byproduct: it brought Nambu into an awareness of the structure of BCS and as a result he introduced the concept of broken symmetry into particle physics: a concept responsible for many of the parts of the "standard model": at least for the "electroweak" theory and for many of the ideas of grand unification etc., now current. It is important to realize that particle physics' borrowing of broken symmetry was repaid by causing condensed matter people to refine and conceptualize their vague notions of broken symmetry, which had been floating around previously and which had been used several times before, as in my antiferromagnetic ground state paper of 1952. The failure of the Higgs mechanism to reappear in particle physics until much later is part of the history of particle physics.

Within condensed matter physics the main opposition to BCS, led basically by B.T. Matthias, with the help of Fröhlich, Bloch and others, was to the phonon mechanism and to the lack of quantitative energetics and of predictive power for the chemical occurrence of superconductivity. At first this focused on the isotope effect, which soon was measured well enough to bring out real or imagined deviations from the BCS $-1/2$. Swihart qualitatively, and Morel and I quantitatively, set out to make a roughly realistic calculation of T_C's from first principles, and soon realized that the BCS equation might be taken fairly literally as an integral equation in the time domain rather than in space, which, using fairly standard dielectric screening theory including phonons, predicted the general run of T_C's and isotope effects, albeit with rather broad limits of errors which by no means satisfied Matthias. As an afterthought, let me note that it was on this work that Cohen and I later based our ideas on the upper limits to T_C in phonon — or for that matter, exciton — based versions of BCS; and I have never seen a serious answer to our arguments.

The quantitative theory of energy gaps which arose from this work came as the result of an informal collaboration between Bob Schrieffer and myself, which, I remember well, had its origins in a long and very boring bus tour we took together of the Polders in Holland, on a rainy day, during the Utrecht many-body theory meeting in 1960. I told Bob of our physical ideas about integration in the time domain, while Bob informed us of his study of the Green's function formalism of the Russians, especially Eliashberg, which was the correct way to do this <u>and</u> to express the tunneling current. At first we went off in different directions, Morel and I with our systematics and Bob with his early work on on-line integration of the gap equation (one of the first on-line scientific calculations ever!) but by summer of 1962 John Rowell had invented differential tunneling spectroscopy and the lines reconverged. Sometime that summer John and I went down to Penn with his data on lead, and a suggested rough model, and the result were the twin letters of Schrieffer, Scalapino and Wilkins, and of Rowell, myself and Thomas which founded the truly quantitative theory of T_C and the coupling parameters, which was later carried to such exquisite precision by McMillan and Rowell. Nowadays, I feel oppressed by that success: it is very hard to break the new orthodoxy based on that work in order to get honest consideration for <u>new</u> mechanisms.

The third objection, which seemed to be best expressed by Felix Bloch and H.B.G. Casimir, was very fundamental: had the basic phenomena of superconductivity really been explained? Felix kept arguing that BCS did not explain the most fundamental observation, that of persistent currents, which Felix himself had shown could not be an equilibrium state. And Casimir kept asking 'how on earth can the voltage be exactly <u>zero</u> along a mile of dirty lead wire ?'

The answer to these questions, in principle, had oddly enough been given essentially simultaneously with the answer to gauge invariance. When I saw Landau in Dec. 1958, he remarked that Gor'kov had derived the Ginsburg-Landau theory from BCS re-expressed in Green's function language and that since Ginsburg-Landau was gauge invariant there was no problem. He was not correct: while Gor'kov's derivation solves <u>Bloch's</u> problem, it does <u>not</u> solve gauge invariance, which involves a correct microscopic, dynamical treatement of collective modes, not basically available via G-L in spite of all the more modern claims to the contrary.

What ensued is one of the most confusing stories in modern science, a story which has permanently cured me of trying to apply logic to history. What should have happened logically is something like the following: Gor'kov should have noticed that his derivation implied that the charge parameter e^* in the Ginsburg-Landau theory was $2e$, unequivocally. Then, Abrikosov should have noticed that <u>his</u> big paper of 1956 on type II superconductivity contained an actual calculation of a quantized flux line and said: 'aha!, the flux quantum is $\frac{hc}{2e}$!' Then everyone should have noticed that in order to kill a persistent current you have to pass flux quanta through the superconductor, which implies an energy barrier $\sim ev/\text{Å}$ of flux line, a number easily derived from Ginsburg-Landau.

None of the above happened. Landau idly speculated that e^* in G-L had to be integer, but nobody understood him. He also noticed the analogy between Abrikosov's lines and Feynman's quantized vortices but Abrikosov's dimensionless numbers concealed the value of the flux quantum. And noone went back and read London's arguments on persistent current. Only Onsager had it all correct, and nobody understood him, because he spoke so cryptically in a deep Norwegian accent.

Meanhile, a race to do the flux quantum experiment involving at least 5 laboratories was won in a dead heat by Döll and Näbauer, and Fairbank and Deaver, in early 1961. Again, at Utrecht prior to that experiment, while planning it: Fairbank asked every theorist he could find what value he would get, if any, and had no unequivocal answers. But oddly enough, of all the papers explaining that value, no Russian pointed out that it had already been derived and the flux line described.

Full understanding came eventually from two new and rather unlikely sources: Brian Josephson and B.B. Goodman, and one could say that by summer 1963 the whole story had become clear and was told at the Colgate meeting at which Kim and I spoke about flux creep and flux flow.

Josephson: as a student in my lectures at Cambridge in 1961-62, he became fascinated by the concept of the phase of the BCS – Ginsburg-Landau order parameter as a manifestation of the quantum theory on a macroscopic scale. He seems always to have hoped that normal causality will somehow break down via quantum mechanics, leaving room for paraphysical phenomena. Playing with the theory of Giaevar tunneling, he found a phase-dependent term in the current which none of us could make go away. Pippard says I first wrote down $J = J_0 \sin \Phi$ but I'm not sure; what I do think is that I may have been the first to say "why not!" Josephson, however, worked out all the consequences in a gorgeous series of papers, private letters, and a privately circulated fellowship thesis.

When I returned from England in June 1962, John Rowell told me that some junctions showed what might be the Josephson effect, and by December we had firm evidence for it. The observational problem was probably the noisy environment of most laboratories at the time. Josephson's equation

$$\hbar \frac{d\varphi}{dt} = 2e\,\Delta\mu$$

which, as he pointed out, is implicit in Gor'kov's theory, then gave us the real and final answer to Casimir's question about the "mile of dirty lead wire". So long as ψ is time-independent, there can be no voltage drop $\Delta\mu$.

Persistent currents had to wait until they were actually shown not to be persistent. Goodman finally reawakened us in the West to Abrikosov's beautiful paper, and we began to understand why such large critical fields were possible in the type II flux quantum array state — but not why they could carry such large currents!

Again in that multiply eventful fall of 1962, Young Kim demonstrated that the critical currents did decay, and I showed that this was a consequence of the pinning of flux lines, their slow release, and Josephson's equation, so we knew how and under what conditions currents persist. Shortly thereafter Kim demonstrated the phenomenon of flux flow: large diamagnetism, a Meissner effect, but plenty of voltage and no persistent current, showing in essence that Casimir and Bloch were correct in separating these phenomena out as independent, in some real sense, from the original BCS theory. They involve topological considerations about the order parameter and the broken symmetry whose most general structure was not understood until much later still, about 1975.

Associated with this story is a bitter little one for Kim. We wanted, when given the opportunity by the A.I.P.'s newsroom, to make a little public relations fuss about the decay of persistent currents, especially because it was to be done at the Seattle APS meeting, and Kim was from Seattle. It may not have been the most important discovery in superconductivity that fall, or even at BTL — the Josephson effect had been found and about that Bell Labs never suggested making any fuss. But it was the first clear solution to the oldest problem in superconductivity and possibly in condensed matter physics. Matthias, however would have none of it and our superiors permitted no release. Bernd felt this result sullied in some way his beautiful materials.

It is in this sense, that the real explanation of the fundamental properties of superconductivity was not explicit at all in the original BCS theory but required all these developments, that it was not until early 1963 that an honest answer could have been given to Bloch and Casimir. Thus we have to say the game was not over in 1957, and that the fat lady really sang "America" roughly in early 1963. Oddly enough, this was also the year that Bloch's presidential address to the APS, published in Physics Today, attacked the thoery of superconductivity as incomplete: a masterpiece of bad timing to which I wrote — and then I believe destroyed — a strong response.

Of course, much advancement followed that point, especially the great work on SLUGS and SQUIDS and other Josephson devices. But until this year, it has seemed that the really vital part of the game was over.

Spin Glass I: A Scaling Law Rescued
Physics Today **41** # 1, 9 (1988)

Spin Glass II: Is There a Phase Transition?
Physics Today **41** # 3, 9 (1988)

Spin Glass III: Theory Raises Its Head
Physics Today **41** # 6, 9 (1988)

Spin Glass IV: Glimmerings of Trouble
Physics Today **41** # 9, 9 (1988)

Spin Glass V: Real Power Brought to Bear
Physics Today **42** # 7, 9 (1989)

Spin Glass VI: Spin Glass as Cornucopia
Physics Today **42** # 9, 9 (1989)

Spin Glass VII: Spin Glass as Paradigm
Physics Today **43** # 3, 9 (1990)

This sequence of brief historical reviews may be seen as a unit, describing the amazing fertility of the spin glass concept and its surprising toughness as a theoretical physics problem.

REFERENCE FRAME

SPIN GLASS I: A SCALING LAW RESCUED

Philip W. Anderson

The history of spin glass may be the best example I know of the dictum that a real scientific mystery is worth pursuing to the ends of the Earth for its own sake, independently of any obvious practical importance or intellectual glamour. If a phenomenon seems likely to contradict the fundamental principles you thought you understood (that's what I mean by a *real* mystery—so long as you also believe the experiments!) you should stick with the phenomenon. This fundamental dictum of good science is increasingly neglected by our masters who provide the money; spin glass is even less popular with them than superconducting materials were before 1987. The pursuit of the spin glass mystery led, *inter alia* and aside from all the good solid-state physics that resulted, to new algorithms for computer optimization, a new statistical mechanics, a new view of protein structure, a new view of evolution and new ideas in neuroscience.

But it all started with some very simple physics. Measuring magnetic resonance and magnetic properties of dilute magnetic ions in insulators (such as Mn in ZnO) to probe magnetic interactions was a very useful game played in the 1950s and early 1960s by John Owen, among others. At about the same time, he and the Berkeley magnetic resonance group of Arthur Kip, Walter Knight, Charles Kittel and others tried diluting manganese into the nonmagnetic metal copper to test the proposed Ruderman–Kittel–Kasuya–Yosida interaction between spins via free electrons, and to see whether the free electrons' exchange interaction with the magnetic ions affected their resonance behavior—that is, to see if there was an electron magnetic resonance equivalent to the

Philip Anderson is a condensed matter theorist whose work has also had impact on field theory, astrophysics, computer science and biology. He is Joseph Henry Professor of Physics at Princeton University.

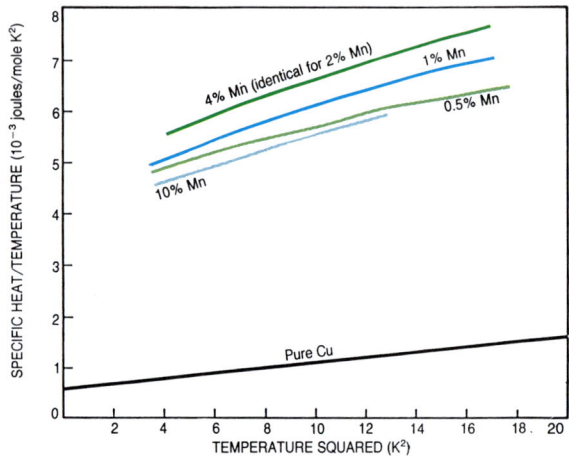

Specific heat of Cu–Mn alloys at low temperatures is independent of the manganese concentration, and it varies linearly with temperature when the contribution due to pure copper is subtracted. In the range of manganese concentrations shown here, the low-temperature properties of the alloy are dominated by the interaction between magnetic moments on manganese atoms, which leads to the spin glass behavior. (Adapted from J. E. Zimmerman, F. E. Hoare, *J. Phys. Chem. Solids* **17**, 52, 1960.)

nmr Knight shift and the Korringa free-electron relaxation of nuclei.

This simple set of experiments opened up at least two cans of worms. The physics of the effect of the free electrons on the individual Mn ions is, of course, closely related to the Kondo problem and the Anderson model, but this is not our worm-receptacle of choice. We are interested in what they saw when the samples contained 0.1–10% manganese and the behavior was dominated by the ion–ion interactions.

The magnetic resonance effects were not very instructive at the time, involving complicated physics that was actually one of the last areas to be clarified. I will discuss this physics in a later column. What first attracted attention were simple thermodynamic measurements such as those of magnetic susceptibility and specific heat.

The extra specific heat at low temperature was perfectly linear in T, with a slope about five times that of the background copper, independently of concentration, but then the extra specific heat peaked at a concentration-dependent temperature T_0 and dropped down at a rate one could imagine to be approximately $1/T^2$. (See the figure above.) The susceptibility χ rose from a constant value to an (apparently) broad peak near T_0, then fell off at a good approximation to the Curie law $\chi = C/T$ (C is the Curie constant) with the right Bohr magneton number for Mn^{++} with its $g = 2$, $S = 5/2$ ground state.

After a bit of to-ing and fro-ing, a

REFERENCE FRAME

group at Grenoble showed that for these "classic spin glasses" (as they much later came to be known) all of these data fit onto a single universal curve with the temperature, energy and entropy scales all linearly proportional to concentration. We know now that this is a good confirmation of the extraordinarily simple underlying physics—that the dominant term in the Hamiltonian for the magnetic moments of solute atoms is the RKKY exchange integrals

$$J(r_{ij}) = J_0 \frac{1}{r^3} \cos(2k_F r_{ij})$$

where k_F is the Fermi momentum. These RKKY exchange integrals fall off slowly with distance and alternate in sign.

In these random, dilute solutions this behavior has the effect of randomizing the exchange integrals (since $1/(2k_F)$ is less than an interatomic distance) and each spin sees quite a few neighbors with each of which it has an exchange interaction that is random in both sign and magnitude. The probability distribution of these exchanges has a scale proportional to $\langle 1/r^3 \rangle_{avg}$, which is in turn equal to the density of solute atoms. The form of the probability distribution, however, is independent of the density except for the scale.

A 1987 postscript to this story was written by Michael Stephen and Elihu Abrahams. The long-range $\cos(2k_F r)/r^3$ form occurs because of the sharp drop in occupation of free-electron states at k_F. Pierre-Gilles de Gennes had proposed, long before, that scattering of the free electrons by impurities reduced the range of this effect by a factor $\exp(-r/l)$, where l is the mean free path of the electrons, by dephasing the electronic wavefunctions at the Fermi surface; this unfortunately would ruin the scaling law for the probability distribution of the interaction. Until recently this argument was universally accepted, but it turns out not to be true, as Stephen and Abrahams have shown: The distribution is almost unchanged by scattering, because scattering merely shifts the phases of the wavefunctions near the Fermi surface without changing their relative phases in any given sample. This is a vital key to the properties of real spin glasses, and in addition restores the scaling laws.

But, as you'll see in my subsequent columns, aside from the scaling law (which has only just been rescued) none of the features of these measurements were to be understood for another 20 years or more, and the real physics is still a problem. ■

REFERENCE FRAME

SPIN GLASS II: IS THERE A PHASE TRANSITION?

Philip W. Anderson

In the late 1950s and early 1960s Jim Kouvel, then at GE, and Paul Beck's group at the University of Illinois spent a lot of time exploring a phenomenon Paul called mictomagnetism. This phenomenon took place in dilute solutions of Mn atoms in Cu (and of other magnetic atoms in other nonmagnetic metals); I discussed these materials in my column in the January issue (page 9). These solutions, as I remarked, seemed to have small linear magnetic susceptibilities of typically paramagnetic magnitude (a few times 10^{-4} in dimensionless units). But Kouvel and Beck showed that the solutions exhibit, at a tiny scale and at very low temperatures, and in addition to the linear susceptibility, many of the phenomena typical of ferromagnetism: hysteresis, remanence and so on. In some ways these solutions are *more* hysteretic than ferromagnets, in that they can remember the sign and direction of the field they were cooled in, even when one applies an opposing field large enough to polarize them in the opposite direction.

Meanwhile Bernd Matthias and the rest of us at Bell Labs were very interested in the possibility that magnetism and superconductivity might coexist. Within the BCS theory the two should be quite incompatible, and in many cases they are; but (some 30 years too soon!) Bernd was determined to show that in at least some cases there would be a close relationship. In some of his dilute solutions of magnetic ions in superconductors (like Gd in CeRu$_2$) he noted the presence of the vague susceptibility peaks and remanences characteristic of spin glasses, and so he said: "Aha!

Philip Anderson is a condensed matter theorist whose work has also had impact on field theory, astrophysics, computer science and biology. He is Joseph Henry Professor of Physics at Princeton University.

Ferromagnetism and superconductivity are *not* incompatible!"

I always tried to listen more carefully to what Bernd's *results* said than what *he* said, since he had little regard for fine distinctions in statistical physics (like that between ferro- and antiferromagnetism, for instance, or between these and some vague bump in the susceptibility), but this is a case where he got to me. I was so certain that the transitions he was talking about were not true ferro- or antiferromagnetism that I failed to note what he had noted, that the transitions seemed remarkably sharp. I was particularly certain that a magnetic transition would involve a significant change in entropy and hence would certainly dominate the tiny energies and entropies of the superconducting state. (This was almost a decade before 1972, when Michael Kosterlitz and David Thouless, following my work on peculiar one-dimensional models, first showed that a phase transition could show no specific heat singularity at all.) Yet these bumps didn't seem to disturb the superconducting transition very much, which I felt meant that they were not phase transitions.

It is a bit ironic that only two or three years later, in 1965, an obscure journal called *Physics*, edited by none other than Bernd and myself, published the first evidence that there really was a spin glass transition, without either of us (or possibly even John C. Wheatley, the author) noticing. Wheatley was interested in testing his then new SQUID magnetometers in an interesting system and chose these same dilute solutions of Mn in Cu. His susceptibilities (measured, perforce, in a tiny magnetic field) followed a very precise Curie law C/T for each solution down to a temperature T_c, which was very exactly proportional to concentration, and then, as abruptly as he could measure, stopped changing with T and became constant. (Note that unlike the older measurements, Wheatley's did not exhibit a peak, because he cooled in a fixed magnetic field; a constant value of the susceptibility is characteristic of spin glasses when they are cooled in such a field.)

It was not until 1970 that the key measurements that woke the rest of us up to this peculiar transition were made—by Vincent D. Canella, John A. Mydosh and Joseph I. Budnick. This group measured ac magnetic susceptibilities with sensitive, but more conventional, methods, and discovered that the key variable is the magnitude of the measuring field. At 1000 gauss, there is only the conventional vague hump; at 1 gauss, a sharp, cusp-like peak appears whose width is less than 1% of T_c. Yet 1 gauss is 10^{-5} the magnitude of the internal field, since T_c is approximately 10 K. This tremendous nonlinearity is the appropriate characterization of the transition; later measurements, by P. Monod and Hélène Bouchiat, for instance, showed that the nonlinear susceptibility $\partial^2\chi/\partial H^2$ diverges as $(T - T_c)^{-P}$, where P is greater than 1. Thus, experimentally there is no doubt that the transition exists and is an equilibrium transition, since the nonlinearity can be measured above the transition point, where no one doubts that equilibrium is established in the system—after all, its natural relaxation frequency should be about 10^{12}–10^{13} sec^{-1}. Nonetheless, no measurement has ever revealed a specific heat singularity at $H = 0$. As we shall see in the next column, the theoretical acceptance of a true phase transition, as well as an understanding of its nature, was much slower to come; and the most striking feature, the nonlinearity, is yet to be calculated, even roughly. ∎

© 1988 American Institute of Physics

PHYSICS TODAY MARCH 1988 9

REFERENCE FRAME

SPIN GLASS III: THEORY RAISES ITS HEAD

Philip W. Anderson

Sam Edwards was finishing out his term as head of the SRC, the British equivalent of the NSF, during the winter of 1974–75, after being appointed to a professorship at Cambridge. (He is now Sir Sam, the Cavendish Professor, successor in that chair to James Clerk Maxwell and four or five Nobel laureates, including Sir Nevill Mott.) Being Sam, he was unfazed by the full-time SRC job and needed research to do on the train back and forth to London, so he dropped in every Saturday at the Cavendish Laboratory for coffee and a chat with me and the theory group. I made a point that year of being there on Saturdays as well as during the week, and we did a lot of talking about localization and the "Fermi glass" (that is, the problem of electrons frozen in place by localization and interactions), the theory of liquids and the glass transition, and other problems of mutual interest.

One of these problems was that of dilute magnetic alloys, which seems to have acquired the name "spin glass" in a 1970 paper I wrote with Wai-Chao Kok (now at Singapore University) for a 65th-birthday festschrift for Mott in the *Materials Research Bulletin*. (See my columns in the January and March issues of PHYSICS TODAY.) I described to Sam the old mystery of continuous, disordered freezing in these alloys, and the new mystery of the sharp cusps and nonlinear behavior that John A. Mydosh had reported. Sam's ears pricked up. He had a notebook full of methods he had been trying on gelation, the glass transition and various polymer questions, but had been frustrated because these are not clean, well-posed problems. (He had also tried the methods on localization; and later, in the hands of Franz Wegner and Shinobu Hikami, they did work reasonably well on that problem—but

Philip Anderson is a condensed matter theorist whose work has also had impact on field theory, astrophysics, computer science and biology. He is Joseph Henry Professor of Physics at Princeton University.

© 1988 American Institute of Physics

some extra frills were needed.) I was convinced that the random Heisenberg Hamiltonian

$$H = \sum_{(ij)} J_{ij} \mathbf{S}_i \cdot \mathbf{S}_j$$

where $\mathbf{S}_i$ is a classical spin vector at site i and J_{ij} is the interaction between the spins on sites i and j, was almost certainly the proper statement of the spin glass problem. Sam was overjoyed when he learned this, because here was a nice clean problem to work his methods on.

The methodology of the resulting, justifiably famous paper was almost entirely his. (That's why I can make such an immodest statement.) But the basic physical concept we worked out together. We decided that the thing to do was to ignore the spatial ordering, that is, to neglect the long-range ordering of spins in space, if any, and instead to look for long-range order in *time*. Richard Palmer later named this concept "nonergodicity" because it means, when present, that the system does not explore all possible states in the course of time. As the measure of long-range order in time, we introduced q, which is the average correlation between a spin $\mathbf{S}_i$ measured at one time and the same spin measured a macroscopic time t later. The equations are

$$q(t) = \langle \mathbf{S}_i(0) \cdot \mathbf{S}_i(t) \rangle_{\text{ave over } i}$$

$$q = \lim_{t \to \infty} q(t)$$

First we did a little physical calculation of the transition temperature below which q became nonzero. One separates out one of the sites, say i, and assumes that all the neighboring sites have a finite value q_0 of the "spin glass order parameter" q. Then one calculates the correlation enforced on site i by the effective fields due to the other sites, and finally one averages it over all sites i in the sample. For self-consistency, the order parameter thus calculated must have the value q_0 initially assumed for sites that are neighbors of i. The order parameter we calculated,

$$q_0 = \langle \langle \mathbf{S}_i(0) \cdot \mathbf{S}_i(\infty) \rangle_{\text{ave}} \rangle_{\text{ave over } i}$$

had a nonzero solution below a certain T_c. (The subscript "ave" stands for the statistical mechanics average over thermal fluctuations.)

Much more devious is Sam's so-called replica method of calculating thermodynamic properties. To do thermodynamics properly, one must average extensive quantities such as free energy and entropy. These are all derivable from the logarithm of the partition function

$$Z = \text{Tr} \exp(e^{-\beta H})$$

which grows exponentially with the size of the system. It is dangerous—in fact wrong—to average the partition function in random systems like the spin glass. This is because the partition function fluctuates too much: Special configurations, such as regions where all the J_{ij}'s are accidentally positive, will dominate its average value.

The key point of principle that makes studies of the spin glass and similar systems difficult problems is this: They are "quenched" random systems, with the values of J_{ij} fixed for all time by the conditions of preparation of the sample. But we want to average over macroscopic samples, in which many different configurations of J_{ij}'s occur, in such a way that the average represents the behavior of a *typical* system and is the proper "extensive" or "intensive" thermodynamic quantity that varies sensibly with the size of the system. Sam recalled the obvious identity

$$\ln Z = \lim_{m \to 0} \frac{(Z)^m - 1}{m}$$

The necessary average is then that of the m-th power of Z, not of its log; but when m is small this is no easier.

Now we do an outrageous thing: We note that the average of Z^m for $m = 1,2,3,4,\ldots$ is calculable because it is the average of an exponential containing J. For m an integer,

$$\langle Z^m \rangle = \underset{(\mathbf{S}^\alpha)}{\text{Tr}} \int \exp\left(-\beta \sum_{\alpha=1}^{m} \sum_{ij} J_{ij} \mathbf{S}_i^\alpha \cdot \mathbf{S}_j^\alpha\right)$$
$$\times P(J_{ij}) d(J_{ij})$$

PHYSICS TODAY JUNE 1988 9

REFERENCE FRAME

which is the average over m identical "replicas" of the system. If

$$P(J_{ij}) \propto \exp(-J_{ij}^2/2J^2)$$

this integration is easily done. It gives rise to the following type of statistical problem:

$$\langle Z^m \rangle = \mathop{\mathrm{Tr}}_{(S^\alpha)} \exp\left[\frac{\beta^2 J^2}{2}\left(\sum_{\alpha=1}^m \mathbf{S}_i^\alpha \cdot \mathbf{S}_j^\alpha\right)^2\right]$$

This is no longer a random problem, but a regular one. It is more difficult, however, because it is biquadratic in spins. It can be solved in mean-field theory—assuming that $\mathbf{S}_i^\alpha$ and $\mathbf{S}_j^\alpha$ are uncorrelated (correlation only gives terms of order m^2) but that $\mathbf{S}_i^\alpha$ and $\mathbf{S}_i^\beta$ *are* correlated. It can also occasionally even be solved by renormalization-group methods. But one has the awful problem of extrapolating from all positive integral values of m to small m. In principle it is not rigorous to take the limit as m goes to 0 when the function is known only for integral values of m. In practice, however, it turns out to be easy since one keeps only terms of order m. What is more, so far none of the real difficulties encountered in the spin glass theory seem to have come from failure of the mathematical extension to $m \to 0$! Recently Haim Sompolinsky (*Phys. Rev. B* **25**, 6860, 1982) and Miguel Virasoro (*Europhys. Lett.* **1**, 77, 1986) have given us some ideas about why that is true.

In the mean-field solution the "Edwards–Anderson order parameter" q reappears in a new guise, as a replica–replica correlation function

$$q_{\alpha\beta} = \langle \mathbf{S}_i^\alpha \cdot \mathbf{S}_i^\beta \rangle$$

Thus in some real sense the different replicas represent very widely separated instants in time at which we choose to look at the same system.

Sure enough, the mean-field theory we worked out showed a nice sharp cusp in the susceptibility, in qualitative agreement with experiment, and weakly nonlinear behavior, qualitatively correct but too small. Unfortunately it also gives a cusp in the specific heat, which to this day has never been seen, and which is surely unphysical for real, finite-dimensional spin glasses. Nonetheless the result, giving a sharp freezing transition and describing a true nonergodicity, seemed sufficiently promising that we felt that the replica methodology was the doorway into the problem and that final solutions were just around the corner.

Little did we know! See next time, when I reveal the Negative-Entropy Catastrophe. ∎

REFERENCE FRAME

SPIN GLASS IV: GLIMMERINGS OF TROUBLE

Philip W. Anderson

In my last column (June, page 9), it seemed as though Sam Edwards's beautiful "replica" scheme had brought us to a highly satisfactory resolution of the old problem of magnetic systems with random exchange interactions—what we now call "spin glasses." (In the replica method, one calculates the partition function of n replicas α of the same random Hamiltonian, averages over the randomness and takes the logarithm by studying the formal limit as $n \to 0$: an indirect, shaky but often useful procedure.) In 1975 David Sherrington, who had been Sam's student and is now at Imperial College, London, tried applying the methods and ideas of the Edwards–Anderson paper to an especially simple model in which the "mean field" version should certainly be exact. In Sherrington's model, every spin in a macroscopic sample of N spins is connected by a random exchange integral $J_{ij} \times 1/\sqrt{N}$ to every other spin. This is precisely the kind of artificial system for which mean-field theory is exact in other magnetic models. Sherrington brought the model with him that summer on a visit to IBM (Yorktown Heights), where he worked with Scott Kirkpatrick. The model is now famous as the SK model. Their conclusion was that the EA method led to a solution that, while superficially plausible, was unequivocally *nonsense*—specifically, they showed that as the temperature approached zero the calculated entropy passed through zero and became negative. Since entropy is the log of an integer (the number of states at energy E), its acquiring a negative value is forbidden in statistical mechanics. The energy near $T = 0$ also seemed to be a little lower (by about 2 percent) in the SK solution than the best that Scott could achieve by simulating the model on a computer.

Naturally, everyone at first assumed that the replica method itself was at fault. In fact David Thouless, Richard Palmer and I set out to produce a solution directly, without the replica method. This so-called TAP theory (1977) adapted the ancient cavity-field method of Lars Onsager and Hans Bethe to include a local-field correction for the response of all the spins affected by the fluctuations of a given spin. This correction, which is absolutely negligible (of order $1/N$) in the corresponding long-range ferromagnet, is finite here, changing T_c by a factor of 2, for instance. But we could "prove" that all further corrections *were* negligible. The results agreed with the SK findings near and above T_c, where in fact we now know that both solutions are right, but they deviated subtly below T_c. One important difference was that we got rid of the negative entropy of the SK solution.

Again we thought we had the answer, and again we were to be disappointed, though the problem surfaced in more subtle ways this time. Central to the TAP solution is a mean-field equation,

$$m_i = \tanh \frac{h_i}{2 k_B T}$$

where

$$h_i = \sum_j J_{ij} m_j - \text{(local field correction)}$$

Here m_i is the mean magnetization at site i. Unlike in the simple case of a ferromagnet, where similar equations are encountered, looking for a nontrivial solution for the magnetization m_i in this equation involves an infinite random-matrix problem at every T. Near T_c, this problem appeared to be just expressible in terms of the known statistical properties of the eigenvalues of the random matrix J, and near zero it depends only on properties of "the" solution at $T = 0$. The former case we solved in lowest order, and it seemed to look OK; and for the latter

"WHAT YOU HAVE DONE, GRUNDIG, IS HELP CONTRIBUTE TO A DIS-UNIFIED THEORY."

Philip **Anderson** is a condensed matter theorist whose work has also had impact on field theory, astrophysics, computer science and biology. He is Joseph Henry Professor of Physics at Princeton University.

REFERENCE FRAME

case, Richard set out to calculate "typical" ground state solutions (that is, solutions at $T = 0$). The limiting form of the above set of random equations at $T = 0$ is

$$m_i = \text{sign}(h_i)$$
$$h_i = \sum_j J_{ij} m_j$$

("Sign" just means that m points in the same direction as h and has the maximum possible magnitude.) Both Richard and Scott had been trying to solve these equations numerically for some time.

Both of them gradually came to the same paradoxical result: They could find no "the" solution to this set of equations. Instead, they found many, many solutions of nearly identical energy. They also noticed that it is very difficult, once one's computer has found one solution, to persuade it to move to another, even if the first has a much higher energy than the optimal one. Incidentally, both Richard and Scott found that the easiest way to find a new solution was to raise the temperature nearly to T_c and come back down again—a procedure that Scott called "simulated annealing."

This peculiar feature, enormously annoying at the time, was the beginning of one of the important discoveries of modern theoretical physics, a discovery comparable to that of chaos in its broad applicability to science. But we didn't quite understand that yet.

Because of this unusual feature, and also for other reasons—Thouless, for instance, was unhappy that our solution near T_c might not be quite stable—the TAP "solution" still did not satisfy. We also needed to know why the replica method had failed. Thouless and a student, Jairo de Almeida, soon discovered the rather unexpected reason. Below a certain line in magnetic field–temperature space, a solution with "replica symmetry"—that is, where every replica has the same correlation $q = q_{\alpha\beta}$ with every other replica—is dynamically unstable to "replica symmetry breaking." This implied that there was some new structure in $q_{\alpha\beta}$ that depended on α and β: According to the ideas that underlay the Edwards–Anderson paper, then, not every time you tried to compare one specimen of a system with another specimen of the *same* system would you get the same answer! Looking back, it seems obvious that this was closely related to the simulation problem, but it was a few years before we caught on to that. In my next column I'll try to explain the final resolution. ∎

REFERENCE FRAME

SPIN GLASS V: REAL POWER BROUGHT TO BEAR

Philip W. Anderson

Gérard Toulouse had always been interested in the spin glass problem. In 1977, subsequent to the work I discussed in my last column (September 1988, page 9), Gérard, then at the Ecole Normale Supérieure in Paris, began to discuss the problem with the powerful "Cargèse" group of theoretical physicists in Paris and Rome: Cyrano de Dominicis, Giorgio Parisi and Miguel Virasoro in particular, and later Bernard Derrida, Nick Sourlas and others. Gérard was the originator of the formal theory of "frustration" (I believe I introduced the *term* originally) as *the* important feature of the spin glass problem. Because the exchange bonds J between spins have random signs in most circuits (loops) of spins returning upon themselves, not all the spins can be made "happy"—hence the "frustration." In a square of four spins, for instance, only if an even number of the J's have the same sign can one satisfy everybody—that is, find a unique minimum-energy configuration of the four spins. Since the world is made up of systems of conflicting desires, from game strategies to a group of people choosing a menu, one begins to see that the spin glass is not that bad a model for many aspects of life.

It was Giorgio Parisi who developed the replica-symmetry-breaking scheme that solved the problems, such as negative entropy, that had been troubling us. You will remember from my column of June 1988 (page 9) that Sam Edwards had introduced the "replica" scheme of making m identical copies $1, \ldots, \alpha, \ldots, m$ of the same system and calculating the average of the product of all m partition functions. One then schematical-

Philip Anderson is a condensed matter theorist whose work has also had impact on field theory, astrophysics, computer science and biology. He is Joseph Henry Professor of Physics at Princeton University.

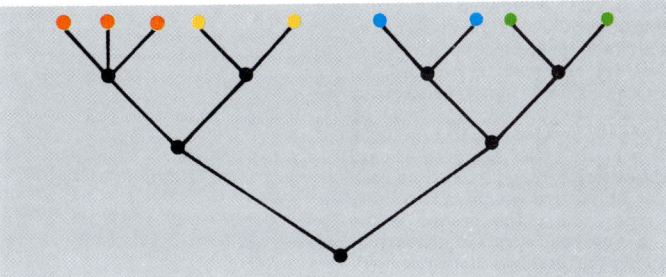

Ultrametric tree. This structure is a convenient way to represent the degree of resemblance between spin glass states (colored dots). The overlap between any pair of states of the same color is q_1; that between "red" and "orange" or between "blue" and "green" states is $q_2 < q_1$. Similarly, the overlap of any of the five states on the left with any of the four states on the right is $q_3 < q_2$. Thus the overlap between two states depends on how deep into the tree one has to go to find a node (black dot) that connects them. One may verify from the figure the amazing property that when any three states are picked at random at least two of the overlaps are equal.

ly takes the limit $m \to 0$ and uses the formula

$$\lim_{m \to 0} \frac{Z^m - 1}{m} = \ln Z$$

to calculate the free energy

$$F = -kT \langle \ln Z \rangle_{\text{ave}}$$

David Thouless and Jairo de Almeida later showed that not all pairs of copies gave the same average correlation

$$q_{\alpha\beta} = \langle S_i^\alpha S_i^\beta \rangle_{\text{ave over } i}$$

This finding was called "replica symmetry breaking." Giorgio was able to produce a form of $q_{\alpha\beta}$ that worked, in the sense that it was a self-consistent and stable solution of the equations in this replica formalism. It would not be wise for me to go through the complicated mathematical structure that evolved from this beautiful and unexpected solution. Instead I shall try to describe what was eventually understood about its implications—with contributions from Toulouse,

Virasoro, and later Haim Sompolinsky, Peter Young, Richard Palmer and many others.

What Giorgio's solution means is that at any temperature below T_c there is no *unique* locally stable thermodynamic state that solves the "TAP" mean-field equations I described in my last column, but rather many such states, which resemble one another to different degrees. Each replica α corresponds to a different solution of the TAP equations; the solutions can be thought of as clusters of states in the N-dimensional configuration space of the N spins. The TAP equations are obtained when the thermodynamic average is restricted to these local clusters of states. The off-diagonal terms in the order parameter $q_{\alpha\beta}$, which represent the average overlap between states in the cluster belonging to replica (solution) α and those belonging to replica β, are a measure of the degree of resemblance between clusters α and β. The diagonal elements $q_{\alpha\alpha}$, which are all the same, are the average overlaps of

REFERENCE FRAME

states within a given replica, or cluster. There is a hierarchy of overlaps, with $q_{\alpha\alpha} \equiv q_0$ being the largest. The next in value is q_1, the overlap between the inequivalent groups of solutions α and β that are closest in the configuration space. Then there is a $q_2 < q_1$ and so on.

At T_c the solutions begin to separate, and the "distance" between them, measured by the deviation of their overlaps from unity, increases until, at $T = 0$, q_0 is 1 but the smallest $q_{\alpha\beta}$ may be nearly zero. Of course, the q's have a continuous distribution in a large ($N \to \infty$) system. Toulouse, Parisi and their collaborators showed that the distances, or overlaps, between different states implied by Parisi's form for the $q_{\alpha\beta}$ could be described by what is called an ultrametric tree. The figure on page 9 shows such a tree, in which no solution in the "red" group is any closer than q_2 to any in the "orange" group.

It is not at all surprising, then, that finding the "best" solution by computer simulations had been impossible: The solutions that separated at T_c became increasingly different as T was lowered. From thermodynamics and the extensive nature of the thermodynamic variables such as entropy and energy, one can show explicitly, as I did, that the only route from one set of solutions to another—through configuration space—passes over energy barriers whose height grows with N, the total number of spins. Thus if you try to get from one solution to another by flipping spins a few at a time, you must make flips that increase the energy by amounts of order N before you can ever get to one of the other solutions or, in particular, to the *best* one. Thus one can represent the solutions as deep valleys connected only by very high passes in a "rugged energy landscape" (to use Stu Kauffman's terms). This is a remarkable result—how truly remarkable and powerful we are only beginning to understand. It implies, among other things, a new thermodynamics—a thermodynamics of systems that are never in thermodynamic equilibrium. Richard Palmer and I called these systems "nonergodic." That one can nonetheless use statistical mechanical methods to get not only the quantitative solutions relevant to such systems but also the structure of the set of solutions is, to say the least, fantastic.

Next time I shall begin my discussion of the implications of this work in fields as far apart as computer science, biology and neuroscience, which normally have been quite outside the purview of physics. ∎

REFERENCE FRAME

SPIN GLASS VI: SPIN GLASS AS CORNUCOPIA

Philip W. Anderson

Some attentive readers will recall a remark I made in my fourth column (September 1988, page 9), to the effect that in the difficulties and annoying features encountered in the study of spin glasses, we were beginning to have an inkling of results that would turn out to be among the most important of modern theoretical physics. I shall now try to make that clear to you. I explained one of the key results last time (July, page 9): the discovery by Gérard Toulouse and his collaborators that there are many inequivalent solutions of the TAP theory of the SK long-range spin glass and that those solutions can be arranged in an "ultrametric tree" whose branches already begin dividing as T is lowered below T_c. To remind you what this jargon means: The TAP theory is the mean-field theory David Thouless, Richard Palmer and I constructed. That theory, we thought, would in principle be exact because fluctuations about it should be negligible in view of the many long-range interactions each spin has in the SK spin glass. "Ultrametric" is an ant's-eye view of a tree, in which the only way to get to another leaf is to climb all the way down to the common branch point and back up (see the illustration in my last column).

Scott Kirkpatrick made a second important connection. Scott observed that finding the lowest-energy state of the SK spin glass—in fact, of almost any spin glass—is a complex optimization problem equivalent to one of the classic examples of what computer theorists call the NP-complete problems. This mysterious class of problems includes a great many mathematical "toys," such as bisecting random graphs, setting up mixed-doubles tournaments and inventing tours of length N for traveling salesmen or Chinese postmen; but it also contains many highly practical problems, such as routing telephone networks to N cities, designing chips with N transistors, connecting N chips together, evolving the fittest animal with N genes and doing almost anything useful with N neurons. Large complex optimization problems are everywhere around us, and almost anything that can be learned about them is of immense importance.

An important branch of computer science is complexity theory, which classifies such large problems according to their "size" N. The size of a complex problem may be thought of as the number of bits necessary to state that problem. For instance, the size of the SK spin glass problem is $N(N-1)/2$, the number of J_{ij}'s. It is strongly conjectured that the number of steps it takes a computer to solve an NP-complete problem cannot be less than a number proportional to an exponential of a positive power of the size. For large N, then, it could take forever. This is clearly the reason why Scott, Richard and others had been unable to find a unique lowest-energy state.

Each instance of the dozens of known NP-complete problems can be converted to an instance of any of the other problems by an algorithm taking only N^p time steps—that is, the number of time steps is a polynomial function of the size of the problem. This suggests that a statistical mechanical "solution" of the spin glass problem might be of general interest for all NP-complete problems. But that is not the case, even if one assumes that the "polynomial" algorithm that maps other problems to the spin glass is not more trouble than it is worth. Our statistical mechanical solution gives *average* answers for an ensemble of examples of the given problem. Such an answer is valid for a generic, or typical, instance of the problem. In the case of the spin glass, the average number describes the generic instance of the problem involving the given distribution of J_{ij}'s. But the mapping algorithm might transform that generic instance into a special case or vice versa. This issue was perhaps somewhat clarified in an exchange between Eric Baum (Princeton), on the one hand, and Daniel Stein (University of Arizona), G. Baskaran (MATSCIENCE, Madras, India) and myself, on the other, about NP-complete problems with "golf course" energy landscapes—landscapes that are flat everywhere except one point! Furthermore, proofs of NP completeness in computer science often refer only to the worst possible case, and some NP-complete problems do not look very hard in generic terms. Finally, the computer scientist discusses—for obvious reasons—the problem of finding the *exact* answer for a particular case, not the average answer correct to order N for the generic case.

Nonetheless, specifying exactly the structure of the landscape of energy values as a function in the 2^N-dimensional space of spins tells us a very great deal about such problems. For instance, the existence of a transition temperature T_c tells us that below some value of energy per site E_c the space bifurcates into regions corresponding to different "solutions," and that as we go lower and lower in energy (or temperature) the space breaks up more and more. This gives us a clear reason why such a problem is "exponentially" hard: If we are in the wrong region, we have to cross an energy and entropy barrier of order N to get a better solution. This kind of

Philip Anderson is a condensed matter theorist whose work has also had impact on field theory, astrophysics, computer science and biology. He is Joseph Henry Professor of Physics at Princeton University.

© 1989 American Institute of Physics

PHYSICS TODAY   SEPTEMBER 1989   9

REFERENCE FRAME

"freezing" phenomenon had been conjectured by computer scientists but never rigorously proved. To counter it, they had evolved a number of heuristic techniques for getting approximate solutions. We now know why this was necessary—namely, to get over the high barriers and sample the entire space of solutions.

Almost the first effect of the kind of thinking developed to understand spin glasses was to provide a *new* heuristic algorithm for the solution of complex optimization problems. That algorithm is called simulated annealing, and it was introduced by Scott and his colleague C. Daniel Gelatt Jr. Kirkpatrick and Gelatt proposed that one imitate the procedure the spin glassers had already been using, of "warming up" the problem above T_c and slowly cooling it back down, or "annealing" it. This could be done by regarding the "cost" for a given problem—say, the cost of connections on a chip—as a "Hamiltonian" function C of the positions to be varied. One plugs this Hamiltonian into a statistical mechanics simulator program, such as the well-known Metropolis algorithm. Then one chooses an appropriately scaled "temperature" T and minimizes $<e^{-C/T}>_{ave}$ for increasingly low temperatures. Simulated annealing, it turns out, is the most effective algorithm only for certain problems, but where it works it is very good indeed, and it is already in regular, profitable commercial use. The question of *why* simulated annealing works as well as it does was approached theoretically by Miguel Virasoro, who showed that, at least for the SK model, the lower the energy of a solution is, the larger is the entropy associated with it near T_c. That is, deeper valleys have bigger basins of attraction near T_c, and so one is more likely to start out in such a valley at T_c.

To me the key result here is the beautiful revelation of the structure of the randomly "rugged landscape" that underlies many complex optimization problems. Physics, however, has its own "nattering nabobs of negativism" (in the immortal phrase of William Safire), and they recently have been decrying the importance of the ultrametric structure, saying that it is a property of the SK model, not of physical spin glasses. Such criticism misses the point: Physical spin glasses and the SK model are only a jumping-off point for an amazing cornucopia of wide-ranging applications of the same kind of thinking. I will write about this in the next—and I hope the last—of these columns. ∎

REFERENCE FRAME

SPIN GLASS VII: SPIN GLASS AS PARADIGM

Philip W. Anderson

In my last column on spin glasses (September 1989, page 9) I tried to show you that the exact solutions of a particular spin glass problem, by Giorgio Parisi and Gérard Toulouse, gave us great insight into the theory of complex optimization problems, as well as an algorithm for solving some of them. One such problem, which has been exhaustively studied by methods of spin glass theory, is the graph partition problem. This is the question of how to divide an arbitrary graph into two pieces, cutting the fewest possible bonds. My student Yao-Tian Fu (now at Washington University, St. Louis) initiated the study of the graph partition problem by replica theory. This classic problem of complexity theory was difficult to solve for sparse graphs by those methods, but another of my students, Wuwell Liao, seems to have done it.

Even more interesting than these applications to complexity theory is the way apparently unrelated areas of science have been stimulated into parallel growth by the spin glass work. John Hopfield (Caltech), who was instrumental in bringing me to Princeton in 1975, became interested in models for neural networks and brain function about 1979–80. It was natural for him to realize that complex, interconnected systems of simple units could have the "rugged landscape," multistable properties of spin glasses. Using, very ingeniously, an ad hoc and apparently unrealistic assumption of symmetric coupling between neurons, John got the following results:

▷ For a given set of coupling synapses (interactions) J_{ij} between neurons (spins) i and j, the conventional McCulloch–Pitts model of neuronal interactions maps onto a "greedy" algorithm for finding the local ground state of a corresponding spin glass. ("Greedy" is the computer scientists'

Philip Anderson is a condensed matter theorist whose work has also had impact on field theory, astrophysics, computer science and biology. He is Joseph Henry Professor of Physics at Princeton University.

self-evident jargon term for jumping directly to the lowest local energy for each spin.)
▷ Modifying the couplings, or choosing the J_{ij}, in such a way as to form the so-called Hebb synapses makes the neural network into a model for "content addressable" memory: a memory like our own, which can reproduce full detail from fragmentary information. A system of N neurons connected by $N(N-1)/2$ symmetric synapses can remember about $N/6$ N-bit messages in the form of locally stable "spin" (that is, neuron firing rate) configurations. (John made this conjecture about the capacity of a neural network on the basis of simulations, but it later turned out to agree with exact analytic theory.) Thus, in exchange for a capacity reduction of a factor of 3 relative to the information-theoretic maximum, one gets the content-addressable feature.
▷ Finally, several other brain functions, such as pattern recognition, could be modeled with the spin glass type of neural network.

Many of you may be aware of the gigantic growth of neural network science in recent years. In 1979, however, when I tried to whip up interest in John's ideas among computer scientists at Bell Labs, there was little response; and he, Alan Gelperin and John Connor got nearly equally short shrift among neuroscientists. Nowadays the neuroscientists and computer scientists like to point to prior claims for each component of John's achievements. I can hardly believe, however, that such further developments in neural networks as the revival of the perceptron would have occurred except as a response to John's beautiful demonstrations that, after all, one such system—the Hopfield neural network—does work and has a rigorous, mathematically respectable basis. In particular, John's work has generated a very healthy trend toward rigorous mathematical demonstrations of limits on capacity, accuracy and so on in neural networks and perceptron-like models, using the statistical mechanics methods provided by Toulouse, Haim Somopolinsky, Miguel Virasoro, John Hertz, Richard Palmer (who was John's associate at Princeton in 1975–78) and many others. It turns out that statistical mechanics can be applied to realistic, asymmetric networks as well, and that there is no real difference between the capabilities of symmetric and asymmetric networks.

I promised I would close my series on spin glasses with this column, so it must be descriptive, not detailed. But I must also mention how the "rugged landscape" of spin glasses relates to theories of biological evolution. In 1981 I visited John Hopfield at Caltech and helped with the course on "physics of information" that he, Richard Feynman and Carver Mead were giving. Stimulated by John's work on neural nets, I came back to Princeton with the realization that I could put my own rugged-landscape ideas into a theory of prebiotic evolution that Daniel Stein (now at the University of Arizona) and I were already working on. The genome of an organism can be thought of as a set of Ising spins—two for each base in the DNA because there are four types of bases and the Ising spin has two possible values. The fitness, or reproductive capacity, of the genome can be modeled by a frustrated, quenched random function of this list of spins, and the simplest random function that satisfies the requisite plus–minus symmetry is a spin glass Hamiltonian function. (The plus–minus symmetry is imposed by the complementarity of base pairing.) With a senior thesis student, Dan Rokhsar, Stein and I made a simple model of primitive evolutionary processes using this idea.

Related ideas, but without the statistical mechanics insights, had already occurred to Gérard Weisbuch at the Ecole Normale in Paris and to Stu Kauffman at the University of Pennsylvania. Nonetheless the extra understanding those insights provide has encouraged us, and especially Stu, to go on and attack all kinds of evolutionary—and other—problems with spin-glass-like random, rugged-landscape models. This approach has become an important part of the program at the Santa Fe Institute, of

REFERENCE FRAME

which Stu and I are members. Unfortunately I cannot discuss here the many other ramifications of this way of thinking. From my point of view, its attractiveness lies in that it allows us to explain simultaneously the contradictory aspects of variety and stability of certain special forms and patterns. Life exhibits these contradictory aspects: Among the countless possible mutations, many lead to stable species. In the language of spin glass theory, there are many "basins of attraction," but each is stable (or at least metastable). It is also clear, as Gérard Weisbuch first pointed out, that evolution in such a landscape will exhibit "punctuated equilibrium," or sudden changes from one deep maximum of fitness to another, a feature that has been emphasized recently as characteristic of much of evolution.

I realize that I never returned to "real" spin glasses, even though it was studies of the low-temperature properties of those dilute magnetic alloys that led to the theoretical ideas I have been discussing. There is a reason: In spite of much beautiful experimental, computational and theoretical work, a complete and consistent understanding of those materials is not yet at hand. Helène Bouchiat and Pierre Monod, among others such as Laurent Levy, have beautifully demonstrated that real spin glasses have divergent nonlinear magnetic susceptibility at T_{SG}, verifying that there is a real phase transition, albeit one without a visible specific heat singularity. Spectacular simulations carried out on special purpose machines by Peter Young (University of California, Santa Cruz) and Andrew Ogielski (AT&T Bell Labs) have also verified the existence of a phase transition in three dimensions. Daniel Fisher (Princeton) and David Huse (Bell Labs), among others, speculate, however, that real spin glasses are really *not* ultrametric or replica-symmetry breaking. The theory is still under development. Some of it was explored in the December 1988 PHYSICS TODAY special issue on disordered solids.

For further information on random landscapes and evolution, Stu Kauffman's forthcoming book *Origins of Order: Self-Organization and Selection in Evolution* (Oxford U. P., New York) is perhaps the best source. John Hopfield has written (with David W. Tank) a *Scientific American* article on his neural network ideas (December 1987, page 104). An article on neural networks by Haim Sompolinsky appeared in the December 1988 PHYSICS TODAY (page 70). ∎

Valence Instabilities and Related Narrow-Band Phenomena
ed. R.D. Parks (Plenum, New York, 1977), pp. 389–396

Present Status of Theory: 1/N Approach
Proc. of 1983 NATO/CAP Inst. "Moment Formation in Solids"
ed. W. J. L. Buyers (Plenum, New York), pp. 313–326

The Problem of Fluctuating Valence in f-Electron Metals
Windsurfing the Fermi Sea, Volume 1,
ed. T.T.S. Kuo and J. Speth (Elsevier Science Publishers, 1987)

These are three of the several reviews I gave of the mixed valence problem in the years from '75–'87. My students made appreciable progress, especially Haldane and Coleman, but I felt and feel the problem has never been sorted out. The first review states the problem; the second gives some vital advice to prospective solvers; and the third represents my final state of confusion.

Reprinted from: VALENCE INSTABILITIES AND RELATED NARROW-BAND
PHENOMENA (1977)
Edited by R. D. Parks
Book available from Plenum Publishing Corporation
227 West 17th Street, New York, New York 10011

EPILOGUE

(transcribed from tapes and edited by R. D. Parks and M. C. Croft)

P. W. Anderson

Bell Laboratories, Murray Hill, NJ 07974 and
Physics Department, Princeton University, NJ 08540

When Ron (Parks) asked me to give a summary of the meeting, I said no, because: (A) I couldn't, (B) I would have had to stay awake through all the talks which I am sure everyone has noticed I haven't and (C) I wasn't really sure I wanted to do it anyhow. Nor do I think you need a summary of what everyone has already said. So we decided on the name epilogue which is something you say afterwards, and as far as I know means nothing else and left me quite free. So here are some of the things I wanted to say. First is a private comment. One name on the organizing committee was mispelled. I don't know how this could have slipped by. The name on the organizing committee was actually C. M. Varma; it was spelled P. W. Anderson. Given that reconstituted organizing committee, there is a second comment I wanted to make, which is (now) a message from the organizing committee. If we are to have another good conference on mixed valence compounds, one way we could get it (probably too) well attended is to point out that among other things almost all good catalysts are mixed valence compounds, and perhaps we should begin to think about that; and, even more important, if we want the next conference to be well attended, we should publicize that fact.

Thinking about the conference in general, it seems to me that the experimentalists have been very successful. They have an enormous number of substances which are undoubtedly undergoing valence fluctuations, and there are many phenomenological and experimental ways of looking at these things. The experimentalists have noted that their theoretical colleagues are scratching their heads and not really getting very far with any of this. And naturally when you have got through with your job, and the other guy's

389

standing there and he seems to be all thumbs, the temptation is to look over his shoulder and offer helpful suggestions. It seemed to me that, during the first couple of days at least, there were a good many helpful suggestions offered from the experimentalists, and as a result there were fewer experiments discussed than some of the theorists might have wished. Even the angels have feared to tread in this field, or at least they're treading very lightly, and perhaps we need even more experiments. On the last day the experimental talks were beautifully devoid of any theories except possibly a right one or two.

After that remark, which I probably shouldn't have made, we come to what the theorists are actually puzzled about; and, a typical object which we could take as one of the main subjects of our concerns here is the resistivity curve of CeAl3 or some such compound (See Fig. 1). I've kind of decorated this curve with all the different kinds of things people discussed. I haven't

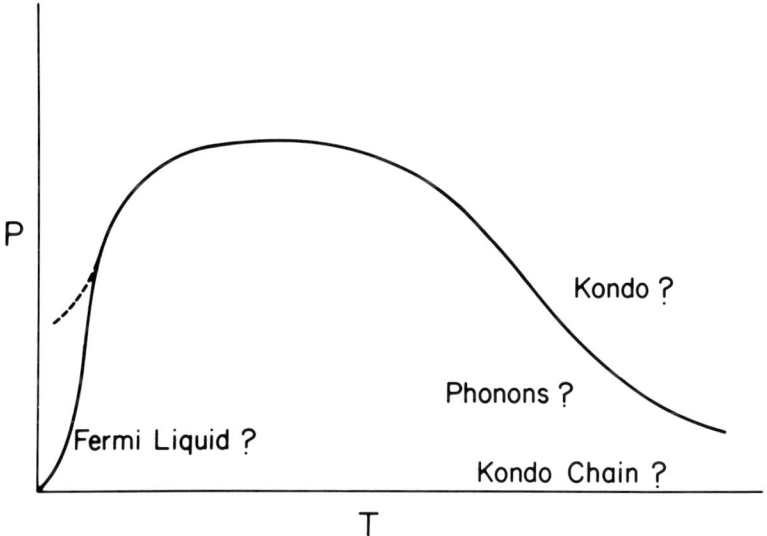

Fig. 1. Resistivity stereotypic of systems such as $CeAl_3$ (see text).

EPILOGUE 391

got everything in there but I have got Kondo over here on the high temperature end and the Fermi liquid at the bottom, and in between either phonons or the Kondo chain or both. Thinking about this and the various other theories, reminded me of the parable (Hindu, I think) about the elephant and the seven blind men. Seven blind men come up to an elephant and one of them pulls his tail and says it's a rope, and another one pulls its trunk and says it's a hose, and a third one says it's a tree, and a fourth one says it's a fan and so on, and it's obvious we have that problem here in this field, so I thought I'd make it explicit (Fig. 2) (extended Laughter). It's obvious that at the low temperature end of the elephant you see a Fermi liquid. If you're standing near the mouth, that's where David Sherrington and Peter Riseborough are listening to the phonons. The rest of it is fairly self-evident. (Question from audience: who is the fellow on the trunk? Answer: Well, it's Kondo, obviously. Even though he isn't there, his name has been invoked often enough.) So, very briefly, I am going to run through various more or less serious questions about this elephant and all the different aspects that you can look at.

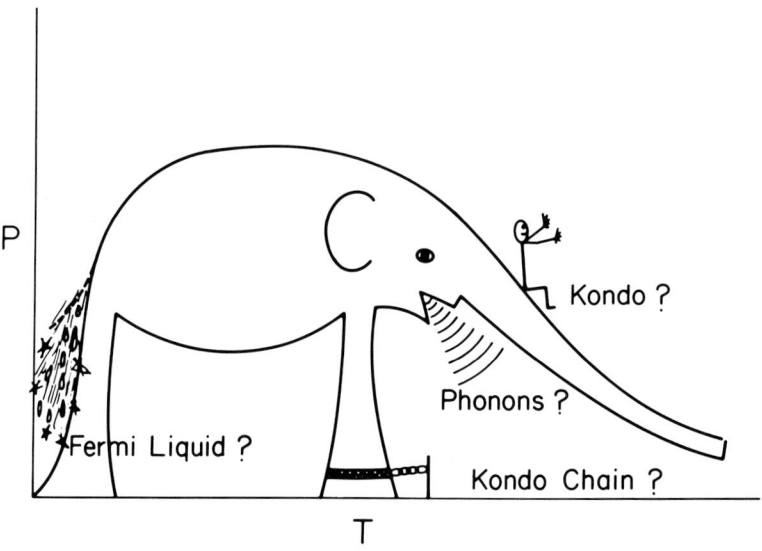

Fig. 2. Elephantine version of Fig. 1 (art work by PWA).

I will now give you one blind theorist's look at the elephant or at least a series of questions I would ask about him. I can't talk very convincingly about what various other people see in this particular elephant. The first question I note is: where is the elephant? And that, we could remark, is a question about phenomenology and quantum chemistry, and more or less under that heading we can include the question of transitions into and out of the mixed valent state. All of these are very interesting and very important questions, particularly for the experimentalist, so that he will know when he is in the presence of this particular elephant and when he isn't. For instance, John Wilson's empirical classification scheme based more or less upon energetics is very useful for understanding to some extent where the different valence states are going to come into rough equality of energy. Basically, one must have as well a quantum chemical picture of how narrow the f bands are and when different configurations are going to come into equilibrium, at which point you will have one or more of the various mixed valence phenomena. It is also interesting to study the transition into and out of the mixed valent state, but from my point of view again, I would put this in the category of approaching the general neighborhood of the elephant as opposed to really studying his intrinsic nature. But, that's one particular blind man and the other people would see that as a much more fundamental thing. So the work on Ce on that kind of thing and the work on the SmS black-to-bronze transitions, which of course are very important, are more or less telling you when you have got into this peculiar state rather than telling you necessarily anything about the state.

Next question: why doesn't he run away. There are a lot of questions about why the elephant doesn't run away, which is a very serious problem for me, at least for the part of the elephant that I see as the core of the problem. The core of the problem is that on a microscopic level you have some atoms at some time which are in one valence and some atoms which are in another valence. They are flipping back and forth at rates that can be either thermally activated, as you heard this morning, or more interestingly, you can have quantum mechanical tunneling back and forth between the different mixed valent states. And yet it is quite clear that when we speak of a mixed valent system we are speaking of a system where the different configurations have in some sense true separate identities. The easy way to see that they have separate idenitities is to look at the kind of XPS data that was presented this morning; that, you understand more or less by the kind of purely theoretical ideas that Hirst has talked about. You get from one configuration into the other by tunneling an electron into or out of the f level, but once you are into it (the mixed valence phase), it is a beast with a lot of internal correlation; and, in particular, it is clear that the two configurations have very different atomic sizes. So, you have this compound, or pure metal, in which objects with different atomic sizes mix at a microscopic level and yet in general do not order. Nor do they segregate. There are two questions in this connection, neither of which has been discussed at this conference. The two questions are: why doesn't the beast run away either (1) by ordering the valences or (2) by segregating the valences. The Friedel, Hume-Rothery theory of alloys would have said that they segregate. The only answer that has been given to that is in a dubious and (I would consider) thoroughly incomplete paper by Chui and myself about phonon effects, which talks about the non-segregation, but it really addresses the question of nonordering to some extent. It's still not obvious why the different valencies don't order. You (almost) obviously gain energy by letting them sit where they want to sit, viz., some ordered way, and if you think hard about the elastic theory, its very tricky that they don't segregate either.

<u>Fundamental</u> <u>model</u> <u>for</u> <u>the</u> <u>elephant</u>. Now we get again to what I consider to be the key aspect of the phenomenon, which is contained in the work of Haldane and similar thoughts by Peter Riseborough (see poster session contributions). Given these ideas you can almost see how the relative position of the two f levels and the fermi level could get pinned in such a way that two stable configurations (stable within conventional Hartree Fock theory) would be pinned to have exactly the same energy within Hartree Fock theory. That I think is an important key to the whole idea of the problem: that within Hartree-Fock theory, looking at things from the point of view of the energetics and time scales of somewhat less than 1 ev, there is a mean field theory which one can think of as Hartree-Fock, which has two separate kinds of states with equal energies on a microscopic level, viz., the non-magnetic state on the one hand and an f level with all the different orbital and spin quantum numbers on the other. At least within Hartree-Fock theory they would be pinned together. Then again you could have various things that you must do with this object in order to understand what really is happening experimentally. This is a fundamental model from which we must all start, and then we must work out a program from this kind of model: various pieces of the program were discussed at this conference. The program's first question is the problem of quenching, which the field theorists might call the idea of <u>confining</u> the f quantum numbers, in some Kondo-like way. One starts doing this of course, with an individual mixed valence impurity, with say one Ce in a background of $LaAl_3$, rather than a whole lot of Ce atoms. But even the one Ce is already a problem, and Wilkins and various other people have thought about this in terms of the asymetrical Anderson model. There is a problem here in even the one impurity problem, for as Duncan Haldane said, "though you do it for one impurity only, its very likely that this is okay in some range of temperatures and in some systems for the concentrated case as well (i.e., for where you have 1 Ce at every site instead of one Ce isolated in a lattice)." The reason being

that at high temperatures the fluctuations will be incoherent; and also, because of the admixture affect or some such effect as Tom Kaplan talked about, the actual number of true renormalized f-electrons (magnetic configurations) may be much less than the nominal number of magnetic centers.

Now we come to the heart and core of the elephant, the part which nobody has really done, which was first mentioned at least as a serious problem here in this conference, namely the Kondo lattice, which Seb Doniach made a start on. What you really have is a lattice full of these objects that fluctuate back and forth from one valence to another. And these are all coupled in lots of ways. There are the phonons, there is the fact that the electrons fluctuate by tossing electrons into the d level on the next site which can then go down into the f levels on yet another site. So the things which toss the valences back and forth are definitely coupled between one site and another. The net result of doing this is something that most of the experiments have told us about: that this probably renormalizes to a very heavy Fermi liquid theory with some kind of strong antimagnetic prejudice in that the f-like objects in the Fermi liquid have somehow lost all of their desire to be magnetic and don't very easily order anymore. This is the extremely hard problem, it's a problem in the same category of problems which are failing to be done in field theory these days, and sure enough it is failing to be done properly in our problem too. Then even when you get down to the bottom (of the elephant) no one has even done the fundamental Fermi liquid theory of what you get there. Does it have a d-band and an f-band, what is the nature of the Fermi surface, how might you calculate it and so on? There is this feeling that when you get over the back of the elephant and start down, at the end you get a Fermi liquid.

In this conference I, at least, began to realize that there is yet one more serious question that we have to consider. Once you get down to this new Fermi liquid, it seems that there is a serious question of then what happens? What does the resulting heavy Fermi liquid do with itself, what further transformation might it undergo? There are several possibilities. One of them which seems to be a very likely one in many cases was mentioned by Kasuya. Namely, this heavy Fermi liquid can localize because the bandwidth is so narrow and because the random potentials can be quite large in these systems. There is no reason at all why it shouldn't localize and maybe there are cases where it localizes. Kasuya gave an argument for one of them. A second possibility a whole series of experiments seem to indicate is that some phase transition takes place in many cases. The question is: what is the nature of these phase transitions? I for one am not ready to accept the idea that they are all simple magnetic phase transitions,

EPILOGUE

e.g., exchange-induced antiferromagnetic phase transitions, as was suggested here by Tournier. He has a theory which shows you in a very concise way how you might get that in a limited range of parameters, but it would (to my mind) be very coincidental that if it does happen, it happens even once, let alone, often. You're getting down to this region where obviously the magnetic character of the f-electrons is dying off; and it's only after you've gotten to a thoroughly non-magnetic region of the f-electrons that all of a sudden the thing decides to order magnetically. They order magnetically in a very nonmagnetic way. If it was going to order magnetically, why didn't it order at some decent temperature with a decent moment? Why would it order at a low temperature with practically no moment? So, maybe it is something else. This heavy Fermi liquid still has some interactions in it other than the ones that were renormalized out in the course of doing the mixed valence job. Maybe there are density wave states in there or maybe there are other kinds of phase transitions. One that nobody mentioned here, fortunately, that I will bring out from under the rug is the possibility of a d to f excitonic phase transition. Maybe there's some kind of d to f excitonic phase transition that either does or does not leave some Fermi surface behind. Maybe there's a density wave. What else?

Another question is brought up by the fact that $CeAl_3$ seems to have in some people's data a phase transition and in other people's data it does not. How many cases are there in which we're quite sure that there is no phase transition and that all we get is the clean renormalization to the state with very low resistance at low temperatures? I mentioned a few possibilities in a slide: $CeAl_3$ perhaps, certainly Yb compounds that weren't much talked about here, etc. An finally, since time is running out, let me mention that last horrid question of what _does_ happen with the phonons: what happens (A) with the optical and (B) with the acoustic phonons? To what extent are the things, which are involved in the screening process that gives the instability in the mixed valence model, phonons and to what extent are they Tomonagons; and how do the phonons and density waves couple? There are lots of fascinating questions here. There are the results of Mook which show that there surely are couplings between phonons and all these phenomena. There's also the questions of whether the Fermi liquid renormalization always take place below the phonon frequency (as it does in many cases) and the Kondo part of it always take place above the phonon frequency. Is that a general rule and is there a reason for it? Those are a number of questions. What is clear is that maybe we have the answer and maybe we're making progress, or maybe the theorists haven't found the elephant at all. So that's my particular view of what is going on (or not going on as the case may be). Clearly there are many more unanswered questions than there are answered ones.

(Note added by PWA: a last slide was omitted in the above which was decorated with a heart and said "Thanks, Ron!")

PRESENT STATUS OF THEORY: 1/N APPROACH

Philip W. Anderson

Bell Laboratories
Murray Hill, N. J. 07974
and
Joseph Henry Laboratories
Princeton University
Princeton, N. J. 08544

When I began to prepare these lectures, and to wonder what I could possibly say that hadn't already been said by all the brilliant lecturers you have already had, I began to realize how many years I have actually been associated with the problem of "Moment Formation". As I have often explained, I don't actually quite go back all the way to the first exploration of this problem, since the virtual state concept was originated in fact by Blandin and Friedel in early 1959[1] or so, and I heard a discussion of it by Blandin in 1959 at Oxford, which stimulated me to develop the model of which so much is made in the mixed valence field.

In all of these years I have learned a number of theorems and quasi-theorems about the moment problem, and developed a rough physical understanding of what is going on, which is usually proof against the great complexity of what we see in real systems. In fact, until recently I had come to think that the mixed valence problem was almost a trivial exercise in physical understanding, but more recently I have come to believe that there are real and very interesting complexities which are much worth discussing, and which do indeed go beyond the primitive, fundamental understanding which one can achieve with simple physical concepts.

Let me list some of these concepts in some kind of order, mostly historical, so that you will get some idea of where I am going and whether or not to go off and play tennis instead of listening.

(1) The Compensation Theorem. Concepts of admixture and of polarization and that admixture ≃ polarization: so net Δm is in localized state.

(2) Friedel sum rule and general idea that chemical energies >> magnetic energies: net occupation is fixed, only dynamics changes. (For "impurity" systems and Coulomb interactions this leads to the Friedel overall sum rule, but in fact mostly each type of electron is conserved: "rule of fixed valence".)

(3) Friedel and Luttinger theorems. Friedel: number of electrons of a given type = (sum of phase shifts/π); Luttinger: Fermi surface volume = number of electrons. General, relating statement: charge of quasiparticles is not renormalized: no e^* (in spite of Falicov).

(4) Scaling ideas.

(A) Ground State Fixed-Point Theorem: Ground states are fixed points of scaling. Three types:

(a) Fermi Liquid
(b) Self-trapped (Ferromagnetic Kondo)
(c) Broken Symmetry. (Magnetic or Superconducting)

(B) Kondo Scaling: From magnetic impurity to Fermi liquid.

(C) Haldane Scaling: Sinking of the singlet and the role of degeneracy.

(5) Finally, the $1/N_f$ idea: weakness of spin-spin interactions and validity of perturbation theory ensured by large degeneracy of the f level.

To complete my preview of what I'm going to say, I should list my more recent problems and thoughts about why it is not all so simple as that:

(1) The Haldane-Kondo crossover and emergence of the "Kondo resonance" in experiment and in 1/N theories.

(2) Absence of phonon effects: is this wavefunction renormalization and the weakness of the Kondo resonance?

(3) Superconductivity: what is the role of finite U and the N = 2 singlet?

----

(1) Let me start out, then, with the compensation theorem and

314

the birth of the Anderson model. The compensation theorem is contained in two abstracts by Al Clogston and myself written in late 1960,[2] trying to understand magnetic resonance data on such magnetic impurity systems as Mn in Cu. Magnetic resonance, both nuclear and electron, was what we had in those days to study magnetic systems, and there was a great puzzlement as to what extent the free electron spins were part of the local moments of the Mn, to what extent they relaxed them (the Korringa relaxation and the Kondo model of free-electron Mn-spin interaction, actually introduced in those days by Yosida), and to what extent the Mn spins polarized the free electrons. These abstracts were the origin of the Anderson model, which was there first studied by simple perturbation theory. But what they emphasized is that a local (say "f") electronic state has, in simple perturbation theory, two effects on the free electron band, not one: first, its wavefunction is admixed into the free electrons' wavefunctions and vice versa, which leads to no net change in moment but a change in the wavefunction or form factor, interchanging spin in the two states; and second, the effective energies of the free electron states are modified, shifting them relative to the Fermi level and causing a net polarization which - since it effectively takes place at $r = \infty$ - is surely only in the free electrons and is antiferromagnetic due to the fact that the filled spin state is below the Fermi energy and repels electrons. It is a common tendency to neglect one or the other of these two effects - in the case of mixed valence, most often admixture, which in many cases gives the whole of the effect - but between the two of them, in the case of a reasonably broad band they approximately cancel and the appearance is of a net decrease in local moment with no effect on the surroundings - a result which usually remains true right through the Kondo and mixed valence regimes, and has confused and disappointed successive generations of NMR specialists. This is why the recent experimental results on $CeSn_3$ are exciting.

(2) The second basic physical concept which is quite important and has not been adequately treated is the question of sum rules and chemical stability. The simplest and most rigorous expression of this idea is the Friedel sum rule, which simply expresses the idea that a metal must always be locally neutral. (Even an insulator normally sustains only slight deviations from local neutrality.) The main "Anderson" and "Kondo" models do not contain charge neutrality automatically and must be carefully handled to do so, which is the first reason which makes "rigorous" solutions of these models somewhat less relevant than is apparent on the surface. The phase shift sum rule of Friedel which I will shortly discuss is absolute. More generally, it is very rare that the valence state of an atom really changes substantively as a function of ordinary temperatures or pressures. The energy differences among valence states, while they can be relatively small, are still expressed in electron volts rather than degrees, and except in very exceptional circumstances, a true "valence transition" by a whole unit will

315

almost never really take place. This is true even in much simpler and less obvious cases: for instance, the valence state at Na in Na metal and in Na^+Cl^- is essentially indistinguishable within a sphere about the Na ion of size equal to the metallic radius, and similarly for Cl.

The distinction between real and nominal valence is like that between "bare electrons" and "quasiparticles" in Fermi liquid theory. The number of "bare" d or f electrons around an atom is very firmly fixed to within at most a couple of tenths of an electron; but the nominal a valence can change quite radically. The most spectacular example of this is the case of transition metal impurities in semiconductors, where magnetic valence can change by several units, while, as Haldane and I suggested[3] and George Watkins proved,[4] the actual number of d electrons changes not at all.

The nominal valence is describable in a number of ways. One way is to ask what the appropriate starting point would be for a perturbation theory which would continuously connect with the actual electronic state of the system. For NaCl, for instance, the appropriate starting point is a set of Wannier functions resembling the Cl^- atomic functions; these will spill over into empty s functions on the Na^+, but one could essentially start putting the solid together from Na^+ and Cl^- ions and never encounter a discontinuity. Na metal, on the other hand, would be best described by starting from Na^+ ions and a free electron gas.

It is in this sense that a substance like SmB_6 or SmS in the insulating state has a nominal valence Sm^{++}, even though when one studies the actual number of f electrons on Sm it is more nearly equal to Sm^{+++}, and the metallic state of SmS has a nominal valence intermediate between the two. In the pressure transition the number of bare f electrons does not actually change very much, as is also the case in Ce metal, because in spite of the large volume change there is not much electronic energy involved.

The final discussion on Friday left the subject of valence in considerably clearer state than my first thoughts above. I think my definitions above remain clear and valid, but it was pointed out that the <u>real</u> valence can change rather radically under some circumstances, as, e.g., in the pressure transition of SmS.

The conclusion we came to is that in normal chemical bonding the matrix elements connecting the various atomic states are much more sensitive to volume or crystal structure than are the atomic states, and hence what will usually happen is a change in the bonding character - in the sense of ionic to covalent, for instance, as in one of the borderline II-VI compounds. A perfect example of this kind of thing was given by Wohlleben, where he shows that the large volume change in Ce metal is accompanied by a very large change in the mixing matrix element and hence in T_F, but little or no net change in valence. That is, the f shell is changing from essentially metallic to ionic bonding character.

On the other hand, where the matrix elements are and stay small, either because of strong intra-f-shell correlations as in Sm or because of a smaller f shell as in Yb, the real valence can change fairly radically. However, it is interesting that in few or none of these compounds does <u>real</u> valence change with T. For example, in SmB_6 there is a big change in nominal valence as defined above, from Sm^{++} at low temperature to 2.6 at high, as determined from the magnetic symmetry character, but no change at all in real valence.

At this point it is perhaps necessary to discuss a side development which has not yet taken its proper place in the mythology or, to be quite honest, in my own mind, the "first order" valence transitions which seem to appear in Hartree-Fock theory when one takes screening properly into account. These were first emphasized by Duncan Haldane[5] and a more realistic, but physically less satisfactory, model has been studied by Schluter and Varma[6]. I think Haldane had the physics correct in his oversimplified model and I will follow him.

Haldane's way of dealing with them was to observe that screening of the "f" charge by the s-d channels could be quite accurately modeled by including a corresponding Falicov-Kimball type of term of form $\lambda n_f n(o)_{free}$ in the Anderson Hamiltonian and then writing $n(o)_{free}$ as the sum of the amplitudes of a set of Tomonaga bosons. In a static Hartree-Fock approximation that just gives you an effective "negative U" contribution $-c(\bar{n}_f)^2$ to the f-electron energy, and this, when balanced against the magnetic energies, makes the valence transition of the Anderson model somewhat first order where, in the simple case, it is second order (I show you, in Figure 1, typical phase boundaries). This is not surprising since

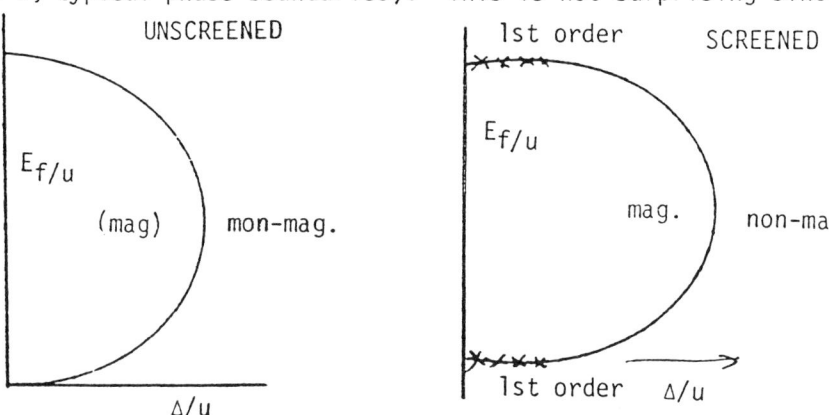

Fig. 1

at the edge of the magnetic region n_f is changing vertically with the parameter E_f and one might expect an effective negative U to recurve it.

I need hardly remind you that much physics is missing here, and specifically that the original "second order" transitions were spurious in that quantum fluctuations spread them out into an entire Kondo-like crossover phenomenon. This will also be true of the first-order jumps. Nonetheless, I believe with Haldane that the screening phenomenon plays a role of considerable but unknown magnitude in at least two ways: in sharpening up the valence transition, so that more physical cases occur near to it, and in reducing the effective f-level width Δ. One may in fact question whether these two roles cannot be combined by simply remarking that all calculations - such as those of Piers Coleman - seem to give f-level mixing integrals V_{ff} and Δ's which are too large, from both points of view, and the renormalization due to screening is most welcome.

When we ask ourselves - as I often did in the early days - why, if the parameter E_f needs to be so very close - within a few hundred ^{0}K - to the central value, so many valence transitions take place, this kind of scale compression is perhaps useful. But actually I now feel this is a non-problem, since the real physics should be much more directly based on the parameter $\bar{n}_f$, which leaves one looking at the Anderson model graph from quite a different perspective (see Figure 2). Screening makes it even flatter.

The confusing f wavefunction transition of Varma and Schluter appears in their model because they do not adequately allow for an f-character free electron channel, which is necessary to accommodate the change in f amplitude in the Haldane first-order transition.

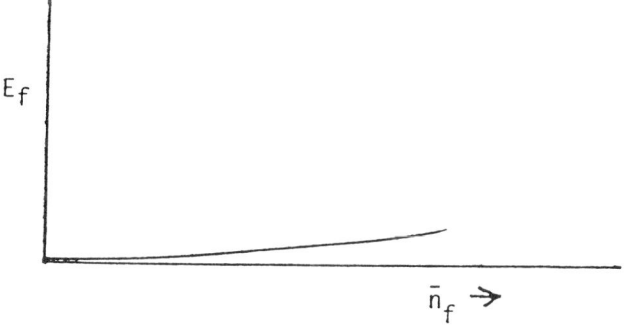

E_f varies slowly with $\bar{n}_f$

Fig. 2

(3) The third basic physical principle is the rigorous theorems which follow from the fact that interactions cannot modify the charge of a quasiparticle, i.e., they conserve numbers of electrons. This means that the "real" state and the "nominal" state must refer to exactly the same number of electrons. For the nominal, or "reference" state, the number of electrons may be obtained essentially by counting nodes of the wavefunctions of the last occupied electronic states, by an ancient theorem of differential equation theory. This is the real source of the Friedel and Luttinger theorems. In the case of an individual atom in a metal, the form such a "node-counting" theorem takes is the Friedel identity for the sum of scattering phase shifts:

$$n = \left(\sum_{\ell,m} \delta_{\ell,m}\right)/\pi$$

Thus, if, as in the Kondo system, there is a transition between a magnetic state at high temperature and a non-magnetic one at low temperature, the sum of the scattering phase shifts for the two signs of spin must remain constant and as a result the two values $\delta_{occ} = \pi - \epsilon$ and $\delta_{empty} = 0 + \epsilon$ (giving $n = 1$) must change to $\delta_\uparrow = \delta_\downarrow = \pi/2$. It follows that the resistance due to magnetic scattering

$$\rho \propto \sum_{\ell,m} \sin^2 \delta_{\ell m},$$

must rise spectacularly and reach the unitarity limit for the appropriate channels. This theorem is especially effective in the mixed valence case, in that, once we recognize that the number of f channels is large (n_f = 6, 8 or 14 depending on spin-orbit coupling) and that we are always interested in accommodating only one electron at most, we recognize that in the non-magnetic state $\delta_f \simeq \pi/n_f$, or that if there is a single "f" scattering resonance, the Fermi level must lie on the edge of that resonance. This is the primary source of the advantage of "large n_f" methods: that the resonance never actually sits right <u>at</u> the Fermi level. (Although, as we will see, in the "Kondo limit" the resonance may be very close by.)

The corresponding theorem for the mixed valence or Kondo lattice problem is the Luttinger theorem: that the volume of the Fermi surface must remain unchanged - the volume of the Fermi surface in k space again being a sophisticated way of counting nodal surfaces of wavefunctions. The most spectacular instance of the operation of this theorem occurs in the mixed valence insulators SmS and SmB_6, where Sm, in the mixed valent ground state, assumes the "nominal" valence 2, even though its real f occupancy is closer

319

to Sm^{+++}. Hence it assumes a non-magnetic, $J = 0$ ground state, and the electron count becomes exactly the equivalent of a simple ionic compound $(X)^{++} S^{--}$ or $(X)^{++} (B_6)^{--}$. Since B_6^{--} and S^{--} are closed shell ions, there can be - and, in fact, should be - a gap at the Fermi surface, which, though extremely small, is observed in both cases. The equivalent, though less spectacularly, will in general hold for all non-magnetic mixed valence ground states: the Fermi surface volume and symmetry will be the equivalent of that for an ordinary non-magnetic atom of the same <u>nominal</u> valence, even though the effective masses can be very much different.

One may very heuristically relate the two phenomena by observing that in the various forms of Green's functions or muffin tin-based methods, and even more in pseudopotential methods, the atom is generally replaced by its effective nonlocal potential, which is simply the phase shifts or quantum defects for the different angular momentum channels. For states precisely <u>at</u> the Fermi level, these must add up to give the appropriate number of f electrons by having the appropriate phase shifts in the f channel.

On the other hand, you will note a slight difference between the single atom version of these theorems, and the coherent one which is going to hold for the ground state of a regular crystal in the Fermi liquid regime. In the incoherent, atomic case, the f resonance actually intersects the Fermi surface and there are necessarily a small number of "nominal" f electrons, i.e., f quasi-particles, although they exist only as the edge of a scattering resonance, admixed into the free electron band. In the Fermi liquid case, recognizable bands of tight-binding, f-like symmetry need not, and in general will not, cross the Fermi surface; instead there will be a strong f-like pseudopotential which affects the ordinary s-d bands of the background metal.

(4) Scaling ideas.

About 1970 we began to understand that the secret to the conceptual understanding of the various Hamiltonians which had been proposed for moment formation problems is the basic idea of scaling or renormalization. This actually was the first entry of the renormalization group into condensed matter physics, quite independently of, and slightly before, the Wilson work on phase transitions. Although this was by no means the way it came about, the great liberating concept which scaling gives us is the idea of states as representing fixed points of a scale transformation, and in particular of ground states as the various kinds of possible stable, low-energy and low-frequency fixed points of a quantum system of the appropriate kind. We know that at high enough temperatures, the magnetic state will try to arrange itself so as to have maximum entropy, by decoupling whatever coupling constant (J in the Kondo case, or V_{mix} in the Anderson model) is involved: but

320

the high-entropy decoupled state usually turns out to be an unstable fixed point and the system goes through a crossover to a stable fixed point.

The appropriate stable fixed point for the single impurity problems, both Kondo and mixed valence, is the so-called Fermi liquid limit. A local spin interacting with a Fermi sea cannot be a low-energy fixed point because of the quantum process of spin-flip scattering. Thus by far the most common and important $T = 0$ fixed point is the Fermi liquid: the replacement of the magnetic spin by an effective non-magnetic center. For visualization purposes, we imagine the local spin binding a free electron into a bound singlet state, which then provides a rather strong scattering center for the remainder of the free electrons, but has no internal dynamics at least at $\omega \cong 0$. The strength of the scattering center at the Fermi surface is controlled by Friedel sum rules and symmetries, as was beautifully explained by Nozieres in what may be the single most important paper ever written on the Kondo effect.[7] For practical purposes, MV centers will have π/N_f scattering.

For the lattice of Kondo or MV centers a possible fixed point is also the unmagnetized Fermi liquid or, in the case where the valence is satisfactory, simply a non-magnetic gapped insulator. We emphasize that in this case the vital sum rule is the Luttinger one - the Friedel sum rules control the pseudopotentials but not the actual Fermi surface - a rather tricky point. Bands do not have pure f or d character, and a valence band which is perfectly normal and effectively non-magnetic can compensate for the extra f-electron phase shift which must lie around say a Ce atom.

Of course, in the lattice case magnetic or superconducting Fermi liquids are also allowed. As we shall see - and as Rama will emphasize for SmB_6 - either external magnetic fields or exchange fields can seriously disrupt the scaling process, and the band version of this is to point out that in the presence of an exchange gap, the magnetic state can of course become an infrared stable fixed point. In fact, any broken symmetry SDW, CDW, or other state can intervene once the Fermi liquid becomes coherent between sites.

To anticipate the 1/N physics which you have been hearing about, it is important to note that the establishment of coherence, and in fact all interatomic interactions, are basically a 1/N effect: the electron goes out in 1 of N channels and has to come back in the same one to be coherent. Thus we hope there is a temperature regime of incoherence where the single impurity is a valid point of view, above a temperature regime where coherence and order sets in.

Finally, there is indeed the possibility of a _decoupled_ ground state - the ferromagnetic Kondo is the canonical case, but we can also imagine a self-trapped small polaron type of state;

321

such states are of importance in other problems and the chemists' mixed valence compounds are usually of that sort - but for the nature of our conference we ignore them.

(B) Kondo Scaling.

The "Kondo" scaling is appropriate when the mean field theory of the Anderson Hamiltonian predicts a stable moment - i.e., well inside the original Hartree-Fock stability region. In the f-series the appropriate spin interaction was derived by Coqblin and Schrieffer. The physics of Kondo is that in the normal case of antiferromagnetic interaction, the spin attracts the electrons with which it can undergo spin-flip scattering, hence increasing the effective interaction indefinitely for the lowest-energy electrons. The simplest way to do the first stages of this scaling is by the "poor man's method" of continually reducing the cutoff D, leading to the famous scaling equations for the two (dimensionless) coupling constants $J_\pm$ and J_z, which are of course equal for isotropic exchange:

$$\frac{dJ_\pm}{d \ln D} = -J_\pm J_z \qquad (+ \text{ higher order})$$

$$\frac{dJ_z}{d \ln D} = -J_\pm^2 \qquad (+ \text{ higher order})$$

These lead to the famous logarithmic Kondo terms in resistivity, etc., when the scaling is carried out using kT as a lower limit on D. It was first recognized by Anderson, Yuval and Hamann that large J is essentially a weak coupling limit in the equivalent space-time scaling problem where one follows the history of the local spin in imaginary time rather than the coupling constant in energy space, and is hence trivial. The crossover from small to large J limits was calculated approximately by Anderson and Yuval and reasonably accurately by Armytage and then, in a famous paper, by Wilson. The space-time approach is an enormous aid to visualization and also, in itself, leads to interesting statistical problems which are of some recent interest (see J. Cardy,[8] and B. G. Kotliar's thesis[9]).

Nozieres outlined the general nature of the ground state of the Kondo system as equivalent to a moderately interactive local non-magnetic center, satisfying a number of symmetry rules, and the important sum rules mentioned above, which often are enough to determine the state nearly exactly. This is a kind of localized "Fermi liquid", and in the corresponding lattice case we of course expect the ground state also to be a highly modified and moderately strongly interacting Fermi liquid.

(C) Haldane Scaling.

F. D. M. Haldane, in his thesis and in some papers based on it,[10] pointed out the key element in the mixed valence case, after identifying the asymmetric Anderson model as the correct model for this problem, and, as we showed earlier, making it plausible that the borderline regime of that model would be fairly common. He observed that in the "poor man's scaling" regime of the asymmetric Anderson model, one encounters a relatively rapid scaling behavior when the electronic transitions carry one between atomic levels with different degrees of degeneracy N_f, and especially so when one of the levels - as in Ce, Sm and Yb '- is non-degenerate and hence non-magnetic. As observed by Varma, the system always scales toward the non-magnetic valence. Haldane showed that the reason was that there are N_f possible electronic transitions <u>out</u> of the non-degenerate state, which lower its energy by perturbation theory, while each of the degenerate states can only hop to the non-degenerate one. Scaling the cutoff D at high energies, we obtain equations such as those given by Coleman:

$$\frac{dE_f^*}{dD} = - \frac{\Delta}{N_f \pi} \frac{1}{D-(E_f-E_b^*)}$$

$$\frac{\delta E_b^*}{\delta D} = \frac{\Delta}{\pi} \frac{1}{D+(E_f-E_b^*)}$$

Note there is no N_f factor in the second, so E_b (the effective singlet energy in this theory) scales N_f times as fast. These scaling equations lead to trajectories as shown in Figure 3 (from Haldane's thesis). Even where the non-degenerate state starts higher, we find that it scales either to a purely non-magnetic case, or to the crossover region into a Kondo limit - which itself then scales to the non-magnetic case, by Kondo scaling.

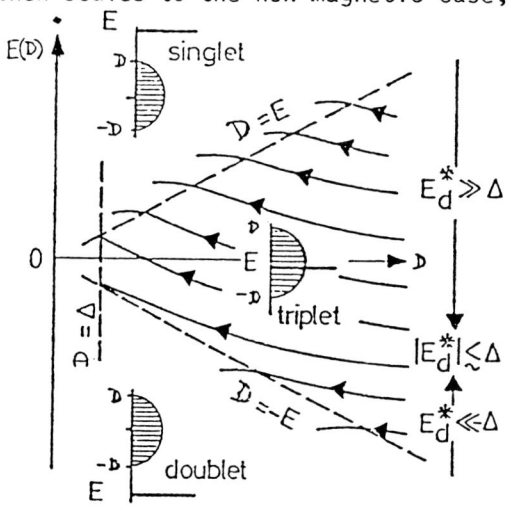

Fig. 3. Scaling trajectories E(D), ending at crossovers (broken lines) to a $<n_d> \simeq 0$ singlet regime for $E_d^* \gg \Delta$, to a $<n_d> \simeq 1$ doublet local moment regime for $E_d^* \ll -\Delta$, and to a mixed valence Fermi liquid regime for $|E_d| \lesssim \Delta$.

323

In poor-man's scaling the crossover region is a difficult one, since here the effective "f" energy level passes out of the band through one of the cutoff energies ±D. One of the energy denominators diverges and the scaling becomes more rapid as we approach the crossover. This is, however, not conceptually difficult in the space-time theory originated by Yuval, Anderson and Hamann. In fact, in his thesis Haldane shows how to carry the two theories in precisely parallel fashion and to control the crossover region. When the scaling crosses over into the Kondo region he avers that T_K is given by

$$T_K \sim \text{const.} \times \sqrt{W\Delta} \exp\frac{-\pi|E_f|}{2\Delta}$$

where W is the overall cutoff. One expects that "2" here is N_f, so the answer in mixed valence terms comes out

$$T_K \sim \sqrt{W\Delta} \exp\frac{-\pi|E_f|}{N_f\Delta}$$

which can often be quite large and is compatible with the observed situation.

The observed behavior of most mixed valence systems is compatible with the idea that the separate atoms undergo this scaling procedure more or less independently. One ends up at T_K or T_f with a set of effective non-magnetic centers, which only then establish coherence and form a band-type Fermi liquid at a lower temperature one might call T_b and might estimate to be $\sim 1/N_f$ smaller.

(5) Usefulness of $1/N_f$.

This brings us to the point of $1/N_f$ theory. In field theory, it has become customary to take advantage of the "large" degeneracy of the quarks and gluons in such theories as color gauge theory to establish a relatively simple limit where various kinds of corrections become small. Since much of the physics is still contained in the limit of large degeneracy number N_d, it is hoped (piously) that qualitative results will be correct for the observed $N_d = 3$. This was not in fact the inspiration for my suggestion of large N_f as a simplification, but it indicates additional value in it. In fact, what I initially suggested was based on the misapprehension that most mixed valence systems were in the top part of the Haldane diagram, with the scaled $E_f > 0$. In that case, many simplifications follow from large E_f. In particular, the sum rules require that the phase shift at E_{fermi} in any one channel be only π/N_f at low temperatures, which is small enough that the resonances can be treated in the original Anderson-Clogston perturbation theory, i.e., the effective scattering potential in any one channel is $\sim 1/N_f$, and one need not worry about the strong resonances which are away from the Fermi energy

324

A second obvious improvement is that the interactions between atoms are considerably reduced. To influence the next atom a scattered electron must travel to it and return <u>in the channel from which it left</u>. (This becomes obvious in any of the possible perturbation theory representations of RKKY etc. type interaction.) This means that there is an effective "dilution factor" of $1/N_f$, and this also represents a factor making incoherent random scattering more effective than coherent Bragg scattering, so that band formation takes place only <u>after</u> the atoms have settled down (the factor of N_f between T_b and T_F already referred to). This appears to possibly have been seen in $CeSn_3$, where the form factor change and χ peak occurs at a temperature a factor of 5 below T_F.

Now for my last thoughts. (1) As I have already remarked, I think some combination of space-time theory and $1/N_f$ techniques may get us through the Haldane-Kondo crossover without too much difficulty. The Kondo lattice problem which then results, however, is undoubtedly a hard one.

(2) The f-like quasiparticles in the Kondo lattice must be expected to have a large wavefunction renormalization factor Z (or actually, it is Z^{-1} which is large). To my knowledge, this has not been calculated but the spectral densities given by the $1/N_f$ calculations show that the density associated with the Kondo peak is very small, probably of order T_K/Δ. The real state of the f site is very close to $n_f = 1$, with the electron typically hopping on and off for brief periods of order $1/E_d$. The phonons cannot follow this actual motion, only the very slight changes in mean occupation which occur on a time scale of order $\hbar/T_K$. Thus the electron-phonon interaction is renormalized away in the Kondo lattice cases. In the more straightforward mixed valence cases of, for instance, Ce metal itself, T_f is higher than the Debye temperature and, again, the occupancy fluctuates too rapidly for phonons to follow. Thus we never expect the local displacements to follow the "valence" fluctuations. To a great extent, here is again a manifestation of the fact that <u>real</u> valence does not actually change much as the <u>nominal</u> valence fluctuates.

Finally, I have promised to say something about superconductivity in the heavy electron systems. The Haldane and Wilson-Nozieres theories leave a repulsive pseudopotential for electron-electron interaction at the effective center, reflected in a (χ/γ) ratio greater than one. It is noteworthy that in UBe_{13} this ratio decreases as we approach T_c, reflecting perhaps some kind of local singlet pairing.

I could guess that what might be occurring is an effective attraction between pairs of f electrons on the same ion <u>in singlet</u>

325

states. This cannot be primarily a phonon interaction, I think, because of the above arguments and because phonons are weaker than the intrinsic Coulomb U once the time-scales of both have been lengthened by Kondo renormalization. A far-fetched idea is the following.

From nuclear shell theory we know that two particles in an open shell (or, in fact, any even number) have a paired-up singlet state which Bohr, Mottelson and Pines showed to be a close analogy of the BCS state. It may be written (for $J = L \pm 1/2$)

$$\Psi_{sing.} = \sum_{M>0} c_M^{J^+} c_{-M}^{J^+} \Psi_0$$

In the nuclear case with attractive interactions, this is the lowest state and is normally seen. (Hence the $I = 0$ value for even-even nuclei.) In atoms, under Hund's rule Coulomb interactions, this is a high-energy state due to repulsive exchange interactions. But this state, like the empty singlet, can be subject to Haldane renormalization by a quantity of order $\Delta \ln W/\Delta$, since it connects with more magnetic substates than vice versa. In the case of UBe_{13}, particularly, we suspect the valence of U to be mixing between 1 and 2, not 0 and 1, and also we expect Δ to be quite large. I have made no calculations but I would expect, under these conditions, the effective Wilson interaction to be attractive. Ce is a more difficult case in $CeCu_2Si_2$, but the photoemission people have found evidence for some $(f)^2$ in several compounds and this one may be the most biased towards $(f)^{1+\epsilon}$ of any. That would be a testable prediction of this theory.

REFERENCES

1. A. Blandin and J. Friedel, J. Phys. Radium 20, 160 (1959).
2. A. M. Clogston and P. W. Anderson, Bull. Am. Phys. Soc. 6, 124 (1960).
3. F. D. M. Haldane and P. W. Anderson, Phys. Rev. B 13, 2553 (1975).
4. G. C. de Leo, G. D. Watkins, W. B. Fowler, Phys. Rev. 23, 1851 (1981).
5. F. D. M. Haldane, Phys. Rev. B 15, 281, 2484 (1977).
6. M. A. Schluter and C. M. Varma, Helv. Phys. Acta 56, 147 (1983).
7. P. Nozieres, J. Low Temp. Phys. 17, 31 (1974).
8. J. Cardy, J. Phys. A 14, 1407 (1981).
9. B. G. Kotliar, Thesis, Princeton University (1983).
10. F. D. M. Haldane, Thesis, Cambridge University, 1977, J. Phys. C 11, 5015 (1978).

5. THE PROBLEM OF FLUCTUATING VALENCE IN f-ELECTRON METALS

P.W. ANDERSON

Joseph Henry Laboratories of Physics, Princeton University, Princeton, NJ 08554, USA

Introduction

It is a pleasure to be here at this tribute to my old friend Gerry Brown, whom I first met, I believe, over a quarter of a century ago, probably at the meeting of the Many-Body community in 1960 in Utrecht when it was still possible essentially to get the whole community in one room, and for all of us to talk, more or less, in a similar language. Meantime, perhaps only Gerry has continued to fight for the great ideal of the unity of this kind of physics against the centripetal forces of specialization; this conference is a delightful vindication of that ideal.

As C.M. Varma will emphasize, the mixed-valence phenomenon is a perhaps somewhat unheralded example of this unity, for almost any branch of theoretical physics can be equally successful and almost any branch of theoretical physics can be equally unsuccessful with the problem [1]. It depends on your taste and judgment what you use. I would like to rely on Chandra and some of the later speakers in the symposium to give you an idea of more of the details than I am going to. My goal here is to be a bit more eclectic and a bit more philosophical about the whole subject. I will focus on two lines of thought:

(1) Some general semi-philosophical ideas about the phenomenon, emphasizing the two points, (a) that most of the physics follows from a few very simple key concepts; and (b), why the rest of the physics makes utter shambles of all the methods which have been tried on it.

(2) Changing gears, I will discuss the question of the chemistry of the phenomenon of mixed-valence and heavy-electron metals, and especially why it is that cerium and uranium can be so similar. In this I am following in the footsteps of a distinguished eclectic physicist, Maria Mayer, who was the first to understand the chemistry of the rare earths as well as of the shell model of the nucleus.

1. Simple ideas

There is a series of key ideas which make the basic physics of mixed valence systems relatively understandable. We start with the Anderson model [2]

$$H = H_{\text{f}} + H_{\text{free}} + H_{\text{hybrid}} + H_{\text{corr}}, \tag{1}$$

$$H_{\text{f}} = E \sum_{m\sigma} n_{m\sigma}. \tag{2}$$

The first term, H_{f}, involves, in the case of the heavy-electron metals, an f resonance near the Fermi level of an otherwise undistinguished metal. Next, there is a hybridizing term mixing the local state with a sea of free electrons, and the sea of free electrons is the third aspect of the problem.

$$H_{\text{hybrid}} = \sum_{k\sigma} V_{km}(a^+_{k\sigma} a_{m\sigma} + \text{h.c.}), \tag{3}$$

$$H_{\text{free}} = \sum_{k\sigma} \epsilon_k n_{k\sigma}. \tag{4}$$

Finally there is the correlation

$$H_{\text{corr}} = U \sum_{m\sigma = m'\sigma'} n_{m\sigma} n_{m'\sigma'}. \tag{5}$$

It is a very simple model, at least if you deal with one atom, and has in principle been solved a number of ways: it was in principle solved about ten or fifteen years ago [3], but recently exact solutions of this model have been found [4]. Exact solutions for the problem of this model in the high concentration of a real metal are completely out of our reach as far as we know, and there are no easy ways of getting at it. Once given this as your model, you have a number of fixed points with which to deal. I will be using the word "fixed point" in both senses. The most important fixed point is the existence of sum rules. First, there is the Friedel sum rule, which says that for each magnetic center, the number of particles in a given channel around such a scattering resonance is related to the phase shift at the Fermi level of that same scattering channel:

$$n = \sum_{l\sigma} (2l+1) \delta_{l\sigma}/\pi. \tag{6}$$

This has been shown by various people; it is clear from Friedel's original derivation [5] that this is remarkably independent of the interaction phenomenon, even independent of putting other complicated resonances close to your original resonance or to putting in neighboring atoms. It's simply a node-counting theorem and as such very solid.

The second basic very important sum rule is derived by Luttinger [6], namely that the volume of the Fermi surface is constant independent of interactions, given by simply counting up the number of electrons in the non-interacting state. In many ways you can think of the mixed-valence phenomenon as a battle of Friedel vs. Luttinger. You have to satisfy them both at once and it's not exactly obvious how to do that. In the case of SmB_6, which is my favorite example of mixed valence, the density of this lattice is such that the samarium wants to have to have exactly 5.4 f electrons, i.e. halfway between Sm^{++} and Sm^{+++}. One way it can do that is to have 0.6 of the atoms be Sm^{++}, a non-magnetic state, and 0.4 of the atoms Sm^{+++}, which has one f electron left over, a spin of $\frac{5}{2}$ and definite magnetic properties. In fact, if you look at high temperatures it is exactly that way: Friedel wins. The valence of any given atom fluctuates rather slowly back and forth from one valence to the other. In looking at any property with a snapshot you will find some of the atoms are Sm^{++} and some of them are Sm^{+++}. If you look at the susceptibility, it is a linear superposition of the two. On the other hand, if you want to have a Fermi surface, Luttinger will say that SmB_6, if you just add up electrons, will have to have an even number of free electrons, it has to have 0 or 2 but certainly not 0.4. But in this high-temperature state it has 0.4 of a free electron Fermi surface. So what does it do? At high temperatures it behaves in the atomic way, and you seem to have some of these electrons hidden in the atomic levels, and 0.6 electrons in a free electron band. At low temperatures, Luttinger wins. There is a kind of a coherent state in which it is a Fermi liquid, there are no magnetic atoms, and it is an insulator. You get a spectacular experimental [7] cross-over phenomenon as you go from the high-temperature to the low-temperature state, as shown in fig. 1. You can think of this dichotomy as a contest between the real valence, which is the real number of f electrons that this particular kind of atom wants to have, namely $Sm^{2.4+}$, and the valence that you deduce that you get by looking at the band structure, which says there are no magnetic atoms there so there is no Sm^{+++}. You can reconcile these two by fixing the phase shifts of all of the atoms at some

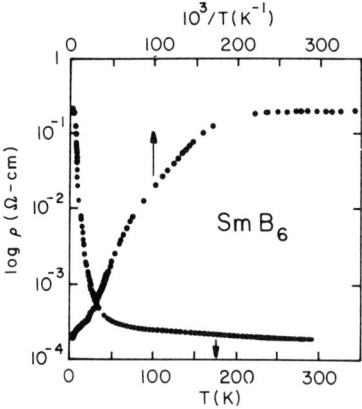

Fig. 1. Temperature dependence of electrical resistivity of of SmB_6 (From ref. [7].)

uniform value which is essentially the valence divided by the number of f states times π. In UBe_{13} and UPt_3 it is even more complicated and spectacular, although the crossover phenomena are not as spectacular: they look rather smooth until you get to the superconducting state which we will talk about later. Uranium at high temperatures definitely appears to be trivalent, whereas at low T, U in UBe_{13} has a nominal valence of zero. It's not magnetic at all.

The next fundamental concept is the idea of scaling and renormalization group: The Anderson–Yuval–Hamann–Wilson [3,8] scaling of the Kondo problem, and the Haldane [9] scaling of the Anderson model. The ground state must be a stable fixed point of one of these kinds of scaling. The Kondo scaling is a much more complex kind of scaling, that you apply to a single magnetic impurity that has a spin exchange interaction with free electrons. You try to scale out the interactions between the spins and the free electrons by reducing a nominal cut-off from some high value, comparable with a bandwidth, to zero. In the original Anderson model which underlies the Kondo model the exchange integral is the width of the resonance divided by the strong-interaction constant U. As you reduce the cut-off the exchange integral actually increases until eventually it is of order unity and all of the energy parameters of the problem become close to unity simultaneously. This identity takes place at a very small value of the energy which is exponentially small in (one over the original cut-off). One can think of it essentially as a renormalization of the resonance width which takes place in the Kondo problem. A later and much simpler kind of scaling was carried out by Duncan Haldane on the Anderson model, with a resonance level whose position starts out near the Fermi level. As you reduce the cut-off you have the effect of leaving the width of the resonance constant and lowering or raising the position of the Fermi level. You eventually raise the position of the Fermi level to the point where you renormalize to a nonmagnetic state, which turns out to be a fixed point of the problem. If you start with the f energy too low compared to the Fermi energy, it ends up doing a crossover into the Kondo problem, but the Kondo problem eventually reduces itself to a resonance which is exactly at the Fermi level. And so no matter how you go, you're stuck and you end up with only one line of possible fixed points, or you have only one fixed point if you adjust your scales properly. For the Kondo problem, the fixed point at absolute zero energy is essentially a nonmagnetic impurity. This can be either a totally noninteracting fixed point or much more often you have the Fermi-liquid type point which was first understood in some kind of detail by Philippe Nozières [10]. The ground state in that case can be very diverse: for example, a simple Fermi liquid ground state with very large mass as in $CeCu_6$, a spin density wave type of ground state, a superconducting ground state, or even some systems which have both spin density waves and superconductivity. It's important to realize that the scaling has the effect of reducing the large interaction parameters to very small values, and whatever happens in the Fermi liquid is essentially on the level with the irrelevant parameters which one does not handle very nicely with the scaling theory. That is what makes it such a terribly hard problem: the really interesting phenomena which happen at low temperatures are mostly at the hard level of the irrelevant parts of the scaling theory.

The final "fixed point" in the physics is to recognize that most of the action is local. Over most of the range, the renormalization which is taking place doesn't really know whether or not there are other mixed-valence atoms present. There are various ways of seeing this, but the most straightforward one is to see the experimental fact that one has an enormous resistance $\rho(T)$ caused by magnetic scattering and indicating that the individual scatterers are acting independently as you go through the so-called f spin fluctuation phenomenon. This is shown in fig. 2. There is still a question as to whether there is a second temperature below the spin fluctuation temperature corresponding to the rate of fluctuations of the valences, i.e., whether there is a coherence temperature below the fluctuation temperature at which the fluctuation of valence and of spin begin to be in phase between the different centers so that one begins to get overall Fermi-liquid behavior. I believe that there is such a lower coherence temperature but none of the theories is good enough to really show you that such a thing happens. Nonetheless there are a number of rough theoretical thoughts that give you this phenomenon. The self-energy of the particles is predominantly local, therefore it is predominantly a function only of the frequency and not of the momentum. There is a residual k dependence. Unfortunately this k dependence, although it is small, is also of the order of the bandwidth in the Kondo lattice. This is another manifestation of the fact that when you scale down to low levels, all the big parameters have become unimportant and the residual k dependence is very important. The residual k dependence may determine the type of ground state that one has.

To summarize, much of the physics is contained in three basic ideas:
(1) the sum rules,
(2) the scaling idea which leads you to a low-temperature Fermi liquid which you

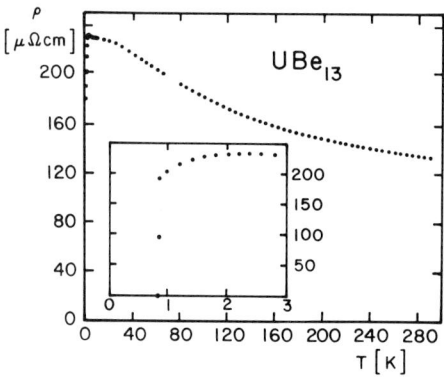

Fig. 2. Temperature dependence of electrical resistivity of UBe_{13}. Inset: low temperature resistivity on an expanded scale (from ref. [11].)

think you understand pretty well because you understand the phase shifts from the Friedel sum rule and you understand the size of the Fermi surfaces from the Luttinger sum rule, and finally

(3) the idea that most of the action is local.

2. Chemistry of mixed valence

Now let me talk about basic physics and chemistry. The heavy-electron materials seem, so far, to appear in three or at most four places in the periodic table: Sm, perhaps Yb, and Ce and U with large amounts of quite neutral, free-electron like solvents: Cu_2Si_2, Cu_6, Al_3, Pt_3, Be_{13}, etc.

The first two are cases in which the Anderson-model width Δ is probably particularly small. But Ce and U often exhibit quite broad f-bands, and in some cases are fairly ordinary metals.

U is particularly anomalous in that one must assume chemically that it has three f-electrons, not one. What I want to propose is that U is like Ce because the first two f-electrons are not particularly magnetic and normally participate in metallic bonding, being in essence part of the free-electron band, not the inner f-shell. Thus I claim that only the third f-electron is really participating.

In this U and its neighbours much more resemble the 3d series than the 4f's. As one studies the series Sc, Ti, V, Cr, Mn (see my book Concepts in Solids) one realizes that Sc and Ti are almost never magnetic although they actually contain 1 and 2 3d-electrons. The interesting properties begin to occur at V, which has all kinds of interesting oxides, while Cr tends to be pretty magnetic.

I pointed out that actually a plane wave contains as much "d" as "s" or "p" character, depending on the ratio of k_f (which is determined by the density of free electrons) and the decay constant κ of the outer parts of the d functions. Only at V does the d-shell begin to become somewhat "inner", primarily because the *third* d-electron cannot behave as a free electron. But occasionally two of them can, even in V.

Much the same is true in the actinides. In Th the two f-electrons are strongly mixed with free plane waves and are essentially free electrons. At U, the f shell is equivocal and may be "inner" or "outer" or partly one or the other.

This fact was observed empirically by Jim Smith in a version of the periodic table he often shows, in which he orders the transition metals 4f, 3d, 5f, 4d, 5d and shows that this aligns the superconducting and magnetic elements much more straightforwardly.

It is also true that the more free-electron like the band, the less the third f-electron in U will hybridize. This may partially explain the tendency of heavy electrons to show up when the solute has a relatively weak pseudopotential, like Be or Cu.

References

[1] P.A. Lee, T.M. Rice, J.W. Serene, L.J. Sham and J.W. Wilkins, Comm. Condens. Matter Phys. 12 (1986) 99.
    G.R. Stewart, Rev. Mod. Phys. 56 (1984) 755.
[2] P.W. Anderson, Phys. Rev. 124 (1961) 41.
[3] P.W. Anderson, G. Yuval and D.R. Hamann, Phys. Rev. B 1 (1970) 4464.
[4] N. Andrei, Phys. Rev. Lett. 45 (1980) 379.
    P.B. Wiegmann, Pis'ma Zh. Eksp. & Teor. Fiz. 31 (1980) 392 [JETP Lett. 31 (1981) 364].
[5] J. Friedel, Adv. Phys. 3 (1954) 446.
[6] J.M. Luttinger, Phys. Rev. 119 (1960) 1153.
[7] A. Menth, E. Buehler and T.H. Geballe, Phys. Rev. Lett. 22 (1969) 295.
[8] K.G. Wilson, Rev. Mod. Phys. 47 (1975) 773.
[9] F.D.M. Haldane, Phys. Rev. Lett. 40 (1978) 416.
[10] P. Nozières, J. Low Temp. Phys. 17 (1974) 31.
[11] H.R. Ott, H. Rudigier, Z. Fisk, and J.L. Smith, Phys. Rev. Lett. 50 (1983) 1595.

Some Ideas on the Aesthetics of Science
Lecture given at the 50th Anniversary Seminar of the Faculty of
Science and Technology, Keio University, Japan, May 1989

Theoretical Paradigms for the Sciences of Complexity
Nishina Memorial Lecture, Department of Physics,
Keio University, Japan, May 1989

These two lectures were given in Japan in 1989 at the behest of Ryogo Kubo. Kubo Sensei was an early soul-mate: he claimed at one time that he "discovered" me, which was certainly in a sense true, in that he arranged for me to be at the Kyoto 1953 meeting where I made many vital contacts; while I certainly have enormous respect for him both as scientist and as organizer.

SOME IDEAS ON THE AESTHETICS OF SCIENCE

P.W. ANDERSON

Joseph Henry Laboratories of Physics
Jadwin Hall, Princeton University
Princeton, NJ 08544

ABSTRACT

The educated layman is used to thinking of science as having aesthetic values in two senses. Often he can recognize the grandeur and sweep of the scientific vision: the cosmological overview of the universe, the long climb of evolution towards complexity, the slow crunch of the tectonic plates, the delicately concentrated energy of the massive accelerator. Also, many visual images from science have aesthetic meaning: images of galaxies, of the complex structures of crystals or of the double helix, the fascinating diversity of organisms and their traces in Nature. What I want to discuss here, however, is the internal, intellectual aesthetic of science, which is often what the scientist himself alludes to when he calls a certain piece of science "sweet" or "beautiful". This is very often a comprehension of internal intellectual connections among diverse phenomena or even fields of science — that the same intellectual structure, for instance, may govern the formation of elementary particles and the flow of electricity in a superconducting wire; another may relate a complex magnetic alloy with the functioning of neuronal circuits. In summary, I will try to describe what the scientist (or, at least, one scientist) finds beautiful in science.

Lecture given at the 50th Anniversary Seminar of the Faculty of Science & Technology, Keio University, Japan, May 1989

During the debate over the Hydrogen Bomb in the early 1950's which eventually led to J.R. Oppenheimer's downfall, he opposed Teller's "crash" efforts to design such a bomb on some combination of technical and moral grounds. But when Stan Ulam, working with Teller, proposed a new configuration, Oppenheimer seems to have withdrawn his opposition, remarking that the new design was "so technically sweet"—*i.e.*, so "beautiful", that it had to be done. This is only a widely publicized incident involving scientists making essentially <u>aesthetic judgments and allowing them to influence their actions</u>; I happen to feel that it is a disgraceful one, but that is beside the point here. All scientists, I think, who are worthy of their calling, have some aesthetic feeling about it, specifically about what is beautiful science and what is not.

It is this aesthetic component of science which I want to discuss here. I am sure that I shall tread on many toes, nor am I absolutely sure that I have got it right in any case; in fact, I would feel that I have done my job if I simply succeed in opening a discussion. In aesthetic matters there is a widespread prejudice summarized in the saying "each to his own taste", but, in fact, I happen to feel there are real criteria both in the arts and in science.

Let me first dispose of some common layman's misconceptions. The most common would surely be that science is not only value-free but without scope for imagination and creativity. It is seen as the application of a systematic "scientific method" involving wearing a white coat and being dull. I feel that too many young people come into science with this view, and that too many fields degenerate into the kind of work which results: automatic crank-turning and data-collecting of the sort which Kuhn calls "normal science" and Rutherford "stamp-collecting". In fact, the creation of new science is a creative act, literally, and people who are not creative are not very good at it. (Equally, one often finds people miscast in scientific careers who do not realize that the second most important skill is <u>communication</u>: this seems to be a special problem for Japanese scientists. Science is the discovery <u>and communication</u> of new knowledge.)

A second layman's problem is the attempt to project his own aesthetic system into science. I have, several times, been asked by artists, for instance, about striking images which can be made from scientific objects, and, of course, in popularizing science every TV program is eagerly hunting for this kind of thing. Science itself contains a fifth column of practitioners—often otherwise respectable—who like to create pretty images, sometimes by computer tricks, or to emphasize the grandeur of the scientific vista by playing games with large numbers. To play pretty games or to inspire awe with large or small magnitudes is perhaps a legitimate, if tricky, way to enhance popular support for science (but what happens to the equally important but unglamorizable subjects?) but it has little or nothing to do with science itself. It is true that different fields of science attract people who are, to some extent, swayed by subject matter: astronomers do like to look at the stars and contemplate deep space, biologists often seem to enjoy the diversity of forms of life, elementary particle physicists are convinced they alone are plumbing the "really" fundamental, etc. But within each science, and across the spectrum of the sciences, it is still possible to distinguish the "sweet" from the ugly.

A third misconception is promulgated by certain sociologists of science, who seem to feel that science is a purely sociological phenomenon, with no intrinsic truth value at all: that scientists' aesthetic and cultural prejudices create the form which science takes, which is otherwise arbitrary. This is mainly refuted by the fact that science works in a real sense: it grows exponentially because it is useful and effective, which means that it produces, one way or another, a true picture of the real world. These sociologists have studied science being done, which is, of course, a confusing set of interactions among highly fallible people with strong prejudices; but they have not enough insight into the subject matter or into the qualitative differences among fields and among people to recognize the rapid disappearance of the shoddy or dishonest result. It is significant that the average scientific paper is cited less than once in the literature, while some are cited thousands of times: some are right, some are wrong, most are meaningless.

To the sociologist of science, observing from the outside, the uncited paper and the "classic" appear equally significant. As we will see, fortunately, the "aesthetic" aspect of science has much to do with values which are also related to its validity and truth, so I am <u>not</u> saying that aesthetics leads scientists to distort the meaning of their work.

I do not deny the regrettable fact that some scientific fields do become detached from the values of the rest of science and lose sight of certain basic reality principles: we have, in the past few months, seen an example of precisely this problem in the field of electrochemistry, which I am told is one of these. But the advancing edge of sciences adhere to unavoidable reality principles.

Having disposed of the negative, let us ask: can we find a theory of aesthetic value which is at all common between the arts as normally understood, and the sciences? The arts, of course, have their equivalent of the facile games I referred to in the sciences: sentimental verse, picture postcard art; there is, of course, a great body of aesthetic theory on which I am certainly not an expert; but I have over a number of years, picked out a number of statements which I think are significant.

In sculpture and painting, the critic Berenson has made much of what he calls "Tactile Values", which seems to mean giving the viewer a sense of personal involvement in the action or motion or scene depicted. A similar, if quite different, statement by a sculptress friend once impressed me strongly: she felt that all successful sculpture, no matter how abstract, referred back to the human body. Finally, also in the visual arts, the use of iconography and symbol is a common bond between ultramodern painters such as Jasper Johns and Frank Stella, and classic painters and sculptors, especially religious art but also classical oriental painting. In the modern paintings, the iconography is self-created by repetition of certain motifs, but it is firmly there. All of these kinds of remarks bring out two theses which I want to put forward and test

(1) That even in abstract art there must be a "content" or "substrate" to

which the viewer is expected to relate. Nothing serious is beautiful in a vacuum; in fact, this is thought now to be a property of the human mind: that it can not think, can not perceive, can not communicate except <u>about</u> something: the mere act of communication requires context.

(2) To be beautiful, a piece of art should have <u>more to it</u> than surface content. It should be enriched by more than one layer of meaning. This brings me to a theory of aesthetics in literature and poetry which very much intrigued me, the ideas of T.S. Eliot and the Cambridge school of critics such as David Daiches. Eliot uses the word "ambiguity" to express himself, perhaps a misleading use of this word which often, in English, means "fuzziness" or lack of clarity; whereas Eliot was always absolutely certain of what he meant. What he really meant was that good poetry should have as many levels of meaning packed into the same words as possible. In his poem "The Wasteland" for instance, there are characters carrying out certain actions on the surface, which is at least clear enough that much of the poem may be read directly as a series of stories. There is a also a surface level of absolutely gorgeous use of language. There is, underneath those two levels, a sense of despair at the moral emptiness of the modern world of the time; and still under that, if we read quite carefully, there are a series of references to myth, especially the Grail legends and those involving the Fisher King. On a more obvious level, his play "The Cocktail Party" has quite obvious Christian symbolism superposed on an apparently clever, brittle drawing-room comedy. But in this, in Japan, I am probably telling you nothing new: in the land of the Haiku, the delicate use of ambiguity and cross-reference needs no explanation.

Leonard Bernstein's Harvard lectures give some beautiful examples of this kind of cross-reference or multiple meaning in, especially, Stravinsky's music; I am not an expert on music and can give you no further examples. But a kind of music I do know well, classic American Jazz, is again a case of multiple-layered meaning and multiple reference. Characteristically, the surface meaning of jazz is a sentimental love song or a naive hymn; this is then overlaid with an ironic twist which pokes fun at its sentimentality or simplicity, and possibly also emphasizes a less respectable meaning of the lyrics; and, finally, there is the contrapuntal improvisation which is a pure, rather abstract musical object, only weakly related to the original tune and often bringing in cross-references to other pieces of music: quotes from Souza marches, bugle calls, or even well-known classical pieces.

As far as I understand the concepts of structuralism and of deconstructionism, my point of view is diametrically opposite to these; I have a feeling that these ideas devalue art and, when applied to science, often have the same effect as the sociological relativism which I have already deplored.

Let me then set out the criteria for beautiful science which I am going to try to abstract from these ideas about beautiful art.

(1) Reality principle: The work must refer to the external world, not just to the contents of the scientist's (artist's) mind. In this I make a real distinction between mathematics and science. Mathematics creates its own world, and because of the long history of mathematics there is a shared substrate of ideas within which cross-reference is possible. But I think any mathematician would agree that beauty in mathematics lies in tying together pre-existing material, rather than in meaninglessly arcane postulational systems.

On the other hand, natural science is the science of nature, not of imaginary worlds; I do not, for instance, feel that cellular automata are part of nature, so that study of their properties must be judged as mathematics, not as science.

I have, myself an aesthetic prejudice in favor of science which takes nature as she is, not that which studies artifacts made by the scientist himself such as gigantic accelerators or fusion machines. I accept that this is personal, not universal, and that clever technology can be beautiful to many people.

(2) Craftsmanship is always an element of beauty, in science as in art. The act of creation must be non-trivial and it must be done well. Much ultra-modern art fails on this score, as a visit to, say, the L.A. County museum can easily convince one. The lucky fellow who happens on a new substance or a new effect may win a

prize, but we, as scientists, do not really value his contribution unless he displays other characteristics: Edison, as scientist, is not a model we really admire. In science, however, one often finds that the discoverer does not necessarily craft his discovery optimally: BCS theory, for instance, was first expressed in its ugliest form, and only refined by Bogoliubov, Nambu and others into a thing of beauty. We accept this as the nature of the beast, and it is perhaps not unknown in art; for instance, the Dutch school discovered counterpoint, but Bach exploited its possibilities beyond their abilities.

(3) Next is the principle of maximal cross-reference, i.e. my "ambiguity" equivalent. This refers both to different levels of meaning and to breadth of reference in the real world. I will talk about examples later, but perhaps I can continue with the BCS theory as a relatively simple one. Once re-expressed in Bogoliubov-Nambu form, it became almost evident that BCS could be a model for a theory of elementary particles as well as of its "surface" meaning, theory of superconductivity. Once expanded by Gor'kov in Green's function form, it not only allowed many new insights into the phenomenology of superconductivity, but acquired a second meaning as not just a "model", parametrized theory but a "microscopic", computable theory. And, finally, Bohr, Mottelson and Pines extended the idea to nuclear matter, and Brueckner, Morel and myself to the anisotropic superfluid ^3He, bringing, in the end, two enormously fertile and unexpected references into the picture.

Where does the beauty reside? Of course, not entirely in the original paper which solved the problem of superconductivity, although indeed that was a well-crafted, very exciting paper. Not in any single object or work: not even any historian of science will be capable of dissecting the entire web of connections brought forth by the phrase "BCS". Perhaps, in some abstract sense, in the citation network: who cites whose paper and why? Science has the almost unique property of collectively building a beautiful edifice: perhaps the best analogue is a medieval cathedral like Ely or Chartres, or a great building like the Katsura detached palace and its garden, where many dedicated artists working with reference to each other's work jointly created a complex of beauty.

(4) I want finally to add one criterion which is surely needed in science and probably so in art: a paradoxical <u>simplicity</u> imposed on all the complexity. There is the famous story of Ezra Pound editing T.S. Eliot's "Wasteland": that he reduced its total length by nearly half, without changing any of the lines that he left in, and greatly improved the poem thereby.

In science, even more than in art, there is a <u>necessity</u> to achieve maximal simplicity, not just an aesthetic preference. The subjects with which we deal, and the overall bulk of scientific studies grow endlessly; if we are to comprehend in any real sense what is going on, we <u>must</u> generalize, abstract, and simplify. Together with the previous criterion, this amounts to a very basic dictum for <u>good</u> science, not just beautiful science. We must describe the <u>maximum</u> amount of information about the real world with the <u>minimum</u> of ideas and concepts. In a way, we can think of this as a variational problem in information space: to classify the maximum amount of data with the minimum of hypotheses. Of course, this is just "Occam's razor" of not unnecessarily multiplying hypotheses, which in fact has been given a mathematical formulation in modern computer learning theory by Baum and others. In this case, our aesthetic concept is severely practical as well.

Again returning to our canonical example of BCS theory, in its original formulation it was not at all clear what the minimal set of hypotheses was: whether the crucial feature was the energy gap, or the zero-momentum pairing idea, or what? With refinement, which came in response to the Russian work and to the Josephson effect, gradually we discarded details and recognized that the one core concept is macroscopic quantum coherence in the pair field, which when coupled with a fermi liquid description of the normal metal leads inevitably to one of the versions of BCS theory. The beauty of the theory lies in the immense variety and complexity of experimental fact which follows from these two concepts. But without the existence of all that variety of experimental fact, and of the painful, exhilarating process of connecting it in to the main mass, the concept alone seems to me to be a meaningless, relatively uninteresting mathematical game. It is in the interplay, the creative tension between theory and experiment, that the beauty of science lies.

Let me give a few examples of beautiful science to try to clarify my ideas further. To begin with, let me hop entirely outside my own specialty and recall an incident from a recent book by Francis Crick. He was describing a dinner meeting at which Jim Watson was to be the feature speaker, and he describes Jim being plied with sherry, wine and after-dinner port, and then struggling with a presentation of their joint work on the double helix. The practical details he got through, but when it came time to summarize the significance, he just pointed at the model and said —"It's so beautiful...so beautiful". And, as Crick says, it was. Why?

As a model of one of the true macromolecules of biology it did, of course, embody brilliant technical advances and insights, and in addition as a structure itself it contains the creative tension of simple repetition yet complex bonding. But of course, he meant far more than that: that with the structure in hand, it was possible to first envisage that the detailed molecular mechanism of heredity, and of the genome determining the phenotype, could eventually be solved. At that point not much further had been solved—one was just at the stage of proving that the obvious mechanism of DNA replication on cell division was really taking place, by quantitative measurements of DNA amounts—but that the original piece of the puzzle lay there in that model was hardly to be doubted. Crick and Watson, to their credit, did see—and did, especially Crick, later participate in and formulate—the whole complex of ideas that was likely to arize from their work. Crick and Brunner called this the Central Dogma, and the role played by macroscopic quantum coherence in BCS theory is played by the Central Dogma in this theory.

The "Central Dogma" is, of course

$$\begin{aligned}&(1) \text{ DNA} \to \text{DNA}\\&(2) \text{ DNA} \to \text{mRNA} \qquad \text{transcription and}\\&(3) \text{ mRNA} \to \text{protein} \quad\text{ gene expression}\\&\qquad\qquad\qquad\qquad\qquad (3) \text{ implies a code}\end{aligned}$$

Some of this was already known in a vague way: that genes determined protein sequence, for instance, so if the gene was DNA, DNA → protein was obvious—but was it? Crick points out that Watson and he were the first to make up the standard list of 20 amino acids, as a response to their realization that a code must exist. Some of the most beautiful—because simple—scientific reasoning in history went into the determination of the code. Enough said—Jim Watson's alcoholic musing was right.

(2) Again, to go outside my specialty, one of the truly beautiful complexes in science is the gauge principle of particle physics: the realization that all four of the known interactions are gauge interactions, in which the form of the forces coupling the particles follows from symmetry and not vice versa. A very nice discussion of this area is to be found in C.N. Yang's scientific autobiography, written as an annotation of his collected papers.

Mathematicians will tell you that they invented gauge theory anywhere from 50 to 100 years before the physicists in the form of something called "fiber bundles". I do not take this seriously—see my remarks about the "reality principle". A theory as a mathematical object is simply a statement about the contents of someone's mind, not about nature. Another point worth noting is that quite often the physicist—or other scientist, as in the case of probability theory—invents his own mathematics which is fairly satisfactory for his purposes, and only later finds the relevant branch of mathematics—as with Einstein and non-Euclidean geometry.

Gauge was first used as a formally symmetric way of writing Maxwell's equations, and formal manipulation with it played a significant role in early attempts to produce a "projective" unification of gravity and electromagnetism. But the gauge idea proper stems from the work of Dirac, Jordan and others in reformulating quantum electrodynamics. What was done was to combine the early ideas of Wigner and Weyl on the role of symmetry in quantum mechanics with the "locality" principle of Einstein's general relativity. Quantum mechanics connects symmetry and conservation laws: time-invariance = energy conservation, rotation-invariance = angular momentum, etc; but from the Einsteinian point of view, the elementary interactions must allow only <u>local</u>, not <u>global</u>, symmetries. The appropriate symmetry principle for charge conservation is phase-invariance of a complex field; but to make phase-invariance <u>local</u> we must introduce the gauge field A and write all derivatives as $i\hbar \nabla - \frac{e}{c}A$. The dynamical theory of the vector potential A is then just electromagnetic theory. This is the message of gauge theory: out of three concepts one gets one. Conservation laws, symmetries, and interactions are not three independent entities but one.

Next—from the physicist's side, this is where the mathematicians make their unjustifiable claims—Yang and Mills realized that the gauge theory was not unique, in that the symmetry involved did not need to be Abelian; but such a theory introduces gauge fields which carry the conserved quantity. After many false starts it became clear that the appropriate theory for strong interactions was color gauge theory, quantum chromodynamics based on the group $SU(3)$. Here yet another thread was brought in by T'Hooft, Gross, Politzer and others: the proof that gauge theories of this sort are <u>asymptotically free</u> and hence renormalizable. To make a long story short, with yet one more beautiful idea, that of <u>broken symmetry</u>, we now contemplate a world in which all four basic interactions are gauge theories: the three-dimensional $SU(3)$, the 2+1 dimensional $SU(2) \times U(1)$ of the electroweak theory, and the 4-dimensional gauge theory which is Einstein's gravity. Whether the fact that the dimensions add up to 10, an interesting number in string theory, is significant is still much under discussion. One can hardly not, even at this stage, sit back and marvel at the beauty and intricacy not just of this simple structure but of its history and its cross-connections to many other ways of thinking.

One could go on following almost any thread of modern science and find an equivalent beauty at the center of it. One more instance will allow me to be a bit self-indulgent: topology, dissipation, and broken symmetry.

This starts with four apparently independent but individually beautiful pieces of work. First, the dislocation theory of strength of materials, when Burgers Taylor and others first invented the concept of the dislocation or line defect of the crystalline order, then—using very modern-sounding topological arguments—proved that it was topologically stable, and finally showed that motion of dislocations was the limiting factor in the strength of most materials.

Second chronologically was the beautiful work of Jaques Friedel's grandfather, G. Friedel, in identifying the defects in liquid crystals—specifically the "nematic" liquid crystal, so-called because the defects appeared threadlike. Third was the domain theory of ferromagnetism, and especially the beautiful sequence of work of Shockley and Williams showing how the motion of domain walls—planar defects where magnetization rotates—accounts for hysteresis loops in magnetic materials. Finally, there are the gorgeous conceptual breakthroughs of Feynman, and then Abrikosov, where Feynman, in particular, invented the superfluid vortex line and showed that it could account for the critical velocity of superfluid helium, while Abrikosov described the vortex state of superconductors and I later pointed out that motion of the vortices implied resistance.

Oddly enough, it was the discovery of superfluidity in ^3He which triggered the realization that these were all the same phenomena. Almost simultaneously, Volovik and Mineev, and Toulouse and Kleman developed the general topological theory of defects in condensed phases, encompassing the physics developed over 100 years prior to 1975 in a single structure, classifying the possible topologies of maps of real space into the space of the order parameter of the condensed phase. For instance, for liquid helium the order parameter has a free phase so one must map space onto a circle; if that map is non-trivial, it implies that at some line in space the order parameter vanishes. This means that the defects are vortex lines. Then Toulouse and I made the general connection between <u>motion</u> of topological defects and the breakdown of a generalized <u>rigidity</u> of the system, implying dissipation: which couples together all these energy dissipation mechanisms. The great generality of this kind of structure has been exploited in the theory of "glitches" in the spinning neutron stars or pulsars: giant slippages of the vortex structure implied by the superfluidity of the neutron matter in such a star, beautifully isomorphic with the "flux jumps" which are the bane of superconducting magnet designers.

More or less at the same time, topology became fashionable in elementary particle physics, with the revival of the "Skyrmion" model of the fermion particle, and the fashion for "θ vacua" and "instantons". This is one of these fascinating cross-connections, although the topological ideas have not yet had their satisfactory resolution in particle theory.

As I already said, pick up almost any thread near the frontiers of modern science and one will find it leading back through some such sequence of connections. For example, an equally glorious story can be made of the separate investigations which, together, make up the present synthesis called "plate tectonics".

But if there is beautiful science, is there also ugly science? I regret to say that this also exists and often flourishes. It does so most commonly when a field falls out of effective communication with the rest of science; one often find fields or subfields which have lost contact with most of science and survive on purely internal criteria of interest or validity. The behaviorist or Skinnerian school of experimental psychology was a notorious example; I suspect that these days we are seeing an exposure of the entire field of electrochemistry to the pitiless light of real science. And certain recent incidents in the field of superconductivity have inclined me to believe that there is an isolated school of electronic structure calculators who have been avoiding contact with reality for some years.

Finally, there is, of course, pseudoscience, which will always be with us: parapsychology, "creation science", "cognitive science", "political science", etc.—

Crick once made the remark that one should always be suspicious of a field with "science"in its title. I leave you with the final thought, that the essentially aesthetic criteria I have tried to describe for you may often be an instant test for scientific validity as well as for beauty.

Theoretical Paradigms for the Sciences of Complexity

P.W. ANDERSON

Joseph Henry Laboratories of Physics

Jadwin Hall, Princeton University

Princeton, NJ 08544

---

* Nishina Memorial Lecture, Dept. of Physics, Keio University, Japan May 1989

I may not be a very appropriate representative for the subject of Materials Science here in a conference focusing on technology and the applications of science to human problems. I am not, strictly speaking, a materials scientist in the narrow sense of these words, and much as I admire and applaud the applications of science in technology, that is not what I do. I am a theoretical physicist much of whose work has involved trying to understand the behavior of more or less complex materials such as metals, magnets, superconductors, superfluids, and the like. I thought that perhaps you would enjoy hearing, in the brief time I have here, not about these investigations or about wonderful materials of the future— as far as I am concerned, from an intellectual point of view, the very impractical and obscure low-temperature phases of the mass-3 isotope of helium are at least as fascinating materials as anything the future is likely to bring—but rather about some of the wider implications of the kind of thing I do. In particular, I have been an active participant for several years in an enterprise called the Santa Fe Institute whose charter involves action in two main directions:

(1) We believe that the growth points of science lie primarily in the gaps between the sciences, so that we believe in fostering cross-disciplinary research in growth areas which are not well served by the conventional structure of the universities or the funding agencies. I use the word cross-disciplinary to emphasize that we are not trying to create new disciplines (like materials science or biomolecular engineering), which often rigidify into new, even narrower intellectual straight-jackets, but that we approach problems by cross-coupling between scientists well grounded in their disciplines but thinking about problems outside or between them—as has often been fertile in materials science, with the coupling of good physicists, good chemists and good engineers.

(2) We believe that there are many common themes in the study of complex systems wherever they occur, from the relatively simple ones which I have encountered in solid state physics or in astrophysical situations, through complex non-linear dynamical systems such as one encounters in hydrodynamics, through biological organization, to complex biological regulatory systems such as the immune system or the nervous system, and on into population biology, ecology, and into human interactions in, for instance, economic systems. Much of our work—not all of it, to be sure—fits under the general rubric of the study of Complex Adaptive Systems: systems which by virtue of their complexity are capable of adapting to the world around them.

It would carry me too far afield to describe all of our activities in Santa Fe; one, for instance, which I have enjoyed very much is a program mixing physical scientists such as myself with a group of theoretical economists in the hope of inventing new directions for the science of economics. Rather, I would like to describe a few of the paradigms for dealing with complex systems, in general, which have come from, or are related to, my science of condensed matter physics and which seem to be generalizable to a great many other types of systems. Let me list three of them here and try to describe each in a few words, relate it to the appropriate part of condensed matter theory, and then show how the idea may be generalized.

(1) The Emergent Property of Broken Symmetry.

(2) The Paradigm of the "Rugged Landscape"

(3) Scale-Free Behavior: Critical Points, Fractals, and $1/f$ noise.

There are several other paradigms—e.g., hierarchical organization, pattern selection, marginal stability, classifier algorithms among others; but surely this is enough ideas for one short talk.

(1) Broken symmetry is actually the basic underlying concept of solid state physics. It seems at first simple and obvious that atoms will want to stack themselves into regular arrays in three dimensions, like cannonballs. Thus one does not recognize that the formation of a crystal lattice is the most-studied and perhaps simplest example of what we call an "emergent property": a property which is manifested only by a sufficiently large and complex system by virtue of that size and complexity. The particles, (electrons and nuclei) of which a crystal lattice is made do not have rigidity, regularity, elasticity—all the characteristic properties of the solid: these are actually only manifest when we get "enough" particles together and cool them to a "low enough" temperature. In fact, there are kinds of particles—atoms of either isotope of helium, or electrons in a metal, for instance—which simply do not normally stack at all and remain fluid right down to absolute zero. This illustrates one of the most important facts about broken symmetry: quantum-mechanical as well as thermal fluctuations are inimical to it.

Why do we call the beautifully symmetric crystalline state "broken" symmetry? Because, symmetrical as it is, the crystal has less symmetry than the atoms of the fluid from which it crystallized: these are, in the ideal case, featureless balls which translate and rotate in any direction, while the crystal has no continuous rotation or translation symmetry.

Mathematically, the properties of the crystal are only to be derived in the so-called "thermodynamic" or "$N \to \infty$" limit of every large system. Of course, for many purposes a very small cluster of atoms, of the order of a few thousand, can behave in somewhat crystalline ways, but the structure at a finite crystal is not really stable against thermal or quantum fluctuations. Thus the characteristic crystalline properties of rigidity, elasticity (as opposed to the shear flow of a viscous fluid) and anisotropy (as e.g., birefringence) are true <u>emergent properties,</u> properties which are <u>only</u> properties of the large and complex systems.

It turns out that many, if not most, of the interesting properties of condensed matter systems are emergent broken symmetry effects. Magnetism is a well-known example; so is superconductivity of metals and the very similar superfluidity of the two forms of helium and of neutrons in a neutron star. The anisotropic properties of liquid crystals, useful in calculator displays, are yet another fascinating example.

Broken symmetry is encountered in several other contexts. One important one is in the theory of the "Big Bang" during which, it is proposed, one or more broken symmetry transitions took place in the state of the vacuum, changing the nature and number of elementary particles available at each one, greatly modifying the energetics of these primeval events, and leaving behind one or more forms of debris—the fashionable one these days being "cosmic strings". The early history of broken symmetry in the vacuum was dominated by the Japanese-American physicist Yoichiro Nambu.

Some scientists have proposed that driven dynamical systems can exhibit broken symmetry effects; I find the analogy between the emergent behavior of equilibrium and of non-equilibrium system less than compelling. Broken symmetry does not generalize in any straightforward way to form a model for the origin of life, for instance; it stands, rather, as an explicit proof of the existence of emergence.

(2) The paradigm of the "rugged landscape" was discussed in connection

with another condensed matter physics problem, the rather obscure phenomenon known as "spin glass". It is almost unnecessary to go into the long and controversial history of the spin glass problem itself, except that it involves the possibility of a phase transition at which the spins in a random magnetic alloy "freeze" into some random configuration. The attempt was initially to find a simple model for the still mysterious behavior of ordinary glass when freezing into a solid-like but disordered state; but it turned out the disordered magnetic alloy had its own different and complex behavior.

The model one uses abstracts the movable spins in the magnetic alloy as having two possible states, up (+) or down (-), like the 0 or 1 of a binary bit. Each spin S_i is presumed to interact in a random fashion "J_{ij}" with many other spins, causing a "frustrated" Hamiltonian

$$\mathcal{H} = \sum J_{ij} S_i S_j$$

in which there are many conflicting terms to optimize and it is not easy to visualize the lowest energy state. J_{ij} is a random variable, equally often positive or negative. It is the values of the energy $\mathcal{H}$–plotted in the multidimensional "configuration space" of the state variable $\{S_i\}$—which constitutes the "rugged landscape". The task of finding a low-energy state is one of seeking for deep valleys in this "rugged landscape", but it can be proved that this task is computationally very difficult, because one gets stuck in one of the many different local minima with no hint as to where to go to find a better state.

In fact, this problem is a prime example of one of the most important classifications of computational complexity, the "NP complete" case. It has already suggested an important new algorithm for solving complex "combinatoric optimization" problems which arise in many emergency situations such as complex chip design: the method of simulated annealing.

Combinatorial optimization problems in the presence of conflicting goals are very common in everyday life: almost every personal or business decision, from ordering from a menu to siting a new factory, is of this nature: so of course information on the basic nature of these problems is of great value. Unique to our approach is the recognition of the "freezing" phenomena, the possibility of being stuck indefinitely in a less than optimum solution.

Two places where the rugged landscape point of view is catching on are in evolutionary biology and in the theory of neural networks—"generalized brains". S. Kaufmann has particularly emphasized the "rugged landscape" approach to problems of molecular evolution, both in the original origins of life and in proposing systems of "directed evolution" to produce organisms with particular traits. In the evolutionary analogy which was pioneered by Stein and myself and picked up by S. Kauffmann, the genetic material is the state-vector $\{S_i\}$ in a multidimensional configuration space. A very important point is that given the "freezing" phenomenon, it may be better to improve by complexifying—adding dimensions to the configuration space—than by optimization within one's obvious capabilities. The reanimation of neural network theory which has recently occurred due to Hopfield's introduction of spin glass like ideas is both so well known and so far afield from my subject that I do not want to go into it more deeply than that.

(3) Scale-free phenomena. Here is a paradigm which has had two joint inputs, initially far apart but growing closer with time.

The first is from mathematics: the fundamental ideas coming from a number of mathematicians such as Hausdorff but the applicability to real world situations (which, to a natural scientist, is far more important) being discovered by Mandelbrot. As Mandelbrot points out, there are many real world objects which have the property of non-trivial scale invariance—properly called anomalous dimensionality, but which he calls fractality. Such objects look the same geometrically no matter what scale they are observed at; in general that is true only in a statistical sense. For instance, he shows that many coastlines have the same geometry at any scale; the same is true of clouds and of many mountain landscapes. This alone is not enough—after all, a simple continuum in any dimension is scale-invariant. The second property is that the size of the object vary with scale in a non-integer way. For instance, he shows that the length of coastlines depends on the length ℓ of the ruler used to measure them as $L \propto \ell^{-P}$ where $P \sim .2$. Mandelbrot gives many beautiful examples, in his books, of fractal objects, and others have discovered additional ones—for instance, the shape of the breakdown paths in a dielectric subjected to too high a voltage, which is an example of Diffusion-Limited Aggregation (DLA), a process of pattern growth common to many systems, and discussed by T. Witten and others. Another important case is the "strange attractor" observed in chaotic, low-dimensional dynamic systems as discussed by Ruelle and co-workers. But in general Mandelbrot has not approached—at least successfully—the question of why fractals are so common and so important in nature.

A second independent observation of scale-invariance is due to condensed matter theorists, specifically Kadanoff, Widom, and Wilson, in the study of "critical points" of phase transition between different thermodynamic phases, such as liquid-gas critical points, superfluid-normal fluid transitions, magnetic critical points, etc. It came to be realized that, at these critical points, again the structure of the substance—in terms of the fluctuations back and forth between liquid and gas—is scale-invariant in the same sense: there are "droplets" of absolutely all sizes from one molecule to comparable to the size of the entire sample. This is the basic nature of critical behavior and the famous "critical fluctuations" which may be beautifully demonstrated experimentally.

It turned out that some of Mandelbrot's fractals are formally equivalent to critical points—e.g., the DLA system, which leads to a critically percolating cluster. It is the recent suggestion of Per Bak, that a great many examples of fractals in nature are systems at or near a critical point, in that he feels that many kinds of systems, when driven hard enough, will maintain themselves at a critical point. He calls this phenomenon "self-organized criticality" and the idea is the center of considerable controversy but also great interest. In particular, Bak points out that many systems in nature exhibit a kind of random fluctuation or "noise" which can also be thought of as scale-free but in time, not space—the ubiquitous "$1/f$ noise" of many different kinds of systems. (Actually, in general $1/f^{1\pm p}$ when p is a small number $\sim.1-.2$.) This kind of noise is technologically very important. Going from the practical and turning to the gigantic, others have proposed that the large-scale structure of the universe may have fractality over some range.

In conclusion, then, let me summarize the point if view I—we, if I may include others at SFI—am taking towards the study of complexity. From the beginning of thought, the system of Pythagoras, the medieval scholasticists, Descartes and his universality of mechanism (and from him stem the ideas of many modern particle physicists, for instance)—it has been a temptation to try to create a universal system—a Theory of Everything. It is precisely in the opposite direction that we search—we try to look at the world and let _it_ tell _us_ what kinds of things it is capable of doing. How, actually, _do_ complex systems behave, and what do these behaviors have in common? The search for the universal _must_ start from the particular.

50 Years of the Mott Phenomenon: Insulators, Magnets, Solids, and Superconductors as Aspects of Strong-Repulsion-Theory

Proceedings of the Enrico International School of Physics, Varenna, July 1987
Frontiers and Borderlines in Many-Particle Physics (North-Holland, 1987)

This was my first full-fledged discussion of the high-T_c phenomenon. The amazing thing to me is not how much is wrong with it but how much is right. At this time the whole RVB idea seemed so straightforward that the gaps in the arguments should fill themselves in the natural course of things. This was not to be the case: several questions which I posed in my final summing up turned up to contain vital parts of the story. In particular, the role of dimensionality was to turn out to be vital: the concept of "confinement" was missing. The second major problem was "finding the holon" which took three years; the simple bosonic holon was a chimera, and in the end the holon turned out to have its own Fermi momentum and Fermi surface.

Nonetheless the basic idea of a projective ground state with the spinon as the $Z = 0$ image of the electron near the Fermi surface was correct.

50 Years of the Mott Phenomenon: Insulators, Magnets, Solids and Superconductors as Aspects of Strong-Repulsion Theory (*).

P. W. ANDERSON

Department of Physics, Joseph Henry Laboratories, Princeton University
Jadwin Hall, P.O. Box 708, Princeton, N.J. 08544

1. – Introduction.

In 1937 R. E. PEIERLS, in a discussion remark on a paper of de Boer and Verwey describing [1] the electrically insulating nature of NiO, introduced the essential physics of the insulating magnetic oxides. He pointed out that the repulsive energy of two Ni^{++} d-shell electrons, which causes the difference between the ionization energies $Ni^{+}\to Ni^{++}$ and $Ni^{++}\to Ni^{+++}$, was larger than the kinetic energy which might be gained in allowing the d electrons to form a band, and that, therefore, the d-shell occupancy of every site would remain d^8 Ni^{++} and the material insulating even though the spins were free to reorient and the occupancy did not correspond to a filled band.

LANDAU seems to have made some similar observations shortly thereafter, and J. H. VAN VLECK [2] used some of the same physics (with HURWITZ) in discussing the magnetism of Ni, but it was two crystal-clear papers by N. F. MOTT, in 1949 and 1956 ([3, 4], see also [5]), which formalized Peierls' remark to apply to the general question of why metals and insulators exist and quite justifiably gave his name to the phenomenon. MOTT also started another fruitful stream of physics by emphasizing the long-range Coulomb contribution to the separation energy of electron and hole, which normally makes the « Mott transition » first order because of screening by the metallic electrons. (This too was informally discussed by PEIERLS.) Much later INKSON and I emphasized the presence of this contribution to the energy gap of every insulator [6]. It is an exchange self-energy, and is a large contribution which evades the illusory « rigor » of the « local-density functional » theory.

(*) This work was supported by NSF Grant No. DMR 8518163.

In 1959 ([7], see also [8]) I used Mott's ideas (in conjunction with the van Vleck-Orgel theory of ligand field effects [9-11]) to understand why the overwhelming majority of insulating magnetic materials were antiferromagnetic [7]. I introduced, for this problem, the simplified model which later came to be known as the «Hubbard Hamiltonian». The process which leads to exchange is the virtual effect of the band-forming «hopping» integrals between sites, frustrated by the repulsive Coulomb energy. I called this process «superexchange» to distinguish it from the more subtle «direct exchange» of ferromagnetic sign between pairs of orthogonal orbitals. (At much the same time HERRING [12] was to give a rigorous analysis of the exchange process for well-separated atoms in which he used the picturesque term «homebase theorem» for the Mott phenomenon.) Since the contribution of the two processes explained magnetic phenomena in insulators in great detail, one might have thought the subject closed, but for some reason the Mott arguments have always infuriated band calculators, beginning with J. C. SLATER. There has been a concerted effort to make them disappear, which has not succeeded, but has had the effect of keeping them out of the undergraduate or first-year graduate texts of solid-state physics, and thus requiring that they be taught anew to each generation of research physicists, at least in the U.S. The most recent round in this one-sided war was fired by TERAKURA et. al. [13], with the laughable contention that (because their band calculation found them so) the insulators FeO, CoO and NiO were metals. This paper had the salutary effect of stimulating SAWATSKY and ALLEN [14] to correct a decade of misinterpreted spectral data and restore the original Hubbard model physics for these oxides almost intact. (A correction, important spectroscopically but not magnetically or statistical mechanically, involves adding in possible further bands based on non-d levels.)

Independently J. HUBBARD [15, 16] had begun to think about magnetism and metal-insulator transitions using a generalized version of the same model, and his sequence of long papers giving approximate solutions to the manybody problem of the metal with strong Coulomb interactions indelibly stamped his name on the model. While giving the field some very useful concepts—especially «upper» and «lower Hubbard» bands, and the Hamiltonian itself—his methods led to results contradicting the fundamental rules of Fermi-liquid theory, and also with some nonphysical aspects. Hubbard's chief purpose was to understand metallic ferromagnetism following the ideas pioneered by VAN VLECK and HURWITZ (and to some extent ZENER [17]), and while that line of investigation has continued (important contributions by KANAMORI, MORIYA, HERTZ, EDWARDS, HEINE and most recently LONZARICH are noteworthy) the properties of magnetic metals are not what we are discussing here. In so far as he approached the metal-insulator transition, his work seemed primarily to show that systematic many-body theory, even applied in unconventional ways, is at a loss when faced with the Hubbard model, the Mott phenomenon and the metal-insulator transition. The strong-interaction problem, and particularly the «Mott insulator» limit, requires methods which deal with the strong coupling from the start.

The most promising method for the Hubbard model was devised by M. GUTZWILLER [18, 19]. Important developments carrying it forward were made by BRINKMAN and RICE [20] with a first plausible theory of the «Mott metal» in the strong-coupling limit, and recently by KAPLAN et al. [21, 22] and RICE et al. [23] in calculating some properties successfully without using Gutzwiller's crude «first approximation». Aside from this work and a growing corpus of numerical simulations, one had the feeling that decades of work on the Hubbard model had produced little understanding of even the basic physics; it remained almost the last hard well-posed problem of many-body theory. This impression received spectacular confirmation in January 1987, when it became clear [24, 25] that the high-T_c superconductors are straightforward cases of the Hubbard model. I shall devote the next two sections to this new aspect of the Mott phenomenon and the Hubbard model—both the basic physics of the relevant materials and new methods and theories which have been brought into play in the past six months to solve this case.

But first I want to continue on with the general discussion of strong-repulsion phenomena and generalized metal-insulator transitions. Another sideline we shall not discuss at length is the idea of the «microcosmic» magnetic state—the question of one or a few magnetic impurities in an otherwise nonmagnetic host background metal. FRIEDEL initiated the many-body resonance way of looking at this problem [26] and I discussed it using the infamous «Anderson model» in 1961 [27]. A local Mott-Hubbard repulsion term is introduced, leading to a behavior of surprising complexity, solved in the end only by introducing the quantum renormalization group method. The fascinating fact here, which puts it out of context with our story, is that in the ultimate end conventional physics wins out and the magnetic state is destroyed by renormalization. The same may be one possible fate of the *periodic* Anderson model, a problem which badly needs to be restudied once we have sharpened our weapons on high T_c.

Let me not leave the electronic problem without referring to some experimental landmarks. I mentioned SAWATSKY and ALLEN; for years there had been attempts to bring into focus by spectroscopy the fundamental parameters of the Hubbard model, which failed partly due to experimental naiveté and partly for reason of misinterpretation of the physical meaning of a model of this sort, which is meant to be expressed in terms of Fermi-liquid-like parameters renormalized not only for the large effects of ligand hybridization, but for dielectric screening and other high-frequency renormalization effects. SAWATSKY and ALLEN seem to have sorted this out.

A second experimental landmark was the work on V_2O_3, one especially favorable case of an experimentally accessible Mott transition [28, 29]. The true experimental problem was not that U is usually too small but that it is in reality so big that the Mott energy scale is above the vaporization temperature; but by happenstance of crystal structure, the relevant electrons in V_2O_3 belong to *pairs* of vanadiums and are thus rather weakly repulsive, forming a spin-one dimer which then «Hubbardizes». In this case the entire antiferromagnetic insulator-paramagnetic insulator-paramagnetic metal phase diagram can be explored, with an indubitably first-order Mott transition. Somewhat more common in this region of the oxides are the dimerization, covalent pairing or «spin-Peierls» transitions such as the titanium oxides exhibit. There are also the magnetite and Magneli-phase valence-density waves, as yet underexplored, which I cannot refer to further; and the two interesting cases $LiTi_2O_4$ and $NaTiO_2$ [30] which will enter our later story. Finally, some «double-exchange» systems occur in the lower d bands where mobile electrons act out the original fantasy of Zener [17], further elaborated by HASEGAWA and ANDERSON [31] and DE GENNES [32], of transmitting ferromagnetic moments from ion to ion by a combination of kinetic energy and Hund's rule direct exchange.

Finally, of great interest in recent years has been the «random Hubbard model» probably represented by the famous Si:P problem. Here we get a *continuous* metal-to-magnetic-insulator transition which exhibits fascinating and accurately measurable exponent anomalies. The recent work of Castellani, Kotliar and Lee [33] may finally have broken this problem—in any case, this gigantic literature can hardly be explored here.

Before immersing us in the high-T_c world I want to bring out the analogues of the Mott phenomenon in other apparently dissimilar problems. These are the problems of quantum solids and fluids and the quantized-Hall-effect problem. The basic ingredients all of these problems have in common are:

1) A basic role of strong repulsive interactions.

2) The presence of a state with an «energy gap», which one can variously think of as zero compressibility (*i.e.* a cusp in the energy as a function of particle density), a «commensurability gap» (an exact state occupancy criterion, as for the Landau levels or the Laughlin states) or an energy gap for free-particle or hole generation, as in true solids or insulators.

3) In general, the energy gap does *not* signal a broken symmetry, and the symmetry of thermally excited «insulating» states is identical with that of metallic states. In fact, the insulating state has an *extra* symmetry, as we have been learning in the past months, a *local* particle conservation or gauge symmetry.

4) Another property of the energy gap state is that it represents a state of special stability (because of the cusp in $E(n)$), so that, although it has no extension in number or occupation space, it may occupy a wide region in

chemical potential or pressure—hence, for instance, the unexpected appearance of Hall-effect quantization.

5) Finally, disorder can have a strong positive *or* negative role relative to interactions; it often favors the insulating state due to localization in electronic systems, but actually favors the superfluid in the Bose superfluid-solid system, and even for superconductors we often encounter localized insulator-to-superconductor transitions.

In a chapter of my book [34], on quantum solids, the concept of the true solid, as opposed to the density wave, is introduced. The true solid is, basically, a density wave with a commensurability condition enforced by a Mott criterion. The overwhelming majority of crystals are true solids, and in fact when we solve a X-ray crystal structure we implicitly assume an integral content of atoms per unit cell, though a surprising number of crystals actually seem to prefer a «defect structure» of some sort. The «superionic conductors» are an example, hence have no Mott-like barrier to ionic conduction. Metals are normally atomic true crystals but electronic density waves, with no true gap for electronic excitations. Finally, smectic liquid crystals are usually merely density waves with no commensurability criterion.

It is when one begins to think about quantum crystals or about charge or spin density waves, their electronic equivalent, that the real problem begins to surface. Classical melting occurs at a high temperature with a high density of defects and the ubiquitous «cusp» quite smeared out, but the quantum solid-fluid transition is often between two ordered states with totally different basic physics and the commensurability question much in evidence. One wishes a mean-field theory starting point for each state, but mean-field theories with «stay-at-home» theorems are not compatible with the quantum fluid description with identical particles—this is the Hubbard model problem surfacing again.

In the relatively easier case of ^{3}He a start has been made by VOLLHARDT, WÖLFLE and myself [35] following VOLLHARDT and BRINKMAN and ANDERSON, the basic idea being to use a Hubbard model on a «fiduciary» lattice which is treated as flexible, and to deal with the Hubbard model by BRINKMAN, RICE and GUTZWILLER. Neither approximation can be considered to be very superior, but prior to January I had planned to use this work and its extension to the Bose case by HSU and myself as the basis for my lectures here. This, and the effect of disorder on the Bose superfluid-solid transition, can both be approached in new ways which I hope to get to in a later section. At least for the regular lattice, one suspects that the «superfluid solid» proposed by LEGGETT [36] is disfavored by the «commensurability gap».

In the Hall effect problem, the «Mott phenomenon» aspect has not been recognized as much in past discussions as it deserves. Even for the integer effect, the spin splitting can be assumed to be predominantly caused by a

Mott-Hubbard repulsive term between identical Landau orbits of opposite spin on the same guiding center, a repulsion which is big compared to the one-electron spin splitting and comparable with the orbital splitting. For the fractional occupations, a similar «commensurability gap» arises, following from the well-known cusp structure for the energy of odd fractional fillings [37]. Transitions into or between fractional fillings are Mott transitions, either first order or continuous, and, as in the Mott transition, there is no true order but a local gauge symmetry, which has been exploited by GIRVIN and MAC DONALD [38] and READ [39] to make a Ginzburg-Landau theory of the phenomenon. Again, fuller discussion may or many not be possible at the end of my lecture series.

Addendum. As I reread the preceding I recognized that I had left out an entire branch of strong-repulsion physics, which has only momentarily and incompletely impinged on the «Mott» situation. This is the aromatic carbon compounds, such as benzene, graphite, and even heme. The magnitudes of the parameters are not far from those in the d-band metals, not as strong coupling as the oxides but highly interacting; and the subject of «ring currents», intensity anomalies and other interaction phenomena is a very old and much-studied one. It may be that RVB physics is not entirely irrelevant to these old problems.

2. – High-T_c superconductors as a magnetic Mott phenomenon.

A recent issue of *Physics Review Letters* published two papers [40, 41] showing that the isotope effect for high-T_c superconductors is, as near as never mind, zero. One of them spent great efforts to show that BCS-MacMillan phonon pairing theory with outrageous parameters applied outside its range of validity might barely give such a result, but did say the parameters were nonphysical. Many theories of these superconductors are less characterized by wishful thinking than that, but almost all have the character of trying to invent some form of attraction between pairs of electrons. This is supposed to give the dynamical screening which is the central mechanism of the BCS theory, a dynamical screening which is necessary because the high-frequency limit of the interelectron potential is just the repulsive Coulomb interaction which can, for the 1s homogeneous gap function of the BCS theory, only be evaded dynamically. All such dynamical theories give isotope shifts of order unity, even though accidental cancellation can occur.

For reasons which are intuitively obvious (electronic parameters depend on ionic mass only via zero-point motion) magnetic-transition temperatures depend on mass only to order θ_D/electronic energies $\sim \sqrt{m/M}$. Thus to those of us who believe these materials are exhibiting a new aspect of an essentially magnetic phenomenon, this news was much easier to take. It was not the first, but the last, of a long series of experimental indications that the rare-earth copper oxides are exceptional among d-band oxides only in the fidelity with which they adhere to the simplest Mott-Hubbard physics.

For some reason, the theories we have been trying to work out for high-T_c superconductivity have been seen as « contentious » and dismissed as « pure poetry » (*Scientific American*, May 1987). In the end, we are forced by the difficulties of the theory to some rather unusual formalisms, which may appear somewhat « poetic » if one does not see their necessity, but at base the theory is the only one so far which is founded in the rather grubby understanding of the underlying microscopic physics of the materials; and it is with a perhaps excessively detailed discussion of that that I want to start.

With very little controversy, I think everyone accepts that all of the electronic action takes place in the copper-oxygen complexed « d » ions. For the record I show figures of the structures of the two materials as yet authenticated (fig. 1), $(La_{1-x}[AE]_x)_2CuO_4$ (*), (La-Sr) and $YBa_2Cu_3O_{7-x}$ (1 2 3). In both there are only slightly distorted square planar layers of CuO_2, with O's between every Cu; and all Cu-O-Cu bonds are equivalent. (Another Y-Ba-Cu material, with planes only, has demonstrated that the chains of Cu-O-Cu-O in 1 2 3 are not relevant.)

Octahedral « ligand » complexes of the d transition metal ions are common to very many magnetic materials: the MnO-NiO series, the ferrites, the garnets, the antiferromagnetic difluorides MnF_2-CuF_2, the perovskite structures $KMnF_3$-$KNiF_3$, etc. They have been studied extensively both by means of band calculations and by local cluster methods, to which they are very suited. VAN VLECK [9] first understood their basic chemistry, which was then extended by ORGEL, BALLHAUSEN and others (see [10, 11]); MATTHIESS [42, 43] made what are now accepted as reasonably accurate band calculations with very similar outcomes.

The primary bonding is ionic, in the sense that a series of bands of symmetry equivalent to a filled O (or F) $2p$ shell is invariably full, and most of them quite far below the Fermi level; while the bands causing the Fermi level are constructed from Wannier functions of metal « d » symmetry (the empty s and d bands of the rare earths and alkaline earths are very high). In all of these materials, however, there is a large stabilization due to « semi-covalent » admixture of ligand (O) $2p$ orbitals with the d's on the transition metal ion. That is, the O $2p$ bands contain a bonding admixture with the Cu d orbitals, and the magnetic orbitals near the Fermi surface are *antibonding* admixtures of Cu $3d$ and O $2p$. As shown by ORGEL, this « ligand field » effect

---

(*) [AE] = alkaline earth.

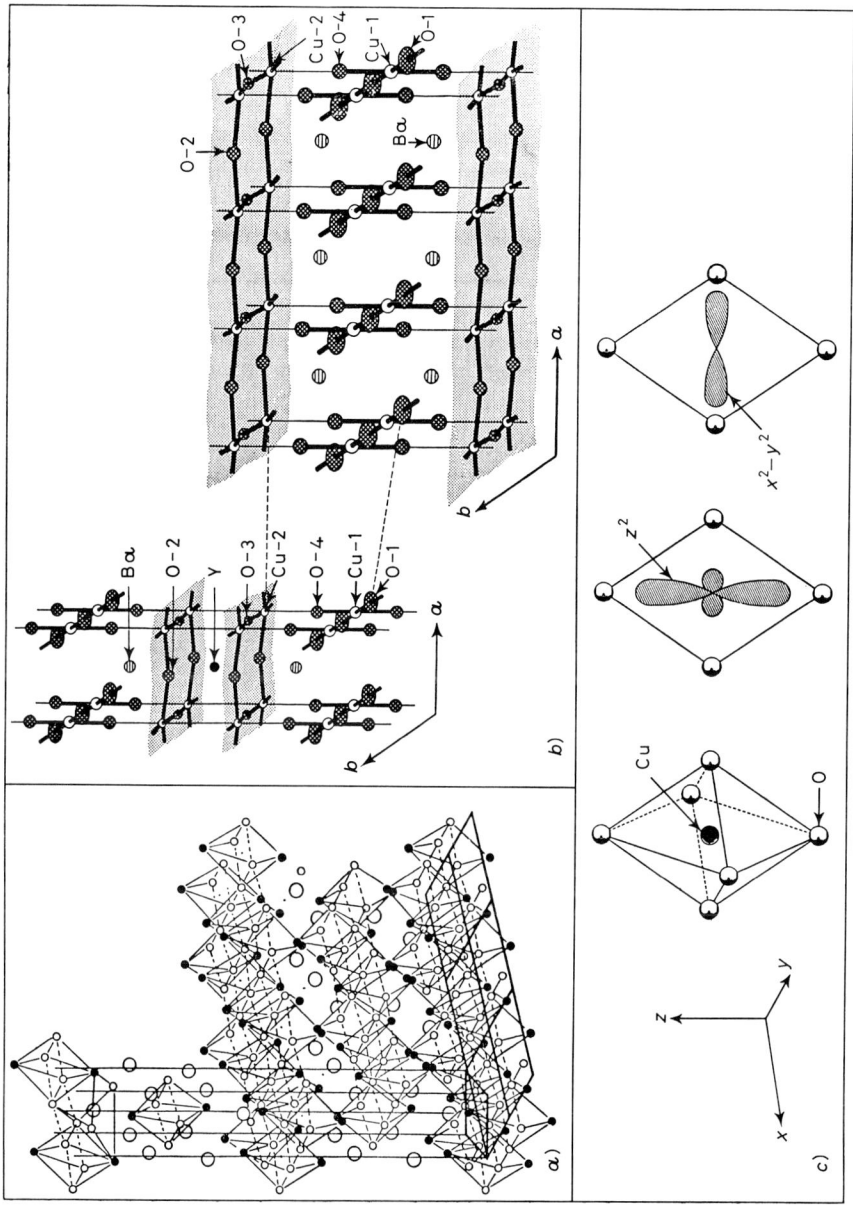

Fig. 1. – a) Crystal structure of La_2CuO_4 (◯ La, ● ○ O), b) crystal structure of $YBa_2Cu_3O_7$, c) Cu-oxygen octahedron.

stabilizes the octahedral ligand shell by energies of the order of the crystal field parameter «10 Dq» ~ 1 eV or more. The $3d_{xy}$ orbitals admix with O $p\pi$'s, more weakly than the $d_\sigma(x^2 - y^2, 3z^2 - r^2)$ admix with the p_σ levels. In the high series members such as Ni and Cu the empty orbitals are d_σ's and the octahedron is especially stable because of the d_σ-d_π splitting 10 Dq, which typically, for a divalent ion, is of order $\sim (10 \div 12\,000)$ cm$^{-1} \simeq 1$ eV. In Cu^{++}, an additional stabilization of at least $\frac{1}{2}$ eV occurs by breaking the pure octahedral symmetry; a so-called «Jahn-Teller» distortion which leaves a band based on the antibonding-hybridized $d_{x^2-y^2}$ orbital above a mass of «spaghetti» due to nonbonding O $2p$'s, filled d orbitals of other symmetries, etc. (Figure 2 gives the band structure as calculated by MATTHIESS [44] and the CuO$_6$ cluster schematic energy level diagram.)

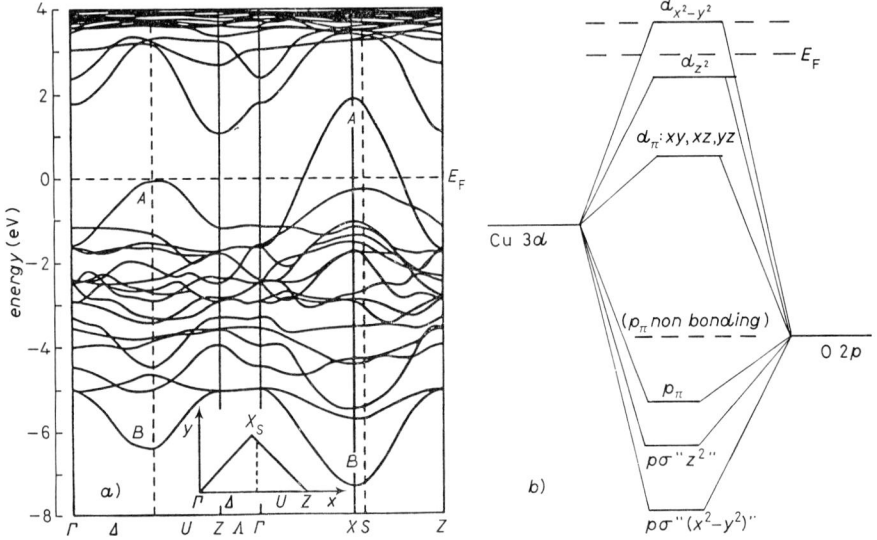

Fig. 2. – a) One-electron band structure [44] for La$_2$CuO$_4$. b) Ligand field theory. Energy levels of a Cu^{++}O$_6$ octahedron.

As we see from either calculation, the levels based on Cu^{++} with one hole in the $d_{x^2-y^2}$ state are nicely isolated in this system. In La$_2$CuO$_4$ this must be the basic electronic state of the system: a singly occupied Mott insulator of a simple type. In a hypothetical ordered YBa$_2$Cu$_3$O$_{6.5}$ with Cu^{++} stoichiometry only, the same would be the case.

The physics of such compounds as La$_2$NiO$_4$, which is insulating and antiferromagnetic, is very simple. There is a «superexchange» interaction through the hybridization of the d levels with the O $2p$ levels which gives a strong antiferromagnetism. The relevant exchange interaction can be roughly estimated by quantifying the hopping integral t by using the ligand field energy;

t is roughly equal to that energy because both involve two steps of d-p transfer. For Cu^{++} we expect rather a large t (of order 2 V as calculated by MATTHIESS) because of the huge Jahn-Teller distortion giving a ligand field splitting nearly twice as big as for Ni^{++}. For Cu^{+++} it will be even bigger. U will be of the conventional size, $(5 \div 10)$ V, which gives a t^2/U of a few tenths of a volt, between 1000 and 5000 K, which must be halved because of the effect of direct exchange and correlated hopping. Hence the antiferromagnetic T_c, if it were not for fluctuation effects which are very serious in 2d and for spin $\frac{1}{2}$, would be extremely high, perhaps as much as 1000 K. (Whatever $T_{Néel}$ is present must, of course, be caused by either the small anisotropy or interlayer forces.)

In the superconducting materials, a fraction of the Cu^{++} has been replaced by Cu^{+++}. The resulting state is less unequivocal *a priori*. Three states are possible: 1) If the Jahn-Teller distortion is large, Cu^{+++} has two holes in $d_{x^2-y^2}$ and is a singlet. This is what we believe happens in reality. Thus in effect the added carriers are simple holes in the pure Cu^{++} network of electrons in $d_{x^2-y^2}$. Note that in fact the extra hole goes into a state heavily hybridized with a d-like linear combination of O ($2p$) electrons so that it is only symmetrywise in a « d » orbital. The actual « d » occupancy changes little between Cu^{++} and Cu^{+++}. 2) A Cu^{+++} could occupy d_{z^2}, if the Hund's rule exchange between the two spins outweighed the energy to be gained from the tetragonal distortion. This leads to a $S = 1$ site, and probably to a large relaxation of the O octahedron around it since the state is orbitally of cubic symmetry. Such a carrier would probably be very immobile—self-trapped, in fact. The corresponding distortion in NiO makes small polarons of the corresponding carriers. Thus these can hardly be the actual metallic carriers in the superconductors, but they may play a role in confusing experimental data as magnetic impurities. For example, if there is a local distortion of the octahedron caused by some defect, it may attract Cu^{+++} $S = 1$ configurations, which would act like magnetic impurities. 3) A final suggestion is that the additional « holes » could occupy one of the « spaghetti » bands and reside in an oxygen-based p-band orthogonal to the Cu d orbitals: some kind of « p_π » band. This is proposed by some theorists, assuming that the interaction leading to superconductivity is via spin fluctuations in the underlying band. Again, *a priori* one cannot rule this possibility out. It can only be eliminated experimentally or by accurate electronic calculations, and such calculations would have to be made in the presence of the background of localized Cu^{++} electrons. What calculations we have seem to isolate the $d_{x^2-y^2}$ band fairly clearly at the Fermi level, but, of course, they are not conclusive. I emphasize that core level spectroscopy does not distinguish very well among these possibilities. A Cu^{+++} will be much more attractive to electrons than Cu^{++}, so the d holes will hybridize much more than in Cu^{++}, and as a result the net amount of d^9 *vs.* d^8 character will hardly change. This phenomenon, that the *actual* valence state of atoms in terms of relative occupancy of the atomic levels usually changes very little even when the magnetic or chemical valence changes a lot, is often deceptive, and has to do with the fact that interaction energies between atoms are smaller ($\leqslant$ a few volt) than the atomic energy level differences of ~ 10 V.

This possibility has become popular and it is essential to show why it is very unlikely. In the first place, it is not reasonable that such bands are based, symmetrywise, on the strongly bonded oxygens of the CuO_2 lattice (see fig. 2b)). The d_σ orbitals of these oxygens are much involved in the strong, rigid semicovalent process which produces the square lattice, while the π bonds are hybridized with the d_{xy} and d_{xz} orbitals and must, therefore, be below these. All these orbitals (in fact, the ones used in these theories) are necessarily below d_{z^2} and will not be the first to empty. (There is *one* genuine nonbonding band of « f » symmetry based on π electrons of the CuO_2 planes; it is clearly as irrelevant as La orbitals.) On the other hand, orbitals on the La or other oxygens, which *a priori* are perfectly possible, are so weakly coupled that the whole phenomenology would be incomprehensible. While these chemical considerations are conclusive, a simple experimental argument is the sensitivity to doping of the antiferromagnetism of La_2CuO_4. A few O $2p$ carriers will not destroy the antiferromagnetism; corresponding carrier densities in other systems do not. « Spin polaron » phenomena in magnetic systems are pretty well understood, and they do not behave like these materials. Second, we might expect effective masses in these bands to be fairly light, and in fact the whole superconductivity phenomenon would look much more like weak-coupling BCS than it actually does, in that the ratio of coupling energies to kinetic energy of the carriers should be small: the binding energy should behave like T_c^2/E_F with $E_F \sim$ volt, for instance, the coherence length $\sim E_F/T_c \cdot$ $\cdot 1/k_F \gg a_0$, etc. This kind of theory seems more suited to the heavy-fermion case, where the T_c's are below 1 K, and where the relevant electrons really do come from substrate bands intermingling with magnetic ones. It is perhaps an alternative to the theory we shall discuss, but has no possibility of explaining the very bizarre behavior of the high-T_c superconductors: such anomalies as the peculiar normal transport properties become incomprehensible.

I would like to make a brief summary of the data up to this point, extremely biased and incomplete, with emphasis on the experimental results I consider especially significant or puzzling.

2˙1. *Phase diagrams.*

La_2CuO_4. Here there is a high-temperature « twitch » transition, orthorhombic to tetragonal, varying from 550 K for pure to $\leqslant 250$ K for doped material. It is controversial whether the displacement decreases near T_c. The displacements are electronically irrelevant: susceptibility and ϱ do not have any anomalies at T_{twitch}; the Hall effect may or may not.

Doping with O vacancies leads to antiferromagnetism with increasing μ

and T_c up to ~ 250 K. (This effect probably is a result of bound Cu^+ defects, and of the O vacancies being in the Cu-O planes which is the natural place chemically and hence disfavoring RVB.) « Pure » La_2CuO_4 seems to contain a small amount of a superconducting phase, $T_c \sim 37$ K, but is possibly antiferromagnetic. $< 0.1\%$ of Sr makes it semiconducting but not antiferromagnetic; but only at 5% does T_c begin to rise, roughly linearly, to 40 K at $(0.15 \div 0.2)$ Sr. More than 0.2 Sr probably does not further alter the number of holes but adds O vacancies (hence we have no proof that T_c goes back down). The changes in magnetic state with small amounts of doping are very unusual and significant. Mixing dopants may raise T_c; in any case there is, in $(0.15 \div 0.2)$ Sr samples, much indication of a second « weak superconductivity » or other phase transition giving a sharp resistivity decrease at $\geqslant 200$ K. This phase also has anomalous sound velocity and attenuation. The normal-state resistivity and thermopower are very anomalous and incomprehensible as conventional electronic transport.

2'2. « 1 2 3 ». – The order-disorder transition for O vacancies forming CuO chains is very high (~ 800 K) (*). No recognizable « twitch » or antiferromagnetism. At $O_{6.5}$ it goes insulating, the absence of carriers being verified by thermopower and Hall effect; T_c falls as we approach 6.5. At least one compound with only planes and no 1-d chains has $T_c \sim 90$ K. Again, many indications of transient or weak truly high-T superconductivity.

Many popular press claims of minority phases with T_c's of 160 K—pretty sound—and higher have been made. These rumors have tended to be right in the past.

Specific heat. All $(La_2CuO_4, (La-Sr)_2CuO_4, 1\ 2\ 3)$ show a linear γT term at low temperature, usually accompanied with a low-T anomaly indicating some magnetic impurity on the 0.1% level [45-48]. We do not know whether the impurity is a Cu^{+++} $S = 1$ near a defect or whether disorder or defects bind spin solitons (see below). The linear specific heat is a key prediction of RVB theory, at least of our « gapless » version (**). The anomaly at T_c is hard to sort out and is fitted by experimentalists to a BCS jump, for no good reason, for a reasonable to large γ value. The coherence lengths derived in any of a number of ways are very short even in very pure material: of order cell size or a factor 2 or 3 larger, *i.e.* basically the distance between active carriers. Of course, there should be some kind of critical behavior and not necessarily a

---

(*) The transition appears to be cooperative with a Jahn-Teller $x^2 - z^2 \to z^2$ electronic transition on the « chain » Cu's. We believe the chain Cu's are irrelevant for more or less obvious chemical-potential reasons, and that with O_7 the valence is 2.5 on the « planes ».

(**) *Note added in proofs.* – Later work by COLEMAN, JOYNTH et al., AFFLECK, KOTHAR and others shows that a gapless « flux phase » also exists with $C \propto T^2$ vs. T.

simple BCS specific-heat jump, which is an artifact of the small parameter a/ξ or $\Delta/E_{\rm F}$. Either from $H_{\rm c}$'s, or from the enormous change in sound velocity [49] ($\sim 0.1\%$), one sees that the binding energy per carrier is about $T_{\rm c}$ and the entropy, therefore, ~ 1, neither the size nor the shape of the BCS specific heat can be expected to fit, since with $S \sim 1$ there will be large fluctuations. A preliminarly look at $H_{\rm c}(T)$ in single crystals suggests $H_{\rm c} \sim t^{\frac{3}{2}}$, which implies a simple specific-heat *cusp*, in not bad agreement with Phillips' data, but other data give $H \sim t$ and the conventional jumps.

GINSBERG shows an anomalous rise in C at $\sim (0.2 \div 0.3)\, T_{\rm c}$, not far from where a typical roton specific heat would begin to show up, and LIANG and I have fitted that to a rotonlike gap of $\sim 3kT_{\rm c}$. Phillips' magnetic-field-dependent C shows a λ-like anomaly which, oddly enough, almost tracks the early data of Ramirez et. al.

2˙3. *Gap measurements*. – Perhaps the worst fiasco is infrared «gap measurements». All of the published values involve fits of data to BCS which do not fit BCS, and there are no direct-gap measurements. These can be totally ignored. The only significant optical feature is a broad, probably electronic, excitation around 0.5 eV, which appears to be proportional to dopant. This may be a transition «spaghetti» $\to Cu^{+++}\, d$ state, or possibly hole soliton $\to$ $\to$ electron hole $+$ spin soliton (again, see below). Preliminary single-crystal data do not show it, or suggest it is polarized out of the plane. There is a dopant-related bump at ~ 250 cm^{-1} which, we suspect, is simply a large polaron TO phonon of the Cu-O planes *vs.* the rest of the lattice.

Tunneling gap measurements vary from $2\Delta = (3 \div 6)\, kT_{\rm c}$, with the higher values favored. Again, clearly BCS should not serve as a model: if nothing else, $T_{\rm c}$ is not the mean-field $T_{\rm c}$ because of fluctuations. Some anomalous very large gaps appear.

In the tunneling curves, the backgrounds are more interesting than the gaps when these appear. For superconductor-superconductor junctions, a gap conductance $\sim |V|^p$ with $p \geqslant 2$ is almost always seen; for normal-superconductor junction, a dependence like $|V|$ is usual. $|V|$ also appears for superconductor-superconductor junctions above $V_{\rm g}$. These behaviors are universal enough that they may represent some new intrinsic process or processes.

A measurement of NMR T_1 is interpreted as a gap $\Delta = 0.65\, T_{\rm c}$ [50]. This is nonsense; the experimentalist has taken no pains to exclude the predicted behavior $T_1 T = \text{const}$. Again, attempted fits to BCS conceal most of the physics. It is also noteworthy that T_1 jumps up at $T_{\rm c}$, not down, by a factor $\lesssim 2$. Do the spinons change here?

I mentioned the sound velocity data—both La-Sr and 1 2 3 stiffen by about 0.1% below $T_{\rm c}$ [49]. This is enormous (stiffness has the dimensions of binding energy and is a rough measure of it, and this means a relative change of order 100 K $\sim T_{\rm c}$!) and also, in itself, proves that the zero isotope effect is not a fluke. A theorem due to CHESTER shows that the normal BCS isotope effect implies softening of the lattice. The stiffening suggests, if not requires, that the lattice enter the problem only by determining the electronic parameters such as t and U. The intrinsic dynamics in t and U is very high frequency. The small isotope effect seen in La-SrCuO$_4$ is shown by BHATT and MILLIS [51] and HSU to be the right order to be simply the effect of zero-point motion on t and U.

A final set of interesting data is the normal-state transport. This shows two behaviors in reasonably good samples, both unusual. The best samples of both La-Sr and 1 2 3 seem to have zero residual resistance extrapolated from above T_c, with a very large T-dependence, often roughly linear but, if interpreted as simple Fermi level electron resistance, implying a $\hbar/\tau$ many times kT [52]. Bosons undergoing elastic scattering have zero residual resistance, but it is not clear that this is the only possible explanation. The Hall effect, as we mentioned, implies roughly one carrier per dopant in La-Sr. The Hall effect for 1 2 3 also counts doping above $O_{6.5}$, but the thermopower vanishes at O_7. This can be a result of all the Cu^{+++} being in the planes, a likely result in fact since the O disorder is in the chains, which must be quite disrupted by it; also there is the compound without chains of CuO at all which has $T_c = 90$ K. Giant thermopower in La-Sr has the strange property of falling *continuously* to zero at T_c.

Another behavior of ϱ which may be significant is a large positive anomaly in the vicinity of T_c which is often seen, and may be of the $T \ln T$ form given by diffusion theory but not necessarily easily fitted to that theory.

3. – RVB theory, such as it is.

I read a recent review of a thriller in which the members of a football squad were being murdered, one by one. If the detectives were to make the assumption that each murder was an independent case, to be solved for its own motive and murderer, one would be justified in putting the book down in disgust. It seems to me that much of the theoretical work on high-T_c superconductivity has ignored the many other crimes being committed in the same vicinity, while the overwhelming probability is that all of them have the same underlying cause. Let us list the peculiarities:

1) High T_c itself.

2) The vicinity of a metal-insulator transition.

The insulator is strange in 3 ways.

3) There is no visible source of the energy gap in the form of antiferromagnetism or a charge density wave.

4) There *is* antiferromagnetism nearby with a peculiar doping sensitivity, but not in the immediate neighborhood of the high-T_c materials.

5) There is a crystal structure («martensitic» or «twitch») transition which emphatically does *not* cause the gap, in fact is itself mysterious.

6) The dimensionality is low.

7) The ions, if magnetic, are isotropic $S = \frac{1}{2}$, and a fair number of $S = \frac{1}{2}$ compounds show strange behavior.

8) The insulator *and* superconductor show $C_{sp} = \gamma T$ around 1 K.

9) Many properties, especially ϱ and tunnel current, have strange dependences in the normal material.

10) Possibly: weak superconductinglike behavior well above T_c.

So we have a strange insulating state, a strange normal state and strange superconductivity. The simplest must be the insulating state and we started with that.

For a long time it had been evident to me that there were very few $S = \frac{1}{2}$ antiferromagnets: the appropriate compounds of Cu^{++} and Ti^{3+} just did not seem to exhibit antiferromagnetism. There are a few: notably $CuCl_2 \cdot 2H_2O$, one of the classic cases of antiferromagnetism, and CuF_2, but the obvious ones like CuO and $KCuF_3$ did not behave normally. I finally began to cast about for a different explanation, and in 1973 ([53], see also [54]) proposed that, at least for some lattices, the Néel state was not a stable ground state of the $S = \frac{1}{2}$ Heisenberg Hamiltonian

$$(3.1) \qquad \mathcal{H} = + \sum_{\langle ij \rangle} J_{ij} \sigma_i \sigma_j$$

to which the Hubbard Hamiltonian leads for a nondegenerate half-filled band. In order to illustrate the point, I used the triangular lattice, but the question is a more general one: can there be a featureless, liquidlike *singlet* ground state for the $S = \frac{1}{2}$ antiferromagnet in 2 or higher dimensions, similar to the 1-d Bethe [55] solution for the antiferromagnetic linear chain? As HULTHÉN [56] described it, the Bethe solution may be described as a linear superposition of states made up of singlet pairs $(\alpha_i \beta_j - \alpha_j \beta_i)/\sqrt{2}$. A nearest-neighbor singlet pair contributes $E = (J/4)(\sigma_i \sigma_j) = -(J/4) \cdot 3$, and the linear chain has $N/2$ pairs, so the zeroth-order energy is $-3/2(NJ/4)$, whereas the naive Néel state has N bonds contributing $\sigma_i \cdot \sigma_j = -1$, or $(-1)(NJ/4)$. The Bethe state is about 15% better than the singlet pair state and its excitations and correlations make much more sense in that context than they do referred to the Néel state.

In order to connect the «spin-Peierls» paired states, paired according to

1 2, 3 4, 5 6, ... to (2 3, 4 5 ...) it is necessary that some number of pairs of very large separation be admixed into the ground state, which also implies that there be low-energy excitations very like free mobile spins, or what we shall call «spinons»: spin solitons behaving like free Fermi particles. These excitations were discovered by DES CLOISEAUX and PEARSON in 1955 [57].

In 1952 I showed that in lowest order of a random-phase approximation the Néel state *could* be stable in two dimensions against zero-point fluctuations of spin waves [58] and later NEVES and PEREZ [59] showed that it *was* stable for not too small a spin value—excluding, however, $S = \frac{1}{2}$. Nonetheless, the frustrated triangular lattice seemed a good case for the «resonating valence bond» state, as I called it, because in this case the triangular Néel structure and the best bond structure had the same energy, $-3/2(NJ/4)$.

We could do nothing but numerical calculations using the techniques developed by HULTHÉN [56] and by RUMER [60] and PAULING [61], who used similar linear combinations of bonding schemes to describe the valence states of aromatic molecular systems—hence my name, «resonating valence bond» (RVB).

Nonetheless, eventually it seems to have been established that at least the triangular lattice is theoretically a RVB state. Recent experimental data ([30] and unpublished) on $NaTiO_2$ are ambiguous on this point, in fact the data on $NaTiO_2$ closely resemble those on La_2CuO_4 in that some samples show RVB and some antiferromagnetism! As we shall see, the exact numerical preference is not a serious issue from the high-T_c point of view. The important point is the possible existence as a locally stable state of this «spin liquid» as a reference frame or «vacuum» state; and the fact that this state is clearly an experimental fact in many samples of La_2CuO_4 and probably in $BaBiO_3$ and $YBa_2Cu_3O_{6.5}$ and $YBaLaCu_3O_7$. When such a state exists, it has very interesting properties, as we shall see and as was not evident back in 1973. In the first place, while remaining insulating it does contain a large amplitude of electron pairing fluctuations, with the pairing energy $\sim J \sim 10^3$ K, which is clearly relevant to high-T_c superconductivity. Even more interesting, it has the possibility of particlelike soliton excitations which will exhibit very nonclassical properties.

It has been understood at least since 1975 that broken-symmetry states have topological soliton excitations, like vortex lines or points, for instance, which arise from a topologically stable configuration of the order parameter and cannot be described directly as a local phenomenon. Some of these, as in the case of defects in charge density waves, can have peculiar spins or charges or statistics. But it has been realized only recently that a featureless liquidlike state can have even more intriguing solitons, with the discovery of the fractional-charge solitons of the Laughlin state—the only other case of a liquidlike state with a commensurability gap [37]. In the RVB state, we have point particlelike solitons with peculiar spin and charge properties. In the original work on RVB [24, 25], I pointed out that the state had uncharged spin excitations obeying Fermi statistics; KIVELSON, ROKHSAR and SETHNA [62] then made the point that in that case there must also be a singly charged boson excitation with no spin. They identify the superconducting transition as a Bose condensation of these particles; with reservations, we accept that such a condensation at least plays a major role.

Let us first give a pictorial view of the state and the excitations, and then discuss a couple of formal ways of expressing the physics. Pictorially, if we have a state made up of a superposition of all possible pair bondings of local spins

$$\Psi = \sum_{p(ij \, kl \, \text{etc.})} (ij)(kl)(mn) \ldots,$$

we can imagine inserting an excitation on site i in three possible ways.

1) The first is the straightforward one of removing the electron from site i leaving its partner j as a free spin. j can, of course, be any site but will tend to be in the neighborhood of i. Neglecting kinetic energy, we will have as our basic Hamiltonian

(3.2) $$\mu \sum_i n_i + U \sum_i n_{i\uparrow} n_{i\downarrow} + \frac{t^2}{4U} \sum_{ij} (\sigma_i \cdot \sigma_j - 1)$$

and the removal of one electron will cost $\mu + 3t^2/4U$ if i and j are nearest neighbors (or, adding one will cost $U - \mu + 3t^2/4U$; the cusp in $\partial E/\partial n$ is U). The resulting excitation has, of course, charge e and spin $\frac{1}{2}$.

2) Second, we can eliminate the site i from the pairing entirely and allow the exchange energy of the background to relax. There will presumably be little change in the state of distant electrons but those nearby—with a decreasing amplitude, actually to ∞—will have to rearrange their pairing. (We emphasize that unlike KRS we firmly suppose that while nearest-neighbor pairs dominate, all pairing distances are allowed and there is no *topological* necessity to modify more than a few bonds.) The cost will now be only μ or $U - \mu + \frac{1}{2}$ (exchange energy) so that *a priori* the « hole soliton » excitation is stabler than the true electron.

3) Finally, we may take the « hole soliton » and refill site i with a free unpaired spin. This still does not break up any unnecessary pairs so only costs half the energy of breaking up a pair, and *no* charge energy: if there really are a number of infinitely long-range bonds, it will not necessarily cost any energy. This is our « spin soliton » or « spinon », the basic excitation from which the whole scheme proceeds. We will try to show that the energy of these excitations vanishes at a surface in k space, the « pseudo Fermi surface ». The spin soliton is the easiest to derive and to demonstrate (via the

2 - *Rendiconti S.I.F.* - CIV

resulting low-T specific heat) but we emphasize that once it exists it can be bound to a real electron to form a hole, so that the hole boson and the spin fermion imply each other.

We have ignored kinetic energy, which is obviously important for the hole and electron excitations; however, it is roughly the same, as we shall see (*). The kinetic energy per se does not affect the spinons.

A final qualitative picture, at least of the spinon excitations, is useful.

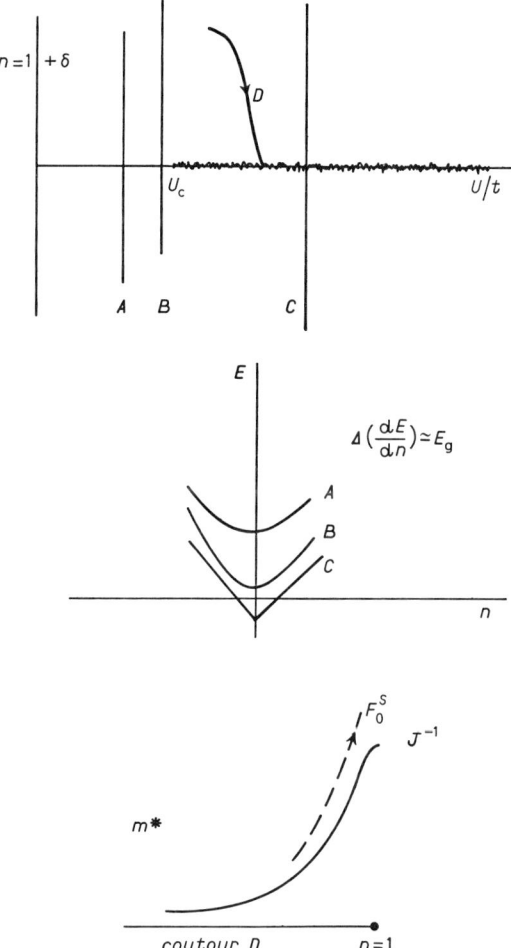

Fig. 3. – Brinkman-Rice phase diagram for the Hubbard model; a path into the insulating state is shown.

(*) *Note added in proofs.* – More recently, we have realized that the hole soliton is much more favorable in kinetic energy.

We imagine gradually moving through the Brinkman-Rice-Gutzwiller phase diagram (see fig. 3) of U/t vs. the occupancy $n = 1 + \delta$, and approaching the insulating state. In general this cannot be really done continuously because the insulating state is antiferromagnetic and because of the long-range Coulomb forces which cause a Mott first-order transition. If, however, we exclude antiferromagnetism, and move through the locally stable or metastable metallic states, according to Fermi-liquid theory the Fermi surface remains *until the system is actually insulating*. The Brinkman-Rice theory [20] assumed that $Z \to 0$ and $m^* \to \infty$ as we approach the insulator, but that was based on the first-approximation solution of Gutzwiller which neglects local antiferromagnetic correlations; in fact, m^* can never become heavier than $m_0 \cdot U/t \sim 1/J$, because of hopping via exchange. The only way the system can become insulating is, then, for m^* to remain finite but the bulk modulus $K \propto F_0^s$ to become infinite, which is merely to say that $K = \partial^2 F / \partial n^2 \to \infty$, which implies a cusp in the energy vs. n, which is an *energy gap* for charged excitations. But the entropy of the spin fluctuations, so long as there is no antiferromagnetism, *must* be manifested in a Fermi surface, which we call the «pseudo Fermi surface» (PFS). One interesting point, however, is that in this limit Z still goes to zero and the fermion excitations have strange properties; one, in particular, is that $s_{i\sigma}$ is not independent of $s_{i-\sigma}^\dagger$, since each actually creates the same excitation: the density of states at the Fermi surface is halved. This comes out formally in the theories which we shall shortly discuss.

The existence of the PFS for the insulating case is only found in one experiment, by PHILLIPS [45] on La_2CuO_4, where a linear specific heat of ~ 1 mJ/mol deg^2 is found, about twice Cu. This implies a rather big J, even accounting for the above factor 2: about 5000 K. One problem is that spinons can clearly become bound to defects, and in fact all data at low T show a high density of free magnetic spins with exchange fields in the $\leqslant 1$ K range, both for pure La_2CuO_4 and for La-Sr and (1 2 3). This is manifest in many of the specific-heat measurements by PHILLIPS, OTT, RAMIREZ [45, 46, 48] and others, and in the unpublished inelastic neutron scattering data of Aeppli (*).

In fact, we have an example of a RVB state with strong disorder in the case of Si:P, which has been understood by BHATT and LEE [63] in the context of pairing of the majority of spins but localized defects in the pairing structure. The specific heat of Si:P is remarkably like that of (1 2 3).

I have «jumped the gun» here by referring to the superconducting state; our best guess is that the PFS remains in the high-T_c superconducting state, and in fact at least seven separate measurements confirm a linear specific heat in the $(1 \div 5)$ K range, often quite a large one: $(3 \div 6)$ mJ/mole K, which implies a J of a more reasonable magnitude.

(*) *Note added in proofs*. – The data of Shirane, Endoh, Birgeneau *et al.* confirm the existence of a PFS.

4. – Formal theories of RVB and high-T_c superconductivity.

4'1. *Introduction; Gutzwiller-BZA theory.* - Modern treatments of the Hubbard model begin with the work of Kaplan *et al.* [21, 22] and of Rice *et al.* [23] on the accurate evaluation of the Gutzwiller wave function. As we have argued, the essential physics is almost all contained in the Hubbard Hamiltonian

$$(4.1) \qquad \mathscr{H} = -t \sum_{\langle ij \rangle} (c^\dagger_{i\sigma} c_{j\sigma} + \text{h.c.}) - \mu \sum_i n_i + U \sum_i n_{i\uparrow} n_{i\downarrow}$$

which can be straightforwardly transformed, to order $1/U$, into the superexchange Hamiltonian

$$(4.2) \qquad \mathscr{H} = -t \sum_{\langle ij \rangle} (1 - n_{i-\sigma}) c^\dagger_{i\sigma} c_{j\sigma} (1 - n_{j-\sigma}) + \mu \sum n_i + \frac{t^2}{U} \sum_{\langle ij \rangle} (\sigma_i \cdot \sigma_j - 1).$$

Before going on, let us mention important corrections. In my original paper, and in KRS, it is emphasized that (4.2) is modified by the possibility of a spin-Peierls coupling of an O breathing mode

$$(4.3) \qquad \lambda \sum_{ij} x_{ij} (\sigma_i \cdot \sigma_j) = \mathscr{H}_\text{s-p}.$$

Barring actual spin-Peierls instability, this probably contributes little to the physics because its major effect will be the second-order-in-λ effect of polarization which is $\propto (\sigma_i \cdot \sigma_j)^2 = -2\sigma_i \cdot \sigma_j + 3$ and changes nothing. Probably, however, $J = t^2/U$ is too large and the RVB frequency is too fast for the spin-Peierls modes. (Though in the disordered RVB they may play an important role in trapping local spin configurations.)

Second is longer-range Coulomb interaction. This seems to play only a minor role in the metal-insulator transition here, primarily because we do not look at the appropriate variables: we do not cross the weak-to-strong coupling boundary, U and t remaining large relative to all temperatures. The Coulomb interaction may play a big role in superconductivity in determining if hole pairs bind; one must not take the Hubbard Hamiltonian too seriously in the details of superconductivity. But we believe it contains the basic physics. U may be estimated at $(5 \div 10)$ V, t, from the calculated band structure or from the crystal field splitting, at $1 \div 2$, giving $J \sim (1000 \div 2000)$ K.

The Gutzwiller method is a variational estimate of the wave function, a rather primitive version of the «Jastrow» wave function often used in other problems. One assumes a given independent-particle wave function (in the classic case, a free-electron determinant) and applies to this a projection operator P_G which eliminates some fraction of the double occupancy inherent in the free-particle wave function. We will be interested here only in the complete elimination of double occupancy where

$$(4.4) \qquad P_G = \prod_i (1 - n_{i\uparrow} n_{i\downarrow}) \,.$$

The above authors applied P_G to the free-particle Fermi sea for half-filled linear chains and found that all but 0.1% of the antiferromagnetic correlations (and hence the energy) of the Bethe solution was recovered. In the one-dimensional case, a rather subtle calculation allows the entire theory to be done exactly except for a final integral.

In two dimensions, an excellent result for the energy was obtained, although it was not anywhere nearly as close to the presumed ground-state energy. The moral is, however, that the free-particle wave function contains a very large fraction of the antiferromagnetic correlations that one might expect in the correct RVB solution. We believe the reason is to be found in the fact that this wave function turns out, by an unexpected identity, to be equivalent to a function based on the RVB idea.

If one wished to construct *a priori* the appropriate RVB state, one would attempt to create $N/2$ nearest-neighbor pair bonds with the operators

$$(4.5) \qquad b_{ij}^\dagger = c_{i\uparrow}^\dagger c_{j\downarrow}^\dagger - c_{i\downarrow}^\dagger c_{j\uparrow}^\dagger \,,$$

place them on the lattice in some array, and add up, in as near equal phase as possible, all the ways of doing this. Since $b_{ij}^\dagger$ is a boson, one suspects that the best way is a Bose-condensed state

$$\Big(\sum_{\langle ij \rangle} b_{ij}^\dagger \Big)^{N/2} \Psi_{\text{vac}} \,,$$

and I showed in my first paper that this is equivalent to a BCS state of a certain type. Such a state, unfortunately, has as many components with site i doubly occupied or empty as does the free-electron state, and it is essential to Gutzwiller project it onto the $d=0$ subspace. But it does have the advantage over the free-electron state that it can be variationally optimized before projection *à la* BCS, to get the best arrangement of pair bonds; although as yet we have no guarantee that optimization and Gutzwiller projection commute.

This optimization process is equivalent to a straightforward BCS theory using the exchange term in (4.2) as an interaction term, and using no kinetic energy at all; this process was carried out by BASKARAN, ZOU and ANDERSON [64]. In the half-filled case, the kinetic energy projects out completely and we have

$$\mathcal{H} = -J \sum_{\langle ij \rangle} b_{ij}^\dagger b_{ij}$$

(with J in this expression $= 4t^2/U$).

The BCS solution for this Hamiltonian is

(4.6) $$\Psi = \prod_k (u_k + v_k c_{k\uparrow}^\dagger c_{-k\downarrow}^\dagger) \Psi_{\text{vac}}$$

with

(4.7) $$\sum_k \frac{v_k}{u_k} = \sum a(k) = 0$$

and $v_k = \pm u_k$ depending on whether k is inside or outside the pseudo Fermi surface determined by the quasi-particle energy

$$E_k = \tfrac{3}{4} J \gamma_k$$

with

$$\gamma_k = \cos k_x a \pm \cos k_y a \,.$$

These quasi-particles are the « spinons » of the theory and, as we see, are gapless. One can get a self-consistent solution in the half-filled case with either the « $2s$ »-like, $\cos k_x + \cos k_y$, « gap function », or the $1d$-like $\cos k_x - \cos k_y$. Apparently, if the slightest trace of kinetic energy intervenes, the latter is preferred, as shown by MIYAKE et al. [65] and by KOTLIAR [66]. It was not an original idea, of course, to use the superexchange Hamiltonian as a BCS interaction energy; this was done by NOGA [67], MIYAKE et al. and probably others for the heavy-fermion problem; but we are discussing this state as an approximation to the insulating RVB state. Although slight indications show that the $1d$ state is physically favored and the electronic gap function may in the end have that structure, we do not believe that the original BZA optimization is close enough to the real physics to show that, although, taking that paper literally, that is what one would conclude (as KOTLIAR has shown). In fact, we shall see that for the half-filled case the two forms are the same wave function.

Completing the story on the BZA version of RVB, BASKARAN then investigated the actual spin correlations induced by this state after projection, and found that they are literally identical to those in a Gutzwiller projection of a free-electron state with the same Fermi surface. The Gutzwiller projection is, as we shall see, equivalent to averaging over a gauge symmetry, and this restricts the possible correlations very severely. On the other hand, the BCS theory does give us a probably quite realistic excitation spectrum, with a Fermi-like spectrum of band width proportional to J and no average charge for the quasi-particles, which are equal mixtures of electrons and holes.

While this is the very best we seem to be able to do, numerically, BASKARAN has been able to write down a more satisfying formal theory which approaches some of the fundamental questions such as the nonexistence of a gap and the rigidity of the state more and which I shall discuss at the end of this section. I should mention also that J. WHEATLEY has attempted to see if this RVB state has any CDW or SDW instabilities other than the $2s$-$1d$ degeneracy, with negative results; the RVB state seems to have a remarkable rigidity against developing a gap or further improvement.

4.2. Zou « slave fermion » theory. – Before discussing gauge theory, let me bring out a first approach to a formal theory of the superconducting state. This is predominantly due to ZOU [68] with many discussions with the rest of us. We believe that this theory is more physical than the modified BCS theory given in BZA and probably represents some approximation to reality. The key idea here is to exploit the soliton structure and to formally separate the charge and kinetic-energy degrees of freedom from the spins, treating their interaction as a perturbation. It is closely related to the « slave boson » theory of Barnes [69, 70], Coleman [71] and Read and Newns [72], used in the mixed-valence problem, but in this theory the bosons are no longer « slaves » but are given physical reality, since they carry charge.

Again we start with the Hubbard Hamiltonian and attempt to canonically transform away the admixture of unpleasantly occupied states with $n_i = 2$ or (alternatively) 0. We exploit a representation of the projection operators given by HUBBARD a number of years ago. We can, of course, write

$$(4.8) \quad 1 = |O_i\rangle\langle O_i| + |\alpha_i\rangle\langle \alpha_i| + |\beta_i\rangle\langle \beta_i| + |\alpha_i\beta_i\rangle\langle \alpha_i\beta_i|,$$

where α_i, β_i are the singly occupied up and down spin states on site i, and O and $\alpha\beta$ have the obvious meanings. HUBBARD pointed out that these projection operators satisfy an algebra which is equivalent to that of a set of two bosons and two fermions $s_{i\sigma}^\dagger, e_i^\dagger, d_i^\dagger$. Using this projection technique, we can express the physical electron operators as products of the projection operators (as, e.g., $c_{i\uparrow}^\dagger = |\alpha\rangle\langle O| + |\alpha\beta\rangle\langle\beta|$) and thence of the « spinons » s and bosons e and d, as, for instance,

$$(4.9) \quad c_{i\sigma}^\dagger = s_{i\sigma}^\dagger e_i + \sigma d_i^\dagger s_{i-\sigma}$$

etc. When these identities are inserted into the Hubbard Hamiltonian we get

$$(4.10) \quad \begin{cases} \mathcal{H} = \mathcal{H}_0 + \mathcal{H}', \\ \mathcal{H}_0 = \sum_{\langle ij \rangle \sigma} (e_i^\dagger e_j - d_i d_j^\dagger) s_{i\sigma}^\dagger s_{j\sigma} + U \sum_i d_i^\dagger d_i + \mu \sum_i (e_i^\dagger e_i - d_i^\dagger d_i), \\ \mathcal{H}' = \sum_{\langle ij \rangle \sigma} e_i^\dagger d_j^\dagger s_{j\sigma} s_{j\sigma} + \text{h.c.} \end{cases}$$

$\mathcal{H}'$ must be eliminated before trying to work with (4.9) because it represents a first-order admixture of the doubly occupied subspace; after canonically transforming it away virtual high-energy processes are removed and our excitations are confined to the low-energy space. The result is simply to reintroduce the famous superexchange term

$$\mathcal{H}_{\text{ex}} = J \sum_{\langle ij \rangle} b^\dagger_{ij} b_{ij}$$

with

(4.11) $$b^\dagger_{ij} = s^\dagger_{i\uparrow} s^\dagger_{j\downarrow} - s^\dagger_{i\downarrow} s^\dagger_{j\uparrow}$$

and to leave

(4.12) $$\mathcal{H} = \mathcal{H}_0 + \mathcal{H}_{\text{ex}}.$$

We found that the charge flows, even after transformation, strictly with the boson degrees of freedom and the spin with the spinons.

The projection condition (4.2) becomes a very severe constraint upon our new degree of freedom:

$$e^\dagger_i e_i + d^\dagger_i d_i + \sum_\sigma s^\dagger_{i\sigma} s_{i\sigma} = 1.$$

In the half-filled case it is consistent to set $e^\dagger = d^\dagger \equiv 0$ and we have only $\mathcal{H}_{\text{ex}}$ expressed in terms of spinons with the «Gutzwiller» constraint $\sum_\sigma s^\dagger_{i\sigma} s_{i\sigma} = 1$. Solution of this by mean-field theory leads directly back to the BZA solution, so in a sense in this case we have gained nothing: the Gutzwiller projection still needs to be done. But this case is helpful in one sense: it tells us a way to think of spinons.

The *physical effect* of $s_{i\downarrow}$, for instance, is identical with that of $s^\dagger_{i\uparrow}$. The two do not cause independent excitations. One cannot, however, identify them from the start, though one is tempted to, since then the constraint is simply the anticommutation relation of the resulting single fermion; one must identify them only in the end after allowing the full Hamiltonian to act on the state. (Otherwise the exchange term loses its x and y components.) In a real sense, the Hamiltonian (4.11) acts in a space in which the fermion degrees of freedom are doubled from the physical one, although in principle its terms, if correctly used, do not take one out of the physical subspace. Projection back removes half of the fermion degrees of freedom, which are duplicated in this space. One may call this, in fact, a «slave fermion» theory rather than a «slave boson» one. It is accurate when e or d is not too large.

The following conjecture needs to be confirmed via more complete study of the gauge theory underlying this physics: it seems likely that it is not correct to change the spinon Fermi surface in accordance with

$$\sum_{k\sigma} n_{k\sigma} = \sum_{i\sigma} n_{i\sigma} = N(1 - |e_i|^2)$$

using a chemical-potential term for spinons, since they are not conserved because of this identity of $s_{k\sigma}$ and $s^\dagger_{-k-\sigma}$. Zou has suggested that one can always describe even the kinetic-energy-like processes proceeding from the $e^\dagger e s^\dagger s$ term in terms of spinon pair creation and destruction only. Thus all of the spinon self-energy can be taken to be of the anomalous type, and with *no chemical-potential term* for the spinons there is always a pseudo Fermi surface as observed, whenever the anomalous self-energy vanishes. Much of this will be clearer in the next subsection dealing with the gauge theory.

With this long discussion, we can now proceed to consider the doped case. When we dope, the chemical potential moves in such a way as to ensure that

(4.13) $$\langle e^\dagger e \rangle = \delta$$

(no chemical potential acts on $s^\dagger s$) and we can assume that at finite temperatures $d \equiv 0$. In mean-field theory, $\langle e_i^\dagger e_j \rangle$ and $\langle s_{i\sigma}^\dagger s_{j\sigma} \rangle$ will be nonvanishing and both the holes and the spinons will acquire mean-field kinetic energy.

We may in fact write

(4.14) $$\mathcal{H} = \sum_k (\varepsilon_k - \mu) e_k^\dagger e_k + \sum_k E_k(s_{k\sigma}^\dagger s_{-k-\sigma}^\dagger + \text{h.c.}) -$$
$$- t \sum_{\langle ij \rangle} (e_i^\dagger e_j - \langle e_i^\dagger e_j \rangle)(s_{i\sigma}^\dagger s_{j\sigma} - \langle \rangle) + J \times (\text{spinon-spinon scattering term}),$$
$$E_k = \gamma_k(\varDelta + t\langle e^* e \rangle).$$

To this Hamiltonian we should append the long-range Coulomb term

$$\frac{e^2}{\varepsilon q^2} \sum_q \varrho_q \varrho_{-q}$$

with

$$\varrho_q = \sum_k e_{k+q}^* e_k.$$

As can be seen, we have boson-fermion field theory with two conserved quantities, charge and spin, and three kind of vertices, of which the hole-hole vertex ϱ^2/q^2 and the hole-spinon vertex t are probably more important.

What can we say about the physics? This can be seen with varying degrees of certainty, and we must keep in mind always that real measurements are made on real impure materials, normally chosen to have the highest possible T_c, and not for our convenience. Our convenience would be served by a very pure single crystal with very low doping, so that the superconductivity is strictly controlled by the Bose condensation of the holes.

One piece of physics is fairly straightforward. In the antiferromagnetic ground state, the mean-field hole kinetic energy vanishes, corresponding nicely to Rice and Brinkman's old calculation which showed that free holes do not move in an antiferromagnet since they leave a trail behind. However, the RVB state does not seem to impede the motion of holes much more than would pure ferromagnetic order, so that, where doping favors double exchange and ferromagnetism for $S > \frac{1}{2}$, it simply favors RVB for $S = \frac{1}{2}$. Oshida's calculations show this for Gutzwiller-Monte Carlo, and it is also evident from small-cluster studies. It is fortunate that, however inaccurately, the slave-fermion theory mimics this effect. BASKARAN and HSU [73] estimated that $< 1\%$ of doping easily destroys antiferromagnetism. This is the reason we are not concerned about the precise stability of RVB for *pure* La_2CuO_4, etc.

Second, given the probably large magnitude of J, we believe that the magnetic state is rather firmly fixed at least up to the «twitch» transition, which occurs below the mean-field BZA T_c. (Note, if this is not already clear, that the BZA T_c of the spinon-BCS theory is an artifact which disappears with phase averaging; RVB and its accompanying spinons appear via a *crossover* near T_c as we shall see.) This means that the hole kinetic energy

$$\varepsilon_k = -t\gamma_k = -t(\cos k_x + \cos k_y)$$

is reasonably well established throughout our temperature range of interest. This means in turn that the holes will Bose condense at a temperature of order $t\delta$, once the doping is big enough to screen out any deep bound states (which appears to be about 5%). Why T_c rises so slowly with δ thereafter is a real mystery which we shall conjecture on shortly. Once Bose-condensed, the spectrum of $e^\dagger$ will take on a characteristic charged Bose-gas form: long-wavelength excitations will mix strongly with plasmons, leaving as lower-energy charged excitations a rotonlike spectrum at finite $k \sim a^{-1} \delta^{\frac{1}{2}}$.

We have, then, a spinon excitation spectrum looking like an uncharged Fermi gas, and a rather high-energy charged boson, plus the condensate: where are the electrons? Electrons in this theory are bound states of e and s, and may or may not really *be* bound states or a continuum, depending on where the hole energies are etc. I tend to believe that the hole-spinon scattering vertex is attractive at momenta near a Fermi surface and that there is a bound state exhibiting a gap spectrum not far from BCS-like. This bound «electron» state will be held together by hole-spinon kinetic-energy terms (and by being wholly transverse it evades the big plasmon terms). The hole comes from the region near $k = 0$, the spinon from near the PFS. It looks like fig. 4a) in terms of diagrams. These «electrons» will see a pairing self-energy due to the hole condensation which provides a two-hole amplitude which, by exchanging a spinon, gives an electron anomalous self-energy as shown in fig. 4b). This gap energy is unrelated to the basic superconductivity mechanism, need not be attractive, and seems to be of order t, but may be much reduced because of having to fight the repulsive Coulomb energy of the two holes, so it may by coincidence be $\sim J$ or T_c.

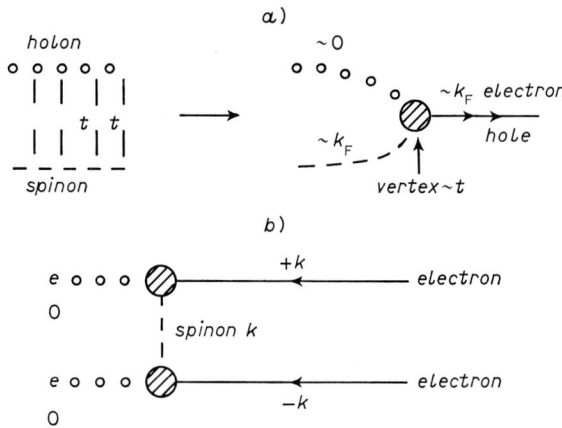

Fig. 4. – Diagrams in the Zou theory: a) binding of hole and spinon to form electron, b) development of pairing amplitude.

We conjecture that as $T \to > T_c$ the hole states cease to be driven up to $\hbar w_p$ and become real particle excitations which can carry current. The electron states may then be relatively unimportant, and one has a continuous spectrum in all channels made up of holes and spinons, and in the charge + + spin channel something roughly like a Fermi gas. The holes, however, will be the least scattered entities (the spinon gas is very weakly excited) and will probably dominate the current, leading to the large T-dependent normal conductivities observed. The holes are very susceptible to scattering by defects and particularly spin defects and there is every reason to believe that we can find explanations for both parts of the anomalous conductivity, and of the thermopower, entirely in terms of holon conduction at moderate temperatures.

One interesting possibility intrigues me. Thinking of the spinon gas as a Fermi liquid with $F_0^s = \infty$, that condition need not hold rigorously at all energies: it can be enforced only up to the charge gap in the $n-1$ case, which is sufficient for all purposes; but in the doped situation there is no true charge gap and the condition $F_0^s = \infty$ may hold only precisely at zero energy. That is, a spinon of finite energy may have a small charge component proportional to some power of that energy. That is, Z will no longer be exactly 0 except right at $T = E_k = 0$. Then the anomalous tunneling below the gap may take place into spinon degrees of freedom which have acquired some charge from contact with the condensate. In fact, the restriction on spin degrees of freedom, plus the requirement that a real electron must do the tunneling, with both spin and charge, can also account for the «background» rise in tunnel conductance well above the «gap» but below the RVB energy J. The point I am trying to make here is not that we can yet calculate any of these processes, including the gap or gaps and the spinon-hole binding energy, but that the possibilities are so various that one can only conceal reality by trying to stuff it into any conventional framework.

Finally, let me say a word about macroscopic coherence in this theory. A macroscopic broken symmetry involving a mean value $\langle e \rangle$ in a whole sample cannot occur, by a theorem of Yang. The two states of the underlying RVB are orthogonal since one has odd spin and the other is a singlet. The constraint firmly ties fermion numbers to boson numbers.

That does not prevent there being a mean value $\langle e^\dagger \rangle$ throughout a single grain or at least a single plane of a superconductor. The question in that case is whether the state with a hole moved from site to site via a path returning on itself can interfere. A hole moving from site to site corresponds to a bond moving half as far, so that in general the returning hole leaves a different state of the substrate after its first excursion. The RVB may be sufficiently flexible that interference can occur nonetheless, in which case we will have true ODLRO within grains in the sense that

$$\langle e(r) e(r') \rangle$$

is a simple product $f(r)f(r')$, though $\langle e(r) \rangle$ cannot exist. If that is the case, a single grain may quantize flux in units of either 1 or 2. Yet another possibility is that $\langle e(r) e(r') \rangle = f(r - r')$ is a function of large but finite range, which would mean an effective coherence length for charge e condensation much shorter than that for $2e$, whatever that means. I favor the ODLRO possibility. (Yet another rather far out conjecture is that e condensation occurs at much higher T than $2e$, but only the latter is seen as true superconductivity!) The measurement of the flux lattice in single-grain films is obviously an important experiment.

4˙3. *Fermion-based mean-field theory.* – The Zou formalism given up to now is an important direction in which to go, but we now feel that there is another, perhaps more straightforward, way to an understanding of the system. We take as an ansatz that the basic RVB state is a projected Bose liquid of singlet pairs—*i.e.* that it has the form of the Gutzwiller-Fulde *et al.*, Rice *et al.* BZA function, with the premise that we imagine that the singlet pairs are allowed virtual interactions to improve their energy without changing the symmetry of the wave function. Thus a suitable model function for the

RVB ground state of the $n = 1$ system is

(4.15) $$\Psi_0 = \prod P_G \Psi_{\text{BZA-BCS}},$$

where

$$\Psi_{\text{BZA-BCS}} = \prod_k (u_k + v_k c_{k\sigma}^\dagger c_{-k-\sigma}^\dagger)$$

with

(4.16) $$\int \frac{v_k}{u_k} \mathrm{d}^d k = 0.$$

Self-consistency requires that, if we define an

(4.17) $$E_k = J(\cos k_x \pm \cos k_y),$$

then

(4.18) $$\frac{v_k}{u_k} = \begin{cases} +1, & E_k > 0, \\ -1, & E_k < 0. \end{cases}$$

However, the form (4.15) is subject to improvement in the same sense as Laughlin's famous trial function for the FQHE—it appears to be variational, but there are no obvious variational parameters which are free, and in particular the sharp Fermi surface is not an artifact but a symmetry requirement. Here we use the exact $n = 1$ projector

$$P_G = \prod_i [n_{i\uparrow}(1 - n_{i\downarrow}) + n_{i\downarrow}(i - n_{i\uparrow})].$$

We now introduce a set of spinon variables to describe fermion excitations which do not violate the projector by the simple procedure

(4.19) $$s_{i\sigma} \Psi_0 = P_G c_{i\sigma} \Psi_{\text{BZA-BCS}},$$

recognizing that this is a general definition which will apply for any suitable ground state. There is one problem: if Φ_0 has an even number of sites, $c_{i\sigma} \Psi_{\text{BZA}}$ has an odd number of electrons and *vice versa*, so these excitations can only be created in pairs. But this a minor restriction as long as there is a pseudo Fermi surface (not in theories with gaps!) so that the second one can be equivalent to a small displacement of that surface.

The state Ψ_0 can be defined in terms of the $s_{i\sigma}$ as the vacuum for the appropriate linear combinations

(4.20) $$\alpha_k = v_k s_{k\uparrow} - u_k s_{-k\downarrow}^\dagger, \quad \text{etc.},$$

which is the best way of defining a BCS state, as the vacuum of the appropriate set of quasi-particles.

The state and the $s_{i\sigma}$ have a $SU(2)$ local gauge symmetry which makes the $s_{i\sigma}$ twofold redundant. This is because, as we said, the physical effects of $s_{i\sigma}$ and $s^\dagger_{i-\sigma}$ are identical, in that they put a spin $-\sigma$ on site i, while leaving the singlet pairing of the rest of the ground state unchanged. Thus we may make any unitary rotation

$$(4.21) \quad \begin{cases} s'_{i\sigma} = u_i s_{i\sigma} + v_i s^\dagger_{i-\sigma}, \\ |u_i|^2 + |v_i|^2 = 1 . \end{cases}$$

This is the symmetry which is responsible for Baskaran's observation that Ψ_{BZA} and $\Psi_{\text{Gutzwiller}}$ are identical, since a permissible operation under it is a uniform rotation in pseudospin space from pseudo x to pseudo z orientation. We shall discuss this gauge theory in the next subsection in attempting to show the state is gapless.

The s's, in spite of their simple construction, are true soliton operators, not constructible as local operators (as we see from the fact that they must be created in pairs). They isolate a single site from the paired background and create a free spin there.

We can, of course, also define a true particle operator which creates a real missing electron,

$$(4.22) \quad c_{i\sigma} \Psi_0 = c_{i\sigma} P_G \Psi_{\text{BZA}} .$$

This operator changes the charge of the wave function as well as its spin, and in the Mott insulator for $n = 1$ it requires an energy $\mu = U/2$. However, we can choose to describe the process of doping as involving the creation of a number $N\delta$ of these particles, hence greatly modifying μ for them. The effect of a $c_{i\sigma}$ is both to create a spin excitation and to empty a site—that is, acting on the completely paired state it leaves both an unpaired spin or spinon *and* an empty site.

One way to make this clear is to introduce the combination

$$(4.23) \quad \begin{cases} c_{i\sigma} s^\dagger_{i\sigma} = e_i , \\ e_i \Psi_0 = c_{i\sigma} P_G c^\dagger_{i\sigma} \Psi_{\text{BZA}} , \end{cases}$$

which removes the spin σ electron which is definitely on site i if we have created a spinon there. This is an unnormalized version of the Kivelson hole soliton which creates a hole at site i, leaving the background paired. It clearly costs less energy than the electron hole itself, by an amount which is of order J, since it leaves no spins unpaired.

It is clear that we could, in turn, define $c_{i\sigma}$ by making it up from s's and e's, and this is the nature of the Zou scheme, to use as basic particles the two forms of soliton; but it makes the physics perhaps a little clearer to start with the two kinds of basic fermions.

We assign the unperturbed kinetic energy $\varepsilon_k - \mu$ to the electron excitations, in the perhaps somewhat optimistic expectation that we will take the interactions into account separately. The chemical potential must be adjusted to give the correct occupation. In contrast to the Zou theory, this theory is aimed at the region where δ is appreciable and, therefore, both hole and particle excitations near the Fermi level, which is well below the $n = 1$ one, are meaningful. Excitations above the original Fermi level are assumed not to be possible, since they are separated by U in energy. This way of treating the strong repulsive interactions is not very satisfactory, but the exercise is useful because it teaches us about the symmetry of the interactions and gives us a model, albeit unrealistic, for calculating the various fundamental vertices in the problem.

The electrons have direct Coulomb interactions of both long and short range, as well as whatever electron-phonon interactions are present; but very likely all such interactions are net repulsive and we may postulate a direct electron-electron interaction

$$(4.24) \qquad V_{ee} = + \sum_q V(q) \varrho_q \varrho_{-q} .$$

The corresponding pair interaction is

$$V_{e\text{-}e} = \sum_{k,k',\sigma} V(k - k') c^{\dagger}_{k\sigma} c^{\dagger}_{-k-\sigma} c_{-k'-\sigma} c_{k\sigma}$$

with $V(k - k') > 0$ and pretty large; approximately,

$$(4.25) \qquad V(q) = \frac{4\pi e^2}{\varepsilon q^2} .$$

While direct electron-electron interaction is dominated by repulsion, the electrons and spinons both feel the superexchange interaction, since both end up actually changing the spin on site i. This is responsible for all of the self-energy of the spinons and for binding spinons and electron holes together into condensate holes.

We postulate, therefore, a spinon-electron interaction

$$\mathcal{H}_{s\text{-}e} = - J \sum_{\langle ij \rangle} (\sigma^e_i \cdot \sigma^s_j) ,$$

and correspondingly for the free electrons, as well as the spinon-spinon interaction. This interaction may be written in the «reduced» form

$$(4.26) \quad \mathcal{H}_{\text{s-e}} = 3J \sum_{\tau} \sum_{k,k'} \exp[i(k-k')\cdot\tau] c_{k\sigma}^\dagger s_{-k-\sigma}^\dagger s_{-k'-\sigma} c_{k'\sigma}.$$

This shows that there is a new type of anomalous self-energy possible which, in practice, is equivalent to a charge-e hole amplitude. That is, we assume that

$$(4.27) \quad \langle e_k^\dagger \rangle = \langle c_{k\sigma}^\dagger s_{-k-\sigma}^\dagger \rangle \neq 0$$

at least in the sense of «ODLRO», that $\langle e^\dagger(r) e^\dagger(r') \rangle = e(r) e(r')$ is asymptotically independent of $r - r'$, within at least the same layer of the same sample.

Conceptually it is easy to express what we intend to do. In the course of doping, we could make up a ground state by multiplying by a number $N\delta$ of $c_{k\sigma}$ operators acting on the $n = 1$ ground state, and then linearly combining all such states. On the other hand, in general that will leave $N\delta$ unpaired spins, which can be remedied by introducing $N\delta$ spinons and pairing them up with the electron spins. (We cannot pair the electron proper since their interactions are repulsive.) Thus the form of ground-state wave function we envisage looks like

$$(4.28) \quad \sum_{k,\varkappa,\sigma} f(k_i, \sigma_i; \varkappa_j, \sigma_j) \times \Pi_{k_i\sigma_i}^{N\delta} c_{k_i\sigma_i} \times P_G \Pi_{k_j\sigma_j}^{N\delta} c_{k_i\sigma_i} \Psi_{\text{BZA}}.$$

That is, the spinon part of the wave function acts entirely under the Gutzwiller $n = 1$ projection and has the characteristic $SU(2)$ symmetry which is responsible for its gaplessness. f satisfies an overall pairing condition in that the total set of k's and $\varkappa$'s are $\pm k\sigma$ pairs, with electron and spinon as well as electron-spinon pairing allowed. The basic scheme can be summarized in diagrams similar to the basic diagram of the BCS theory expressing the formation of a bound state from pairs and the consequent development of a condensate, but in this case there are three condensates of charges zero, e and $2e$, and these lead to three anomalous self-energies D_k, Δ_k and Γ_k (see fig. 5).

We proceed in a manner similar to conventional BCS theory by introducing the anomalous self-energies in an effective Hamiltonian

$$(4.29) \quad \mathcal{H}_{\text{eff}} = \sum_{k\sigma} (\varepsilon_k + \mu) c_{k\sigma}^\dagger c_{k\sigma} + $$
$$+ \sum_{k\sigma} \{D_k^\dagger c_{-k-\sigma} c_{k\sigma} + \Delta_k^\dagger c_{-k-\sigma} s_{k\sigma} + \Gamma_k^\dagger s_{-k-\sigma} s_{k\sigma}\} + \text{h.c.},$$

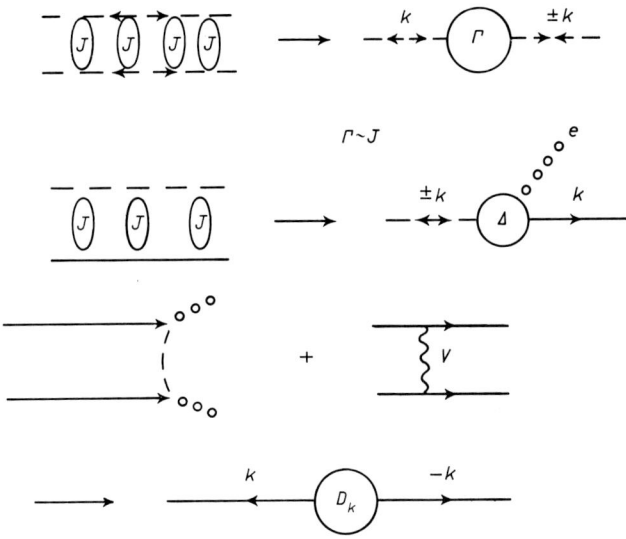

Fig. 5. – Diagrams showing the three anomalous self-energies in the « all-fermion » theory.

where D, Δ and Γ are to be determined self-consistently via their definitions in fig. 5 and the propagators determined from (4.29):

(4.30)
$$\begin{cases} \Gamma_k = J r_k \sum_{k'} \gamma_{k'} \langle s_{-k-\sigma} s_{k\sigma} \rangle, \\ \Delta_k = J \gamma_k \sum_{k'} \gamma_{k'} \langle s_{-k-\sigma} c_{k\sigma} \rangle, \\ \text{etc.}, \\ D_k = \sum_{k'} V_{kk'} \langle c_{-k-\sigma} c_{k\sigma} \rangle. \end{cases}$$

The corresponding secular equation is

(4.31)
$$\begin{matrix} c_{k\sigma}^\dagger \\ c_{-k-\sigma} \\ s_{k\sigma}^\dagger \\ s_{-k-\sigma} \end{matrix} \begin{pmatrix} \varepsilon_k - \mu - \omega & D_k & & \\ D_k & -\varepsilon_k + \mu - \omega & & (cc) \\ \Delta_k^a & \Delta_k^{b*} & \omega & \Gamma_k \\ \Delta_k^b & -\Delta_k^a & \Gamma_k & \omega \end{pmatrix} = 0.$$

The components Δ_k and the $SU(2)$ symmetry must be carefully thought out. By means of a $SU(2)$ transformation on the amplitudes of s, the quantities Δ and Γ are transformed, it turns out in just such a way as to multiply (4.31) by a unitary matrix, which leaves its eigenvalues the same:

$$U = \begin{pmatrix} 1 & 0 & 0 \\ 0 & 1 & \\ 0 & & U \end{pmatrix}.$$

In the representation (4.31), where we have written the spinon self-energy in anomalous form, it seems inconsistent to allow a Hartree-like electron-spinon self-energy Δ_k^a, and we set

(4.32) $$\Delta_k^a = 0,$$

but I see no proof for this. It is a choice which is consistent with the idea that only electrons have true kinetic energy.

When we write the spinon self-energy conventionally, the Δ sectors transform differently and we get

(4.33) $$\mathcal{H} = \begin{pmatrix} \varepsilon_k - \mu & D & & (cc) \\ D & -\varepsilon_k + \mu & & \\ \Delta_1 & \Delta_2 & \Gamma & 0 \\ \Delta_2 & -\Delta_1 & 0 & -\Gamma \end{pmatrix}$$

and, if (4.32) is satisfied, $\Delta_1 = \Delta_2$. If Δ_2 or Δ_1 were zero, the secular equation would factorize in this form and holes and electrons would never mix; this unphysical result is one of our reason for choosing (4.32).

For simplicity let me neglect the direct-pairing energy D_k; since V is probably repulsive, this cannot arise self-consistently on its own but only as a consequence of indirect pairing via Δ. D will obviously be nonzero and will contain whatever phonon effects are relevant; clearly it can weakly affect T_c. But the essence of the results is the effect of Δ. After these simplifications, the eigenvalue equation becomes

(4.34) $$[\omega^2 - (\varepsilon_k - \mu)^2](\omega^2 - \Gamma_k^2) - 2\omega^2 |\Delta|^2 + |\Delta|^4 = 0$$

with the solutions (omitting the k and setting $\varepsilon_k - \mu = \varepsilon$)

$$\omega^2 = \frac{\varepsilon^2 + \Gamma^2}{2} + \Delta^2 \pm \frac{1}{2}\sqrt{(\Gamma^2 + \varepsilon^2)^2 + 4\Delta^2[\Gamma^2 + \varepsilon^2]}.$$

There are only two interesting cases: The neighborhood of the PFS, where $\omega^2 \sim \Gamma^2 \sim \Delta^2 \simeq 0$, and of the «real» Fermi surface, if any, $\varepsilon_k - \mu \simeq 0$. In the former case we have

(4.35) $$\text{PFS:} \quad \omega^2 \simeq \Gamma^2 + \frac{\Delta^4 - 2\Gamma^2\Delta^2}{\mu^2}$$

(and $\omega^2 \simeq (\varepsilon_k - \mu)^2$).

Thus the PFS is intact. There is some mixing of spinons with charged excitations—this is the reason for the higher-order terms—but it is rather weak and there is some cancellation, so it may not be a big effect in all situations. The second « electron » branch unmixed with spinons may be strongly scattered by them and may only be a broad smear in the Green's function.

In the region of the gap for charged excitations, we have to use the full solution to get it quite right, but the lowest energy will be

(4.36) $$\omega_G = \sqrt{\Delta_k^2 + \frac{\Gamma_k^2}{4}} - \frac{\Gamma_k}{2},$$

which is, fortunately, positive and of order Γ. (The second is $\sqrt{\Delta^2 + \Gamma^2/4} + \Gamma/4$.) The gap has a very complex structure even neglecting the anisotropy of Γ and Δ. Nonetheless, there is a gap and it is induced by the process of the diagram, namely spinon exchange with emission of condensed holons via Δ.

It is abundantly clear why tunneling has not clarified the gap situation in these materials. Our own feeling is that the Bose condensation of holons is the major phenomenon accounting for T_c, and that the pair condensation is in a way an epiphenomenon, although it is absolutely necessary for the observation of superconductivity, since condensed holons cannot tunnel from plane to plane, and the currents in the c direction are entirely due to the relatively weak pair condensation. It may be that even the Bose condensation is a weak 3-dimensionality effect. It may be very difficult to force fluxons to penetrate the planes, and it may be hard to see a simple, regular fluxon lattice whether of hc/e or $hc/2e$ fluxons.

4'4. *Gauge theory*. – We can make a start in putting a rather rigorous base behind some of the above with a development initially due to BASKARAN [74]. (AFFLECK also developed a similar theory, and some of the ideas are contained in [62].) For the RVB insulating state proper we can get quite far, but the extension to superconductors is still programmatic. We attempt to understand the overcompleteness induced by the Gutzwiller projection in terms of the local $SU(2)$ symmetry induced on the $s_{i\sigma}$, $s^\dagger_{i-\sigma}$ operators. For the RVB, the Heisenberg Hamiltonian

$$\mathcal{H} = J \sum_{\langle ij \rangle} b^\dagger_{ij} b_{ij} = J \sum (S_i \cdot S_j - \tfrac{1}{4})$$

has the $SU(2)$ gauge symmetry when expressed in terms of spinons, and

Elitzur's [75] theorem shows that this cannot be spontaneously broken, and hence

$$\langle b_{ij}^\dagger \rangle \equiv 0 \tag{4.37}$$

at all T, as originally pointed out by BZA.

This is, of course, not an exact symmetry of all states of the original Hubbard Hamiltonian; it is true only so long as we stay below the energy gap U. But it is a symmetry which controls the subspace containing the ground state and low-order excitations, and in this subspace it can be made exact. It is intriguing that a similar gauge symmetry appears in the Girvin-Reed theory of the FQHE. It also has a *global* spin symmetry which *can* be broken (v. the Néel state).

The essence of what we do is similar to field-theoretic transcriptions of other problems in which the algebra and/or the constraints of the physical variables are complicated and difficult to handle, such as the spin glass replica problem. We introduce a classical field Γ_{ij} coupled to $b_{ij}^\dagger$ which is, however, a 2×2 matrix and not a simple number; it can be thought of as the local equivalent of the Γ's in our secular equations. To make a proper theory we would have to introduce a nonlinear σ-model–type constraint, which carries me beyond my depth, but I think the following is schematically correct. We write

$$Z = \exp[-\beta \mathscr{H}] = \operatorname{Tr} \exp\left[\frac{J}{T} \sum b_{ij}^\dagger b_{ij}\right] = \tag{4.38}$$
$$= \prod_{ij} \int d\Gamma_{ij} \int d\Gamma_{ij}^* \operatorname{Tr}_s \exp[-T/J \Gamma_{ij}^* \Gamma_{ij} + \Gamma_{ij} b_{ij}^* + \Gamma_{ij}^* b_{ij}].$$

The fermions can now in principle be traced out because *a)* it is only a linear problem for them, *b)* the gauge symmetry average takes care of the projection.

Fortunately, much of that problem is solved by the BZA mean-field theory, which tells us the nature of the energy as a function of *uniform* Γ_{ij}.

BASKARAN has evaluated, roughly, the first nonuniform terms, and we find a Ginsburg-Landau form

$$Z_{\text{eff}}(\Gamma) = a \sum_{ij} |\Gamma_{ij}|^2 + b \sum_{ij} |\Gamma_{ij}|^4 + (--)_1 + \tag{4.39}$$
$$+ c \sum_{\text{plaquets}} \operatorname{Tr} \Gamma_{12} \Gamma_{23} \Gamma_{34} \Gamma_{41} + (--)_2$$

with

$$a \sim \frac{k_B}{J}(T - T_{\text{MF}}),$$
$$b \sim c \sim 1,$$

$(--)_2$ can be guessed to fall off rather fast with size of plaquet. Only the plaquet term gives coupling of the phases of different order parameters, and this type of term controls any symmetry breaking which may occur. As we see, at $kT = J/2$ there is a cross-over at which $\langle |\varDelta|^2 \rangle$ grows to a large value, signaling the BZA mean-field T_c.

The plaquet term is not, according to accepted ideas about mean-field theory, capable by itself of causing broken symmetry at all [75]. The only kind of transition it can lead to involves the Wilson-Wegner loop correlation function defined on a closed loop in the lattice as

(4.40) $$W(c) = \langle b_{ij}^\dagger b_{jk} b_{kl}^\dagger b_{lm} \dots b_{ni} \rangle \propto \langle \varGamma_{ij}^\dagger \dots \varGamma_{ni} \rangle \,.$$

In the «confining» phase which is continuous with the high-T_c limit, this falls off exponentially with the area of the loop; it can make a transition to a so-called «free» phase where it falls off only with the perimeter, but only in dimension higher than 2 or 3. This would imply an infinite correlation length and a gap for excitations, since as $T \to 0$ the correlation length no longer changes. But in fact the system stays in the «confining phase» and there is no gap. Further work on this gauge theory is obviously the best way to solve the mysteries of the RVB, and it is by no means an intractable one: much is known. In particular, at low T it is clear that $|\varGamma|^2$ takes on a roughly constant amplitude $\varGamma_0^2$—the potential for $\varGamma$ is the famous «Mexican hat»—and the plaquette term is just expanded around a chosen uniform state

$$\varGamma_0 \cos(\theta_{ij} - \theta_{jk} + \theta_{kl} - \theta_{li})\,,$$

where $\varGamma = \varGamma_0 \exp[i\theta_{ij}]$. The resulting correlations among the θ_{ij} may, as suggested in ABZH, affect various lattice modes of the oxygens via the overlap charges on the valence band oxygens. We are no closer than that statement to an understanding of the «twitch» transition on the La_2CuO_4 lattice.

This gauge theory is very preliminary and I hesitate to discuss it at length here, especially since we have not completely correlated its treatment of the superconducting case with that of the more explicit «slave fermion» theory.

Our best guess is that the correct way to deal with it is to leave the spinons defined as acting under projectors and hence always with the symmetry. Then a holon e is thought of as a bound state of a c_σ and a s_σ and its amplitude is not $SU(2)$ invariant, but its *charge* follows the c part, which is *not* infected by $SU(2)$. Essentially, we always pick a particular $SU(2)$ gauge and find the saddle points, then average, and $SU(2)$ does not infect the rest of the world.

5. – Conclusion and loose ends.

As I look back on the above, I view the most important open questions as the following:

1) Why is T_c so low? I conjecture at least three possible reasons. *a*) We

are, especially in La-Sr, operating in the neighborhood of a mobility edge. The hole kinetic energy can be thereby seriously reduced. *b)* Antiferromagnetic spin fluctuations are not good but bad: they too will raise the mass of the hole degrees of freedom, while possibly lowering that of the spinons, both trends in the direction of experimental fact. *c)* The effect of dimensionality is always an unknown. To what extent is the 3d superconducting T_c lowered by weak interplanar coupling (which has to occur via hole pairs, we recognize, so must not occur until there is a good strong pair amplitude). In fact, is all of the above *only* a theory of the «normal» state and is 3-dimensionality vital to superconductivity?

2) To what extent does the soliton description break down at finite energy? Is $Z \equiv 0$ at zero energy only, and do the spinons acquire charge gradually rather than binding a full-scale hole and becoming real electrons? This is favored by all kinds of IR and tunneling data, which suggest a background density of charged excitations forming with a power of energy. It also solves the very knotty problem of whether we can really live with two separate Fermi surfaces, one for spinons and one for electrons.

3) What is the twitch and in fact how does the RVB couple to phonons? Why is the 200 cm^{-1} phonon so prominent?

Nonetheless, the idea that the explanation of high T_c lies in these general ideas is hardly to be doubted.

* * *

I shall emphasize, as I have again and again in the text, that the above work is to a massive extent influenced by the ideas of G. BASKARAN and Z. ZOU. The entire Princeton group has contributed in more ways than I can possibly enumerate and I list them in alphabetical order: E. ABRAHAMS, I. A. AFFLECK, S. COPPERSMITH, T. HSU, R. KAN, S. D. LIANG, J. SAULS, T. WEN, J. WHEATLEY. Conversations and preprints were important from P. FAZEKAS, J. HIRSCH, W. KOHN, S. KIVELSON, D. ROKHSAR, T. Y. RAMAKRISHNAN, T. M. RICE, G. KOTLIAR, N. READ, P. A. LEE, and no doubt many others. Many experimentalists have been most helpful, especially N. E. PHILLIPS, T. H. GEBALLE, L. GREENE, G. AEPPLI, B. BATTLOGG, G. A. THOMAS, D. BISHOP, H. R. OTT, D. MCLACHLAN, C. W. CHU. The hospitality of the Fairchild program at Cal Tech should be mentioned, as well as of the Indian Institute of Physics.

REFERENCES

[1] J. H. DE BOER and E. J. W. VERWEY: *Proc. Phys. Soc. London Sect. A*, **49**, 59 (1937).

[2] J. H. VAN VLECK: *Rev. Mod. Phys.*, **25**, 220 (1953). Also H. HURWITZ: Thesis (1941).
[3] N. F. MOTT: *Proc. Phys. Soc. London Sect. A*, **62**, 416 (1949).
[4] N. F. MOTT: *Can. J. Phys.*, **34**, 1356 (1956).
[5] N. F. MOTT: *Adv. Mat. Phys.*, **3**, 76 (1952).
[6] J. INKSON: *J. Phys. C*, **5**, 2599 (1972). See also P. W. ANDERSON: in *Elastic Excitation in Solids*, edited by A. DE VREESE (Plenum Press, New York, N. Y., 1974), p. 1.
[7] P. W. ANDERSON: *Phys. Rev.*, **115**, 2 (1959).
[8] P. W. ANDERSON: *Solid State Phys. Adv.*, **14**, 99 (1963).
[9] J. H. VAN VLECK: *J. Chem. Phys.*, **3**, 843 (1935).
[10] J. D. DUNITZ and L. E. ORGEL: *J. Phys. Chem. Solids*, **3**, 20 (1957).
[11] J. S. GRIFFITH and L. E. ORGEL: *Q. Rev. (London)*, **11**, 381 (1957).
[12] C. HERRING: *Magnetism*, Vol. II B, edited by G. RADO and H. SUHL (1966), p. **1**.
[13] K. TERAKURA, A. R. WILLIAMS, J. OGUCHI and J. KUBLER: *Phys. Rev. Lett.*, **52**, 1830 (1983).
[14] G. A. SAWATSKY and J. W. ALLEN: *Phys. Rev. Lett.*, **53**, 2339 (1984).
[15] J. HUBBARD: *Proc. R. Soc. London, Ser. A*, **276**, 238 (1963).
[16] J. HUBBARD: *Proc. R. Soc. London, Ser. A*, **277**, 237 (1964).
[17] C. ZENER: *Phys. Rev.*, **82**, 403 (1951).
[18] M. GUTZWILLER: *Phys. Rev.*, **134**, A 923 (1964).
[19] M. GUTZWILLER: *Phys. Rev.*, **137**, A 1726 (1965).
[20] W. F. BRINKMAN and T. M. RICE: *Phys. Rev. B*, **2**, 4302 (1970).
[21] T. A. KAPLAN, P. HORSCH and P. FULDE: *Phys. Rev. Lett.*, **49**, 889 (1982).
[22] T. A. KAPLAN and P. HORSCH: *J. Phys. C*, **16**, 4203 (1983).
[23] T. M. RICE: this volume, p. 171.
[24] P. W. ANDERSON: *Proceedings of the International Conference on Mixed Valence, January 1987* (Bangalore, to be published).
[25] P. W. ANDERSON: *Science*, **235**, 1196 (1987).
[26] J. FRIEDEL and A. BLANDIN: *J. Phys. Radium*, **19**, 573 (1958).
[27] P. W. ANDERSON: *Phys. Rev.*, **124**, 41 (1961).
[28] D. B. McWHAN, J. P. REMEIKA, T. M. RICE, W. F. BRINKMAN, J. MAITA and A. MENTH: *Phys. Rev. Lett.*, **27**, 941 (1971).
[29] D. B. McWHAN, J. P. REMEIKA, T. M. RICE, W. F. BRINKMAN, J. MAITA and A. MENTH: *Phys. Rev. B*, **5**, 2252 (1972).
[30] K. HIRAKAWA, H. KADOWAKI and K. UBIKOSHI: *J. Phys. Soc. Jpn.*, **54**, 3526 (1985).
[31] P. W. ANDERSON and H. HASEGAWA: *Phys. Rev.*, **100**, 675 (1955).
[32] P.-G. DE GENNES: *Phys. Rev.*, **188**, 141 (1960).
[33] C. CASTELLANI, G. B. KOTLIAR and P. A. LEE: *Phys. Rev. Lett.*, **59**, 323 (1987).
[34] P. W. ANDERSON: *Basic Notions in Condensed Matter Physics* (J. Wiley, New York, N. Y., 1984), Chap. 4.
[35] D. WOLLHARDT, P. WÖLFLE and P. W. ANDERSON: *Phys. Rev. B*, **35**, 6703 (1987).
[36] A. J. LEGGETT: *Phys. Rev. Lett.*, **25**, 1543 (1971).
[37] R. B. LAUGHLIN: *Phys. Rev. Lett.*, **50**, 1395 (1983).
[38] S. M. GIRVIN and A. H. MacDONALD: *Phys. Rev. Lett.*, **58**, 1252 (1987).
[39] N. READ: in preparation (1987).
[40] B. BATLOGG, R. J. CAVA, A. JAYARAMAN, R. B. VAN DOVER, G. A. KOUROUKLIS, S. SUNSHINE, D. W. MURPHY, L. W. RUPP, H. S. CHEN, A. WHITE, K. T. SHORT, A. M. MUJSCE and E. A. RIETMAN: *Phys. Rev. Lett.*, **58**, 2333 (1987).
[41] L. C. BOURNE, M. F. CROMMIE, A. ZETTL, H.-C. ZUR LOYE, S. W. KELLER, K. L. LEARY, A. M. STACY, K. J. CHANG, M. L. COHEN and D. E. MORRIS: *Phys. Rev. Lett.*, **58**, 2337 (1987).

[42] L. F. MATTHIESS: *Phys. Rev. B*, **2**, 3918 (1970).
[43] L. F. MATTHIESS: *Phys. Rev. B*, **6**, 4718 (1972).
[44] L. F. MATTHIESS: *Phys. Rev. Lett.*, **58**, 1028 (1987).
[45] N. E. PHILLIPS et al.: in *Novel Superconductivity*, edited by S. A. WOLF and V. Z. KRESIN (Plenum Press, New York, N. Y., 1987), p. 739.
[46] H.-R. OTT: private communications.
[47] M. E. REEVES, T. H. FRIEDMAN and D. M. GINSBERG: *Phys. Rev. B*, **35**, 7207 (1987).
[48] A. P. RAMIREZ: *Phys. Rev. B*, **35**, 5234 (1987).
[49] D. J. BISHOP, P. L. GAMMEL, A. P. RAMIREZ, R. J. CAVA, B. BATLOGG and E. A. RIETMAN: *Phys. Rev. B*, **35**, 8788 (1987); also **36**, 2408 (1987).
[50] M. LEE, M. YUDKOWSKY, W. P. HALPERIN, J. THIEL, S.-J. HWU and K. R. POEPPLEMEIER: *Phys. Rev. B*, **36**, 2378 (1987).
[51] R. N. BHATT and A. MILLIS: private communication.
[52] P. A. LEE and N. READ: *Phys. Rev. Lett.*, **58**, 2691 (1987).
[53] P. W. ANDERSON: *Mater. Res. Bull.*, **8**, 153 (1973).
[54] P. FAZEKAS and P. W. ANDERSON: *Philos. Mag.*, **30**, 474 (1974).
[55] H. A. BETHE: *Z. Phys.*, **71**, 205 (1931).
[56] L. HULTHÉN: *Ark. Mat. Astr. Fys. A*, **26**, No. 11 (1938).
[57] J. DES CLOISEAUX and J. J. PEARSON: *Phys. Rev.*, **128**, 2131 (1962).
[58] P. W. ANDERSON: *Phys. Rev.*, **86**, 693 (1952).
[59] F. J. NEVES and J. F. PEREZ: *Phys. Lett. A*, **114**, 331 (1986).
[60] G. RUMER: *Nachr. Akad. Wiss. Gött., Math.-Phys. Kl.*, 337 (1932).
[61] L. PAULING: *J. Chem. Phys.*, **1**, 280 (1933).
[62] S. KIVELSON, J. SETHNA and D. ROKHSAR: *Phys. Rev. B*, **35**, 8865 (1987).
[63] R. N. BHATT and P. A. LEE: *Phys. Rev. Lett.*, **48**, 344 (1982).
[64] G. BASKARAN, Z. ZOU and P. W. ANDERSON: *Solid State Commun.*, **63**, 973 (1987).
[65] K. MIYAKE, S. SCHMITT-RINK and C. M. VARMA: *Phys. Rev. B*, **34**, 6554 (1986).
[66] G. B. KOTLIAR: *Phys. Rev. B*, **37**, 3664 (1988).
[67] M. NOGA: preprint (1986).
[68] Z. ZOU and P. W. ANDERSON: *Phys. Rev. B*, **37**, 627 (1988).
[69] S. E. BARNES: *J. Phys. F*, **6**, 1375 (1976).
[70] S. E. BARNES: *J. Phys. F*, **7**, 2631 (1977).
[71] P. COLEMAN: *Phys. Rev. B*, **29**, 3035 (1984).
[72] N. READ and D. NEWNS: *J. Phys. C*, **16**, 3473 (1983).
[73] P. W. ANDERSON, G. BASKARAN, Z. ZOU and T. HSU: *Phys. Rev. Lett.*, **58**, 2790 (1987).
[74] G. BASKARAN and P. W. ANDERSON: to be submitted (1987).
[75] S. ELITZUR: *Phys. Rev. D*, **52**, 3978 (1975).

Reprinted From
Frontiers and Borderlines
in Many-Particle Physics
© 1988 CIV Corso
Soc. Italiana di Fisica - Bologna - Italy

The Central Dogmas

Sir Francis Crick is a model for me of how to go about real discovery in science, as opposed to stamp collecting. His Central Dogma device as explained here was immensely valuable in biology: find what you are sure of an fill in the gaps. First stated at Cargese in 1989 where they were the substance of my lectures, these principles are still completely valid. The gaps are mostly now filled in.

CHAPTER II
The "Central Dogmas"

At a certain point in the process of unravelling the "Secret of Life"—for which read the mechanisms of reproduction and transcription of biological information—F.C. Crick propounded what he called "The Central Dogma" which constrained the overall structure of any description of the actual mechanism. The Central Dogma was determined by logical deduction from the overall experimental facts of biology. The very important conceptual function which was played by the "Central Dogma" was to limit serious discussion of mechanisms and theories to those which were consistent with logic and with the overall burden of experimental fact, while allowing a great deal of freedom in working out specific mechanisms, and leaving the overall structure of the theory immune to changes in specific processes. For instance, it was "dogma" that a genetic code existed, but the theory was independent of details of that code.

The main function of this chapter is to convince the reader that such a system of "dogmas" is useful for the field of high T_c superconductivity, a field which has the same kind of complexity and confusion as microbiology had at that time. As in molecular biology, there is enough irrelevant complexity that an unwitting theorist may never reach the neighborhood of the actual problem, even though he is working along a line which is widely represented in the literature. Understanding high T_c involves not one, but a multiplicity of steps, and it is vital to provide a map through the maze of alternative paths, almost all of which can be eliminated by simple logic using simple and well-founded experimental or theoretical reasoning, of a sort which should be immediately persuasive.

A second reason for propounding these "dogmas" is to correct the general misapprehension that there is no viable theory of high T_c superconductivity. There is in fact an evident path indicated through all the complications. The problem is not the lack of a theory but its complexity and the fact that it consists of a number of steps of widely different completeness, involving different types of arguments, published in a bewildering variety of places and versions, and embedded in a literature of great complexity which is beset with controversy. A problem is that few theorists nowadays are familiar with the process of rigorous deduction from theoretical concepts combined with a broad range of experimental facts, the process which is the primary source of the picture we now have. (As it was the primary source of the Fermi liquid theory of real metals which it replaces.) Relatively few theorists now active have been through the process of actually solving a puzzle like superconductivity, or the Kondo effect.

The problem is not primarily finding a theory, but clearing away the underbrush of too many theories, some of which contain germs of truth because of tying in to a few valid experimental facts or theoretical concepts, but which do not take into account the key requirement of overall consistency with the complete picture. This is the key requirement because of the mature state of condensed matter physics, which leaves theoretical problems extremely <u>over</u>determined: finding a theory is a redundantly posed problem. Any correct theory must be consistent with anomalous behavior of a bewildering variety of experimental probes, in addition to the very basic requirement of being internally consistent and embodying the entities which are really there: the known crystal structure, the outermost electronic bands and their interactions, the lattice vibrations, and nothing <u>else</u>. The theory must touch its base in the quantum theory of these entities. There remains very little flexibility, and one would feel there was no hope of finding such a theory except for one's knowledge that physics actually <u>does</u> work and there has to be one—and only one—solution. This then, is the final methodological clue—that one must retain one's faith that the solution exists. Thus when one has found <u>a</u> way through the maze of conflicting requirements, that is certain to be <u>the</u> way, no matter how many deep-seated prejudices it may violate and no matter how unlikely it may seem to those trained in the conventional wisdom.

Aside from theoretical papers which have some germs of truth in them, there is a larger group which are completely inconsistent with the basic realities of the subject. Many of these belong sociologically in the mode of particle theory, where there is no solid a priori foundation and speculations about the underlying physical model are acceptable, but many others come from naive or careless thinking in which extraneous independent entities like "anyons" or "spin fluctuations" or "bipolarons" are introduced but not tied down to the actual physical model and/or experimental observations. . In general, many of the papers written in this field are not just easily falsified, in the Popperian sense, but actually falsified before they start. I think the problem is one of a new kind of scientific sociology: physics has become so specialized and fragmented that an attempt at overall consistency with the observed facts and fundamental restrictions is not seen as a necessary precondition for publication. Pure speculation, or at the other extreme simple-minded applications of standard formulas, with no discussion of validity, seem to be seen as acceptable work.

We feel the best presentation of the complete picture is to give the overall view, step by step, postponing as far as possible the details, such as precise justifications, alternative techniques, and detailed critiques. We give the resulting steps in the reasoning in the form of "dogmas", which take us, step by step, through the process of solving a typical, if somewhat difficult, problem in quantum condensed matter physics. Typically, one has to go through several stages of "renormalization", which is a fancy word for abstracting the relevant parts of the problem and eliminating the irrelevant high-energy degrees of freedom. This is why it is a canard that such a problem can be solved by enough computing power: it is hopeless until one has gone through this process. The most important terms by far are those which open up gaps in the spectrum, because states above any such gap can always be eliminated exactly and replaced by effective interactions among the remaining low-energy, low-frequency degrees of freedom: a process discovered by Van Vleck many decades ago. The whole problem of high T_c is a lesson—almost a poem—in restricting Hilbert space.

To summarize what we shall do, the first three of the six dogmas have to do with such eliminations. The first two restrict our attention to a single band of the one-electron spectrum, the antibonding, "$d_{x^2-y^2}$" symmetry band on the CuO_2 planes. Other bands are separated by large energy gaps from the relevant degrees of freedom near the Fermi level, and can be eliminated: the only relevant Hilbert space is this band. The third dogma tells us that only one of the interactions of the electrons in this band is so large as to open up yet another gap and further restrict Hilbert space. The next two dogmas are descriptions of the state of the "normal" non-superconducting metal which we encounter when we enforce these restrictions, both of which are almost equally supported by experimental and theoretical arguments. The first states that the resulting metal is in an unconventional state which we call a "Luttinger liquid", possessing a Fermi surface but no conventional electron quasiparticles; and the second that this state is strictly two-dimensional in this case. Only the last dogma, then, has to do with the superconducting state: it tells us which of the residual interactions which are left can be strong enough to give us the unconventional high transition temperatures which are observed. Again much of the argument is from experiment.

Let us then set out this list of "dogmas" with some discussion of the basis behind each and of alternatives which have been proposed.

Dogma I: All the relevant carriers of <u>both</u> spin and electricity reside in the CuO_2 planes and derive from the hybridized $O_{2p} - d_{x^2-y^2}$ orbital which dominates the binding in these compounds.

The main alternative was the "chain" school, but now the one compound, YBCO, which has chain coppers, is in a tiny minority among some dozen compounds, none of the rest of which have chains and all of which behave with remarkable similarity. The infrared data on single untwinned crystals of Schlesinger and Collins show that chain conductivity is qualitatively different and relatively little affected by superconductivity. A more persistent and subtle fallacy, which this dogma excludes, is the literal acceptance of band theory results which often give bits of Fermi surface attached to other parts of the structure: the notorious "bismuth pockets" in BISCO, for instance, predicted by Freeman et al and discussed in ARPES papers. The c-axis resistivity in BISCO, 10^5 times that in the planes, means unequivocally that no essentially 3-dimensional pockets of carriers can exist in that case. The band calculations, based as they are on an idealized, stoichiometric structure, are also incompatible with the graphite-like cleavage between the Bi layers, which shows that no bands near the Fermi surface are occupied at Bi. No band calculations are to be trusted at this kind of level of accuracy, and the real band structure must be deduced from experiment. Whether there are ever pockets of other carriers, or chain carriers, is not important but may help explain some results.

The most commonsensical approach to Dogma I is to recognize the many anomalous properties of the cuprate layer materials, which are unique to those materials; and the rather unique chemistry of the cuprate layers, involving an unusual valence of Cu and very strong Jahn-Teller distortion and semicovalent bonds. The "Anderson mystery story principle", which is the original source of much of the dogma, then operates: this principle is that we must associate all truly unusual events with each other and with the basic problem.

Excitations outside of the planes, and probably even non-bonding or bonding bands in the planes, play very little role and can be ignored except for minor renormalizations. There is a remarkable similarity in behavior of materials with widely different chemistry outside the planes: for instance, the ab plane normal state resistivity per plane in optimally doped materials doesn't vary by more than a factor 2 among 5 or 6 compounds as Battlogg observes. Normal state infrared and tunneling data are also very similar. Only T_c varies widely, a fact which we will discuss later.

Dogma I in summary: look at the planes only (a great and welcome simplification.)

Dogma II: Magnetism and high T_c superconductivity are closely related, in a very specific sense: i.e., the electrons which exhibit magnetism are the same as the charge carriers.

The initial source of this was the "generalized phase diagrams" of state vs. doping, which can be traced out in several compounds. $\delta = 0$ (pure Cu^{++} oxidation state) is an antiferromagnetic insulator, with relatively high T_{Neel} if not frustrated: it is a straightforward Mott-Hubbard insulator with at least a 2 volt charge-transfer gap. The present view is that a relatively sharp transition (which is sometimes first order) occurs at $\delta \sim .1$ to a metallic state which almost always is superconducting, with usually a T_c which is initially finite. T_c seemed originally to rise continuously with δ from the insulator but this has not been demonstrated clearly in most cases. The metallic state is always peculiar as we shall later discuss; when it turns into a more normal metal, with excessive doping, T_c goes down.

Since in (almost) all other substances low carrier number, metal-insulator transitions, and antiferromagnetism decrease T_c, the mystery story principle requires the association of the magnetism with the phenomenon of high T_c. A theoretical point: the effect of doping on Mott insulators has been a controversial and unsolved problem for decades; again the overwhelming temptation is to associate difficulties.

More straightforward, and equally logically compelling, are optical, photoelectron spectroscopy, and NMR data. From optics and PES it is clear that the carriers appear in the Mott-Hubbard gap in proportion to the doping. NMR data show that the hyperfine couplings of the metallic carriers are identical with those of the spins responsible for magnetism.

Theory—so long as optics, PES, and other probes confirm the presence of the new carriers in the same orbitals as the magnetism—is equally compelling. The strength of the semicovalent bond due to $O_{p\sigma} - d_{x^2-y^2}(Cu)$ hybridization is responsible for the great integrity of the square planar configuration of CuO_2 planes. This suggests that, next to the Cu^{++} "U" repulsion, the second largest parameter is the $2p\sigma - d\sigma$ hybridization t_{dp} which must be of order $2 - 3\,ev$, leading to the observed $\sim 6ev$ splitting of bonding and antibonding bands. Between these two levels are a spaghetti of weakly-bonding $Cu - O$ hybrids which form a very large hump $\gtrsim 1 - 2ev$ below the Fermi level. The antibonding band has only one state per Cu ion, and therefore the appropriate Hilbert space for leaving the magnetic state intact and introducing a new set of carriers does not exist.

A considerable body of reliable electronic structure calculations by Schluter and others confirms this picture and gives us reasonably reliable values for the Hubbard "U" and "t" parameters.

The most persistent fallacy evading Dogma II is the "extended Hubbard Model" and various variants thereof, which are at least formally correct in that they can be reduced to the right model unless they have the wrong parameter values, but are an unnecessary detour of no physical value. More naive theories simply don't question where, electronically, the magnetism comes from, and use coupled magnetism-carrier physics. Various probes show that the magnetic and charge form-factors of the carriers differ somewhat: they are $\gtrsim 60\% \, d$, at least, magnetically, and $\gtrsim 60\% \, p$ electrically. This can be understood as different polarization of the background bands by exchange and Coulomb interactions, by a single band of renormalized carriers. Thus the conclusion is:

II: We must solve the old problem of doping a single Mott-Hubbard band before we can begin the problem of high T_c. After renormalizing away high-energy excitations, the physical particles live in a single band. The problem is the very old problem of reconciling their magnetic structure and their charge transport.

Dogma III: The dominant interactions are repulsive and their energy scales are all large. Clearly the existence of a Mott half-filled band insulator implies large repulsive interactions whose scale may be bounded below by the Mott-Hubbard gap for charge transfer. A second scale is set by the exchange parameter which, by various accurate experimental measurements, especially spin wave velocities and Raman spectra, is at least 1200°K, leading to spin wave bands $.2 - .3 \, ev$ wide and a spin wave velocity comparable ($\sim 1/4$ at least) to Fermi velocities. In Hubbard model terms, we are in the case of large but not infinite U.

Many clear indications place the intrinsic electron-phonon coupling at normal to moderately strong. Most striking are the shifts of optical phonons associated with the gap, shown in Fano resonances with the anomalous electronic background in the Raman effect. There is no reason to doubt various direct electronic calculations of these couplings. What is striking is that they have so little effect. We believe the resistivity which would have been caused by phonons and by static lattice distortions and defects in the (non-stoichiometric) normal metal, is larger, at least at low T, than the observed ρ_{ab}, which, again, does not vary from substance to substance as much as a phonon resistivity would. Equally, "phonon bumps" do not show up strongly in tunneling, infrared, and ARPES spectra. This situation reveals one of the crucial anomalies of the high T_c materials. We can and will describe the physics by saying that the strong repulsions dominate and restrict the response of the charge and spin density fluctuations to phonon and static potentials.

A little more may be said. We are accustomed to the fact that collective, bose-like modes (phonons, plasmons, spin waves) are much less easily scattered at low frequencies and long wavelengths than particle wave functions. In the "Luttinger liquid" type of theory, for charge transport the particle modes have been replaced by the collective motions, by the simple construction dating back to Tomonaga: these experimental facts support this kind of electronic theory. For instance, the heat conductivity suggests that the phonons are strongly scattered, if not vice versa. How can this happen? Thus we have, to the lowest order,

III: Restrict your attention to a <u>single band, repulsive</u> (not too big) U Hubbard Model.

Dogma IV. The "normal" metal above T_c is the solution of the planar one-band problem resulting from Dogma III, and is not a Fermi liquid, in the sense that $Z = 0$. (Z being the quasiparticle wave-function renormalization constant.) But it retains a Fermi surface satisfying Luttinger's theorem at least in the highest T_c materials. We call this a Luttinger Liquid.

This has several sound experimental and theoretical bases. The most vital experimental evidence lies in the giant anisotropy of resistivity ρ_{ab}. The resistivity perpendicular to the planes extrapolates to ∞ at T=0 with an exponent $T^{-1\sim 2}$ and is in all cases well above the Mott limit $\frac{h}{e^2 k_F}$. This means that there is no coherent electronic transport in the c-direction: all motions are inelastic. Fermi liquids cannot localize in one direction and extend in a second, because simple localization is a question of coherence: are the electrons coherent in extended or in localized states? Thus a fortiori the normal state is a two-dimensional metal with only inelastic processes connecting the layers.

In these experimental considerations and also in later ones a theorem due to Schrieffer (as far as I know) is very important: single-particle tunneling, and transport are not renormalized by the wave-function renormalization constant "Z", in any conventional—or even unconventional, as in the case of weak localization—Fermi liquid theory. This was discovered in relation to superconducting tunneling, and in the verification of phonon interactions in BCS theory it played a vital role, but it was equally important in the hands of Mott, Thouless, and other early workers in localization theory, who recognized that the conductance e^2/h is a universal boundary between metallic and insulating states, independently of dynamic effects, because conductivity is not renormalized. The modern ideas on conductivity using the S-matrix, pioneered by Landauer, give us considerable understanding of the universality of this theorem. The essential idea is that conductivity contains no <u>dynamical</u> corrections: it may be written entirely in geometrical terms as

$$\sigma_a = \frac{e^2}{h}(c_D) A_F^a \cdot \ell_a \tag{1}$$

where c_D is a dimensionless, D-dependent constant of order unity, A_F is the Fermi surface area perpendicular to the direction a, and ℓ_a is the mean free path in direction a. (1) is proved by essentially the same technique as Schrieffer's proof for tunneling conductivity, which gives it as an integral over the entire quasiparticle spectrum which reduces to a contour integral around the pole with no "Z" correction.

Several authenticated cases of "insulating" transverse conductivity exist in the literature in other systems (e.g., TaS_2) and we would suggest that these be re-examined, since this behavior will not normally occur in a Fermi liquid. If the matrix elements are genuinely tiny, it is possible that inelastically assisted conductivity could dominate, but in all high T_c materials but BISCO the observed $3d$ superconductivity rules this option out (see later).

A second very strong argument is the small value of ρ_{ab} and the relative sharpness of the features in the ARPES spectrum. Reasonable estimates of impurity scattering by the large non-stoichiometry in (214) or BISCO leads to a mean free path $l \sim$ a few $xa_0 \sim 20$Å while ρ_{ab} corresponds to roughly $l = 50 - 100$Å at T=100°K. Any reasonable estimate of phonon scattering also lead to a bigger ρ_{ab} than observed. The ARPES peak widths are in good agreement with ρ_{ab} while the feature sharpness is even smaller. Hence, giant concentrations of charged impurities have no effect on ρ_{ab}; tiny percentages of substitutional uncharged impurities in the planes, on the other hand, which carry free bound spins according to several measurements, lead to reasonable residual resistances and to T_c lowering. This point of view is carefully worked out in my Science article.

We can only conclude that current is carried by some collective or soliton excitation whose motion is controlled by uncharged entities, i.e., $Z = 0$, $F_{OS} = \infty$, Fermions as we shall shortly discuss. A good model for such a state and such excitations is given by appropriately reinterpreting the exact Lieb-Wu solutions of the one-dimensional Hubbard model.

$Z = 0$ is confirmed by the several measurements, all of which agree, of inelastic scattering τ's: infrared, Raman background, ARPES. All tell us that $\frac{1}{\tau} \propto \omega$, and if we treat the carriers as quasiparticles with a self-energy $\sum$, $\sum_{im} \propto \omega$ and $\sum_{re} \propto \omega \ln \omega$ by Kramers-Kronig transform, hence $\frac{\partial \Sigma}{\partial \omega}$ diverges.

In fact, it is likely that the true ω-dependence is not $\omega \ln \omega$ but $\omega^{1-\alpha}$ and both real and imaginary parts are power laws. In either case $\frac{\partial \Sigma}{\partial \omega}|_{\omega=0} \to \infty$ and $Z = (1 - \frac{\partial \Sigma}{\partial \omega})^{-1} \to 0$. One's first response to $Z = 0$ is to abandon entirely the Fermi liquid quasiparticle theory on which this derivation is based, as well as the Fermi surface. However, we shall see that both the one-dimensional model and the experimental facts show that the Fermi surface and the Fermion-like excitations may remain even when electron-like quasiparticles are absent.

There is a strong theoretical argument and motivation for this peculiar non-Fermi liquid normal state. The one-band Hubbard model has the property of having the "upper Hubbard band", a separated band of states thought of as comprising the motions of electrons on doubly occupied sites. This can be given a precise meaning in at least two ways: as a band of separated "anti-bound" particle-particle scattering states, or by the Rice-Kohn-Anderson canonical transformation procedure to the "t-J" model in which the kinetic energy term is exactly projected onto an equivalent singly-occupied subspace. The key operative word is "projective": the Hilbert space of the new low-energy problem is smaller than that of the corresponding Fermi liquid states, because projection operators have zero eigenvalues. This change of Hilbert space means that the states of the $N+1$ body problem live in a new Hilbert space and are necessarily orthogonal to those of the N body problem, hence Z—which is a ground state to ground state overlap integral— is zero.

Intrigued by this problem of doping of the Mott-Hubbard insulator, theorists have rather ingeniously found at least three viable alternative approaches, and we take it as an optional premise that one should follow only one of these—but as dogma what the final result must be like.

Two reasonable but somewhat indirect approaches are twisted antiferromagnetic order parameters and "flux" or "anyon" phases. A few carriers doped into the antiferromagnetic insulator can be shown rigorously to generate a co-moving distortion ("twist") of the antiferromagnetic order parameter, and the resulting soliton may be a model for our $Z = 0$ object—it is an excellent one in 1-dimension. A second concept is to model the Mott insulator with an "RVB" liquid of short-range singlet pairs, and to study solitons in this; again, a limiting process starting this way is a way of thinking about 1d. The short-range RVB may be based on a "flux phase" and give solitons which are anyons. The resulting particles carry gauge fields in order to implement the projective transformations. But experiment seems firmly to reject the short-range picture. Yet another method is to follow the original suggestion of Anderson and Zou modeled on the short-range RVB ideas and to implement the projection with a "slave boson" (or "slave Fermion") theory with spinons, holons, and gauge fields, calculating directly using the full gauge theory. Lee and Nagaosa, and Yoffe and Wiegmann, calculate in this way without assuming short-range RVB, and in particular Lee assumes a Fermi surface for his spinons. Wilczek and Zou carry this system still further. Many properties can be approximately calculated in this way, but as yet it is not accurate, as far as one can tell. A fourth concept, closest to the original picture of Anderson and BZA, is a $Z = 0$ liquid of spin 1/2 Fermi-like particles, and charged S=0 excitations or holons. Perhaps the most straightforward way to describe such a fluid is as a limiting case of the Fermi liquid, such a system as one might find if—as in 3 dimensions—there is a U_{crit} dividing Fermi liquid from Mott insulating states. As one approaches U_{crit} $Z \to 0$ implying $\frac{\partial \Sigma}{\partial \omega} \to \infty$. However, we recognize that there remains a finite spin velocity in this limit, hence the Fermion mass does not go to zero. This is only possible if $\frac{\partial \Sigma}{\partial \omega} / \frac{\partial \Sigma}{\partial k} = \frac{1}{v}$ remains finite, hence the compressibility which is proportional to $(\frac{\partial \Sigma}{\partial k})^{-1} \to 0$ and the Landau parameter $F_0^S \to \infty$. Thus charge cannot be carried by the Fermions, any Fermion being perfectly surrounded by compensating charge in the medium. It must reside in collective sliding motions of the Fermion liquid, or bose like charged excitations which we think of as a second, charged soliton or "holon". All of this picture is precisely modeled on the one-dimensional case which is exactly soluble.

Several experimental facts drive us to this type of theory. The Pauli-like spin susceptibility (except in the anomalous case of $Y(Ba)_2Cu_3O_{6.65}$ which has many unique properties) is that of an equivalent spinon liquid. Most strikingly, the existence of a Fermi surface in the ARPES measurements on the normal state can only be understood this way (the interpretation of ARPES energy distribution data certainly does not contradict, and possibly strongly supports, the "Luttinger liquid" picture).

One of the methodological strengths of condensed matter physics is the overdetermination by data. It is often the strongest evidence for a theory that none of the wide variety of possible probes contradicts it; let us assure you this is the case here.

The main alternative theory, which is excluded by the evidence for Dogma IV, is any of the many versions of conventional or unconventional Fermi liquid theory. The data require that the electron excitation be composite, not elementary, and, in fact, that it decay at a rate given by the available phase space. This type of behavior of ρ_c is seen elsewhere only in conjunction with some kind of "unusual" condensed state such as SDW's or CDW's in dichalcogenides or organics, where the charge carriers are coupled to an order parameter. And the existence of the upper Hubbard band, which drives the Z=0 process, is clear in optical data.

The "marginal Fermi liquid" theory is an alternative path which uses much of the above experimental argument but attempts to distance itself from the Hubbard model. It is true that attractive models can also have varieties of non-Fermi liquid behavior in 1 and 2d systems. The primary distinction is the rejection of the concept of spinons and of the spinon Fermi surface without any consistent alternative being proposed. The theory as it is normally given rejects our arguemnt that v_F and m^* remain finite, so has no second excitation for charge. The evidence which is relevant is that for Dogma II, and for the upper Hubbard band. Our frank assessment is that the most valid results of MFLT are those which are equally well explained by the main line of reasoning. There is one interesting question: assuming, as we must, that $Z = 0$ implies charge-spin separation and the spinon Fermi surface, is it still necessarily true that there is a true holon excitation, or could the charge be carried by a collective resonance near the $2k_F$ edge of the spinon pair spectrum? Calculations aimed at proving the existence of a bound $2k_F$ charged collective mode of this sort have not succeeded. We tentatively reject such an alternative, but this is a serious question for study.

We summarize Dogma IV by the statement: the normal metal is a two-dimensional Luttinger liquid: i.e., Fermion-like spin excitations—spinons—establish a Luttinger Theorem Fermi surface. Charge is carried by an alternative excitation which necessarily itself exhibits a Fermi surface. Charged excitations cannot be thought of as having fixed statistics and none of the three: electron, spinon, holon, is a simple bound state of the other two (in contradiction to short-range, KRS RVB and to PWA's Varenna notes!)

Dogma V : The above state is strictly two-dimensional and coherent transport in the third dimension is blocked. We feel it is also, but less, evident that this two-dimensional state is not superconducting and has no major interactions tending to make it so, at least near the usual T_c. For the Dogma, the often cited c-axis resistivity would be argument enough, (the most common, data being those in the infrared, see Uchida et al and Tanner et al), but there is a second important experimental one. The ARPES data on BISCO 2212 resolve a very sharp cusp feature near the Fermi surface, and an even-sharper quasiparticle peak below T_c in a variety of directions presumably moving through the Fermi surface. Because of the pair of close CuO_2 planes in BISCO 2212, the Fermi surface in the planes should split in two by an amount equal to the effective interplane hopping integral $t_\perp$, and this effect is indeed to be seen in the calculated band structures. The relevant splitting is a couple of tenths of an ev. At some symmetry points, the splitting is small because the hopping effectively takes place via the Sr^{++} ions between the CuO_2 planes, and hence the effective matrix element can be frustrated. Nonetheless, enough directions have been probed to indicate strongly that this odd-even splitting of the CuO_2 planar states doesn't exist.

Theoretically—here we have only the one-dimensional model to use as an analogy. Several workers in the heyday of one-dimensional physics pointed out that interchain hopping, if weak, renormalizes to irrelevant at $\omega, k \to O$. (More recently, D.-H. Lee and ourselves have come to the same conclusion.) Thus there exists a critical value of the hopping, $t_\perp$, between CuO_2 layers, below which the physics of the Luttinger Liquid can remain strictly lower-dimensional. [More recently, this seems to have become controversial, but as far as I can see this is purely a semantic misunderstanding (see JETP Letters)]

This is, however, strictly a one-dimensional result depending on the small exponent α in the Green's function, while we feel that the physical effect is larger and more obvious. If the Luttinger liquid dogma is correct, the electron is not a stable excitation in the plane any more than a quark is in free space: its charge and spin move off immediately at different velocities. Thus it has no $\omega = 0$ amplitude: it cannot find a stable eigenstate with the same quantum numbers of charge, spin and momentum, which is the prerequisite for coherent motion. Thus theoretically dogma IV seems to lead automatically to dogma V.

The impact of Dogma V, then, is that the two-dimensional state has separation of charge and spin into excitations which are meaningful only within their two-dimensional substrate; to hop coherently as an electron to another plane is not possible, since the electron is a composite object, not an elementary excitation.

The assumption that the two-dimensional system is not superconducting is based on the scale and the variability of T_c. T_c varies from 0 to 6 to 130° depending on the overall 3-dimensional structure: except for ρ_c, the only highly variable experimental parameter. This is a scale which is much smaller than the scale at which the Luttinger Liquid properties set in, which is not less than $\sim 1000°$ K. The properties of the planes have become reasonably uniform, and dominated by strong interaction effects, from this temperature downward. The question is—what could possibly make T_c be of an entirely different, and widely varying, scale? It seems almost required that T_c itself depends radically either on interactions between planes or with the substrate, and is not a purely 2 dimensional effect. When superconductivity in fact ensues, the properties seem indeed to become more isotropic, and the penetration depth λ_c^2 which is basically a measure of the c-axis plasma frequency changes to a value which is much too high vis a vis the c-axis resistivity.

There are simply no indications of a unique, purely two-dimensional type of superconductivity. We go into this point more fully, since it is true that 2d fluctuation effects of a fairly conventional type exist above T_c in YBCO and in fact in all the "multilayer" systems such as Bi and Tl 2212. We propose that pairs or triplets of layers do become superconducting by dint of their interlayer coupling and behave like a single conventional superconducting layer. This is quite different physically from "anyon superconductivity" within a single layer. The Kosterlitz-Thouless behavior often seen is consistent with this idea.

It is possible that the various 2-dimensional superconducting states proposed by others are in fact nothing but the $Z = 0$ Luttinger Liquids of Dogma IV in a new guise. No attempts at a fluctuation or thermodynamic theory of "anyon superconductivity", for instance, have been made, only a ground state is discussed. Thus a real possibility is that T_c is zero or the state is sensitive to impurity scattering of one type or another. The anomalous ρ_{ab} conductivity seen in experiments extrapolates to zero and hence, in some sense, the state really is "superconducting"; with $T_c = 0$; but some other T_c mechanism intervenes, probably that of Dogma VI.

Dogma VI Interlayer hopping together with the "confinement" of Dogma V is either the mechanism of or at least a major contributor to superconducting condensation energy.

There is better than a rough identity between the conductivity per electron in a given direction and the total kinetic energy: each are proportional to a velocity-velocity correlation function. Kohn has shown that in a single tight-binding band the frequency integral of the conductivity is proportional to the mean kinetic energy per electron

$$\int \sigma(\omega)d\omega \propto < t_{ij} c_i^+ c_j >$$

Along the c-axis there is a great defect in conductivity: there is no coherent motion of electrons in the c-direction. This means that there is, in the normal state, a missing energy of order $t_\perp^2/t_{||}$, which is regained in the superconducting state (since, experimentally, the superconducting response function in this direction appears to be consistent with band structure, and this measures the restoration of the low-frequency part of $\int \sigma(\omega)d\omega$).

There is, therefore, a contribution to the condensation energy which is not a theoretical but an experimental fact and which comes from the interlayer tunneling energy. It is compelling to identify this as the source of the condensation energy, but, of course, pairing has other consequences and may well gain energy from phonon and other modes as well.

The heuristics of transition temperatures lend strong support to this view. We have published a study of relative T_c's of Bi and Tl 2 and 3-layer materials on this basis. Pressure coefficients are very favorable: these are large of the right sign in all 1-layer materials, but for two or more strongly-coupled layers one cannot be sure what pressure will do. The one mystery is Tl one-layer, which has a very mysterious chemistry, fluctuating from zero to 60° with physical state. We suspect that Tl orbitals are close to the Fermi level. The fact that this T_c is so sensitive to interlayer chemistry is in itself mysterious if T_c is intralayer. We also point out (stimulated by a remark of S. Trugman) that the fact that the top layer of an YBCO crystal seems not to be superconducting is very strange on any interlayer theory. We suspect that YBCO surface problems are caused by the lack of a satisfactory cleavage in this crystal separating the pairs of layers at the chain lattice planes.

The mathematics of "confinement"—the blockage of interlayer coherent motion—is not complete but is becoming so.

On mechanism, there are no plausible competing theories. It is hardly necessary to detail this. It may be necessary to point out that the "BCS mechanism" has quite a subtle heuristics which was well understood by Morel and Anderson following Bogobliubov within 3 years of BCS. No other "theory" of high T_c is beyond the level of the Cooper model calculation, using vaguely described "attractive forces" of some hypothecated kind with no chemical or structural heuristic whatever. What is more, no hint of these mechanisms appears in well-designed experiments.

The mystery story principle requires that a unique consequence—high T_c—be associated with the only other truly unique properties: $Z = 0$ Fermi liquids, and exact two-dimensionality. In summary, then, Dogma VI tells us that whatever the normal state physics, we can count on the mechanism of interlayer tunneling deconfinement to account for T_c.

A parenthetical remark: the anomalous case of $YBa_2Cu_3O_{6.65}$ (the "60°" superconductor") has attracted a lot of attention. It may well be that this spin liquid is a real example of one of the other alternative states, with spirals or fluxes. However the tendency to a reduction of all spin fluctuations with temperature suggests that we are actually getting a singlet-paired CDW forming, not an SDW. All such liquids have the $Z = 0$ syndrome and charge-spin separation, so all are equally subject to the same interlayer mechanism.

These six dogmas lead us through a rather intricate maze of alternatives to a consistent, coherent view of the whole fascinating phenomenon of the superconducting cuprates. The data-overdetermination of condensed matter physics tells us, actually, that if a crucial countervailing experiment existed we would surely not be able to sustain such a heavy structure.

There is, however, a set of experiments which can confirm it beyond the shadow of a doubt: and, in preliminary fashion, partially they have done so. These are of three kinds. **(1)** Careful infrared and superconducting studies of the c-axis electromagnetic responses, especially in 214 or other one-layer materials. If (VI) is right, the conductivity which is missing in the normal state should partially reappear in the superconducting response $1/\lambda^2$. (This has been beautifully done in 2 1 4 by Uchida.) Infrared can confirm the absence of an appropriate $\sigma_c(\omega)$ in the whole relevant range in the normal state.

An experiment which can be quite interesting would be to search for the "$\sigma - \pi$" interband absorption in 2-layer materials in the c direction. There is, in every band calculation, a splitting of the Fermi surface, due to $t_\perp$, of $\sim .1ev$, essentially between bands even and odd in the Y or other ion between the close CuO_2 planes. This should, in Fermi liquid theory, cause a very strong c-active infrared absorption, which should be easily visible. In fact, it should also have a very strong tail towards low frequencies because of the vanishing of this splitting in symmetry directions. If it is not there our diagnosis of "confinement" is confirmed very strongly. The absence of ρ_c can be thought of, naively, as some kind of localization effect, but the strong reduction of electronic conductivity in a wider frequency range would be very telling.

As with the other experiments we will discuss, these electromagnetic anomalies will manifest themselves as an apparent <u>failure of a sum rule</u>: the superconducting response $1/\lambda^2$ will not be equal to the integral up to the gap frequency of the normal conductivity.

$$(\frac{1}{\lambda^2})_c \neq \int_0^\Delta \sigma^c_{norm}(\omega)\, d\omega$$

This is a striking prediction which may already have been verified.

(2) Careful experiments on tunneling along the c axis. "H I H" junctions consisting either of single crystals clamped together, relying on a surface inactive layer, or true tunnel junctions, for instance created by MBE, should manifest extremely interesting properties. The same arguments needed to explain $\sigma_c \simeq 0$ seem to lead to weak tunnel currents as well, in this case. We should see a strong contrast between the normal and superconducting states. In the superconducting state Andreev scattering allows quasiparticles to move freely between planes; spinons pick up charge from the condensate and turn back into electron quasi particles. Thus tunneling, at least in the energy gap region, will be relatively normal. (Of course, energies above the gap are not affected). Thus we can expect to see a rather clean superconducting tunneling spectrum complete with gap and Josephson supercurrent in junctions which appear more or less insulating in the normal state. This is a tricky but easy experiment. I suspect that such junctions have already been observed and discarded by a number of experimentalists.

It also may be relevant to reexamine really clean c-oriented normal to superconductor contacts—why are they so bad? Do they also show changes below T_c?

(3) $G_1(k,\omega)$ as measured by ARPES, like the electromagnetic response, will probably not obey the usual sum rules. When the sample is normal, the quasiparticles are composite and G_1 is spread over a broad spectrum. We have shown elsewhere that when the system becomes superconducting the quasiparticle amplitude becomes finite and peaks reappear in the Green's function. Again, the effect is an apparent failure of the sum rule for quasiparticle amplitude: the peaks appear to arise from nowhere, because they are borrowed from a wide range of energies. There is a difference, however, in that in this case the original sum rule is still satisfied if all frequencies are summed over.

In the above we have given a logically consistent overall picture of the physics of high T_c. At the very last stage we find that the picture is susceptible to several clear experimental checks, involving phenomena inexplicable on other grounds. Some other vital areas of experimental study can also be suggested: clearly high-resolution, careful ARPES will continue to be of great value. A less obvious but important area is doping the planes with spin impurities, which by smearing the spinon Fermi surface will rapidly destroy the unique Luttinger liquid properties, reintroducing residual resistance and metallic transport along the c axis as well as reducing T_c. In subsequent chapters we will flesh out these ideas. We will focus on transport, experiment and theory; on spectroscopy, Raman, IR, and PES; and then on theoretical concepts: the 1d Hubbard model, Luttinger liquids in higher dimensions, and finally the interlayer mechanism for superconductivity.

The Infrared Catastrophe: When Does it Trash Fermi Liquid Theory?

Yet another attempt to clarify why I believe I have proved the failure of perturbative renormalization for Fermi liquids of various kinds. It provokes a special fury in those trained in quantum field theory methods, compounded in equal parts of fear of the unknown and of conviction that nothing simple enough for a mere solid state physicist to find could mess up their beautiful theory so thoroughly. My respect for mathematical physics does not extend to the majority of its practitioners. That experimental fact is unequivocal seems not to concern them. When did physics cease to be an experimental science?

The "Infrared Catastrophe:" When Does It Trash Fermi Liquid theory?

P.W. ANDERSON[*]

Joseph Henry Laboratories of Physics
Jadwin Hall, Princeton University
Princeton, NJ 08544

ABSTRACT

We give an historical discussion of the "infrared catastrophe" and the "x-ray edge anomalies" of Mahan associated with scatterers in a Fermi sea of electrons. The infrared catastrophe provides a perspicuous way into understanding the difficulties with many-body perturbation theory which have recently been discovered as a result of a study of high T_c superconductivity, and we show how this "catastrophe" is avoided in some cases, but cannot be avoided in the one and 2-dimensional electron gas systems. Finally, we indicate the new type of theory which is necessary in the event of such a breakdown.

---

[*] This work was supported by the NSF, Grant # DMR-9104873

Nearly 30 years ago G.D. Mahan[1] conjectured a fundamental anomaly associated with a Fermi sea of electrons interacting with a local potential, which he supposed to be responsible for experimentally observed power law anomalies in the x-ray absorption and emission edges of metallic systems. This conjecture, based on extrapolating the results of a perturbation treatment, was turned into a proof in Ref. (2) and into a full theory of the x-ray edge effect by Nozieres and de Dominicis in 1969.[3]

Mahan's original conjecture encountered a great deal of resistance at the time, in fact his paper was delayed in the refereeing process; and the general community did not accept the reality of the effect until NDD's formal theory appeared (or even after). This is somewhat understandable: it is easy, and was done from time to time, to produce false Fermi surface anomalies, such as false "bound states", caused by the sharp edge in the density of states at the Fermi level; and Kohn and Majunder[4] had apparently given a general refutation of such efforts a year or two earlier. The proofs of Refs. (2) and (3) showed that the effect did exist and did not not contradict any fundamental principles, however, including those discussed in ref. 4.

The effect is so simple in the form described in Ref. (2) that it is worth demonstrating here. We suppose ourselves to have introduced a local, finite scattering potential (representing the effect of the inner-shell electron in the x-ray process) $V(r - r_0)$ which is so short-range that its matrix element for scattering a free electron from state k to state k' is essentially a constant for all states near the Fermi surface:

$$V_{kk'} \simeq V \qquad (1)$$

We calculate the probability that state k below the Fermi surface is scattered into state k' above it ($\epsilon_{k'} > 0$) which is, to lowest order

$$P_{k \to k'} = \frac{|V|^2}{|\epsilon_{k'} - \epsilon_k|^2}$$

The sum of all such scattering probabilities diverges logarithmically:

$$\sum_{k,k'} P_{k \to k'} = |V|^2 \sum_{\epsilon_{k'}>0} \sum_{\epsilon_k>0} \frac{1}{|\epsilon_{k'} - \epsilon_k|^2} \qquad (2)$$

$$\sim |N(E_F)V|^2 \ln \frac{(\epsilon_{k'} - \epsilon_k)_{\max}}{(\epsilon_{k'} - \epsilon_k)_{\min}} \tag{3}$$

(where $N(\epsilon_k)$ is the density of states at the Fermi level) and the minimum possible $\epsilon_{k'} - \epsilon_k$ is of order (size of system Ω)$^{-\frac{1}{2}}$, so the divergence is as $\ln \Omega$.

Such a divergence was shown, in Ref (2), to mean that the ground state of the system <u>with</u> the scattering potential V (a Slater determinant of scattered wave functions) is orthogonal to the original free electron Slater determinant in the limit $\Omega \to \infty$. The overlap contains a singular power law

$$\left(\psi_0(V),\ \psi_0(0)\right) \propto (\Omega)^{-\frac{1}{2}|N(E_F)V|^2} \tag{4}$$

which can also be shown to appear as a function of ω, in various response functions as well, and causes the "x-ray edge anomalies".

The property of V which is crucial is that it leads to at least one finite scattering phase shift η (in this case, as the isotropic $\ell = 0$ channel). The wave functions in the asymptotic region, far from the scatterer, are, for free electrons,

$$\varphi_{\ell=0}^{\text{free}} \sim N \frac{\sin kr}{r} \tag{5}$$

and when the scatterer is introduced,

$$\varphi_{\ell=0}^{\text{scatt}} \sim N \frac{\sin(\tilde{k}r + \eta)}{r} \tag{6}$$

where N is a normalization constant and $\tilde{k}$ must be modified to fit boundary conditions at the edge R of the sample. The expansion of the scattered wave functions (6) in terms of free wave functions (5), when we take into account the boundary condition which makes $\tilde{k}$ different from k, gives exactly the overlaps which one obtains as transition probabilities from perturbation theory:

$$S_{kk'} \propto \frac{\eta}{k - k' + \frac{\eta}{R}}$$

This way of thinking of the effect of the scatterer is, however, more precise and generalizable: one must find the scattered wave functions for a given set of boundary conditions and calculate the overlap of the appropriate Slater determinants.

This effect, apparently simple in old-fashioned Brillouin-Wigner perturbation theory or in conventional scattering theory, is not easy to deal with in modern many-body theory. Among many other reasons, boundary conditions, which play a vital role, are absent in many-body theory; and it is assumed in that theory that the set of wave vectors [k] cannot change. Nozieres and co-workers[5] labored through two long, difficult papers before getting the effect, and Ref. (3) uses a radically different method from the usual techniques. Many-body perturbation theory is based on the Feynman diagram technique, which in turn depends crucially on the fact that a hole (a "positron") is simply the time-reverse of an electron, with the Dirac Sea having no dynamical consequences (this is the famous "Z" diagram of Feynman, describing pair creation as simply reversing the space-time path of the electron.) In these phenomena which we treat here, the "Sea" is no longer a dead, meaningless object but—as we see from the fact that the logarithm in (3) contains both the lower and upper cutoffs in energy—plays a vital role. In metals, we may have to distinguish holes from electrons—we paint the "hole" branch of the space-time path a different color from the "electron" one, in a manner of speaking.

The "orthogonality catastrophe" is closely tied in with two other important Fermi Sea concepts, again not natural to diagram techniques: the Friedel theorem and bosonization. The Friedel theorem is that in the presence of a scattering potential, the region near the scatterer contains an excess of particles given by

$$\delta n = \sum_\ell \delta n_\ell = \sum_\ell (2\ell + 1)\frac{\eta_\ell(E_F)}{\pi} \qquad (7)$$

(This is related to certain theorems of scattering theory due to Wigner: the object on the right is the imaginary part of $\ln S$, S the general scattering matrix at the Fermi energy.) In fact, the exponent in the overlap is proportional to the sum of the squares of the δn_ℓ's. This demonstrates that a phase shift of π is exactly equivalent to the formation of a locally bound state, 2π is 2 bound states, etc.,, as far as the Ω-dependence of the overlaps is concerned. This fact that $\eta = n\pi$—which corresponds to a zero scattering cross-section—has this profound effect, more than anything else demonstrates that the properties of a Fermi sea cannot really be understood without taking into account the "counting theorems" of Friedel and Luttinger, in addition to the simple diagrammatic perturbation theory of electrons at the Fermi surface.

Schotte and Schotte[6] showed that the overlap could be calculated by "bosonizing" the variables in the $\ell = 0$ channel, which is basically equivalent to treating the position of the Fermi surface in the relevant channel in momentum space as the appropriate dynamical variable to describe the Fermi sea, rather than using particle coordinates and momenta. This very important and useful concept has been recently generalized by Haldane[7]; but the rather formidable mathematics involved is not necessary to our story.

The existence of the orthogonality catastrophe of course called into question immediately the whole structure of many-body perturbation theory as it is applied to metals. If an electron added to the Fermi sea acted as a scatterer for the other electrons, with a finite scattering phase shift in some channel, the new state of the "sea" electrons would be orthogonal to the original state. The obvious consequence is that the quasiparticle renormalization constant "Z" would vanish since this is just the overlap between the relaxed and "bare" particle states. This would mean among other things, that the Fermi surface discontinuity of occupancy of states in momentum space would vanish. The basic point of a vanishing "Z" is even deeper than that: Z represents the connection between the "bare" particle states and the exact low-energy eigen-excitations of the interacting system. A finite Z means that these remain in one-to-one correspondence with each other, while a vanishing Z means that the exact excitations can—and do—have totally different character from the original excitations. If "Z" were zero in quantum electrodynamics, that would mean that the underlying theory contained no "bare" entities resembling the physical electrons which we see in nature. Such a phenomenon actually occurs in the transformation from bare quarks to physical nucleons.

Even for electrons this is not impossible at all: in fact, the bewildering complexity of the various exact or asymptotically exact theories of one-dimensional interacting electrons (for a review see Solyom[8]) can be traced to just this fact: that in one dimension an added electron has a finite scattering phase-shift for the electrons near the Fermi surface. As a result these theories have no discontinuity at the Fermi surface and the eigen-excitations are not electrons but bosons representing vibrations of the local Fermi surface in k-space.[9] Metzner and di Castro,[10] as well as Haldane[9], have recently emphasized that for real electron systems there is really only a single free parameter—representing this phase shift—in the known solutions of the one-dimensional interacting electron gas, except at special points where commensurability with the lattice plays a role, and aside from a trivial mean field (RPA) response.

But many Fermi systems of ordinary dimensions (2 or 3) behave more or less like "Fermi liquids", which is the shorthand for a system described by quasiparticle excitations, presumably the result of applying a convergent perturbative renormalization to the bare Hamiltonian, leading to quasiparticles with effective masses and Fermi velocities, and restricted types of interaction, but not to totally new entities. D.R. Hamann,[11] to my knowledge, was the first to think about and to solve the problem posed by the "orthogonality catastrophe" for Fermi liquid theory. He pointed out that in dimensionality ≥ 2, the Fermi surface scattering is much reduced by the dynamic recoil of the electron, i.e., simply by the fact that when the particle of momentum k' is scattered to a momentum $k' + q$, the original particle of momentum k is scattered to $k - q$. In general, this removes the vanishing energy denominators in the crucial sum (2). This argument is entirely valid for foreign light particle scatterers such as positrons or μ mesons, for which there are never any infrared problems. The inner shell electrons which are involved in the x-ray edge process are, however, so narrow-band that they cannot recoil appreciably.

The step which can avoid Fermi-edge divergences is the replacement of the scattering matrix element $V_{kk'}$ by a "pseudopotential" or "scattering vertex" $T_{kk'}$. This idea was one of the seminal ideas behind many-body theory, in its early form of "multiple-scattering theory", which was aimed at dealing with strong individual scattering potentials and was first applied (to nuclear matter) by Brueckner and collaborators.[12] One attempts to solve the problem of repeated scattering of two particles and to insert the result of the repeated scattering as a formal renormalized effective or "pseudo"-potential. The calculation of repeated scatterings is essentially the same as solving the Schrödinger equation for the two particles and deriving from this their scattering amplitude. This renormalizes the large effect of strong repulsive interactions or "hard cores" but it also has the effect of allowing the scatterer to recoil. In diagrammatic terms the idea is to replace the individual scattering vertex $V_{kk'}$ by the sum of "ladder diagrams" (see Fig. 1) in which repeated scatterings $k \to k+q \to k+q' \ldots \to k$ and $k' \to k'-q \to k'-q' \ldots \to k'$ are allowed. The other particles are assumed not to be involved in this process. This procedure is therefore exact in the limit of low density, and in fact was the procedure used in the treatment of the low density Bose and Fermi systems by Yang[13] and co-workers and by Gor'kov and Galitskii,[14] and later on by Bloom[15] for the 2D electron gas. But it is actually essentially correct even at finite densities for the purposes of our problem, because for particles close enough to the Fermi surface or within it real scatterings by other particles are restricted severely by the exclusion principle, and virtual scatterings by other particles can be renormalized away. The only important correction is that the ladder diagram sum must take into account the fact that states $k'-q$ etc., may be excluded because of being already occupied. So it is literally exact, in principle, to consider an electron of up spin as a foreign particle scatterer for down spins exactly like the inner shell hole of the x-ray edge problem. Like that hole, it acquires a singular self-energy due to the resulting modification of all of the energies and wave-functions of the down-spin electrons—this is exactly the method of calculation used by Nozieres and de Dominicis, in fact, for the x-ray problem.

In many cases this "recoil" calculation actually works to reduce "Z" to a finite value. These are the cases in which the scattering can be described for small scattering vectors Q, by an "effective range" or "scattering length" theory, as was done in the works by Yang et al,[13,14,15] referred to above and as is assumed without proof in AGD.[16] This seems to be correct in the limit of low densities or weak scattering in 3 or more dimensions.

The "effective range" theory assumes that the scattering is like that from a hard sphere of radius a, with "a" a parameter which can be determined by solving the radial wave equation for two free particles. The wave function in relative coordinates, near the origin, becomes

$$\varphi(r) \sim \frac{\sin Q(R-a)}{r} \tag{8}$$

so that the phase-shift η is obviously Qa, which vanishes as Q, the relative momentum, goes to zero. This is sufficient to restore Fermi liquid theory, for reasons which require a little subtlety to explain.

We consider the state $(k\uparrow)$ to be the added particle whose self-energy Σ_k and renormalization constant $Z_k = \left(1 - \frac{\partial \Sigma}{\partial \omega}\right)^{-1}$ we are seeking. The state $k'_\downarrow = (k+Q)_\downarrow$ is the scattered state whose energy and wavelength shift due to the existence of $k\uparrow$ we want to calculate. In order to fit the new s-wave state to the boundary condition at a radius $\sim R$ ($R^3 \sim \Omega$), we must shift Q by

$$\delta Q R = Qa$$
$$\delta Q = \frac{aQ}{R} \tag{9}$$

But plane wave states k and k' are only fractionally in an s-wave state with respect to each other, the number of angular momenta possible for a given $k - k'$ being

$$N_L \simeq QR \tag{10}$$

only one or, at most, a small finite number of these angular momenta are affected by the scattering so that

$$\delta k' \propto \frac{\delta Q}{QR} \propto \frac{a}{R^2} . \tag{11}$$

The separation between k-values $\Delta k'$ is $= \frac{\pi}{R}$ so that

$$\frac{\delta k'}{\Delta k'} \simeq \frac{a}{R} <<< 1 , \tag{12}$$

vanishing in the $\Omega \to \infty$ limit. This is the condition that no orthogonality catastrophe takes place; this slight shift can be compensated, as extra particles are added, by a shift in mean potential, adjusting the chemical potential so that the density of k-values in momentum space remains constant. Any higher power-law dependence of η on Q is a fortiori even more harmless: the key condition is that the forward scattering phase shift vanishes or not as $Q \to 0$.

It is often stated that Fermi liquid theory is based on a convergent perturbation theory in which no resummation of divergent terms has taken place. This is obviously not the case in view of the above derivation: the "ladder" summation which reduced $\eta(Q = 0)$ to zero and gave us effective range theory is precisely such a summation. The rather indirect way in which this summation is argued around in the standard derivations is the subject of a short paper which has been submitted,[17] but that argument is not necessary to the present discussion. Suffice it to say that the summation of formally divergent ladder diagrams as $Q \to 0$ is an integral part of the theory.

In dimensions lower than three the recoil argument becomes more difficult. In fact, in one dimension $\eta(Q = 0)$ does not vanish even for free particles and, as we already pointed out, no interacting system has a finite Z. In two dimensions a free scatterer can be shown to obey

$$\eta_{2D}(Q) = \frac{1}{|\ln Qa|}$$

which, as Bloom showed, is just adequate to give a convergent first few terms of perturbation theory in the formal zero-density limit of the 2d hard core gas.

In 2 or 3 dimensions there are certain theorems ("Levinson theorems") which demonstrate that in the absence of bound states the scattering phase shifts go to zero for free particles at zero relative energy. However, there is no proof of such theorems for scattering of two particles which live in a Fermi sea and must obey the exclusion principle. The effect of the exclusion principle is to make it much harder for the two particles to recoil, especially at very low energies: essentially, "soft" recoils become impossible, because the states involved are occupied. Levinson's arguments have no relevance in this case.

It is very important to understand precisely the scattering problem we are considering, and why it is done in this way. We are adding a particle of $k_\uparrow$ near (but below) or on the Fermi surface: i.e., we visualize having added the particles one by one and space has been left for this one. We now consider each particle $k'_\downarrow$ in the down-spin Fermi sea and allow it to adjust its wave-function to the new scattering by $k_\uparrow$. The scattering problem for each k' is the "on-shell" problem in which we study scattering which is precisely elastic. Lifetimes, near the Fermi surface, of the holes that these particles will fill are $\propto (k-k_F)^2$ and can be made negligible by well-known arguments.

Both particles we are taking as renormalized quasiparticles as far as high-energy virtual interactions with the Fermi sea electrons are concerned. These various statements can be justified quite rigorously by lengthy, fussy arguments whose general character is quite obvious and irrelevant to the present considerations. This is a slightly unfamiliar scattering problem, not the same as the one which would be done to produce the conventional "scattering vertex" for two added particles or holes in perturbation theory, because we are concerned with the wave functions of electrons actually present in the Fermi sea, not with added excitations over the Fermi sea as vacuum. As in the x-ray problem, we are looking for the response of the "sea" considered as a physical fluid to the addition of a perturbing particle.

The diagrammatic description of this process is that we sum all ladders which scatter k, k' to $k+Q, k'-Q$, eliminating all occupied states for both particles with a factor

$$F(Q) = (1-f_{k+Q})(1-f_{k'-Q}) \tag{13}$$

but allowing the particles to return to their initial state $Q = 0$. The result is a simple Schrödinger equation which, using a Hubbard-model local potential U, reads

$$1 = U \sum_Q \frac{1}{E-E_Q} F^*(Q)$$

where E_Q is $\epsilon(k+Q) + \epsilon(k'-Q)$ and F^* is (13) for all states Q except those for which $E-E_{Q_0}$ is of order ΔE_{Q_0}, the spacing between levels of different Q_0.

These are the states which must be treated separately to allow the scattered wave function to satisfy the boundary condition enforced by the scattering. Essentially, these states must be allowed to move in k-space in order to let the wave function modify itself in order to satisfy Schrödinger's equation both at ∞ and at the origin. Formally, we do this by writing

$$\frac{1}{U} = \sum_Q \frac{F^*(Q)}{E - E_Q}$$
$$= \sum_{Q \simeq Q_0} \frac{1}{E - E_{Q_0}} + P \int \frac{dN}{dE_{Q'}} \frac{dE_Q}{E - E_{Q'}} F(Q')$$
$$= N(E) \cot \pi \eta + P \int \frac{dN}{dE_Q} \frac{dE_{Q'}}{E - E_{Q'}} F(Q')$$

This, in 2D, leads in general to a finite η as $Q \to 0$ and $k \to k_F$, because $F(Q)$ vanishes as Q^2 and the principal part integral is convergent as $Q \to 0$.

This computation is trivial, and in fact has been checked in much more complete calculations by Fukuyama et al.[18] and by Engelbrecht et al.[19] Both find that <u>on the energy shell</u> the forward scattering phase shift is finite. What they find is that the phase of the scattering matrix $T(Q,\omega)$ is finite exactly along the line $E_Q = \omega$ but not in either the limit $\omega \to 0$ or $Q \to 0$. Thus less careful calculations miss this "on-shell" effect. Where we differ is in the use of this information: I point out that this phase shift implies a shift of the momentum values δk which is turn leads to the orthogonality catastrophe. These authors, however, do not draw the obvious conclusion that the orthogonality catastrophe occurs, because they are working with conventional perturbation theory, in which the orthogonality catastrophe does not appear and the consequences of a finite phase shift are not evident. As we showed earlier, the orthogonality phenomenon shows up as a divergence when a finite system is taken to $\Omega = \infty$, but perturbation theory assumes that the infinite volume limit is approached uniformly. Special stratagems are necessary to uncover the orthogonality phenomenon without recourse to taking the limit from finite systems.

One of the interesting side implications of this result is that it may not be possible in general to draw conclusions about the states of finite systems using many-body perturbation theory as it now exists. In finite systems, the modification of wave functions upon adding a particle is not small and must be taken into account. This may be one of several reasons for difficulties with the LDA approach to such problems. Genuine Hartree-Fock, which does take the change in wave functions into account, may often give qualitatively preferable answers.

It is not yet clear when or whether 3D systems also are liable to the orthogonality catastrophe and the breakdown of perturbation theory. The simple forward scattering calculation (14) does not lead to finite η because $N(E_Q) \to 0$ at $Q = 0$. General arguments suggest that two conditions may need to be satisfied in 3D:

(1) The relevant band must be isolated by an energy gap from higher, free-electron like bands.

(2) U must be stronger than a critical value.

These are the conditions for validity of the "t-J" model transformation of the interacting Hubbard model to a projective model in which the exact "double occupancy" restriction is enforced. Such a restriction can be removed with the addition of a local gauge symmetry, which leads in general to a major modification in the elementary excitation spectrum.

CONCLUSION

A great deal of work has been done involved in fitting this breakdown of perturbation theory into a possible scheme for dealing with the 2-dimensional electron gas. This scheme involves using the multidimensional bosonization scheme of Luther and Haldane, which they presented as a transcription of Landau's Fermi Liquid Theory. As they show, the hydrodynamics of the Fermi liquid are not adequately described by the random-phase approximation; however, in the Fermi liquid theory there are no interesting collective modes other than at $Q \simeq 0$, where RPA is satisfactory. In their picture, the hydrodynamics of the Fermi sea is that of an incompressible liquid blob in k-space, with the low-frequency modes being vibrations of this blob at each point of its Fermi surface, with conservation laws for every angle around that surface.

The singularity due to forward scattering fits neatly into this hydrodynamics and can be expressed by modifying the bosons representing the fluctuations of the Fermi surface, giving separate Fermi velocities for charge and spin bosons. This seems to be not only theoretically sound but also vindicated by a wide variety of experimental phenomena in the cuprate metals, some of which were predicted prior to observation. The mathematics of this approach is, however, beyond the scope of this article.

This approach leaves open the next order corrections: coupling of bosons on different patches of the Fermi surface, etc. It also has, so far, only been applied in the 2D case. The conventional many-body theory has been noteworthy in its failure to deal adequately with strongly-interacting and low-dimensional systems; we now have the first breakthroughs which demonstrate that this is not a mere problem of difficulty in application but a failure in principle. It is likely to be essential to completely revise our conceptual structure before going further.

REFERENCES

1. G.D. Mahan, Phys. Rev. **163**, 612 (1967); Phys. Rev. **153**, 882 (1967)
2. P.W. Anderson, Phys. Rev. Lett. **18**, 1049 (1967)
3. C.J. DeDominicis and P. Nozieres, Phys. Rev. **178**, 1097 (1969)
4. W. Kohn and C. Majumdar, Phys. Rev. A **138**, 1617 (1965)
5. P. Nozieres, O. Raulet, J. Gavoret, Phys. Rev. **178**, 1072, 1084 (1969)
6. U. Schotte and D. Schotte, Phys. Rev. **182**, 429 (1969)
7. F.D.M. Haldane, Unpublished Notes, Les Houches Summer School, 1992.
8. J. Solyom, Adv. in Physics, **28**, 201 (1979)
9. F.D.M. Haldane, J. Phys C **14**, 2585 (1981)
10. W. Metzner and C. di Castro, preprint.
11. D.R. Hamann, Private Communication.
12. K. Brueckner, et al, Phys. Rev., **92**, 1023 (1953); Phys. Rev. **95**, 217 (1954), etc.
13. T.D. Lee, K. Huang, C.N. Yang, Phys. Rev. **106**, 1135 (1958)
14. L.P. Gorkov and G. Galitskii, as quoted in Ref. 17
15. P. Bloom, Phys. Rev., **B12**, 125 (1975)
16. A.A. Abrikosov, L.P. Gorkov, I. Dzialoshinskii, Methods of Quantum Field Theory, Prentice Hall, N.J. (1963).
17. P.W. Anderson, submitted to Phys. Rev. Lett.
18. H. Fukuyama and O. Narikiyo, J. P. Soc. Japan **60**, 372 (1991); **60**, 1032 (1991)
19. J.R. Engelbrecht and M. Randeria, Phys. Rev. Lett **65**, 1032 (1990).

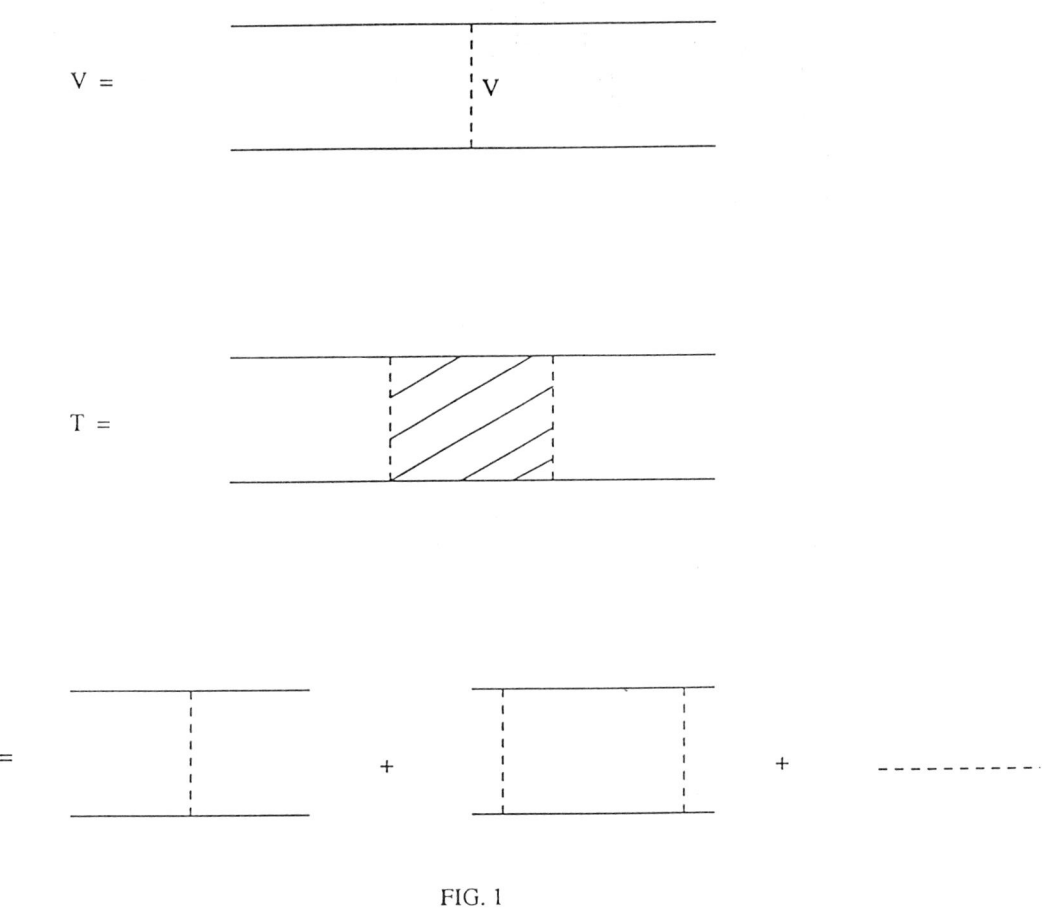

FIG. 1

THE LADDER SUM FOR THE EFFECTIVE SCATTERING POTENTIAL

Experimental Constraints on the Theory of High-T_c Superconductivity

P. W. Anderson

Analysis of the many experiments on high-temperature superconductivity indicate several essential aspects of any theory. The conductivity and other transport properties as a function of disorder, temperature, and frequency point to a non–Fermi liquid–like behavior, whereas photoemission experiments and magnetic properties indicate the presence of a Fermi surface in momentum space. To reconcile this apparent contradiction, a new type of electron liquid, called a Luttinger liquid, has been postulated, and the present article aims to show the need for this postulate. Theory and experiment indicate that the suitable phenomenological electronic structure model of the CuO layers is that of the one-band Hubbard model. It is also argued that experiment clearly indicates that interlayer interactions strongly affect the superconducting transition temperature, T_c, consistent with the fact that no theoretical calculations on two-dimensional Hubbard models have resulted in the prediction of high transition temperatures, and that anyon models are not favored by experiment.

A common misapprehension is that the theory of high-temperature superconductivity is in disarray. It may be true that theorists are in disarray, perhaps understandably, in that the situation genuinely is complex, many of the most crucial arguments are quite subtle and involve many esoteric and unfamiliar branches of physics, and people have committed themselves to diverse points of view that seemed tenable at the outset. I have considerable understanding of some of these theories, having tried them myself over many years even before 1987 to explain puzzling phenomena in many of the older superconductors. I think it would be useful to summarize the reasons for discarding many of these kinds of ideas, for the sake of not only theorists but also bewildered experimentalists and, most important of all, the curious outsider.

It is possible to produce a reasonably complete and consistent theoretical viewpoint that does not encounter any serious theoretical or experimental problems (1). This theory involves many elements that are esoteric even to the average many-body theorist, and it might therefore best be discussed in a book or a series of specialized articles, and this is being done. A more generally understandable review, which does not require special technical expertise, on the other hand, may be written about a number of serious constraints on any viable theory, most of which come from simple experimental measurements and simple but quite rigorous theoretical interpretations.

In a widely circulated but unpublished paper about a year ago, I expressed some of the constraints (2). It is now possible to do so even more clearly.

There are four major areas in which one can give clean results. These are (i) deductions from transport of electrons in the ab plane, (ii) deductions from transport along the c axis, (iii) growing consensus on the model: results from electronic calculations and from optical and photoelectronic probes, and (iv) deductions from the heuristics of the transition temperature, T_c.

Many theoretical approaches start from the premise that the real problem is to find a novel mechanism leading to enhanced attractive interactions of some sort, pairing the electrons of the normal metal that is supposed to exist above T_c. As a first step theorists do conventional band calculations to derive the "electronic structures" of the materials. The experimental experts on the various kinds of probes for studying Fermi surfaces and band structures—angle-resolved photoemission, angle-resolved positron annihilation, and de Haas van Alphen studies—sometimes take such structures very literally and use them as a guide. That such structures must not be taken too literally is, for instance, shown by the fact that experimentally the c-axis mean free path is less than the unit cell size so that it is not physically meaningful to attempt to determine departures of the Fermi surface from precise two dimensionality, since c-axis momentum cannot be defined. It is of course natural to continue to use conventional tools in conventional ways for lack of anything else, and the experimentalists especially express themselves with some caution, but as in this instance data interpretation may be influenced by theoretical preconceptions.

In the third section on models I will sort out the evidence that the fundamental model must transcend band theory, a point of view that is indeed beginning to be a consensus among theorists and that was well covered in the popular article by Schrieffer and Anderson (3). In the first two sections, however, I want to focus on the deeper and more precise question that was stated but not answered in that article: Is the normal metal above T_c in the cuprates a Fermi liquid in the precise meaning of that term? This question is the clear focus of the present discussion. It is my view that the experimental answer is unequivocally negative. If the answer is that the normal metal above T_c is not a Fermi liquid, the conventional approach sketched above is not likely to be fruitful but must be supplemented.

Fermi Liquids

Let me explain what the Fermi liquid question means and why it is important. Not since Wigner's work of the 1930s have theorists felt they could ignore the strong Coulomb interactions among electrons and of electrons and lattice vibrations in metals, and use a pure free-electron Sommerfeld-Drude-Bloch model. But that model worked with great precision as to qualitative behaviors; the general understanding of the reasons, in the operation of the exclusion principle, dates to the early 1950s or even earlier; but it was the Russian group around Landau and Migdal who codified what is now known as the Landau Fermi liquid theory, which expresses the precise sense in which this is true. In this theory an exact meaning is given to the concept of "quasiparticle," an exact low-energy elementary excitation of the Fermi liquid that has all the properties of a real free electron but can have a modified velocity, mass, and so forth and contains all of the high-energy effects of the interactions. In perturbation theory the quasiparticle excitation is connected to the real bare particles by a finite "wave function renormalization factor," Z. Z is also the fraction of the amplitude in a photoemission or positron experiment, for instance, that appears in the peak associated with the quasiparticle; the rest is incoherent many-quasiparticle superpositions.

One should not get the impression that a

The author is at the Joseph Henry Laboratories of Physics, Jadwin Hall, Princeton University, Princeton, NJ 08544.

small Z means that the quasiparticles are not the whole story: they are, in that the Landau theory shows that all sufficiently low-frequency dynamical properties follow from quasiparticle motions alone, because they are the only low-frequency electronic excitations.

When we find such a quasi-exact low-energy theory, we now know, from the work of Wilson (4) and others, that it is certainly what we call a "fixed point of a renormalization group." In this case this means that we start at high energy and successively eliminate (renormalize away) the interactions and states, leading to an effective theory at low energy and long wavelength that in general will be simpler and simpler as we renormalize: it will converge to a "fixed point." This means that, even though we may not have access to the lowest energies, the renormalization group will have come a long way from the high-energy phenomena where it started and the resulting description of the state will be very far from the various alternatives. Thus, although it may seem like nitpicking to argue, as we do, about whether Z is really exactly zero or only 1%, it makes a qualitative difference, because it means we are near a qualitatively different fixed point and have a qualitatively different set of elementary excitations.

In summary, if we find that many of the electronic properties have settled down to characteristic temperature dependences as we go to the lowest available temperatures (and it is a serious problem in some cases that T_c is so high that we do not really have very low temperatures available), we should assume that we have reached the neighborhood of a fixed point, whether Fermi liquid or some other new possibility.

Properties of High-T_c Materials

One of the obvious facts about the cuprates is how similar the "normal" metal properties are among dozens of chemically very different materials (see Table 1). For most of these materials, if we are reasonably close to the optimum doping for T_c, we have very similar resistivity per plane, for instance, in the direction parallel to the planes (5); the spin susceptibility (6), nuclear magnetic relaxation (7), Hall effects (8), photoelectric (9) and infrared spectra (10), and tunneling curves (11), just to give a few examples, each measured on some subset of the few materials that can be prepared in single-crystal or at least well-characterized form, always have striking resemblances to each other and striking differences from other metals. Most observers accept that such details as the "chains" in YBCO or the "apical oxygens" of La_2CuO_4 are not controlling the properties and that the electronic properties are dominated by the CuO_2 planes.

Here I must appeal to a point of logic. The common response, when one makes a firm statement that all of these materials are not Fermi liquids because of one or another observation, is to say that the observation encounters exceptions among these many materials. But that is not the point: if they are all at the same fixed point—and they clearly are—it will be non–Fermi liquid for all if it is not for any one: it is necessary only to prove the negative in one instance. Exceptions are logically irrelevant.

Characteristic of Fermi liquids are a number of typical behaviors, with well-understood modifications due to special effects such as strong random scattering or magnetic impurities. One strong invariant is the volume within the Fermi surface, a theorem attributed to Luttinger, which, it turns out, is even stronger than Fermi liquid theory (FLT) because it still holds in many non-FLT one-dimensional models and is likely to hold in the cuprates. One can sketch a general proof that, if there is no new broken symmetry, any surface of low-energy excitations in k-space will enclose the appropriate volume for the density of electrons and hence will have a "large," "Luttinger" Fermi surface (12). This is a point on which there is some confusion (in which at one time I shared): the one-dimensional solutions clearly show that the Fermi surface may exist in a non–Fermi liquid.

Transport and response functions for a Fermi liquid are as follows: a constant density of states is reflected in finite T-independent spin susceptibility, a specific heat of γT, and $KT_1(NMR)T = $ const. (where T_1 is the time constant for relaxation of spins). The resistivity is $\rho = \rho_{res} + AT^2 + \rho_{phonon}$; Matthieson's rule of additivity is good if ρ_{res} is not too high, and the deviations due to weak localization are well understood. Heat conductivity $K = $ const. $\times T$ obeys the Wiedemann-Franz law of proportionality to σ approximately. The Hall effect ideally is independent of T both at very low T and anywhere above $1/4$ T^{Debye}.

A second strong rule is that $\rho_{res}^{(T=\omega=0)}$ does not renormalize with Z (where ω is frequency): it is a purely geometrical, wave-function effect of scatterers on the phases of quasiparticle wave functions [Mott's theorem, an early version having been proved by Schrieffer (13)]. It does not participate in the dynamic effects that cause Z. This is also true (in Migdal theory) of phonon resistivity. This is why the strongest arguments against FLT are based on resistivity data. Let me sketch the best of these.

Resistivity in the ab plane. The characteristic observation is that ρ_{ab} is not exceptionally high and that it has two unusual

Table 1. Materials.

Nickname	Composition at (approximately) highest T_c	T_c (K)	Structural features	Single crystals?	Well-characterized
214	$La_{1.85}Sr_{0.15}CuO_4$ (also many analogous)	~40	Octahedron; all planes identical	Yes	Yes
$YBCO_7$	$YBa_2Cu_3O_7$	~95	Chains and pyramidal plane pairs	Yes	Yes
$YBCO_{6.6}$	$YBa_2Cu_3O_{6.6-0.7}$	~60	Chains (O-deficient) + plane pairs (pyramids)	Yes	Yes
BISCO					
2212 (32)	"$Bi_2Sr_2Ca_1Cu_2O_8$"	~80	Pyramidal plane pairs	Yes	Yes
2201	"$Bi_2Sr_2CuO_6$"	~10	Single tetrahedral planes	Yes	Yes?
2223	"$(Bi,Pb)_2Sr_2Ca_2Cu_3O_{10}$"	~110	Two pyramids + one simple plane	No	?
	All nonstoichiometric				
TlCOs					
2212	"$Tl_2Ba_2Ca_1Cu_2O_8$"	100	Same as BISCOs	No	Yes
2201	"$Tl_2Ba_2CuO_6$"	10–70	Same as BISCOs	No	?
2223	"$Tl_2Ba_2Ca_2Cu_3O_{10}$"	125	Same as BISCOs	No	Yes
"Electron doped" (33)	$Nd_{2-x}Ce_xCuO_4$	38	Ce	Yes	?
YBCO "124" (34)	$Y_1Ba_2Cu_4O_8$	81	Two chain layer + two planes	No	Yes
YBCO "247" (35)	$Y_2Ba_4Cu_7O_{15}$	95	Alternate two + one chains	No	Yes
"∞-layer" n (36)	$Sr_{1-y}Nd_yCaO_2$	40	Only planes, no apices: electrons	No	No
"∞-layer" p (37)	$Sr_{1-x}Ba_xCuO_{2+y}$	60–90	Same, holes	No	No

characteristics: usually there is no residual resistivity, and the T dependence is roughly linear (5):

$$\rho_{ab} = 0 + AT^{1\pm\epsilon} \quad (1)$$

(The correction to linear is weak: $\epsilon \sim 0.1$, or the correction could be logarithmic.) The conductivity for planes doped to have maximum T_c (15 to 30% hole doping) is not very different per plane from one substance to another. There are exceptions to these rules but exceptions from a generic behavior cannot restore the validity of the Fermi liquid fixed point.

This behavior is particularly striking in that Eq. 1 holds both for the pure, stoichiometric single crystal Y Ba$_2$Cu$_3$O$_7$ and for the very nonstoichiometric materials BISCO (2212) and (La$_{1.84}$ − Sr$_{0.16}$)CuO$_4$. There is little effect, therefore, of random scattering by the doping ions. In BISCO, for instance, the best estimates are that ∼25% hole doping occurs entirely through the presence of off-stoichiometric distributions of the various ions. One may calculate rather easily the residual resistance that would result from a 25% off-stoichiometric concentration of ionic charges near the planes.

Another theorem that does not renormalize with many-body effects is the "Friedel sum rule" (14), which states that the screening charge around an ion in a metal is related to the scattering phase shifts by (in two dimensions)

$$n = 2 \sum_m \frac{\delta_m}{\pi} \quad (2)$$

where $m = 0, \pm 1, \pm 2$, and so forth (the rough approximation of circular symmetry is inessential; the numbers do not depend on it). The screened ion is not more than 4 Å from the plane, so it is screened by electrons in about one unit cell of the square lattice, which implies that $\delta_{m=2}$ is probably small. Setting $\delta_{m=0} = \delta_{m=1} = \pi/6$, I estimated that the transport cross section for a single ion is

$$S = \frac{1}{k_F} \frac{1}{2\pi} \int_0^{2\pi} d\theta \sum_m |(1 - e^{2i\delta_m}) \cos m\theta|^2 \quad (3)$$

$$\simeq \frac{3}{k_F}$$

(where k_F is the Fermi wave vector and θ is an integration variable) and so the mean free path is

$$\ell^{-1} = nS \simeq 0.075 k_F \quad (4)$$

If we use the simple, unrenormalizing formula

$$\sigma = \frac{e^2}{2\pi\hbar}(k_F \ell) \quad (5)$$

we get (taking a typical plane spacing of 6 Å) a residual resistivity in the neighborhood of 100 microohm-cm. This resistivity is characteristically comparable to that per plane of BISCO, YBCO, 124, or Tl2212 in the region of 100 to 200 K. In fact, when these are successfully doped in the planes with scatterers that are strong enough to cause residual resistivity, this is a fairly usual value of ρ_{res} (8). What is, however, strikingly characteristic is that this resistivity due to the off-stoichiometric doping is never seen, nor is there any resistivity visible that is clearly caused by conventional phonon scattering, a resistivity that can be calculated to be of a similar order of magnitude although with a more complicated temperature dependence.

In summary, residual resistivity can be induced by certain kinds of doping, and, when it exists, it is additive to the mysterious linear term AT (A depends very little on purity). But the easily calculable residual resistivity of the dopant ions and the phonon resistivity are missing.

There is a great deal of information on the infrared conductivity of the ab plane that is confirmatory of the conclusions reached here, especially that the phonons (which are clearly visible in infrared and Raman spectra) have surprisingly little influence on the electronic response, and that the strong linear T dependence precisely corresponds to a linear ω dependence of relaxation rate τ^{-1} which extends smoothly out to frequencies near and beyond 1000 cm^{-1}, where it is unthinkable that phonon scattering could be involved (10).

As I emphasized, the absence of residual resistivity means that FLT has failed. This is independently confirmed strongly by the less rigorous argument that the ω dependence implies a mean scattering rate τ^{-1} that would lead to a divergence to zero of Z, as calculated from the perturbation theoretic self-energy of the quasiparticles. [This is the "marginal Fermi liquid" argument (15).] It hardly needs to be said that the entire apparatus of the conventional theory of metals, including the phonon or other boson-coupling pairing mechanisms as embodied in the "Eliashberg equations" that we used so effectively for conventional superconductors, might not be usable in such a radically modified environment. Thus, the great bulk of the literature speculating on one or another "pairing mechanism" may be irrelevant because it does not confront these deeper questions. The best description of the normal state from a purely empirical point of view is that the observed conductivity is that of a material that is going to be a superconductor at $T = 0$, and that has a depairing mechanism which prevents that superconductivity from being manifest, yet conventional scattering mechanisms do not cause resistance. (A similar behavior, although with different T dependence, is that of σ_{xx} in the quantum Hall effect.)

Resistivity along the c axis. A second, equally rigorous argument against the validity of FLT can be constructed around the same set of theorems applied to the resistivity in the c-axis direction, perpendicular to the CuO$_2$ planes. Again relying on the basic Mott theorem (13) that conductivity is not renormalized by the dynamic quasiparticle renormalizations, we observe that for conventional metals, almost no matter how anisotropic the mass, a conductivity greater than the Mott minimum metallic conductivity

$$\sigma_{min} \simeq \frac{e^2}{\hbar} k_F \quad (6)$$

can be expected in all directions. The factor k_F can in fact be anisotropic if the Fermi surface is very cylindrical, having low dispersion in the c direction; k_F then represents the average magnitude of this dispersion in k-space and might be an order of magnitude lower relative to the ab direction. A common misapprehension is that localization is possible in one direction only, not in all; this is untenable because localization is a coherent backscattering phenomenon that requires the electron to retain its local coherence in all three directions in order to observe any localization in one.

A more serious problem arises in the very anisotropic material BISCO and perhaps elsewhere. Here it is reasonable to suppose that the dispersion in the c direction, "$t_\perp$" in energy units, is less than the inelastic scattering rate $\hbar/\tau$. In this case we will have incoherent Giaever tunneling in the c direction, not metallic conduction. However, in all such materials there are close pairs or triplets of CuO$_2$ planes that would give large infrared conductivity owing to the infrared-active interplanar hybrids of odd and even symmetry, a conductivity, again, at worst an order of magnitude less than the conductivity along the ab plane.

The general observation on c-axis conductivity is that it (i) is usually lower than, and often much below, the minimum Mott conductivity and (ii) often decreases rapidly with decreasing T. Early measurements often showed a roughly inverse behavior to the ab plane conductivity, that is, $\sigma_c \propto T$, $\rho_c \propto 1/T$ (16), at least at low temperature. The key argument is a logical one: that if even a few of the cuprates are clearly violating Fermi liquid inequalities, the overwhelming implication is that all are; demonstrating that "pure enough" or "adequately oxygenated" crystals show metallic behavior in the sense of a positive temperative coefficient fails to answer the question of what was going on in the original materials, which certainly comprise the great majority. In no sense does the Mott limit restrict itself to "pure" or "sufficiently oxygenated" materials. In view of the anisotropy ratios (17) from 300 to 10^4 observed in many single crystals, it is clear that defects of all sorts—spiral dislocations, for example, which are common, as well as a-axis twins, grain boundaries, and so forth—may account for some of the observed "normal" T dependences. Again, the logic is that, if any one cuprate is not a Fermi liquid, none is.

The linear T dependence, with a very large coefficient of resistance at high temperature, is sufficiently common that one comes to suspect that it may be a generic behavior of Giaever tunneling for these materials. No theory of Giaever tunneling for non–Fermi liquids has been developed. The fact that this linear resistivity often adds in series with a $1/T^P$ term (where P is the power law constant) is significant in showing that this is not a simple Drude or Giaever conductivity.

The infrared conductivity polarized along the c axis is quite hard to measure, but what measurements exist show no sign of a very strong σ-π absorption in the region from ~500 to 1000 cm^{-1} (18). These measurements seem to confirm another puzzling paradox: the conductivity that in conventional superconductors determines the square of the inverse penetration depth λ^{-2} via the relation derived from a sum rule:

$$\int_0^\Delta \sigma(\omega)d\omega = \frac{1}{\lambda^2}\frac{8}{c} \quad (7)$$

(c is the speed of light) is not adequate to explain the observed penetration depth (19): that is, the normal metal is in some cases strictly two dimensional as a metal yet becomes a three-dimensional superconductor. Very few of the candidate theories can cope with this fact.

There are many other unconventional aspects of transport, as, for example, a strongly and anomalously T-dependent Hall effect (8), but I have focused here on rigorous, quantitative, inescapably logical deductions that force one to a non–Fermi liquid fixed point.

The One-Band Hubbard Model

A consensus has grown up among many theoretical students of the high-T_c problem as to the appropriate model that must be solved to understand all this anomalous behavior. A semipopular exposition on this subject was given by Schrieffer and Anderson in *Physics Today*, June 1991 (3).

Underlying this model are a number of quite solid deductions from optical and photoemission data, electronic calculations and simulations, and other observations. To give the answer first, the equivalent model Hamiltonian—not, now, at the "transport" scale of ~100 K but at the 1-eV scale of the fundamental electronic interactions—is a "one-band Hubbard model" (20).

There is only one band of electrons that plays a role in the low-energy properties (less than ~1 to 1.5 eV) of these substances on the CuO_2 planes. This band is a hybrid (antibonding in character) between $Cu d_{x^2-y^2}$ orbitals and $O_{2p\sigma}$ orbitals, which overlap strongly and whose wave functions are even relative to the CuO_2 plane. There is only one "Wannier function," local orbital, per Cu site. Charge and spin polarization effects cause the apparent degree of hybridization to vary somewhat depending on which experiment we study, but not outside of reasonable limits. The basic strong interaction between these electrons is repulsive, opening up a "Hubbard gap" U between states containing, respectively, two holes or one hole per Cu.

One should understand that the above picture is a model, albeit a quite accurate one. The "upper Hubbard band" is really above a "charge transfer" gap, not a pure Coulomb gap; the added electron is to some extent in an s orbital in the surrounding Cu ions; and the band is a fairly complicated antibonding hybrid. But through careful photoemission analysis Sawatsky (21) has been able to demonstrate how the Hubbard band forms above the high density of states peak of the nonbonding O orbitals. Optical studies show the rather well developed Mott-Hubbard gap of about 1.5 eV in the antiferromagnetic, insulating cuprates. As holes are doped into the "lower Hubbard band," intensity disappears in both optical and BIS bands, which can be clearly identified as the upper Hubbard band. The rate of disappearance agrees reasonably well with that calculated for a one-dimensional Hubbard model (22).

This picture is strongly supported by various computations, which, on this energy scale, can be done quite accurately. Careful electronic calculations by Schluter and Hybertsen (20) show that energy levels of clusters of several CuO_2 units in the appropriate background can be closely matched to those of a one-band Hubbard model; also, a fairly good account of the overall spectra can be produced from direct simulations [as, for example, in Horsch's work (23)] from a one-band Hubbard. Finally, nuclear magnetic resonance (NMR) coupling constants have been shown by the Illinois, Los Alamos, and Zürich groups (24) to be compatible with this band picture and not with one in which the magnetic electrons are distinct from those carrying the charge. Moreover, neutron (25) as well as Raman and NMR studies (7) have shown that antiferromagnetic interactions and antiferromagnetic correlations persist well into the superconductivity, or "strange metallic," regime and exhibit the superconducting gap. These magnetic couplings follow only from a basic repulsive electron-electron interaction such as the Mott-Hubbard U, of fairly strong magnitude.

Two popular types of theories are excluded by this basic model information. Both rely on large phonon coupling: the "negative U" or bipolaron scheme and the "density of states peak" idea. In the negative U scheme it is supposed that preformed pairs of electrons are bound together by large phonon displacements and their Bose condensation is T_c. Aside from the well-known problem of such a theory that the Franck-Condon displacements make metallic conductivity impossible and the Bose temperature negligible, the clear evidence that the magnetic and superconducting electrons are the same—in the neutron data of Rossat-Mignod *et al.* (25) they show the same energy gap—rules this out. The fundamental interactions of the superconducting electrons are repulsive.

The "Van Hove peak" idea of a high density of states peak coincident with some strong phonon coupling again fails to account for the clear evidence of an upper Hubbard band (26). Such speculations go back to the older "high T_c's" of the A15 and Chevrel structures, where also they were not successful. It is not consistent to ignore the dominant effect of the inescapable repulsive interactions when dealing with narrow bands or narrow-band features, because both theoretically and experimentally such features enhance repulsive effects (27).

T_c Is Controlled by Interlayer Interactions

A great many attempts at parametrizing the T_c data have been made, with greater or less success, but, in general, there is a tendency for such studies to be somewhat selective as well as very uncritical of sample characterization issues. A rough outline of what is really there, as far as I can see, follows.

1) For every material there is an optimum (but not necessarily very sharply defined) degree of doping for high T_c: below this level of doping, the material tends to be too insulating; above this level, almost always there is a crossover in the normal state toward conventional Fermi liquid behavior. It is significant that Fermi liquid–

like behavior contraindicates high T_c.

2) At and near the optimum doping there is rather little difference in normal state properties, especially the linear T resistivity per plane, suggesting that these are controlled by the planes themselves. Unfortunately, only a restricted number of single-crystal samples are available to verify this estimate, but so far it holds.

Very large differences in T_c's are caused by the "reservoir" layers between the planes. Bismuth materials show this very clearly: single planes have $T_c < 10$ K; double planes, ~80 K; triple planes, 110 K; $(La - Sr)_2CuO_4$ has T_c near 40 K, but the per plane properties are practically identical to 95 K YBCO, or to 80 K BISCO. Single-plane Tl cuprates may be optimized to nearly 80 K but also can drop to 10 K without enormous change in planar properties, while double and triple planes go to 105 and 125 K (28).

By careful selection, some of these data points can be made to fit on various universal curves but not in a way that carries conviction. To me, it seems an inescapable conclusion from the overwhelming tendency of the data that T_c is not a single-plane property. Many theorists have been trying to find a T_c mechanism within the single-band, one-layer Hubbard model, yet even a casual glance at the experimental facts convinces one that superconductivity is caused by effects outside that simple model and does not occur with high T_c in an isolated cuprate plane. (It can and does occur in a single unit cell with more than one plane.)

Many other structural features have been postulated as vital, but fortunately all have now been excluded by the finding of high T_cs without them. A list of a few of these follows (see Table 1 for supporting evidence).

1) The "chain" layer of YBCO, missing in almost all other high-T_c materials.

2) The "apical oxygen," the oxygen completing either a bipyramid or square pyramid with the CuO_2 square planar group. Materials with T_c's of 40 to 90 K have been found with no apices at all.

3) Hole-particle asymmetry. Electron-doped materials have lower T_c's and do not show all the anomalies as clearly but are definitely part of the picture.

4) Order versus disorder. YBCO is a stoichiometric crystal, BISCO is as disordered as you can get, with T_c's within 15 K of each other. In fact, disordering YBCO can change its planar properties, anisotropy, for instance, sharply without changing T_c at all.

5) Tight groups of planes. $(La - Sr)_2CuO_4$ and the newer "∞-plane" materials show that, although tight groups (pairs or triplets) of planes are good for highest T_c's, they are not essential.

If there is a generalization that so far has not failed, it is that superconductivity is always a two- to three-dimensional crossover, experimentally; it is this generalization which our theory exploits (1).

Anyon Superconductivity?

One of the types of theories proposed for high-T_c superconductivity is vaguely described as "anyons": theories in which the superconducting or normal state or both are described by spontaneous time-reversal (or parity) breaking and an excitation spectrum consisting of vortex-like solitons. I have not specifically discussed these theories above but, in view of the interest they attract, they may deserve separate mention.

As a description of the normal state, they suffer primarily from the evidence that the normal fixed point has a true Fermi surface [evinced by photoemission data and the Korringa NMR relaxation of many nuclei, demonstrating a large constant density of states of Fermi-like spin fluctuations (29)]. Anyons are motivated by, and require, a gapped or pseudogapped spectrum.

If we, ignoring the evidence that the normal state is a distinct fixed point, confine ourselves to the superconducting state, one experiment seems to mitigate against anyon states. This is the very clean IBM demonstration of persistent currents in a loop composed partly of high-T_c and partly of ordinary superconducting material (30). This experiment shows that the phase of a singlet pair wave function is correlated macroscopically and seems only explicable with a conventional Bardeen-Cooper-Schrieffer pair order parameter. This, with many other less conclusive experiments, shows that conventional Ginsburg-Landau theory with a true order parameter describing singlet pairs is the phenomenology of high T_c. The highly sophisticated theoretical demonstrations of anyon superconductivity do not demonstrate the existence of a suitable order parameter and are not completely convincing as to Ginsburg-Landau behavior. The optical experiments, which seemed, momentarily, to support T or P noninvariance, seem now to be inconclusive at best, having been contradicted by more sensitive tests (31).

Finally, much is made in several papers of "spin gaps": actually, these are pseudogaps involving considerable loss of density of states for a few tens of degrees above T_c. Some materials with the highest T_c's do not show much hint of such gaps, which brings into play our "single fixed point" argument. Also, attempts to fit data with such a gap leave it relatively small even compared with the known superconducting gaps, and certainly out of scale with the "fixed point" physics that occurs throughout the range up to 1000 to 2000 K. It must be an additional, mostly irrelevant quirk of the very complex physics.

Summary

The theoretical picture of high T_c becomes very much less confusing when one examines experimental results critically with certain minimal theoretical results in mind. The careful reader may wonder if any candidate highly regarded theory survives, but he should not despair; in my opinion there is at least an existence proof that there is one theory (1) that is both internally consistent and compatible with all these experimental constraints. Whether there is another time will tell.

REFERENCES AND NOTES

1. P. W. Anderson, in *Proceedings of the "Materials and Mechanisms of Superconductivity, High Temperature Superconductors III" Conference*, Kanazawa, 22 to 26 July 1991, M. Tachiki, Y. Muto, Y. Syono, Eds. (North-Holland, Amsterdam, 1991), p. 11; *Physica C* 185–189, 11 (1991).
2. ———, in preparation.
3. J. R. Schrieffer and P. W. Anderson, *Phys. Today* 44, 54 (June 1991).
4. K. G. Wilson, *Phys. Rev.* 179, 149 (1969). That the Fermi liquid is a fixed point was perhaps first expressed in P. W. Anderson, in *Proceedings of Nobel Symposium 24*, S. Lundqvist and B. Lundqvist, Eds. (Nobel Foundation, Uppsala, 1973), p. 266.
5. For overviews, for instance, see: Y. Iye, in *Studies of High Temperature Superconductors: Advances in Research and Applications*, A. Narlikar, Ed. (Nova Science, New York, 1991), p. 199; N.-P. Ong, in *Physical Properties of High Temperature Superconductors*, D. M. Ginsberg, Ed. (World Scientific, Singapore, 1989), vol. 2, p. 459.
6. D. B. Mitzi *et al.*, *Phys. Rev. B* 41, 6564 (1990); M. Miljak *et al.*, *Europhys. Lett.* 9, 723 (1989); M. Miljak, *Phys. Rev. B* 42, 10742 (1990).
7. H. Alloul *et al.*, *Phys. Rev. Lett.* 63, 1700 (1989); Y. Kitaoka *et al.*, in *Proceedings of Mt. Fuji Conference, Strong Correlations and Superconductivity*, H. Fukuyama, S. Maekawa, A. P. Malozemoff, Eds. (Springer, Tokyo, 1985), p. 262; S. E. Barrett *et al.*, *Phys. Rev. B* 41, 6283 (1990).
8. N.-P. Ong *et al.*, *Phys. Rev. Lett.* 67, 2088 (1991).
9. C. G. Olson *et al.*, *Phys. Rev. B* 42, 381 (1990).
10. R. J. Collins *et al.*, in *Proceedings of Mt. Fuji Conference, Strong Correlations and Superconductivity*, H. Fukuyama, S. Maekawa, A. P. Malozemoff, Eds. (Springer, Tokyo, 1985), p. 289; *Phys. Rev. B* 41, 11237 (1980); *Phys. Rev. Lett.* 65, 801 (1990); *ibid.* 63, 422 (1989); *Phys. Rev. B* 43, 8201 (1991). Striking confirmation of the phenomenology of Collins *et al.* is found in I. Bosovic *et al.*, *ibid.* 42, 1969 (1990); *ibid.* 43, 1169 (1991); *Physica C* 174, 435 (1991).
11. R. C. Dynes, in *Proceedings of the "Materials and Mechanisms of Superconductivity, High Temperature Superconductors III" Conference*, Kanazawa, 22 to 26 July 1991, M. Tachiki, Y. Muto, Y. Syono, Eds. (North-Holland, Amsterdam, 1991), p. 234; *Physica C* 185–189, 234 (1991).
12. The method would be a generalization of the proof of J. Friedel's very similar theorem on the relation between scattering phase shifts and number density. One would embed the interacting gas in a suitably chosen noninteracting one with the same Fermi surface and study the scattering of Fermi surface electrons. See P. W. Anderson, in *Proceedings of the International School of Physics*, Varenna, Italy, June 1966 [*Nuovo Cimento Suppl.* 37, 50 (1967)]; *Phys. Rev. Lett.* 17, 95 (1966).

13. N. F. Mott and E. H. Davis, *Electronic Properties of Non-Crystalline Materials* (Taylor and Francis, London, 1975), p. 79; J. R. Schrieffer et al., *Phys. Rev. Lett.* **10**, 336 (1963). Mott observed that the conductivity may be expressed entirely in terms of scattering cross section and wavelength (or k_F), with mass and velocity, the dynamic quantities, canceling out.
14. J. Friedel, *Adv. Phys.* **3**, 336 (1954).
15. As presented by C. M. Varma et al., *Phys. Rev. Lett.* **63**, 19969 (1989).
16. S. W. Tozer et al., *ibid.* **59**, 1768 (1987).
17. S. Martin, A. J. Fiory, P. Fleming, L. Schneemeyer, J. V. Woczak, *Phys. Rev. B* **41**, 846 (1990).
18. T. Timusk and D. B. Tamura, in *Physical Properties of High T_c Superconductors*, D. M. Ginsberg, Ed. (World Scientific, Singapore, 1989), p. 339; A. V. Bazhenov et al., *Physica C* **169**, 381 (1990). Recent measurements by T. Timusk and by L. D. Rotter (private communication) on YBCO show a very low featureless infrared conductivity σ_c from $\omega = 0$ out to 2000 to 3000 cm^{-1}, that is, no conventional Drude term at all, and certainly no hint of the interplanar term at 500 to 1000 cm^{-1}, which would be there in a simple band theory.
19. A. P. Malozemoff, in *Physical Properties of High T_c Superconductors*, D. M. Ginsberg, Ed. (World Scientific, Singapore, 1989), p. 71.
20. M. Schluter and M. S. Hybertsen, *Physica C* **162–164**, 583 (1989).
21. G. A. Sawatsky, in *Proceedings of the Los Alamos Symposium on High Temperature Superconductivity*, K. S. Bedell et al., Eds. (Addison-Wesley, Redwood City, CA, 1990), p. 297; S. Uchida, *Physica C* **185**, 28 (1991).
22. C. Stafford, thesis, Princeton University (1992).
23. P. Horsch, *Helv. Phys. Acta* **63**, 345 (1990).
24. Experiments as quoted in (6), interpreted in: F. Mila and T. M. Rice, *Physica C* **153**, 561 (1989); *Phys. Rev. B* **40**, 1382 (1990); A. Millis et al., *ibid.* **42**, 167 (1990).
25. J. Rossat-Mignod et al., in *Proceedings of the "Materials and Mechanisms of Superconductivity, High Temperature Superconductors III" Conference*, Kanazawa, 22 to 26 July 1991, M. Tachiki, Y. Muto, Y. Syono, Eds. (North-Holland, Amsterdam, 1991), p. 86; *Physica C* **185–189**, 86 (1991); *Physica B* **169**, 58 (1991).
26. S. Tajima, S. Tanaka, J. Ido, S. Uchida, in *Proceedings of the Second International Symposium on Superconductivity (ISS '89)*, 14 to 17 November 1989, Tsukuba, in *Advances in Superconductivity II*, T. Ishiguro and K. Kajimura, Eds. (Springer-Verlag, Tokyo, 1990), p. 569; S. Uchida, in *Proceedings of the "Materials and Mechanisms of Superconductivity, High Temperature Superconductors III" Conference*, Kanazawa, 22 to 26 July 1991, M. Tachiki, Y. Muto, Y. Syono, Eds. (North-Holland, Amsterdam, 1991), p. 28; *Physica C* **185–189**, 28 (1991).
27. N. F. Berk and J. R. Schrieffer, *Phys. Rev. Lett.* **17**, 433 (1966).
28. T. C. Hsu and P. W. Anderson, *Physica C* **162**, 1445 (1989).
29. This point is exhaustively demonstrated in a recent review by W. E. Pickett, H. Krakauer, R. E. Cohen, D. J. Singh, *Science* **255**, 46 (1992).
30. P. Chaudhari et al., *IBM J. Res. Dev.* **33**, 299 (1989).
31. S. Spielman et al., *Phys. Rev. Lett.* **65**, 123 (1990); further confirmation by the same group (A. Kapitulnik et al.) and R. B. Laughlin et al. has been widely circulated.
32. D. B. Mitzi, L. W. Lombardo, A. Kapitulnik, S. S. Laderman, R. D. Jacowitz, *Phys. Rev. B* **41**, 6564 (1990).
33. T. Tokura, H. Takagi, S. Uchida, *Nature* **337**, 345 (1989); J. M. Tarascon et al., in *Proceedings of "Materials and Mechanisms of Superconductivity, High Temperature Superconductors" Conference*, Stanford, 1989; *Physica C* **162–164**, 285 (1990).
34. J. Karpinski, S. Ruzieki, E. Kaldis, E. Bucher, E. Jilek, *Physica C* **160**, 449 (1990).
35. J. Y. Genau et al., *Physica C* (Suppl.), in press.
36. M. G. Smith, A. Manthiram, J. Zhou, J. B. Good-